International Conference on Process Engineering and Advanced Materials (ICPEAM2018)

IOP Conference Series: Materials Science and Engineering Volume 458

Kuala Lumpur, Malaysia
13 – 14 April 2018

Editor:

Mohamed Mahmoud Nasef

ISBN: 978-1-5108-8004-7
ISSN: 1757-8981

Printed from e-media with permission by:

Curran Associates, Inc.
57 Morehouse Lane
Red Hook, NY 12571

proceedings.com

Some format issues inherent in the e-media version may also appear in this print version.

Printed by Curran Associates, Inc. (2019)

Institute of Physics
Dirac House, Temple Back
Bristol BS1 6BE UK

Phone: 44 1 17 929 7481
Fax: 44 1 17 920 0979

techtracking@iop.org

Additional copies of this publication are available from:

Curran Associates, Inc.
57 Morehouse Lane
Red Hook, NY 12571 USA
Phone: 845-758-0400
Fax: 845-758-2633
Email: curran@proceedings.com
Web: www.proceedings.com

TABLE OF CONTENTS

ADVANCED MATERIALS AND ENGINEERING

ENVIRONMENTAL ENGINEERING

PROCESS AND SYSTEM ENGINEERING

REACTION ENGINEERING AND KINETICS AND OTHERS

Preface

The 5th International Conference on Process Engineering and Advanced Materials (ICPEAM 2018) is biannual event organized by Chemical Engineering Department of Universiti Teknologi PETRONAS, under the banner of World Engineering, Science & Technology Congress (ESTCON 2018) that was successfully held in Kuala Lumpur, Malaysia in August 13-14, 2018. The conference provides a platform to promote interaction between international experts, academicians and industrialists to discuss contemporary industrial issues and review cutting-edge research in a wide range of interdisciplinary chemical engineering areas including advanced materials and engineering, environmental engineering, process and system engineering and reaction engineering & kinetics and more. The conference also provides great opportunities for participants to meet experts, exchange ideas, share experiences and establish the ground for future research and business relations with local and global partners. This ICPEAM 2018 Proceedings contains 86 selected papers that have been accepted for presentation in the conference. All articles were subjected to revisions by a minimum of two expert reviewers appointed by the conference technical committee. The open access mode of these proceedings is meant to allow wider access to readers around the world to benefit from the outcomes of the conference. The guest editor would like to sincerely thank the publication team for their efforts in preparing this proceeding. Finally, I would like to acknowledge the invaluable support by the conference organizing committee chaired by Dr. Suriati Sufian towards the publication of this proceedings in IOP Conference Series: Materials Science and Engineering. Last but not least, the publication committee wishes to thank all the authors for their contributions to ICPEAM 2018 proceedings and we look forward to welcoming more of your contributions to the ICPEAM 2020.

Editorial
Guest Editor: Mohamed Mahmoud Nasef
Editorial members:
Suzana Yusup
Zakaria Man
Marappagounder Ramasamy
Muhammad Moniruzzaman
Kiki Adi Kurina
Raihan Mahirah Ramli

Guest Editor Email: mohamed.nasef @utp.edu.my
Phone No. +605-3687561

IOP Conf. Series: Materials Science and Engineering **458** (2018) 011002 doi:10.1088/1757-899X/458/1/011002

Peer review statement

All papers published in this volume of *IOP Conference Series: Materials Science and Engineering* have been peer reviewed through processes administered by the proceedings Editors. Reviews were conducted by expert referees to the professional and scientific standards expected of a proceedings journal published by IOP Publishing.

Zeolite NaY Synthesis by using Sodium Silicate and Colloidal Silica as Silica Source

M Aimen Isa[1,2], Thiam Leng Chew[2], Yin Fong Yeong[2]

[1] PETRONAS Research Sdn. Bhd., Kawasan Institusi Bangi, Kajang, 43000 Selangor, Malaysia.
[2] Department of Chemical Engineering, Universiti Teknologi PETRONAS (UTP), Seri Iskandar, 32610 Perak, Malaysia.

Email: thiamleng.chew@utp.edu.my

Abstract. Zeolite NaY (FAU) has been prepared through direct hydrothermal method by using Sodium Silicate and Colloidal Silica as the silica source. Aging time was carried out at 24 hours while hydrothermal temperature was performed at 90°C for 24 hours. Detailed comparisons and characterizations were performed by using several methods onto the synthesized zeolite NaY (FAU) by using powder X-Ray Diffraction (XRD), Field Emission Scanning Electron Microscopy (FESEM) and Energy dispersive X-Ray spectroscopy (EDX). The results showed that the products obtained from the starting gel were highly dependent on the type of raw material used- silica source which affects the structure of the products.

1. Introduction

Zeolites are crystalline aluminosilicates with open-framework structures and have been widely used for various applications including separations, ion exchange and catalysis. Furthermore, the existence of uniformly distributed cavities and channels in molecular dimensions endows zeolites with unique molecular recognition, discriminations and organization properties [1]. One of it is molecular sieve zeolite from Faujasite (FAU) family with large three-dimensional pore structure and being perpendicular to each other of x-y-z (111) co-ordinations with pore openings at about 0.74nm [2–5]. In general, FAU zeolites are separated into two classes which are zeolite NaX and zeolite NaY which zeolite NaY has silicon to aluminium Si/Al molar ratio above 1.5 [2,6–8]. This value of Si/Al ratio is particularly important as it will determine the hydrothermal stabilities of the zeolite NaY. Previous investigations on influential parameters of zeolites synthesis revealed that the chemical composition of the reaction mixture was one of the most crucial factors in controlling the properties of the zeolites during the synthesis process [9]. Synthesis of zeolites will follow three main stages which are: (1) Mixing of chemicals in order to obtain aluminosilicate precursor solution, (2) Aging of the precursor solution and (3) Crystallization of the precursor solution which is conducted at higher temperature [10]. Normally, zeolites are prepared hydrothermally for a certain period of times ranging from within hours to a few days. As for zeolite NaY, it is prepared from precursor solution which mainly contains the mixture of the silica source and aluminium source to form concentrated aluminosilicate solutions which contain amorphous gel [4,8,11]. During the growth process under the alkaline state with suitable aging and crystallization time, this amorphous precursor solution serves as a reservoir for the crystals to grow hence to be deposited at the surface of the gel. Hence, the source of the precursor solution is one of the important parameters that need to be investigated extensively in order to produce zeolite NaY with desired properties.

In the current study, two different of silica sources (Sodium Silicate and Colloidal Silica solution) were used separately in preparing the precursor solution for zeolite NaY synthesis. The effect of different silica sources on the formation of zeolite NaY was investigated. The synthesized samples were characterized by using different analytical techniques including X-Ray Diffraction (XRD), Field Emission Scanning Electron Microscopy (FESEM) with Energy dispersive X-Ray spectroscopy (EDX).

The method applied during the synthesis is based on a method reported in previous studies [12–16].

2. Experimental

2.1. Materials

All chemicals were directly used as received without any further purifications. The reactants used were Sodium Hydroxide (NaOH), Sodium Aluminate (Fischer, 53wt% Al_2O_3; 42.5wt% Na_2O), Sodium Silicate (Fischer, Na_2O 51 wt%; SiO_2 45.5 wt%), Colloidal Silica (Ludox, AS-40), commercial zeolite NaY powder (0.3 lm crystal size, Si/Al = 2.5, Tosoh) and deionized water (DI).

2.2. Synthesis of Zeolite NaY by Using Sodium Silicate and Colloidal Silica

Precursor solution with a molar composition of Al_2O_3: 12.8 SiO_2: 17 Na_2O: 975 H_2O was used for the hydrothermal synthesis of zeolite NaY in this study. During this procedure, identified amount of NaOH, Sodium Aluminate, and DI water were mixed. The resultant solution was heated at 70°C for 1h until a clear solution was obtained. After cooling down to room temperature, Sodium Silicate solution (as silica source) was added drop-wise and stirred until completely dissolved. After aging for 24 hours, the mixture was transferred to a Polypropylene (PP) bottle and then kept inside a heated oven at 90 °C for 24 hours. After cooled down, the solid product was vacuum filtered and washed with DI water repeatedly in order to remove any unreacted chemicals until pH 7-8. Similar procedure was repeated by using Colloidal Silica solution as the other silica source.

Besides that, commercial zeolite NaY powder (Tosoh) was also added as seeds into several samples of the precursor solution prior to heating in an oven at 90°C for 24 hours. On the other hand, several runs of the synthesis were conducted by varying the amount of NaOH used in the precursor solution with respect to its the molar ratio (Na_2O/Al_2O_3) under similar synthesis conditions (aging time for 24 hours at room temperature and heating in over at 90 °C for 24 hours) in order to study the effect of Na^+ ions towards formation of zeolite NaY. Table 1 shows all the samples synthesized with different synthesis conditions in the current study.

Table 1. Samples synthesized at different conditions.

No	Sample Name	Silica Source	Seeded	Na_2O/Al_2O_3
1	S1	Ludox-AS-40	No	17
2	S2	Ludox-AS-40	Yes	17
3	S3	Sodium Silicate	No	17
4	S4	Sodium Silicate	Yes	17
5	S5	Sodium Silicate	No	20
6	S6	Sodium Silicate	No	25

2.3. Characterizations of the Zeolites

The synthesized samples and commercial zeolite NaY (Tosoh) were characterized by several different techniques including X-Ray Diffraction (XRD), Field Emission Scanning Electron Microscopy (FESEM) with Energy dispersive X-Ray spectroscopy (EDX). X-ray Diffraction (XRD) was conducted by using Cu-Kα radiation with 40kV and 30mA in the measuring range of 2θ: 5-50° at a scanning rate of 4 min^{-1}. XRD was used to analyze the crystallinity of the samples with respect to the reference sample (commercial NaY from Tosoh). The morphology of the samples was observed through FESEM at room temperature. The images of the samples are captured by the FESEM equipment from the electrons that are ejected over the surface of the samples at an acceleration voltage of 15kV. For Energy Dispersive X-ray Spectrometer (EDX) attached to the FESEM, it was used to conduct elemental analysis of the samples. The final Si/Al ratios for each sample were obtained by taking the average of at least 3 measurements at different spots.

3. Results and discussions

The XRD peaks for each sample synthesized were compared with the XRD of commercial zeolite NaY (Tosoh). It was reported that the main diffraction peaks should appear approximately at 2θ of 6.31°, 10.31°, 12.10°, 15.92°, 20.71°, 24.06°, and 31.95° [9]. Figure 1 shows the XRD pattern for the synthesized sample and commercial zeolite NaY (Tosoh). When Colloidal Silica solution was used as the silica source, with NaY seeds (S2 sample) or without NaY seeds (sample S1), zeolite NaY was successfully formed as the XRD pattern agreed well with the XRD pattern of the commercial zeolite NaY (Tosoh). However, different results were observed whereby zeolite NaY was successfully formed only when NaY seeds (S4 sample) was added into the precursor solution as compared to sample S3 (without the addition of NaY seeds) which displayed amorphousity, indicating failure in forming zeolite NaY by using Sodium Silicate as the silica source. These results are in agreement with those reported by previous studies which used Sodium Silicate as its silica source that require NaY seeds in the starting gel [17,18]. Possibly, due to the availability of NaY seeds which acts as nuclei hence to assist in the crystallization process of NaY zeolite as compared to the samples without addition of seeds.

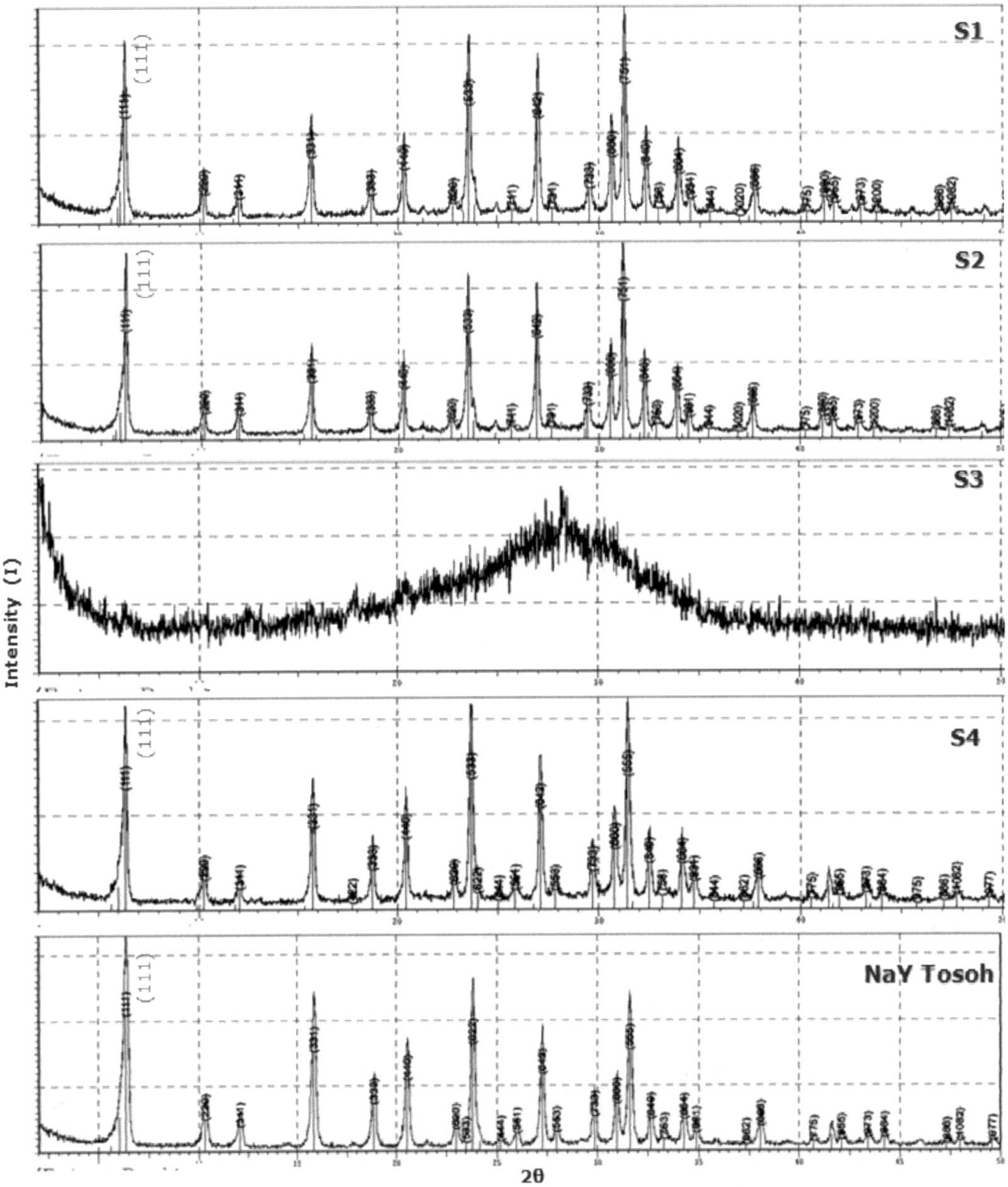

Figure 1. XRD patterns of synthesized samples and commercial NaY (Tosoh).

The crystal structure of the synthesized zeolite NaY was reported to be in octahedral shape [11,19–21]. The FESEM images of the synthesized samples are shown in Figure 2. For the samples synthesized using Colloidal Silica and Sodium Silicate as silica source with the addition of NaY seeds (Tosoh), the FESEM images showed fine octahedral shape with particle sizes around 500-800nm. Meanwhile, for zeolite NaY samples synthesized with Sodium Silicate used as its silica source without the addition of NaY seed (Tosoh), only amorphous phase could be observed. For EDX results, all samples synthesized from Colloidal Silica and Sodium Silicate with the addition of NaY seed (TOSOH) displayed Si/Al molar ratio of more than 1.5 which represents zeolite NaY, which is in agreement with the value reported in the previous studies [2,6–8].

Figure 2. FESEM of synthesized samples and commercial NaY (Tosoh).

It was showed that even without the addition of NaY (Tosoh) seeds zeolite NaY were successfully synthesized by using Colloidal Silica as its silica source. This could be results from the availability of Na^+ ions which are solely from NaOH that are more easily to dissociate into ions as compared to Na^+ ions from Sodium Silicate which are only slightly soluble in water. High availability of Na^+ ions in the colloidal solutions helps to stabilize the highly open FAU structures [22]. Furthermore, a higher amount of Na^+ ions also resulted in the increase of alkalinity of the colloidal solution which permit higher solubility of aluminosilicate in order to form crystalline NaY zeolite which interrelated to the crystallization rate of the zeolite itself [2,19,23]. Apart from that, due to lower solubility of Sodium Silicate in water, less amount silicalite anions could be released into the surrounding solution in order to react with aluminate anions hence to produce aluminosilicate solids. These aluminosilicate solids which are rich in Al^- and hydrated Na^+ cations are crucial in the formation of zeolite NaY (FAU) during crystallization process [22]. An EDX analysis was performed on the solid products synthesized from Colloidal Silica and Sodium Silicate used as the silica source together with or without zeolite NaY seeds (Tosoh). Higher amount of Na element was found in the samples synthesized by using Colloidal Silica as compared the one synthesized by using Sodium Silicate used as the Silica source. This shows the importance of Na^+ ions in the precursor solution during the aging process which allow on the stabilization of the highly open zeolite structures and to form zeolite NaY [22]. This also indicates that the zeolite NaY is strongly dependent on Na^+ ions concentration in the synthesis mixture [11]. Table 2 showed the EDX results of the samples.

Table 2. EDX results on samples synthesized at different conditions.

No	Sample Name	Silica Source	Seeded	EDX Na (wt%)	Phase
1	S1	Ludox AS-40	No	12.36	Zeolite NaY
2	S2	Ludox AS-40	No	11.87	Zeolite NaY
3	S3	Sodium Silicate	No	9.15	Amorphous
5	S4	Sodium Silicate	Yes	11.72	Zeolite NaY

Figure 3 shows XRD micrographs of the samples synthesized with different Na_2O/Al_2O_3 molar ratio in the precursor solution. It was observed that as the amount of Na_2O/Al_2O_3 molar ratio was not varied, no definite diffraction peaks were observed from the XRD indicating the sample was still in an amorphous phase. However, as the amount of Na_2O/Al_2O_3 molar ratio was increased, phase transformation started to occur from amorphous to a crystal phase. Nonetheless, an increase in Na_2O/Al_2O_3 molar ratio assists in the crystallization rates of the samples as well as resulted in the formation of other impurities i.e. NaP (GIS) instead of NaY (FAU) zeolite in the synthesized samples. This could be seen from the' spacing (2θ) as well as miller indices at (301) co-ordinations which indicates the existence of zeolite NaP (GIS) zeolite in the synthesized samples i.e. sample S6[24]. This could also be the results of too high sodium concentration in the synthesis solution which favors the condensation reactions on the surface of particles hence contributes to their growth and irregular shapes as well [19]. This result is in agreement with previous studies indicating zeolite NaP (GIS) which is in more stable phase is the most competing phases with zeolite NaY (FAU) [6,25–27]. Figure 4 showed the FESEM images for each of the samples synthesized at different Na_2O/Al_2O_3 molar ratio whereby Zeolite NaP (GIS) could be identified by its spherical/ globular shape which is different from zeolite NaY (FAU) which is in octahedral shape [7,19,21].

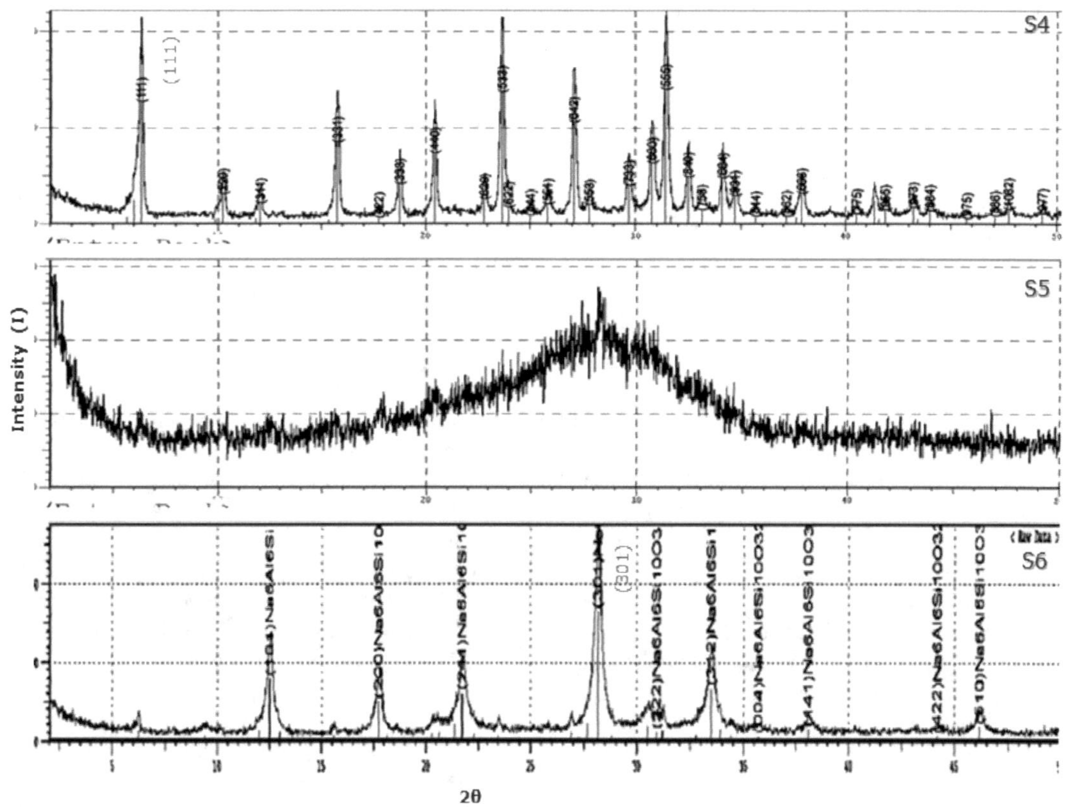

Figure 3. XRD patterns of samples synthesized with different Na_2O/Al_2O_3 molar ratio in the precursor solution.

Figure 4. FESEM of samples synthesized with different Na_2O/Al_2O_3 ratio in the precursor solution

4. Conclusions

By using Colloidal Silica as the silica source without NaY seed (Tosoh) in the precursor solution, zeolite NaY was successfully obtained with octahedral crystal structure as confirmed by XRD patterns and FESEM micrographs after aging time 24 hours followed by heating at 90°C for 24 hours. Higher amount Na^+ ions are readily available when Colloidal Silica was used in the precursor solution in order to form aluminosilicates which are crucial in the formation of zeolite NaY (FAU). This was confirmed by the studies on variations of Na_2O/Al_2O_3 ratio which indicates the effects of readily available Na^+ ions in the precursor solution which helps to stabilize the highly open zeolite NaY (FAU) structures. Thus, it could be seen by having Colloidal Silica as an alternative to Sodium Silicate as the silica source in the synthesis solution is effective in synthesizing NaY (FAU) zeolite with the octahedral shape.

5. Acknowledgment

The financial support from GR&T PETRONAS Research Grant (cost center: 0153C-025) is greatly appreciated. The authors appreciate the technical support from the Institute of Contaminant Management, UTP and CO2 Research Centre (CO2RES), UTP.

6. References.

[1] Huang Y, Wang K, Dong D, Li D, Hill M R, Hill A J and Wang H 2010 Microporous and Mesoporous Materials Synthesis of hierarchical porous zeolite NaY particles with controllable particle sizes *Microporous Mesoporous Mater.* **127** 167–75

[2] Liu Z, Shi C, Wu D, He S and Ren B 2016 A Simple Method of Preparation of High Silica Zeolite Y and Its Performance in the Catalytic Cracking of Cumene **2016**

[3] Ozdemir O D and Pişkin S 2013 Zeolite X Synthesis with Different Sources *Int. J. Chem. Environ. Biol. Sci.* **1** 229–32

[4] Ahmedzeki N S and Al-tabbakh B A 2016 Synthesis and Characterization of Nanocrystalline Zeolite Y Synthesis and Characterization of Nanocrystalline Zeolite Y

[5] Larsen S C 2005 Development of Improved Materials for Environmental Applications: Nanocrystalline NaY Zeolites **39** 1214–20

[6] Lang W Z, Ouyang J X, Guo Y J and Chu L F 2011 Synthesis of tubular faujasite X-type membranes with mullite supports and their gas permeances for N2/CO2 mixtures *Sep. Sci. Technol.* **46** 1716–25

[7] Wang Z, Kumakiri I, Tanaka K, Chen X and Kita H 2013 Microporous and Mesoporous Materials NaY zeolite membranes with high performance prepared by a variable-temperature synthesis q *Microporous Mesoporous Mater.* **182** 250–8

[8] Karami D and Rohani S 2009 A Novel Approach for the Synthesis of Zeolite Y 4837–43

[9] Motazedi K, Mahinpey N and Karami D 2016 Preparation and Application of Faujasite-Type Y Zeolite-based Catalysts for Coal Pyrolysis using Sodium Silicate Solution and Colloidal Silica as Silicon Source *Chem. Eng. Commun.* **203** 300–17

[10] Don J 2010 The Influence of Alumosilicate Gel Aging on the Synthesis of NaX Zeolite **22** 30–4
[11] Li Q, Creaser D and Sterte J 2002 An Investigation of the Nucleation / Crystallization Kinetics of Nanosized Colloidal Faujasite Zeolites 1319–24
[12] Jeong B, Hasegawa Y, Sotowa K, Kusakabe K and Morooka S 2002 Vapor Permeation Properties of an NaY-Type Zeolite Membrane for Normal and Branched Hexanes 1768–73
[13] Kusakabe K, Kuroda T, Uchino K, Hasegawa Y and Morooka S 1999 Gas Permeation Properties of Ion-Exchanged Faujasite-Type Zeolite Membranes **45** 1220–6
[14] Hasegawa Y, Watanabe K, Kusakabe K and Morooka S 2002 Influence of alkali cations on permeation properties of Y-type zeolite membranes *J. Memb. Sci.* **208** 415–8
[15] Hasegawa Y, Watanabe K, Kusakabe K and Morooka S 2001 The separation of CO 2 using Y-type zeolite membranes ion-exchanged with alkali metal cations **23** 319–25
[16] Hasegawa Y, Tanaka T, Watanabe K and Jeong B 2002 Separation of CO 2 -CH 4 and CO 2 -N 2 Systems Using Ion-exchanged FAU-type Zeolite Membranes with Different Si / Al Ratios **19** 309–13
[17] Materials A 1992 A physicochemical study of the aging of colloidal silica gels used in zeolite Y synthesis **12** 733–41
[18] Carl Vance McDaniel, Laurel, Md., Heyman Clarke Duecker M 1971 Process for preparing high silica Faujasite
[19] Chaves T F, Pastore H O and Cardoso D 2016 Synthesis and properties of zeolite FAU with nanocrystals *Microporous Mesoporous Mater.* **161** 67–75
[20] Zhu G, Qiu S, Yu J, Sakamoto Y, Xiao F, Xu R, Terasaki O, Al S, Na O T M A O and O O H 1998 Synthesis and Characterization of High-Quality Zeolite LTA and FAU Single Nanocrystals Since Milton synthesized zeolites by means of hydro- thermal treatment of alkaline aluminosilicate gels in the 1940s , 1 more than a hundred different kinds of zeolites **4756** 1483–6
[21] Sato K 2007 Application of FAU-type zeolite membranes to vapor / gas separation under high pressure and high temperature up to 5 MPa and 180 ° C **101** 312–8
[22] Anon 1992 The effects of gel aging on the synthesis of NaY zeolite from colloidal silica **12**
[23] Larlus O, Mintova S and Bein T 2006 Environmental syntheses of nanosized zeolites with high yield and monomodal particle size distribution **96** 405–12
[24] Treacy M M J 2001 Collection of Simulated XRD Powder Patterns for Zeolites Editors :
[25] Qiang L, Ying Z, Zhijun C, Wei G and Lishan C 2010 In fl uence of synthesis parameters on the crystallinity and Si / Al ratio of NaY zeolite synthesized from kaolin 403–9
[26] Anon 2010 A115 2010
[27] Gu X, Dong J and Nenoff T M 2005 Synthesis of Defect-Free FAU-Type Zeolite Membranes and Separation for Dry and Moist CO 2 / N 2 Mixtures 937–44

Formation of pure NaX zeolite: Effect of ageing and hydrothermal synthesis parameters

Asma' Ahmadon[1], Liyana Salwa Mohd Nazir[1, 2], Yin Fong Yeong[2], Shamsina Sabdin[1]

[1]PETRONAS Research Sdn. Bhd., Lot 3288&3289, Off Jalan Ayer Itam, Kawasan Institusi Bangi, Kajang, 43000, Selangor, Malaysia
[2]Chemical Engineering Department, Faculty of Engineering, Universiti Teknologi PETRONAS, Bandar Seri Iskandar, 32610, Perak, Malaysia
liyana_salwa@petronas.com

Abstract. Among the challenges of synthesizing NaX zeolite is the formation of competing phases during the hydrothermal process which results with impurities in the product. The competing phases of NaX zeolite may include NaP zeolite, hydroxylsodalite and NaA zeolite. In this paper, relationships between the preparation parameters and hydrothermal synthesis parameters towards the formation of high purity of NaX zeolite are studied. The studied parameters include ageing duration, hydrothermal synthesis temperature and hydrothermal synthesis duration. Sodium silicate, sodium aluminate, sodium hydroxide and distilled water are used as the main reactant sources at a fixed molar ratio. The morphology, Si/Al ratio and zeolite phase of the resultant particles were characterized using field emission scanning electron microscope (FESEM), energy-dispersive x-ray spectroscopy (EDS) and X-ray diffraction analyser (XRD), respectively. Hydrothermal synthesis temperature of 90°C and synthesis duration of 16 hours were identified as the optimum synthesis parameters for producing high crystalline, single phase NaX zeolite. In addition, the relationships between synthesis parameters and NaX zeolite produced were also studied. Longer ageing duration and longer hydrothermal synthesis duration results in high crystalline product with potential impurities phases, while higher hydrothermal synthesis temperature produced pure NaX zeolite at shorter hydrothermal synthesis duration.

1. Introduction

Zeolites have superior thermal and chemical stability and good shape-selectivity, which are suitable for various applications, particularly the oil and gas industry. Their polycrystalline structure with well-defined pores allows the permeation of selective gas molecules through molecular sieving and competitive adsorption. The size-exclusion mechanism happens when the zeolite pore size falls between the kinetic diameters of the desired gas molecule with those of the unintended species in the feed [1]. FAU framework zeolites, namely NaX (Si/Al ratio of 1 to 1.5) and NaY (Si/Al ratio larger than 1.5), are being studied for CO_2 and N_2 separation from natural gas. The pore size of these zeolites are 0.74 nm, allowing the permeation of certain molecules in natural gas. The selective permeation through the zeolite pore is determined by the surface diffusion and pore constriction [2].

In comparison with polymeric membranes for gas separation, inorganic membranes have the potential of CO_2 and N_2 removal from natural gas with better performance, such as CO_2/CH_4 selectivity of 200 and N_2/CH_4 selectivity of 10 [3]. The current available membranes reported in literature have yet to be developed for commercial application due to the difficulty in scaling up. Recently, inorganic membranes with zeolite separation layer have shown significant separation and promising results. One of the challenges in fabrication of NaX zeolite membrane is the formation of impurities phases during hydrothermal synthesis. The growth of impurities, such as NaP zeolite, NaA zeolite and hydroxylsodalite, on the membrane surface hinders the permeation of desired gas through the membrane, thus, reducing the performance of the membrane [4]. This essential factor is among the reasons of the study to develop high purity NaX zeolite. Their nature as metastable materials causes the zeolites to transform into other more stable materials if left in their aluminosilicate gel solution. Interestingly, other factors can also impact zeolite crystallization even when the gel solution is stable, such as hydrothermal synthesis duration and temperature, types of raw materials used, the order of mixing these materials, stirrer method and aging technique [5,6].

In this study, the synthesis of high purity NaX zeolite via hydrothermal method is reported. The ageing duration and hydrothermal synthesis parameters and their effect towards the synthesis of NaX zeolite are investigated.

2. Experimental Section

2.1. Materials and Methods

For the NaX zeolite synthesis, the materials used were as follows: (1) sodium silicate (100% (w/w) Na_2SiO_3; Fisher Chemical, UK); (2) sodium aluminate (100% (w/w) $NaAlO_2$. Fisher Chemical, UK); (3) sodium hydroxide pellets (90-100% (w/w) NaOH; Avantor Performance Materials, USA) and (4) Deionized water. No further purification was performed and the raw materials were used as received. Sodium silicate and sodium aluminate were used as the silica source and alumina source, respectively.

The NaX zeolite synthesis batches were prepared by mixing the silica source solution into the alumina source solution and stirred at 600 rpm until a homogeneous mixture is formed. The gel mixture was stirred with varying durations (4 – 72 hours) at room temperature using a fixed stirrer speed in order to age the mixture. Then, the synthesis gel was filled into an autoclave and hydrothermal synthesis was carried out at temperatures of 80 to 100 °C and from 8 to 24 hours without stirring under autogenous pressure. Upon completion of hydrothermal synthesis, the gel was cooled at room temperature, filtered and washed until the filtrate was at pH 7. The crystals collected was then dried at 90 °C. Throughout the experiment, the final composition of the aluminosilicate gel mixture of NaX zeolite batches was maintained at $Al_2O_3:SiO_2:Na_2O:H_2O$ = 1:4.8:17:975 in a molar composition, according to the method reported by Hasegawa et al [7].

2.2. Characterization

The zeolite powder samples were characterized by using powder X-ray diffraction (XRD; Shimadzu XRD-7000) with Cu-Kα radiation, operated at 40.0 kV and 30.0 mA in order to identify the X-ray diffraction patterns of the crystal structure. The zeolite morphology was observed using field emission scanning electron microscope (FESEM; Hitachi SU8020) while the elemental Si/Al ratios were determined using energy dispersive X-ray spectroscopy (EDS; Horiba EMAX).

3. Results and Discussion

3.1. Effects of Ageing Duration

Figure 1 shows the XRD patterns of NaX zeolite powder samples obtained using synthesis solution aged at room temperature for 24 and 72 hours, respectively. Both samples were synthesized at 80 °C for 24 hours. Referring to Figure 1, a mixture of low crystallinity and amorphous phase can be seen for the sample aged for 24 hours with all the weak peaks corresponding to NaX zeolite, as reported by Treacy and Higgins [8]. Further aging of the synthesis solution of up to 72 hours resulted in a highly crystalline phase, indicated by the sharp and intense XRD peaks and low background radiation line. However, the sample contained both zeolite phases of NaX and NaP.

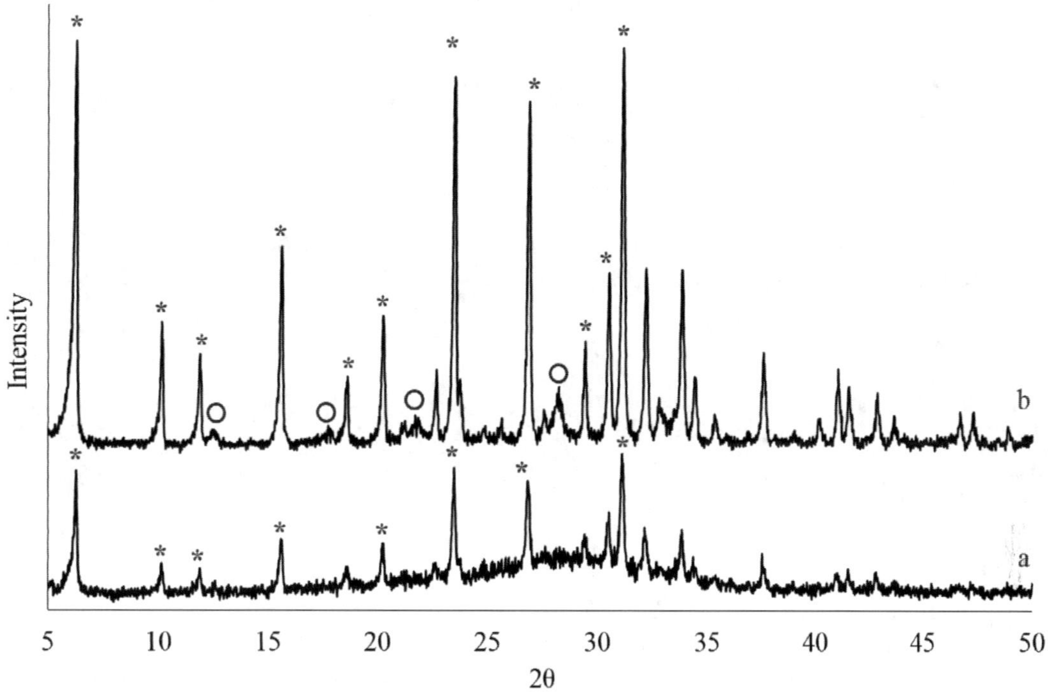

Figure 1. XRD patterns of zeolite samples synthesized from solution aged for: (a) 24 hours and (b) 72 hours. (o: NaP zeolite, *: NaX zeolite).

These results are consistent with previous studies reported by Oyinade et al, which suggest that ageing is a significant factor in the nucleation process of zeolite NaX [9]. The longer the ageing duration, the higher the crystallinity of the sample. The ageing process enables the control of crystal size, acceleration of crystallization and reduction of the formation of impurity phases [10–14]. It was observed that sufficient ageing leads to rapid crystallization of FAU zeolite and to attain this, a shorter hydrothermal synthesis duration is necessary to maintain the purity and avoid the formation of NaP zeolite [13]. During the ageing process, more siliceous species are formed which consequently merges with the aluminosilicate species to form FAU phase precursors. At longer ageing duration, the dissolution of the sodium silicate is accelerated and incorporated into the high Al-rich aluminosilicate gel to form a pure phase of NaX zeolite [15–17]. Ageing encourages the self-generation of structured elements (crystal proto-nuclei) in comparison to gels, which were prepared and immediately heated for hydrothermal synthesis without ageing [11].

The ageing process gradually releases the siliceous species to evolve and transform the secondary building units (SBUs) from zeolites with small membered rings to larger structures [13,14]. Bondareva reported that longer initiation time is needed for the formation of NaP zeolite due to its slow rate of nucleation as compared to NaX zeolite [18]. This explains the formation of NaP zeolite at 72 hours ageing duration.

The FESEM images of the samples are shown in Figure 2. Based on Figure 2, both samples demonstrate crystals in the shape of octahedral, resembling FAU zeolite framework. For the sample synthesized using synthesis solution aged at 24 hours (Figure 2(a)), both amorphous and crystalline phased were observed, where the NaX crystals were between 4.7 to 5.5 μm in size. On the other hand, referring to Figure 2(b), NaP zeolite crystals in the form of spherelutic particles and the NaX crystals with 3.0 – 4.5 μm in size [19] were obtained in the sample synthesized using synthesis solution aged at 72 hours. This finding explains that longer ageing duration produces smaller particle sizes and is in agreement with the ideas of Ogura et al and Ginter et al [13,16]. The FESEM results are in consistent with the XRD results shown in Figure 1.

Figure 2. FESEM images of samples synthesized using synthesis solution aged for: (a) 24 hours and (b) 72 hours.

3.2. Effects of Hydrothermal Synthesis Parameters

In order to study the effect of hydrothermal synthesis temperature on the formation of NaX zeolite, the synthesis solution was aged for 4 hours and hydrothermal synthesis duration was fixed at 12 hours. Meanwhile, the hydrothermal synthesis temperatures used were varied from 80 to 100 °C. Referring to Figure 3(a-b), it can be observed that the samples synthesized at 80 and 90 °C for 12 hours exhibit NaX zeolite structure with low crystallinity and presence of amorphous phase. Due to the resulting amorphous phase, further investigation was done where zeolite samples were prepared at 90 °C and 100 °C with longer hydrothermal synthesis duration (24 hours). Figure 3(c) shows that the crystallinity of the sample improving significantly and the XRD peaks show high crystallinity and pure NaX zeolite for 90 °C. Elemental analysis for the sample synthesized at 90 °C for 24 hours was done using EDS. The Si/Al ratio was 1.45, which is within the range of NaX zeolite. Figure 3(d) shows that the samples obtained from hydrothermal synthesis conducted at 100 °C resulted in a mixture with NaX zeolite as the main phase and NaP zeolite as the minor phase.

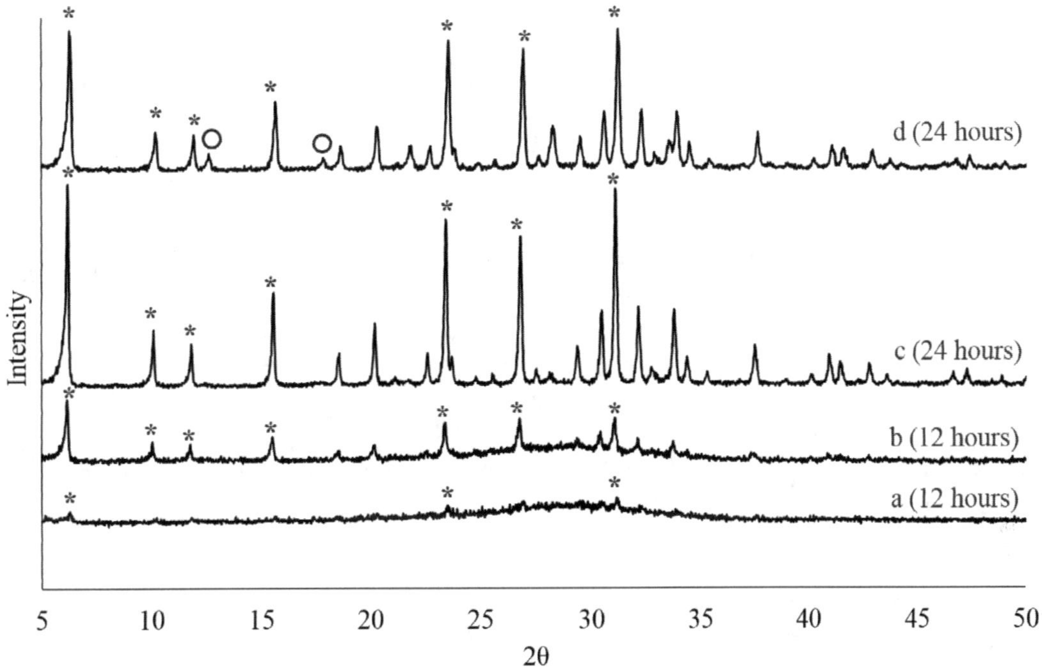

Figure 3. XRD patterns of samples synthesized at different hydrothermal synthesis temperatures of: (a) 80 °C, (b) 90 °C, (c) 90 °C and (d) 100 °C (o: NaP zeolite, *: NaX zeolite).

The effect of hydrothermal synthesis temperature on zeolite formation is observed in this study with the gradual increment in crystallization phase, which agrees with reported literature. The amorphous phase of a zeolite sample signifies that the shorter hydrothermal synthesis duration was insufficient to form full crystalline phases. At higher hydrothermal synthesis temperature, the crystallization rate of NaX zeolite increases and the structure tend to form dense zeolite structures, such as NaP zeolite [1,20]. The hydrothermal synthesis temperature majorly affects the nucleation process and crystallization process of zeolite formation. The energy contained from the high reaction temperature increases the nucleation and crystallization rate, thus shorter hydrothermal synthesis duration is required [21].

To study the effect of hydrothermal synthesis duration, samples were aged at room temperature for 4 hours and treated at 90 °C. The XRD patterns of samples synthesized at 8 to 24 hours are shown in Figure 4. From the patterns, it is apparent that the samples obtained below the synthesis duration of 14 hours demonstrate low crystallinity, which signifies that the hydrothermal synthesis duration to achieve NaX zeolite was insufficient. Hydrothermal synthesis duration of 8 hours produced an amorphous phase zeolite sample. Weak peaks relating to the NaX zeolite structure start to emerge at 12 hours of hydrothermal synthesis. At 14 hours of hydrothermal synthesis durations, the percentage crystallinity of the samples increased and sharp XRD peaks was achieved. Upon longer hydrothermal synthesis duration (20 to 24 hours), the gel crystallized into a mixture of zeolites: NaX and NaP as minor peaks of zeolite NaP are observed. These results are consistent with the results reported by Zhang et al [21].

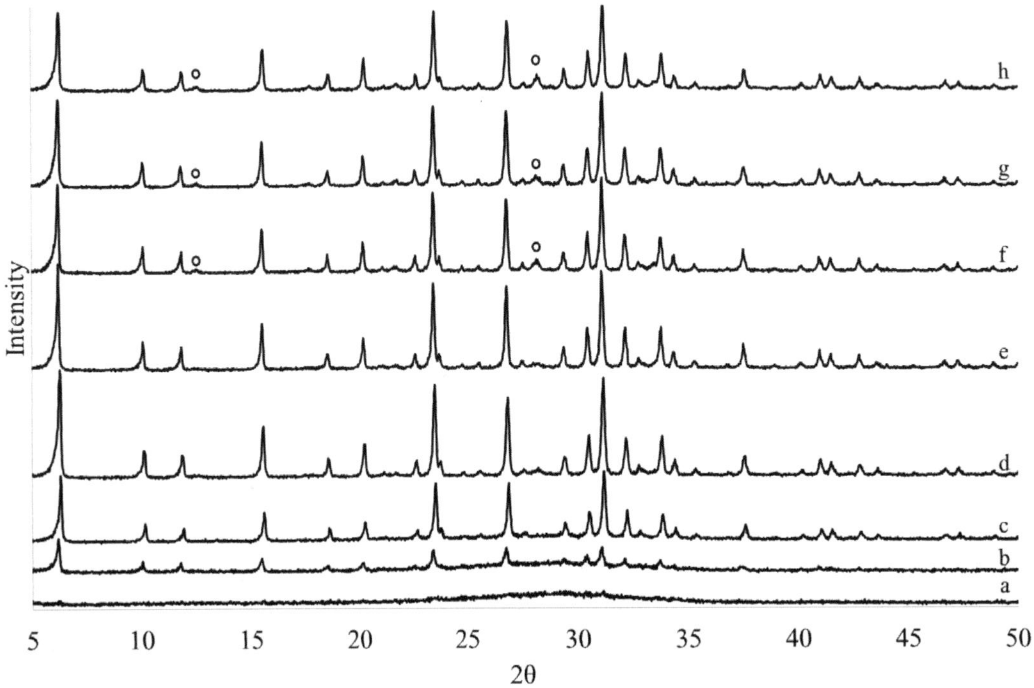

Figure 4. XRD patterns of samples synthesized at different hydrothermal synthesis duration (a) 8 hrs, (b) 12 hrs, (c) 14 hrs, (d) 15 hrs, (e) 16 hrs, (f) 20 hrs, (g) 22 hrs and (h) 24 hrs at synthesis temperature of 90 °C (o: NaP zeolite,)

In this study, the hydrothermal synthesis duration of 16 hours resulted in high purity NaX zeolite. From elemental analysis using EDS, the Si/Al ratio for the samples synthesized at 16 hour hydrothermal synthesis duration was 1.46, confirming the formation of NaX zeolite. Overall, these results indicate that exposing the zeolite samples at longer durations higher than 16 hours resulted in the formation of impurities (NaP zeolite).

The morphology of the zeolite samples synthesized at different durations were further analysed using field emission scanning electron method, and their FESEM images are shown in Figure 5. Referring to Figure 5(a), samples obtained at 14 hours demonstrates loosely packed crystals with the presence of amorphous phase. Formation of octahedral crystals of NaX zeolite was visible for the sample synthesized at 15 and 16 hours, as shown in Figure 5(b) and (c). Further synthesis duration beyond 16 hours ignited the growth of NaP zeolite, which resemble the shape of spherelutic particles (Figures 5(d), (e) and (f)). Beyond the hydrothermal synthesis duration of 16 hours, the zeolite morphology of the samples shows mixture of spherical aggregates and octahedral crystals [19].

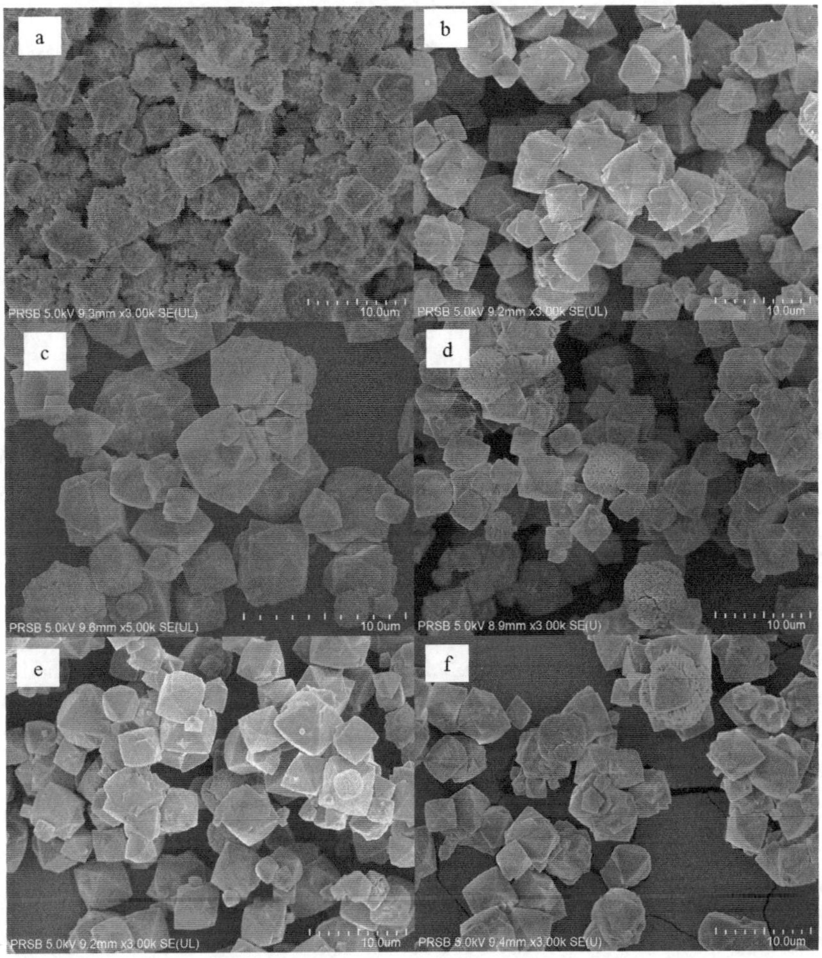

Figure 5. FESEM images of samples obtained at different hydrothermal synthesis durations of: (a) 14 hours, (b) 15 hours, (c) 16 hours, (d) 20 hours, (e) 22 hours and (f) 24 hours at synthesis temperature of 90 °C

4. Conclusions

In this work, pure NaX zeolite can be synthesized using sodium silicate and sodium aluminate by adjusting the ageing and hydrothermal synthesis parameters. Hydrothermal synthesis temperature of 90°C and synthesis duration of 16 hours were identified as the optimum synthesis parameters for producing high crystalline, single phase NaX zeolite. Longer ageing duration and longer hydrothermal synthesis duration results in high crystalline product with potential impurities phases, while higher hydrothermal synthesis temperature produced pure NaX zeolite at shorter hydrothermal synthesis duration. Further work will be focusing on the growth of zeolite membrane for the separation of CO_2 and N_2 from natural gas using NaX zeolite seed prepared using the optimized parameters in this study.

Acknowledgements

The financial support provided by GR&T PETRONAS Research Grant (cost center: 0153C-026) and the technical support from CO_2 Research Centre (CO2RES), Institute of Contaminant Management, Universiti Teknologi PETRONAS are highly appreciated.

References

[1] Gu X, Dong J, Nenoff T M, Gu X, Dong J and Nenoff T M 2005 Synthesis of Defect-Free FAU-Type Zeolite Membranes and Separation for Dry and Moist CO2/N2 Mixtures *Ind. Eng. Chem. Res.* **44** 937–44

[2] White J C, Dutta P K, Shqau K and Verweij H 2010 Synthesis of ultrathin zeolite Y membranes and their application for separation of carbon dioxide and nitrogen gases. SI *Langmuir* **26** 10287–93

[3] Rufford T E, Smart S, Watson G C Y, Graham B F, Boxall J, Diniz da Costa J C and May E F 2012 The removal of CO2 and N2 from natural gas: A review of conventional and emerging process technologies *J. Pet. Sci. Eng.* **94–95** 123–54

[4] Kita H, Fuchida K, Horita T, Asamura H and Okamoto K 2001 Preparation of Faujasite membranes and their permeation properties *Sep. Purif. Technol.* **25** 261–8

[5] Breck D W 1973 *Zeolite Molecular Sieves: structure, chemistry and use* (New York: John Wiley & Sons)

[6] Grand J, Awala H and Mintova S 2016 Mechanism of zeolites crystal growth: new findings and open questions *CrystEngComm* **18** 650–64

[7] Hasegawa Y, Tanaka T, Watanabe K and Jeong B 2002 Separation of CO2-CH4 and CO2-N2 Systems Using Ion-exchanged FAU-type Zeolite Membranes with Different Si/Al Ratios *Korean J. Chem. Eng.* **19** 309–13

[8] Treacy M M J and Higgins J B 2001 *Collection of simulated XRD powder patterns for zeolites* (Elsevier)

[9] Oyinade A, Kovo A S and Hill P 2016 Synthesis, characterization and ion exchange isotherm of zeolite y using Box-Behnken design *Adv. Powder Technol.* **27** 750–5

[10] Karami D and Rohani S 2009 Synthesis of pure zeolite Y using soluble silicate, a two-level factorial experimental design *Chem. Eng. Process. Process Intensif.* **48** 1288–92

[11] Cundy C S and Cox P A 2003 The hydrothermal synthesis of zeolites: History and development from the earliest days to the present time *Chem. Rev.* **103** 663–701

[12] Bo W and Hongzhu M 1998 Factors affecting the synthesis of microsized NaY zeolite *Microporous Mesoporous Mater.* **25** 131–6

[13] Ogura M, Kawazu Y, Takahashi H and Okubo T 2003 Aluminosilicate species in the hydrogel phase formed during the aging process for the crystallization of FAU zeolite *Chem. Mater.* **15** 2661–7

[14] Fan W, Shirato S, Gao F, Ogura M and Okubo T 2006 Phase selection of FAU and LTA zeolites by controlling synthesis parameters *Microporous Mesoporous Mater.* **89** 227–34

[15] Zhu G, Li Y, Chen H, Liu J and Yang W 2008 An in situ approach to synthesize pure phase FAU-type zeolite membranes: Effect of aging and formation mechanism *J. Mater. Sci.* **43** 3279–88

[16] Ginter D M, Bell A T and Radke C J 1992 The effects of gel aging on the synthesis of NaY zeolite from colloidal silica *Zeolites* **12** 742–9

[17] Lutz W 2014 Zeolite Y: Synthesis, modification, and properties - A case revisited *Adv. Mater. Sci. Eng.* **2014**

[18] Bondareva G V., Rat'ko A I and Azarov S M 2003 Hydrothermal synthesis of zeolite NaX on porous ceramic supports *Inorg. Mater.* **39** 716–21

[19] Rees L V C 2001 Zeolite P *Verified Syntheses of Zeolitic Materials* ed H Robson pp 169–70

[20] Yang S, Vlessidis A G and Evmiridis N P 1997 Influence of Gel Composition and Crystallization Conditions on the Conventional Synthesis of Zeolites *Ind. Eng. Chem. Res.* **36** 1622–31

[21] Zhang X, Tang D, Zhang M and Yang R 2013 Synthesis of NaX zeolite: Influence of crystallization time, temperature and batch molar ratio SiO2/Al2O3 on the particulate properties of zeolite crystals *Powder Technol.* **235** 322–8

IOP Conf. Series: Materials Science and Engineering **458** (2018) 012003 doi:10.1088/1757-899X/458/1/012003

Chemical structure, water absorbency and thermal properties of poly(acrylamide-*co*-acrylic acid)-*grafted*-poly(styrene-*co*-methyl methacrylate) "raspberry" – shape like structure microgels

Ros Azlinawati Ramli[1], Shahrir Hashim[2], Waham Ashaier Laftah[3]

[1] Material Technology Program, Faculty of Industrial Sciences & Technology, Universiti Malaysia Pahang (UMP), 26300 Gambang, Kuantan, Pahang.
[2] Department of Polymer Engineering, Faculty of Chemical Engineering, Universiti Teknologi Malaysia (UTM), 81310 Skudai, Johor, Malaysia.
[3] College of Oil and Gas Engineering, Basrah University for Oil and Gas, 61004 Basrah, Iraq.

Email: azlinawati@ump.edu.my

Abstract. This contribution presents the characterization of poly(acrylamide-*co*-acrylic acid)-*grafted*-poly(styrene-*co*-methyl methacrylate) [poly(AAm-*co*-AAc)-*g*-poly(St-*co*-MMA)] The chemical structure of poly(AAm-*co*-AAc)-*g*-poly(St-*co*-MMA) microgels was proposed. The effect of MBA contents on the degree of cross-linking (%), water absorbency (EWC, %), glass transition temperature, T_g and thermal degradation were studied. The degree of cross-linking (%) and EWC, % increased at 1.5 to 1.675 wt.% of MBA might be due to mono cross-links of MBA. The fluctuation of T_g and melting temperature (T_m) from 1.85 to 2.2 wt.% of MBA might be due to di, poly and cyclic cross-links of MBA. Decomposition of poly(AAm-*co*-AAc)-*g*-poly(St-*co*-MMA) was slightly affected by increasing in MBA contents from 1.5 to 2.2 wt.%. T_g of poly(AAm-*co*-AAc)-*g*-poly(St-*co*-MMA) was decreased at 1.5 to 1.675 wt.% of MBA due to increase in ordered structure and crystallinity content, which led to an increase in T_m.

Keywords. Microgels; Cross-linking; Emulsion; Core-shell; Network

1. Introduction

Microgels have become a very attractive research area combining at the same time fundamental and applied topics of great interest. Interest in microgels comes from their unique blend of properties with consistent multi useful aspects of conventional macrogels and colloidal dispersion [1]. Because of the versatility of these materials, there has been a growing interest in a development of more advanced architecture, hence are multifunctional. The advanced architecture is core-shell morphology, which offers potential applications in many fields, including drug delivery, controlled release, and sensors [2]. The microgels systems might combine the characteristics and properties of both shell and core. Furthermore, the surface properties of the hydrogels shell are translated to the core, imparting new functionality to the microgels [3].

Consequently, a fundamental understanding of the mechanism leading to the formation of the polymeric chains is very important to develop product design and manufacture. Cross-linker is multifunctional monomer normally used in free-radical polymerization to construct network structure [4]. The concentration of cross-linker is mainly depending on polymerization technique with lower cross-linker content need to be used in emulsion polymerization than in bulk and solution polymerization. Furthermore, the semi-batch process used a lower concentration of cross-linkers than a batch process. In emulsion polymerization, cross-linker is used to control the particle morphologies and enhancement the product mechanical properties [5]. At present, the network structure within these microspheres is not well understood, and a precise control of the cross-linked network structure seems to be a formidable task except controlling the overall cross-linking density by changing the amount of cross-linking agent [6]. Thus, it is important to measure the optimum amount of cross-linker used and study the structure-properties relationship.

Degradation can affect the deterioration in polymer properties where it can be caused by heat, stress, and radiation due to a breakdown of its chemical structure [7]. The thermal gravimetric analysis is a simple and accurate method to determine polymers thermal stability. Knowledge of degradation and mode of decomposition under the influence of heat is highly recommendable in the processing and fabrication procedures. The threshold decomposition temperature indicates the highest fabrication temperature. Cross-linked networks have better thermal properties than corresponding homopolymers. This is usually, attributed to the networking of the phases [8].

Previously, the research of synthesis, characterization and morphology study of poly(AAm-*co*-AAc)-*g*-poly(St-*co*-MMA) "raspberry"- shape like structure microgels have been carried out. The evidence of the "raspberry"- shape like structure core microgels via electron microscopy have been shown [9]. In the present work, the effect of the cross-linker amount on cross-linked network structure, water absorbency and thermal properties of poly(AAm-*co*-AAc)-*g*-poly(St-*co*-MMA) microgels were investigated.

2. Materials and Methods

2.1. Materials

Acrylamide (AAm, 99%), methyl methacrylate (MMA, 99%), acrylic acid (AAc, 99%) and styrene (St, 99%) were used as monomers. *N,N'*-methylenebisacrylamide (MBA, 99%) was used as cross-linker. sodium dodecyl sulfate, (SDS, 99%), sorbitan monooleate (span 80) and octylphenol ethylene oxide, (Triton X-100) were used as surfactants. Potassium persulphate, (KPS, 99%) was used as the initiator. All of the materials were purchased from MERCK, Germany and used as received.

2.2. Microgels preparation

Poly(AAm-*co*-AAc)-*g*-poly(St-*co*-MMA) microgels were prepared via pre-emulsified semi-batch emulsion polymerization as reported in previous work [9].

2.3. Characterization

2.3.1. Degree of cross-linking (%)

The degree of cross-linking (%) during the polymerization was carried out using an extraction method based on the polymer mass that remained after extraction with dimethylformamide (DMF). A pre-determined weight of dried microgel samples (0.3 ± 0.05 g) were immersed in DMF for a period of 15 hours at room temperature. The samples were filtered and dried in an oven at 80°C until a constant weight was obtained to ensure that all of the solvents was vaporized. The degree of cross-linking of the microgels films based on the total emulsion solids was calculated using Equation 3.1:

$$\text{Degree of cross-linking (\%)} = (W_1 / W_0) \times 100 \qquad (3.1)$$

Where, W_1 and W_0 are the weight of dry microgels film after and before extraction, respectively [10].

2.3.2. Water Absorbency

The dried sample (0.3 ± 0.05 g) was immersed in 200 ml of distilled water at room temperature for a period of 24 hours. The equilibrium water content (EWC) was calculated using equation 3.2 [11]:

$$\text{EWC (\%)} \quad = \quad [(w_s - w_d) / w_s)] \times 100 \qquad (3.2)$$

Where w_s, is the weight of the swollen sample and w_d is the weight of the dry sample after swelling. All of the experiments were carried out five times repeatedly and the average values were reported.

2.3.3. Differential Scanning Calorimeter (DSC)

DSC was used to determine glass transition temperature, T_g of the microgels. DSC assessments were carried out according to ASTM D 3418 using Perkin Elmer DSC 7 (USA), at a heating rate of 10°C/min in the range of 30 to 500°C. Dried microgels film with the weight of 7-10 mg was placed in a sealed aluminum pan, and an empty aluminum pan was used as a reference.

2.3.4. Thermogravimetric Analysis (TGA)

TGA was used to determine the thermal degradation of the microgels film. TGA assessments were carried out according to ASTM E 1131 using Perkin Elmer TGA 7 (USA). The dried microgels film was scanned from 30-950°C, at a heating rate of 10°C/min. Each film was placed in an open platinum pan where nitrogen was used together as the purge gas at a flow rate of 20 mL/min. Onset temperature of a 10% and 75% weight loss deviation from the baseline of the thermogravimetric (TG) curve were used as an indicator for resulted products thermal stability.

3. Results and discussion

3.1. Proposed route

Figure 1 shows the proposed route to the formation of cross-linked network structure of poly (AAm-*co*-AAc)-g-poly (St-*co*-MMA) microgels based on mono cross-links of MBA. The polyfunctionality of MBA cross-linker which has four reactive sites that can be linked both sides to the poly(AAm-*co*-AAc)-*g*-poly(St-*co*-MMA) to form three dimensional (3D) network [12]. During emulsion polymerization persulfate radicals anion were formed and react with monomers dissolved in the aqueous phase to produce charged oligomers [13]. The polymerization was proceed in micelles and poly (AAm-*co*-AAc) and poly (St-*co*-MMA) were formed. Poly (AAm-*co*-AAc) is hydrophilic and poly (St-*co*-MMA) is hydrophobic. The difference in hydrophilicity between poly (AAm-*co*-AAc) and poly (St-*co*-MMA) led to form poly (AAm-*co*-AAc) seeds at an early stage of polymerization. After that, poly (AAm-*co*-AAc) particles were grafted with poly (St-*co*-MMA) core, adhere, and then coagulate to form the shell. The original seeds of poly (AAm-*co*-AAc) deformed from its original shape of a sphere into an outer shell then uniformly cross-linked with MBA. Thus, at a certain weight ratio of poly (AAm-*co*-AAc)-g-poly (St-*co*-MMA), microdomains have grown in size and joined together to form a ''raspberry''-shape like structure core. Finally, these microdomains coalesce together to complete the shell formation of the ''raspberry'' particles [9].

Cross-linker acting as a bridge between the polymer chains by forming an equivalent bond from its reactive sites. Mono, di, poly, and cyclic cross-links might be formed inside the network structures of poly (AAm-*co*-AAc)-g-poly (St-*co*-MMA) microgels as illustrated in figure 2 Mono, di and poly cross-links occur between two different chains (intermolecular cross-linking). Furthermore, cyclizations occur in the single chain (intramolecular cross-linking). Primary cyclization (Mz_1), occurring between the same growing primary chains, creating loops or cycles. Secondary cyclization (Mz_2) occurring between the same macromolecule but belonging to a different primary chain [5].

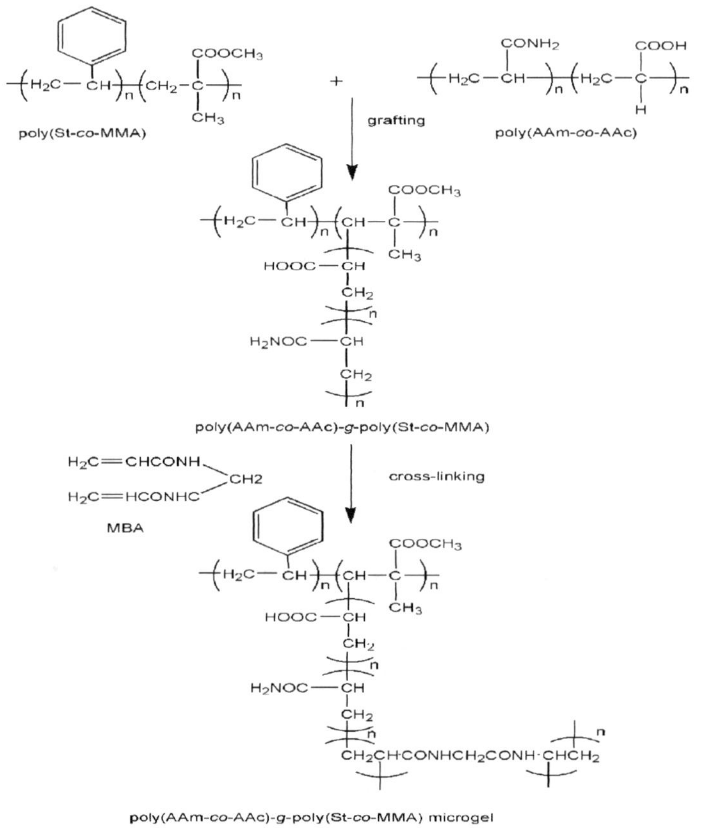

Figure 1. The route to the formation of cross-linked network structure of poly (AAm-*co*-AAc)-*g*-poly (St-*co*-MMA) microgels.

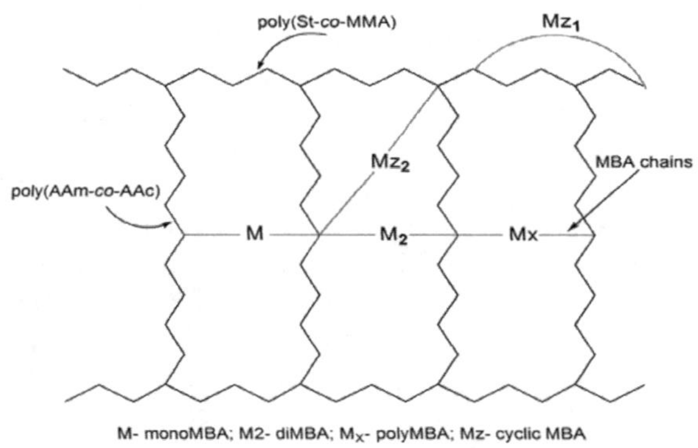

Figure 2. Cross-linked network structure formed by mono, di, poly and cyclic MBA cross-links.

3.2. Degree of cross-linking (%) and Water Absorbency

Figure 3 shows the degree of cross-linking and EWC, % as a function of MBA amount for every batch of poly(AAm-*co*-AAc)-*g*-poly(St-*co*-MMA) microgels. It is clear that the cross-linking degree shared the same trend with EWC, %. Maximum EWC, 72.6 % was obtained at 1.675 wt.% of MBA, and it is decreased to 63.9 % at 1.85 wt.% of MBA. Decreasing in EWC, % at 1.85 wt.% of MBA might be as a result of the formation of di and poly cross-links, network space gets diminished and less water can trap inside the microgels [14]. Moreover, further increases of MBA amount at 2.025 wt.% led to increasing EWC to 69 % due to the enhancement of the 3D network form which increase the amount of trapped water. However, further increases of MBA amount at 2.2 wt.%, produced tighter polymer network with decreasing the pores size and preventing the water to diffuse into the network structure and thus reduces the EWC, %. The water absorbency of the microgels film is the evidence that the hydrogels properties of poly(AAm-*co*-AAc) shell were translated to the poly(St-*co*-MMA) hydrophobic core.

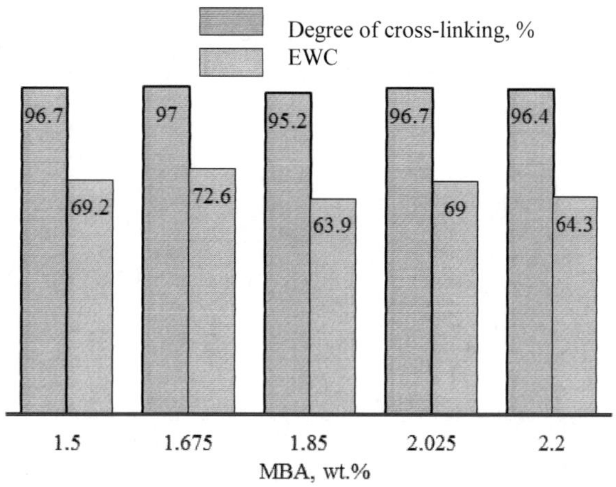

Figure 3. Cross-linking degree (%) and equilibrium water content (EWC, %) of poly(AAm-*co*-AAc)-*g*-poly(St-*co*-MMA) as a function of MBA contents.

3.3. *Differential Scanning Calorimeter (DSC) Analysis*

Figure 4 shows DSC curves of poly(AAm-*co*-AAc)-*g*-poly(St-*co*-MMA). Two distinct glass transitions temperature, T_g peaks were observed, and that the curve returns essentially to baseline values in between the two peaks is the evidence that the two polymers are highly phase separated copolymers, this result are similar to work reported by Stubbs [15]. The first transition representing the core of the microgels, poly(St-*co*-MMA) and the second (T_{g2}) transition for the poly(AAm-*co*-AAc) shell. The results show that the of T_g of poly(AAm-AAc) (161°C) is higher than poly(St-*co*-MMA) (103°C). The value of both T_g is higher than theoretical T_g due to the cross-links structure of the microgels particles. It is well known that introducing cross-linkages into a polymer raises the glass temperature [16]. In addition, there is a decrease in the specific volume, since introducing cross-linkages involves the exchange of Van der Waals' bonds for shorter, more compact primary bonds. Thus, the process of cross-linking should affect the segmental mobility and the related properties in a manner similar to the effect of an increase in the molecular weight of the primary chains [17]. Furthermore, high T_g of poly(St-*co*-MMA) might be due to the thermal properties of poly(AAm-*co*-AAc) hydrogels shell were translated to the core. For each polymer phase, the change in the heat capacity across the T_g is representative of the amount of that phase [18]. The third transition shows the melting points (T_m) of the film sample. The results in Table 1 indicated that T_{g1} and T_{g2} decreased and T_m increased with the increase in MBA amount from 1.5 wt.% to 1.675 wt.%. This might be due to a homogenous network of pores formed by mono cross-links led to enhance the ordered structure and crystallinity content consequently increases the T_m from 390 to 430°C. However, further increases of MBA amount from 1.85 to 2.2 wt.%, which results in fluctuation of T_g and T_m due to inefficient cross-linking which affected the 3D network. Furthermore, the formation of di, poly, and cyclic cross-links caused to produce heterogeneous network of pores which affected the thermal properties.

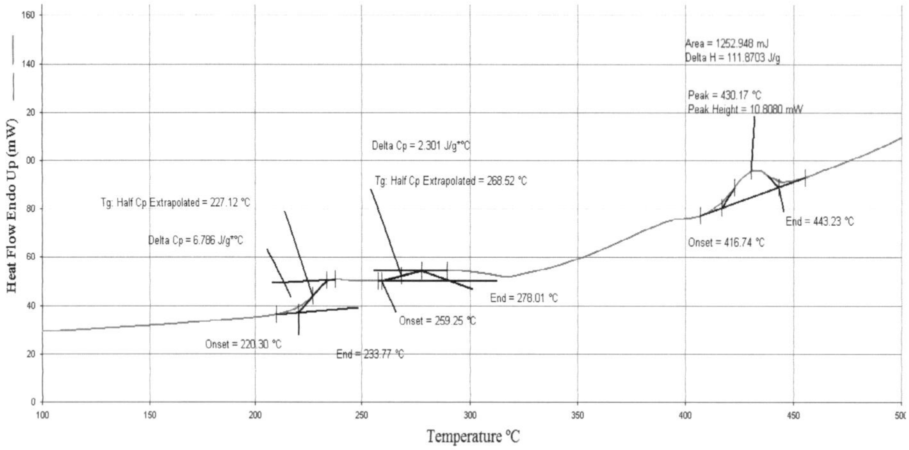

Figure 4. DSC trace for poly(AAm-*co*-AAc) hydrogels shell and poly(St-*co*-MMA) hydrophobic core.

Table 1. Glass transition temperatures, T_g and crystalline melting points, T_m of microgels with varying amount of MBA cross-linker.

MBA (pphm)	Exothermic		
	T_{g1} (°C)	T_{g2} (°C)	T_m (°C)
1.500	235	270	390
1.675	227	269	430
1.850	238	266	430
2.025	228	297	385
2.200	226	292	431

3.4. Thermal Gravimetric Analysis (TGA)

TGA data of poly(AAm-*co*-AAc)-g-poly(St-*co*-MMA) microgels with varying amount of MBA cross-linker are shown in Table 2. The data showed 10% (T_d), 75% ($T_{75\%}$) and the maximum ($T_{\max}$) weight loss temperature and their char residue at 600 °C. Generally, the results indicated high temperatures of $T_{10\%}$ for all samples. Increasing in $T_{10\%}$ weight loss is an indication of enhancement in thermal stability [19]. Increasing in $T_{10\%}$ from 219 °C to 288 °C is due to the increase in MBA amount. When MBA amount was increased, more cross-linking points were generated to bond with polymer molecules. Furthermore, the intramolecular hydrogen bond within MBA units also increased consequently more heat was needed to break the structures.

In all cases, poly(AAm-*co*-AAc)-*g*-poly(St-*co*-MMA) microgels showed four steps of decomposition as shown in derivative thermal gravimetric (DTG) curves in figure 5 The first step below 100 °C is due to the removal of adsorbed water molecules [20]. The second step is about 249 to 260 °C might be due to the destruction of carbonyl groups (e.g., amide, carboxyl) [20, 21]. The third step is attributed to the main chain pyrolysis beginning at about 290 to 330 °C with the evolutions of aromatics from the degradation of styrene. Polystyrene degrades into a mixture of styrene (40%), toluene (2.4%), methyl styrene, etc. above 300°C [8, 20]. The last step weight loss was corresponding to degradation of the polymer backbones at about 404 to 445 °C [21]. At 600 °C, residue of poly(AAm-*co*-AAc)-*g*-poly(St-*co*-MMA) with varying contents of MBA cross-linker is about 10.12 to 17.55%. Depending on the mentioned results can be concluded that increasing in MBA amount slightly affected the decomposition of poly(AAm-*co*-AAc)-*g*-poly(St-*co*-MMA) microgels.

Table 2. TGA data of the poly(AAm-*co*-AAc)-*g*-poly(St-*co*-MMA) with varying amount of MBA.

MBA (pphm)	Temperature (°C)		Max wt. loss	T_{max}	Residue at 600 °C
	$T_{10\%}$	$T_{75\%}$	(%)	(°C)	(%)
1.500	219	458	71.06	385.16	17.55
1.675	248	454	74.65	377.43	14.47
1.850	255	449	76.14	410.82	10.12
2.025	258	455	75.15	387.40	15.64
2.200	288	458	77.53	399.16	16.91

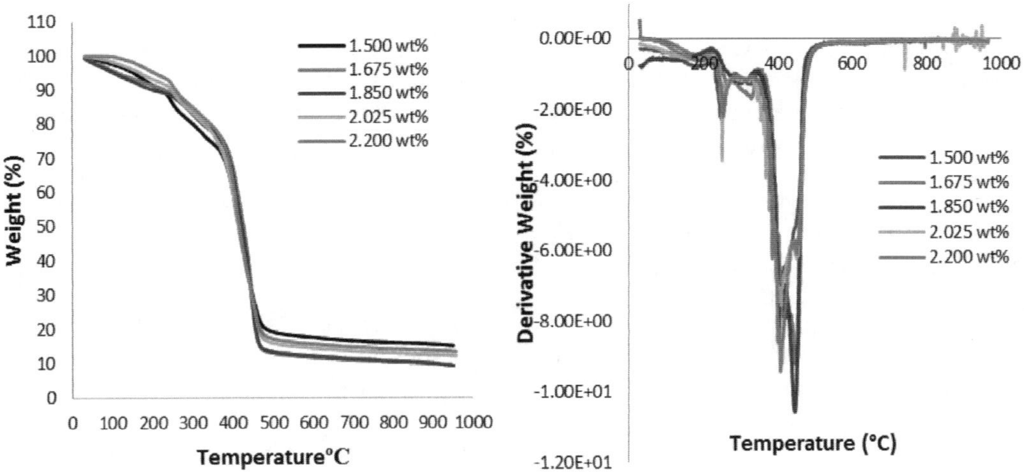

Figure 5. (A) TG Curves, and (B) DTG curves of poly(AAm-*co*-AAc)-*g*-poly(St-*co*-MMA) microgels showed four steps of decomposition.

4. Conclusions

Poly(AAm-*co*-AAc)-*g*-poly(St-*co*-MMA) "raspberry"- shape like structure microgels with varying contents of MBA cross-linker were synthesized via pre-emulsified semi-batch emulsion polymerization. Variation of the cross-linker concentration provides good understanding relationships between the cross-linked structure and properties. The formation of cross-linked network structure of poly(AAm-*co*-AAc)-*g*-poly(St-*co*-MMA) microgels was proposed based on mono cross-links of MBA. 1.675 wt.% of MBA was suggested as the optimum amount of the cross-linker used. The maximum degree of cross-linking, % and EWC, % was achieved at 97 and 72.6% respectively. The water absorbency of the microgels film and high T_g of poly(St-*co*-MMA) core is the evidence that the properties of the hydrogels shell were translated to the hydrophobic core. Such structure-properties relationships are important for the design of future systems to use in potential applications such as medical devices coating.

Acknowledgment
The authors would like to thank to Universiti Malaysia Pahang and Universiti Teknologi Malaysia for the financial support under University Grant (No. RDU1703286).

References.

[1] Ramli, R.A., W.A. Laftah, and S. Hashim, Core–shell polymers: a review. Rsc Advances, 2013. 3(36): p. 15543-15565.
[2] Ramli, R.A., Hollow polymer particles: a review. RSC Advances, 2017. 7(83): p. 52632-52650.
[3] Hendrickson, G.R., et al., Design of Multiresponsive Hydrogel Particles and Assemblies. Advanced Functional Materials, 2010. 20(11): p. 1697-1712.
[4] Laftah, W.A., S. Hashim, and A.N. Ibrahim, Polymer Hydrogels: A Review. Polymer-Plastics Technology and Engineering, 2011. 50(14): p. 1475-1486.
[5] Bouvier-Fontes, L., et al., Seeded Semicontinuous Emulsion Copolymerization of Butyl Acrylate with Cross-Linkers†. Macromolecules, 2005. 38(4): p. 1164-1171.
[6] Tobita, H., Crosslinking kinetics in emulsion copolymerization. Macromolecules, 1992. 25(10): p. 2671-2678.
[7] Shah, A.A., et al., Biological degradation of plastics: A comprehensive review. Biotechnology Advances, 2008. 26(3): p. 246-265.
[8] Mathew, A.P., S. Packirisamy, and S. Thomas, Studies on the thermal stability of natural rubber/polystyrene interpenetrating polymer networks: thermogravimetric analysis. Polymer Degradation and Stability, 2001. 72(3): p. 423-439.
[9] Ramli, R.A., S. Hashim, and W.A. Laftah, Synthesis, characterization, and morphology study of poly(acrylamide-co-acrylic acid)-grafted-poly(styrene-co-methyl methacrylate) "raspberry"-shape like structure microgels by pre-emulsified semi-batch emulsion polymerization. Journal of Colloid and Interface Science, 2013. 391(0): p. 86-94.
[10] Xie, F., Z.-H. Liu, and D.-Q. Wei, Curing Kinetic and Properties of Acrylic Resin Cured with Aziridine Crosslinker. Chinese Journal of Polymer Science, 2002. 20(1): p. 65-70.
[11] Rodkate, N. and M. Rutnakornpituk, Multi-responsive magnetic microsphere of poly (N-isopropylacrylamide)/carboxymethylchitosan hydrogel for drug controlled release. Carbohydrate Polymers, 2016.
[12] Singh, B. and N. Chauhan, Modification of psyllium polysaccharides for use in oral insulin delivery. Food Hydrocolloids, 2009. 23(3): p. 928-935.
[13] Young, R.J., and Lovell, P. A. , Introduction to Polymers. 1991: Chapman and Hall: Kluwer Academic Publisher.
[14] Wu, J., et al., Study on starch-graft-acrylamide/mineral powder superabsorbent composite. Polymer, 2003. 44(21): p. 6513-6520.
[15] Stubbs, J.M. and D.C. Sundberg, A round robin study for the characterization of latex particle morphology—multiple analytical techniques to probe specific structural features. Polymer, 2005. 46(4): p. 1125-1138.
[16] Loshaek, S., Crosslinked polymers. II. Glass temperatures of copolymers of methyl methacrylate and glycol dimethacrylates. Journal of Polymer Science, 1955. 15(80): p. 391
[17] Fox, T.G. and S. Loshaek, Influence of molecular weight and degree of crosslinking on the specific volume and glass temperature of polymers. Journal of Polymer Science, 1955. 15(80): p. 371-390.
[18] Sundberg, D.C. and Y.G. Durant, Latex Particle Morphology, Fundamental Aspects: A Review. Polymer Reaction Engineering, 2003. 11(3): p. 379-432.
[19] Yu, Y.-Y. and W.-C. Chen, Transparent organic–inorganic hybrid thin films prepared from acrylic polymer and aqueous monodispersed colloidal silica. Materials Chemistry and Physics, 2003. 82(2): p. 388-395.
[20] Ding, R., et al., Preparation and characterization of polystyrene/graphite oxide nanocomposite by emulsion polymerization. Polymer Degradation and Stability, 2003. 81(3): p. 473-476.
[21] Liu, Y.-L., C.-Y. Hsu, and K.-Y. Hsu, Poly(methylmethacrylate)-silica nanocomposites films from surface-functionalized silica nanoparticles. Polymer, 2005. 46(6): p. 1851-1856.

IOP Conf. Series: Materials Science and Engineering **458** (2018) 012004 doi:10.1088/1757-899X/458/1/012004

Study on the plasticiser type for biobased microcrystalline cellulose filled polylactic acid composite

M M Pang[1], S C Koay[1], J H Low[1], Y F Buys[2], K Y Tshai[3]

[1] School of Engineering, Taylor's University, Lakeside campus, No.1, Jalan Taylor's, 47500 Subang Jaya, Selangor, Malaysia.
[2] Faculty of Engineering, International Islamic University Malaysia, Jalan Gombak, Selangor, Malaysia.
[3] Faculty of Engineering, University of Nottingham Malaysia Campus, Jalan Broga, 43500 Semenyih, Selangor, Malaysia.

E-mail: mingmeng.pang@taylors.edu.my

Abstract. The aim of this work was to study the effect of different type of plasticisers, i.e., polyethylene glycol (PEG) and coconut oil (CO), on the properties of microcrystalline cellulose (MCC) filled polylactic acid (PLA) composite. The sample compounding was carried out via melt blending method. The properties of the plasticised MCC filled PLA composites were characterized by tensile, thermal gravimetric analysis (TGA) and scanning electron microscopy (SEM). It was found that CO showed a better plasticising effect, with improvement in elongation at break by 12% as compared to neat PLA and higher decomposition temperature at 302°C.

1. Introduction

Polylactic acid (PLA) is a biobased polymer derived from a renewable resource and has widely used in many applications such as packaging, consumer items and medical implants [1]. PLA has many advantages such as low toxicity, high mechanical strength, biocompatibility and biodegradable. However, its main drawbacks are its brittleness and poor thermal stability [2]. Many methods have been investigated, such as adding a reinforcement agent or additives, irradiation using electron beam etc to enhance the performance of PLA [3-4].

Microcrystalline cellulose (MCC) is a biodegradable polymer derived from renewable origin such as wood and is often used as reinforcing agent in composites, particularly to improve the stiffness and thermal stability of the material [5-6]. However, previous work showed that the tensile strength and elongation at break decreased with the addition of MCC [6]. Thus, plasticiser is added into PLA composite in order to improve the flexibility and reduce the friction during processing. Apart from it, it helps to enhance its elongation at break [7].

In this study, two types of plasticiser were evaluated: polyethylene glycol (PEG) and coconut oil (CO) were added at the dose of 10 wt%, respectively into the biobased MCC filled PLA composites. The plasticiser was expected to provide plasticising effect via interaction with the polymer chains and enhance the ductility of the composite. Biobased content of the composite was determined via the ASTM D6866. It was reported that generally a minimum 25% biobased content must be fulfilled in order to certify as a biobased product [8].

2. Experimental

2.1. Materials
The PLA 3051D in pellet form was obtained from NatureWork LLC. PEG 400(brand R&M) and MCC (brand Sigma Aldrich), CO (brand ChemSoln) were purchased from Synertec Enterprise Kuala Lumpur.

2.2. Sample preparation
The melt blending process was carried out by using an internal mixer (Haake Polylab System) at 180°C and 50 rpm, with a retention time of 10 min. Prior to melt blending, PLA and MCC were dried in an oven maintained at 70°C for 24 hr to remove the moisture. The dosage of MCC and plasticiser (PEG or CO) were set at 10 wt% each and the composites were designated as PLA, PLA/MCC/PEG, PLA/MCC/CO as shown in table 1. Rectangular composite sheets (12 x 15 cm) with a thickness of 1 mm were produced by using compression molding. The compound was filled in a mold and being compressed by a hot press machine (model: Moore E53) at 185°C. The hot press operation steps included a pre-heating for 8 min, the compression process for 3 min at a pressure of 150 bar and cooling for 8 min. The sheets were cut into dumbbell shape (type IV) for the tensile test.

Table 1. Formulation for composite

Sample	PLA(wt%)	MCC(wt%)	PEG(wt%)	CO(wt%)
PLA	100	0	0	0
PLA/MCC/PEG	80	10	10	0
PLA/MCC/CO	80	10	0	10

2.3. Characterisation

2.3.1. Thermal gravimetric analysis (TGA). The thermal stability of the sample was studied by using the Perkin Elmer Pyris 1(USA). The sample, weighing between 6-8 mg, was heated from 30 to 600°C at a rate of 20°C/min, under nitrogen atmosphere. The onset decomposition temperature was recorded.

2.3.2. Scanning electron microscope (SEM). Field emission SEM (model FEI QUANTA 400 F FESEM) was used to observe the surface and fracture morphology of the specimens. Specimens were coated with a thin layer of gold to prevent electrical charging during the examination.

2.3.3. Biobased carbon content. The biobased carbon content of the sheet sample was determined according to ASTM D6866-16 Method B using the accelerator mass spectrometry method. The required solid sample size was approximately 1 g.

2.4. Polymer Testing

Tensile test. The tensile properties were determined with an Instron model 3366 tester according to ASTM D638 at a cross head speed of 5 mm/min. The average of 5 measurements for the tensile strength (TS), elongation at break (EB) and Young's modulus (YM) were collected.

2.4.1. The TS was calculated by using the force divided by the initial cross-sectional area as shown in Eq.1.

$$TS = \frac{\text{Force}}{\text{Area}} \tag{1}$$

2.4.2. The EB was calculated by dividing the extension at break point by the original length as shown in Eq.2

$$\mathrm{EB} = \frac{\mathrm{Extension}}{\mathrm{Length}} \times 100\% \tag{2}$$

2.4.3. The YM was the ratio of stress and strain as shown in Eq.3

$$\mathrm{YM} = \frac{\text{tensile stress}}{\text{tensile strain}} \tag{3}$$

3. Results and discussion

3.1. Tensile test

Table 2 shows the tensile properties of the tested specimens. The PLA/MCC/CO composite showed deterioration in TS and YM, but improvement in EB. This finding is in accordance with the result reported by Mohapatra et al. [9]. The addition of CO plasticiser reduced the intermolecular forces along the polymer chains, which led to reduction in TS and YM, and at the same time, increased the flexibility as reflected in higher EB. In this study, PEG did not show any improvement as a plasticiser. This could be attributed to the insufficient interaction between the PEG and the MCC filled PLA composite. The previous study by Taib et al [10] reported the presence of PEG might have disturbed the fiber-matrix interaction, which caused the unplasticised PLA/fiber composite exhibited better tensile properties than the plasticised composite. However, CO has shown its great potential in plasticising the MCC filled PLA composite, the EB improved by 12% as compared to neat PLA.

Table 2. Tensile properties

Sample	TS(MPa)	EB(%)	YM(MPa)
PLA	56.2±1.37	22.6±3.78	393±14.1
PLA/MCC/PEG	26.8±2.07	12.2±1.53	349±9.88
PLA/MCC/CO	22.2±0.47	25.4±5.34	346±5.12

3.2. TGA

The thermal degradation data are tabulated in table 3 and the TGA curves are shown in figure 1. The CO plasticised MCC filled PLA exhibited higher thermal stability than its counterpart, PEG. The thermal degradation for PLA/MCC/CO began at 302°C, it also showed improvement compared to neat PLA at 290° C. The improvement showed by CO could be attributed to its higher onset decomposition temperature, 257°C against the PEG at 150°C [11-12]. Lu et al reported high stability was found in virgin CO due to its high content of saturated fatty acids [13].

Table 3. Thermal properties

Sample	Onset decomposition temperature (°C)	Residual weight% at 600°C
PLA	290	0
PLA/MCC/PEG	280	8.59
PLA/MCC/CO	302	5.24

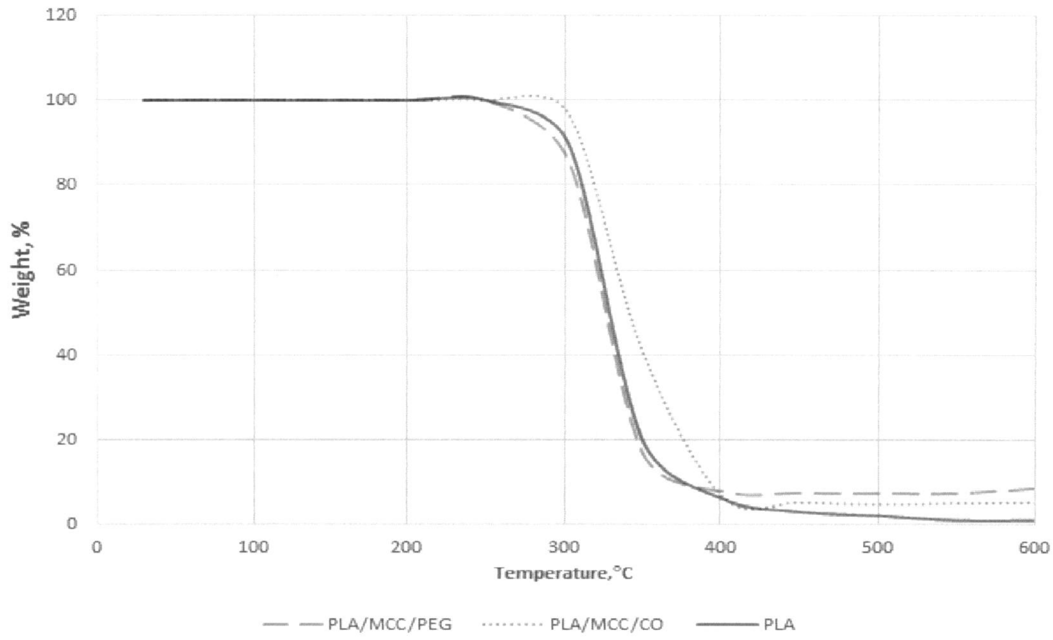

Figure 1. TGA curves

3.3. Morphology study

The SEM micrograph of the plasticised MCC filled PLA is showed in figure 2 (a) for CO and (b) PEG. Refer to figure 2(a), it can clearly see the homogenous fracture surface, indicating CO plasticiser and MCC particle dispersed uniformly in the PLA matrix. However, for the PEG plasticised sample can see the MCC fibril embedded in the PLA matrix (white arrow) and an aggregate of the MCC (white circle). The poor adhesion allows the explanation of poorer performance in the tensile properties for PLA/MCC/PEG. The previous study by Bijarimi et al. [14] revealed that PEG was dispersed as droplets in the PLA matrix with a clear boundary between the PLA and PEG phases, which led to a reduction in tensile strength.

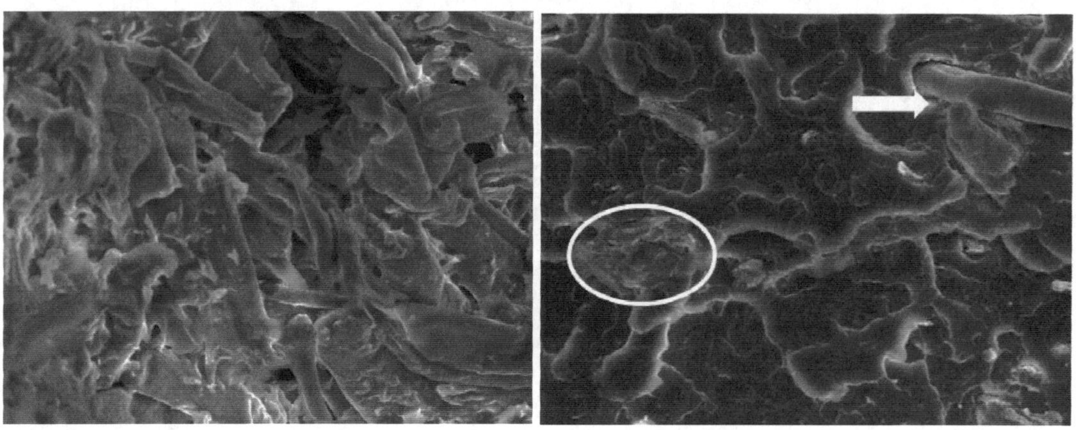

(a) (b)

IOP Conf. Series: Materials Science and Engineering **458** (2018) 012004 doi:10.1088/1757-899X/458/1/012004

Figure 2. SEM fracture surface (a) PLA/MCC/CO (b) PLA/MCC/PEG

3.4. Biobased carbon content

The biobased carbon content for the PLA/MCC/CO was reported as 95±3% as shown in figure 3. This means of all the organic carbons present in the PLA/MCC/CO, 95% is biobased and 5% is petroleum based. Theoretically, all the 3 components (PLA, MCC, and CO) were derived from plant (i.e., sugar cane, wood pulp, coconut) and supposed to have zero fossil carbon in it. The present of non-renewable carbon (~5%) in the PLA/MCC/CO could be attributed to the impurities during the compounding process or fossil-based chemical used to produce the MCC. Anyway, the biobased content up to 95% was sufficient to certify as a biobased product under the intermediates-plastic resins category as per requirement of U.S. Department of Agriculture (USDA) BioPreferred Program.

RESULT:	95 % Biobased Carbon Content (as a fraction of total organic carbon)
Laboratory Number	Beta-487628
Percent modern carbon (pMC)	106.09 +/- 0.31 pMC
Atmospheric adjustment factor (REF)	100.5; = pMC/[1/(100.5/112)]

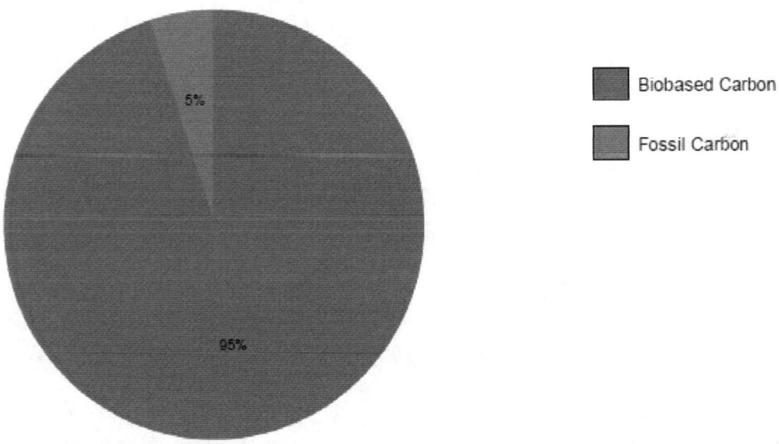

Figure 3. Biobased carbon content for PLA/MCC/CO

4. Conclusions

Plasticiser CO was better than PEG in providing the plasticity effect on the MCC filled PLA. The CO plasticised sample showed improvement in EB by 12% and recorded higher onset decomposition temperature at 302°C as compared to neat PLA at 290°C. SEM micrograph revealed a homogenous mixture of the PLA/MCC/CO. The PLA/MCC/CO biobased carbon content tested was 95% which was meeting the requirement of USDA to be certified as a biobased product.

Acknowledgments

This research was financially supported by the Taylor's University, Malaysia (Grant: TRGS/ERFS/1/2017/SOE/031).

References

[1] Saini P, Arora M and RaviKumar M N V 2016 Adv. Drug Deliv. Rev 107 47-59
[2] Chen Y, Geever L M, Killion J A, Lyons J G, Higginbotham C L and Devine D M 2016 Polym Plast Technol Eng 55 1057-1075
[3] Koay S C, Husseinsyah S and Osman H 2013 Polym Eng Sci 53 1109-1116
[4] Shin B Y, Han D H and Narayan R J 2010 J Polym Environ 18 558-566
[5] Mathew A P, Oksman C and Sain M 2004 J Appl Polym Sci 97 2014–2025
[6] Haafiz M K M, Hassan A, Zakaria Z, Inuwa I M, Islam M S and Jawaid M 2013 Carbohydr Polym 98 139-145
[7] Rahman M and Brazel C S 2004 Prog Polym Sci 29 1223-1248
[8] Clark J H, Hunt A, Topi C, Paggiola G and Sherwood J 2017 Sustainable Solvents: Perspectives from Research, Business and International Policy (UK:Royal Society of Chemisty)
[9] Mohapatra A K, Mohanty S and Nayak S K 2014 Polym Compos 35 283-293
[10] Taib R M, Ramarad S, Ishak Z A M and Todo M 2010 Polym Compos 31 1213-1222
[11] Jayadas N H and Nair K P 2006 Tribol Int 39 873-878
[12] Wang S, Wang Q, Yang X, Wang L and Z H 2003Chin J Polym Sci 21 87-91
[13] Lu H F S and Tan P P 2009 Int Food Res J 16 343-354
[14] Bijarimi M, Ahmad S, Rasid R, Khushairi M A and Zakir M 2016 AIP Conf Proc 1727 020002

Electrochemical behaviour and analysis of Zn-Ni alloy anti-corrosive coatings deposited from citrate baths

S Anwar[1], Y Zhang[1] and *F Khan[1]

[1]Centre for Risk, Integrity, and Safety Engineering Group (C-RISE)
Faculty of Engineering & Applied Science, Memorial University
St. John's, Newfoundland and Labrador, A1B3X5, Canada.

*Corresponding Author E-mail: fikhan@mun.ca; Tel: +1 709 864 8939

Abstract. Zinc-Nickel alloy coatings have been widely used to improve the corrosion resistance of steels. Electrodeposition of zinc-nickel alloy films on a steel plate under various deposition conditions is studied. The effects of plating variables such as bath composition, pH and current density on the coating composition, morphology and corrosion property have been systematically investigated. The zinc and nickel ions in plating baths are stabilized using complexing agent potassium citrate. Increasing the molar concentration of potassium citrate extends the pH of nickel hydroxide and zinc oxide precipitation. The morphological properties and crystalline phase structure of the deposited alloy coatings are examined by scanning electron microscopy (SEM) and X-ray diffraction (XRD). The analyses reveal that the coatings deposited from citrate bath contain γ-$NiZn_3$ and γ-Ni_2Zn_{11} alloy phases. These phases improve the corrosion resistance of the steel plate. Electrochemical behaviour analyses of the Zn-Ni film illustrated by electrochemical impedance spectroscopy (EIS) measurement revealed that the sample (a) deposited from 0.0326M citrate bath at acidic pH shown higher impedance modulus than non-citrate bath coated samples, which improved anti-corrosive properties of the coatings. This study confirms that uniform Zn-Ni coating with smaller grain sizes has better anti-corrosive properties.

1. Introduction

Corrosion is one of the major causes of equipment failure and safety problems in offshore and marine operations, automobiles and aircraft industries[1,2]. The anti-corrosive coating is an effective approach for protecting the steel structures/machinery against corrosion [2]. The zinc-nickel alloy coatings are extensively employed in anti-corrosion purposes. Electrodeposition is one of the most widely applied technologies for the fabrication of corrosion resistant metallic coatings due to high technological feasibility and economic viability[3]. Zinc alloy coatings such as Zn-Ni, Zn-Cu and Zn-Fe alloy films are often used to provide excellent corrosion resistance and maintain good mechanical properties for steel parts or equipment. They have better corrosion resistance compared to pure zinc coating[4,5]. Electrodeposition is one of the most economical and conventional technologies for producing metallic coatings. The challenges for Zn and Zn alloy electrodeposition are hydrogen evolution and bath stability. The Zn alloy coatings deposited from citrate bath have been studied by Silva et al. (2010). They deposited Cu-Zn alloy coating on mild steel substrate from a bath containing sodium citrate as a complexing agent and benzotriazole and cysteine as additives. The results demonstrated that the coatings obtained with additives were brighter with small grain size. Cystine in the electrodeposition bath increases the potential for hydrogen evolution reaction (HER) and also prevents direct zinc deposition, and the deposited film exhibits high corrosion resistance[6].

Nowadays, Zn-Ni alloy coatings are the crucial eco-friendly substitute for toxic cadmium coating [7] and have attracted significant attention due to providing better corrosion resistance and mechanical properties at higher temperatures and severe environmental conditions, compared to pure zinc and other zinc alloy coatings [8]. Several analyses have been undertaken to understand the characteristics of the electrodeposition process of Zn-Ni alloys. It is found that the features of the deposited coating depend on the current density, pH, bath composition, applied voltage, additives and temperature[6].

The chemical composition, phase formation and microstructure of the deposited Zn-Ni alloy are essential characteristics which control corrosion resistance and other mechanical properties[7, 8].

As seen from the previous research, there are several experimental studies have been done on the Zn-Ni deposited from citrate bath at various operating parameters. To the best of the author's knowledge, there is no systematic study specifically Zn-Ni alloy at higher concentration of citrate, optimum parameter condition and challenge to suppress the hydrogen evolution reaction (HER) to improve corrosion resistant properties of the coating. Therefore, in order to fulfil this gap, the present research is to utilize the incremental study on the Zn-Ni alloy deposited at higher level operating conditions is investigated and analyzed.

The aim of this work is to electrodeposit Zn-Ni alloy films on steel substrates to improve corrosion resistance properties using potassium citrate as a complexing agent. The effects of plating variables such as bath composition, pH and current density on coating composition, morphology and corrosion property are investigated. The stability of Zn-Ni plating bath and suppression of HER are examined. The electrochemical behaviour of the deposited Zn-Ni alloy coatings is also analyzed.

2. Experimental methods

The electrodeposition experiments were carried out in a bath cell with the 200ml solution at room temperature. A rectangular steel plate of 10 cm^2 was used as the working electrode (cathode), and a graphite rod was used as the counter electrode (anode). The anode and cathode were connected to the D.C power supply via a multi-meter. There were two types of baths prepared as shown in Table 1. The pH of the bath was adjusted to acidic using hydrochloric acid. The agitation speed was 350 rpm and the electroplating time was 10 min. The steel plates for electroplating were polished by 400, 600, and 800 grit papers. They were degreased with 10 wt% sodium hydroxide solution for 5 min, then rinsed with 10 wt% hydrochloric acid and alcohol for a few seconds. The conditions for electrodeposition are listed in Table 2. After deposition, the Zn–Ni alloy coatings were washed with distilled water and dried in air. The stability of the baths was calculated using OLI software. Electrochemical analyses were performed using a ZAHNER IM6 electrochemical workstation produced by ZAHNER-Elektrik GmbH & Co.KG, Germany. The electrochemical analysis experiments were carried out in a conventional three-electrode cell with the 200ml solution. A 10 cm^2 rectangular steel plate was used as the cathode. A graphite rod was used as the anode. Silver/silver chloride electrode with saturated potassium chloride salt bridge (Ag/AgCl/KClsat) was used as the reference electrode. The morphology of the deposited samples was investigated by scanning electron microscopy (FEI MLA 650F). A Rigaku Ultima IV X-ray diffractometer with a copper x-ray source and a scintillation counter detector was used to study the crystal phase structures of deposits, and the databases used to identify the peaks were pdf# 03-065-5310 and 00-004-0831 RDB Minerals, International Centre for Diffraction Database (ICDD).

Table 1. Bath compositions for the electrodeposition of Zn-Ni alloy coatings

Bath Composition	Molar Concentrations	
	Bath-1	Bath-2
Zinc Chloride ($ZnCl_2$)	0.550M	0.550M
Nickel (II) chloride hexahydrate ($NiCl_2.6H_2O$)	0.105M	0.105M
Boric Acid (H_3BO_3)	-	0.161M
Sodium dodecyl sulfate (SDS)	0.000017M	0.000017M
Potassium Citrate ($K_3(C_6H_5O_7)$)	-	0.0326M
Sodium chloride (NaCl)	0.171M	0.171M

IOP Conf. Series: Materials Science and Engineering **458** (2018) 012005 doi:10.1088/1757-899X/458/1/012005

Table 2. Conditions for depositing the Zn-Ni alloy coatings

Bath conditions	Samples	pH range of bath
	(a)	Acidic (pH=2.5)
Bath-1 (with0.0326M potassium citrate)	(c)	Natural (pH=5.5)
	(b)	Acidic (pH=2.5)
Bath-2 (without potassium citrate)	(d)	Natural (pH=5.5)

3. Results and Discussion

3.1. Effect of complexing agent on the stability of plating bath

The stability of an electroplating bath can be studied through the stability diagram or Pourbaix diagram. Complexing agents such as potassium citrate were employed to stabilize the metal and alloy plating baths. Figures 1 (a) and (b) are the pourbaix diagrams calculated for the Zn-Ni alloy plating baths without the addition of potassium citrate and with the addition of 0.0326M potassium citrate.

It can be seen that, thermodynamically, the stability of a Zn-Ni alloy plating bath is dominated by the precipitation of $Ni(OH)_2$ and ZnO at pH 6.8 and 8.6, respectively (Figure 1a). After the addition of 0.0326M of potassium citrate, the Zn-Ni alloy plating bath is thermodynamically stable until pH 7.8 for $Ni(OH)_2$ precipitation due to the formation of stable complexing species $Ni[C_6H_5O_7]^-$, and ZnO precipitates out at pH over 8.7 with the given concentration of metal ions (Figure 1b).

The calculated stability diagrams demonstrate that, thermodynamically, citrate can efficiently stabilize Zn-Ni alloy plating baths, preventing the precipitation of metal hydroxides at a higher pH.

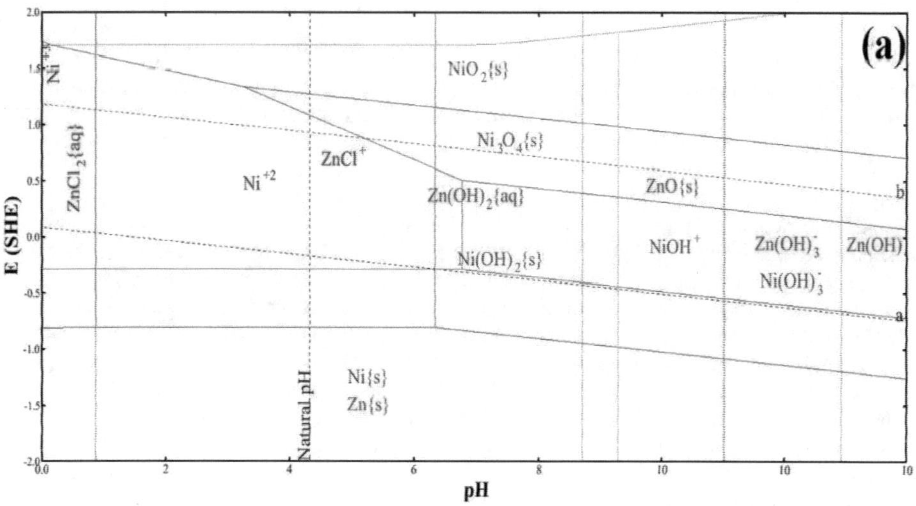

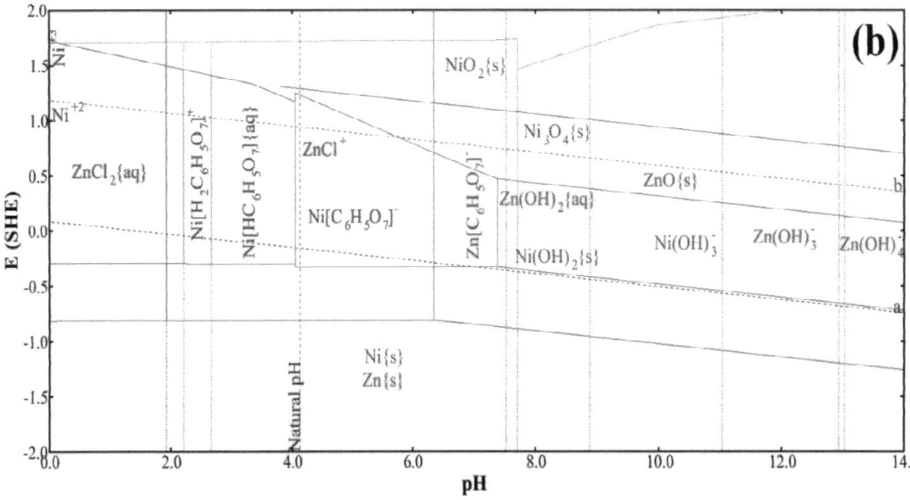

Figure 1 (a). The Pourbaix diagram for Zn-Ni alloy plating bath without the addition of $K_3(C_6H_5O_7)$ (Bath-1). The dashed lines 'a' and 'b' refer to the equilibrium lines for H^+/H_2 and $(O_2+H_2O)/OH^-$ respectively, as they do in Figure 1b. **(b).** The Pourbaix diagram for Zn-Ni alloy plating bath with the addition of 0.0326M potassium citrate $K_3(C_6H_5O_7)$ (Bath-2).

3.2. Electrochemical Impedance Spectroscopy (EIS) analysis

The Electrochemical Impedance Spectroscopy (EIS) measurement is used as a corrosion resistance analysis technique to evaluate the characteristics and kinetics of the electrochemical process prevailing at the electrode/solution interface in corrosive solutions[11,12]. The EIS measurements are carried out for the films deposited from citrate and non-citrate baths in a 3.5% NaCl solution to analyze the corrosion resistance behaviour of Zn-Ni coatings. The frequency range of open circuit potential (OCP) is 100kHz to 100mHz. The measured EIS data displayed as Nyquist and Bode plot are shown in Figures 3 and 4. In the Nyquist plots, the polarization resistance resembles the shape of a semicircle[13][14]. The EIS measurement of sample (a) deposited from 0.0326M citrate bath at acidic pH shown higher impedance modulus than the non-citrate bath coated samples. In the view of impedance vs frequency, impedance value increases by adding the potassium citrate. As for bode plots of frequency vs phase angle, phase angle also increases from citrate bath as shown in Figure 4.

The measuring setup for corrosion resistance of the presented Zn-Ni alloy coating can be explicated by an equivalent electrical circuit. The calculated fit data of the equivalent circuit for films deposited from citrate and non-citrate baths at acidic pH are listed in Table 3, where R_{sol} represents the solution/electrolyte resistance between the reference and working electrode (i.e., Zn-Ni coated specimen). R_3 and CPE_3 correspond to the resistance and capacitance of the formation of the thin oxide film that is reinforced by the ionic conduction through its pores. R_2 and CPE_2 correspond to the resistance and capacitance of the Zn-Ni alloy films. R_1 represents the charge transfer resistance associated with the reaction of zinc oxide and reduction of oxygen. CPE_1 corresponds the electric double layer capacitance at the interface of coating/electrolyte solution. All the measured impedance values were fitted using Zahner Thales software. The charge transfer resistance (R_1) of the Zn-Ni coating is 3.6 $K\Omega cm^2$ from 0.0326M citrate bath, which illustrates that the higher oxide reduction and zinc oxidation occurring on the coated surface; while, for the non-citrate Zn-Ni alloy sample, the R_1 value is 688 Ωcm^2 due to the increase of the active surface. It is well known that higher polarization resistance (R_2) values indicate better corrosion resistance. The decrease of R_2 from 879 Ωcm^2 to 264 Ωcm^2 suggests the easier access of the electrolyte to the Zn-Ni coated surface deposited from non-citrate baths. Moreover, Zn-Ni coating deposited from non-citrate bath had lower R_1 and R_2, which means that electrolyte diffused into the coating surface exhibit less corrosion resistance. Zn-Ni alloy from citrate bath had maximum coating resistance (R_2) with minimum capacitance (CPE_2) of 879 Ωcm^2 and 28.2 $\mu F cm^2$. This behaviour suggests that the addition of the complexing agent ($K_3(C_6H_5O_7)$) in a bath is a practical approach to increase the corrosion resistance of Zn-Ni alloy coating. Similar electrical equivalent circuit have been reported by Feng et al. (2016) prepared Zn-Ni alloy coating from a new DMH-based bath as a replacement of Zn and Cd coatings[15].

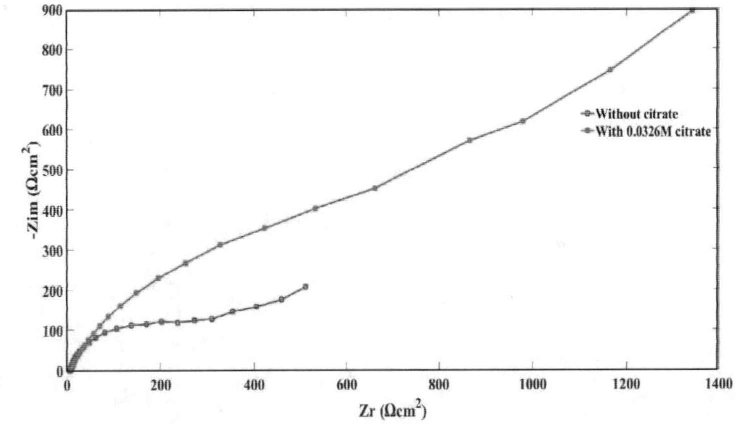

Figure 2. Nyquist plot for the Zn-Ni alloy coatings deposited from citrate and non-citrate baths at pH= acidic

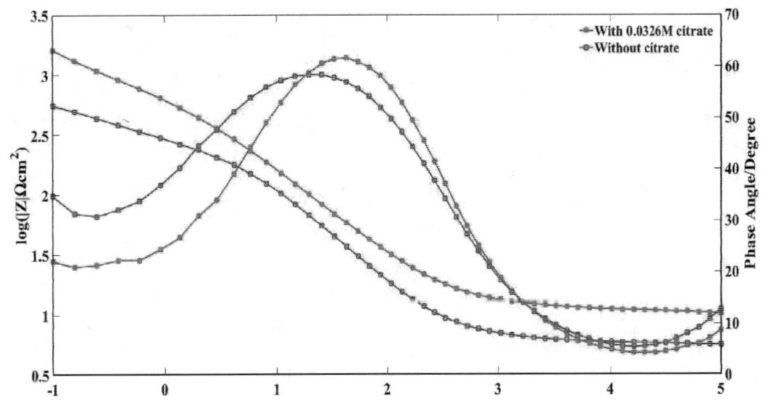

Figure 3. Bode plot for the Zn-Ni alloy coatings deposited from citrate and non-citrate baths at pH= acidic

Table 3. Electrochemical parameters determined by equivalent circuit modelling

Bath condition	R_{sol} (Ωcm^2)	R_1 (Ωcm^2)	R_2 (Ωcm^2)	R_3 (Ωcm^2)	CPE_1 (μFcm^2)	CPE_2 (μFcm^2)	CPE_3 ($nFcm^2$)
From citrate	3	3.6K	879	4.2	40	28	98
Non-citrate	2.6	688	264	3.3	106	64	104

3.3. Surface morphology and chemical compositions

The surface morphology of the Zn-Ni coatings deposited from citrate bath and non-citrate bath at different pH level was investigated using scanning electron microscopy (SEM) as shown in Figure 5. The deposited Zn-Ni alloy coating obtains different uniformity, porosity and grain size by varying bath compositions, current density and pH. All the images are taken at the same magnification. The sample (a) Zn-Ni alloy coating at a plating current density equal to 60 mA/cm² exhibits smaller and more uniform grain sizes, and no swelling is found on the coatings. Rahman et al. (2009) reported that from 40-60mA/cm^2 plating current density grain of the deposit is smaller, uniform and no porosity is found whereas, with increasing plating, current density after 60mA/cm^2 have to produce non-uniform crystals and coarse-grained coatings[9].Thus, the nature of the deposits is uniformly adherent and dense (Figure 4(a)). The sample (b) coating at same conditions reveals non-uniformity, swelling with larger and coarser grain sizes are found due to the HER. The sample (c) coating has bigger grain size and rough surfaces, the sample (d) coating has more massive coarse structure with some cracks.

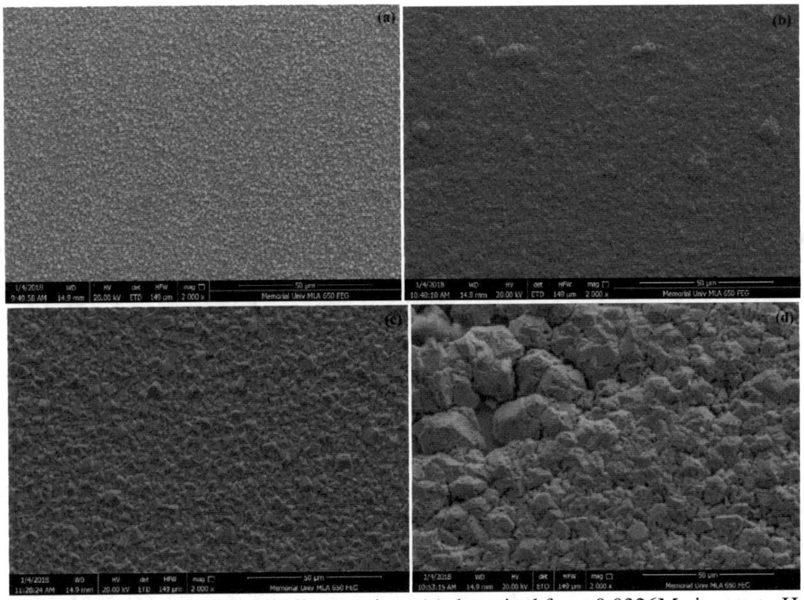

Figure 4. The SEM images of Zn-Ni alloy coatings (a) deposited from 0.0326M citrate at pH=acidic, (b) deposited from non-citrate at pH= acidic, (c) deposited from 0.0326M citrate at pH=natural, (d) deposited from non-citrate at pH=natural

3.4. X-Ray Diffraction (XRD) analysis

The XRD patterns distinguish the crystalline phases and orientation of the Zn-Ni coatings deposited at different conditions, as shown in Figure 6. The peaks around 38^0, 43^0 and 73^0 correspond to γ-$NiZn_3$ (831) and γ-Ni_2Zn_{11} (330) (631) crystal planes. The other peaks are at 51^0, 69^0 and 82^0 matching the Zn-Ni (111) and Zn (110) (200) planes. The Zn-Ni coating deposited from an acidic citrate bath has the highest intensity of the γ-phase (γ-Ni_2Zn_{11}) with (330) plane orientation, as shown in Figure 6 (b). Moreover, the peak value of γ-phase (330) plane is higher than those of other γ-phase orientations such as γ-$NiZn_3$, demonstrating that γ-phase with (330) plane orientation is preferentially formed in the Zn-Ni alloy coatings[16][17][18]. The additives or complexing gents are adsorbed on the Zn-Ni matrix preferentially and inhibits Zn-Ni alloy depositions, resulting in the smaller reduction of Zn^{+2} and Ni^{+2} on these matrices, therefore smooth and homogeneous deposits can be observed[19].The intensity of the γ-phase (γ-Ni_2Zn_{11}) of Sample (b) is high in comparison to other samples. In sample (a), the γ-Ni_2Zn_{11} peak is replaced by γ-$NiZn_3$ (815). Therefore, the coating with high-intensity γ-phase with (330) (815) plane orientation exhibits better corrosion resistance.

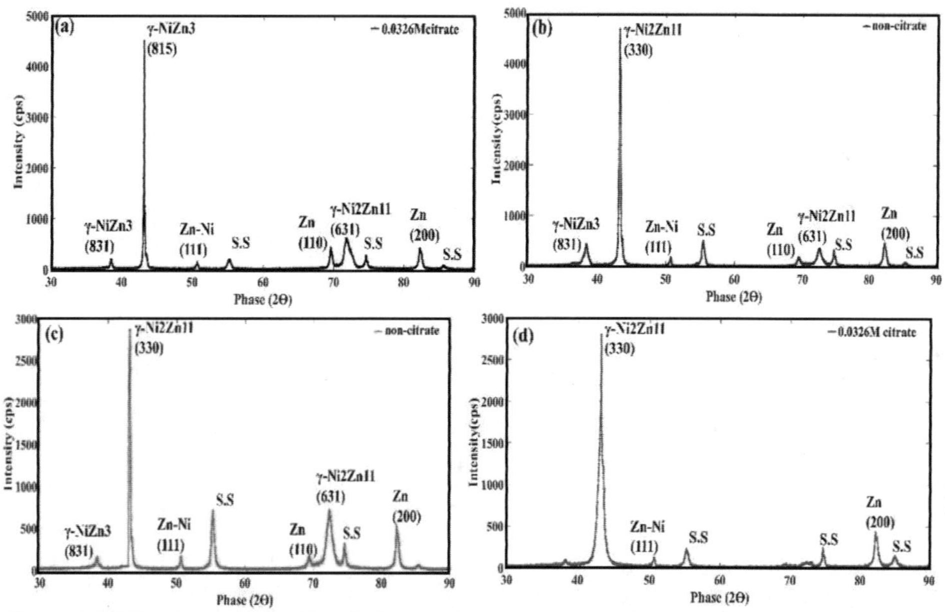

Figure 5. XRD patterns for the Zn-Ni alloy coatings (a) deposited from 0.0326M citrate at pH=acidic, (b) deposited from non-citrate bath at pH= acidic, (c) deposited from 0.0326M citrate at pH=natural, (d) deposited from non-citrate bath at pH=natural

4. Conclusions

The electrochemical behaviour of Zn-Ni alloy coating from citrate and non-citrate bath are concluded as follows:

- Stabilized plating baths and complexing agents such as potassium citrate are employed for Zn-Ni alloy electrodeposition. This extends the precipitation of nickel hydroxide ($Ni(OH)_2$), and zinc monoxide (ZnO) to a higher pH and the hydrogen deposition is also suppressed.
- The EIS measurement of sample (a) deposited from 0.0326M citrate bath at acidic pH shown higher impedance modulus than the non-citrate bath coated samples.
- The electrical equivalent circuit modelling data best fit the generated Nyquist plot demonstrated that the Zn-Ni alloy coating on steel substrate deposited from citrate bath had maximum coating resistance (R_2) with minimum capacitance (CPE_2) of 879 Ωcm^2 and 28.2 μFcm^2. This behaviour indicates that the addition of the complexing agent ($K_3(C_6H_5O_7)$) in a bath is a simple and effective process to improve the corrosion resistance of Zn-Ni alloy coating.
- The equivalent electrical circuits obtained from the experimental fitted data also demonstrated the drastic change in charge transfer (R_1) and polarization resistance (R_2) occurring on the Zn-Ni alloy coated surface deposited from citrate in comparison to a non-citrate bath.
- The sample (a) Zn-Ni alloy coating at a plating current density equal to 60 mA/cm^2 exhibits smaller and more uniform grain sizes, and no swelling is found on the coatings.
- The phase structure of the Zn-Ni deposit obtained from an acidic citrate bath having the highest intensity of γ-$NiZn_3$ (815) and γ-Ni_2Zn_{11} (330) (631) planes exhibits better corrosion resistance.

References

[1] Venkatesha T V., Rajappa S K and Praveen B M 2008 Chemical treatment of zinc surface and its corrosion inhibition studies *Bull. Mater. Sci.* **31** 37–41

[2] Lin C C and Huang C M 2006 Zinc-nickel alloy coatings electrodeposited by pulse current and their corrosion behavior *J. Coatings Technol. Res.* **3** 99–104

[3] Schwarzacher W 2006 Electrodeposition: A Technology for the Future *Electrochem. Soc. Interface* **15** 32–5

[4] Fratesi R, Roventi G, Giuliani G and Tomachuk C R 1997 Zinc-cobalt alloy electrodeposition from chloride baths *J. Appl. Electrochem.* **27** 1088–94

[5] Karahan I H and Güder H S 2009 Electrodeposition and properties of Zn, Zn–Ni, Zn–Fe and Zn–Fe–Ni alloys from acidic chloride–sulphate electrolytes *Trans. IMF* **87** 155–8

[6] Silva F L G, Do Lago D C B, D'Elia E and Senna L F 2010 Electrodeposition of Cu-Zn alloy coatings from citrate baths containing benzotriazole and cysteine as additives *J. Appl. Electrochem.* **40** 2013–22

[7] Sriraman K R, Brahimi S, Szpunar J A, Osborne J H and Yue S 2013 Characterization of corrosion resistance of electrodeposited Zn-Ni Zn and Cd coatings *Electrochim. Acta* **105** 314–23

[8] Gnanamuthu R, Mohan S, Saravanan G and Lee C W 2012 Comparative study on structure, corrosion and hardness of Zn-Ni alloy deposition on AISI 347 steel aircraft material *J. Alloys Compd.* **513** 449–54

[9] Rahman M J, Sen S R, Moniruzzaman M and Shorowordi K M 2009 Morphology and Properties of Electrodeposited Zn-Ni Alloy Coatings on Mild Steel *J. Mech. Eng. Mech. Eng. Div* **40**

[10] Ebrahimi F and Li H . 2003 Structure and properties of electrodeposited nanocrystalline FCC Ni-Fe alloys *Mater. Sci. Eng.* **A**, **347** 93

[11] Punith Kumar M K, Venkatesha T V., Pavithra M K and Shetty A N 2012 A Study on Corrosion Behavior of Electrodeposited Zn-Rutile TiO_2 Composite Coatings *Synth. React. Inorganic, Met. Nano-Metal Chem.* **42** 1426–34

[12] Fashu S, Gu C D, Zhang J L, Huang M L, Wang X L and Tu J P 2015 Effect of EDTA and NH_4Cl additives on electrodeposition of Zn-Ni films from choline chloride-based ionic liquid *Trans. Nonferrous Met. Soc. China (English Ed.* **25** 2054–64

[13] Mansfeld F, Shih H, C.H.Tsai and Greene H 1993 Analysis of EIS Data for Common Corrosion Processes *Am. Soc. Test. Mater.* **1188** 37–53

[14] Tozar A and Karahan I H 2014 Structural and corrosion protection properties of electrochemically deposited nano-sized Zn-Ni alloy coatings *Appl. Surf. Sci.* **318** 15–23

[15] Feng Z, An M, Ren L, Zhang J, Yang P and Chen Z 2016 Corrosion mechanism of nanocrystalline Zn–Ni alloys obtained from a new DMH-based bath as a replacement for Zn and Cd coatings *RSC Adv.* **6** 64726–40

[16] Boonyongmaneerat Y and Saenapitak, S Saengkiettiyut K 2009 Reverse pulse electrodeposition of Zn-Ni alloys from a chloride bath *J. Alloy. Compd.* **487** 479–82

[17] Feng Z, Li Q, Zhang J, Yang P and An M 2015 Studies on the enhanced properties of nanocrystalline Zn–Ni coatings from a new alkaline bath due to electrolyte additives *RSC Adv.* **5** 58199–210

[18] Feng Z, Li Q, Zhang J, Yang P and An M 2015 Electrochemical Behaviors and Properties of Zn-Ni Alloys Obtained from Alkaline Non-Cyanide Bath Using 5,5'-Dimethylhydantoin as Complexing Agent *J. Electrochem. Soc.* **162** D412–22

[19] Hosseini M G, Ashassi-Sorkhabi H and Ghiasvand H A Y 2008 Electrochemical studies of Zn-Ni alloy coatings from non-cyanide alkaline bath containing tartrate as complexing agent *Surf. Coatings Technol.* **202** 2897–904.

Development and characterization of glycerol coating on the PAN/PVDF composite membranes

***M K Chan[a], C S Ong and P Kumaran**

[a] Department of Chemical Engineering, SEGi University, Jalan Teknologi, Kota Damansara, 47810, Selangor, Malaysia

*Corresponding Author E-mail: mkchan@segi.edu.my

Abstract. Studies on investigating the effects of glycerol coating on the intrinsic and permeation properties of polyacrylonitrile (PAN) - polyvinylidene fluoride (PVDF) membranes were carried out. The surface properties of the membranes were determined by using field-emission scanning electron microscopy (FESEM), atomic force microscopy (AFM), fourier transform infrared spectroscopy (FTIR) and contact angle goniometer. The percentage of shrinkage, pore size, porosity and the water permeation of the membranes were also studied. The presence of new -OH and C-O bands in FTIR spectra showed that glycerol was successfully coated on the surface of the membranes. It was found that with increasing glycerol concentration from 0 to 30 vol%, the membranes exhibited higher water flux due to lower contact angle and larger pore size. However, the water flux was decreased when more than 20 vol% of glycerol was used as the coating material. It was due to the pores blocking on membrane surface and decreased surface roughness as shown by the FESEM and AFM images. Overall, glycerol coating can be a promising coating material to improve the surface properties of the membrane and its permeability performance.

1. Introduction

Polymeric membrane is widely used for water treatment due to its easy forming properties, higher flexibility, smaller space footprint and low operating cost [1]. Polyacrylonitrile (PAN) and polyvinylidene fluoride (PVDF) are widely studied due to their beneficial characteristics as host polymers. PVDF is well-known for its excellent thermal stability, chemical resistance and mechanical strength [2,3]. Meanwhile, PAN exhibits good chemical stability and good solubility in common organic solvents [4,5]. PAN also shows higher degree of hydrophilicity and better fouling resistance compared with other polymeric materials such as polyethersulfone (PES), polysulfone (PSf), and polyethylene (PE) [6]. The polymeric membrane prepared from the composites of PVDF and PAN may have the synergistic advantages of both PVDF and PAN. However, PVDF membrane would shrink after the drying process.[7, 8] Additionally, it is difficult to prepare a miscible blend of PVDF and PAN. Yin et al. [9] and Liu et al. [10] demonstrated that PAN/PVDF composites have rough surfaces, which are undesirable as it is prone to fouling. Hence, it is necessary to improve the surface properties of the PAN/PVDF membranes and minimises the shrinkage effect.

Surface modification on membrane affects its hydrophilicity, chemicals resistance and separation performance significantly [11]. An extra layer coated on the membrane surface can profoundly affect these characteristics and its propensity to foul. The modification can be done by using different coating or grafting methods in order to improve the hydrophilicity of the membrane and its anti-fouling property [12,13]. Compared to blending method, Mojtahedi et al. [14] reported that membranes surface modified via coating technique was more hydrophilic, which resulted in a higher water flux. Glycerol is known as non-solvent additive for preparing porous membranes. Hess et al. [15] found that membranes prepared by increasing the glycerol concentration in the dope solution exhibited an increase in their mean pore size and porosity. Similar findings were also reported by other researchers [16,17]. However, the use of glycerol as the PAN/PVDF membrane surface coating material has not been studied extensively by the researchers.

In this study, an attempt was made to investigate the role of glycerol in improving its surface properties as well as reducing the effect of shrinkage on the PAN/PVDF membranes. The physical properties of the membranes were evaluated by using contact angle goniometer, Fourier transform infrared spectroscopy (FTIR), atomic force microscopy (AFM) and field emission scanning electron microscopy (FESEM). Besides, the pore sizes, porosities and the water permeability of the membranes were also reported in this study.

2. Materials

Polyvinylidene fluoride (Shanghai ofluorine, PVDF T-1) and polyacrylonitrile (Sigma-Aldrich, 150kDa) were used as membrane forming materials. N-Methyl-2-pyrrolidone (Merck, 99.5%) was used as solvent to dissolve polymers. Glycerol (QReC, 99.5%) was used as the coating agent. All the chemicals were used without further purification.

3. Methodology

The polymer solution was prepared by dissolving 16 wt.% of polymer, with a ratio of PAN/PVDF: 80:20 in NMP at 65°C. The solution was allowed to stir at 400 rpm for four hours to achieve homogeneity. The solution was stored at room temperature for 24 hours to remove the bubbles. Membranes with approximate 200 μm were prepared by using the semi-automated casting machine at 7 cm/s. [18] The cast polymer solutions were immersed in reverse osmosis (RO) water to complete the phase inversion process. The membranes were then pre-treated in microwave at 100W for 30 minutes to remove excess NMP. The membranes were immersed in glycerol solution, ranged from 5, 10, 20 and 30 vol% for 1 hour followed by drying at room temperature. The membranes were labeled according to the volume percentage (vol %) of glycerol used; for instance, PAN/PVDF5 indicates 5 vol % of glycerol was coated on PAN/PVDF based composite membrane.

The shrinkage of the membrane was measured by comparing the length of the membranes before (L_i) and after (L_f) under different concentrations of glycerol. The equation to calculate the percentage of shrinkage is as below:-

$$\frac{L_i - L_f}{L_i} \times 100\% \quad (1)$$

The porosity (ε) of the membranes was measured using the gravimetric method and the porosity was calculated by using the following equation (2):

$$\varepsilon = \frac{(w_a - w_b)/\rho_{water}}{(w_a - w_b)/\rho_{water} + w_b/\rho_{membrane}} \quad (2)$$

Where w_a = Weight of wet membrane (g), w_b = Weight of dry membrane (g), ρ_{water} = Density of water (g/cm^3), $\rho_{membrane}$ = Density of membrane (g/cm^3).

The contact angle of the membrane was measured by using the contact angle goniometer (Attension/KSV Theta, Japan). Approximate 3μL of RO water was dropped on the surface of the sample. The contact angle was measured from the water-membrane interface. At least 10 locations were arbitrarily chosen on the membrane surface in order to obtain an average value.

The water permeation of the membranes was determined by using the stirred ultrafiltration cell (Millipore, model 8050) at 2.5 bars. The flux(J) is calculated from the following equation:

$$J = \frac{Q}{AP} \tag{3}$$

Where Q the water flow rate in L/h is, A is the effective membrane surface area which is 13.4 $cm^2$and P is the operating pressure (bar).

The surface and cross sectional images of the membranes were obtained using FESEM (Model: SU8020, Hitachi). The dry membrane samples were immersed in liquid nitrogen and fractured, followed by sputter-coating with platinum using a sputtering device. The effect of glycerol concentrations on the membrane surface roughness was studied by using non-contact AFM mode (Park SystemXE-100). FTIR spectra of the membranes were recorded in the mid IR region between 700cm^{-1} and 4000 cm^{-1} by using Perkin Elmer Spectrum One FTIR Spectrophotometer.

The pore radius (r_p) of the membranes were determined by using the Guerout-Elford-Ferry equation.[19]

$$r_p = \sqrt{\frac{(2.90 - 1.75\varepsilon)8V\mu\delta}{\varepsilon PA}} \tag{4}$$

Where V is the water flow rate in m^3/s, μ is the dynamic viscosity of water at 25°C (0.891 mPa.s), and δ is the thickness of the membrane.

4. Results & Discussion

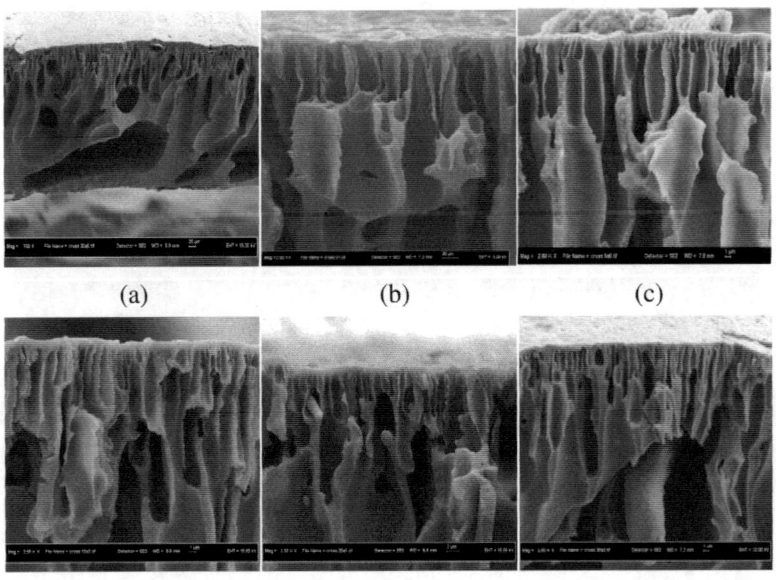

(a) (b) (c)

(d) (e) (f)

Figure 1. FESEM cross-sectional image of (a) PAN/PVDF (150x) (b) PAN/PVDF0 (2.5Kx) (c) PAN/PVDF5 (2.5Kx) (d) PAN/PVDF10 (2.5Kx) (e) PAN/PVDF20 (2.5Kx) and (f) PAN/PVDF30 (2.5Kx).

The overall cross-sectional image of the asymmetric PAN/PVDF is shown in Figure 1(a). A finger-like structure was observed under the skin layer and a porous macro voids structure was found in the support layer of the membrane. The formation of this structure was due to fast exchange of solvent and non-solvent during the phase inversion process [17,20]. This FESEM image was similar to the blended

PAN/PVDF membranes fabricated by Yang et al [20]. Figures 1 (b) to (f) showed the structure of the membranes after glycerol coating. Although different concentrations of glycerol were coated on the PAN/PVDF membranes, the cross sectional images were similar, where the finger like structures were found under the skin layer. This showed that the coating technique did not affect the intrinsic structure of the membrane, which was formed during the phase inversion process.

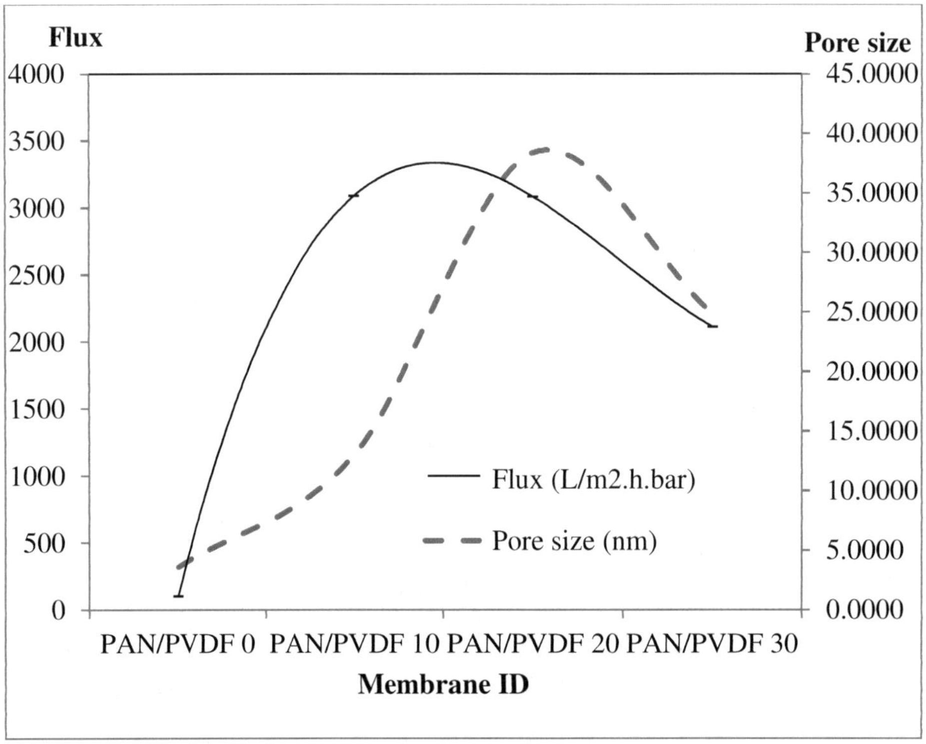

Figure 2. The flux and the pore size data for PAN/PVDF membranes.

Figure 2 showed that the mean pore size of the membrane was the smallest for PAN/PVDF0 membrane, which was 3.58 nm and the biggest mean pore size was 38.37 nm for PAN/PVDF20 membrane. This was due to the minimum shrinkage in the latter membrane, as shown in Table 1.Without the glycerol coating, the shrinkage of PAN/PVDF0 membrane was ~20%. When the concentration of glycerol increased from 5 vol% to 20 vol%, the effect of shrinkage reduced from ~20% to ~3%. This showed that shrinkage might lead to pore collapse and hence reduced the membrane pore size. Further increase of glycerol beyond this concentration did not improve the shrinkage of the membrane but it blocked the pores on the membrane. This explains the smaller pore size for PAN/PVDF30 membrane compared to PAN/PVDF20 membrane. It is worth to mention that an increase of 0.07 μm membrane thickness was measured from PAN/PVDF0 to PAN/PVDF30 membranes, revealing that glycerol coating does to some extent governs membrane performance in this study. Experimental work conducted by Bilad et al. [21] showed that no pore size larger than 0.05 μm was found for the membrane which experienced severe shrinkage effect. Comparatively, for the modified membrane which possesses a high degree of resistance to shrinkage, maximum pore size with ~0.2 micron was reported.

Table 1. The properties of PAN/PVDF membranes.

Membrane ID	Shrinkages (%)	Contact angle (°)	Porosity (%)
PAN/PVDF0	19.9004 ± 2.5806	80.7042 ± 2.6456	35.0321 ± 3.0301
PAN/PVDF5	19.1525 ± 2.0781	70.8447 ± 2.8890	48.0192 ± 3.4825
PAN/PVDF10	19.4514 ± 2.0276	55.9529 ± 2.8548	53.8582 ± 2.9260
PAN/PVDF20	3.3323 ± 2.7832	59.2653 ± 2.6851	22.0000 ± 2.8284
PAN/PVDF30	3.9683 ± 2.3662	58.7473 ± 2.9418	20.0998 ± 4.9661

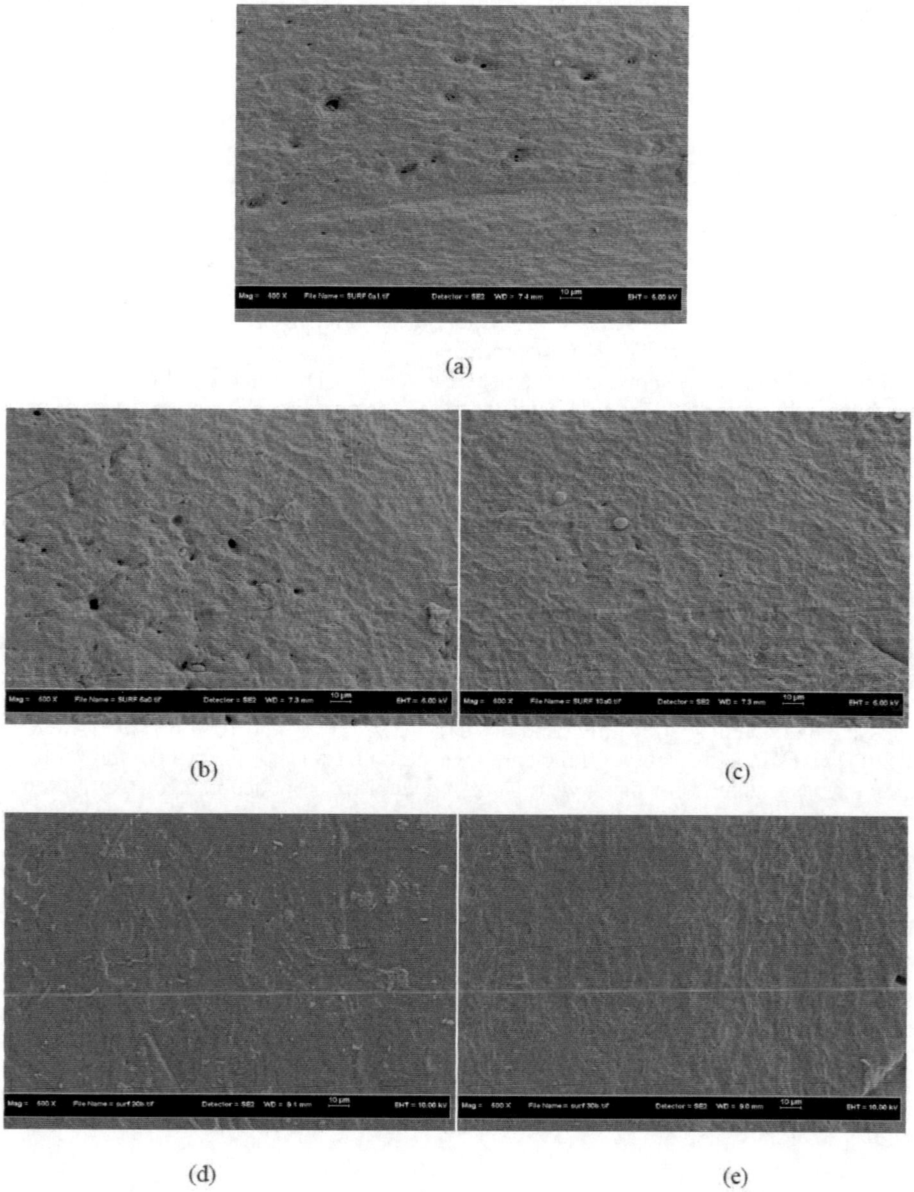

Figure 3. FESEM surface images of (a) PAN/PVDF0 (b) PAN/PVDF5 (c) PAN/PVDF10 (d) PAN/PVDF20 (e) and PAN/PVDF30.

Figure 2 also illustrated the influence of glycerol coating on the pure water permeation of the membranes. The lowest flux was observed in PAN/PVDF0 membrane due to its small mean pore size and hydrophobic surface as shown in Table 1. The contact angle of this PAN/PVDF0 membrane was 80.70°, which indicates the nature hydrophobicity of pristine PAN/PVDF membrane. When the concentration of glycerol in the coating solution was increased to 10 vol% and 20 vol%, the membranes pore size and the hydrophilicity were increased. Thus ~3100 L/m^2.h.bar of flux was found. Based on the data in Table 1, although PAN/PVDF20 membrane was slightly hydrophobic and less porous compared to PAN/PVDF10, the flux of the membrane was maintained at ~3100 L/m^2.h.bar. It is due to the pore size of PAN/PVDF20, which is about 25 nm larger compared to PAN/PVDF10 membrane. When 30 vol% was coated on PAN/PVDF membrane surface, the flux of the PAN/PVDF30 membrane reduced to ~2100 L/m^2.h.bar due to its smaller pore size compared to PAN/PVDF20 membrane.

During the coating process, glycerol which carries three hydroxyl groups; is attached physically on the surface of the membrane. Thus, the hydrophilicity of the membrane is improved from 80.70° to 55.95°as the amount of glycerol is increased from 0 vol% to 10 vol% in the coating solution, as shown in Table 1. However, beyond this point, the PAN/PVDF20 and PAN/PVDF30 membranes became slightly hydrophobic compared to PAN/PVDF10. Similar trend is observed for the porosity of the membrane, where the PAN/PVDF 10 membrane was more porous compared to PAN/PVDF5 and PAN/PVDF0. Further increase in the concentration of glycerol in the coating solution reduces the porosity from ~54% for PAN/PVDF 10 membrane to ~20% for PAN/PVDF10 and PAN/PVDF20 membranes. Zhao et al. [22,23] reported that contact angle of a membrane is affected by its porosity. Hence, the low porosity in membranes PAN/PVDF20 and PAN/PVDF30 contributed to their high contact angle, compared to PAN/PVDF10 membrane.

FESEM images in Figure 3 shows that the use of glycerol coating improved the surface roughness of the membranes. In Figure 3 (a)-(c), defects due to the shrinkage effect was observed on the surface of the membranes. This is in agreement with the shrinkage data in Table 1, where approximately 20% of shrinkage was measured for these membranes. When the amount of glycerol in the coating solution is increased to 20 vol% and 30 vol%, defects free and smooth surface are seen in Figure 3 (d) and (e). The AFM images in Figure 4 also illustrated the similar finding where the roughness average (Ra) value of PAN/PVDF0 is 35.958 nm and this value is reduced to 23.822 nm for PAN/PVDF30 membrane.

Figure 5 shows the FTIR spectra of PAN/PVDF membranes. The characteristic peaks for the PAN/PVDF0 membrane are observed at 1,451.61 cm^{-1} (stretching band of CH_2), 2,919.58 cm^{-1} (C-H) and 1249.39 cm^{-1} (CN); which are consistent with the inherent characteristic of PVDF and PAN materials [24]. In addition to these, the peak at 1403.47 cm^{-1} could be attributed to β-phase of PVDF [25,26]. Compared to the control membrane, all the glycerol coated PAN/PVDF membranes showed broad peaks at 3,424.43 cm^{-1} (stretching band of OH), 1179.17 cm^{-1} and 1097.64 cm^{-1} (stretching band of C-O) [20]. This is in agreement with the data presented in Table 1, where the hydrophilicity of the membranes increase along with the glycerol coating due to the attachment of –OH group on the membranes surface.

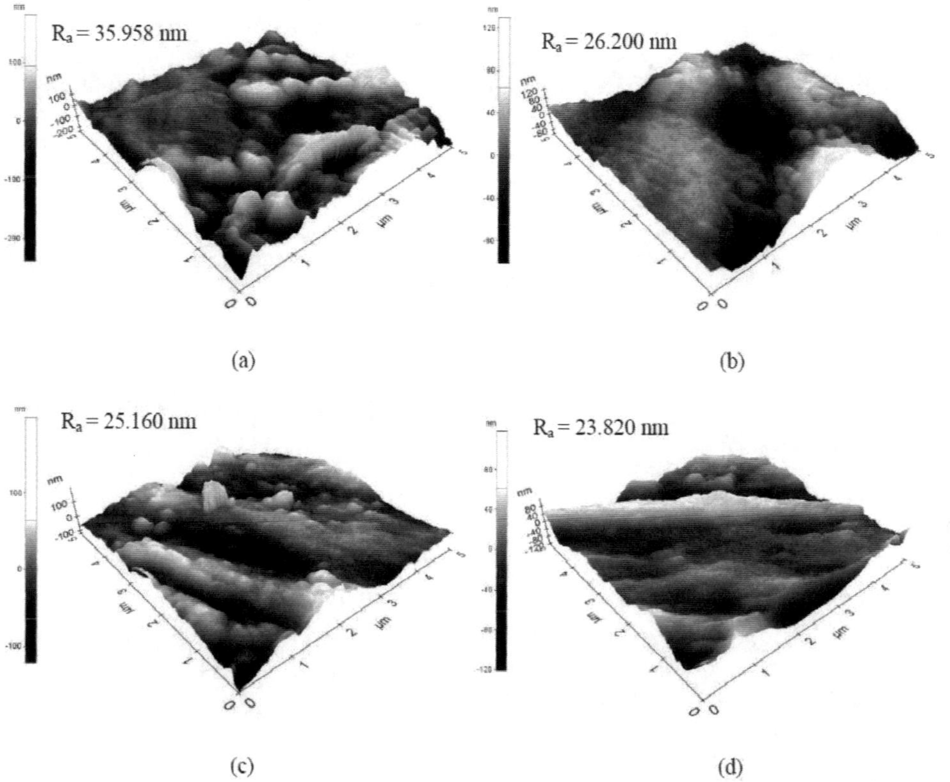

Figure 4. AFM surface images for (a) PAN/PVDF 0 (b) PAN/PVDF 10 (c) PAN/PVDF 20 and (d) PAN/PVDF 30 membranes.

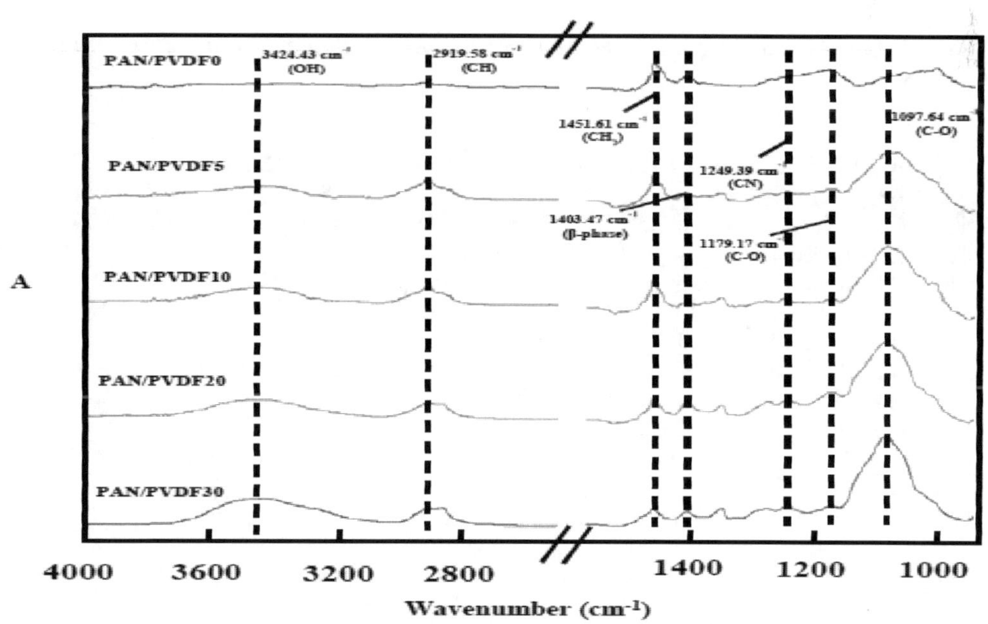

Figure 5. FTIR Spectra of PAN/PVDF membranes.

5. Conclusion

The intrinsic and permeation properties of PAN/PVDF membranes coated with varied vol% of glycerol have been studied. Glycerol was successfully coated on the membranes as new -OH and C-O stretching bands are found in the FTIR spectra. Experimental results revealed that at least 20 vol% of glycerol is required to minimize the shrinkage effect on the membranes. It also minimized the collapse of pores and resulted in the largest membrane surface pore size. The glycerol coating increased the water flux of the membrane as the membrane became more hydrophilic and porous. FESEM and AFM images showed that glycerol coating improved the surface roughness of the membranes. Based on the results obtained from this study, it shows that glycerol can be a very promising coating material to improve the PAN/PVDF membrane flux performance and surface properties, which might extend membrane application in industrial water and wastewater treatment process.

Acknowledgement

The authors gratefully acknowledge the financial support from SEGi Research Fund (Project number: SEGi/2015-01/FOEBE-15/84)

References

[1] Ng L Y, Mohammad A W, Leo C P and Hilal N 2013 *Desalination* **308** 15–33
[2] Ji J, Liu F, Hashim N A, Abed M R M and Li K 2015 *React. Funct. Polymers* **86** 134–153
[3] Hong G, Xin L, Ziyi W, Bao'an L, Jixiao W and Shichang W 2018 *Chinese Journal Of Chemical Engineering* 1-6
[4] Nazri N A M, Lau W J, Ismail A F and Veerasamy D 2015 *Desalination Water Treat* **55**, 1771-85
[5] Jian C, Jian Z, Liling Z, Gregory S. O and Tai S C 2017 *J. Membr. Sci* **539** 295–304
[6] Jung B 2004 *J. Membr. Sci* **229** 129–136
[7] Jian K and Pintauro E N 1997 *J. Membr. Sci* **135** 41-53
[8] Wang D, Li K and Teo W K 2000 *J. Membr. Sci* **178** 13–23
[9] Yin X, Cheng H, Wang X and Yao Y 1998 *J. Memb. Sci* **146** 179-184
[10] Liu T Y, Lin W C, Huang L Y, Chen S Y and Yang M C 2005 *Polym. Adv. Technol* **16** 413-9
[11] Yin J and Deng B 2015 *J. Membr. Sci* **479** 256–75
[12] Rana D and Matsuura T 2010 *Chem. Rev* **110** 2448-71
[13] Sun J X, Gui E C and Zhen L X 2018 *Journal of Industrial and Engineering Chemistry* **58** 179–188
[14] Mojtahedi Y M, Mehrnia M R and Homayoonfal M 2013 *Desalination and Water Treat* **51** 6736-42
[15] Hess S C, Kohll A X, Raso R A, Schumacher C M, Grass R N and Stark W J 2015 *ACS Appl. Mater. Interfaces* **7** 611-7
[16] Shim H C, Yeh Y S and Yasuda H 1990 *J. Membr. Sci* **50** 299-317
[17] Lee J, Park B, Kim J and Park S B 2015 *Macromol. Res* **23** 291-9
[18] Chan M K, Raoo P, Eu Y K and Lim J M Y 2015 *Mater. Manuf. Processes* 1941-7
[19] Ye L, Fan Z and Zhou J 1981 *Membrane Science and Technology* **2** 24-30
[20] Yang M C and Liu T Y 2003 *J. Membr. Sci* **226** 119–130
[21] Bilad M R, Burrieza E G, Mavukkandy M O, Al Marzooqi F A and Arafat H A 2015 *Desalination* **376** 62–72
[22] Zhao Y H, Zhu B K, Ma X T and Xu Y Y 2007 *J. Membr. Sci* **290** 222
[23] Chan M K, Ng S C, Noorkhaniza S and Choo C M 2016 *Innovation and Analytics Conference & Exhibition (IACE)* 1-5
[24] Hashim N A, Liu Y and Li K 2011 *Chem. Eng. Sci* **66** 1565–75
[25] Socrates G, 2001 Wiley, New York
[26] Manjula D D, Anand P A and Kim J K 2018 Vibrational Spectroscopy **94** 74–82

Removal of 2,4-dinitrophenol (2,4-DNP) by using magnetic nanoparticles (MNPs) coated with polypyrrole (PPy)

Nor Azean Nadhirah Mohamed Sani[1], Nurul Aqilla Mazlan[1], Ahmad Husaini Mohamed[1], Kavirajaa Pandian Sambasevam[1], Khairil Anuar Jantan[2], Muggundha Raoov Ramachandran[3,4] and *Jamil Mohamed Sapari[1]

[1]School of Chemistry and Environment, Faculty of Applied Sciences, Universiti Teknologi MARA (UiTM) Cawangan Negeri Sembilan, Kampus Kuala Pilah, Pekan Parit Tinggi, 72000 Kuala Pilah, Negeri Sembilan

[2]Faculty of Applied Sciences, Universiti Teknologi MARA (UiTM) Cawangan Selangor, Shah Alam, 40450 Selangor, Malaysia

[3]University of Malaya Centre for Ionic Liquids, Department of Chemistry, Faculty of Science, Universiti Malaya, Kuala Lumpur 50603, Malaysia.

[4]Department of Chemistry, Faculty of Science, Universiti Malaya, Kuala Lumpur 50603, Malaysia.

*Corresponding Author Email: jmohamedsapari@gmail.com or kavirajaa@live.com

Abstract. The research represented here is oriented towards developing a new approach to functionalise a surface of Magnetic Nanoparticles with Polypyrrole (MNPs-PPy). This development will then applied to remove 2,4 dinitrophenol (2,4-DNP) in aqueous solution. A various analytical technique such as Fourier Transform Infrared Spectroscopy (FTIR) analysis, Scanning Electron Microscope (SEM), X-ray Diffraction (XRD) was employed for complete characterisation of MNPs-PPy. FT-IR spectra have shown the peak of Fe-O at 580 cm-1 while; the SEM images illustrated the globular structure of the surface of MNPs-PPy. Also, XRD result has shown the MNPs-PPy is in the crystalline form. The ultimate aims would then be the optimisation of the superlative condition of several parameters such as effect pH, temperature and concentration, contact time, amount of adsorbent, interfering ions and reusability for the removal of (2,4-DNP) in an aqueous solution by MNPs-PPy. The optimum conditions were observed at pH 6, 318 K, 10 ppm of analyte, 75 minutes and 50 mg of adsorbents which shown 98.67 % removal of analyte.

1. Introduction

According to the US Environmental Agency (EPA) and the European Union, phenol and its derivative become the main importance and apprehension which are comprised of their list of the aquatic environment. The manufactured of plastic, dyes, drugs, pesticides, antioxidants and paper in the petrochemical industries are created from the phenolic compound. The mutual disposal of all these above production products such as 4-nitrophenol (4-NP), 2,4-dichlorophenol (2,4-DCP) and 2,4,6-

trichlorophenol (2,4,6-TCP) are formed in water sources. Thus, the removal of phenol groups is crucial compared to the determination in water samples [1].

Since the discovery of magnetic nanoparticles (MNPs), specifically iron oxides, it becomes one of the advantageous resources in various uses such as magnetic fluids catalysis and magnetic resonance imaging. MNPs have high surface area because it is in nano-sized particles. So, the adsorption size and performance is increasing [2].

Tahmasebi et al., [3] claim that polymer coating is another way to enhance the stability of the particles and permit the alteration of the useful surface for well-sensitivity and selectivity. One of the most explored coating polymers is Polypyrrole (PPy) that displays advantageous characteristics such as ability to coat on a unique element such as MNPs, simply prepared in liquid and solid medium whether chemical and electrochemical process, high porosity (quick ion exchange with circumstances), high electrical conductivity, good stability, harmless and inexpensive.

Therefore, preparation of MNPs functionalised with PPy will be synthesised to analyse its performance towards removal of 2,4-dinitrophenol in water samples with different pH, temperature and concentration, contact time, amount of adsorbents, the effect of interfering ions and its reusability.

2. Experimental

2.1. Materials.

All experiments were carried out in the air, and the products obtained appear indefinitely stable towards the atmosphere, whether in solution or the solid state. Reagents and solvents were used as received from commercial sources. Magnetic nanoparticles and 2,4-dinitrophenol that act as analyte was purchased from Bendosen company, Selangor. High purity solvent (99%) of hydrochloric acid, ethanol and acetonitrile was supplied by R&M, Shah Alam. Potassium PeroxydiSulfate (KPS) and pyrrole (99%) were bought from Merck, Shah Alam.

2.2. Characterization Technique.

Magnetic nanoparticle (MNP) and magnetic nanoparticle-Polypyrrole (MNPs-PPy) were analysed using the various analytical technique. Infrared data were obtained using a Perkin-Elmer Spectrum 100 FT-IR spectrometer. Phase identification of crystalline material was analysed using X-Ray Diffraction (XRD). The surface topography and composition of nanoparticles were obtained using Scanning Electron Microscopy (SEM) at high-resolution (80-200 kV), Ultra Violet Visible (UV-Vis) spectrophotometer analysis was conducted using PG-instrument T80/T80+ type equipped with 1 cm quartz cell.

2.3. Synthesis of magnetic nanoparticle coated with Polypyrrole (MNPs-PPy).

A solution of Potassium Peroxydisulfate (KPS) (5.0 g) in a hydrochloric acid solution (50 mL, 0.1 M) was treated with magnetic nanoparticles (1.25 g) in pyrrole (1ml, 0.01 mol). The solution was stirred continuously at 250 ppm for approximately 2h. The solution then was left for 24 hours. The resulting solid product was washed several times with ethanol, methanol and deionised water to remove any impurities [4] and dried in the oven for 24 hours at 45 °C. The dried product was transferred into a clean vial and kept in a desiccator and away from any external sources of the magnetic field.

2.4. Removal and Determination of Phenolic Compound.

MNPs-PPy were weighed (20 mg) in three separate glass vials. A stock solution of 2,4-dinitrophenol (10 mL of 10 ppm) was pipetted into the glass vials. The mixture was shaken for 75 minutes using at 250 rpm at room temperature. The supernatant solution was then filtered and transferred to the centrifuge tubes with the aid of an external magnet. The effect of varying different parameter pH (2-12), temperature (10 – 50 °C), concentration (10 – 100 ppm), amount of adsorbent (5 - 50 mg) and interfering ions were investigated. The reusability of the synthesised MNPs-PPy was used to measure the concentration of the analyte by using UV-Vis Spectrophotometer.

2.5. Effect of pH.
The concentration of analyte was adjusted using HCl (0.1 M) and of NaOH (0.1 M). The determination of phenolic compound was done by transferring the solution into the glass vials containing MNPs-PPy

2.6. Effect of Temperature and Concentration.
Different concentration of analyte (10, 20, 40, 60, 80 and 100 ppm) was prepared and analyse at a different temperature (25, 35 and 45 °C).

2.7. Effect of Interfering Ions.
There could be some other ions present in the water sample that will interrupt the adsorption of phenolic compound towards the MNPs-PPy. Some of the interfering ions were Cl-, NO3- and SO42-. About 10 ppm of the analyte was added with 10% w/v of the Cl- solution. Next, determination of the phenol anion was continued. All the steps were repeated using NO3- and SO42- solution and its triplicate measurements.

2.8. Reusabilities.
The residue of the adsorbent was washed three times with deionized water and dried in the oven. Then, the overall steps were repeated more than three times to check for the adsorbent performance. The percentage of removal, removal percentage (% R) will be calculated.

3. Result and Discussion

3.1. Fourier Transform Infrared Spectroscopy (FTIR).
The spectrum in figure 1, shows both MNPs and MNPs-PPy shows peak at 580 cm^{-1} indicate the Fe-O bond while the peak at 772.58 cm^{-1} indicate the C-H out of plane ring deformation bond. C-H pyrrole in rings has observed in the spectrum at 1034.30 cm^{-1} and C-N stretching band was observed at 1314 cm^{-1}. Both peak at 1537.20 cm^{-1} and 1451.70 cm^{-1} were assigned to the C-C asymmetric and symmetric stretching vibrations of pyrrole rings [5]. Most of the spectrum for MNPs-PPy indicated the peak for polypyrrole which indicates that the MNP has been successfully coated with polypyrrole.

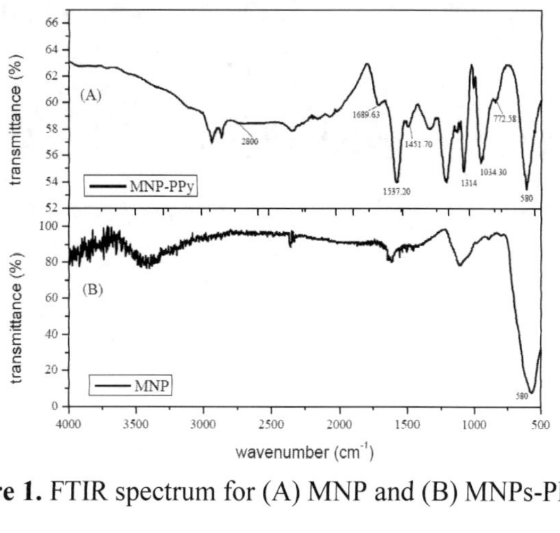

Figure 1. FTIR spectrum for (A) MNP and (B) MNPs-PPy

3.2. X-Ray Diffraction (XRD)
The result obtained as in figure 2 for XRD diffractogram has proven that the MNPs-PPy was in crystal structure. This is due to the highest peak observed at $2\theta=35°$ which indicated to the degree of

crystallinity of PPy. Therefore, the XRD results has shown that the amorphous PPy present on the surface of MNPs [6]. This is because the intensity of MNPs-PPy is lower than MNPs because the crystinallity of MNPs has been decreased due to the existing amorphous surface of PPy.

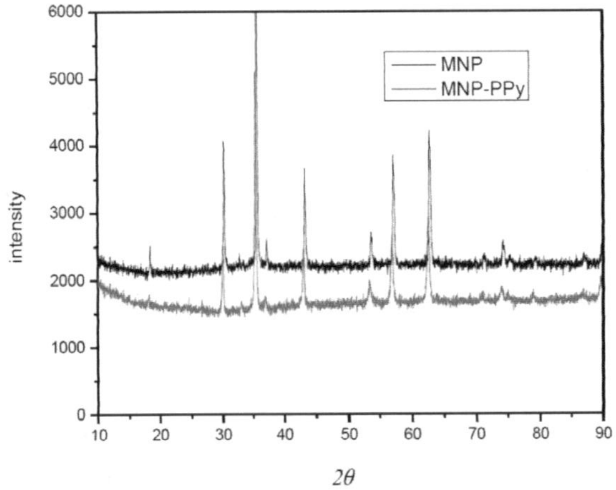

Figure 2. X-ray Diffraction (XRD) diffractogram of MNP and MNPs-PPy

3.3. Scanning Electron Microscopy (SEM)
Figure 3, the surface of the MNPs is not uniform. The results obtained for the MNPs-PPy has the smooth and uniform surface which has been proven that the PPy has successfully coated the MNPs surface. The MNPs-PPy also has a globular structure. The results was compared with the studies of Shanehsaz et al., [7] and Tahmasebi [3]. MNPs-PPy has the porosity surface so that it can absorb more analyte compared to MNPs that has less porosity surface.

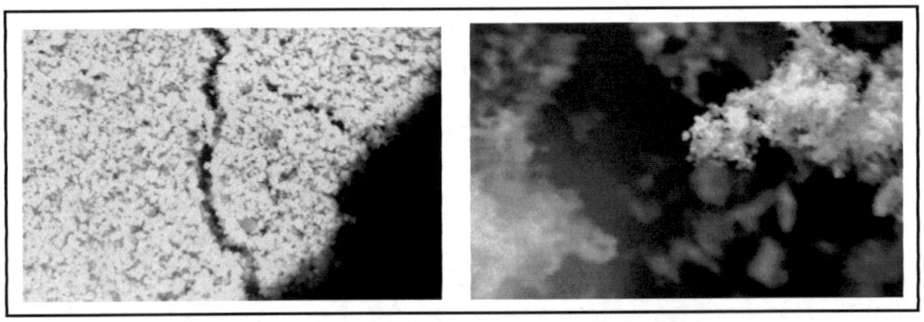

Figure 3. Morphology structure of MNP and MNPs-PPy

3.4. Scanning Analyte
Ultraviolet Visible Spectrophotometer (UV-Vis) was used to check the ⅄max of the analyte which is 2,4-dinitrophenol (2,4-DNP). The highest absorbance for the 2,4-DNP is 360 nm. The value 360 nm was chosen due to the molar absorptivity of the analyte where it shows the highest peak at 360 nm as shown in figure 4.

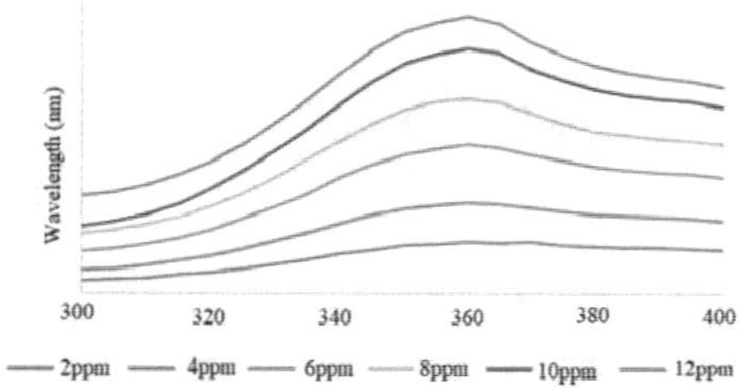

Figure 4. Scanning of 2,4-Dinitrophenol (2,4-DNP)

3.5. Effect of pH

Based on the figure 5, the best PH results obtained for the removal of 2,4-dinitrophenol (2,4-DNP) was at PH 6 which is in an acidic medium. The percentage removal almost achieved 95 %. Studies from Shaneshaz et al., [7] states that in an acidic condition, the nanoparticles have a positive surface charge which can act one upon another with anion dye molecules. The positive charge density on the surface of adsorbent can be formed under lower PH. So, the adsorbent can attract more of the negative charged analyte and causing the removal efficiency is high. Meanwhile the adsorption process between adsorbent and analytes are affected by the PH of the solution.

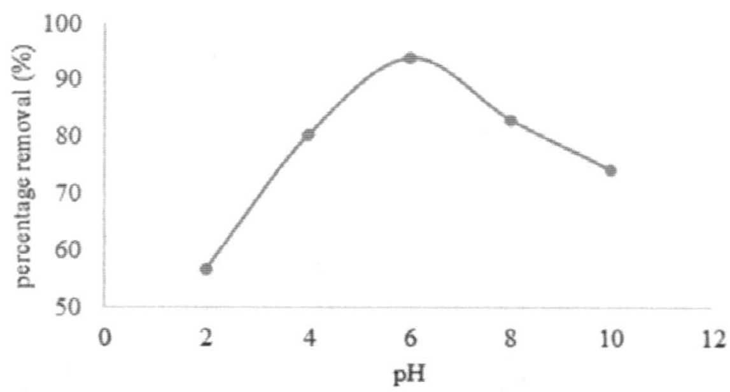

Figure 5. Removal percentage of 2,4-DNP at different pH

3.6. Effect of Temperature and Concentration

Based on the figure 6, the results obtained for the effect of temperature and concentration, the best removal achieved at temperature 318.15 K and concentration of analyte is 10 ppm. The removal is done at the acidic condition which is at pH 4 at all temperatures. The results is backup by the studies from which states that higher temperature will increase the removal of 2,4-DNP because of better diffusion rate. So, it will increase the tendency for the analyte to be absorbed by the adsorbent. Moreover, as the temperature higher, the adsorbent may increase its number of active surface centre for the absorption process.

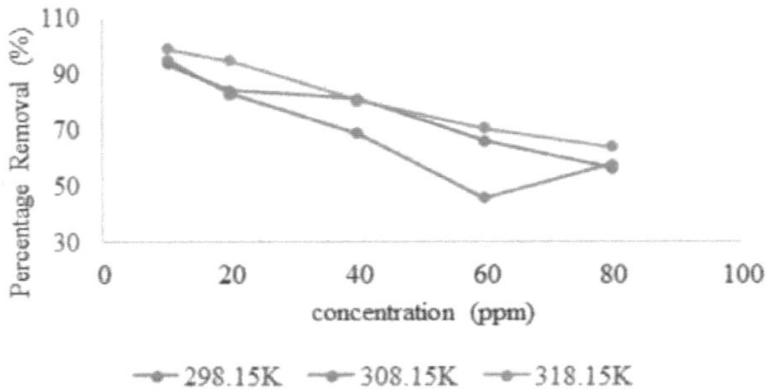

Figure 6. Removal percentage of 2,4-DNP based on Concentration and Temperature

3.7 Contact in Time

Figure 7 shows the effect of contact time on the removal of 2,4-DNP by the MNPs-PPy. The MNPs-PPy showed high percentage removal (95.77 %) at 75th minutes. After 75 minutes, the percentage remained around 95 %. The increasing of contact time until 2 hours give the result of removal percentage analyte reached 93.1 % up until 95.6 % which it has reached equilibrium. When the contact time is extend after 2 hours, the percentage of removal is remained because there is no available adsorption site on the MNPs-PPy surface [8].

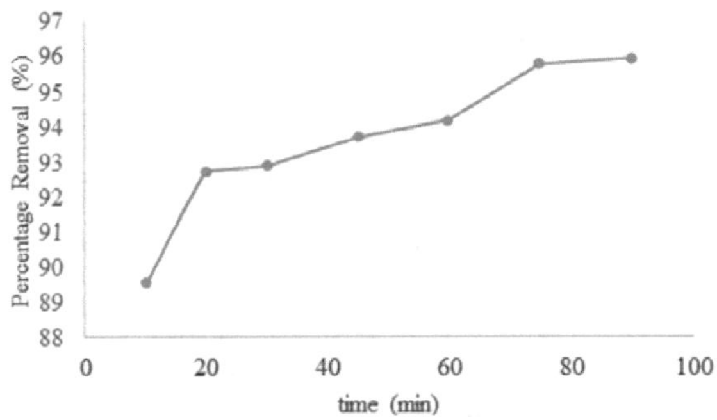

Figure 7. Removal percentage of 2,4-DNP based on Contact Time

3.8. Effect on amount of adsorbent

Based on the figure 8, the best amount of adsorbent to remove 10 ppm of 2,4-DNP is 50 mg with the percentage removal is 98.67 % at 318.15 K and 75 minutes. Shahneshaz *et al.,* [7] stated in their studies that as the amount of adsorbent increase, the surface area also increase and there will be more adsorption surface exists to absorb the analyte.

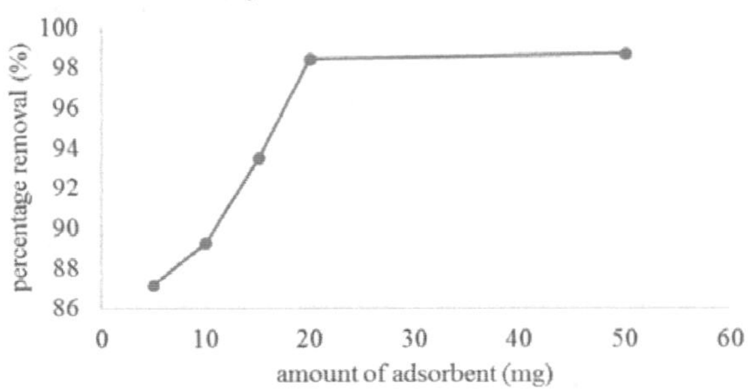

Figure 8. Removal percentage of 2,4-DNP based on Contact Time

3.9. Effect of Interfering Ions

From the result obtained on figure 9, it shows that the addition of salt will decrease the percentage removal of 2,4-DNP in the presence of respective interfering ions when the %w/v of ions increase, the removal efficiency also decrease. The ions that had been investigated were from salt which were Cl^- from NaCl, NO_3^- from KNO_3 and SO_4^{2-} from Na_2SO_4. This condition can be described as the salting out effect where the addition of salt may change the physical properties of the analyte, reduce the diffusion rates of analytes towards the adsorbent and thus decrease the removal ability [9]. Besides, studies from Thavanugul [10] stated that when the concentration of ionic strength increases, the solubility of analyte decreased. This is due to the increasing viscosity of the solution. Therefore, diffusion rate of analyte towards the adsorbent decreases and hence, the removal efficiency also decrease.

3.10 Reusability

Based on the figure 10, it shows that the more the adsorbent is used, the efficiency of the adsorbent will decrease. The MNPs-PPy can be used up until 5 times only. The more it reused, the lower its efficiency. The reusability of adsorbent becomes less efficient because the surface area has been reduced due to the surface alteration of the adsorbent until the fifth times.

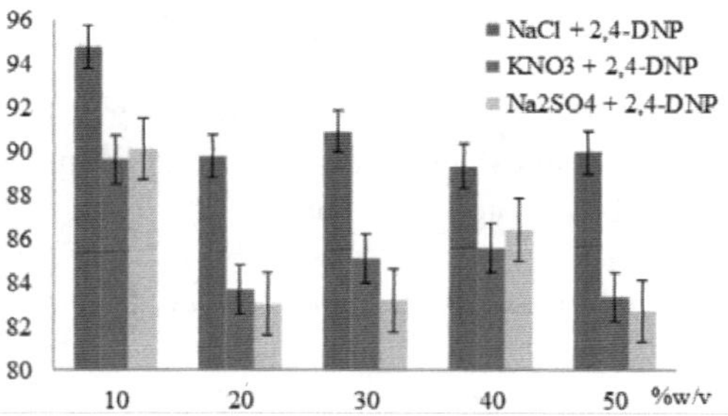

Figure 9. Removal percentage of 2,4-DNP based on amount of Adsorbent.

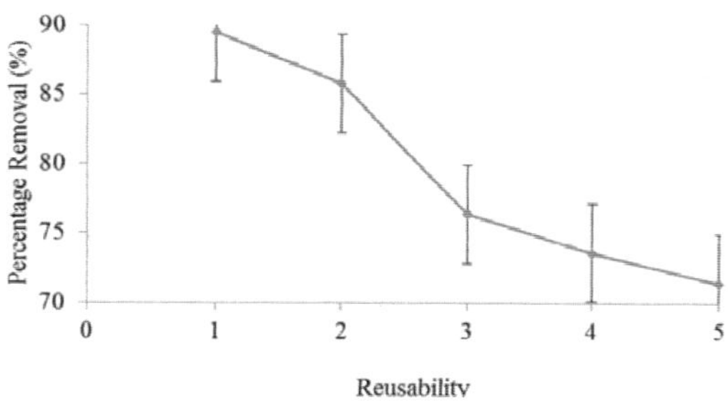

Figure 10. Removal percentage of 2,4 DNP by the effect of Interfering Ions

4. Conclusions

As a conclusion, the magnetic nanoparticles coated with polypyrrole was successfully removed the 2,4-Dinitrophenol (2,4-DNP) by using several parameters. The best condition for the 2,4-DNP to be remove by MNPs-PPy were at pH 6, temperature at 318.15 K and concentration at 10 ppm, 75 minutes for the time removal and 50 mg amount of adsorbent. The MNPs-PPy can be used five times only. FTIR spectrum has proved that the MNP has been successfully coated with polypyrrole where the major peak for Fe-O bond at 580 cm-1. The XRD result had shown the MNPs-PPy was in crystal form where the highest peak exist at $2\Theta=35°$. The MNPs-PPy has globular structure which has been proven by using SEM.

Acknowledgment

Special thanks to UiTM Cawangan Negeri Sembilan for supporting this research.

References

[1] Amirah Ahmad, Mohd Hasmizam Razali, Mazidah Mamat, Faizatul Shimal Binti Mehamod and Khairul Anuar Mat Amin (2017). Adsorption of methyl orange by synthesized and functionalized-CNTswith 3-aminopropyltriethoxysilane loaded TiO2 nanocomposites. Chemosphere. Volume 168, 474 – 482.

[2] Baharin, S. N. A., Muhamad Sarih, N., Mohamad, S., Shahabuddin, S., Sulaiman, K., and Ma'Amor, A. (2016). Removal of endocrine disruptor di-(2-ethylhexyl) phthalate by

modified polythiophene-coated magnetic nanoparticles: Characterization, adsorption isotherm, kinetic study, thermodynamics. *RSC Advances*, *6*(50), 44655–44667.
[3] Tahmasebi, E., Yamini, Y., Seidi, S., and Rezazadeh, M. (2013). Extraction of three nitrophenols using polypyrrole-coated magnetic nanoparticles based on anion exchange process. *Journal of Chromatography A*, *1314*, 15–23.
[4] Muliwa, A. M., Leswifi, T. Y., Onyango, M. S., and Maity, A. (2016). Magnetic adsorption separation (MAS) process : An alternative method of extracting Cr (VI) from aqueous solution using polypyrrole coated Fe_3O_4 nanocomposites. *Separation and Purification Technology*, *158*, 250–258.
[5] Bagher, M., Yamini, Y., Dayeni, M., and Seidi, S. (2015). Journal of Environmental Chemical Engineering Adsorptive removal of alizarin red-S and alizarin yellow GG from aqueous solutions using polypyrrole-coated magnetic nanoparticles. *Biochemical Pharmacology*, *3*(1), 529–540.
[6] Jiao, Z., Zhang, Y., and Fan, H. (2016). Ultrasonic-microwave method in preparation of polypyrrole-coated magnetic particles for vitamin D extraction in milk. *Journal of Chromatography A*, *1457*, 7–13.
[7] Shanehsaz, M., Mohammad, S., Shoja, R., and Poursaberi, T. (2016). Removal of Reactive Red195 Synthetic Textile Dye using Polypyrrole-coated Magnetic Nanoparticles as an Efficient Adsorbent, *96*, 85–96.
[8] Fan, L., Wei, C., Xu, Q., and Xu, J. (2017). Polypyrrole-coated cotton fabrics used for removal of methylene blue from aqueous solution, *5000*(March).
[9] Ebrahimpour, B., Yamini, Y., Seidi, S., and Tajik, M. (2015). Analytica Chimica Acta Nano polypyrrole-coated magnetic solid phase extraction followed by dispersive liquid phase microextraction for trace determination of megestrol acetate and levonorgestrel. *Analytica Chimica Acta*.
[10] Thavarungkul, P. (n.d.). Polypyrrole / silica / magnetite nanoparticles as a sorbent for the extraction of sulfonamides from water samples.

IOP Conf. Series: Materials Science and Engineering **458** (2018) 012008 doi:10.1088/1757-899X/458/1/012008

Concatenation of carbonaceous nanofillers for mixed matrix membrane development

P S Murugiah[1], *P C Oh[1] and K K Lau[1]

[1]CO_2 Research Centre (CO_2RES), Institute of Contaminant Management, Chemical Engineering Department, Universiti Teknologi PETRONAS, 32610 Seri Iskandar, Perak, Malaysia.

*Corresponding Author Email: peiching.oh@utp.edu.my

Abstract. Two different carbonaceous nanofillers, carbon nanofiber (CNF) and graphene oxide (GO) were synergistically incorporated into glassy poly (2,6-dimethyl-1,4-phenylene oxide) (PPO_{dm}) polymer to synthesize a novel mixed matrix membrane (MMM) for CO_2/CH_4 separation. The combined effect of binary filler on the filler dispersion and gas separation performance of resultant membrane were analysed. The binary filler-based MMM rendered homogenous filler dispersion quality, as compared to single filler-based MMM. The CNF particles dispersed in between GO nanosheets, inhibited the restacking of GO whilst the strong steric effect imposed by GO nanosheets improved the dispersion of CNF particles in the polymer matrix. The optimum gas separation performance was recorded at 2 wt% CNF/3 wt% GO loading with ideal CO_2/CH_4 selectivity of 59.70 and CO_2 permeability of 62.32 Barrer at 3.5 bar feed pressure. The CNF particles with smooth wall and larger diameter allowed gases to diffuse through faster by creating a gas pathway and eventually improved the gas permeability. Meanwhile, GO nanosheets with surface hydroxyl and carboxyl groups facilitated the transport mechanism of CO_2 molecules, acted as a selective barrier and rendered high selectivity. Therefore, both permeability and selectivity of resultant membrane were enhanced.

1. Introduction

Mixed matrix membrane (MMM) is a heterogeneous membrane consisting of inorganic filler embedded in polymer matrix. It combines the desirable properties of both polymeric and inorganic material to form high performance gas separating membrane. Generally, the addition of inorganic fillers is expected to alter the transport properties of gases, modify the properties of the neighbour polymer phase and alter the packing, dynamics or conformation of polymer chain near its surface [1]. According to Ismail *et.al.*, [2] the integration of inorganic fillers enhances the gas separation properties of polymer matrix via two ways. Firstly, the interaction between the polymer-filler particles disrupt the polymer chain packing, increases the polymer matrix free-volume and eventually increases the gas diffusion parameter. Thereafter, the functional groups attached on the filler surface interact with some gases and enhances the gas solubility parameter. Over the past decades, various inorganic particles have been incorporated in the polymer matrix to form MMM. Excessive works have been carried out especially with the conventional fillers including zeolites, silicas, metal oxides and carbon molecular sieves as they possess some unique and individualistic properties which contributed positively to the robust performance of resultant membrane [3]. Various modifications have been performed on the filler particles and the performance of resultant membrane were well documented. However, Goh *et.al.*, [4] found that the development of MMMs using these conventional fillers has reached its bottleneck whereby the performance of resultant membrane could not be further improved. Meanwhile, very few works have been reported on the utilization of alternative fillers in MMM fabrication, thus far. Alternative fillers are relatively new for gas separation membrane synthesis and its full potential is yet to be explored.

Besides, the recent practice of developing MMM using nano-sized inorganic filler is found to encounter severe agglomeration issues in polymer matrix due to their high specific surface area and strong intermolecular van der Waals interactions. Moreover, the addition of single filler was often insufficient to improve both permeability and selectivity of resultant membranes [5]. Therefore, in order to improve the filler dispersion properties and to enhance the gas separation efficiency, a new technique of combining two different fillers with distinct properties, morphologies, nature and dimension in the polymer matrix is proposed [6-10]. In this technique, two fillers are synergistically incorporated into a polymer matrix at different weight loading to engineer the performance of resultant MMM. The complementary interaction between the two fillers with distinct surface chemistry promote polymer-filler interaction, thereby reducing the possibilities of filler agglomeration, whilst retaining the native properties of the filler particles. On top of that, it is hypothesized that the addition of binary fillers could enhance both permeability and selectivity of MMM. One of these binary fillers could improve the gas permeability across the membrane while the other could enhance the selectivity. Therefore, this work focuses on the development of a novel MMM with enhanced filler dispersion and gas separation performance using two different alternative carbonaceous nanofillers, namely carbon nanofiber (CNF) and graphene oxide (GO). To the best of author's knowledge, these filler combinations has not been attempted in gas separation membrane development.

2. Materials and Method

2.1. Material

Poly (2,6-dimethyl-1,4-phenylene oxide) (PPO_{dm}) polymer powder and chloroform solvent with ≥ 99.5 purity were purchased from Sigma Aldrich. Carbon nanofiber (CNF, > 70% purity) and graphene oxide (GO, > 99% purity) with an average diameter of 200 nm and 500 nm, respectively were purchased from Suzhou Hengqiu Graphene Technology Co. Ltd, China. Carbon dioxide (CO_2) and methane (CH_4) gases with > 99.995% purity were supplied by Gas Walker, Malaysia.

2.2. Membrane fabrication

Five membranes with different filler loadings (Table 1) were synthesized via dry-phase inversion technique, as depicted in Figure 1. A specified amount of filler was ultra-sonicated for 30 min in chloroform to form well-dispersed homogenous suspension. Subsequently, 22 wt% of preheated polymer powder was added to the filler-solvent suspension gradually with continuous stirring for 24 h at 60°C. The dope solution was degassed for 4 h and allowed to stand at room temperature overnight. The solution was cast on a clean glass plate using a film applicator at a gap setting of 150 µm and allowed to dry at room temperature for the first 24 h, followed by drying at 80°C for 72 h. For binary filler-based MMM, similar procedures were used in which CNF particles were sonicated for the first 30 min and further sonicated for another 30 min after the adding GO in the suspension. Meanwhile, pristine PPO_{dm} membrane was fabricated via the same procedure without the addition of filler particles.

2.3. Characterization

Powder X-Ray Diffractometer (XRD, X'Pert[3] Powder & Empyrean, PANalytical) was used to study the atomic arrangement of polymer chain upon the incorporation of filler particles. The analysis was carried out in the 2θ range of 5-90° with scanning speed of 10°/min. The spacing between two adjacent planes (*d*-spacing) was calculated using Bragg's equation 1.0:

$$\lambda = 2d \sin \theta \tag{1.0}$$

where λ is the wavelength of CuKα radiation (λ=0.1540 nm), d is the basal spacing and sin θ is the scattering angle.

The surface and cross sectional morphology of membranes were analyzed using Variable Pressure Field Emission Scanning Electron Microscope (VP-FESEM, Zeiss SupraTM 55 VP) at an accelerating voltage of 15 kV. All membrane samples were sputter-coated with gold using Polaron Range SC 7640. For cross-sectional analysis, the membrane samples were cryofractured with liquid N_2 prior to scanning.

The dispersion and distribution properties of filler particles in the PPO_{dm} polymer matrix were further examined using Optical Microscope (OM, OLYMPUS Model BX53M) at 500 X magnification. For clearer visual inspection, the images were converted to binary images wherein for single filler-based MMM the filler particles were depicted in red, while for binary filler-based MMM, GO and CNF were portrayed in yellow and blue, respectively.

2.4. Gas permeation test

The gas permeation properties of resultant membrane were measured using an in-house gas permeation test rig at fixed feed pressure of 3.5 bar and room temperature. Membrane with an effective surface area of 1.77 cm^2 were dried for at least 1 h and secured on a metal plate of a flat sheet membrane module. The permeate flow rate across the membrane was measured using a soap bubble flow meter after the system reached equilibrium. Five measurements were taken for each sample and the average value was used to calculate the gas permeability. Gas permeability (P_i, Barrer) and ideal CO_2/CH_4 selectivity (α_{CO_2/CH_4}) were calculated using the method described in our previous work [11].

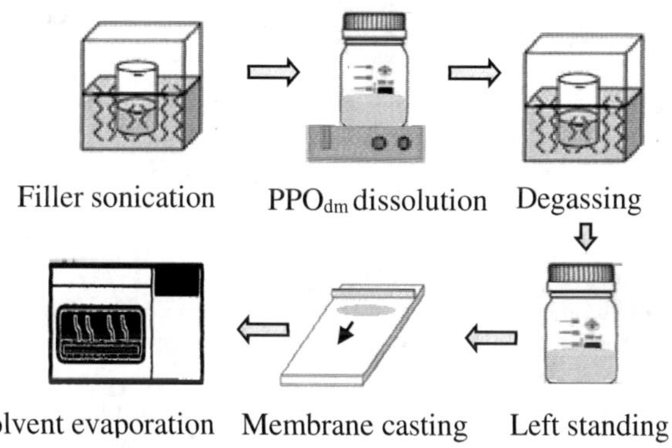

Figure 1: Visualization of MMMs development

Table 1: Filler compositions

Membrane	Filler loading (wt %)	
	CNF	GO
P	-	-
P-5G	-	5
P-2C/3G	2	3
P-3C/2G	3	2
P-5C	5	-

3. Results and Discussion

3.1. X-Ray diffraction
Figure 2 depicts the XRD spectrum of resultant membranes at various filler composition. Initially, the pristine PPO_{dm} membrane (Figure 2 a) exhibited a single amorphous peak at 2θ = 14.10° which was in good agreement with spectra reported by Khulbe *et.al.*, [12]. Upon the addition of single filler, P-5G (Figure 2 b) and P-5C (Figure 2 e), additional low intensity peak at 2θ = 10.24° and 2θ = 26.25° which corresponded to the main reflection peak of GO and CNF were observed. The presence of these additional peaks in MMM illustrated that the filler's crystallinity was not altered during membrane fabrication process. It also validated the dispersion of filler particles in the polymer matrix as well the polymer-filler interaction. In addition to that, the broad peak of PPO_{dm} polymer matrix experienced a minor shift to lower angles with the addition of single filler and to higher angles with the addition of binary fillers as tabulated in Table 2.

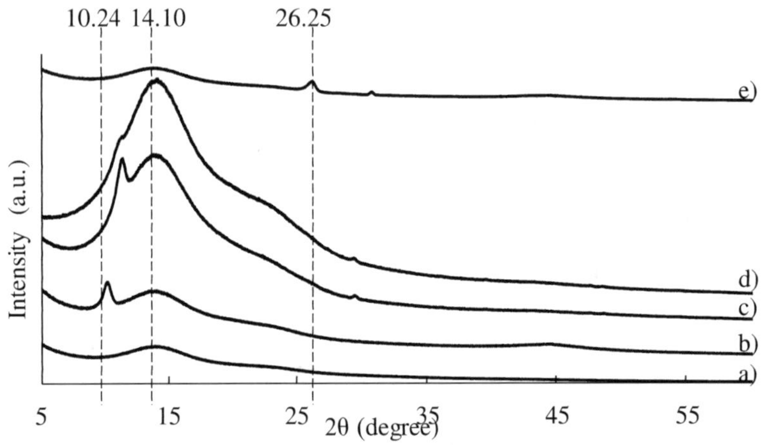

Figure 2: XRD spectra of a) P, b) P-5G, c) P-2C/3G, d) P-3C/2G and e)P-5C MMM

From Table 2, the addition of single filler was found to increase the *d*-spacing of polymer chains and eventually produced more free volume of polymer chains into the membranes which tend to allow the gases to permeate through faster [13]. Nevertheless, upon the addition of binary fillers a slight reduction in the polymer chain distance were observed due to polymer-filler strong interaction. Sarfraz, M., & Ba-Shammakh [14] reported similar trend upon the addition of binary nanofillers (ZIF-302 and GO) into PSf polymer matrix. Moreover, the intensity of GO characteristic peak at 3C/2G filler loading was found to be very weak and insignificant as compared to 2C/3G and 5G filler loading. This indicated the exfoliation and homogenous dispersion of GO nanosheets in the PPO_{dm} polymer matrix with the presence of CNF. Therefore, the presence of co-existing nanofillers, wherein CNF dispersed between GO nanosheets and inhibited the restacking of GO in PPO_{dm} polymer matrix [15].

3.2. Field emission scanning electron microscope
The internal microstructure of the membranes were analysed from the FESEM morphological images, as shown in Figure 3. Pristine PPO_{dm} membrane shows dense, non-porous and defect-free structure. From Figure 3 a) no voids or pinholes were observed. Furthermore, no phase separation was observed, in which the polymeric membrane was regularly packed. Upon the addition of fillers, the dense and non-porous structure of membrane was maintained with continuous and defect-free interconnected network morphology. However, the filler particles were observed to disrupt the polymer intermolecular chain

Table 2: d-spacing analysis of PPO_{dm} main characteristic peak at $2\theta = 14.10^o$

Sample	Angle (degree °)	*d*-Spacing (Å)
P	14.10	6.27
P-5G	14.07	6.29
P-2C/3G	14.24	6.18
P-3C/2G	14.31	6.17
P-5C	13.78	6.42

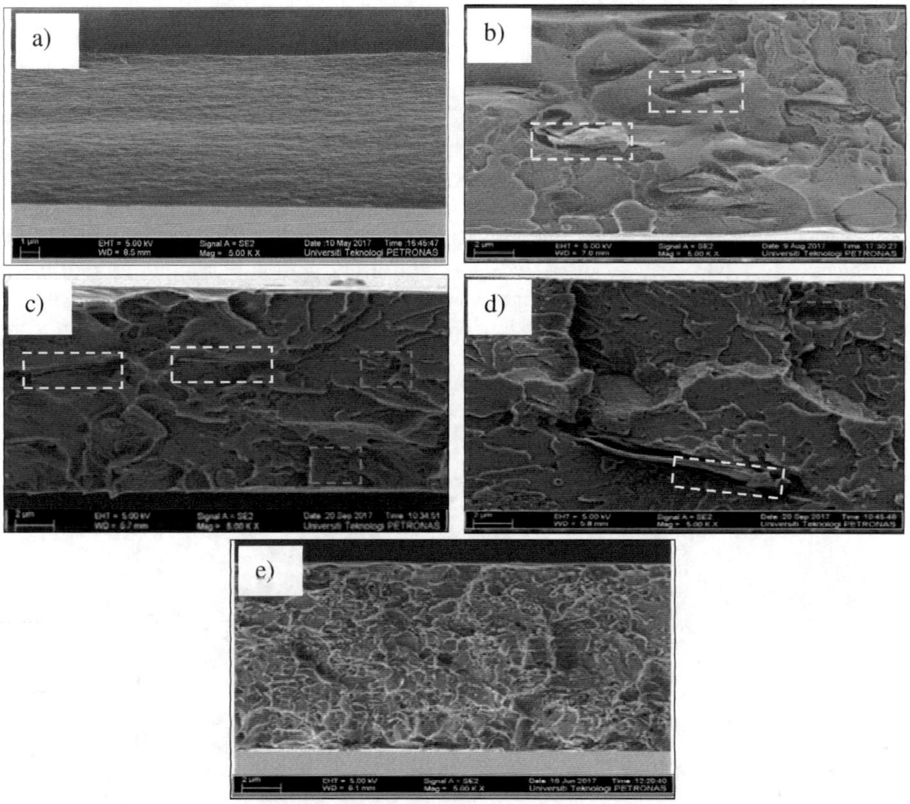

Figure 3: Cross-sectional morphology of a) P, b) P-5G, c) P-2C/3G, d) P-3C/2G and e) P-5C. Yellow and red boxes show the location of GO and CNF, respectively

packing which eventually induced some wrinkles and groves to the cross-sectional morphology of the resultant membrane. In addition to that, fracturing the MMMs in liquid N_2 caused the thin layer to fragment with slight plastic deformation due to the presence of filler particles. At 5 wt% CNF loading (Figure 3 e), the CNF particles were noticed to form small clusters due to their hydrophobic nature. At higher loading, the cohesive forces between the CNF particles dominates and eventually leads to particles agglomeration and undesirable polymer-filler interfacial defects. On the other hand, at 5 wt% of GO loading (Figure 3 b), GO nanosheets were found to disperse regularly with preferential horizontal orientation to the membrane surface. Interestingly, the filler particles were dispersed more uniformly in the polymer matrix upon the addition of binary filler. No filler agglomeration or polymer-filler interfacial defects were noticed. The sheeted morphology of GO with horizontal orientation provides significant steric effect and prevented the aggregation of CNF particles in the polymer matrix [16]. Hence, the CNF particles dispersed among GO nanosheets and loosely entangled in the polymer matrix without forming agglomerates as observed at 5C filler loading. Therefore as hypothesized earlier, the complementary interaction between these two nanofillers inhibits the formation of particle agglomeration and rendered homogenous filler dispersion properties in PPO_{dm} polymer matrix.

3.3. Optical microscopy

Optical microscope was used to analyze the relative dispersion properties of filler particles in PPO_{dm} polymer matrix. Figure 4 (a&b) depicts the optical micrographs of P-5G and P-5C MMMs. From visual inspection, no major agglomeration was observed with the exception of several small clusters. However, the distribution was less uniform wherein some larger non-reinforced polymeric domains could be observed especially for P-5G MMM. Upon the addition of binary fillers Figure 4 (b&c), the dispersion properties of fillers were relatively improved with minimum number of filler aggregates. At 2C/3G and 3C/2G loadings, the size of aggregates specifically GO nanosheets (yellow) were noticed to be reduced as compared to P-5G MMM (Figure 4 a). It might be resulted from the addition of CNF which dispersed in between the GO nanosheets and reduced the tendency of GO restacking and agglomeration in polymer matrix. Meanwhile the CNF particles dispersed more uniformly without forming any clusters or stress convergence points at the presence of GO nanosheets. Therefore, the synergistic effect of combining CNF and GO evidently improved the dispersion properties of filler particles in the polymer matrix. Furthermore, the area of filler reinforcement were enhanced with the addition of binary fillers.

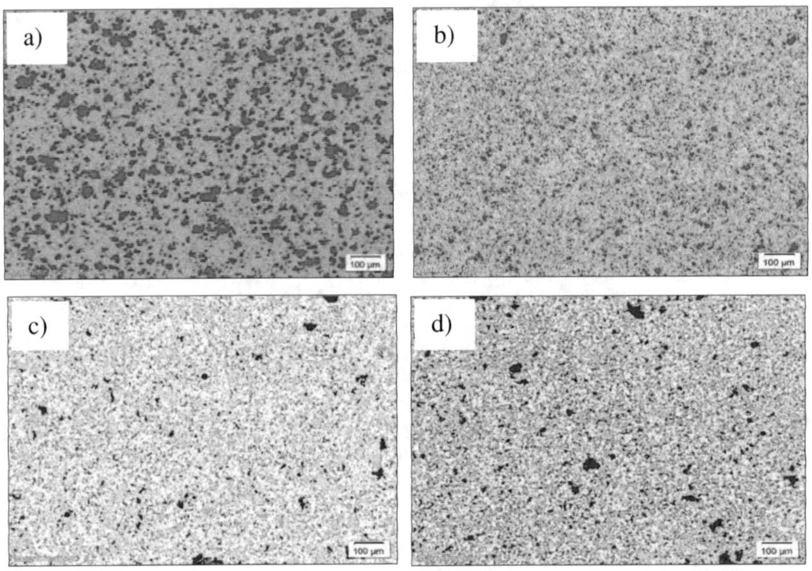

Figure 4: Optical micrographs of a) P-5G, b) P-5C, c) P-2C/3G and P-3C/2G MMM

3.4. Gas separation performance

At a feed pressure of 3.5 bar, the pristine PPO_{dm} membrane exhibited high CO_2 permeability of 50.64 Barrer with low CO_2/CH_4 selectivity of 7.35. Upon the addition of 5 wt% GO, the CO_2 permeability reduced to 45.16 Barrer while selectivity increased drastically to 40.65. GO nanosheets acted as a barrier that impedes the direct diffusion of gases through the membranes. In addition, GO nanosheets caused the polymer chain to become more discontinuous and tortuous. At this condition, the gases diffused through the tortuous path created by GO nanosheets, which resulted in lower permeability. Simultaneously, the addition of GO enhanced the CO_2/CH_4 selectivity of pristine PPO_{dm} membrane by 453%. Since, GO nanosheets increases the tortuous diffusion pathway in the polymer matrix, the polymer chain mobility was reduced while generating rigidified polymer-filler interface. The rigidified interface impedes the diffusion of larger gas molecules (CH_4) while favors the diffusion of small molecules (CO_2) with less resistance [13-14]. Moreover, the favorable interaction between the polar groups especially carboxyl (-COOH) and hydroxyl (-OH) on GO and the polar individual C-O bonds of CO_2 accelerates the transport of CO_2 through the membrane. Meanwhile, the addition of 5wt% CNF improves the CO_2 permeability to 129.08 Barrer with little improvement in ideal CO_2/CH_4 selectivity, 8.23 . The addition of CNF disrupted the polymer chain packing and increased the polymer matrix free volume, resulting in improved gas permeability. This was confirmed by the increment in *d*-spacing value of polymer chain upon the addition of CNF in PPO_{dm} polymer matrix in XRD analysis (Table 2). Moreover, the smooth walls of CNF allows the gases to diffuse through faster than the dense layer of polymer matrix by creating a gas pathway. Therefore it could be concluded that, P-C MMM displayed high permeability of CO_2 with moderate CO_2/CH_4 selectivity, while P-G MMM exhibited low CO_2 permeability but high selectivity for CO_2/C_4 separation. On the other hand, the addition of binary fillers enhanced the overall gas separation efficiency of MMM. At 2C/3G filler loading, the CO_2/CH_4 selectivity value reached 59.70, which was 712%, 625% and 46% higher than the selectivity reported for pristine PPO_{dm} membrane, P-3C MMM and P-5C MMM, respectively. Meanwhile, the CO_2 permeability was measured at 62.32 Barrer which was 23% higher than pristine PPO_{dm} membrane. These values were higher than the performance reported for single-filler based MMM, in which CNF acted as the pathway to increase the gas permeability, while GO nanosheets acted as selective barrier to accelerate the separation efficiency of the resultant membrane [15]. The CNF and GO dispersed regularly, disrupted the polymer chain packing and enhanced the gases diffusivity properties. Meanwhile, the hydroxyl and carboxyl group on GO had a stronger affinity with polar gas, CO_2 and increases the CO_2 solubility across the membrane. Thus, the resultant P-C/G MMM provides selective solubility of CO_2 polar gas and restricted CH_4 diffusion. CO_2 molecules permeated via facilitated transport mechanism combined with solution-diffusion mechanism, while CH_4 permeated via solution-diffusion mechanism only. Hence, the combination of 2 wt% CNF and 3 wt% GO displayed favourable synergistic effect on the gas separation performance of resultant membrane. Nevertheless, the barrier effect of GO nanosheets predominated the performance of resultant membrane, in which the superior gas permeability achieved by P-C MMM could not be obtained for P-C/G MMM.

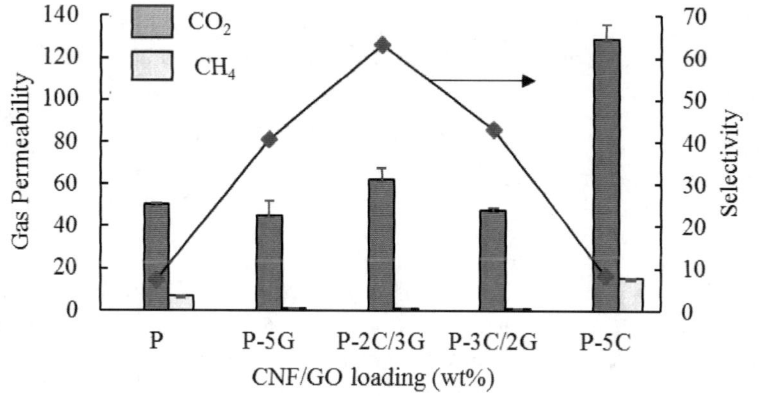

Figure 5: The effect of filler loading on membrane gas separation performance

4. Conclusion and Recommendation

Binary fillers-based MMMs were successfully developed using two different carbonaceous nanofillers via dry phase inversion technique. The resultant membranes remained mainly amorphous with two additional filler characteristic peaks which corroborated the dispersion of CNF and GO in PPO_{dm} polymer matrix. The MMMs exhibited dense structure with no significant filler agglomeration and polymer-filler interfacial defects. The filler dispersion quality relatively improved for P-C/G MMMs, as compared to P-C and P-G MMMs. Moreover, significant improvement could be noticed in the gas separation performance of P-2C/3G MMM, in which the CO_2/CH_4 selectivity and CO_2 permeability were enhanced by 712% and 23%, respectively. Therefore, this technique of combining two carbonaceous nanofillers is deemed effective for efficient CO_2/CH_4 gas separation MMM fabrication.

Acknowledgment

This work was financially supported by Universiti Teknologi PETRONAS (UTP) under YUTP-FRG grant No. 0153AA-E08.

References

[1] Houshmand, A., Wan Daud, W. M. A., & Shafeeyan, M. S. (2011). Exploring potential methods for anchoring amine groups on the surface of activated carbon for CO_2 adsorption. *Separation Science and Technology*,*46*(7), 1098-1112.

[2] Ismail, A. F., Goh, P. S., Sanip, S. M., & Aziz, M. (2009). Transport and separation properties of carbon nanotube-mixed matrix membrane. *Separation and Purification Technology*, *70*(1), 12-26.

[3] Julian, H., & Wenten, I. G. (2012). Polysulfone membranes for CO_2/CH_4 separation: State of the art. *IOSR Journal of Engineering*, *2*(3), 484-495.

[4] Goh, P. S., Ismail, A. F., Sanip, S. M., Ng, B. C., & Aziz, M. (2011). Recent advances of inorganic fillers in mixed matrix membrane for gas separation. *Separation and Purification Technology*, *81*(3), 243-264.

[5] Rezakazemi, M., Amooghin, A. E., Montazer-Rahmati, M. M., Ismail, A. F., & Matsuura, T. (2014). State-of-the-art membrane based CO_2 separation using mixed matrix membranes (MMMs): an overview on current status and future directions. *Progress in Polymer Science*, *39*(5), 817-861.

[6] Galve, A., Sieffert, D., Staudt, C., Ferrando, M., Güell, C., Tellez, C., & Coronas, J. (2013). Combination of ordered mesoporous silica MCM-41 and layered titanosilicate JDF-L1 fillers for 6FDA-based copolyimide mixed matrix membranes. *Journal of Membrane Science*, *431*, 163-170.

[7] Valero, M., Zornoza, B., Téllez, C., & Coronas, J. (2014). Mixed matrix membranes for gas separation by combination of silica MCM-41 and MOF NH2-MIL-53 (Al) in glassy polymers. *Microporous and Mesoporous Materials*, *192*, 23-28.

[8] Zornoza, B., Seoane, B., Zamaro, J. M., Téllez, C., & Coronas, J. (2011). Combination of MOFs and zeolites for mixed-matrix membranes. *ChemPhysChem*, *12*(15), 2781-2785.

[9] Wang, T., Yang, C. H., Man, C. L., Wu, L. G., Xue, W. L., Shen, J. N., ... & Yi, Z. (2017). Enhanced Separation Performance for CO_2 Gas of Mixed-Matrix Membranes Incorporated with TiO_2/Graphene Oxide: Synergistic Effect of Graphene Oxide and Small TiO_2 Particles on Gas Permeability of Membranes. *Industrial & Engineering Chemistry Research*, *56*(31), 8981-8990.

[10] Xue, Q., Pan, X., Li, X., Zhang, J., & Guo, Q. (2017). Effective enhancement of gas separation performance in mixed matrix membranes using core/shell structured multi-walled carbon nanotube/graphene oxide nanoribbons. *Nanotechnology*, *28*(6), 065702.

[11] Jamil, A., Ching, O. P., & Shariff, A. M. (2017). Mixed matrix hollow fibre membrane comprising

polyetherimide and modified montmorillonite with improved filler dispersion and CO_2/CH_4 separation performance. *Applied Clay Science*, *143*, 115-124.

[12] Khulbe, K. C., Matsuura, T., Lamarche, G., & Lamarche, A. M. (2000). X-ray diffraction analysis of dense PPO membranes. *Journal of Membrane Science*, *170*(1), 81-89.

[13] Li, X., Cheng, Y., Zhang, H., Wang, S., Jiang, Z., Guo, R., & Wu, H. (2015). Efficient CO_2 capture by functionalized graphene oxide nanosheets as fillers to fabricate multi-permselective mixed matrix membranes. *ACS Applied Materials & Interfaces*, *7*(9), 5528-5537.

[14] Sarfraz, M., & Ba-Shammakh, M. (2016). Synergistic effect of incorporating ZIF-302 and graphene oxide to polysulfone to develop highly selective mixed-matrix membranes for carbon dioxide separation from wet post-combustion flue gases. *Journal of Industrial and Engineering Chemistry*, *36*, 154-162.

[15] Li, X., Ma, L., Zhang, H., Wang, S., Jiang, Z., Guo, R., ... & Wang, B. (2015). Synergistic effect of combining carbon nanotubes and graphene oxide in mixed matrix membranes for efficient CO_2 separation. *Journal of Membrane Science*, *479*, 1-10.

[16] Qiu, L., Yang, X., Gou, X., Yang, W., Ma, Z. F., Wallace, G. G., & Li, D. (2010). Dispersing carbon nanotubes with graphene oxide in water and synergistic effects between graphene derivatives. *Chemistry-A European Journal*, *16*(35), 10653-10658.

Synthesis and characterization of ionic liquid polymer composite with zeolite and its application for carbon dioxide capture

Noorshahidah Ahmad[1], A. Vijaya Bhaskar Reddy[2], Bidyut Baran Saha[3], Nasruddin[4], Muhammad Moniruzzaman[1,2*]

[1]Department of Chemical Engineering, Universiti Teknologi PETRONAS, 32610 Seri Iskandar, Perak, Malaysia
[2]Centre of Research in Ionic Liquids (CORIL), Universiti Teknologi PETRONAS, 32610 Seri Iskandar, Perak, Malaysia
[3]International Institute for Carbon-Neutral Energy Research (WPI-I2CNER), Kyushu University, 744 Motooka, Nishi-Ku, Fukuoka 819-0395, Japan
[4]Department of Mechanical Engineering, Faculty of Engineering, University of Indonesia, Indonesia

m.moniruzzaman@utp.edu.my

Abstract. Among the different ways to capture carbon dioxide (CO_2), ionic liquid technology is one of the prominent approaches and gained the interest of researchers to improve this technology from time to time. In this study, we have prepared and characterized a novel ionic liquid polymer composite with zeolite and tested its efficiency to capture CO_2. The ionic liquid monomer, 1-vinyl-3-ethylimidazolium bis(trifluoromethylsulfonyl)imide [Veim][Tf2N] was polymerized through free radial polymerization process adding zeolite to form poly[Veim][Tf2N] zeolite polymer composite. The synthesized polymer ionic liquid zeolite composite was characterized by FT-IR, FE-SEM and EDX. The formation of polymer ionic liquid zeolite composite was indicated by the presence of zeolite peaks in FT-IR spectra. Additionally, the EDX results were confirmed the presence of zeolite in the prepared composite. Finally, to prove the efficiency of prepared polymer ionic liquid zeolite composite, the adsorption studies have been conducted using CO_2 adsorption cell. The results revealed the capability of ionic liquid zeolite polymer composite in adsorbing high amount of CO_2 and recorded the highest pressure drop compared to other polymer ionic liquid and monomer ionic liquid materials.

1. Introduction

In recent years, the development of industrial revolution has created higher demand for energy that is being supplied by thermal power plants. The higher production of energy to meet the demand of industrial needs is primarily achieved by the combustion of fossil fuels (i.e., coal, oil and gas), whose combustion creates a negative impact on the environment and producing excess of greenhouse gases that are mainly consist of carbon dioxide (CO_2). According to United States Environmental Protection Agency (2015), CO_2 is contributing for the largest percentage of greenhouse emissions, which is roughly 82% from 6587 million metric tons in 2015 compared to other greenhouse gases [1].

NOAA's Earth System Research Laboratory and Scripps Institution of Oceanography (May 2016) were responsible to measure the emission of greenhouse gases in the atmosphere, and both the agencies have reported that concentration of CO_2 in the atmosphere has increased by more than 2.0 ppm every year from 2012 to 2015. They were also reported that, the estimated average amount of CO_2 in 2016 (404.16 ppm) was increased by almost 4.0 ppm when compared with the reported value of average amount of CO_2 in 2015 (400.30 ppm) [2].

The above data clearly indicates that, the concentration of CO_2 in the atmosphere is increasing every year. Therefore, many researchers and scientists have paid attention to discover the new technologies and methods that control the concentration levels of CO_2 in the atmosphere, which is the prime contributor of greenhouse gases. Several methods have been discovered to capture and separate CO_2 physically and chemically including sorption processes and membrane separations [3]. In recent years, ionic liquids have been proposed as promising alternatives to the conventional volatile solvents, which possess high CO_2 solubility. Basically, ionic liquids are low melting point salts with negligible vapour pressure, thermal stability with high CO_2 solubility. The trivial negative impact of ionic liquids to the environment has created the advantage and opportunity for their use in the CO_2 capture [4].

The major advantage of ionic liquids is their excellent solvation property for a variety of organic and inorganic compounds and their ability to undergo modification according to specific objectives and requirements with the marginal effect to environment is sufficient for them to be known as 'greener solvents' [5, 6]. Even though, ionic liquids have the advantage for the absorption of CO_2, the higher practical Henry's Law constant of ionic liquids does not help in the effective application of ionic liquids. Additionally, the liquid structure of ionic liquids has lower thermal stability and makes the process of CO_2 separation difficult to be done. Therefore, researchers were attempted to find the new ways to enhance the efficiency of CO_2 sorption performance by polymerization of ionic liquids [7]. Ionic liquid polymers were found to have higher sorption capacities than the room temperature ionic liquids. For instance, a study reported that the performance of poly(ionic liquid)s i.e., P[VBTMA][BF4] and P[MATMA][BF4] was increased over 7.6 capacities with 6.0 times of the efficiency compared to their monomer [bmim][BF4] [8]. Additionally, a large variety of adsorbents have been incorporated with ionic liquids to improve the CO_2 adsorption including activated carbon, zeolite, metal organic framework (MOF), alkali metal carbonate and amine based solids [9].

Therefore, in the present study, zeolite has been considered as the adsorbent/solid phase because of its abundant quantity on the earth and its low cost gives the advantage for the process to become more economical similar to activated carbon [10]. Additionally, zeolite has the crystalline molecular structure of aluminium and silicon atoms surrounded by four oxygen atoms, or it can be aluminosilicate consists of uniform porous surface enable the undesired molecule to be adsorbed by penetrating it. Therefore, it is proposed that the incorporation of zeolite into ionic liquid polymer may enhance the adsorption capacity of polymer ionic liquid [11-13]. Hence, the objective of this study is to synthesize and characterize the polymer ionic liquid zeolite composite and to investigate the efficiency of prepared ionic liquid zeolite polymer composite for the sorption of CO_2. The results of this study were also composed with the existed literature to prove the efficiency of synthesized ionic liquid zeolite polymer composite.

2. Experimental

2.1. Materials

The chemicals used to synthesize the ionic liquids including ethanol, bromoethane and ethyl acetate were obtained from Merck, New Jersey, USA. Lithium bis(trifluoromethylsulfonyl)imide, 1-vinylimidazole, zeolite 13X, trimethylolpropane trimetacrylate were purchased from Sigma-Aldrich, St. Louis, Missouri, USA. The chemicals used as supplied without further purification. Appropriate amount (about 1.0%) of salts were dissolved in ethanol to prepare ionic liquid zeolite polymer composite.

2.2. *Synthesis of ionic liquid monomers*

Bromoethane and 1-vinylimidazole were used as primary materials for the synthesis of 1-vinyl-3-ethylimidazolium bromide [Veim][Br] ionic liquid. For this, 33 g of 0.3 mol bromoethane was added to 28 g of 0.3 mol 1-vinylimidazole under the vigorous stirring in ice bath. Then, the solution was heated to 40 °C for 2.0 hrs to ensure the reaction of total component. Thereafter, the sample was dried in rotary evaporator for 5.0 hrs at 70 °C and drops the pressure until 50 mmHg to ensure that the prepared ionic liquid was completely dry.

For the preparation of second ionic liquid monomer i.e., 1-vinyl-3-ethylimidazolium bis(trifluoromethylsulfonyl)imide [Veim][Tf2N], ionic liquid [Veim][Br] and Li[Tf2N] were used as starting materials. For this, 40 g of 0.197 mol [Veim][Br] was poured and mixed into 20 mL of distilled water. Then, 62 g of 0.216 mol Li[Tf2N] was added to [Veim][Br] in 1:1 (v/v) ratio by drop wise to avoid the excess heat generation. Two phases of liquids were formed on mixing at room temperature under constant stirring, indicating [Veim][Tf2N] formation at bottom phase and aqueous Li[Br] at top phase. The mixture was then separated using a separating funnel and the desired product [Veim][Tf2N] was washed thoroughly with distilled water and the process was repeated few times to ensure that there is no Br- remained. Distilled water containing Br- was removed using separating funnel and using rotary evaporator to ensure the removal of water completely.

2.3. *Synthesis of ionic liquid polymer composite zeolite*

The ionic liquid polymer, 1-vinyl-3-ethylimidazolium bis(trifluoromethylsulfonyl)imide was prepared by conventional free-radical polymerization. Initially, 6.0 g of 0.015 mol ionic liquid monomer i.e., [Veim][Tf2N] was added to 6.0 g of 0.13 mol ethanol and then the solution was purged with nitrogen and stirred well at 300 rpm for 30 minutes. Next, 120 mg of 0.72 mmol 2,2'-azobis isobutyronitrile (AIBN) and 300 mg of 0.9 mmol trimethylolpropane trimetacrylate (TRIM) were added to the above solution and purged again with nitrogen for another 10 minutes. Thereafter, the mixture was heated at 65 °C for 30 minutes without any purging under nitrogen. A white gel-elastic solid was formed at the end of stirring, then it has been diluted with ethanol to remove any unreacted monomer. The solution was poured into distilled water to precipitate the polymer composite. The samples were washed carefully with ethanol for few times prior its filtration using filter funnel and filter paper. After the filtration, the samples were dried using rotary evaporator to remove solvents at 55 °C with minimum pressure of 80 mmHg for 15 minutes. Likewise, the preparation of ionic liquid polymer zeolite composite is similar to the preparation of poly[Veim][Tf2N] except the addition of zeolite 13X along with the ionic liquid monomer as well as ethanol with mass 180 mg with 6.0 g [Veim][Tf2N]. The resulted poly[Veim][Tf2N] composite zeolite was appeared to be white elastic-solid which contain white powder of zeolite, and it is quite transparent elastic-solid compared to poly[Veim][Tf2N] (Fig. 1).

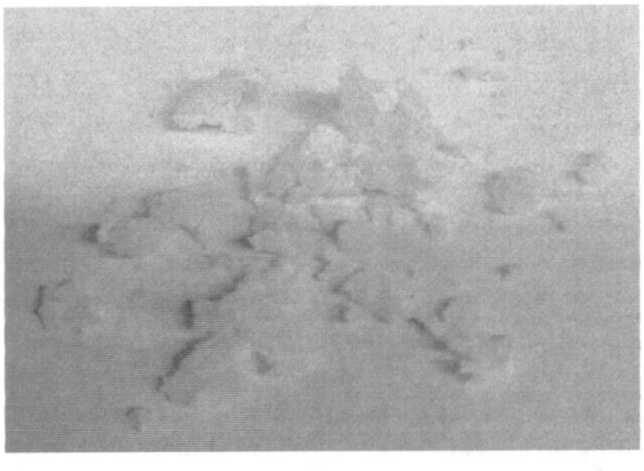

Figure 1. Poly[Veim][Tf2N] zeolite composite

2.4. *Characterization of samples*
The fabricated polymer composite was characterized by fourier transform infrared spectroscopy (FT-IR) (Perkin Elmer Spectrum One) to evaluate the incorporation of zeolite into polymer ionic liquid. Next, the field emission scanning electron microscopy (FE-SEM) with energy dispersive X-ray spectroscopy (EDX) (Zeiss Supra 55 VP) was used to observe the crystalline structure and morphology of the sample, as well as to analyze the elemental composition in the prepared polymer composite. Finally, the ionic liquids were characterized using ^{1}H NMR, ^{13}C NMR (Bruker Avance 500 MHz).

2.5. *Application of samples to CO_2 capture*
Ionic liquid monomer [Veim][Tf2N] and its polymer composite materials have been tested for CO_2 adsorption using CO_2 adsorption cell at room temperature 298 K with the initial pressure exerted at 10 bars. This equipment measured the pressure drop of CO_2 gas after passing through the sample. Amount of CO_2 adsorbed by the samples has been analysed using material balance before and after the test and compressibility factor was calculated using Soave-Redlich-Kwong (SRK) equation of state as shown below.

$$P = \frac{RT}{(Vm - b)} - \frac{a}{\sqrt{T}\, Vm(Vm + b)}$$

Where, P is the gas pressure; R is the gas constant; T is temperature; Vm is the molar volume (V/n); a & b are the constants to correct the attractive potential of molecules and volume respectively.

3. Result and discussion

3.1. *Characterization of samples*
The ionic liquid monomers namely [Veim][Br] and [Veim][Tf2N] were characterized using H^1 NMR to verify their purity. Polymeric materials including poly[Veim][Tf2N] and poly[Veim][Tf2N] zeolite composites were characterized by FT-IR and compared the respective peaks of monomer, polymer and zeolite. Fig. 2 exhibited some difference among the peaks of monomer and polymer as the intensity at the range 1600-1680 cm^{-1} is very weak for poly[Veim][Tf2N] and higher for the monomer. Basically, this range indicates the presence of non-aromatic alkenes. Therefore, the results expressed that alkene (C=C) from the monomer was successfully bonded and connected to same ionic liquid molecule to form polymer structure. At the peak 1729 cm^{-1}, its intensity for polymer was higher compared with monomer, which indicates the presence of carbonyl group (C=O) from TRIM. Poly(Veim][Tf2N] zeolite composite was characterized by FT-IR and compared with poly[Veim][Tf2N] ionic liquid. In the same figure, the difference has been displayed in the black box indicating the higher intensity peak at 950-1000 cm^{-1} and the presence of peak at 420-500 cm^{-1} for zeolite polymer composite compared to non-composited polymer. Actually, both the peaks were common for zeolite which represents the tetrahedral characteristic of the internal QO4 where Q is Si or Al [14]. Based on this different, it is confirmed that zeolite is completely incorporated into polymer and resulted to form a composite.

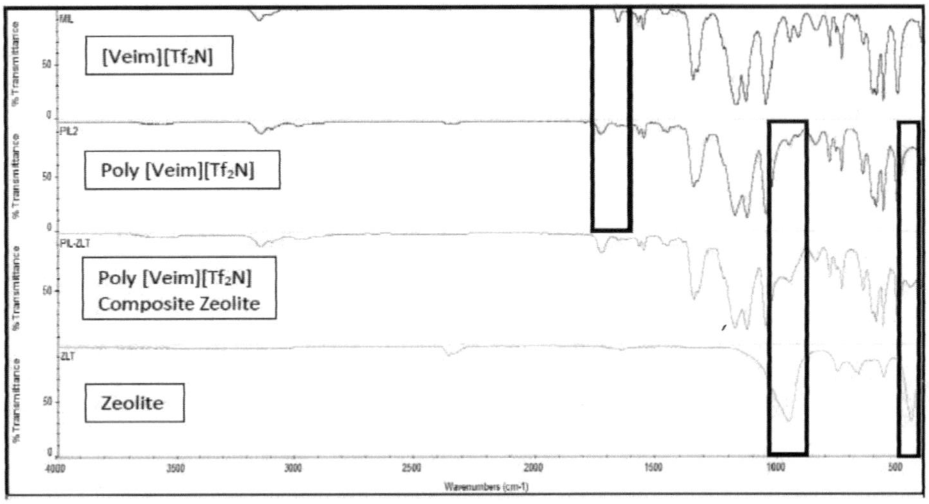

Figure 2. FTIR spectra of [Veim][Tf_2N], poly[Veim][Tf_2N], poly[Veim][Tf_2N] zeolite composite and zeolite.

Additionally, the polymer composite material was characterized by Energy Dispersive X-ray (EDX) and represented in Fig. 3 and Fig. 4. The EDX analysis showed that the major elements in both the polymers were carbon, oxygen, fluorine and sulphur. Hence, based on Fig. 4, it is revealed that the presence of zeolite components such as aluminium and silicon in the polymer indicates that the zeolite was successfully integrated into polymer structure.

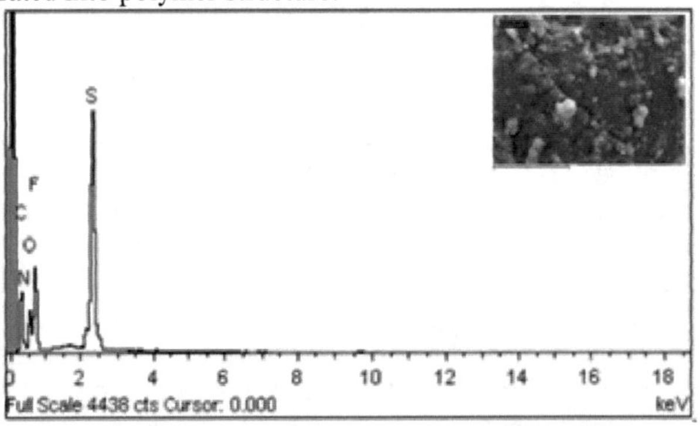

Figure 3. EDX analysis for Poly [Veim][Tf2N]

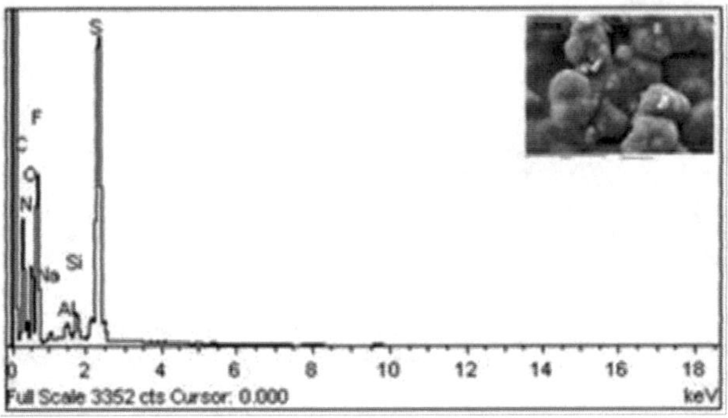

Figure 4. EDX analysis of Poly[Veim][Tf2N] zeolite composite

Besides, the surface morphology of the polymeric materials was determined using field emission scanning electron microscopy to observe the porosity and structure of polymers. Figure 5 indicates that the porosity of polymer was reduced after zeolite introduction into the polymer structure. Even though there is reduction in the porous structure, it indicates that zeolite if fully mixed with the polymer and the structure of zeolite itself is microporous which would increase the efficiency of CO_2 adsorption process.

Figure 5. FESEM morphology at magnification 10000;
Left: Poly[Veim][Tf2N], Right: Poly[Veim][Tf2N] zeolite composite

3.2. Application of samples for CO_2 Capture

A CO_2 adsorption cell was used to evaluate the capability of prepared polymer composite samples for CO_2 capture. This equipment measured the equilibrium pressure of CO_2 after passing through the sample. The same initial pressure was exerted for all the samples i.e., 10 bars and the adsorption process was conducted for about 48 hrs at ambient temperature. The value of pressure was being recorded for every 60 sec. CO_2 concentration has been calculated using material balance before and after the adsorption experiment using the following Peng-Robinson equation of state.

$$CO_2 \text{ uptake} = n = \frac{P_{ini}.V_{tot}}{Z_{ini}.R.T_{ini}} - \frac{P_{eq}(V_{tot} - V_{sample})}{Z_{eq}.R.T_{eq}}$$

where n = moles of CO_2 captured; P_{ini} = initial pressure obtained in the absence of sample after expansion of the gas sample from the bomb into the whole system; V_{tot} = total volume of system; Z_{ini} = compressibility factor ($P_{ini} \times T_{ini}$); Z_{eq} = compressibility factor ($P_{eq} \times T_{eq}$); P_{eq} = pressure at equilibrium; V_{sample} = volume of sample; R is the ideal gas constant; T_{ini} = initial temperature and T_{eq} = temperature at equilibrium.

The SRK equation of state has determined the compressibility factor of CO_2, which was found to be 0.894. Based on the results in Fig. 6, it is proved that poly [Veim][Tf2N] zeolite composite is capable to provide higher adsorption capacity compare to the poly [Veim][Tf2N] and [Veim][Tf2N] ionic liquid monomer. Further, the efficiency of the prepared poly[Veim][Tf2N] zeolite was compared with the reported methods, which indicated the clear improvement in CO_2 adsorption when zeolite is incorporated into ionic liquid polymer material [15-17].

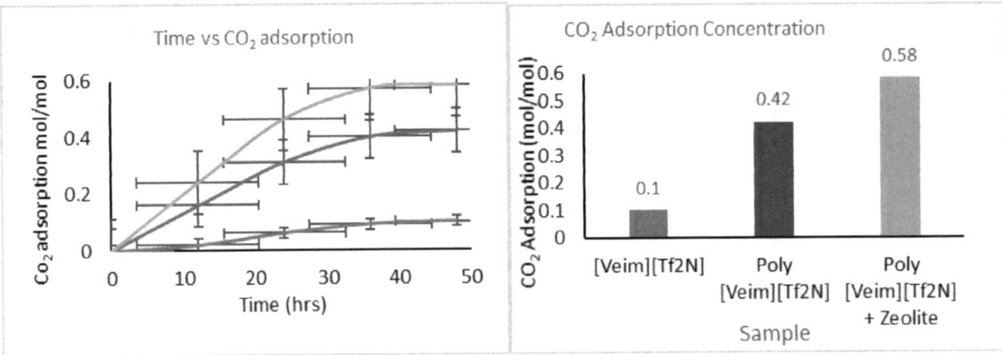

Figure 6. CO_2 adsorption for [Veim][Tf2N], poly[Veim][Tf2N], and poly[Veim][Tf2N] zeolite composite evaluated by CO_2 adsorption cell

4. Conclusion

Ionic liquid polymers namely poly[Veim][Tf2N] and poly[Veim][Tf2N] zeolite composite was successfully synthesized through free radical polymerization process from its monomer [Veim][Tf2N]. During the polymer composite formation 2,2'-azobis(isobutyronitrile) (AIBN) and trimethylolpropane trimethacrylate (TRIM) were added as initiator and cross linker respectively. The synthesized polymer material composites have been characterized using FT-IR, EDX to evaluate the composition of zeolite in the polymer structure. The morphology of the polymer surface was also analysed using FESEM to observe porosity of the polymer. The results showed that zeolite was successfully incorporated into the polymer.

The performance of synthesized ionic liquid materials has been tested using CO_2 adsorption cell to evaluate the adsorption capacity between the polymer material either composited or non-composited and also the ionic liquid monomer. From the results, it is proved that ionic liquid composite polymer gave the best performance as expected by providing higher adsorption capacity compare to non-composited and ionic liquid monomer. The composition of polymer ionic liquid composite i.e., combination between external adsorbent (zeolite) and enhancement of ionic liquid into polymer was believed to influence the efficiency of CO_2 capture. Future work is also recommended to analyse the best desorption process for CO_2 from polymer material to separate CO_2 and to reuse the adsorbent material. Also, proper optimisation between the amount of polymer ionic liquid and external adsorbent is very necessary to increase its efficiency with lower cost.

Acknowledgements

The present work is supported by an international project (Cost center: 0153AB-M32) from Faculty of Engineering, University of Indonesia, Indonesia.

References

[1] Shaun A R, Creason J, Jameel A, Sara O, Jeffrey E P and Robert H B 2015 Global mitigation of non-CO_2 greenhouse gases: marginal abatement costs curves and abatement potential through 2030 *J. Integr. Environ. Sci.* **12** 155-168

[2] Min L, Jiabing W, Zhu X, He H, Jia W and Xiang W 2015 Evolution and variation of atmospheric carbon dioxide concentration over terrestrial ecosystems as derived from eddy covariance measurements *Atmos. Environ.* **114** 75-82

[3] Roda B, Nancy L, Denis R, Denis R, CécileV and Eric F 2006 Membrane processes for post-combustion carbon dioxide capture: A parametric study *Energy* **31** 2556-2570

[4] Magaret S, Moniruzzaman M and Masahiro G 2016 Recent advances in exploiting ionic liquids for biomolecules: solubility, stability and applications *Biotechnol. J.* **11** 1000-1013

[5] Han D and Row K H 2010 Recent applications of ionic liquids in separation technology *Molecules* **15** 2405-2426
[6] Animesh P, Maisara S R S, Moniruzzaman M, Cecilia W, Sourav M, Kyaw T and Bidyut B S 2017 Ionic liquid as a new binder for activated carbon based consolidated composite adsorbents *Chem. Eng. J.* **326** 980-986
[7] Cheng H, Ping W, Luo J, Jan F, Dirk E D V and Luo Z H 2015 Poly (ionic liquid)-based nanocomposites and their performance in CO_2 capture *Ind. Eng. Chem. Res.* **54** 3107-3115
[8] Jianbin T, Huadong T, Weilin S, Henry P, Maciej R and Shen Y 2005 Poly(ionic liquid)s: A new material with enhanced and fast CO_2 absorption *Chem. Commun.* **0** 3325-3327
[9] Hamayoun M, Hafiz M A S, Moniruzzaman M and Suzana Y 2016 Synthesis of ionic liquid polymer incorporating activated carbon for carbon dioxide capture and separation *Adv. Mater. Res.* **1133** 566-570
[10] Siriwardane R V, Shen M S and Edward P F 2005 Adsorption of CO_2 on zeolites at moderate temperatures *Energ. Fuel.* **19** 1153-1159
[11] Petrov I and Michalev T 2012 Synthesis of zeolite A: A review. Zlatarov University 1-5
[12] Yang H., Xu Z, Fan M, Gupta R, Slimane R B, Alan B E and Wright I 2008 Progress in carbon dioxide separation and capture: A review *J. Environ. Sci.* **20** 14-27
[13] Jadhav P, Chatti R V, Biniwale R B, Labhsetwar N K, Devotta S and Rayalu S S 2007 Monoethanol amine modified zeolite 13X for CO_2 adsorption at different temperatures *Energ. Fuel.* **21** 3555-3559
[14] Ojumu T V, Du Plessis P W and Petrik L F 2016 Synthesis of zeolite A from coal fly ash using ultrasonic treatment-A replacement for fusion step *Ultrason. Sonochem.* **31** 342-349
[15] Mojtaba M, Babak M, Alireza B and Ali S 2018 Solubility of carbon dioxide and methane in 1-hexyl-3-methylimidazolium nitrate ionic liquid, experimental and thermodynamic modeling *The J. Chem. Thermodyn.* **122** 31-37
[16] Franciele L B, Rafael B D, Marisol F R, Maria C C F, Luz A C, Vitaly V C, Felipe D V and Sandra E 2018 Cellulose based poly(ionic liquids): Tuning cation-anion interaction to improve carbon dioxide sorption *Fuel* **211** 76-86
[17] Andre B, Jianbin T, Hu X, Sugata P T, Youqing S and Maciej R 2007 Carbon dioxide solubility in polymerized ionic liquids containing ammonium and imidazolium cations from magnetic suspension balance: P[VBTMA][BF4] and P[VBMI][BF4] *Ind. Eng. Chem. Res.* **46** 5542-5547

Density Measurement of Aqueous solution of Tetrapropylammonium Hydroxide at different Temperatures and Concentrations: Data and Correlation

Nur Ani kartikawati; Bhajan Lal; Muhammad Ibrahim Bin Abdul Mutalib; Suriati Bt Sufian; Azmi Mohd Shariff.

Chemical Engineering Department Universiti Teknologi PETRONAS, 32610, Bandar Seri Iskandar Perak, Malaysia.

Email: bhajan.lal@utp.edu.my

Abstract. Densities of aqueous Tetrapropylammonium hydroxide solution (5,10, 15, 20 and 25 wt %) were invistigated over a temperature range of 298.15 to 333.15 K and at atmospheric pressure. Derived properties, molar volume (V_m) and excess molar volumes (V_m^E) were estimated and correlated with temperature using the least square method. The predicted values from correlations are in good agreement with measured values. The obtained results from derived properties for a binary system, V_m^E are interpreted based on intermolecular interactions.

Keywords: Density, Thermal expansion, Excess molar volume, and TPAOH

1. Introduction

The emission of greenhouse gases, particularly CO2 has been considered as a major concern for environment. It causes incresing of global temperature which araise climate change fear. CO2 separation by absorption using aqueous amines solutions is the most extensively used method due to its wide operating experience in the industry and technical maturity. Even have many limitation of problems faced in CO_2 absorption such as corrosion, foaming, solvent degradation and volatility[1]. Ionic Liquids have attracted considerable interest recently, due to their unique physical and chemical properties include low vapour pressure, high thermal stability, and good slovent for wide range of organic and inorganic compounds [2].

Thermo physical properties of absorbent are necessary to characterize the solvents used in industrial applications. These properties are also essential in the design of treatment equipment of the acid gas removal. [3], [4]. The density parameter is crucial for physical solubility of CO_2. Anabella, et al., (2012) Have measured density with speed of sound for aqueous imidazolium-based ionic liquids at different temperatures [5]. Their density result found to decrease with an increase in temperatures. Govinda et al., (2015), has done about the effect of the alkyl chain length of the cation on the interactions between water and ammonium-based ionic liquids at 25 °C [6]. These obtained results declared that the ammonium-based ILs significantly affect the intermolecular interactions between the solvent molecules. Further Kartikawati et al., (2018) have measured density along with other physicochemical properties for aqueous solution of Ammonium based ionic liquids over a wide range of temperatures (295.15 to 333.15 K) [7]. The studied property is found to decrease with an increase in temperatures. Zhang et al., (2013) investigated the densities were increased linearly with a decrease in temperature and be higher than the density of water [8]. Machanova et al., (2012) reported the effect of temperature and the alkyl chain length on involved properties. They found that the density and viscosity decrease with temperature as opposed to conductivity [9]. Given that the density is crucial in absorption process, due the aqueous solution of Tetrapropylammonium hydroxide is potensial solvents for CO_2 capture, as well as the density data for aqueous TPAOH (Tetrapropylammonium Hydroxide) solution and comparison with conventional solvent at different temperatures is limited, it is very interesting to investigating about density of aqueous solution of TPAOH.

In the present study, the density of TPAOH solutions over temperature range (298.15 to 333.15 K) and concentration 5,10, 15, 20 and 25 wt %) have been studied and reported. Density data is also used to calculate other properties such as excess molar volume, and coefficient of thermal expansion. From the data obtained can be used to propose an accurate estimate of the temperature and concentration solvent for CO_2 capture.

2. Experimental Section

2.1. Material

TPAOH (purity ≥99.00%) was acquired from Sigma-Aldrich. Deionized water is used to prepare different (5,10, 15, 20 and 25 wt %) concentrations of TPAOH solution. Samples were prepared and weighed using an analytical balance with the accuracy of ± 0.1mg. All the solvent were kept in airtight bottles.

2.2. Density Procedure measurement

The densities of the TPAOH aqueous solutions are measured at temperatures ranged from 298.15 to 333.15 K using Anton Paar DMA-5000 digital vibrating tube density meter with an accuracy of ± 0.0006 g·cm^{-3}. The measuring cell temperature is adjusted using a built-in solid state thermostat, which maintains a temperature having an exactness of ±0.01 K. The densities reported using an average of three measurements. Before used, the equipment was calibrated with a standard water of Millipore quality.

3. Result and discussion

The density of TPAOH aqueous solution prepared with different concentrations (5,10, 15, 20 and 25 wt %) at a temperature range between of 298.15 to 333.15 K was measured.

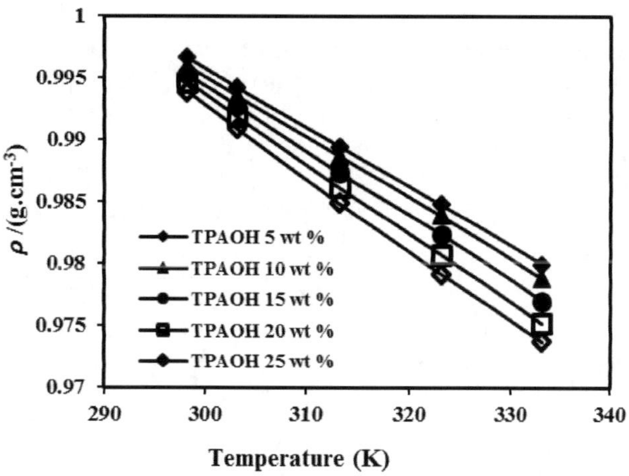

Figure 1. Graph of experimental data of density of aqueous TPAOH solution as a function of temperature for different concentration

The value of densities measurement of aqueous TPAOH solution is shown in Fig. 1. An inverse relationship is observed as the density is seen to increase with the decrease of temperature. The same trend has been reported in the literature [10]–[12]. By increasing the temperature, the molecules tend to migrate further apart as the kinetic energy within each molecule increases with an increase in the temperature [13]. It later was found that the experimental data regarding the density of aqueous TPAOH solutions are in the range of 0.965–0.995 g.cm^{-3} at various concentrations (5,10, 15, 20 and 25 wt %)

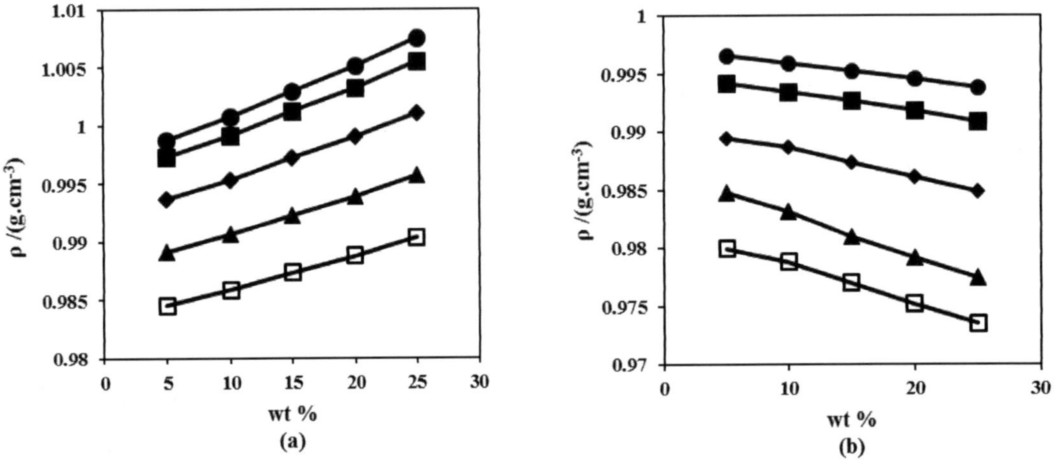

Figure 2. Comparison the density of literature data [maham 1993) for aqueous solution of MEA (a) with the experiment data of aqueous solution of TPAOH (b). and●, 298.15 K, ■, 303.15 K; ♦, 313.15 K;▲, 323.15 K;□, 33.15 K.

From the Fig.2, it can be seen that MEA and TPAOH aqueous solutions have different behavior for different concentration. For MEA the density increase with the increase of concentration while TPAOH showed different behavior. The results can be demonstrated based on the high interaction molecule between MEA and water. MEA has strong attractive forces to water molecules which reduce the intermolecular space, and the total volume of the solution can be smaller than the combined volume of its components, leading to an extra increase in density [14], [15]. For TPAOH solutions, the increasing of concentration means a higher amount of bulk molecules with hydrophobic alkyl chain are incorporated into the system. Thus, density is decreased [1].

Table 1. Correlation equation parameters and SD for densities ρ /(g·cm^{-3}) of TPAOH aqueous solution.

Cons (wt%)	ρ (exp)	T(K)	A0	A1	R2	ρ (predic)	SD
	0.99663	298.15	1.1383	-0.0005	1	0.989225	0.0077753
	0.99418	303.15				0.986725	
5	0.989448	313.15				0.981725	
	0.984774	323.15				0.976725	
	0.979937	333.15				0.971725	
10	0.99596	298.15	1.1412	-0.0005	0.9998	0.992125	0.0035824
	0.99349	303.15				0.989625	

	0.98868	313.15				0.984625	
	0.98393	323.15				0.984625	
	0.97882	333.15				0.974625	
	0.995288	298.15	1.1504	-0.0005	0.9998	1.001325	0.0076447
	0.99271	303.15				0.998825	
15	0.987361	313.15				0.993825	
	0.98241	323.15				0.993825	
	0.977022	333.15				0.983825	
20	0.99463	298.15	1.1601	-0.0006	0.9999	0.98121	0.014122
	0.99187	303.15				0.97821	
	0.98614	313.15				0.97221	
	0.98075	323.15				0.96621	
	0.97521	333.15				0.96021	
	0.993873	298.15	1.1659	-0.0006	0.9996	0.98701	0.0071
	0.990891	303.15				0.98401	
25	0.984867	313.15				0.97801	
	0.979172	323.15				0.97201	
	0.97371	333.15				0.96601	

The density correlated with temperature using Eq (1). The values of densities and the corresponding coefficients are presented in Table 2.

$$\rho = A_0 + A_1 (T) \tag{1}$$

Where ρ is density, A_0 and A_1 are correlation parameters, and T is the temperature in Kelvin. The standard deviation (SD) of predicted density by equation one is calculated using Eq 2.

$$SD = \left[\frac{\sum_1^n (X_{exp} - X_{cal})^2}{n}\right]^{0.5} \tag{2}$$

Where SD is standard deviation, X_{exp} is the value of experiment, X_{cal} is the estimated value, and n is some measure points. It can be observed from Table 2, R^2 value is higher than 0.95 which refer to the good agreement between experimental and predicted density.The density data and correlation parameter are used to calculate the coefficient of thermal expansion α_p according to Eq. 3. This property quantifies the expansion of the material by temperature. Thermal expansion is also necessary for designing acid gas separation system [18].

$$\alpha_p = -\frac{1}{\rho}\left[\frac{\partial \rho}{\partial T}\right] = \frac{A_1}{A_0 + A_1 T} \tag{3}$$

Where α_p stand for the thermal expansion coefficient, T is temperature, ρ is density and A_0 and A_1 are correlation parameter in Eq 1. From the result shown in Table 3, α_p increased with increasing temperature and concentration. It can be noted from Table 3 that the change in thermal expansion coefficients values affected slightly by temperature. Therefore, the variation of volume expansion of the studied solutions could be considered as independent of temperature. Meanwhile, adding more TPAOH sound to have higher thermal expansion due to weak bonding with water molecules compared with the water-water system.

Table 2. Coefficient of thermal expansion of aqueous solution of TBAOH

T(K)	α_p				
	5 wt %	10 wt %	15 wt %	20 wt %	25 wt %
298.15	0.1495143	0.1495131	0.1495096	0.1794072	0.1794046
303.15	0.1520143	0.1520131	0.1520096	0.1824072	0.1824046
313.15	0.1570143	0.1570131	0.1570096	0.1884072	0.1884046
323.15	0.1620143	0.1620131	0.1620096	0.1944072	0.1944046
333.15	0.1670143	0.1670131	0.1670096	0.2004072	0.2004046

The same trend has been investigated in the literature [15,19]. This property also is an advantage of TPAOH to be used as an absorbent for CO_2 capture system at a higher temperature so the thermal expansion in the absorbent can be avoided. The excess molar volume (V_m^E) of the solution is defined as the amount of volume which is in excess to that an ideal solution at the same conditions of the composition. For a binary system, V_m^E can be calculated according to the following equation (Eq. 4) using mole fractions, molar masses, and densities to represent the volumes of solution and simple components [18].

$$V_m^E = \left[x_2\, M_2(\frac{1}{\rho} + \frac{1}{\rho 2}) + (1 - x_2)M_1(\frac{1}{\rho} - \frac{1}{\rho 2})\right] \quad (4)$$

Where V_m^E is excess molar volume; ρ is the density of the aqueous solution, ρ_2 is the density of deionized water, M_1 is a molar mass of water, M_2 is a molar mass of solvent and x_2 is mole fraction of solvents. The excess molar volume results are tabulated in Table 4

Tabel 3. Excess molar volume of aqueous solution of TPAOH

T(K)	V_m^E(cm3.mol-1)				
	5 wt %	10 wt %	15 wt %	20 wt %	25 wt %
298.15	9.958458	9.950181	9.941879	15.978	15.91262
303.15	9.921438	9.912889	9.903224	15.84434	15.75944
313.15	9.846037	9.836455	9.819995	15.60283	15.49145
323.15	9.766976	9.756355	9.737225	15.44546	15.30607
333.15	9.681880	9.950181	9.644832	15.32206	15.18817

The results show a positive excess molar volume values in the presence of TPAOH. That can be attributed to physical properties, which are nonspecific interactions between the real species present in the mixture. The increase of excess molar volume is due to increasing dipole–dipole interactions which results strong in higher hydrogen bonding of components [19], [20]

Conclusion

In this paper, the densities of aqueous Tetrapropylammonium hydroxide solution (5,10, 15, 20 and 25 wt %) were invistigated over a temperature range of 298.15 to 333.15 K and at atmospheric pressure. The experimental data for density measurement revealed that as the temperature of the solution increased the value of the density value decreased. Moreover, the density of the solution also decreased with the increase of TPAOH concentration. The behavior is opposite of MEA solution. Predicted values of density are shown good agreement with the experimental value. Increasing the temperatures and concentrations is increasing the thermal expansion. The negative value of excess molar volume for the TPAOH aqueous solution decreased with increasing the temperature. Each physical property is correlated with temperature by the least square method, and the corresponding coefficients for each property are reported and used to predict the density.

Acknowledgment
The authors gratefully acknowledge Universiti Teknologi PETRONAS for providing the technical support to complete the present work

References

[1] O. Nashed, J. C. H. Koh, and B. Lal, "Physical-chemical Properties of Aqueous TBAOH Solution for Gas Hydrates Promotion," Procedia Eng., vol. 148, pp. 1351–1356, 2016.
[2] S. Garg, A. M. Shariff, M. S. Shaikh, B. Lal, A. Aftab, and N. Faiqa, "Selected physical properties of aqueous potassium salt of L-phenylalanine as a solvent for CO2capture," Chem. Eng. Res. Des., vol. 113, pp. 169–181, 2016.
[3] S. N. Khan, S. M. Hailegiorgis, Z. Man, A. M. Shariff, and S. Garg, "Thermophysical properties of aqueous N-methyldiethanolamine (MDEA) and ionic liquids 1-butyl-3-methylimidazolium trifluoromethanesulfonate [bmim][OTf], 1-butyl-3-methylimidazolium acetate [bmim][Ac] hybrid solvents for CO2capture," Chem. Eng. Res. Des., vol. 121, pp. 69–80, 2017.
[4] R. Umapathi, P. Attri, and P. Venkatesu, "Thermophysical Properties of Aqueous Solution of Ammonium-Based Ionic Liquids," J. Phys. Chem. B, vol. 118, no. 22, pp. 5971–5982, 2014.
[5] A. J. L. Costa et al., "Density , Thermal Expansion and Viscosity of Cholinium-Derived Ionic Liquids," pp. 1–9, 2012.
[6] V. Govinda, P. Venkatesu, and I. Bahadur, Molecular interactions between ammonium-based ionic liquids and molecular solvents: current progress and challenges, vol. 18, no. 12. Royal Society of Chemistry, 2016.
[7] N. A. Kartikawati, R. Safdar, B. Lal, M. I. B. A. Mutalib, and A. M. Shariff, "Measurement and correlation of the physical properties of aqueous solutions of ammonium based ionic liquids," J. Mol. Liq., vol. 253, no. 2017, pp. 250–258, 2018.
[8] M. Hou et al., "Thermodynamic properties of aqueous solutions of two ammonium-based protic ionic liquids at 298.15 K," J. Mol. Liq., vol. 178, pp. 149–155, 2013.
[9] J. Jacquemin, "Thermophysical Properties of Ammonium-Based Bis{(tri fl uoromethyl)sulfonyl}imide Ionic Liquids: Volumetric and Transport Properties," 2012.
[10] K. A. Kurnia, F. Harris, C. D. Wilfred, M. I. Abdul Mutalib, and T. Murugesan, "Thermodynamic properties of CO2 absorption in hydroxyl ammonium ionic liquids at pressures of (100-1600) kPa," Journal of Chemical Thermodynamics, vol. 41, no. 10. pp. 1069–1073, 2009.
[11] M. G. Montalbán, C. L. Bolívar, F. G. Díaz Baños, and G. Víllora, "Effect of Temperature, Anion, and Alkyl Chain Length on the Density and Refractive Index of 1-Alkyl-3-methylimidazolium-Based Ionic Liquids," J. Chem. Eng. Data, vol. 60, no. 7, pp. 1986–1996, 2015.
[12] R. L. Gardas, D. H. Dagade, J. a P. Coutinho, and K. J. Patil, "Thermodynamic studies of ionic interactions in aqueous solutions of imidazolium-based ionic liquids [Emim][Br] and [Bmim][Cl].," J. Phys. Chem. B, vol. 112, no. 11, pp. 3380–3389, 2008.
[13] P. Kilaru, G. A. Baker, and P. Scovazzo, "Density and surface tension measurements of imidazolium-, quaternary phosphonium-, and ammonium-based room-temperature ionic liquids: Data and correlations," J. Chem. Eng. Data, vol. 52, no. 6, pp. 2306–2314, 2007.

[14] T. G. Amundsen, L. E. Øi, and D. A. Eimer, "Density and Viscosity of Monoethanolamine + Water + Carbon Dioxide from (25 to 80) ° C," pp. 3096–3100, 2009.

[15] A. B. Pereiro, J. L. Legido, and A. Rodríguez, "Physical properties of ionic liquids based on 1-alkyl-3-methylimidazolium cation and hexafluorophosphate as anion and temperature dependence," J. Chem. Thermodyn., vol. 39, no. 8, pp. 1168–1175, 2007.

[16] R. Safdar, A. A. Omar, L. B. Ismail, A. Bari, and B. Lal, "Measurement and correlation of physical properties of aqueous solutions of tetrabutylammonium hydroxide, piperazine and their aqueous blends," Chinese J. Chem. Eng., vol. 23, no. 11, pp. 1811–1818, 2015.

[17] A. Muhammad, M. I. A. Mutalib, C. D. Wilfred, T. Murugesan, and A. Shafeeq, "Viscosity, refractive index, surface tension, and thermal decomposition of aqueous N-methyldiethanolamine solutions from (298.15 to 338.15) K," J. Chem. Eng. Data, vol. 53, no. 9, pp. 2226–2229, 2008.

[18] S. Md Nayeem, M. Kondaiah, K. Sreekanth, and D. Krishna Rao, "Thermoacoustic, volumetric, and viscometric investigations in binary liquid system of cyclohexanone with benzyl benzoate at t = 308.15, 313.15, and 318.15 k," J. Thermodyn., vol. 2014, 2014.

[19] N. Deenadayalu, S. Kumar, and P. Bhujrajh, "Liquid densities and excess molar volumes for (ionic liquids + methanol + water) ternary system at atmospheric pressure and at various temperatures," J. Chem. Thermodyn., vol. 39, no. 9, pp. 1318–1324, 2007.

[20] G. H. Y. T. L. Chen, "Density , Viscosity and Excess Molar Volume of the Aqueous Ionic Liquid Tris (monoethanolamine)," pp. 2117–2132, 2014.

Effects of Na: Al and water: solid ratios on the mechanical properties of fly ash based geopolymer

***Sarini Mat Yaakob[1], Nurul Ekmi Rabat[1], Suriati Sufian[1]**

[1]Department of Chemical Engineering, Universiti Teknologi PETRONAS, 32610, Perak.

*Corresponding Author Email: sarinimatyaakob@yahoo.com

[Abstract]. Every segment industry infrastructure like onshore/offshore setup, power generation and water management is part of a complex system which involves large investment. Unfortunately, these infrastructures easily susceptible to be attacked by corrosion, resulting in hundreds of millions of dollars in losses. Conventionally, steel is protected by anticorrosive chromate coatings. But these coatings contain carcinogenic hexavalent chromium compound which may cause to environment and health related issues. To reduce those impacts, conventional coatings are replaced by water-based alternatives and inorganic coatings. Geopolymer as a green inorganic material with superior properties is in great research to be explored as a new coating material. The present paper outlines briefly the potential of geopolymer as a coating material through the mechanical performance test of adhesion and flexural strength. Scanning electron microscopy (SEM) was used to examine the morphology of geopolymer and unreacted fly ash matrix. The test was conducted by varying two aspects of Na: Al and water: solid ratios. The findings show both the alkali and water content plays a asignificant role in geopolymerization. The increase in water content at low Na/Al ratio causes dilution of activator thus resulted in decreased of adhesion and flexural strength.

1. Introduction

Estimates of the annual cost of corrosion around the world lead to a huge number. In any case, corrosion represents a tremendous economic loss and much can be done to reduce it. Corrosion of bridges is a major problem as they age and require replacement, which costs billions. A large chemical company spent more than $400,000 per year for corrosion maintenance in its sulfuric acid plants. The petroleum industry spends a million dollars per day to protect underground pipelines [1].

In order to maintain the durability of systems, coatings are applied on metal surface. There are several types of coatings that available in the market such as organic coatings, metallic coatings, zinc coatings and ceramic coatings, in which, different coatings serve for different application [2]. When the coatings itself is subjected into certain failure or damage, consequently it will cause a serious corrosion issue occurs. Basically, there is a gradual formation of defect such as the presence of void or a pore inside the coatings, which may cause detachment occurred. Later, it will followed by the penetration of the corrosive species likes chlorine ions that promote adhesion loss and indirectly attacks the underlying metal. Thus, in order to solve that such issue, a major operation process is required which caused big economic loses [3] - [6].

There are several types of coatings that have been used as surface protection materials such as acrylic, polyurethane, epoxy, etc. However, these coatings usually cover the material surface by physical absorption. In fact, to achieve an excellent protection, coatings should be coated the surface by both physical and chemical absorption. On top of that, most of the available coating materials in the market nowadays are being derived from the compound that is not only harmful to the humans but also to the environment. The content of volatile organic compounds (VOCs) in coatings may lead into serious health related issues such as asthma, bronchitis and sometimes skin and lung cancer [7]. Meanwhile, it may also affect to the depletion of the ozone layer and greenhouse effect.

Geopolymers have been materialized as a potentially new inorganic material with interesting properties such as high compressive strength, good thermal stability, high flexural strength and good fire resistance. Geopolymers are synthetic green polymeric materials prepared by the chemical interaction of raw aluminosilicates obtained from power plants such as coal fly ash, mineral clays (metakaolin), and slag form metallurgy, with an aqueous alkaline solution of hydroxide and/or silicate as reaction process accelerators resulted in the development of solid, insoluble binding material [8] - [9]. Chemically geopolymers are tetrahedrally cross linked alumino silicates with Si^{+4} and Al^{+4} units and the structure is charge stabilized by alkali cation. The content of Si and Al will much influence the chemistry of geopolymers [10]. Geopolymers chemistry is synonymous to zeolite with the exceptions that geopolymers are amorphous and the unreacted starting material is also intact with geopolymer gel.

Early research on the alkali activation works dating back to 1908, when Kuhl *et al* patented their first formulation called slag cement by mixing the blast furnace slag (BFS) with alkaline components [11]. Later on, in 1930-1950, work by Prudon et al was conducted in a lab scale project to fabricate cement from steel slag and alkalis. Based on the findings, the idea and technology were adopted in 1950 to commercialize the materials in Belgium. Noticeably, the idea was implemented in the con struction buildings at that period [12]. Glukhovskii from Russia had developed building materials using alkali activation of low calcium aluminosilicate precursor by calling the material as soil cement and soil silicates [13]. The term of 'geopolymer' was initially introduced by Davidovits in 1970s for alkali activated metakaolin. The term of geopolymer was used by him as to resemblances the geopolymers with organic thermoset resins [14]. The findings from Davidovits' work become a direction for researchers to develop new material for building structure. A hybrid concrete as called Pyrament with high early strength was commercialized in 1980-1990 in North America [15].

The application of fly ash in geopolymer based was first reported in 199G0s by Wastiels *et al*. The binder was prepared for the work by chemical reaction of alkaline solution with blast furnace slag and fly ash [16]. A superior properties of geopolymer such as controlled setting time, low water permeability, resist to corrosion, heat and fire attack make it a new potential material that can be used as a coating. Temuujin *et al* has proven thin coatings from geopolymer had shown a good resistance towards the heat and fire. High adhesion strength of >3.5MPa was obtained from the work and good thermal stability as only 5-8% mass loss was recorded during heating from 50-800°C [17]. It was found that different parameters significantly influenced the performances of geopolymer coatings, including chemical composition, water content, and particle sizes of the precursors [30]-[32]. Besides, the ratio of Na: Al, Si:Al and water: solid plays a significant role towards geopolymer properties. In another study by Temuujin *et al* reported the adhesion strength of geopolymer coating was much affected by Si: Al and water: solid ratio. Noticeable, an increasing in strength was directly proportional with increasing of Si: Al and water: solid ratio [18].

Reflected to the available knowledge in literature, current research was conducted on the fly ash based geopolymers as to expend more the mechanical properties especially in adhesion strength and flexural strength. In order to find the optimum performance of geopolymers, there were two parameters that are chosen to be tested on synthesization of geopolymers. The effect of Na: Al and water: solid ratios were tested in the current study. The results were comparatively reported.

2. Experimental section

2.1 Raw materials

Coal fly ash was procured from Malaysia power plant. To minimize the variation in the particle size distribution, and to spate larger particles, the fly ash was sieved through a 45μm sieve using a vibratory sieve shaker (AS200 Basic, RETSCH, Germany). Sodium hydroxide (99%) was procured from Merck Millipore, Malaysia. Hydroxide solution was prepared individually by dissolving NaOH pellets in demineralized water, according to the Na: Al and water: solid ratio. The solution was stored in air tight Pyrex container to prevent changes in the composition of the solution due to contamination. A carbon steel plate with the dimension of (3 x 50 x 127 mm) was used for adhesion testing. Prior to adhesion testing, the plate was gone for surface finishing using abrasive paper at specific grade. Then, the plate was degreased with acetone and finally washed with distilled water to remove all the impurities on the plate surface.

2.2 Geopolymer synthesis process

Fly ash was mixed with an alkali activator solution of NaOH using an overhead mechanical stirrer at a stirring speed of 1000 rpm for 5 minutes. Prior to adhesion and flexural testing, samples were cured in an oven at 60°C. The ratios of water: solid and Na: Al were varied to find a better and stable geopolymers property. The Si: Al ratio of 1.78 was the actual composition of fly ash, whereas the Na: Al ratio was altered by use of alkalis of different concentrations.

Twelve different geopolymers formulations were produced with varying Na: Al molar ratio (0.60, 0.80, 1.00 and 1.20) and water to solid ratios of 0.30, 0.33, and 0.36. The samples were nominated using a general formula Na/AlxW/Sy where x and y represents the ratio values as shown in table 1.

Table 1. Mix design of fly ash based geopolymers coatings.

Sample Id	Abbreviation	Si:Al	Na:Al	Water: solid
Na/Al $_{0.60}$W/S$_{0.30}$	AD1	1.78	0.60	0.30
Na/Al $_{0.60}$W/S$_{0.33}$	AD2	1.78	0.60	0.33
Na/Al $_{0.60}$W/S$_{0.36}$	AD3	1.78	0.60	0.36
Na/Al $_{0.80}$W/S$_{0.30}$	AD4	1.78	0.80	0.30
Na/Al $_{0.80}$W/S$_{0.33}$	AD5	1.78	0.80	0.33
Na/Al $_{0.80}$W/S$_{0.36}$	AD6	1.78	0.80	0.36
Na/Al $_{1.00}$W/S$_{0.30}$	AD7	1.78	1.00	0.30
Na/Al $_{1.00}$W/S$_{0.33}$	AD8	1.78	1.00	0.33
Na/Al $_{1.00}$W/S$_{0.36}$	AD9	1.78	1.00	0.36
Na/Al $_{1.20}$W/S$_{0.30}$	AD10	1.78	1.20	0.30
Na/Al $_{1.20}$W/S$_{0.33}$	AD11	1.78	1.20	0.33
Na/Al $_{1.20}$W/S$_{0.36}$	AD12	1.78	1.20	0.36

2.3 Adhesion strength testing

Prior to adhesion testing, a carbon steel plate was gone for surface finishing using abrasive paper at specific grade. The plate was degreased with acetone and washed with distilled water to remove all the impurities on the plate surface. The adhesion strength of the coating material was determined using pull-off adhesion tester called Elcometer 108 following the ASTM D-4541 [19]. This adhesion tester employs a pull-off method to determine the force required to pull an area of coatings from the steel substrate. Before application, the surface of the Dolly (2mm) was abraded with an abrasive paper and was fixed to the coating surface using epoxy adhesive (Araldite regular, Belgium) and isothermally cured at 60°C for 24 h in an electric oven. The sides of the dolly were cut around to minimize the effect of the surrounding coatings. The dolly was carefully pulled off by Elcometer 108, and the adhesion strength was documented for 3 days cured samples. The detached surfaces were examined for any adhesion failure. If the coating was completely removed with dolly, the test was considered 100% valid. Only 100% valid test was considered in this study.

2.4 Flexural strength testing

Flexural strength measurement was conducted on specimens (3.2 mm x 12.7 mm x 125 mm) using a three-point bending flexural test on an Instron Universal Testing Machine, following ASTM standard D790 B [20]. The purpose of this test is to determine the flexural properties of geopolymer utilizing a three-point loading system to apply a load to a simply supported specimen.

3. Result and Discussion

3.1 Adhesion strength analysis

Figure 1 shows the adhesion strength analysis of cured geopolymers samples coated on steel plate for 3 days. There was a gradual increase in adhesion strength with increasing in Na:Al ratio from 0.60 to 1.20 with the value varied in the range of 1.80 – 4.00 MPa.

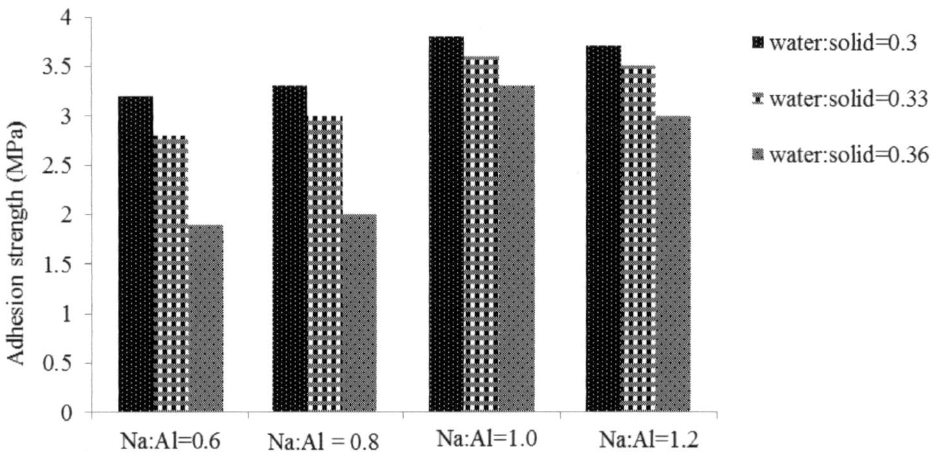

Figure 1. Adhesion strength of geopolymers at different Na: Al and water: solid ratios

Noticeable, the effect of water: solid ratio plays a significant role in adhesion strength of geopolymers. The adhesion strength of all geopolymers decreased with the increasing of water: solid ratio. There are several factors that affect the adhesion of geopolymer to metal surface include the type of raw aluminosilicate material, its mineralogical composition, particle size, geopolymer formulation, test matrix design and nature of metal substrate. The presence of water will function as a reactant in geopolymerization process. High volume of water in the system will caused the activator become more diluted, thus resulting in decreasing of adhesion strength [21].

On the other side of the effect of Na: Al ratio to the adhesion strength, surprisingly the strength was increased with the increasing of ratio. The highest adhesion strength was obtained at 4.0 MPa and 3.5 MPa when the ratio of Na: Al are at 1.0 and 1.2 respectively. While, the lowest adhesion strength were recorded for Na: Al = 0.6 and 0.8 because of the weakness of alkali and lesser dissolution rate of fly ash. The process of geopolymerization started with the dissolution of fly ash, whereby the dissolution rate is dependent to the alkali content [22]. When the amount of Na^+ cations is increased, the more aluminium will leach out from the fly ash, thus the formation of geopolymers will be more [23]. By definition, geopolymers has negatively charged aluminum that was stabilized by metal cations, therefore at least one mole of Na^+ cation was needed for the formation of stable geopolymers. This explanation is in agreement with the strength of systems at Na: Al =1.0 and 1.2. Surprisingly, inconsistent behavior of the system with Na: Al =1.2 can be related to the unreacted sodium hydroxide [24] – [25].

3.2 Flexural strength analysis

The flexural strength of geopolymers samples was shown in Figure 2. It can be seen that, the effects of Na: Al and water: solid ratios have significantly influenced the strength properties. Highest flexural strength of 6.7 MPa and 6.3 MPa were obtained when Na: Al ratio = 1.0 and 1.2 respectively. Both systems achieved almost 50% increment in strength as compared to the 0.6 and 0.8 Na: Al ratios. This could be explained by the large amount of Al was leached out from the fly ash, resulted in more stable geopolymers. As a result, higher flexural strength was recorded [26]. A similar finding was observed in the study by Nur Ain *et al* [27], in which formation of homogeneous and denser geopolymers structure was obtained by increasing the concentration of sodium hydroxide.

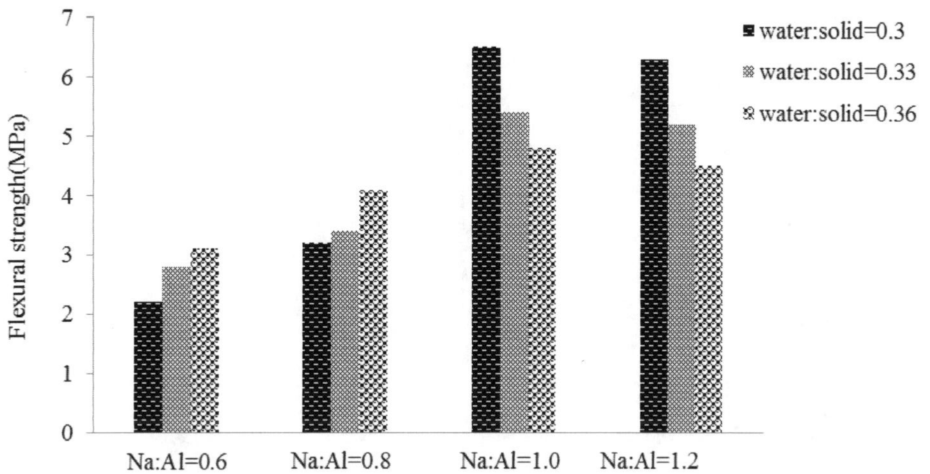

Figure 2. Flexural strength of geopolymers at different Na: Al and water: solid ratios

3.3 Scanning Electron Microscopy (SEM) analysis

SEM images of geopolymer of AD 1 and AD 8 were chosen as to observe the effect of Na: Al and water: solid ratio on morphological of samples. Figure 3 shows there were two types of fly ash particles found in the geopolymer, i.e. unreacted and reacted particles. Larger particles remained unreacted in AD 1(Figure 3(a) might be due to the low water content and less alkalinity. Meanwhile, reacted particles was observed at AD 8 as shown in Figure 3(b). Higher water content and high alkali in system might caused to the morphology obtained which directly contributed to the high strength of adhesion and flexural. This can be supported by EDS analysis as shown in Table 2, in which the amount of Al was increased in AD 8 system indicating more Al was leached out from the fly ash.

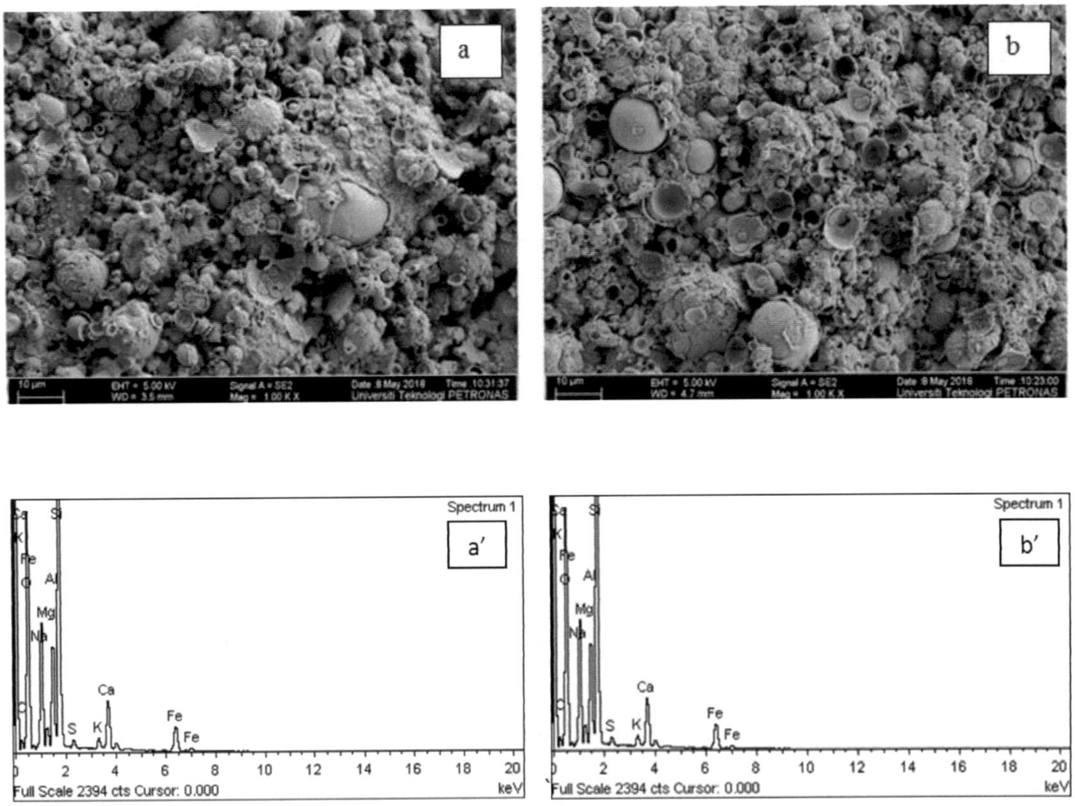

Figure 3. SEM image and EDX analysis of geopolymer of AD 1(a, a') and AD 8(b, b')

Table 2. EDX of AD1 and AD 8

	Weight %	
Elements	AD 1	AD 8
O K	50	51.75
Na K	25.4	22.64
Al K	9.61	18.77
Si K	4.36	3.4
K K	1.02	0.44
Ca K	4.48	2.21
Fe K	4.89	0.79

3.4 Fourier Transform Infra-red (FTIR) analysis

An infrared spectrum of geopolymer is presented in Figure 4. The spectra are consisted of different peaks, and to better define the position and shape of these peaks the spectra are divided into Figure 4(a) and (b) representing wave number region of 4000-850 cm^{-1} and 850-450 cm^{-1}, respectively. In Figure 5 (a) , the peaks at 3451 cm^{-1} and 1631 cm^{-1} are assigned to the asymmetric stretching and bending vibrations of O-H groups, respectively. Peak at 1091 cm^{-1} originated due the Si-O-Si asymmetric stretching vibrations [28]. The presence of peaks due to hydroxyl group can be linked with presence of water in the raw material and KBr. Besides, it also represents the presence of silanol functional group (Si-O-H) [29].

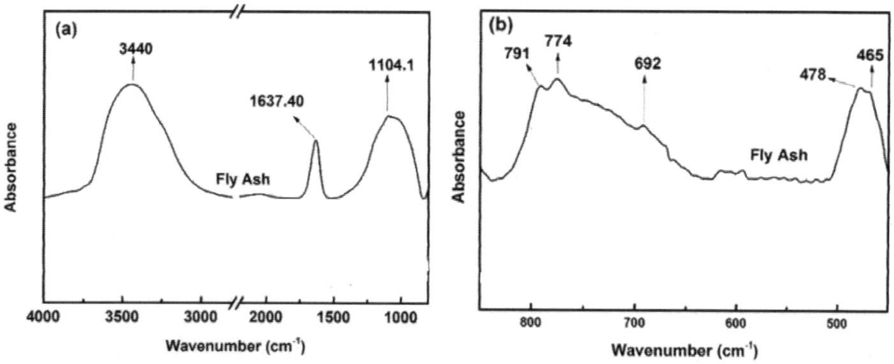

Figure 4. FTiR spectra of fly ash (a) 4000-900 cm^{-1} (b) 900-400 cm^{-1}

4. Conclusion

In this study, fly ash based geopolymers with different Na/Al and water /solid ratios were successfully prepared and synthesized. The samples were then examined under adhesion and flexural analysis. Adhesion strength was much affected with the Na: Al ratio where, there was an increment in adhesion strength up to 4.0 MPA when the ratio increased from 0.60 to 1.20. A similar trend was also observed for flexural strength analysis, in which the strength was much affected by Na: Al ratio. Almost 50% of the increment in flexural strength was obtained by increasing Na: Al ratio to 1.0 and 1.2.

References

[1] Kunze K 2003 T-duality and Penrose limits of spatially homogeneous and inhomogeneous cosmologies *Phys. Rev.* D **68** 063517 (*Preprint* gr-qc/0303038)
[1] Ahmad Z 2006 Principle of corrosion engineering and corrosion control *Elsevier Ltd*
[2] Fedrizzi L and Bonora P L 1997 Organic and inorganic coatings for corrosion prevention EUROCORR'96 *European federation of Corrosion*
[3] Cathodic disbondment *http://www.corrosionpedia.com/defination/228/cathodic-disbondment*
[4] Leidheiser H 1981 Electrical and electrochemical measurements as predictor of corrosion at the metal-organic-interface *Corros. Control by Org. Coating, NACE, Houston, Texas*
[5] Armstrong R D, Wright J D, and Handyside T M 1992 Impedance studies into the corrosion protective performance of a commercial epoxy acrylic coating formed upon tin plated steel *J. Appl. Electrochem* **22** no. 9 pp 795–800
[6] Armstrong R D, Lindholm B, and Sharp M 1986 Impedance characteristics of a modified electrode *J. Electroanal. Chem. Interfacial Electrochem* **202** no. 1–2 pp 69–74
[7] Armstrong R D, Johnson B W, and Wright J D 1991 An Investigation into the cathodic disbondment of epoxy-polyamine protective coatings *Electrochim Acta* **36** pp 1915–1923
[8] Provis J L and Bernal S A 2014 Geopolymers and related alkali-activated materials *Annu. Rev. Mater. Res.* **44** pp 299–327
[9] Provis J L 2013 Geopolymers and other alkali activated materials:why, how, and what? *Material Struct*ure **47**, pp 11–25
[10] Palomo A, Krivenko P, Garcia-Lodeiro I, Kavalerova E, Maltseva O, and Fernandez-Jimeneza A 2014 A review on alkaline activation: new analytical perspectives Material Construction **64**
[11] Kuhli H 1908 Slag cement and process of making the same
[12] Prudon A O 1940 The action of alkalis on blast-furnace slag *J. Soc. Chem. Ind.* **59** pp 191–202

[13] Glukhovskii V D, Pashkov I A, Starchevskaya E, and Rostovskaya G S 1967 Soil silicate concrete for hydraulic and irrigation structures *Hydrotechnical Constr* **1** pp 120–124
[14] Davidovits J 982 Mineral polymers and methods of making them Google Pat
[15] Davidovits J 2008 Geopolymer chemistry and applications
[16] Wastiels J, Wu X, Faignet S, and Patfoort G 1994 Mineral polymer based on fly ash J. *Resour. Manag. Technol* **22** pp 135–141
[17] Temuujin J, Minjigmaa A, Rickard W, Lee M, Williams I, and van Riessen A 2010 Fly ash based geopolymer thin coatings on metal substrates and its thermal evaluation *J. Hazard. Mater* **180** pp 748–752
[18] Temuujin J, Minjigmaa A, Rickard W, and van Riessen A 2011 Thermal properties of spray-coated geopolymer-type compositions *J. Therm. Anal. Calorim* **107** pp 287–292
[19] ASTM D-4541 Standard test method for pull-off strength of coatings using portable dhesion Testers".
[20] ASTM standard D790 B, "Standard Test Methods for Flexural Properties of Unreinforced and reinforced plastics and electrical insulating materials
[21] Khan M I, Azizli K, Sufian S , Man Z 2015 Sodium silicate-free geopolymers as coating materials: Effects of Na/Al and water/solid ratios on adhesion strength *Ceramics International* **41** pp. 2794-2805
[22] Yusuf M O, Johari M A, Ahmad Z A, and Maslehuddin M 2014 Effects of H_2O/Na_2O molar ratio on the strength of alkaline activated ground blast furnace slag-ultrafine palm oil fuel ash based concrete *Materials and Design* **56** pp 158-164
[23] Criado M, Fernandez-Jimenez A, and Paloma A 2007 Alkali activation of fly ash: Effect of the SiO_2/Na_2O ratio *Microporous and Mesoprous Materials* **106** pp 180-191
[24] Haha M B, Lothenbach B, Le Sout G, and Winnefeld F 2011 Influence of slag chemistry on the hydration of alkali activated blast furnace slag-Part I: Effect of MgO *Cement and Concrete Research* **41** pp 955-963
[25] O'Connor S J, Mackenzie K J D, Smith M E, and Hanna J V 2010 Ion exchange in the charge-balancing sites of aluminosilicate inorganic polymers *Journal of Materials Chemistry* 20 10234
[26] Palomo A, Krivenko P, Garcia-Lodeiro I, Kavalerova E, Maltseva O, and Fernandez-Jimeneza A 2014 A review on alkaline activation: new analytical perspectivs *Materiales De Construction*
[27] Jaya N A, Al Bakri Abdullah M M, Ghazali C M R, Hussain M, Hussin K and Ahmad R 2010 Effect of NaOH concentration on flexural strength, phase formation and microstructural development of kaolin geopolymer ceramic *Materials Science Forum* **857** pp. 405-411
[28] Hajimohammadi A, Provis J L, and van Deventer J S 2011 Time-resolved and spatially-resolved infrared spectroscopic observation of seeded nucleation controlling geopolymer gel formation *Journal of colloid and interface science* **357** pp. 384-392
[29] Rees C A, John L, Grant C, Jannie S 2007 In situ ATR-FTIR study of the early stages of fly ash geopolymer gel formation *Langmuir : the ACS journal of surfaces and colloids* **23** pp. 9076-9082
[30] Ilkentapar S, Atis C D, Karahan O, Görür Avsarog˘lu E B 2017Influence of duration of heat curing and extra rest period after heat curing on the strength and transport characteristic of alkali activated class F fly ash geopolymer mortar *Constr. Build. Mater.* **151** pp. 363–369
[31] Nguyen K T, Ahn N, Le T A, Lee K 2016 Theoretical and experimental study on mechanical properties and flexural strength of fly ash-geopolymer concrete *Constr. Build. Mater.* **106** pp. 65–77
[32] Soutsos M , Boyle A P, Vinai R, Hadjierakleous A, Barnett S J 2016 Factors influencing the compressive strength of fly ash based geopolymers, *Constr. Build. Mater.* **110** pp. 355–368

IOP Conf. Series: Materials Science and Engineering **458** (2018) 012012 doi:10.1088/1757-899X/458/1/012012

Theoretical and experimental study of valine molecular imprinted polymer using sol-gel process on silica microparticles surface

***M N Azalina, K L Chua and I Noorhidayah**

Faculty of Engineering Technology, Universiti Malaysia Perlis, Kampus Uniciti Alam Sg. Chuchuh, 02100 Padang Besar, Perlis, Malaysia.

*Corresponding Author E-mail: azalina@unimap.edu.my

Abstract. Molecular imprinted polymer using sol-gel process on silica microparticles surface has been prepared using valine as a template, 3-aminopropyltriethoxysilane (APTES) as the functional monomer, and tetraethoxysilane (TEOS) as the cross-linker. The theoretical results of molar ratio 1:2 (Valine:APTES) showed highest interaction energy with -3.8479 kcal/mol. The molecular imprinted polymer (MIP) and non-imprinted polymer (NIP) were prepared followed by characterization and evaluation of the absorption binding performance of imprinted valine. Observation of MIP morphology on Scanning Electron Microscopy (SEM) showed that the surface of silica were coated with imprinted polymer. In Fourier-Transform Infra-Red Spectroscopy (FT-IR), the strong intensity peaks were observed at 2921.71cm^{-1} and 1500cm^{-1} represent the functional group of alkane due to presence of the C-H stretching and C-N vibration from APTES of functional monomer, respectively. Observations in MIP and NIP samples confirmed the existence of a monomer and cross-linker on the silica surface. However, the experimental binding capacity result contradict the theoretical computation, where the experimental binding capacity showed that the maximum absorption binding capacity was at molar ratio 1:1 instead of 1:2. In conclusion, this novel molecular imprinted technology might efficiently be used in the identification and selection of amino acid fragments such as valine on its biological action.

1. Introduction

Valine is a branched chain amino acid and classified as a group 1 non-polar amino acid. Valine is usually used to enhance normal growth, repair tissues, control blood sugar, stimulate the central nervous system and provide energy to the body. One of the method used to produce L-Valine is microbial fermentation, however, microbial fermentations produced impurities. Hence, extraction and purification are needed to reduce the by-products formed. Valine can be extracted using various methods, such as the enantioselective liquid-liquid phase extraction and distillation [1], however, there is no report on the extraction of valine using molecular imprinted polymer.

Molecular imprinting is a method used to make selective binding sites in synthetic polymers [2]. Molecular imprinting has high affinity and selectivity towards its template because the presence of the template mold formed in the polymer on recognition sites. A selected functional monomer are attached to the template and undergo polymerisation process in the presence of cross-linker. After imprinted polymer had polymerised, the template will be removed from the polymeric structure and a cavity complementary in shape and chemical properties is present in the structure. The imprinted polymer becomes available to bind molecules identical to the template [3]. Molecular imprinted polymers are inexpensive, strong, easy to prepare and have high resistant towards high temperature and pressure. There are several known methods used in the polymerisation process including the sol-gel process, hydro-gel process, free-radical process and aerogel process [4,5]. Different techniques can be applied in each process such as the bulk polymerisation, precipitation polymerisation, surface imprinting, dispersion polymerisation and suspension polymerisation [6].

In this research, the polymerization process and technique employed were sol-gel process and surface imprinting, respectively, in which valine was used as the template, while 3-aminopropyltriethoxysilane (APTES) and triethoxysilane (TEOS) were used as the functional monomer and cross-linker, respectively, in the molecular imprinting process. Valine molecular imprinted polymer will be synthesized by applying sol-gel polymerization method. The research scope was to evaluate the theoretical and experimental criterion for molecular imprinted polymer design based on the interaction between the functional monomer and the template using computational modelling. The computational study will be implemented to optimize the ratio between the functional monomer and the template within prepolymerization complex. Next, the valine imprinted polymer was synthesized and characterized by batch sorption experiments in order to validate the computational study.

2. Materials and Methods

2.1. Materials

The silica-gel (230-400 mesh, 0.040-0.063 mm in size) used as the support material for surface imprinted polymer was purchased from Merck, Germany. Valine, 3 aminopropyltrimethoxysilane (APTES) and DMSO were obtained from Sigma Aldrich, Germany. Tetrarthoxysilane (TEOS) was purchased from Acros (Sigma, Germany) and hydrochloric acid was obtained from Fisher Scientific, USA. The analytical grade methanol was used as washer solution supplied from HmbG, Germany.

2.2. Experimental Procedure

2.2.1. Computational Design of Polymer for Valine. The simulations was carried out by using a molecular modelling software program known as Hyperchem software (Hypercube Inc., Gainesville). Valine and APTES were drawn in 2D and converted to 3D structures. The geometry opyimization of each molecular structure was carried out using Polak Ribiene algorithm until the energy gradient was below 0.01 kcal/mol. Computational study has been suggested as an alternative approach to seek rational MIP design, which is based on calculation of interaction energy (ΔE) between template molecule and the functional monomer. The interaction energy was calculated using the equation (1):

$$\Delta E = E_{template-monomer} - (E_{template} + \sum E_{monomer}) \quad (1)$$

Where, ΔE is the interaction energy between the template and functional monomer (kcal/mol), $E_{template-monomer}$ is the energy between the template and functional monomer (kcal/mol), $E_{template}$ is the template energy (kcal/mol), $E_{monomer}$ is the monomer energy (kcal/mol).

2.2.2. Activation of silica microparticles surface (silica gel). Before the molecular imprinting process was carried out, the surface of silica-gel was activated. The mixture of silica-gel and 6M of HCl were first refluxed for 10 hours. Then, the resulting mixture was filtered and washed thoroughly with deionized water. The solid activated silica gel and water were saperated using centrifuge at 5,000 rpm and the process were continued until neutral. Lastly, the particles were dried overnight at temperature 70 °C in an oven. The purpose of silica activation was to enhanced the content of the silanol group. Moreover, by washing, it is able to eliminate nitrogen-containing impurities and metal oxide [7, 8].

2.2.3. Preparation of molecularly imprinted polymer. Valine, 3-aminopropyltriethoxysialne (APTES) and tetraethoxysilane (TEOS) were used as a template molecule, functional monomer and cross-linker, respectively. The data obtained from computational study was used in the preparation of valine-molecularly imprinted polymer. In order to prepare template:monomer with the ratio of 1:1, 0.1172 g of valine template was dissolved in 15 ml dimethyl sulfoxide (DMSO) together with 5 ml of water. After dissolving the template in the solvent completely, 0.234 ml of APTES was added follow by 2.238 ml of TEOS. Next, 2 g of activated silica gel and 0.2 ml of 0.012 M HCl were subsequently added into the dissolved valine template mixture under continuous stirring at temperature 40 °C. The mixture was then stirred continuously for 24 hours. Then, the imprinted polymer was filtered out and the imprinted polymer was washed using 50 ml of ethanol and 50 ml of water until it pH reaches neutral. The imprinted polymer was left to dry overnight at temperature 80°C. The steps were repeated in preparing template: monomer with the ratio of 1:2 where 0.468 ml APTES was used. Then, the template were extracted with the mixture of 50 ml MeOH and HCl (10% in water) in the ratio of 1:1, followed by methanol and distilled water. The mixture were shaken using an orbital shaker at 100 rpm for 30 minutes to remove the template on the imprinted polymer. The supernatant at every washing solution was tested using UV-Vis at 265nm^{-1} until no longer template be detected in the washing solution. Finally, the imprinted polymer was collected by filtration and dried at 80°C overnight in oven. The NIP was also prepared using identical procedure without template.

2.3. Batch Rebinding.

Fifty miligrams of imprinted polymer was placed in a 50 ml centrifuge tube. Then, 5 ml of valine in water at different concentration ranged between 0.0625 mg/ml-0.25 mg/ml was added. The centrifuge tubes were placed horizontally in an orbital shaker at 100 rpm for 4 hours at room temperature. Then the centrifuge tubes were placed in the centrifuge and spin at 5000 rpm for 10 minutes to remove the imprinted residual from the supernatant. The supernatant was filtered using nylon syringe filter (0.45 μm), and was measured via UV-visible spectroscopy at 265 nm^{-1}. The binding test was tested by calculating the binding energy, Q using equation (2):

$$Q = \frac{V(C_i - C_o)}{W} \tag{2}$$

Where, Q is the rebinding energy (mg/g), V is the volumes of standard valine solution (ml), C_i is the initial concentration (mg/ml), C_o is the final concentration (mg/ml), W is the weight of the imprinted polymer used (mg). Furthermore, the imprinting factor (IF) was used to evaluate the selectivity properties of molecular imprinted polymer and non-imprinted polymer towards valine the template. The imprinting factor, IF was calculated based on Q_{MIP} over Q_{NIP}.

3. Results and Discussion

3.1. Theoretical Optimization of Template: Functional Monomer Ratio

In this study, Hyperchem software was used for computational modelling on the study of valine as template while APTES as monomer. The imprinted molecule structures between valine and APTES with different molar ratio were optimized and an interaction energies were calculated. The interaction positions of the template-monomer was shown as well. The interaction energy between valine and APTES was calculated using equation (1). Molar ratio of 1:1 and 1:2 were chosen and to be used in this experiment because valine has only two active sites, which are carbonyl group and amine group that allow it to bind with the monomer.

Table 1 showed the energy of the imprinted molecules and the interaction energies, ΔE for both molar ratio of template: monomer at 1:1 and 1:2. The interaction energy, ΔE for molar ratio 1:2 shown higher ΔE -3.8479 kcal/mol compared to molar ratio 1:1 which, the value was calculated as - 1.4404 kcal/mol. Complex polymer with higher interaction energy is more stable compare to the polymer that has lower interaction energy [9]. This interaction energy also indicates that the molar ratio of 1:2 has greater affinity toward the template (valine) during rebinding test. However in a polymerization process, excess number of mole of functional monomer may cause non-specific binding and a reduction in the selectivity properties of the polymer [10].

Table 1. Energy and Interaction energies of valine with different ratio of 3-aminopropyltriethoxysilane (APTES).

Compound	Energy (kcal/mol)	Molar ratio (Template:Monomer)	Energy (kcal/mol)	Interaction Energy, ΔE (kcal/mol)
Valine	-1769.2398	1:1	-5171.0855	-1.4404
APTES	-3400.4053	1:2	-8573.8983	-3.8479

The optimized potential interactions between template and functional monomers with different ratio complexes were illustrated in Figure 1. The geometrie hydrogen bondings between valine and APTES at molar ratio 1:1 were presented in Figure 1 (A), where hydrogen atom from the carbonyl group on template was hydrogen-bonded with the nitrogen atom from amine group on monomer APTES.

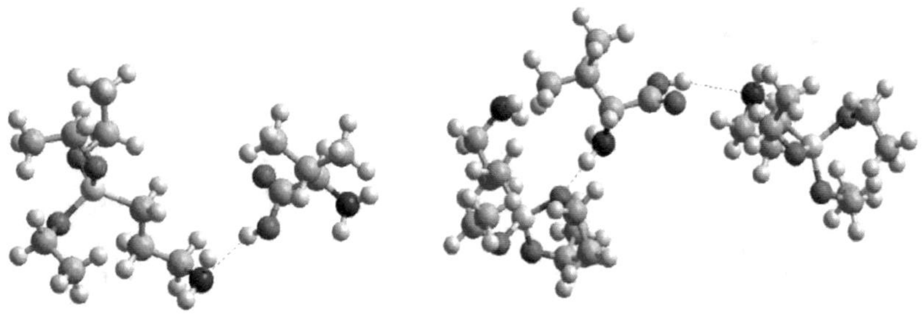

(A) (B)

Figure 1. The optimized structures of valine and APTES with molar ratio 1:1 (A) and molar ratio 1:2 (B) . Nitrogen [dark blue ball], Carbon [light blue ball], Oxygen [red ball], Silicone [white big ball] and Hydrogen [white small ball]. Hydrogen bond shown as dotted line.

Figure 1(B) showed the interaction of molar ratio 1:2, where two hydrogen bonds were obtained. The first hydrogen bonding was formed between hydrogen atom from carbonyl group on valine with nitrogen atom from amine group on monomer. In which, same hydrogen bonding as formed in molar ratio 1:1. The second hydrogen bonding was formed between hydrogen atom from amine group on valine with oxygen atom from monomer. Therefore, from the structure obtained, the valine and APTES in molar ratio 1:2 is more stable as compares to the structure of valine and APTES in 1:1 molar ratio. This is because molar ratio 1:2 had a greater molecule structure than molar ratio 1:1.

3.2. Characterisation of polymer

3.2.1. Fourier transform infrared spectroscopy (FT-IR). Perkin Elmer FT-IR Spectroscopy 100 was used to determine the possible functional group present in the polymer using KBr pellet method. The wave numbers of FT-IR analysis were collected at one data point per 2 cm^{-1} and ranged from 400-4000 cm^{-1}. Figure 2 showed the FT-IR spectrum of silica gel, molecular imprinted polymer (MIP) and non-imprinted polymer (NIP).

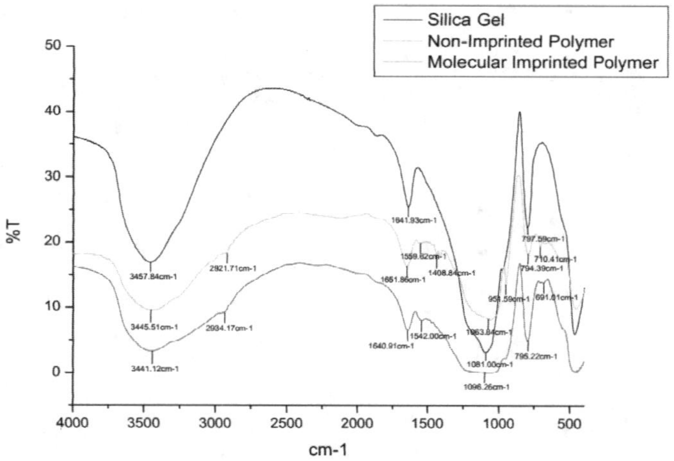

Figure 2. FTIR spectroscopy of silica gel, non-imprinted polymer and molecularly imprinted polymer for molar ratio 1:2.

The IR spectra (Figure 2) for both NIP and MIP showed similar peak. This observation may be due to the same functional group present in both polymers. The similar peaks for silica gel, MIP and NIP were observed at 3457.84 cm^{-1}. These strong and broad intensity peaks correspond to the presence of H-N-H bond and alcohol (-OH) stretching. However, the stretching vibration slightly shift to 3445.51 cm^{-1} and 3441.12 cm^{-1} due to the removal process. The peaks observed at 2921.71 cm^{-1} and 2934.17 cm^{-1} represent the functional group of alkane due to the presence of the C-H stretching from the monomer that have strong intensity. Meanwhile, between 1400 cm^{-1} and 1500 cm^{-1} represent the presence of the C-N vibration from APTES of functional monomer. However, there was no peak on the silica gel curve, therefore, the template and monomer were successfully imprinted on the activated silica gel. The peaks of stretching vibration Si-O-Si observed at 1063.84 cm^{-1}, 1081.00 cm^{-1}, and 1096.26 cm^{-1} were due to silica particles, whereas Si-O peak observed at 797.59 cm^{-1}, 794.39 cm^{-1}, and 795.22 cm^{-1} were due to the vibration from the interaction with the silica surface and APTES.

3.2.2. Scanning Electron Microscopy (SEM). Figure 3 showed the image captured using SEM (JEOL Scanning electron microscopy) for the activated silica gel and valine MIP. The activated silica gel exhibit a smooth surface as shown in Figure 3(A) while surface of silica coated with imprinted polymer (Figure 3(B)) exhibited a rougher surface morphology, which indicate that valine MIP have been successfully prepared.

Figure 3 (C) exhibited small agglomerate at the surface of the silica which may be due to the impurities from insufficient washing [7,8]. However, the overall surface was smooth compared to MIP in figure 3(D). The SEM scans verified that the imprinting polymer of siloxane leads to the formation of agglomerates with multiporous surfaces. The roughness on the surface of valine MIP can be clearly seen compared to activated silica gel. Moreover, the imprinting were spread homogenously around the silica microparticle surface, which proved that the silica gel had been fully coated with imprinting polymer [11].

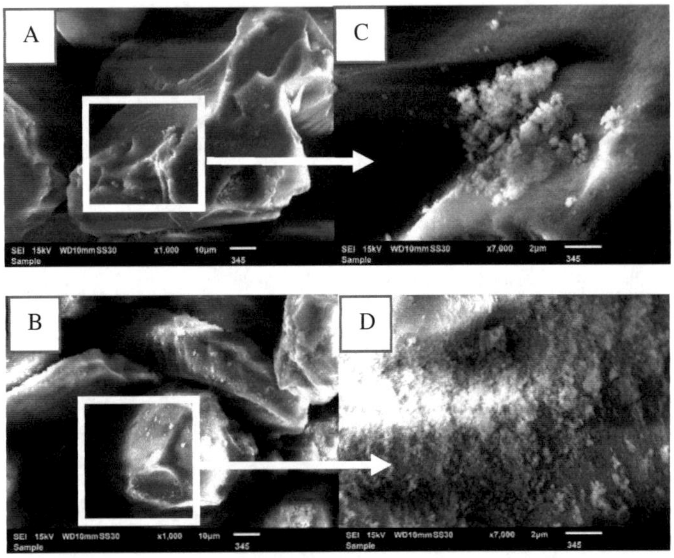

Figure 3. SEM image of (A) activated silica gel and (B) valine MIP at 1000 magnification while, (C) activated silica gel and (D) valine MIP at 7000 magnification respectively.

3.2.3 Batch Rebinding Experiment

The adsorption binding capacities for MIP of molar ratio 1:1 was 35.06 ± 0.0007 mg/g higher than NIP in which, was 26.85 ± 0.0005 mg/g (Figure 4). The situation could be explained by the existence of recognition sites at the surface of silica microparticles which enhance the surface area of adsorption. The amount of valine bound to both MIP and NIP at molar ratio 1:1 increased sharply with increasing concentration of valine but the adsorption ability of NIP was lower than MIP. This showed that MIP has higher loading ability towards the template as compared to NIP. The results showed that the existence of imprinting cavities on MIP differentiated the binding ability of NIP. Similar observation was also observed in both MIP and NIP at molar ratio 1:2. The amount of valine bound to MIP and NIP at ratio 1:2 increased with increasing concentration of valine and then, the curve reached equilibrium 0.25 mg/ml (Figure 4). However, the adsorption ability on NIP (26.64 ± 0.0012 mg/g) was lower than MIP (28.39 ± 0.0018 mg/g) (Figure 4).

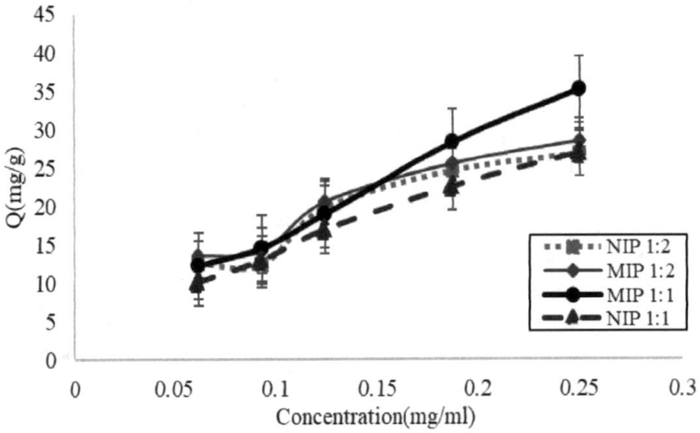

Figure 4. Comparison for the binding test using valine on NIP and MIP of molar ratio 1:1 and molar ratio 1:2.

MIPs of both molar ratio 1:1 and 1:2 had higher binding capacity, attributed to the presence of specific binding sites formed in the process of molecular imprinting and enhance the binding performance of template fits into the imprinted polymer cavity. The difference in binding capacity between both molar ratios showed that molar ratio 1:1 had higher affinity binding sites while molar ratio 1:2 had lower affinity sites. Whereas in NIPs, although functional monomer were present in the polymer, the monomer was ineffective for binding with the template due to the lack of recognition cavities [12]. Therefore, the binding capacity of NIPs for both molar ratio 1:1 and 1:2 were lower than MIPs. However, the template-monomer complexation was influenced after polymerization by the number of monomers attached and the availability of the functional groups on the monomers [13].

The efficiency of the molecular imprinting polymer could be measured by the imprinting factor (IF). The imprinting factor is a measure of the imprinted polymer towards the template molecule [14]. Imprinting factor is the ratio of the substrate obtained from the molecular imprinted polymer and the recovery yield obtained using the same substrate on non-imprinted polymer with the same monomer. Our calculation indicated that the imprinting factor for molar ratio 1:1 (1.306) was higher compared to the imprinting factor for molar ratio 1:2 (1.066), indicating molar ratio 1:1 having highest selectivity towards valine compared to molar ratio 1:2.

In term of stability, the structure of the polymeric complex for molar ratio 1:2 should be greater affinity toward the tamplate compare to the imprinted polymer structure for molar ratio 1:1. Therefore, higher adsorption binding capacity should obtained. However, a contradictory observation was obtained. This may be due to the recognition sites were dependent only to the distribution of functional groups. Moreover, in solution only one valine interact with one of the monomer and after polymerization the oxygen atom attached to template may not be available for interacting with APTES, due to steric hindrance [13]. Another possibility could be due to damaged monomer on the cavities after rewashing, which lead to the decrease in the affinity for the template [14] and also an incomplete removal of the valine during washing step [15].

The binding capacity is dependent on the end application. MIP for environmental analysis or sensing application does not require a high capacity, but for long-term controlled release, higher polymer loading capacities may be required [16].

4. Conclusion

The theoretical study using Hyperchem software was performed to obtain the interaction energy between the template and the functional monomer using AM1 semi empirical. The study showed that molar ratio 1:2 (-3.8479 kcal/mol) have highest stability compare to molar ratio 1:1 (-1.4404 kcal/mol). The synthesized valine imprinted polymer according to the result from computational studies was characterization. Results from the characterization proved that animo silane was successfully imprint on the activated silica gel. The adsorption studies were performed by batch rebinding and the results indicated that for both molar ratio 1:1 and 1:2 of MIP have highest adsorption results compare to NIP. However, both adsorption binding capacity and IF for MIP on molar ratio 1:1 was higher than the molar ratio 1:2. The results showed that the molar ratio 1:1 has higher selectivity towards valine. The contradicting result found between theoretical and experimental were due to several factors incuding the distribution of functional group on the recognition site, steric hindrance and the damaged functional monomer on the cavities. Overall, our study revealed that high selectivity MIP can be prepared using a simple and fast sol-gel method.

Acknowledgments
The author would like to thank to Ministry of Higher Education of Malaysia the support for providing financial support through fundamental research grant No. FRGS/1/2016/TK05/UNIMAP/03/1.

References

[1] Tang K W, Fu T, and Zhang P L, 2013 Enantioselective liquid-liquid extraction of (D,L)-valine using metal-BINAP comples as chiral extractant *J. Chem. Technal. Biotechnol.* vol. **88**, pp. 1920-1929

[2] Małgorzata R, Justyna P-W, Calum M, Piotr P W, Jacek N, Mariusz M, 2018 Application of molecularly imprinted polymers in analytical chiral separations and analysis *Trends Anal. Chem.* vol. 102, pp. 91-102

[3] Gama M R and Bottoli C B G 2017 Molecularly imprinted polymers for bioanalytical sample preparation *J. Chromatogr. B,* vol. **1043**, pp. 107-121

[4] Mujahid A, Lieberzeit P A, and Dickert F L, 2010 Chemical sensors based on molecularly imprinted sol-gel materials *Materials,* vol. **3**, pp. 2196-2217

[5] Takuya K, Koji O, 2016 Recent progress for the selective pharmaceutical analyses using molecularly imprinted adsorbents and their related techniques: A review *J. Pharm. Biomed. Anal.* vol. 130, pp. 68-80

[6] Boutevin B, Boyer C, Csetneki I, David G, Ferguson J S, and Filipcsei G, 2007 *Oligomers-Polymer Composites-Molecular Imprinting* vol. **206** (Berlin: Springer Science)

[7] Ismail A R, and Vejayakumaran P, 2012 Synthesis of silica nanoparticles by sol-gel: Size-dependent properties, surface modification, and applications in silica-polymer nanocomposites-A review *J. Nanomater.*, vol. **2012**, pp. 1-15

[8] Syed N S, Kallidanthiyil C L, Abdul M I M, and R M P, 2015 Extraction and Recovery of Naphthenic Acid from Acidic Oil Using Supported Ionic Liquid Phases (SILPs) *Chem. Prod. Process Model.*, vol **10**, pp. 221–228

[9] Riahi S, Eynollahi S, Ganjali M, and Norouzi P, 2010 Computational approach to investigation of template/monomer complex in imprinted polymers; dinitrobenzene sensor *Int. J. Electrochem. Sci.,* vol. **5**, pp. 509-516

[10] Ishak N, Ahmad M N, Nasir A M, and Islam A K M S, 2015 Computational Modelling And Synthesis Of Molecular Imprinted Polymer For Recognition Of Nitrate Ion *Malaysian J. Anal. Sci.,* vol. **19**, pp. 866-873

[11] Liao S, Zhang W, Long W, Hou D, Yang X, and Tan N, 2016 Adsorption characteristics, recognition properties, and preliminary application of nordihydroguaiaretic acid molecularly imprinted polymers prepared by sol–gel surface imprinting technology *Appl. Surf. Sci.,* vol. **364**, pp. 579-588

[12] Zhao S, Yang X, Zhao H, Dong A, Wang J, Zhang M, Huang W, 2015 Water-compatible surface imprinting of 'Saccharin sodium' on silica surface for selective recognition and detection in aqueous solution *Talanta,* vol. **144**, pp. 717-725

[13] Fu X, Yang Q, Zhou Q, Lin Q, and Wang C, 2015 Template-Monomer Interaction in Molecular Imprinting: Is the straongest the best?," *Open J. Org. Poly. Mater.,* vol. **5**, pp. 58-68

[14] Chen L, 2015 *Synthesis, characterization and theoretical study of amino acid-based magnetic molecularly imprinted polymer for recognition of sulfonamide drugs* (Malaysia:University of Malaya)

[15] Puzio K, Delépée R, Vidal R, and Agrofoglio L A, 2013 Combination of computational methods, adsorption isotherms and selectivity tests for the conception of a mixed non-covalent–semi-covalent molecularly imprinted polymer of vanillin *Anal. Chim. Acta,* vol. **790**, pp. 47-55

[16] Fitzhenry L, 2011 *Development of Molecularly Imprinted Polymers for Corticosteroids* (Republic of Ireland:Waterford Institute of Technology)

Potential of Mining Waste from Metallic Mineral Industry for Carbon Sequestration

Sharifah Nur Munirah Syed Hasan and Faradiella Mohd Kusin

Department of Environmental Sciences, Faculty of Environmental Studies, Universiti Putra Malaysia, 43400 UPM Serdang, Selangor, Malaysia.

Email: faradiella@upm.edu.my

Abstract. Anthropogenic carbon dioxide emission mainly from metallic mining industry is the major gases that increase greenhouse gases. Mining wastes that are discarded from mining operation might have potential as a solid buffer in storing atmospheric carbon dioxide for long term by mineral carbonation. Therefore, this study was conducted to identify the major minerals in the waste rocks at waste dump and stockpile of a gold mine, and to determine the potential of silicate minerals from gold mining wastes for mineral carbonation process. Waste rocks sample from the waste dump and stockpile were obtained and were analyzed using X-ray diffractometer. Findings indicated that quartz, mica, dolomite, and calcite are the major minerals found in the waste rocks. Results also demonstrated that carbonate minerals such as dolomite, calcite, and siderite are in stable form that are considered a reservoir for carbon storage. Silicates minerals such as chlorite, pyroxene, mica, and amphiboles that are present at both waste dump and stockpile contain magnesium, calcium, and iron silicates ions, which can facilitate the formation of carbonate minerals. Thus, the presence of potential minerals from gold mining wastes provides a great possibility for carbon sequestration that can encourage evaluation of carbon footprint in mining industry.

1. Introduction

The removal of soil and rocks through extraction and processing of materials from mining industry to gain access to buried ore, includes the leftover materials such as solids, water, and gases that are called waste materials or residues. This process releases carbon dioxide (CO_2) which contributes to high accumulation of anthropogenic greenhouse gases (GHGs) in the atmosphere [1,2], and yields large amount of waste that has little or no economic value. However, mining wastes in a positive approach can help to mitigate environmental pollution due to its potential for CO_2 storage in reducing anthropogenic GHG emission to the atmosphere. Therefore, the process of passive CO_2 sequestration or carbon capture and storage (CCS) is required to stabilize CO_2 in the atmosphere which involves trapping CO_2 from the gas streams into geological formations such as soil and rocks in carbonate form by natural carbonation reaction [3,4,5]. Carbonate minerals that form in the soil include formation of stable carbonate or bicarbonate to capture carbon through weathering of primary minerals that release ions such as hydrogen (H^+), magnesium (Mg^{2+}), and calcium (Ca^{2+}), reacting with CO_2 [1,4,5,6,7,8]. This natural process of carbonation provides a great potential for carbon storage of mining residue in reducing GHG emission. Furthermore, mining waste from different types of mining production might have potential for CO_2 sequestration and also provide economic benefits that involve minimal cost technology [3,6,9].

1.1. Theory of calculating carbon sequestration

Waste rock might have potential to sequester CO_2 s via mineral weathering in order to mitigate climate change. Waste rock is a coarse material that consists of different sizes of particles which is produced from mineral excavation alongside overburden that frequently has low mineral value [2,10]. This material is often stored in heaps or dumps and disposed of in stockpiles on the mine site, where the sizes of dumps or stockpiles are very huge and signify complex geotechnical constructions [11]. Mine waste rocks from ultramafic-hosted ore deposits are commonly preferred for carbonation mineralization due to high amount of magnesium silicates from final product of mining process and contains primary ultramafic minerals such as olivine (Mg_2SiO_4) and weathered minerals such as serpentine [$Mg_3(OH)_4(Si_3O_5)$] [12]. This minerals release magnesium ions and react with carbonate and bicarbonate ions to form carbonate minerals via carbonation reaction as shown in the following reactions (1) [8,13]:

$$2Mg_2SiO_4 \text{ (olivine)} + H_2CO_3 \rightarrow Mg_3Si_2O_5(OH)_4 \text{ (lizardite)} + MgCO_3 \text{ (magnesium carbonate)} \quad (1)$$

Moreover, ultramafic rocks in mine tailings also allow carbon sequestration to occur through formation of magnesium and/or calcium carbonates from natural weathering process [14]. For instance, during rock weathering, carbonation reaction of calcium silicates in metamorphic rocks produces wollastonite ($CaSiO_3$) minerals (reaction 2), which increase carbon uptake [8,13,15]. As a result of rock weathering, it accelerates carbonation of calcium silicates, which occur naturally and promote carbon sequestration.

$$CaSiO_3 \text{ (wollastonite)} + H_2CO_3 + H_2O \rightarrow CaCO_3 \text{ (calcium carbonate)} + H_4SiO_4 \quad (2)$$

Mineral carbonation has the potential to sequester large amounts of carbon in stable carbonate minerals through natural weathering process of silicate such as magnesium and calcium silicates. However, Mg-silicates such as olivine and serpentine are preferred to be used in mineral carbonation in waste rock due to more reactive and accessible [11]. Estimation of carbon sequestration potential from mining waste is measured from cation content of magnesium oxide (MgO) and calcium oxide (CaO) in weight percent (wt%) which represents percentage of waste rock to be produced at mine site. 100% conversion rate is assumed where one tonne of MgO or CaO wt% can sequester one tonne of CO_2 based on the following equation 3, R is the total amount of CO_2 sequestered [11]:

$$R = (\text{MgO/CaO wt\%}) \times (\text{waste rock tonnage}) \quad (3)$$

The maximum carbonation potential is calculated by assuming that all of Mg and/or Ca available in mining waste are converted to $CaCO_3$. For example, 1 mol of cations captures 1 mol of carbon from $MgCO_3$ mineral [8]. Therefore, 1 mole of cation can capture 1 mole of carbon dioxide from mining waste.

Thus, the objectives of this study were to: (1) to identify the major minerals in the waste rocks at waste dump and stockpile of a gold mine; and (2) determine the potential of silicate minerals from gold mining wastes for mineral carbonation process. Mineral compositions of other types of metallic mining waste are also discussed for their potential in carbon sequestration.

2. Materials and methods

2.1. X-ray diffraction analysis

Samples of rocks were obtained from a gold mine for field trial assessment. The rock samples were crushed and were grounded into fine particles using agate mortar and pestle. The fine samples were sieved to 1 mm size fraction using particle sieves because grain size of samples is a critical consideration that can affect results of X-ray diffraction (XRD) analysis [16]. Then, sieved samples were grounded in a very finely powder form using a special mortar and about 1 ± 0.5 g of fine powder samples were placed on special sample holder before being attached on X-ray machine [17]. Powder samples were analyzed using XRD instrument (model Philips X'Pert Pro Pranalytical-PW3440/60, Netherlands) at 1°/min rate (0.02° step size) over the 5–50° scattering angle range [2,6], where the range of detection limit was between 1 to 2% [18]. This analysis was used to classify mineral characterization at the waste dump and stockpile within the gold mine [2,17]. Integrated intensity of peak areas from the single peak function was determined using Diffrac AT EVA software (v.9.0) and the d-spacing was identified using OriginPro 8 software (Originlab Corporation, Northampton, UK).

3. Results and discussion

3.1. Mineralogy of waste rocks and potential of silicate minerals from gold mine

Example from the case study of a gold mining is presented here. Potential feedstocks for mineral carbonation found for this case study were the carbonate and silicate minerals as shown in Table 1. There were 15 types of minerals that were identified from XRD analysis of the waste rock samples at the waste dump and stockpile of the gold mine. There were four major minerals found in both the waste dump and stockpile samples which are quartz, mica, dolomite, and calcite, followed by other minerals such as kaolinite, pyroxene, goethite, anatase, chlorite, illite, larosite, rutile, siderite, amphiboles, and opal. Major minerals were discovered widespread at both mining wastes due to parent rocks from the waste rock samples were originated from volcanic, sedimentary (e.g., limestone, carbonaceous shale, and conglomerate), and metamorphic rocks (e.g., phylite) [10]. Figure 1 and 2 indicate that the highest peak of quartz and mica were present in both types of mining wastes because these minerals were resistant to weathering [19]. Thus, quartz and mica minerals were considered as the common minerals found in all types of rocks.

From the XRD results, majority of minerals were present in the waste dump and stockpile except for minerals such as siderite, amphiboles, and rutile that were found at stockpile only, whereas opal mineral was only present at the waste dump (Table 1). Data in the table shows that silicate minerals such as chlorite, pyroxene, mica, and amphiboles that were present at the waste dump and stockpile are the potential feedstocks for mineral carbonation. These silicate minerals contain magnesium and/or calcium silicates ions, which can react with CO_2 to form stable carbonate mineral [7,11,13]. For instance, reaction of chlorite mineral with CO_2 will produce magnesite as shown in following reaction (4):

$$Mg_6Si_4O_{10}(OH)_8\,\text{(chlorite)} + 6CO_2 \rightarrow 6MgCO_3\,\text{(magnesite)} + 4SiO_2 + 4H_2O \quad (4)$$

The presence of magnesium and calcium silicates from chlorite, pyroxene, mica, and amphiboles minerals at the mining wastes are favorable for mineral carbonation processes. Therefore, this shows that the mineral components of the waste dump and stockpile of the gold mine have valuable potential of mineral carbonation for carbon sequestration.

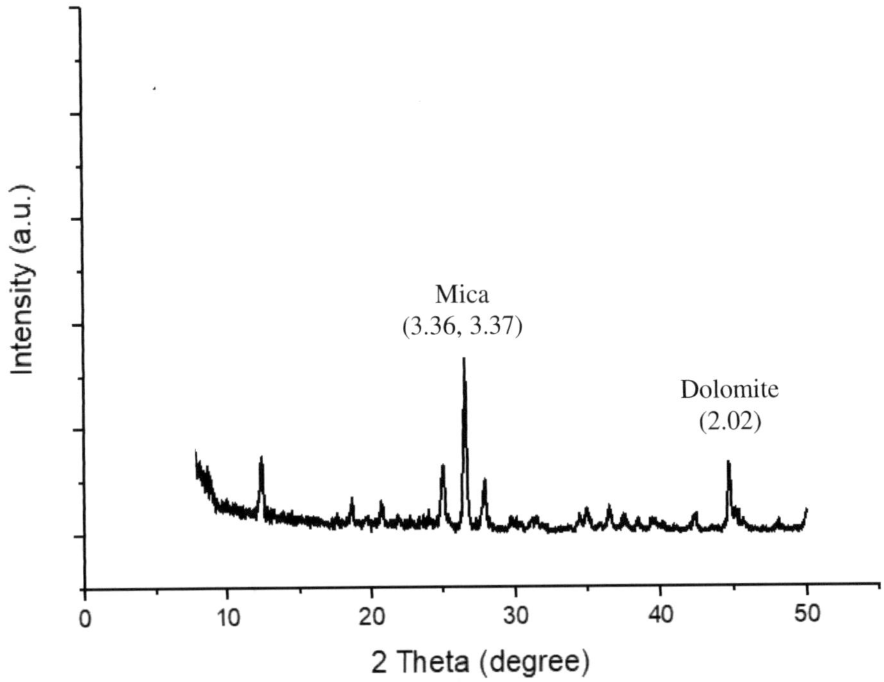

Figure 1. Major minerals with d-spacing value from XRD analysis of waste rock at the waste dump.

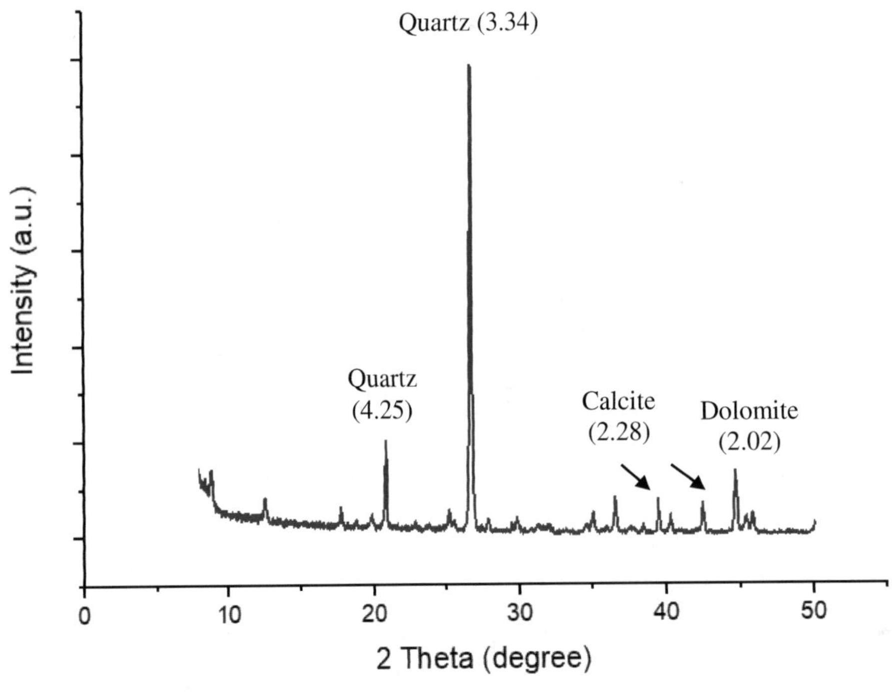

Figure 2. Major minerals with d-spacing value from XRD analysis of waste rock at the stockpile.

Table 1. Summary of composition of minerals in waste rocks of a gold mine. Symbol of (√) is referring to the presence of minerals in waste dump and stockpile.

Minerals	Chemical formula	Waste dump	Stockpile
Primary minerals			
1. Quartz [a]	SiO_2	√	√
2. Mica [a]	$K(Mg,Fe)_3(AlSi_3O_{10})(OH)_2$	√	√
3. Pyroxene [a]	$(Ca,Na)(Mg,Fe,Al)(Al,Si)_2O_6$	√	√
4. Amphiboles [a]	$NaCa_2(Mg,Fe,Al)_5(Al,Si)_8O_{22}(OH)_2$		√
Secondary minerals			
5. Dolomite [b]	$CaMg(CO_3)_2$	√	√
6. Calcite [b]	$CaCO_3$	√	√
7. Siderite [b]	$FeCO_3$		√
8. Kaolinite [a]	$Al_2Si_2O_5(OH)_4$	√	√
9. Chlorite [a]	$Mg_6Si_4O_{10}(OH)_8$	√	√
10. Goethite	$HFeO_2$	√	√
11. Illite [a]	$(K,H_3O)(Al,Mg,Fe)_2(Si,Al)_4O_{10}[(OH)_2,(H_2O)$	√	√
12. Anatase	TiO_2	√	√
13. Larosite	$(Cu,Ag)_{21}(Pb,Bi)_2S_{13}$	√	√
14. Rutile	TiO_2		√
15. Opal	$SiO_2.nH_2O$	√	

[a] Silicates minerals, [b] Carbonate minerals.

3.2. Mineral composition of metal mining waste for carbon sequestration

Minerals are the largest components in soils approximately 45% to 49% of mineral portion, consisting of primary and secondary minerals. Primary minerals such as sand and silt are similar to parent minerals, while secondary minerals is produced from weathering of primary minerals that released ion such as hydrogen, magnesium and calcium ion and react with CO_2 to form stable carbonate mineral such as silicates [6,7]. Table 2 shows the mineral compositions and its percentage of waste rock materials, which have the potential for carbonation reaction to increase carbon capture. In terms of mineral composition, waste rocks contain magnesium minerals that are potential for mineral carbonation such as brucite, nesquehonite, hydromagnesite, and landsfordite, which are hydrated during natural weathering process from the rocks [4,20]. In addition, alkaline earth silicate type of waste rocks such as ultramafic rock (e.g., peridotite, amphibolite, pyroxenite, and serpentinite) and mafic rock (e.g., basalt and gabbro) [21,22,23], are more preferable for mineral carbonation reaction due to abundance of magnesium and calcium minerals that can bind with CO_2 and form a carbonate, which can reduce CO_2 in the atmosphere. Generally, the waste rocks have the potential feedstock for carbon sequestration due to high amount of magnesium and calcium content that can form a stable carbonate during mineral carbonation process.

Various metal mining wastes are potential to sequester more CO_2 permanently. For instance, metal mining waste from bauxite has a potential for carbon storage due to high concentration of divalent cations such as calcium oxide (CaO = 40%) from calcite ($CaCO_3$) and gypsum ($CaSO_4.2H_2O$) minerals [24] and magnesium oxide (MgO = 39.50%) from brucite [$Mg(OH)_2$], chrysotile [$Mg_3(Si_2O_5)(OH)_4$], talc [$Mg_3(SiO_5)_2(OH)_2$], and chlorite [$Mg_6Si_4O_{10}(OH)_8$] minerals [8] (Table 3). These minerals have potential to form carbonates such as $CaCO_3$ and $MgCO_3$ through mineral carbonation reactions. Bauxite also has high concentration of aluminum oxide (Al_2O_3 = 33.10%) which is from boehmite ($Al_2O_3.H_2O$), gibbsite ($Al_2O_3.3H_2O$), and kaolinite ($Al_2O_3.2SiO_2.2H_2O$) minerals [25]; silica (SiO_2 = 23.80%) [2], and ferric oxide (Fe_2O_3 = 71.9%) from wustite (FeO), hematite (Fe_2O_3), magnetite (Fe_3O_4), and goethite ($HFeO_2$) minerals [6]. Other minerals composition such as chantalite [$CaAl_2SiO_4(OH)_4$], chamosite [$(Mg,Fe,Al)_6(Si,Al)_{414}(OH)_8$], hydrogarnet [$Ca_3AlFe(SiO_4)(OH)_8$], and albite [$(Na,Ca)Al(Si,Al)_3O_8$ /$NaAlSi_3O_8$)] also exists in bauxite mining residue.

Table 2. Characteristics and compositions of waste rock from mining industry for carbonation reaction.

	Waste rock	References
Percentage of mineral present	Forsterite = 64.7% Lizardite = 29.6% Magnetite = 4.2% Quartz = 0.3% Brucite = 2.5%	[12]
Mineral composition	*Mg-rich minerals* Brucite Nesquehonite Hydromagnesite Lansfordite *Mg-silicates* Olivine Chrysotile Talc Chlorite Phlogopite *Fe minerals* Magnetite *Ca-rich* Albite	[4,20,21]
Type of rock	*Ultramafic rock* Peridotite Amphibolite Pyroxenite Serpentinite Mafic rock Basalt Gabbro	[21,22,23]

Iron ore also has high possibility to be used for carbon sequestration due to the presence of secondary mineral phases, such as iron oxides (FeO), which can form stable iron carbonate ($FeCO_3$) from mineral carbonation process. For example, mineral composition from iron ore residue consists of 41.10% of MgO, 5% of CaO [22], and 16.04% of FeO from minerals such as siderite ($FeCO_3$) and goethite ($HFeO_2$) [26] (Table 3). Further mineral compositions of iron contain 67.33% of ferric oxide (Fe_2O_3) from both hematite (Fe_2O_3), and magnetite (Fe_3O_4) minerals [27]; 5.79% of quartz (SiO_2) mineral [23], and other minerals including pyrite (FeS_2), ilmenite ($FeTiO_3$), and chamosite $(Mg,Fe,Al)_6(Si,Al)_{414}(OH)_8$. Therefore, high concentration of carbonate minerals in iron and bauxite can increase carbon capture and storage, thus reducing carbon dioxide emission to the atmosphere. Other metal mining waste from gold, tin, and manganese mining might also have potential for CO_2 sequestration but the composition of the divalent cations are not known from the available literature.

Table 3. Percentages of estimated divalent cation content of metal mining wastes expressed in oxides.

	Estimated divalent cation contents (%)		
Metallic mineral	MgO	CaO	FeO
Bauxite	39.50[a]	40[c]	unknown
Iron	41.10[b]	5[b]	16.04[d]
Gold	unknown	unknown	unknown
Tin	unknown	unknown	unknown
Manganese	unknown	unknown	Unknown

[a][8], [b][22], [c][24], [d][26].

4. Conclusions

Metallic mineral from mining industry is highlighted in this study due to their potential to mitigate the release of CO_2 to the atmosphere through mineral carbonation processes. Extraction of gold for instance, are mined from mafic and ultramafic rocks for economic purposes, however natural weathering of the waste rocks results in the formation of carbonates that encourage carbon sequestration processes. Gold mining wastes have a great potential for carbon sequestration in reducing CO_2 in the atmosphere. The major minerals that were discovered at the waste dump and stockpile of the gold mine include the quartz, mica, dolomite, and calcite. Calcite, dolomite, and siderite are considered a reservoir for carbon capture and storage because these minerals were in stable carbonate forms. The presence of silicate minerals such as chlorite, pyroxene, mica, and amphiboles at the waste dump and stockpile are the potential feedstocks for carbon sequestration because these minerals are rich in calcium and magnesium that are essential for mineral carbonation processes.

Acknowledgments

The authors would like to thank Department of Mineral and Geoscience Malaysia, Ministry of Natural Resources and Environment Malaysia for providing information on mining industry in Malaysia throughout this study. Funding for the research was provided by the Ministry of Higher Education Malaysia.

References

[1] Assima G P, Larachi F, Molson J and Beaudoin G 2014a Can. J. Chem. Eng. 92 2029–2038
[2] Harrison A L, Power I M and Dipple G M 2013a Environ. Sci.Technol. 47 126–134
[3] Assima G P, Larachi F, Beaudoin G and Molson J 2013 Int. J. Greenhouse Gas Control 12 124–135
[4] Wilson S A, Dipple G M, Power I M, Thom J M, Anderson R G, Raudsepp M, Gabites J E and Southam G 2009 Econ. Geol. 104 95–112
[5] Assima G P, Larachi F, Molson J and Beaudoin G 2014b Chem. Eng. J. 245 56–64
[6] Lechat K, Jean-Michel L, Molson J, Beaudoin G and Hébert R 2016 Int. J. Greenhouse Gas Control 47 110–121
[7] Power I M, Harrison A L and Dipple G M 2013 Rev. Mineral. Geochem. 77 305–360
[8] Renforth P 2011 Mineral Carbonation in Soils: Engineering the Soil Carbon Sink (United Kingdom: Newcastle University) p 315
[9] Assima G P, Larachi F, Beaudoin G and Molson J 2012 Ind. Eng. Chem. Res. 5 8726–8734
[10] Makoundi C, Zawa K, Large R R, Meffre S, Chun-Kit L and Hoe T G 2014 Gondwana Res. 26 241–261
[11] Hitch M, Ballantyne S M and Hindle S R 2010 Int. J. Min. Reclam. Environ. 24(1) 64–79
[12] Jia-jie L and Hitch M 2015 Int. J. Miner. Metal. Mater. 22(10) 1005–1016
[13] Manning D A C and Renforth P 2012 Environ. Sci. Technol. 1–7
[14] Gras A, Beaudoin G, Molson J, Plante B, Bussière B, Lemieux J M and Dupont P P 2017 Int. J.

Greenh. Gas Control 2017 60 10–23
[15] Qafoku N P 2014 Overview of different aspects of climate change effects on soils. Report Prepared for the U.S. Department of Energy p 52
[16] Saat A, Hamzah Z and Bakar Z A 2009 Journal of Nuclear and Related Technologies 6(1) 230–236
[17] Arce G L A F, Neto T G S, Ávila I, Luna C M R, José C S and João A C J 2017 Hydrometallurgy 169 142–151
[18] Kandji, E H B, Plante B, Bussière B, Beaudoin G and Dupont P P 2017 Environ. Sci. Pollut. Res. 24(12) 11734–11751
[19] Shamshuddin J 2011 Methods in Soil Mineralogy (Malaysia: Universiti Putra Malaysia Press) pp 14–42
[20] Renforth P, Manning D A C and Lopez-Capel E 2009 Appl. Geochem. 24(9) 1757–1764
[21] Assima G P, Larachi F, Molson J and Beaudoin G 2013b Thermochim Acta 566 281–291
[22] Bodénan F, Bourgeois F, Petiot C, Augé T, Bonfils B, Julcour-Lebigue C, Guyot F, Boukary A, Tremosa J, Lassin A, Gaucher E C and Chiquet P 2014 Miner. Eng. 59 52–63
[23] Sarvaramini A, Assima G P, Beaudoin G and Larachi F 2014 Fuel 115 749–757
[24] Gräfe F and Klauber C 2011 Hydrometallurgy 108 46–59
[25] Swain R, Padhy L N and Rao R B 2011 Iran. J. Mater. Sci. Eng. 8(3) 37–49
[26] Sara K, Javad G, Alireza Y and Hosein M 2010 Min. Sci. Technol. 20 0855–0860
[27] Assima G P, Larachi F, Molson J and Beaudoin G 2014c Miner. Eng. 59 39–44

Synergistic effect of binary fillers (MCM-41/C20) hybrid membrane for CO_2/CH_4 separation

Wong Yean Sang[1], *Oh Pei Ching[1], Lau Kok Keong[1]

[1]CO_2 Research Centre (CO_2RES), Institute of Contaminant Management, Chemical Engineering Department, Universiti Teknologi PETRONAS, 32610 Seri Iskandar, Perak, Malaysia.

*Corresponding Author E-mail: peiching.oh@utp.edu.my

Abstract. A novel binary fillers hybrid membrane was formed by incorporating Mobil Composition of Matter-41 (MCM-41) nanoparticles and Cloisite 20 nanoclay layers (C20) into polysulfone (PSf) polymer matrix in order to resolve common agglomerations issue in hybrid membrane and to enhance gas separation performance. Enhanced dispersion degree of MCM-41 nanoparticles and enhanced exfoliation of C20 layers were achieved in the binary fillers hybrid membrane. The resultant membrane showed excellent gas separation performance compared to pristine PSf membrane and single filler hybrid membrane. A 25% improvement in CO_2/CH_4 ideal selectivity (α = 28.7) was observed in the gas permeation of binary fillers hybrid membrane without compromising CO_2 permeability (P_{CO_2} = 6).

1. Introduction

In recent decades, membrane technology has successfully emerged as one of the most important separation methods in industries [1]. Hybrid membrane (HM), which consists of inorganic fillers dispersed in polymer matrix, is of great interest to scientists due to its excellent mechanical and thermal properties [2]. However, achieving uniform dispersion and distribution of inorganic fillers in polymer matrix is often challenging as inorganic particles with identical nature tends to agglomerate and cause poor dispersion, leading to uneven properties of membrane [3]. Furthermore, it is often noticed that the resultant hybrid membranes showed either improvement in terms of gas permeability or selectivity, but not both [4-6]. An ideal membrane should possess high gas permeability and selectivity, hence this led to the development of binary fillers HM. Binary fillers dispersion method was recently introduced to homogeneously disperse nanoparticles in HM and to enhance overall membrane gas separation performance.

Zornoza et al. investigated the synergistic effect of binary fillers by combining metal-organic framework (MOF) and silicalite-1 (S1C) into polysulfone (PSf) for gas separation [3]. Two MOFs, ZIF-8 and HKUST-1, with partially organic nature were also incorporated, since they ensure the chemical affinity with polymer matrix. Due to the increase in polymer matrix free volume caused by inorganic filler disruption, 88% increment in gas permeability was achieved. Furthermore, the combination of HKUST-1/S1C shows optimum performance with 50% improvement in selectivity compared to single filler HM. This is due to enhancement in particles dispersion as particles of different nature were incorporated into the matrix. Recently, Li *et al.* adapted the binary fillers technology from waste water treatment by combining carbon nanotube (CNT) and graphene oxide (GO) to form HM for CO_2/CH_4 separation [7]. They found that HM with 5wt% of CNT and 5wt% of

GO (Matrimid® CNT/GO-5/5) gives highest gas permeability and selectivity. The excellent separation performance of Matrimid® CNT/GO-5/5 was attributed to homogeneous dispersion of CNT and GO in the polymer matrix as well as good adhesion between fillers and polymer. GO which is in layered form, possesses strong steric effect and prevented CNT from agglomeration. The well dispersed CNT, on the other hand, inhibited the stacking of GO nanolayer in the membrane. In another study, Mahdavi and Moradi-Garakani [8] also discovered improved gas separation performance through incorporation of MOF and fumed silica in polyethersulfone HM. However, minimal work has been done to correlate the dispersion degree to gas separation performance.

In the present work, a novel binary fillers HM was formed by incorporating mesoporous silica nanoparticles (Mobil Composition of Matter-41, MCM-41) and layered-silicate (Cloisite 20, C20) into polysulfone (PSf) polymer matrix. The objective of this research was to enhance fillers dispersion degree and CO_2/CH_4 gas separation performance. The effects of binary fillers on MCM-41 nanoparticles dispersion, exfoliations of C20 nanoclay layers, and gas separation performance were investigated through comparison with pristine and respective single filler HM. Furthermore, image processing toolbox from MATLAB was employed to quantitatively analyse the blob area fraction (agglomerates area fraction) and size of agglomerates (L_a) for particles dispersion degree study.

2. Experimental Method

MCM-41, C20 and PSf loadings were pre-determined and optimized at 0.5 wt%, 1 wt% and 26 wt%, respectively. C20 was first dispersed in tetrahydrofuran (THF) for 30 minutes using IKA T25 Digital Ultra-Turrax disperser. MCM-41 was slowly added and dispersed in the mixture for another 30 minutes. PSf was added batch-wise into the mixture and stirred until the pellets were completely dissolved. The solution was cast on a clean and flat glass plate using casting knife with 150 µm gap setting. The resultant film was placed in a covered tray to slow evaporation. The formed membrane was dried for 3 days at room temperature. Pristine PSf membrane and single filler HM at similar loadings (PSf/MCM-41 and PSf/C20) were fabricated via the same method. Images of the membrane were obtained using scanning electron microscope (SEM, Hitachi TM3030) and transmission electron microscope (TEM, FEI Tecnai F20S). The blob area fraction and size of largest agglomerates (L_a) of resultant HM were estimated through developed MATLAB coding. Gas permeation properties of resultant membranes with 1.8 cm^2 effective area, were measured using CO_2 and CH_4 (99.995% purity) at feed side pressure of 8 bar. Permeability and ideal selectivity of gases were calculated based on previously published equations [9, 10].

3. Results and Discussion

3.1. Effect of binary fillers on nanoparticles agglomeration

Blob area fraction of nanoparticles (MCM-41 and C20) in resultant binary fillers HM was obtained and compared with the respective single filler HM, as depicted in figure 1. It was noticed that binary fillers HM demonstrated the smallest blob area fraction, which indicates that better dispersion and distribution of nanoparticles were obtained. A small blob area fraction also suggested that lesser agglomerates were formed in the HM. In addition, L_a of binary fillers HM was estimated from SEM images. It was discovered that the L_a of binary fillers HM was the smallest, as shown in figure 2. This again verifies that the dispersion and distribution of nanoparticles were enhanced in binary fillers HM. This improvement can be attributed to the low cohesive strength of nanoparticles. According to Khare and Burris [11], nanoparticles of the same surface nature (or the same type of nanoparticles) possess strong cohesive strength and weak adhesive force. These particles usually exhibit high specific surface area and energy, which caused them to "stick" to one another and were difficult to separate. On the contrary, nanoparticles of different nature demonstrated an antagonistic effect and caused them to separate from each other. Furthermore, the C20 nanoclay layers in binary fillers HM acted as a sheet barrier that hindered the agglomeration of MCM-41. This steric hindrance effect caused MCM-41 nanoparticles to disperse well in PSf matrix, hence reduced the size of agglomerates. On the other hand, the well dispersed MCM-41 nanoparticles had also inhibited the restacking of C20 nanoclay layers, resulting in increased basal spacing of the layers, as illustrated in figure 3. Compared to C20 single filler HM, which demonstrated a basal spacing of 17.9 nm, binary fillers HM displayed increased basal spacing at 28.55 nm. Thus, it can be concluded that enhancement in nanoparticles dispersion and clay exfoliations were achieved by incorporating binary fillers in HM.

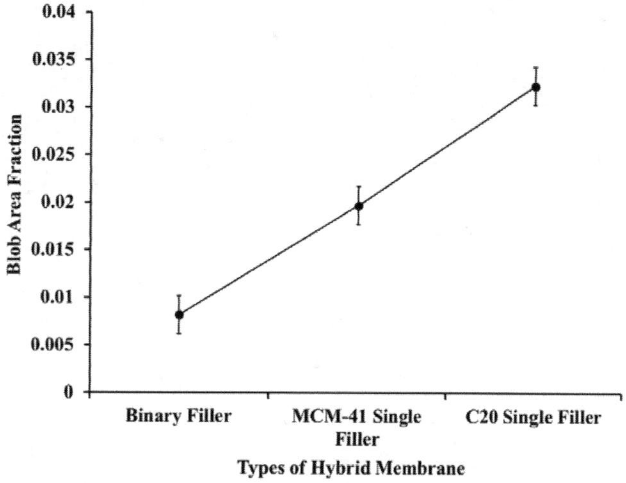

Figure 1. Blob area fraction for different types of HM.

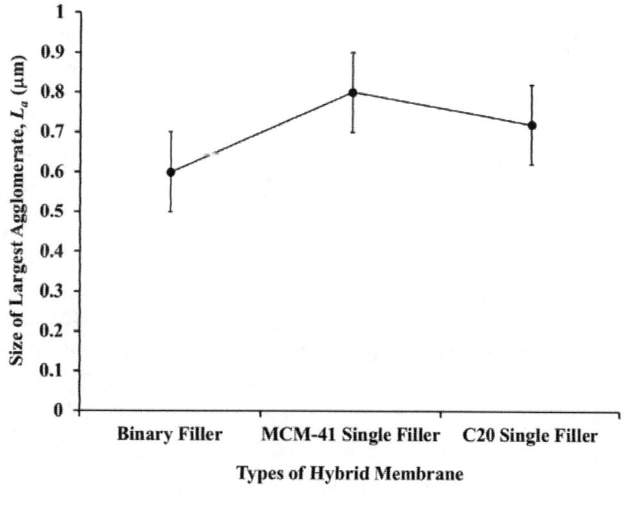

Figure 2. Size of largest agglomerates (L_a) for different types of HM.

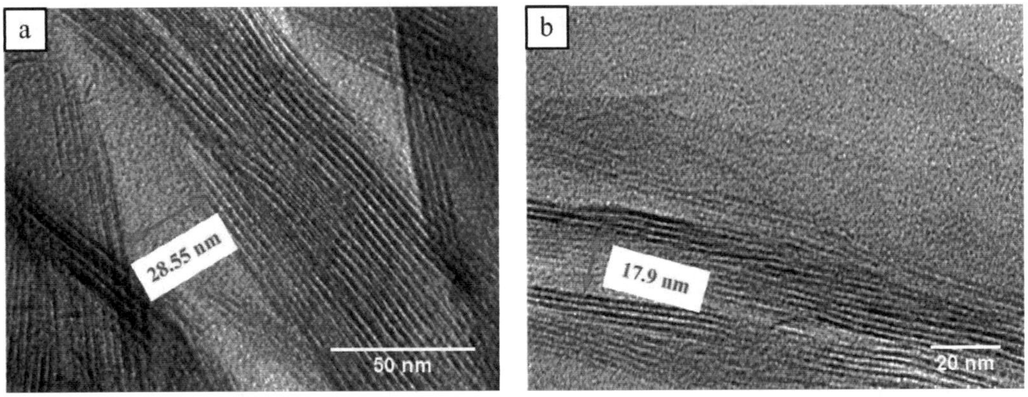

Figure 3. TEM micrographs of (a) binary fillers HM and (b) C20 single filler.

3.2. Effect of binary fillers on gas separation performance

The CO_2/CH_4 ideal selectivity against CO_2 permeability of resultant membranes were plotted, as illustrated in figure 4. Binary fillers HM demonstrated improved gas separation performance. The resultant binary fillers HM showed 25% improvement in CO_2/CH_4 selectivity (α = 28.7) compared to pristine PSf (α = 22.9) membrane without compromising the gas permeability (P_{CO_2} = 6). This increase in selectivity can be attributed to the well exfoliated non-porous C20 nanoclay layers. The non-porous structure of C20 acts as a gas barrier that restricted the direct diffusion of gases through the membrane by forcing gases to diffuse through tortuous route created by the exfoliated layers [12, 13]. CH_4 that has larger kinetic diameter ($d_k(CH_4)$= 3.8 Å) than CO_2 ($d_k(CO_2)$= 3.3 Å) experienced higher restriction in the diffusion through the membrane with well exfoliated and randomly oriented C20 nanoclay layers. Reduction in permeability of CH_4 caused an increase in CO_2/CH_4 selectivity of the membrane. In addition, the mesoporous structure of MCM-41 facilitated gas permeation by allowing gases to pass through the pores with minimal resistance. Thus, a reduction in CO_2 permeability was not observed even with the incorporation of C20 in binary fillers HM. Besides, it was also interesting to note that MCM-41 single filler HM demonstrated highest CO_2 permeability (P_{CO_2} = 7.5) and lowest CO_2/CH_4 selectivity (α = 22.1) among all the membranes. This could be due to the formation of large MCM-41 mesoporous agglomerate in MCM-41 single filler HM, as shown in figure 2. The mesoporous structure of MCM-41 agglomerates allow both CO_2 and CH_4 diffuse through the pores with minimal resistance, hence, low CO_2/CH_4 selectivity was observed. In contrast, MCM-41 agglomerates were reduced in binary fillers HM. Thus, reduction in CO_2/CH_4 selectivity was not observed in binary fillers HM. On the other hand, PSf/C20 single filler HM showed increase in CO_2/CH_4 selectivity (α = 27.7) but decrease in the permeability of CO_2 (P_{CO_2} = 4.5) compared to pristine PSf membrane due to the tortuous path created by the exfoliated C20 nanoclay layers. However, the decrease in permeability of CO_2 was compensated by the introduction of MCM-41 in the binary fillers HM. Thus, this proved that the synergistic incorporation of binary fillers in the fabrication of HM enhanced the gas separation performance.

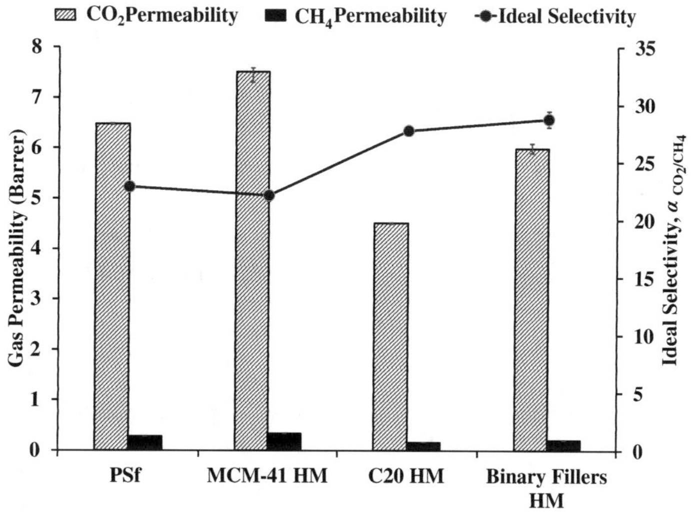

Figure 4. CO_2 permeability and CO_2/CH_4 ideal selectivity versus types of membrane

4. Conclusion

Synergistic incorporation of binary fillers in the fabrication of HM enhanced MCM-41 nanoparticles' dispersion, C20 nanoclay layers exfoliation and gas separation performance. MCM-41 nanoparticles were well dispersed in binary fillers HM due to the steric hindrance effect caused by exfoliated C20 nanoclay layers. On the other hand, the restacking of C20 nanoclay layers was inhibited by the well dispersed MCM-41 nanoparticles. 25% improvement in CO_2/CH_4 ideal selectivity was achieved by the binary fillers HM without compromising CO_2 permeability.

Acknowledgment

This research work was supported by Universiti Teknologi PETRONAS and YUTP-FRG Grant No. 0153AA-E08.

References

[1] Rybak A, Kaszuwara W, Awietjan S, Sysel P, Grzywna ZJ 2017 *Mater. Lett.* **208** 14-18
[2] Chung TS, Jiang LY, Li Y, Kulprathipanja S 2007 *Prog. Polym. Sci.* **32** 483-507
[3] Zornozav B, Seoane B, Zamaro JM, Téllez C, Coronas J 2011 *Chem. Phys. Chem.* **12** 2781-85
[4] Zhao D, Ren J, Li H, Li X, Deng M 2014 *J. Membr. Sci.* **467** 41-47
[5] Wu H, Li X, Li Y, Wang S, Guo R, Jiang Z 2014 *J. Membr. Sci.* **465** 78-90
[6] Khan AL, Klaysom C, Gahlaut A, Khan AU, Vankelecom IFJ 2013 *J. Membr. Sci.* **447** 73-79
[7] Li X, Ma L, Zhang H, Wang S, Jiang Z, Guo R 2015 *J. Membr. Sci.* **479** 1-10
[8] Mahdavi H and Mordadi-Garakani F 2017 *Chem. Eng. Res. Des.* **125** 156-165
[9] Jamil A, Oh PC, Shariff AM 2017 *Appl. Clay. Sci.* **143** 115–124
[10] Qin G, Zhang Y, Wei W 2017 *Mater. Lett.* **209** 75-77
[11] Khare HS, Burris DL 2010 *Polym.* **51** 719-729
[12] Choudalakis G, Gotsis A 2009 *Eur. Polym. J.* **45** 967-984
[13] Choudalakis G, Gotsis A 2012 *Curr. Opin. Sci.* **17** 132-140

IOP Conf. Series: Materials Science and Engineering **458** (2018) 012015 doi:10.1088/1757-899X/458/1/012015

Fabrication and analysis of HNBR-Silica / Carbon Phenolic ablative composites as an effective thermal protection system (TPS)

*S Farrukh[1], N A Tahir[2], A Hussain[1], M Ayoub[3], M Rafiq[2], M H Khan[2]

[1]School of Chemical & Material Engineering, National University of Science &Technology, H-12, Islamabad – Pakistan
[2]National Engineering and Scientific Commission, Islamabad – Pakistan.
[3]Chemical Engineering Department, Universiti Teknologi PETRONAS, Bandar Sri Islander, Perak, Malaysia

*Corresponding Author Email: Sarah.farrukh@scme.nust.edu.pk

Abstract:
The hydrogenated nitrile butadiene rubber (HNBR) based thermal protection systems (TPS) were successfully developed using silica phenolic and carbon phenolic as fillers by filler particle size variation and fixing filler amount in different formulations. Both the TPS were characterized for their confirmations and functional properties using various analytical techniques. The fabricated composites demonstrated improved mechanical, thermal and ablative properties that can meet requirements of aerospace industry for specific applications. A comparative study was also carried out to perceive the effects of particle size of the filler on thermo-mechanical and ablative properties showing that with decrease of the particle size of the filler, mechanical properties and thermal stability increased, while ablation rate decreased. Furthermore, silica phenolic based composites had better mechanical, thermal and ablative properties as compared to its counterpart carbon phenolic TPS.

1. Introduction

The increasing demands of thermal protection system (TPS) in aerospace and defense industry focused the scientists and engineers to develop thermal protections / insulations that can withstand at high temperature and pressure environments [1, 2]. Usually, these thermal protections are composed of special materials having ablative properties, shielding the space vehicles and defense ammunition from thermal destruction during exposure at high temperatures and flow of flame [3, 4, 6].

Literature has revealed that ablative materials are used to protect certain structural parts or equipment from the intense hypothermal environmental conditions [2, 5, 6]. Mostly, such composites are used in aerospace industry [7]. Thermal protection system (TPS) of space vehicles and long range ballistic missiles is to protect them from hypersonic atmospheres, encountered during their missions, are composed of ablative materials that are highly endothermic sacrificial materials [8, 9].

Rubber itself cannot survive the severe environments such as high temperature, pressure and high-speed flow erosion. These properties can be improved by incorporation of some fillers. Silica and carbon phenolic fillers may be incorporated into the rubber to improve its ablative as well as mechanical properties [9-10]. Elastomeric ablative materials with a rubber matrix are widely used in heat shields. Thermal insulation by ablation is achieved through the degradation of the material itself and the formation of a refractory char on the material surface, thereby hindering the transfer of heat into the interior of material [11-12].

The above-mentioned description, convinced us to study the effects of carbon and silica phenolic fillers in hydrogenated nitrile butadiene rubber (HNBR) for thermal protection applications. It was significant part of research to explore, whether these fillers can improve the thermal protection at high temperature or not. Other than this a comparative study was also carried out to perceive the effects of particle size of the filler on thermo-mechanical and ablative properties of both fabricated composites.

2. Experimental

2.1 Materials

The material utilized in this work includes Hydrogenated Acrylonitrile Butadiene Rubber (HNBR) (having CAN content 29% and density 1.1 gm/cm^3), Silica Phenolic and carbon Phenolic (as fillers, mesh size 30 to 70 mesh), Sulphur (curing agent), ZnO, Steric acid (activators), Tetramethylthiuram disulfide (TMTD) , Dibenzothiazyle Disulphide (MBTS) (accelerator), Dioctyl Phthalate (DOP) (Platicizer) and Phenyl Naphtylamine (PBN) (anti-oxidant).

2.2 Fabrication of HNBR based ablative composites

The fabrication of HNBR based ablative composites was done by mixing the ingredients in optimized quantities as shown in Table 1. 11 different recipes were prepared, 1 neat sample without filler incorporation and 5 samples for each filler, fixing its amount at 20 phr with mesh size variation from 30 mesh to 70 mesh.

After mixing of all ingredients, curing of the vulcanized HNBR was carried out by using compression molding machine at 165 °C and pressure 10 MPa. After curing, the sample sheets of fabricated ablative composite were prepared for characterization.

Table 1: Ingredient quantities for compounding of HNBR and other ingredients

Material	A/U	Amount
HNBR	Phr*	100
S		3
ZnO		10
Stearic Acid		2
Fillers		20
MBT		2
DOP		5
TMTD		1
PBN		2

*Parts per hundred rubber

3. Characterization

Fabricated ablative composites were characterized by using following instruments:

3.1 Mechanical Properties

Mechanical properties like tensile strength and elongation at break was carried out by using INSTRON-4465, UK according to ASTM D412-98a standards.

3.2 Thermal Conductivity

Thermal conductivity was carried out using guarded hot plat according to ASTM C177-97.

3.3 Thermal Stability

Thermal stability was characterized by using TGA Q50, TA Instruments, and USA according to ASTM D6370-99. Linear ablation rate was found using oxyacetylene torch method according to ASTM E-285-80.

4. Results and Discussion

4.1 Physical Properties

4.1.1 Density. Density of fabricated ablative composites increases with decrease in filler particle size for both fillers, the result is shown in figure 1. This increase in density is due to increase in surface area and good compaction of filler with matrix material as the particle size decreases. Moreover, density of the fabricated composite is also dependent on the specific gravity of additives and since specific gravity of silica Phenolic (1.23) is higher that neat sample of HNBR having specific gravity (1.035).It can also be concluded that silica phenolic gives slightly higher values of the density as compared to its counterpart.

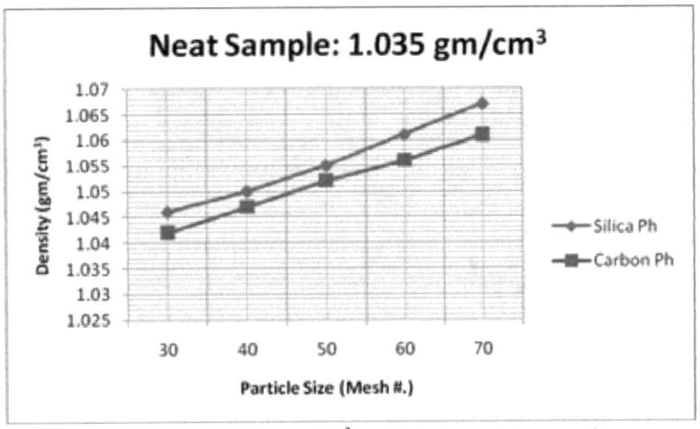

Figure. 1: Density (g/cm^3) of fabricated composites

4.1.2 Hardness. The effect of particle size variation on the hardness of the composite is shown in the figure 2. It is quite evident that hardness of ablative composites increases as the particle size of the filler decrease for both types of the fillers. This increase in hardness is because more rigid filler particles replace the relative softer part of the rubber which also lowers the elasticity of the polymeric chains of the rubber. It can also be concluded that silica Phenolic filler gives slightly higher values of the hardness as compared to carbon Phenolic.

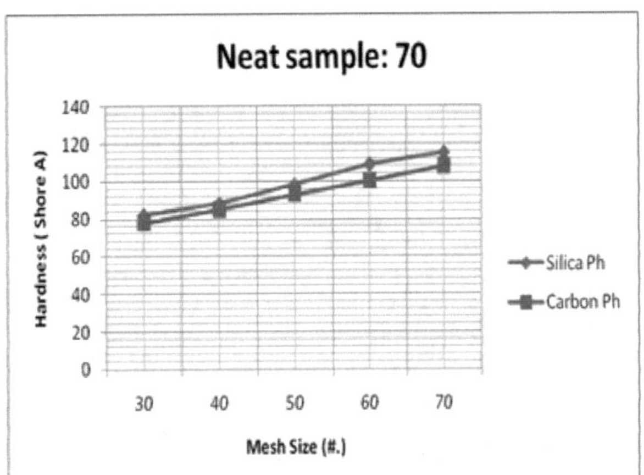

Figure. 2: Hardness of fabricated composites

4.2 Mechanical Properties

4.2.1 Tensile Strength. The effect of particle size variation on the tensile strength of the fabricated ablative composite is shown in the figure 3. It is evident that tensile strength of composites increases as the particle size of the filler decrease for the both types of the fillers. This increase in tensile strength can be explained by the fact that rubber filler interaction improves as the particle size of the filler decreases i.e. smaller the particle size higher will be interaction. It can also be concluded that silica Phenolic gives slightly higher values of the hardness.

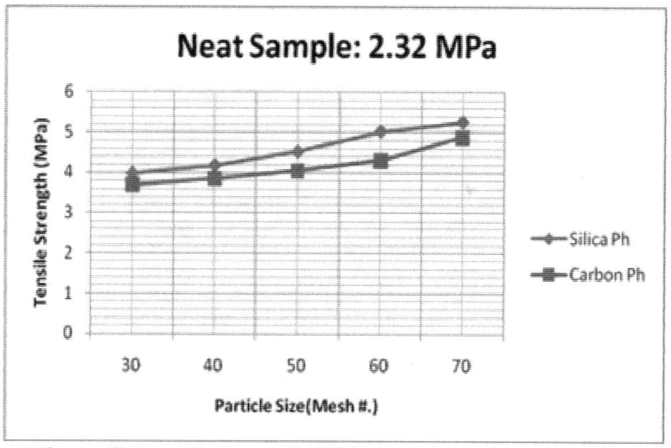

Figure 3: Tensile strength of fabricated ablative composites

4.2.2 Elongation at break. The effect of particle size variation on the elongation at break of the composite is shown in the figure 4. It is clear that elongation at break of fabricated ablative composites decreases as the particle size of the filler decreases for the both types of the fillers. Since elongation at break measures the elastic behavior of the material, this decrease in elongation at break is because more rigid filler particles replace the relative softer part of the rubber which also lowers the elasticity of the polymeric chains of the rubber. Moreover, Silica phenolic filler gives lower values of the elongation at break in comparison to its counterpart Carbon Phenolic.

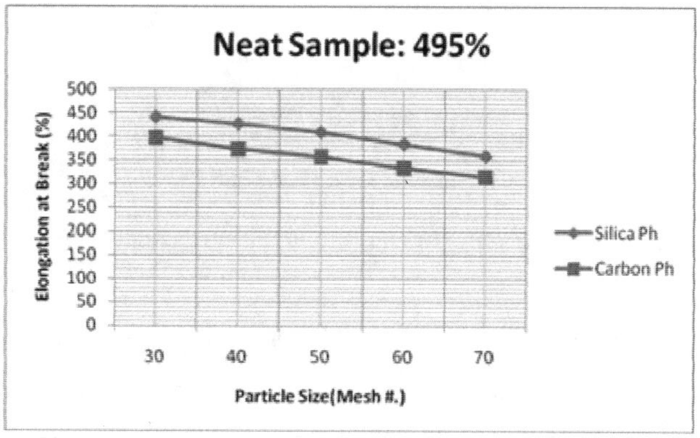

Figure 4: Elongation at break (%) of fabricated composites

4.3 Thermal Properties

4.3.1 Thermal stability analysis using TGA. Thermal stability comparison of fabricated ablative composite using silica / carbon phenolic as filler is shown in figure 5 and 6, respectively. Results obtained by TGA indicate that composites had superior thermal stability than the neat matrix of HNBR by filler particle size variation for both fillers.

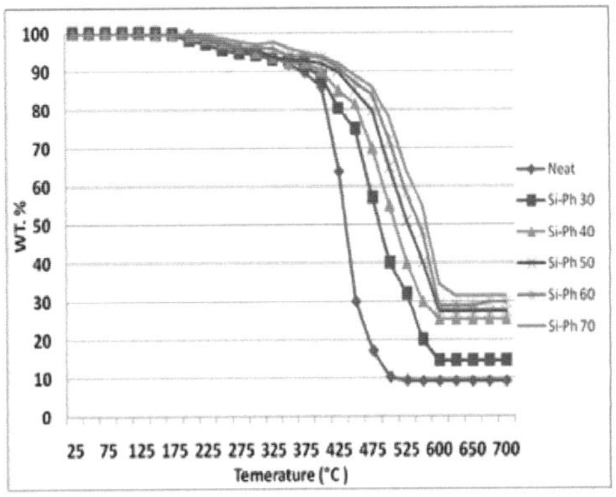

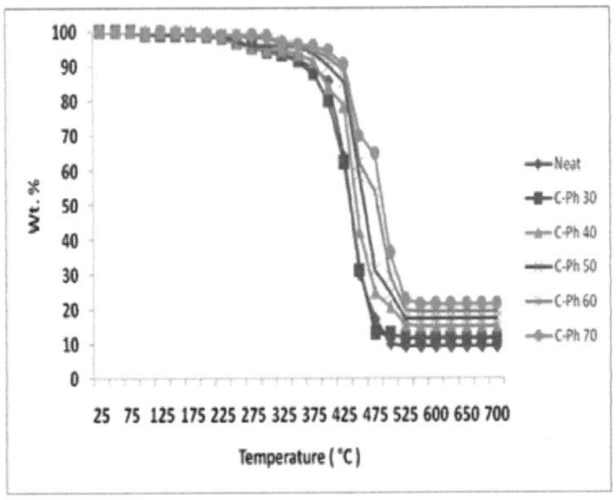

4.3.2 Thermal conductivity. The effect of particle size variation on the thermal conductivity of the composite is shown in the figure 7. It is clear that thermal conductivity of fabricated ablative composites decreases as the particle size of the filler decreases for the both types of the fillers. Moreover, Silica phenolic filler gives lower values of the elongation at break in comparison to its counterpart Carbon Phenolic. Decrease in thermal conductivity was due incorporation of non conducting fillers which actively blocks the path of thermal energy flow as compared to neat HNBR.

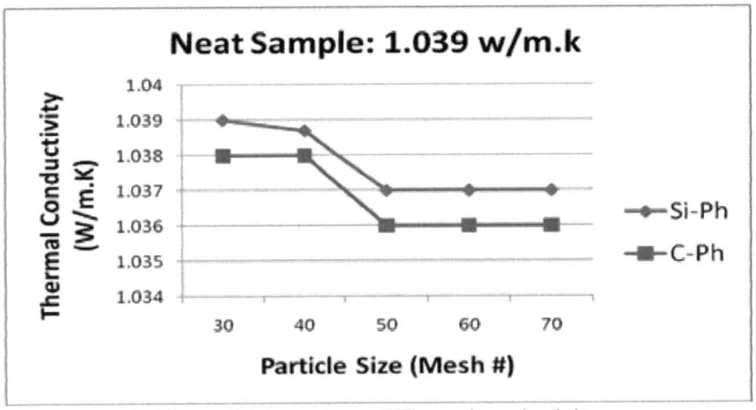

Figure 7. Comparison of Thermal conductivity

4.4 Linear Ablation Rate

The figure 8 shows the effect of filler particle size variation on the linear ablation rate on ablative composites. It gives clear picture that as the particle size of the filler decreases, linear ablation rate also decreases. Results showed incorporation of silica/carbon phenolic fillers into HNBR helps to considerably reduce erosion rate because as the fabricated composites melt, they tend to produce a more viscous layer of residue that binds to substrate. It can also be concluded that Silica phenolic filler gives good ablation results as compared to its counterpart carbon phenolic.

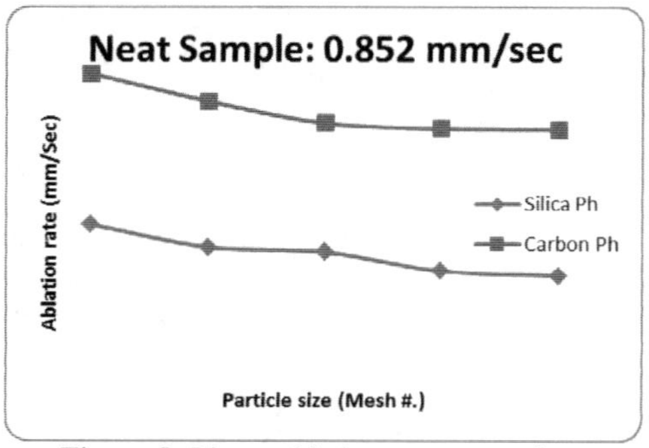

Figure 8: Linear ablation rate comparison

5. Conclusions

It can be depicted from experimental study that tensile strength of fabricated composites increases with the decrease the particle size of both fillers. However, silica phenolic gives better strength as compared to its counterpart carbon phenolic. The elongation at break decreases with the decreases of the fillers size, while an increase in the values of initial modulus of the composite was found as filler particles became coarse to fine. The general increasing trend among density and hardness was also found as filler particles became coarse to fine.

Results of thermal conductivity of composites showed a gradual decrease when size of filler particles is decreased. TGA results indicated that thermal stability of newly fabricated ablative composites was improved as compared to neat sample. Similarly, results of linear ablation rate also showed the effectiveness of HNBR composites as compared to blank HNBR.

From the test results, it can be concluded that in general, physico-mechanical, thermal and ablative properties of HNBR based composites can be improved effectively for optimal and desired results by decreasing particle size of the filler. Furthermore, silica phenolic filler has presented better results as compared to its counterpart carbon phenolic.

6. Acknowledgement

This work was supported by School of Chemical and Materials Engineering, National University of Sciences and Technology, Islamabad, Pakistan and NESCOM Islamabad.

References

[1] Stewart RR, In Flora of West Pakistan, An Annotated Catalogue of the Vascular Plants of West Pakistan and Kashmir 1972 *Fakhri Press, Karachi*, p. 223.

[2] Saleem A B, Ali M K and Khan C H, Natural Inhibitors for Inhibiting the Production of Aflatoxins 2000 *J. Nat. Prod.*, **57**, 1422 .

[3] Ahmed E, Ph.D. Thesis, Phytochemical Investigation on Haloxylon recurvum, Silybum marianum and Aerva persica 2006, University of Karachi.

[4] Almendro J L, Martín J, Sánchez A and Nozal F, Elektromagnetisches Signalhorn, Patent-DE29702195U 1998.

[5] Saleem A B Ali *M K and* Khan C H, Natural Inhibitors for Inhibiting *t*he Production of Aflatoxins 2000 *J. Nat. Prod.*, **57,** 1422.

[6] Shen X T, at el, Ablation behavior of C/C–ZrC composites in a solid rocket motor environment 2015 *Ceram Int.*, **41**, 11793-11803.

[7] Wang Z J, Kwon D K, Ablative and mechanical evaluation of CNT/phenolic composites by thermal and microstructural analyses 2014 *Compos Part B.*, **60** 597-602.

[8] Eslami Z,et al. Thermal and mechanical properties of phenolic-based composites reinforced by carbon fibres and multiwall carbon nanotubes. 2015 *Compos Part A.*, **72** 22-31.

[9] Pulci G, at el. Carbon–phenolic ablative materials for re-entry space vehicles 2010 *J Manufacturing and properties,* **41** 1483-1490.

[10] Chu HT, Chou JH. Experimental study of ablative materials by oxyacetylene flame flow system. In: 2012 International Conference on Industrial and Intelligent Information (ICIII 2012) Singapur, 2012 **31**.
[11] Deuri AS, et al. Thermal and Ablative Properties of Rocket Insulator Compound Based on EPDM 1998 *J.Polym Degrad Stab.*, **21** 21-28.
[12] Bhuvaneswar CM, et al. Filled Ethylene-propylene Diene Terpolymer Elastomer as Thermal Insulator for Case-bonded Solid Rocket Motors 2008 *J.Defense science journal*, **58** 94-102.
[13] Hassan H H, et al. Effect of filler concentration on the physico-mechanical properties of super abrasion furnace black and silica loaded styrene butadiene rubber 2012 *J Materials and Design.*,**34** 533.
[14] Natali M, et al. A nanostructured ablative bulk molding compound: Development and characterization 2011 *J.Appl Sci Manuf.*, **42**:1197–1204.
[15] Jovanovic V, et al. Composites based on carbon black reinforced NBR/EPDM rubber blends 2013 *Compos Part B Eng.*, **4** 333-340.
[16] Eslami, et al. Thermal and mechanical properties of phenolic-based composites reinforced by carbon fibres and multiwall carbon nanotubes 2015 *Compos Part A.*, **72** 22-31.
[17] Srikanth I, et al. Mechanical, thermal and ablative properties of zirconia, CNT modified carbon/phenolic composite 2013 Compos Sci Technol., **80** 1-7.
[18] ASTM Standards (Mechanical properties ASTM D412-98a, Thermal conductivity ASTM C177-97, TGA ASTM D6370-99 & ablation rate ASTM E-285-80.

The Effect of Acidic Gases and Thermodynamic Inhibitors on the Hydrates Phase Boundary of Synthetic Malaysia Natural Gas

Cornelius B. Bavoh[1,2], Muhammad Saad Khan[1,2], Vennesa Johnny Ting[1], Bhajan Lal [1,2], Titus Ntow Ofei[3], Joel Ben-Awuah[4], Muhammad Ayoub[1], Azmi Bin Mohd Shariff[1,2]

[1] Chemical Engineering Department, Universiti Teknologi of PETRONAS, Bandar Seri Iskandar, 32610, Perak, Malaysia.
[2] CO_2 Research Centre (CO2RES), Bandar Seri Iskandar, 32610, Perak, Malaysia.
[3] Petroleum Engineering Department, Universiti Teknologi of PETRONAS, Bandar Seri Iskandar, 32610, Perak, Malaysia.
[4] Petroleum Geoscience Department, Cutin University, Sarawak, Malaysia.

Email: bhajan.lal@utp.edu.my

Abstract. In this work, PVTSim was used to predict the hydrate phase equilibrium boundaries of a Malaysia synthetic natural gas (SNG) composition with increasing CO_2 and H_2S compositions in the presence of commonly used hydrate thermodynamic inhibitors such as methanol (MeOH), ethanol (EtOH), mono-ethylene glycol (MEG), diethylene glycol (DEG), tri-ethylene glycol (TEG). The increasing CO_2 and H_2S concentrations studied ranged from 5 – 50 mol%, while the inhibitors were tested at 10 wt%. To validate the accuracy of PVTSim, a validation test was conducted by predicting the phase behavior of a natural gas in the open literature. The predicted results were in good agreement with the literature data with a percentage deviation of 0.08% from the experimental data. From the predicted hydrate phase equilibrium for pure SNG field, it was found that, when the SNG composition was modified, the hydrate formation region shrank at increasing CO_2 composition, reducing the hydrate formation risk whereas the hydrate formation region expanded at increasing H_2S composition, increasing the hydrate formation threat. TEG was found to best inhibit hydrate formation in the studied gas compositions.

1. Introduction

As exploration and production move further into the deeper sea, flow assurance risk management in subsea pipelines becomes crucial due to problems such as gas hydrates, slugging, corrosion, wax, and scales [1], [2]. Among these flow risks, gas hydrate formation is of major concern to which millions of dollars are spent annually combating its effects [3], [4]. Gas hydrates are solid-crystalline compounds formed by hydrogen-bonded water molecules (host molecules) forming solid crystalline lattice structure by the engaging of gas molecules (guest molecules) such as methane, ethane, propane and etc. at low temperature and high pressure conditions [5], [6], [7], [8]. In Malaysia, a natural gas field situated in the South China Sea offshore of the Sarawak Province holds more than 13.2 trillion cubic feet (tcf) of natural gas with 70% of CO_2 [9]. The high content of CO_2 not only makes it difficult for the operator to separate CO_2 from the natural gas and condensate but also subjects the field to hydrate formation threat [10]. On the other hand, gas hydrates also potentially form inside production facilities whenever the formation conditions are fulfilled.

Although gas hydrates may have potential both as a source of hydrocarbon energy and as a medium for storing and transmitting natural gas, they create major threats. Gas hydrates plug subsea flowlines and can rupture the pipeline which leads to spills of containments causing environmental disasters, high costs of repair and maintenance or even loss of assets. [11]–[15]. Mechanical methods such as insulation and active heating are some existing control measures. However, these methods are impractical and ineffective under some conditions [16]–[20]. For this reason, kinetic and thermodynamic inhibitors are introduced to prevent hydrate formation. A kinetic inhibitor prolongs the induction time of hydrate formation more than the residence time of free water in pipeline while the widely used technique; thermodynamic inhibition shifts the hydrate envelope to the left side of the Pressure-Temperature Diagram to displaces the hydrate formation region to a lower temperature and/ or higher pressure conditions as shown in figure 1 [17], [21].

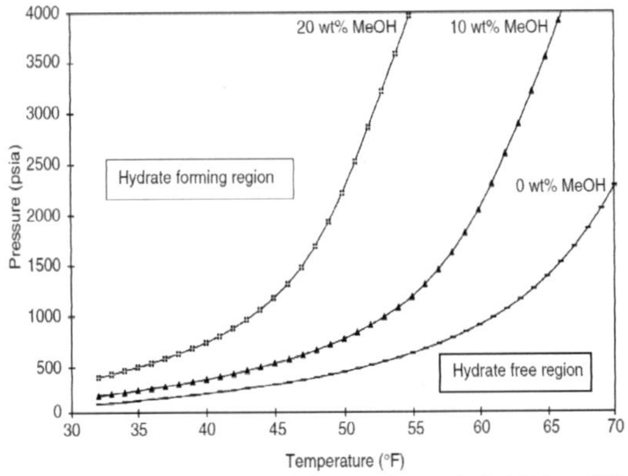

Figure 1. Effect of methanol thermodynamic inhibitor [5]

In order to successively inhibit gas hydrates formation, prediction of hydrate phase equilibrium becomes critical in industrial application. Four available methods are employed for predicting hydrate phase equilibrium in the natural gas systems: (1) hand calculation- suitable for single gas component but limited applicability, (2) empirical correlations- though can be applied to gas mixture, but also limited applicability, (3) experimental method- costly, time consuming and limited apparatus available currently and (4) simulation- easily accessible, fast and reliable using software such as PVTSim, HYSYS, EQUIphase, PIPEphase and etc. [11], [22], [23]. Due to these reasons, the simulation method is nowadays more preferred than other methods.

There are various types of hydrate simulation software such as CSMGem and PVTSim. However, PVTSim is the widely used software. Ward et al. [24] studied the hydrate equilibrium phase of H_2S experimentally and validated his work using PVTSim and reported that PVTSim could be accurately used for H_2S hydrate equilibrium phase studies. In addition, Sule and Rahman [25] and Bavoh et al [26] predicted the hydrate equilibrium phase of synthetic natural gas in Canada and Ghana in the presence of H_2S and inhibitor (methanol) using PVTSim. Their results suggested that PVTSim gives an accurate prediction. More so, the presence of H_2S increased the hydrate formation threat in the studied natural gas systems.

Several studies [27], [28] have to use PVTSim to study the hydrate phase behave is some regions around the world for safe field operation. However, such studies have not been reported on Malaysia field in the open literature. Therefore, this paper seeks to implement simulation method to predict the hydrate phase equilibrium of a synthetic natural gas system in Malaysia using PVTSim and to study the resulting effects on the hydrate phase equilibrium when the gas composition is modified and/ or when commercial gas hydrate inhibitors such as methanol (MeOH), ethanol (EtOH), mono-ethylene glycol (MEG), diethylene glycol (DEG), tri-ethylene glycol (TEG) and glycerol is present.

2. Methodology

2.1. Materials

A set of experimental data as shown in Table 1 was taken from reference [5] to validate the accuracy of PVTSim in predicting hydrate phase equilibrium. Table 2 shows the main data used in this work.

Table 1. Experimental data from literature [5]

CH_4	C_2H_6	C_3H_8	iC_4H_{10}	nC_4H_{10}	nC_5H_{12}	N_2
8.41	6.47	3.57	0.99	1.14	0.78	0.64

Table 2. Synthetic natural gas composition (SNG)

CO_2	H_2S	N_2	CH_4	C_2H_6	C_3H_8	iC_4H_{10}	nC_4H_{10}	iC_5H_{12}	nC_5H_{12}	H_2O
2.13	-	3.91	88.92	4.42	0.50	0.05	0.05	0.01	0.01	2.59

2.2. Applied Software Model
PVTSim owned and developed by Calsep is an adaptable equation of state (EOS) modeling software used for the simulation of fluid properties and experimental PVT data. PVTSim is designed to suit a variety of processes and problems in the industry as it includes features such as advanced flash and regression algorithms, reliable and predictive reservoir fluid characterization, self-regulating fluid regression for minimization of manual tuning, user-friendly interface designed for fast and efficient workflow, etc. [29] PVTSim is divided into a few modules such as characterized fluid, flash and unit operations, asphaltene, wax, hydrate, hydrate open structure, etc. The hydrate and hydrate open structure modules were employed in this work due to its specialized functions in the simulation of hydrate formation conditions, with and without inhibitors as well as the effect of impurities (such as acidic gases) on hydrate formation. [29] A suitable fluid package, Peng Robinson (PR) Peneloux fluid package was selected for the prediction of hydrate phase equilibrium for its capability to cater for a wide range of operating conditions and to simulate result with higher accuracy compared to the Soave-Redlich-Kwong (SRK) Peneloux fluid package [28].The PVTSim predicted results were tabulated and analyzed using Microsoft Excel.

2.3. Simulation Contents and Methods
To set up the modeling environment in PVTSim, every gas composition to be used in this work was registered and stored in the fluid database. Next, the hydrate module was selected in order to employ the hydrate calculations.

For validation analysis, the natural gas composition of the experimental data from open literature [5] was selected from the fluid database. In the hydrate open structure module, the hydrate formation temperature function was selected, and the phase equilibrium pressures of the experimental data were inputted to predict the corresponding phase equilibrium temperatures. The predicted temperatures were compared to the experimental temperatures in Microsoft Excel and the average absolute percentage error (APE) between the predicted result and the experimental result was calculated.

In the first part of the main study, pure SNG composition was selected from the fluid database. The hydrate formation PT curve function in the hydrate open structure module was selected to predict the hydrate phase equilibrium at a pressure range of 16 bar – 200 bar and a temperature range of 2 °C – 20 °C, with reference to the good design pressure, 49.63bar, and temperature, 37.76°C. From this prediction, the hydrate-forming region and hydrate free region can be determined for the safe operation of the field.

The second part of the main work consists of two case studies: The effect of increasing CO_2 and H_2S composition in the synthetic natural gas composition and the effect of thermodynamic inhibitors for effective hydrate inhibition. In case study 1, the CO_2 composition in SNG was increased by 5mol%, 10mol%, 15mol%, 20mol%, 30mol%, 40mol% and 50mol%. Firstly, SNG with additional 5mol% of CO_2 was selected from the fluid database and the hydrate phase equilibrium was predicted at a pressure range of 16 bar – 200 bar using the hydrate formation temperature function. Under the same pressure range, the hydrate phase equilibrium for the increasing CO_2 compositions in the SNG was predicted and compared to that of pure SNG. Next, the average temperature deviation of the hydrate phase equilibrium curves between pure SNG and SNG with increasing CO_2 composition was calculated for each increment and was plotted in Excel. Finally, the CO_2 effect on the hydrate phase equilibrium were analyzed and discussed. The same procedure was repeated for increasing H_2S composition under the same pressure and temperature conditions.

In case study 2, the composition of pure SNG with the addition of methanol (MeOH) at 10wt% concentration was selected from the fluid database. The hydrate phase equilibrium was then predicted under the same pressure range used in case study 1. This step was repeated with ethanol (EtOH), mono-ethylene glycol (MEG), diethylene glycol (DEG), tri-ethylene glycol (TEG) and glycerol. The hydrate phase equilibrium curves generated for each inhibitor were plotted in the same graph in Excel to analyze and discuss the inhibition effect for the pure SNG so as to determine the best inhibitor for the particular composition. The same procedure was repeated for SNG + 30mol% CO_2 and SNG + 30mol% H_2S to study the effect of the inhibitors incorporating the effect of change in gas composition.

3. Results and Discussions

Prior to the simulation, the accuracy of PVTSim in predicting the hydrate phase equilibrium curve was validated by predicting the experimental hydrate equilibrium phase data of the natural gas composition taken from reference [5]. Fig. 2 shows the comparison of the hydrate phase equilibrium curves between the experimental and PVTSim predicted results. A close agreement is observed between the two curves with an average absolute percentage error of 0.08%, which verifies the accuracy of PVTSim in the prediction of hydrate phase equilibrium conditions.

Figure. 3 shows the hydrate phase equilibrium curve of pure SNG composition predicted by PVTSim. In Fig. 3, the area above the curve (to the left) shows the hydrate-forming region whereas the area below the curve (to the right) shows the hydrate free region. Based on the predicted hydrate phase equilibrium data of the SNG system, the gas field flow assurance operations must be conducted within the temperatures and pressures conditions in Fig. 3 to avoid flow assurance problems such as hydrate formation in the pipelines.

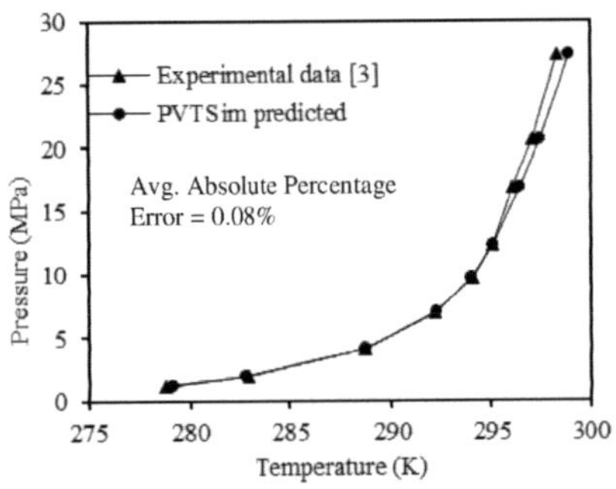

Figure 2. Comparison of PVTSim predicted result and experimental result

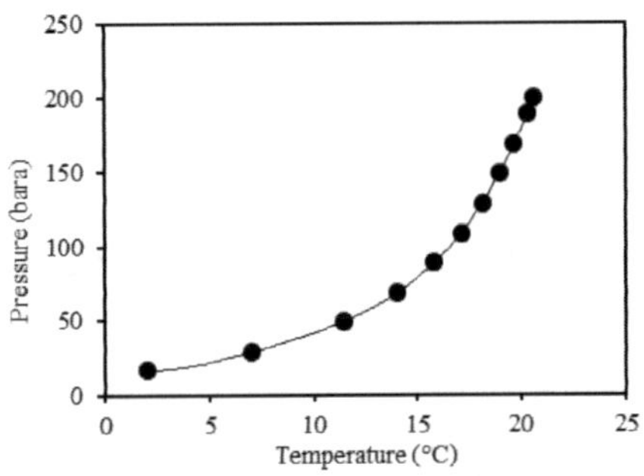

Figure 3. PVTSim predicted hydrate phase equilibrium of pure synthetic natural gas (SNG) composition

3.1 Case Study 1: The effect of CO_2 and H_2S on SNG composition

The addition of acidic gases such as CO_2 and H_2S makes the gas mixture sour. In the life of a field, the gas composition may change which can affect the hydrate equilibrium phase curve of the mixed gas system. Hence, the prediction of the hydrate equilibrium phase curve of the SNG system when some of its gas composition change is necessary for gas hydrate formation prevention strategies in natural gas field flow assurance operations. Fig. 4 shows the effect of increasing CO_2 composition in the SNG system at CO_2 increasing concentrations of 5mol%, 10mol%, 15mol%, 20mol%, 30mol%, 40mol% and 50mol%. From Fig. 4, it is observed that the pure SNG hydrate phase equilibrium curves shift to the left as CO_2 concentration increases. This shows that CO_2 behaves like an inhibitor by shrinking the hydrate formation zone, decreasing the risk of hydrate formation in the system. However, the inhibiting effect is observed to be minimal with an average temperature depression of 1°C at all studied increasing CO_2 concentration. In addition, the SNG hydrate phase equilibrium shifts to the left more at higher pressures.

To further scrutinize the effect of CO_2 increasing concentration on the SNG hydrate equilibrium phase, the average temperature deviations of the SNG hydrate equilibrium phase in the presence of increasing CO_2 concentrations are calculated and presented in Fig. 5. The negative value of the average temperature deviation is an indication that the hydrate phase equilibrium curves are shifted to the left.

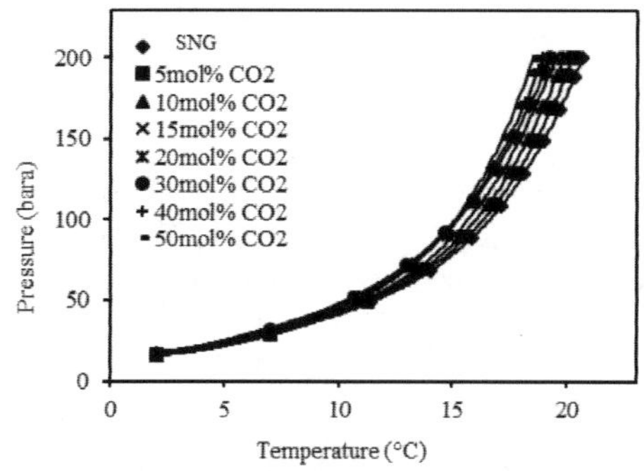

Figure 4. PVTSim predicted hydrate equilibrium phase curve of pure SNG composition and SNG + different CO_2 concentration ranging from 5mol% to 50mol%

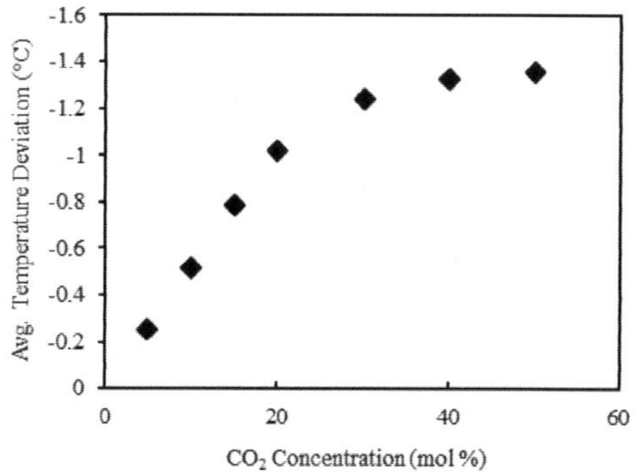

Figure 5. Average temperature deviation between pure SNG composition and SNG + different CO_2 concentration ranging from 5mol% to 50mol%

However, it is noticed that the inhibiting effect of increasing CO_2 composition in the SNG system stabilizes at 30mol%. Increasing CO_2 concentration above 30mol% exhibits small or negligible temperature deviation compared to CO_2 concentrations lower than 30mol%. This means that adding more CO_2 to the SNG system will not have any further significant effect on its hydrate equilibrium phase curves. Therefore, it is deduced that hydrate formation risk in the SNG system relatively / slightly reduces significantly with increasing CO_2 concentration until 30mol% CO_2 concentration. Fig. 6. shows the effect of increasing H_2S composition on the SNG system at concentrations of 5mol%, 10mol%, 15mol%, 20mol%, 30mol%, 40mol% and 50mol% predicted by PVTSim. In contrary to the increasing effect of CO_2 concentration in the SNG system, increasing H_2S concentration further expands the hydrate formation region by shifting the SNG hydrate phase equilibrium curve to higher temperatures and lower pressures region. This signifies that, the presence of H_2S increases in gas hydrate formation threat in the system as H_2S concentration increases as shown in Fig. 6. Surprisingly, it is seen that H_2S gives a more significant effect than CO_2 when added to the SNG system. For example, with 50mol% of H_2S in the SNG mixture, the hydrate phase equilibrium is shifted to the right of the PT diagram for an average temperature difference of 10°C, which is about ten times more than the shift that occurs in the case of 50mol% of CO_2 added to the gas mixture that gives only an average temperature difference of 1°C (see Fig. 4).

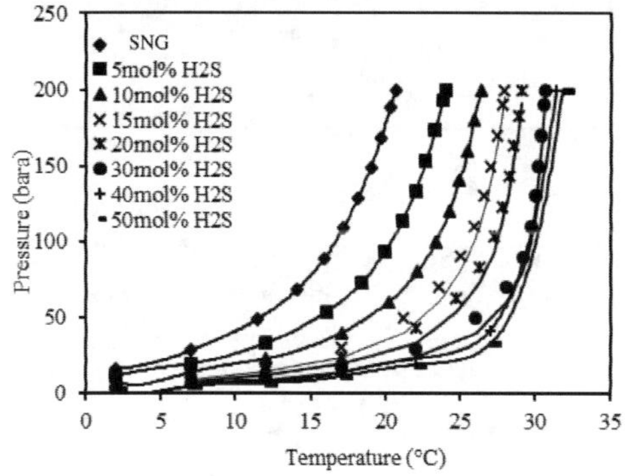

Figure 6. PVTSim predicted hydrate equilibrium phase curve of pure SNG composition and SNG +

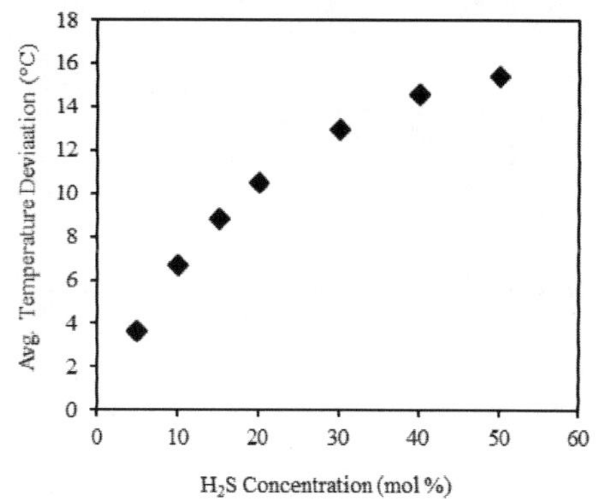

Figure 7. Average temperature deviation between pure SNG composition and SNG + different H_2S concentration ranging from 5mol%

different H_2S concentration ranging from 5mol% to 50mol%

to 50mol%

Furthermore, it can be observed from the trend of the hydrate phase equilibrium curves of each H_2S concentration that the curve shifts drastically to the right of the PT diagram when pressure is below 100 bar and the shifting starts to retard as pressure increases causing the curves to incline more vertically. Fig. 7 shows the average temperature deviation of the SNG composition and SNG + different H_2S concentration ranging from 5mol% to 50mol%. The positive value of the average temperature deviation represents the right-hand side shifting of the hydrate phase equilibrium curves. The deviation in temperatures increases consistently and is observed to decrease at 40mol% H_2S, indicating the stabilization of the H_2S effect on the SNG hydrate equilibrium curve. However, the concentration of H_2S above 50mol% is seen to have a negligible effect in shifting the hydrate phase equilibrium.

In case study 1, it can be inferred that SNG + CO_2 or H_2S hydrate phase equilibrium curves tends to shift towards the hydrate equilibrium curve of pure CO_2 or H_2S component as CO_2 or H_2S concentration increases in the system, agreeing to similar finding reported by Lee and Kang in their study of the hydrate-phase equilibria for gas mixture $(CH_4 + C_2H_4)$ and $(C_2H_6 + C_2H_4)$ with various concentrations along with the equilibrium curves of the pure components. [30] Also, explained in their work that the shifting of the hydrate phase equilibrium is affected by the cavity filling of mix gases present in the system. On the other hand, Ripmeester and others also studied on the $H_2S - CH_4$ hydrates cage occupancy and found out that occupancy of cavities of Structure I by H_2S increases as the concentration of H_2S increases, which is also represented by the shifting of the hydrate phase equilibrium to the right of the $H_2S - CH_4$ hydrates PT diagram [31].

3.2. Case Study 2: The effect of thermodynamic inhibitors

Commonly, thermodynamic inhibitors such as methanol, ethylene glycol, glycerol and etc. are used in the industry to prevent hydrate formation. Thermodynamic inhibitors generally inhibit hydrate formation by distracting the activity of water in hydrate formation by forming hydrogen bonding with water molecules which results in shifting the hydrate phase equilibrium curve to lower temperatures and higher pressures conditions to keep the system out of the hydrate formation zone.

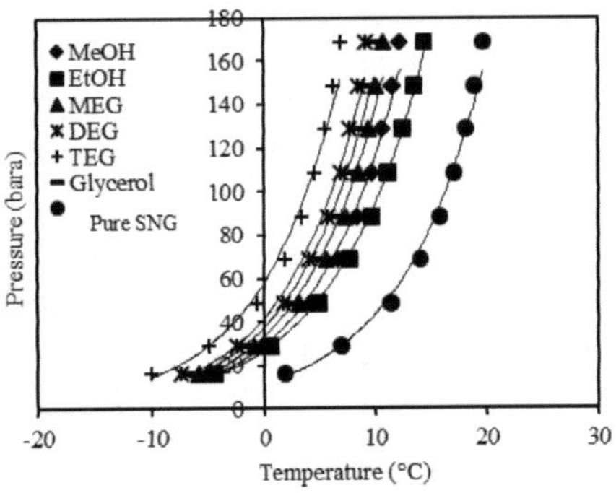

Figure 8. PVTSim prediction of hydrate phase equilibrium of SNG composition + different types of thermodynamic inhibitors at 10wt%

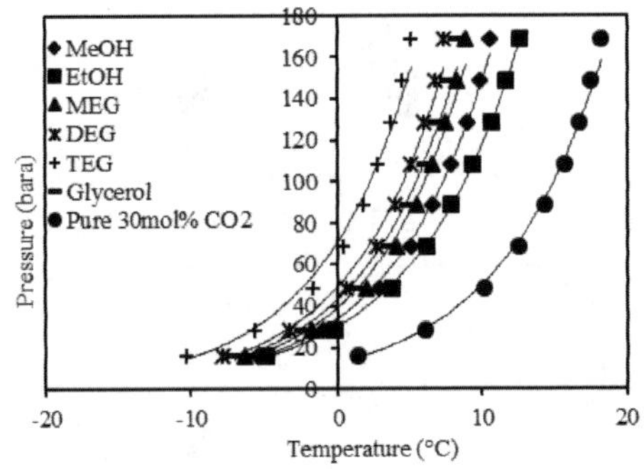

Figure 9. PVTSim prediction of hydrate phase equilibrium of SNG + 30mol% of CO_2 in the presence of commonly used gas hydrate thermodynamic inhibitors at 10wt%

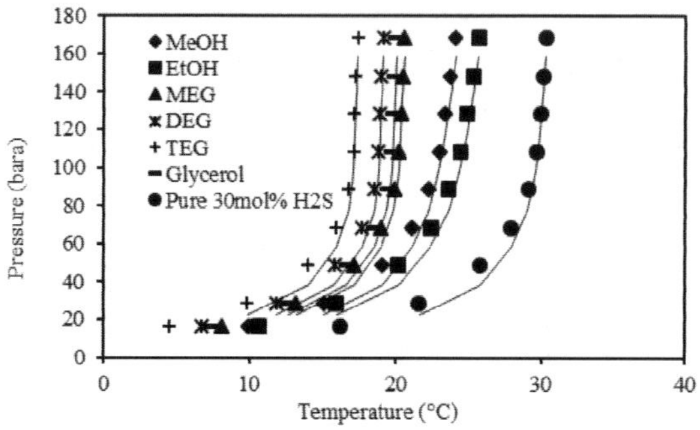

Figure 10. PVTSim prediction of hydrate phase equilibrium of SNG + 30mol% of H_2S + in the presence of commonly used gas hydrate thermodynamic inhibitors at 10w%

The PVTSim predicted effect of six commonly used thermodynamic inhibitors namely: methanol, ethanol, mono ethylene glycol (MEG), diethylene glycol (DEG), tri-ethylene glycol (TEG) and glycerol on pure SNG composition, SNG + 30mol% of CO_2 and SNG + 30mol% of H_2S at a concentration of 10wt% are as shown in Fig. 9, Fig. 10 and Fig. 11 respectively. In Fig. 9, it is seen that all inhibitors shift the SNG hydrate phase equilibrium to the left. It is also observed that the inhibitors increase in inhibition in an ascending order of magnitude of EtOH, MeOH, MEG, Glycerol, DEG, and TEG. Therefore, TEG can best inhibit hydrate formation as it incurs the greatest shift of the hydrate phase equilibrium to the left of the PT diagram. The presence of different gas in the SNG composition can affect the inhibition outcome. For optimum results, the effect of thermodynamic inhibitors on the hydrate equilibrium curves of SNG + 30mol% CO_2 and SNG + 30mol% H_2S were also studied and presented in Fig. 10 and 11 respectively. From Fig. 10 and 11, similar inhibiting performances as presented in Figure 4.7 are observed in which TEG can best reduce the hydrate formation zone. However, some slight changes of inhibition in the SNG +30mol% H_2S composition + inhibitor equilibrium curve, suggest that the performance of a thermodynamic inhibitor can be affected by the type of gas composition.

4. Conclusions

PVTSim was applied to predict the hydrate phase equilibrium conditions of a synthetic natural gas composition of a field in Malaysia. Based on the study, the hydrate formation risk in the SNG system relatively reduces significantly with increasing CO_2 concentration until 30mol% CO_2 concentration. In contrary, increasing H_2S expands the hydrate formation region in a greater extent but starts to stabilize at 40mol% H_2S concentration and shows negligible phase boundary increasing effect at 50mol% H_2S concentration. Furthermore, it was found that TEG can best inhibit hydrate formation in pure SNG, SNG + 30mol% CO_2 and SNG + 30mol% H_2S compared to other inhibitors. This study is significance to ensure the safe operation of natural gas wells.

5. Acknowledgments

Authors wishes to acknowledge the YUTP funding under grant no. 0153AA-E75 for providing financial support. Authors also would like to thank the Chemical Engineering Department, Universiti Teknologi PETRONAS for providing laboratory facilities throughout the study.

6. References

[1] I. S. Author, "Phase Behaviour of Hydrate Formations in Oil and Gas Production in Subsea Conditions," 2014.

[2] C. B. Bavoh, B. Lal, O. Nashed, M. S. Khan, K. K. Lau, and M. A. Bustam, "COSMO-RS: An ionic liquid prescreening tool for gas hydrate mitigation," Chinese J. Chem. Eng., vol. 24, no. 11, pp. 1619–1624, 2016.

[3] K. K. Salam, A. O. Arinkoola, and Y. E. Ayansola, "Prediction of Hydrate Formation Conditions in Gas Pipelines," Int. J. Eng. Sci., vol. 2, no. August, pp. 327–331, 2013.

[4] O. Nashed, D. Dadebayev, M. S. Khan, C. B. Bavoh, B. Lal, and A. M. Shariff, "Experimental

and modelling studies on thermodynamic methane hydrate inhibition in the presence of ionic liquids," J. Mol. Liq., vol. 249, pp. 886–891, 2018.
[5] E. D. S. Jr. and C. Koh, "Clathrate Hydrate of Natural Gases." 2007.
[6] C. B. Bavoh, B. Partoon, B. Lal, and L. Kok Keong, "Methane hydrate-liquid-vapour-equilibrium phase condition measurements in the presence of natural amino acids," J. Nat. Gas Sci. Eng., vol. 37, pp. 425–434, 2017.
[7] "Can gas hydrates be an answer to India's energy woes?," Market Wrap, 2015. .
[8] M. S. Khan, C. S. Liew, K. A. Kurnia, B. Cornelius, and B. Lal, "Application of COSMO-RS in Investigating Ionic Liquid as Thermodynamic Hydrate Inhibitor for Methane Hydrate," Procedia Eng., vol. 148, no. 2016, pp. 862–869, 2016.
[9] M. S. Khan, C. B. Bavoh, B. Partoon, O. Nashed, B. Lal, and N. B. Mellon, "Impacts of ammonium based ionic liquids alkyl chain on thermodynamic hydrate inhibition for carbon dioxide rich binary gas," J. Mol. Liq., vol. 261, pp. 283–290, 2018.
[10] "Petronas to develop K5 project on its own offshore Malaysia Sarawak," 2B1st Consulting. .
[11] A. Bahadori, "A simple mathematical predictive tool for estimation of a hydrate inhibitor injection rate," Nafta, vol. 62, no. 7–8, pp. 213–223, 2011.
[12] C. B. Bavoh, B. Partoon, L. K. Keong, B. Lal, and C. D. Wilfred, "Effect Of 1-Ethyl-3-Methylimidazolium Chloride and Polyvinylpyrrolidone on Kineticsof Carbon Dioxide Hydrates," Int. J. Appl. Chem., vol. 12, no. 1, pp. 6–11, 2016.
[13] C. B. Bavoh, B. Partoon, B. Lal, G. Gonfa, S. Foo Khor, and A. M. Sharif, "Inhibition Effect of Amino Acids on Carbon Dioxide Hydrate," Chem. Eng. Sci., 2017.
[14] C. B. Bavoh, O. Nashed, M. Saad Khan, B. Partoon, B. Lal, and A. M. Sharif, "The Impact of Amino Acids on Methane Hydrate Phase Boundary and Formation kinetics," J. Chem. Thermodyn., vol. 117, no. February 2018, pp. 48–53, 2018.
[15] M. S. Khan, B. Lal, C. B. Bavoh, L. K. Keong, and A. Bustam, "Influence of Ammonium based Compounds for Gas Hydrate Mitigation : A Short Review," Indian J. Sci. Technol., vol. 10, no. 5, pp. 1–6, 2017.
[16] O. Wilfred and D. Appah, "Analyzing Thermal Insulation for Effective Hydrate Prevention in Conceptual Subsea Pipeline Design Analyzing Thermal Insulation for Effective Hydrate Prevention in Conceptual Subsea Pipeline Design," no. October, 2015.
[17] H. E. Christiansen, "Rate of Hydrate Inhibitor in Long Subsea Pipelines," no. June, 2012.
[18] M. S. Khan, B. Partoon, C. B. Bavoh, B. Lal, and N. B. Mellon, "Influence of tetramethylammonium hydroxide on methane and carbon dioxide gas hydrate phase equilibrium conditions," Fluid Phase Equilib., vol. Volume 440, no. May 2017, pp. 1–8, 2017.
[19] M. S. Khan, C. B. Bavoh, B. Partoon, B. Lal, M. A. Bustam, and A. M. Shariff, "Thermodynamic effect of ammonium based ionic liquids on CO2 hydrates phase boundary," J. Mol. Liq., vol. 238, no. July, pp. 533–539, 2017.
[20] C. B. Bavoh, B. Lal, L. K. Keong, M. binti Jasamai, and M. binti Idress, "Synergic Kinetic Inhibition Effect of EMIM-Cl + PVP on CO2 Hydrate Formation," Procedia Eng., vol. 148, pp. 1232–1238, 2016.
[21] A. H. Jonassen, "Distribution of Hydrate Inhibitors in Processing of Gas-Condensate," no. June, 2013.
[22] A. Bahadori and H. B. Vuthaluru, "A novel correlation for estimation of hydrate forming condition of natural gases," J. Nat. Gas Chem., vol. 18, no. 4, pp. 453–457, 2009.
[23] R. Davarnejad, J. Azizi, and J. Azizi, "International Journal of Engineering Prediction of Gas Hydrate Formation using HYSYS Software," vol. 27, no. 9, pp. 1325–1330, 2014.
[24] Z. T. Ward, C. E. Deering, R. A. Marriott, A. K. Sum, E. D. Sloan, and C. a Koh, "Phase Equilibrium Data and Model Comparisons for H2S Hydrates," J. Chem. Eng. Data, vol. 60, pp. 403–408, 2014.
[25] I. Sule and A. Rahman, "Phase behaviour of hydrate formations in oil and gas production in subsea conditions," 2014 Ocean. - St. John's, Ocean. 2014, 2015.
[26] E. Broni-Bediako, R. Amorin, and C. B. Bavoh, "Gas Hydrate Formation Phase Boundary Behaviour of Synthetic Natural Gas System of the Keta Basin of Ghana," Open Pet. Eng. J., vol. 10, no. 1, pp. 64–72, 2017.
[27] G. Gupta and S. K. Singh, "Hydrate inhibition - Optimization in deep water field," Soc. Pet.

Eng. - SPE Oil Gas India Conf. Exhib. 2012, OGIC - Further, Deep. Tougher Quest Contin., no. March, pp. 311–317, 2012.

[28] O. G. Avaldsnes, T. Barth, and P. Fotland, "An Analysis of CO2 , CH4 and Mixed CO2 -CH4 Gas Hydrates: Experimental Phase Equilibria Measurements and Simulations with State-of-the-Art Software," no. June, 2014.

[29] Calsep, "PVTsim Technical Overview," 2016.

[30] J. Lee and S. Kang, "Formation Behaviours of Mixed Gas Hydrates Including Olefin Compounds," vol. 32, no. 1997, pp. 1921–1926, 2013.

[31] J. A. Ripmeester, "The Characteristics of Gas Hydrates Formed from H2S and CH4 under Various Conditions," no. March 2016.

Comparative Study On Water Uptake And Ionic Transport Properties Of Pre- And Post Sulfonated Chitosan/PVA polymer Exchange Membrane

Chun Yik Wong[1], Wai Yin Wong[1], Kee Shyuan Loh[1], Wan Ramli Wan Daud[1,2], Kean Long Lim[1], Keh Shyuan Loh[1], Rashmi Walvekar[3], Mohamad Khalid[4]

[1] Fuel Cell Institute, Universiti Kebangsaan Malaysia, 43600 UKM Bangi, Selangor, Malaysia.

[2] Department of Chemical and Process Engineering, Faculty of Engineering and Built Environment, Universiti Kebangsaan Malaysia, 43600 UKM Bangi, Selangor, Malaysia.

[3] School of Engineering, Taylor's University Lakeside Campus, 47500 Subang Jaya, Selangor, Malaysia.

[4] Graphene & Advanced 2D Materials Research Group (GAMRG), Research Centre for Nanomaterials and Energy Technology (RCNMET), School of Science and Technology, Sunway University, 47500 Subang Jaya, Selangor, Malaysia.

E-mail: nickwcy72@gmail.com; waiyin.wong@ukm.edu.my

Abstract: Chitosan/poly(vinyl alcohol) (PVA) blend composite was prepared through two strategies of chemical modification, namely pre- and post-sulfonation. The sulfonation was carried out by using 4-sulfopthalic acid (sPTA) as the sulfonating agent. The modified chitosan/PVA blend composite was prepared under a range of chitosan content (10:90, 25:75, 50:50, 75:25 and 90:10). Water uptake and ion exchange capacity (IEC) of modified membranes were evaluated by titration and gravimetry methods respectively. Accordingly, both pre- and post-sulfonated composite showed a decrease in water uptakes and IEC values with an increase in chitosan content from 10 to 50 vol.%, attributed to the greater number of hydrogen bond pairs between the two polymers. At 75 vol.%, the composite was predicted to be dominant by the hydrophilic nature of chitosan, in which IEC values and water uptakes were shown to increase. The composite with 90 wt.% chitosan was found to be excessively hydrophilic with tremendously high water uptake, hence not suitable for fuel cell application. Besides that, the post-sulfonated composite showed a trend of increase in the IEC values and water uptakes with a decrease in chitosan content from 10 to 50 vol.%, and bounced back at 75 vol.%. Despite both pre- and post-sulfonation methods demonstrated similar trends in the results, it was notable that post-sulfonation method emerged with higher water uptake and ionic conductivity was found more favorable, attributed to the possibility that sulfonation took place on both PVA and chitosan which has caused a significant increase in sulfonic groups that purportedly exhibited higher ion transport mobility.

Keywords: Chitosan/poly(vinyl alcohol) blend composite, fuel cell, functionalisation of polymers, membranes.

1. Introduction

In the past few years, biopolymers have been appraised as the potential substituents for synthetic polymers used in low and intermediate temperature polymer electrolyte-based fuel cells [1]. Biopolymers have demonstrated excellent properties such as biocompatibility, biodegradability, nontoxicity, naturally abundant and lower cost compared to synthetic polymers. Despite that, an apparent obstacle of low ionic conductivity and high crystallinity is observed on pristine biopolymers such as chitosan and cellulose, potentially hinder the performance of a solid polymer electrolyte (SPE) [2].

In correspondence, diverse strategies have been made to comprehend the aforementioned challenges, for instance, performing copolymerization, grafting, polymer blend, crosslinking, chemical modification, the addition of fillers and plasticisers [3-6]. Among all, chemical modification excelled as an exceptional strategy to tailor the polymer properties mainly contributed to the available functional groups in the polymer that allow functionalization while providing multitude variations and specific chemical functionalities [7, 8]. Hence, the scope of study in this work constraint on investigating the effect of pre- and post-sulfonation through the blending of chitosan and poly(vinyl alcohol) (PVA) as a potential SPE for fuel cell application.

Chitosan is regarded as the second most abundant natural polymers in the world which can be obtained through deacetylation of chitin by alkaline hydrolysis or enzymatic method. It has been widely considered for a range of applications owing to its excellent properties and its constructive role as a polymer backbone that is readily for complexation and modification [9, 10]. In contrary, PVA is a synthetic polymer produced from the hydrolysis of poly(vinyl acetate), which is well-suited for blending with natural polymers due to its highly polar characteristic [11]. According to Buraidah and co-workers [12], the blending of chitosan and PVA can form complexation through strong hydrogen bonding from hydroxyl groups of PVA with amino and hydroxyl groups of chitosan that results in greater conductivity and mechanical properties. The desirable properties of mechanical and conductivity are shown improved with an enhancement through blending of chitosan and PVA, however the effect is not immense as ion exchange capacity (IEC) is reported with 0.0960, 0.1200 and 0.2530 mequiv/g for pristine PVA, chitosan and chitosan/PVA composite respectively, compared to 0.9500 mequiv/g of Nafion membrane [13, 14].

Herewith, the work adopted two modification methods as mentioned above: (1) pre- and (2) post-sulfonation of the polymer membranes. The concept of post-sulfonation is first done by cross-linking the polymer membrane and followed by exposing to a sulfonation agent. This sulfonation process is considerably slow and not uniform across the membrane [15-18]. As for the latter, namely, pre-sulfonation is the opposite of post-sulfonation whereby cross-linking of the polymer membrane is done after sulfonation. This was shown with a better control over the degree of sulfonation and deactivated sites in the polymer backbone, which further avoids side reactions such as decomposition and hydrolysis of polymers that resulted from post-sulfonation method [19]. Recently, Chanthad and co-workers [20] had reported a work involving sulfonated PVA composite with 4-sulfopthalic acid (sPTA) as sulfonation agent. The study has shown that PVA was found to be chemically modified at the hydroxyl groups of PVA molecules with the acid. With the addition of sulfonic acid groups, the electrical properties of the sulfonated-PVA composite membrane had demonstrated a vast improvement in comparison to pristine PVA membrane. With respect to that, this objective of this work is to focus on the comparative parameters obtained through pre- and post-sulfonation including water uptake, ion exchange capacity (IEC), and ionic conductivity.

2. Methodology

2.1 Materials

Chitosan (low molecular weight, M_w = 50,000–190,000 g/mol, Sigma-Aldrich), poly(vinyl alcohol) (PVA, 99.5% hydrolysed, M_w = 85,000–124,000 g/mol, Sigma-Aldrich), 4-sulfopthalic acid (sPTA, 50% aqueous solution, Sigma-Aldrich), acetic acid glacial (CH_3COOH, 99.8%, AR Grade, Friendemann Schmidt), sodium hydroxide, pellets (NaOH, AR Grade, Macron Fine Chemicals) and sodium chloride (NaCl, AR Grade, R&M Chemicals) were used. Throughout the study, deionized water was used to perform cleaning and analytical testing.

2.2 Preparation of chitosan and PVA solutions
The polymeric blend involving both chitosan and PVA were prepared as described by Alhosseini and co-workers [21]. Dissolution of chitosan and PVA were performed separately under respective solvents. 1 wt.% of chitosan solution was prepared under the dissolution in 0.1 M acetic acid glacial at room temperature for 24 hours under continuous stirring; whereas 5 wt.% of PVA was dissolved in deionized water at 120°C for approximately 6 hours under continuously stirring with a speed of 1000 rpm.

2.3 Pre-sulfonation of chitosan/PVA composite membrane
The PVA solution was firstly exposed to sPTA as described by Seeponkai and co-workers [22]. It was added drop-wise with a dosage level of 10 wt.% to the polymer weight of PVA solution and was further stirred at room temperature for 24 hours. The procedure for preparation of blend composite was conducted accordingly as described by Kuzmina and co-workers [23] with varying ratio of chitosan to the sulfonated-PVA solution. The resulting blend solution of 20 ml under different proportion will be cast on a glass petri dish and left evaporate overnight prior to further drying in vacuum oven at 60 °C for 24 hours.

2.4 Post-sulfonation of chitosan/PVA composite membrane
PVA and chitosan solutions prepared were blended under different ratio prior to sulfonation. The dosage level of sPTA was added drop-wise with a level of 10 wt.% to the polymer weight of the solution (PVA and chitosan in this case) and was further stirred at room temperature for 24 hours. The resulting blend solution of 20 ml with different proportions will be cast on a glass petri dish and left evaporated in a fume hood overnight prior to further drying in vacuum oven at 60 °C for 24 hours.

Table 1. Test samples performed with vary in chitosan: PVA content under vol.%

Pre-sulfonation (Chitosan/sulfonated-PVA) and post-sulfonation (Sulfonated-chitosan/ PVA)	10:90
	25:75
	50:50
	75:25
	90:10

2.5 Water uptake
Membranes were first placed in a vacuum oven for 24 hours at 60 °C to eliminate any moisture content prior to water uptake test. Then the measurement for dried specimens was carried out at room temperature, using a sorption apparatus equipped with microbalance [24]. The blended composite is equilibrated with water vapor for 24 hours and mass change was recorded by weighing before and after equilibration. Equation 1 shows the calculation of water uptake:

$$Water\ uptake, \% = \frac{M_{wet}-M_{dry}}{M_{dry}} \times 100\% \qquad (1)$$

where M_{eet} is the weight of wet membrane, g; and M_{dry} is the weight of dry membrane, g

2.6 Electrochemical impedance spectroscopy (EIS)
The ionic conductivities of the membranes were measured using Autolab PGSTAT128N potentiostat. The membrane was sandwiched between two stainless steel electrodes (two-electrode setup) in a measurement cell under alternative current (AC) mode. A setting of frequency ranged from 1000 kHz to 1 Hz and signal amplitude 0.1 V was applied [25]. The diameter of the measurement cell 19 mm and all samples was tested under hydrated state at room temperature.

In order to generate informative data from the potentiostat, NOVA software was used. Data was collected in a form of Nyquist plot, in which electrochemical circle fit was then applied to determine its electrolyte resistance, R_s and obtained a convergence curve. The ionic conductivities of the samples were calculated with Equation 2.

$$\lambda = \frac{t}{A \times R_s} \quad (2)$$

where λ is the ionic conductivity, S/cm; *t* is the thickness of the blended composite membrane, cm; *A* is the area of the blended composite membrane covered by measurement cell; and R_s is the electrolyte resistance, Ω.

2.7 Ion exchange capacity (IEC)

IEC of the membrane was measure by titration technique as described by Chanthad and co-workers [20]. The membrane was firstly prepared in a size of 3×3 cm^2 before weighing the dry sample weight. Secondly, the dry sample was immersed in 0.1 M sodium chloride (NaCl) solution for 24 hours to allow the exchange of hydrogen and sodium ions. After that, the sample inside the NaCl solution will be removed before performing titration with 0.1 M sodium hydroxide (NaOH) solution to evaluate the amount of HCl generated from the exchange process. Based upon the data collected, IEC of the sample will be calculated with the following equation:

$$IEC = \frac{M_{NaOH}}{W_{dry}} \quad (3)$$

where M_{NaOH} is the mol equivalent (mequiv) of NaOH; and W_{dry} is the weight of the dry membrane.

3. Result And Discussion

3.1 Water uptake of the membrane

Water uptake is an important attribute to all solid polymer electrolytes (SPE), whereby it reflects the balance between ionic conductivity and mechanical stability [26]. Table 2 shows that the water uptake of PVA membrane chemically modified with sPTA was slightly lower than pure PVA. This effect could be attributed to the stronger ionic interaction that resulted in an increase in cross-linking density of the sulfonated membrane and consequently allow only limited water uptake [27].

The blend composite membrane of chitosan and PVA under pre-sulfonation demonstrated similar trend to post-sulfonation, water uptake of the chitosan/PVA composite decreases with higher chitosan content and noticeable bounce was shown at 90 vol.% chitosan containing composite with the values of 280.6% and 578.8 %, respectively. According to our understanding, PVA imbedded a partially ionic proton-donor group and a collinear proton-acceptor atom with asymmetric lone-pair orbitals in hydroxyl groups that is feasible for the formation of hydrogen-bonding [28]. Owing to this, the water uptake of chitosan/PVA composite displayed a reduction with increasing chitosan content. At 50 vol.% chitosan containing composite, the blended composite had the lowest water uptake of 116.4 % probably attribute to that all lone-pair orbitals were filled as we hypothetically plotted an equally balanced on mole ratio according to the calculation. Theoretically, chitosan consists of 3 lone-pair electrons on hydroxyl and amine functional groups and 2 ether groups as the backbone of polymer that are ready for complexation through hydrogen bonds formation [29]. In this study, chitosan prepared in 1 wt. % consists of 5 functional groups, namely 2 hydroxyls groups, 1 amine group, and 2 ether groups; while PVA prepared in 5 wt.% consists of 5 hydroxyls. With 50 wt.% of chitosan, the complexation was likely attributed to the formation of hydrogen bonds between all five pairs of functional groups in PVA and chitosan that has resulted in a drastic reduce in the available hydrophilic groups ready for water uptake, hence, recorded the lowest value among all composites.

On the other hand, the pre-sulfonated membrane showed lower water uptakes as compared to post-sulfonated membranes. This might be due to the sulfonating agent being successfully crosslinked with PVA and thus leaving less hydrophilic groups available when blended with chitosan to form an SPE. Following the same argument, the post-sulfonation composite membrane with dominant chitosan content showcased possible hydrolytic degradation with a severe water uptake of 578.8 % as opposed to similar ratio composite membrane under pre-sulfonation [19].

Table 2. Data result of membranes for both pre- and post-sulfonation

Sample	Water uptake, %		Ionic conductivity, S/cm		IEC value, mequiv/g	
	Pre-sulfonation	Post-sulfonation	Pre-sulfonation	Post-sulfonation	Pre-sulfonation	Post-sulfonation
PVA	127.3	-	0.0662	-	0.0440	-
Sulfonated-PVA	124.5	-	0.2613	-	0.0890	-
CS:PVA 10:90	130.4	173.9	0.1225	0.2280	0.3430	0.4000
CS:PVA 25:75	119.6	125.1	0.0805	0.1538	0.2906	0.3450
CS:PVA 50:50	104.1	116.4	0.0702	0.0883	0.1080	0.2209
CS:PVA 75:25	127.8	122.5	0.0916	0.2183	0.1370	0.2304
CS:PVA 90:10	280.6	578.8	0.1874	0.6081	0.3015	0.3212

3.2 The ionic conductivity of the membrane

Table 2 shows the results of ionic conductivity for all tested samples under the hydrated state. A remarkable increase from 6.620×10^{-5} to 2.613×10^{-4} S/cm on the sulfonated PVA membrane as a comparison to pure PVA membrane was demonstrated. This notable enhancement in ionic conductivity reflects the positive contribution of sulfonation, likely attributed to the reaction at the hydroxyl groups of PVA and carboxylic groups of sPTA to form sulfonic groups that allowed the formation of water clusters in the membrane and subsequently increased the ion transport mobility [30].

Aside from that, the ionic conductivities of both pre- and post-sulfonation were shown to decrease with higher chitosan content. For membrane with 10 vol.% containing chitosan, the behavior was controlled by PVA that was reflected with greater water uptake and ionic conductivity due to its higher affinity towards water. Further, increase in the chitosan content beyond 50 vol. % has shown to increase the ionic conductivity, which is in a similar trend as water uptake results. In view of this, chitosan being an established natural hydrophilic swellable polymer was prime to indulge higher water uptake and ionic conductivity. However, the exaggeration of 578.8 % in water uptake for 90 vol.% containing chitosan composites was not favorable due to the fact of being excessively hydrophilic and would cause undesired excessive swelling and hydrolytic degradation when immersed into water.

In a comparison between both pre- and post-sulfonation, the ionic conductivities for pre-sulfonation were found lower. This was possible due to lesser proton transport sites in the pre-sulfonation composite membrane as aforementioned and thus resulted in lower proton conductivity. Another explanation is possibly attributed to the non-uniform distribution of proton transport sites that resulted in improper balance of proton conduct mechanisms and subsequently lead to non-linear effects in the membrane resistance [31]. Conclusively, the post-sulfonation method exhibited higher ionic conductivity due to the likelihood of sulfonation taken place on both PVA and chitosan. As a result, the blended composite could potentially possess a significant increase in sulfonic acid groups that are responsible for ionic transport across the membrane.

3.3 Ion exchange capacity (IEC) of the membrane

Table 2 shows the IEC values of samples including both pre- and post-sulfonation. The pure chitosan was found dissociated over time and subsequently dissolved fully in the 0.1 M NaCl and deionized water solution after 24 hours, in which no data was reported. Besides that, the sulfonated PVA membrane showed an enhancement of the IEC value from 0.0440 to 0.0890 mequiv/g. In such circumstances, the prediction stated above with the potential of hydroxyl groups of PVA being reacted with carboxylic groups of sPTA was valid.

Notably, the IEC values reported were shown aligned with the water uptake on both pre- and post-sulfonation methods. In regards, IEC presented an indirect and reliable root to investigate the ionic conductivity as described by Palani and co-workers [32]. Among all the samples, 10 vol.% containing chitosan composite under post-sulfonation demonstrated the highest IEC value with 0.4000 mequiv/g. Nevertheless, despite that the 90 vol.% containing chitosan composite displayed highest water uptake and ionic conductivity, the possible reason of lower IEC value obtained could be due to the lesser ion-exchangeable sites presented in the blended composite by mean of amino and hydroxyl functional groups interaction from chitosan and PVA [32]. In contrary, the same content of chitosan under pre-sulfonation emerged with lower IEC value, 0.3015 mequiv/g that may indicatively to the lower water uptake which interprets better water management in comparison to post-sulfonation.

4. Conclusion

The effect of pre- and post-sulfonation in chitosan/PVA blend composite was studied. Results tabulated for pre-sulfonation was shown identical to post-sulfonation on the trend for the proton conductivity, water uptake and IEC results. It was found that at 10 vol.% containing chitosan composite, it exerted a balance between water management (water uptake) and active proton transport sites (IEC value). This was followed by a decrease in both properties with increase in chitosan content to 50 vol.%. As the chitosan content being a further increase, the ion transport properties took off with a bounce that shown with an enhancement in all analysis tested. However, it was remarkable at 90 vol.% whereby the water management falls sharply with potential hydrolytic degradation occurrence on the composite. Overall, the post-sulfonation method had led to higher ionic transport properties likely due to sulfonation took place on both PVA and chitosan which induced with a significant increase of sulfonic groups that purportedly exhibited higher ion transport mobility.

Acknowledgement

This work is financially supported by the Universiti Kebangsaan Malaysia research grant (GGPM-2017-043).

References

[1] Muthumeenal A, Pethaiah S S and Nagendran A 2017 Biopolymer Composites in Fuel Cells Biopolymer Composites in Electronic vol 1, ed K K Sadasivuni, D Ponnamma, J Kim, J-J Babibihan and M A Almaadeed (Amsterdam: Elsevier) chapter 6 pp 185–217
[2] Wan Y, Creber K, Peppley B and Tam Bui V 2004 J. Polym. Sci. B: Polym. Phys. 42 1379
[3] Ding J, Chuy C and Holdcroft S 2001 Chem. Mater. 13 2231
[4] Ding J, Chuy C and Holdcroft S 2002 Macromolecules 35 1348
[5] Huang J, Zhang K, Wang K, Xie Z, Ladewig B and Wang H 2012 J. Membr. Sci. 423 362
[6] Díaz M, Ortiz A and Ortiz I 2014 J. Membr. Sci. 469 379
[7] Ma J and Sahai Y 2013 Carbohydr. Polym. 92 955
[8] Ye Y-S, Rick J and Hwang B-J 2012 Polymers 4 913
[9] Varshney P and Gupta S 2011 Ionics 17 479
[10] Dash M, Chiellini F, Ottenbrite R M and Chiellini E 2011 Prog. Polym. Sci. 36 981

[11] Cinelli P, Chiellini E and Imam S H 2008 J. Appl. Polym. Sci. 109 1684
[12] Buraidah M H and Arof A K 2011 J. Non-Cryst. Solids 357 3261
[13] García-Cruz L, Casado-Coterillo C, Iniesta J, Montiel V and Irabien A 2016 J. Membr. Sci. 498 395
[14] Slade S M, Ralph T R, Ponce de León C, Campbell S A and Walsh F C 2010 Fuel Cells 10 567
[15] Brack H P, Buchi F N, Rota M and Scherer G G 1997 Abstr. Pap. Am. Chem. Soc. 214 343
[16] Brack H P, Buehrer H G and Scherer G G 1997 Abstr. Pap. Am. Chem. Soc. 214 192
[17] Brack H P, Bonorand L, Buehrer H G and Scherer G G 1998 Abstr. Pap. Am. Chem. Soc. 216 U154
[18] Lehtinen T, Sundholm G and Sundholm F 1999 J. Appl. Electrochem. 29 677
[19] Zhu X, Liu Y and Zhu L 2011 Polymer Composites for High-temperature Proton-Exchange Membrane Fuel Cells Polymer Membrane for Fuel Cells vol 1, Zaidi J and Matsuura T (New York: Springer US) chapter 7 pp 159–184
[20] Chanthad C and Woothikanokkhan J 2006 J. Appl. Polym. Sci. 101 1931
[21] Alhosseini S N, Moztarzadeh F, Mozafari M, Asgari S, Dodel M, Samadikuchaksaraei A, Kargozar S and Jalali N 2012 Int. J. Nanomedicine 7 25
[22] Seeponkai N and Woothikanokkhan J 2007 J. Appl. Polym. Sci. 105 838
[23] Kuzmina O, Heinze T and Wawro D 2012 ISRN Polym. Sci. 2012 1
[24] Rouquerol J and Davy L 1978 Thermochimica Acta 24 391
[25] Aziz S and Abidin Z H Z 2013 J. Soft. Mat. 2013 1
[26] Linkous C A, Rhoden S L, Linkous W G, Pearman B P and Mohajeri N 2008 ECS Trans. 16 705
[27] Eisenberg A and Kim J S 1998 Introduction to Ionomers (New York: Wiley)
[28] Goodman I 1963 Synthetic Fibre-forming Polymers and Co-polymers Fibre Structure vol 1, Hearle J W S and Peters R H (Amsterdam: Elsevier) chapter 4 pp 111–180
[29] Winie T and Arof A K Biopolymer Electrolytes for Energy Devices Nanostructured Polymer Membranes vol 2, Visakh P M and Nazarenko O (Weinheim: Wiley-VCH) chapter 8 pp 311–356
[30] Bahlakeh G, Nikazar M, Hafezi M-J, Dashtimoghadam E and Hasani-Sadrabadi M 2012 Int. J. Hydrogen Energy 37 10256
[31] Eikerling M H, Malek K and Wang Q 2008 Catalyst Layer Modeling: Structure, Properties and Performance PEM Fuel Cell Electrocatalyst and Catalyst Layers vol 1, Zhang J (London: Springer-Verlag) chapter 8 pp 381–446
[32] Palani P B, Abidin K S, Kannan R, Sivakumar M, Wang F-M, Rajashabala S and Velraj G 2014 RSC Adv.4 61781

IOP Conf. Series: Materials Science and Engineering **458** (2018) 012018 doi:10.1088/1757-899X/458/1/012018

Formaldehyde production via partial oxidation of methanol over oxides of Cr, Mo and W supported on ceria-zirconia

†H Inokawa[1], *S F Zaman[2], H Driss[2], M Daous[2], A Al-Zahrani[2], H Miyaoka[3], T Ichikawa[4], Y Kojima[3] and L A Petrov[2]

[1]Department of Nanoscience, Faculty of Engineering, Sojo University,4-22-1 Ikeda, Nishi-ku, Kumamoto 860-0082, Japan

[2]King Abdulaziz University, P.O.Box 80204, Jeddah 21589, Kingdom of Saudi Arabia

[3]Advanced Materials Division, N-BARD, Hiroshima University, 1-3-1 Kagamiyama, Higashi-Hiroshima,739-8530, Japan

[4]Graduate School of Engineering, Hiroshima University, 1-3-1 Kagamiyama, Higashi-Hiroshima,739-8530, Japan

* Corresponding Author E-mail: zfsharif@gmail.com, sfzaman@kau.edu.sa

† Co-corresponding Author E-mail: inokawa@nano.sojo-u.ac.jp

Abstract: Oxides of chromium (Cr), molybdenum (Mo), or tungsten (W) were supported on mixed oxides of cerium and zirconium (CeO_2-ZrO_2) by an incipient wetness impregnation process, and used to investigate their activities for partial oxidation of methanol. High Resolusion Transmission Electron Microscope (HR-TEM) revealed that nanoparticles of Cr_2O_3, MoO_3 or WO_3 were formed on the CeO_2-ZrO_2 support. The particle sizes of Cr_2O_3, MoO_3 and WO_3 were 2-4, 3-6 and 4-5 nm, respectively. As a result of partial oxidation of methanol at 275 – 450 °C, formaldehyde (CH_2O) was obtained as the major product over all these catalysts. Cr_2O_3 supported on CeO_2-ZrO_2 produced both CO and CO_2 along with H_2 production, while MoO_3 and WO_3 on CeO_2-ZrO_2 support generated only CO as a by-product. It was found that MoO_3 supported on CeO_2-ZrO_2 showed the highest methanol conversion and CH_2O yield among these three catalysts.

1. Introduction

Partial oxidation of methanol (POM) has been investigated in view of hydrogen production [1-3] and formation of formaldehyde (CH_2O) via oxidative dehydrogenation (ODH) [4-6]. Transition metal elements in the 6B group, chromium (Cr), molybdenum (Mo) and tungsten (W), are highly related with these kinds of reactions involving methanol in the literature. For example, Cr_2O_3 supported on zinc oxide was used for methanol synthesis from CO and H_2 at high pressure [7, 8]. Because catalysts have possibility to be effective for the reverse reaction, it can be assumed that Cr_2O_3 based catalysts can be also used for H_2 production from methanol, which is a reverse reaction of the methanol synthesis. Similarly, Mo based catalysts, which are MoO_3 and $Fe_2(MoO_4)_3$, have been investigated for formaldehyde production via ODH of methanol [9, 10], and tungsten oxide (WO_x) supported on TiO_2 was used for dehydration of methanol to produce dimethyl ether [11].

Recently, mixed oxides of CeO_2-ZrO_2 (CZ) have attracted attention due to their high oxygen mobility property [12]. Bake et al. compared the activity of gold (Au) catalysts supported on ZrO_2, CeO_2-ZrO_2, or CeO_2-ZrO_2-TiO_2, and found that Au supported on CeO_2-ZrO_2 exhibited the highest activity toward hydrogen generation via POM [1]. They suggested that the contribution of oxidized gold particles and redox property of CeO_2-ZrO_2 brought about the high conversion of methanol and the high selectivity of H_2 production. In this study, oxides of Cr, Mo and W were synthesized on a CeO_2-ZrO_2 (CZ) mixed support and their catalytic performance for methanol partial oxidation have been investigated in order to clarify the activities of 6B group elements supported on CeO_2-ZrO_2 (CZ) toward POM and ODH reaction.

$$\text{POM reaction: } CH_3OH + \tfrac{1}{2}O_2 \rightarrow CO_2 + 2H_2 \quad (1)$$

$$\text{ODH reaction: } CH_3OH + \tfrac{1}{2}O_2 \rightarrow CH_2O + H_2O \quad (2)$$

2. Experimental

2.1. Catalyst Preparation

CeO_2-ZrO_2 mixed support was prepared by a simple mechanical mixing of powders of CeO_2 and ZrO_2 [1]. Powders of CeO_2 (Acros) and ZrO_2 (Aldrich) with a weight ratio of 3: 1 were ultrasonically homogenized in water, and then dried at 120 °C in an oven overnight. Cr, Mo or W (1 mass% of each transition metal) was supported on CeO_2-ZrO_2 by an incipient wetness impregnation process using aqueous solution of $Cr(NO_3)_2{\cdot}9H_2O$ (Aldrich), $(NH_4)_6Mo_7O_{24}{\cdot}4H_2O$ (Aldrich) or $(NH_4)_6H_2W_{12}O_{40}{\cdot}xH_2O$ (Aldrich). CeO_2-ZrO_2 support was added into the aqueous solutions of individual salts, and water was removed by rotary evaporator at 40 °C and around 50 mbar. After further drying at 110 °C for 2 h in an oven, the powders were calcined in the air at 500 °C for 5 h at a ramping rate of 5 °C min^{-1}. CeO_2-ZrO_2 supporting oxides of Cr, Mo and W are named CrO_x/CZ, MoO_x/CZ and WO_x/CZ, respectively, throughout this manuscript.

2.1.1. Characterization. Crystal structures of the samples were characterized by X-ray diffraction (XRD) and High Resolution Transmission Electron Microscope (HR-TEM). TEM was also used to characterize size and shape of particles. Content of Cr, Mo or W was analyzed by X-ray Fluorescence (XRF). Oxidation states of Cr, Mo or W were anayzed by X-ray Photoelectron Spectroscopy (XPS). Adsorption and desorption isotherm of N_2 was analyzed at 77 K after degassing at 200 °C for 2 h under vacuum, and analyzed by Brunauere Emmette Teller (BET) method in order to evaluate specific surface area.

2.1.2. Catalyst activity measurement setup. Partial oxidation of methanol was carried out in a system (PID Eng & Tech, Spain) with a heat box and a continuous fixed bed reactor of quartz (o.d. = 6 mm), as shown in Figure 1. The catalytic reactions were examined at atmospheric pressure in a temperature interval between 275 and 450 °C according to conditions of previous experiments done by Bake et al [1]. The reactor was charged with 100 mg of catalyst. The heat box was kept at 150 °C to avoid inline condensation. Methanol and oxygen were fed to the reactor while being dilutied with helium. O_2/Methanol molar ratio and gas hourly space velocity (GHSV) were 0.5 and 30,000 ml g^{-1} h^{-1}, respectively, for all reactions. Residuals of methanol (CH_3OH) and oxygen (O_2) and products, which were formaldehyde (CH_2O), carbon monoxide (CO), carbon dioxide (CO_2) and hydrogen (H_2) were quantitatively analyzed by a gas chromatograph (GC, Hewlett Packard HP6890) with separation columns of Molecular seive 5A and Porapak QS.

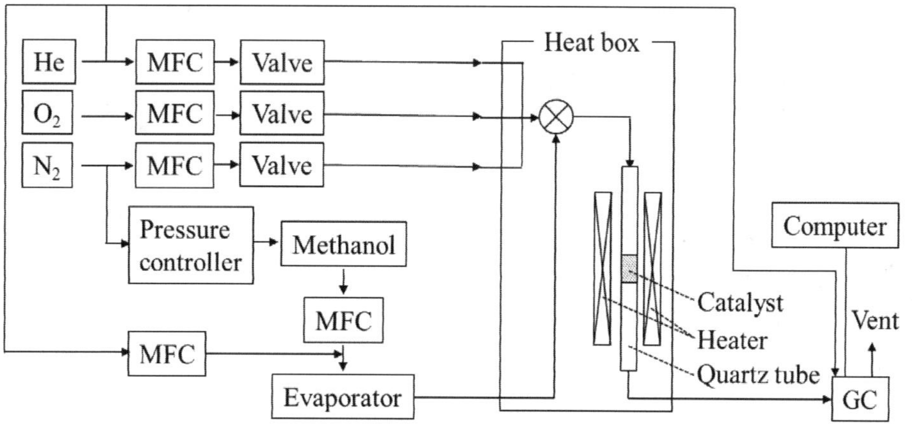

Figure 1. Schematic image of partial oxidation of methanol (POM) reaction system.

Conversion of reactant and selectivity of products were calculated as follows.

$$CH_3OH\ conversion\ (\%) = \frac{mol.flow\ of\ CH_3OH_{in} - mol.flow\ of\ CH_3OH_{out}}{mol.flow\ of\ CH_3OH_{in}} \times 100 \quad (3)$$

$$O_2\ conversion\ (\%) = \frac{mol.flow\ of\ O_{2in} - mol.flow\ of\ O_{2out}}{mol.flow\ of\ O_{2in}} \times 100 \quad (4)$$

$$CH_2O\ selectivity\ (\%) = \frac{mol.flow\ of\ CH_2O}{mol.of\ CH_3OH_{consumed}} \times 100 \quad (5)$$

$$CO\ selectivity\ (\%) = \frac{mol.flow\ of\ CO}{mol.of\ CH_3OH_{consumed}} \times 100 \quad (6)$$

$$CO_2\ selectivity\ (\%) = \frac{mol.flow\ of\ CO_2}{mol.of\ CH_3OH_{consumed}} \times 100 \quad (7)$$

$$H_2\ selectivity\ (\%) = \frac{mol.flow\ of\ H_2}{2 \times mol.of\ CH_3OH_{consumed}} \times 100 \quad (8)$$

3. Results and Discussion

3.1. Characterization

XRD patterns of CeO_2-ZrO_2 (CZ) support, CrO_x/CZ, MoO_x/CZ and WO_x/CZ are shown in Figure 2. XRD patterns of the catalysts do not show any characteristic peaks assigned to the oxides of Cr, Mo or W, which were impregnated on the support. This may arise from the fact that the mass% of each transition metal is 1%, very low concentration which may cause formation of very small crystallites, and that may be under the detection limit of the XRD equipment. And/or the peaks are so small that it is shaded under the ceria and zirconia peaks. Later on HR-TEM images confirmd the formation of small nano particles (< 5 nm) of transition metals over CZ support.

In order to identify crystal structures of CrO_x, MoO_x and WO_x on the support, their particles were observed by HR-TEM as shown in Figure 3. HR-TEM of CrO_x/CZ, as shown in Figure 3(a), revealed that CrO_x has d space of 3.64 Å corresponding to (0 1 2) plane of Cr_2O_3. HR-TEM also showed that the particle size of Cr_2O_3 was smaller than 5 nm. HR-TEM of MoO_x/CZ showed that MoO_3 was formed on the support because d space of the MoO_x was 6.9 Å corresponding to MoO_3 (0 0 1). The particle size of MoO_3 was around 5 nm. HR-TEM and Fast Fourier Transform (FFT) image of WO_x/CZ also revealed that WO_3 was formed as shown in Figure 3(c). The particle size of WO_3 was 4 - 5 nm.

XPS profiles shown in Figure 4 are consistent with the identification by HR-TEM, which shows the formation of Cr_2O_3, MoO_3 and WO_3 over CrO_x/CZ, MoO_x/CZ and WO_x/CZ, respectively. From above results, it is clarified that fine and well dispersed nano particles of Cr_2O_3, MoO_3 and WO_3 were formed on CrO_x/CZ, MoO_x/CZ and WO_x/CZ catalysts, respectively, via the catalyst preparation process.

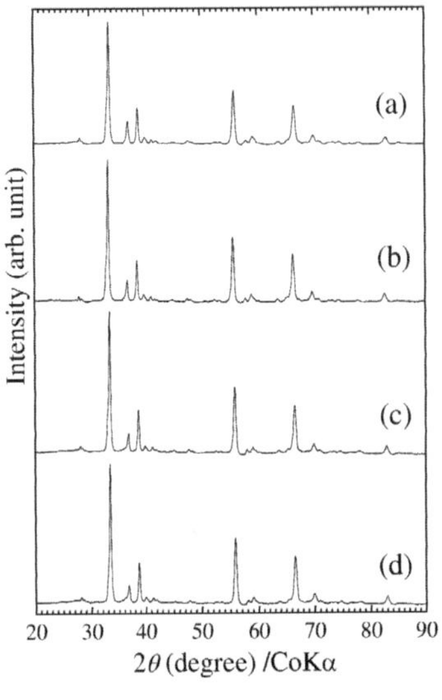

Figure 2. XRD patterns of (a) CeO_2-ZrO_2 support, (b) CrO_x/CZ, (c) MoO_x/CZ and (d) WO_x/CZ.

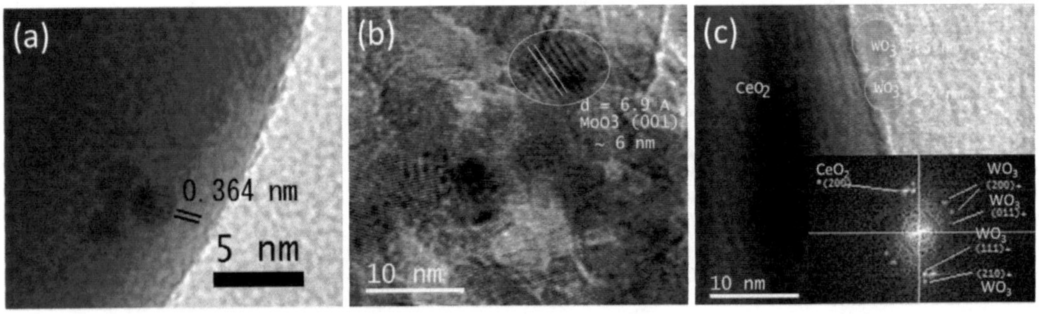

Figure 3. HR-TEM images of (a) CrO_x/CZ, (b) MoO_x/CZ and (c) WO_x/CZ with Fast Fourier Transform (FFT) image.

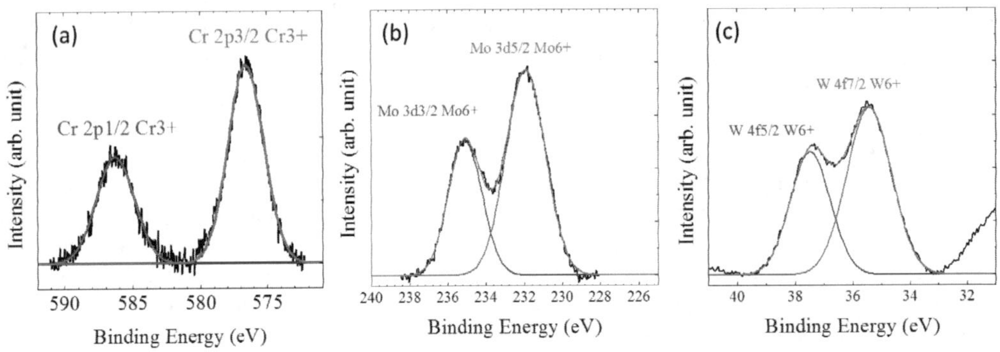

Figure 4. XPS profiles of (a)CrO_x/CZ, (b)MoO_x/CZ and (d)WO_x/CZ with peak separation analysis.

From the results of XRF, the contents of Cr, Mo and W in each sample were estimated as shown in Table 1. The content of the active material, which was Cr, Mo or W, in the catalysts was approximately 1 mass%, which means the catalysts were prepared successfully according to the desired mass% of the active metal on the support. Specific Surface Area (SSA) of the samples are also shown in Table 1, where all the catalysts have similar values, which are close to the support CeO_2-ZrO_2 SSA value.

Table 1. Content of M (M = Cr, Mo or W) and specific surface area (SSA) of catalysts.

Sample (M)	CrO_x/CZ (Cr)	MoO_x/CZ (Mo)	WO_x/CZ (W)	CeO_2-ZrO_2 support
M content (mass%)	0.95	1.08	1.09	0
Surface area (m^2/g)	6.3	4.3	5.5	7.4

3.2. Partial Oxidation of Methanol (POM)

Temeperature dependency of CH_3OH and O_2 conversion and selectivity of formaldehyde (CH_2O), CO_2, CO and H_2 over three catalysts are shown in Figure 5. Over all catalysts, formaldehyde (CH_2O) was obtained as the major product, suggesting that oxidative dehydrogenation of methanol (reaction 2) is mainly catalyzed over these oxides of 6B group elements supported on CeO_2 and ZrO_2.

CO was also generated as a by-product over these catalysts with the increase of reaction temperature. CO might be generated by thermal decomposition of CH_3OH to CO and H_2 (reaction 9) or oxidation of CH_2O to CO and H_2O (reaction 10). If thermal decomposition of CH_3OH (reaction 9) took place, H_2 selectivity must have same value of CO selectivity. The observed H_2 selectivity is, however, lower than CO selectivity, suggesting that H_2 was generated by other reactions. So it can be considered that CO was generated by oxidation of CH_2O (reaction 10). In particular, behaviors of CH_2O selectivity (decreasing) and CO selectivity (increasing) are showing opposite trends on CrO_x/CZ with the increase of temperature, as shown in Figure 5(a), which suggests that the oxidation decomposition of the CH_2O is happening over the surface to produce CO according to (reaction 10).

On both MoO_x/CZ and WO_x/CZ, CO selectivity increased with temperature, while very little H_2 and CO_2 was generated. The behavior of CO and H_2 generation on MoO_x/CZ and WO_x/CZ is consistent with oxidation of CH_2O (reaction 10). Only CrO_x/CZ produced considerable amount of both CO_2 and H_2 as by-products, indicating that partial oxidation of methanol (reaction 1) or water gas shift reaction (reaction 11) took place over CrO_x/CZ.

Mole of generated H_2, CO and CO_2 were almost same among them on the CrO_x/CZ. Because the amount of H_2 generated via the partial oxidation of methanol (reaction 1) is theoretically double of the amount of CO_2, the partial oxidation reaction can be assumed not to have taken place on CrO_x/CZ. Therefore, it can be considered that water gas shift reaction (reaction 11) took place from CO and H_2O generated by oxidation of CH_2O. When about a half of the generated CO reacts with H_2O, the final product have same amount of CO, CO_2 and H_2, and the selectivity of CO, CO_2 and H_2 on CrO_x/CZ can be explained over this catalyst.

In regard to formaldehyde production, CH_3OH conversions and CH_2O yeilds of the three catalysts were compared as shown in Figure 6. Among the three catalysts, MoO_x/CZ showed the highest methanol conversion and CH_2O yield at all the investigated temperature range. The maximum yield of formaldehyde was observed at 375 °C for MoO_x/CZ.

The catalysis of Cr_2O_3 on CeO_2-ZrO_2 is similar to that of CrO_3 supported on SiO_2, which obtained formaldehyde as the main product and acetic acid, CO and CO_2 as by-products [13]. The activity toward ODH reaction might be due to the nature of chromium oxides.

We have observed that MoO_x/CZ showed the highest activity toward ODH of CH_3OH to produce CH_2O (reaction 2) among these three catalysts under investigation as well as reported Mo based catalysts with high activities for formaldehyde production in literature [9, 10, 14-17].

Although tungsten oxide (WO_x) has been reported as a catalyst for dehydration of CH_3OH (feed pure CH_3OH) to obtain dimethyl ether [11], in this study CH_2O was obtained as a main product via

ODH of CH_3OH over WO_x/CZ. The difference of the activity and selectivity of WO_x/CZ catalyst might be caused by the use of CZ support and the presence of oxygen to perform partial oxidation.

$$\text{Methanol thermal decomposition: } CH_3OH \rightarrow CO + 2H_2 \quad (9)$$

$$\text{Oxidation of formaldehyde: } 2CH_2O + O_2 \rightarrow 2CO + 2H_2O \quad (10)$$

$$\text{Water gas shift: } CO + H_2O \rightarrow CO_2 + H_2 \quad (11)$$

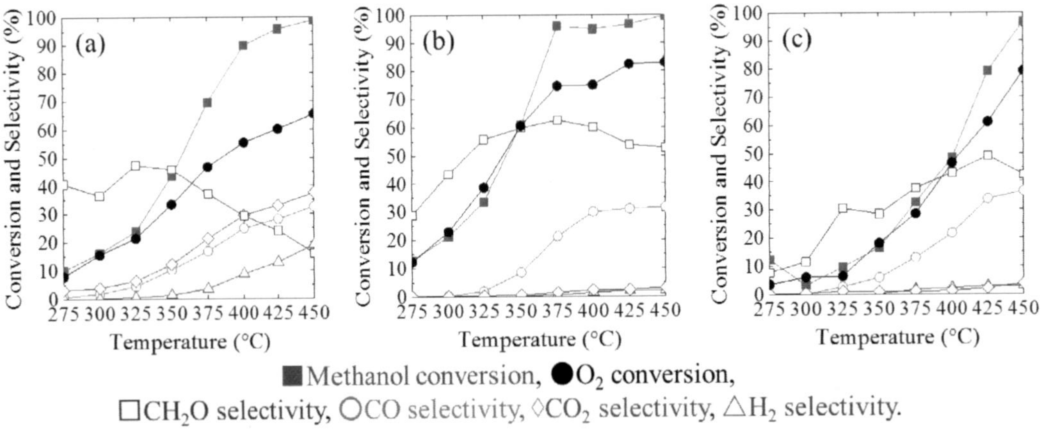

Figure 5. Conversion and selectivity of partial oxidation of methanol over (a) CrO_x/CZ, (b) MoO_x/CZ and (c) WO_x/CZ.

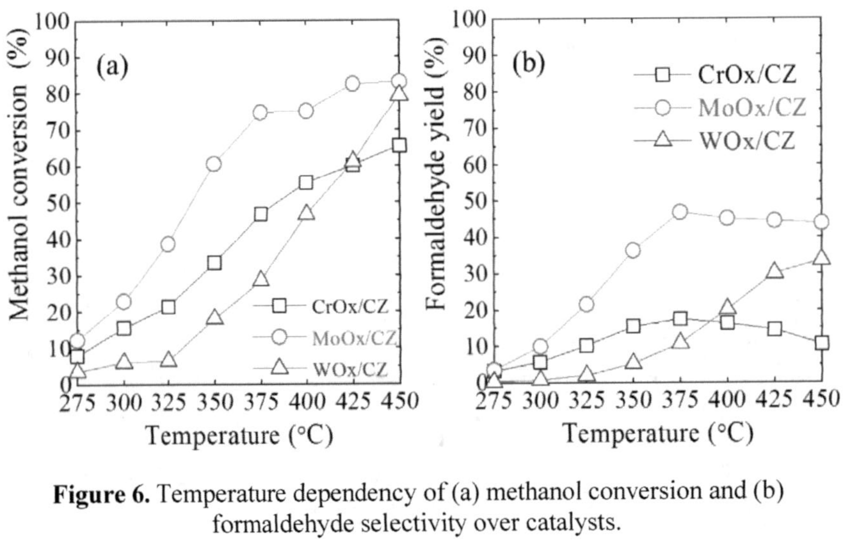

Figure 6. Temperature dependency of (a) methanol conversion and (b) formaldehyde selectivity over catalysts.

4. Conclusions

Oxides of 6B group elements, Cr, Mo or W, were supported on mixed oxides of Ce and Zr in order to investigate their catalytic activity toward partial oxidation of methanol. Nanoparticles of Cr_2O_3, MoO_3 or WO_3 were successfully synthesized on the support and all these catalysts resulted in the selective formation of formaldehyde (CH_2O). The MoO_3 supported on CeO_2-ZrO_2 showed the highest methanol conversion and CH_2O yield among these three catalysts.

Acknowledgement
The authors are grateful for the research support given by Chemical and Materials Engineering Department, Faculty of Engineering at King Abdulaziz University, Jeddah, Saudi Arabia and Department of Nano science, Faculty of Engineering at Sojo University, Kumamoto, Japan.

References

[1] Bake H, Zaman S F, Alhamed Y A, Al-Zahrani A A, Daous M A, Rather S U, Driss H and Petrov L A 2016 *RSC Advances* **6** 22555
[2] Qi A, Wang S, Hong X, Fu G and Wu D 2000 *Xiandai Huagong/Modern Chemical Industry* **20** 36
[3] Wang Z, Xi J, Wang W and Lu G 2003 *Journal of Molecular Catalysis A: Chemical* **191** 123
[4] Bowker M, Brookes C, Carley A F, House M P, Kosif M, Sankar G, Wawata I, Wells P P and Yaseneva P 2013 *Physical Chemistry Chemical Physics* **15** 12056
[5] Deo G and Wachs I E 1994 *Journal of Catalysis* **146** 323
[6] Whiting G T, Bartley J K, Dummer N F, Hutchings G J and Taylor S H 2014 *Applied Catalysis A: General* **485** 51
[7] Bradford M C J, Konduru M V and Fuentes D X 2003 *Fuel Processing Technology* **83** 11
[9] Dias A P S, Montemor F, Portela M F and Kiennemann A 2015 *Journal of Molecular Catalysis A: Chemical* **397** 93
[10] Edwards J, Nicolaidis J, Cutlip M B and Bennett C O 1977 *Journal of Catalysis* **50** 24
[11] Ladera R, Finocchio E, Rojas S, Busca G, Fierro J L G and Ojeda M 2013 *Fuel* **113** 1
[12] Jen H W, Graham G W, Chun W, McCabe R W, Cuif J P, Deutsch S E and Touret O 1999 *Catalysis Today* **50** 309
[13] Kim D S, Tatibouet J-M and Wachs I E 1992 *Journal of Catalysis* **136** 209
[14] Estévez Sánchez A M, Fernández Tena A and Márquez Moreno M C 1989 *Reaction Kinetics and Catalysis Letters* **38** 193
[15] Babichev I V, Ilyin A A, Rumyantsev R N, Ilyin A P and Kuryleva K V 2015 *Russian Journal of Applied Chemistry* **88** 590
[16] Lafyatis D S, Creten G and Froment G F 1994 *Applied Catalysis A: General* **120** 85
[17] Yeo B R, Pudge G J F, Bugler K G, Rushby A V, Kondrat S, Bartley J, Golunski S, Taylor S H, Gibson E, Wells P P, Brookes C, Bowker M and Hutchings G J 2016 *Surface Science* **648** 163

IOP Conf. Series: Materials Science and Engineering **458** (2018) 012019 doi:10.1088/1757-899X/458/1/012019

Heck reaction using palladium-benzimidazole catalyst: synthesis, characterisation and catalytic activity

N R Said, M A Mustakim, N N Md Sani and S N A Baharin

School of Chemistry and Environment, Faculty of Applied Sciences, Universiti Teknologi MARA (UiTM) Cawangan Negeri Sembilan, Kampus Kuala Pilah, Pekan Parit Tinggi, 72000 Kuala Pilah, Negeri Sembilan, Malaysia.

Email: nurra1435@ns.uitm.edu.my

Abstract. Palladium is used as a catalyst for carbon-carbon couplings such as in the Heck, Stille, and Suzuki reactions. This research reports on two palladium catalysts for the Heck reaction, namely bi-1-3-bis(phenylmethyl) benzimidazolium bromo palladium(II) (Pd-NHC-H) and bi-1,3-bis[(4-chlorophenyl)methyl]-1-benzimidazolium bromo palladium(II) (Pd-NHC-Cl). The catalyst preparation involved two steps of synthesis. First was the synthesis of N-heterocyclic carbene (NHC) ligands by reacting benzimidazole with benzyl bromide and 4-chlorobenzylbromide. It was then followed by the reacting of both synthesized ligands with palladium chloride. The synthesized NHC ligands and catalysts were characterized using Fourier Transform Infrared (FT-IR), ^{1}H and ^{13}C Nuclear Magnetic Resonance (NMR), and UV-Vis spectroscopy. All results gave evidence of the successfulness of the synthesized compounds. A preliminary complexation study between the Pd^{2+} cation and N-heterocyclic carbene (NHC) ligand showed stoichiometric of 1 Pd^{2+} cation to 2 NHC ligand. The catalytic activity of the Pd-NHC-H and Pd-NHC-Cl was studied on a Heck carbon-carbon coupling reaction between 1-bromo-4-nitrobenzene and styrene at different catalyst loadings and reaction times. The catalytic performance was analysed using GC-FID. Results showed that excellent catalytic activity was achieved using Pd-NHC-Cl with a catalyst loading of 0.5 mmol% and 60 minutes of reaction time.

1. Introduction

Palladium catalysts are efficient and powerful tools for carbon-carbon bond formations as the reactions have a high reaction rate, good yield percentage and high turnover number. The Heck reaction, Suzuki reaction and Stille reaction are examples of reactions that use palladium as a catalyst. The reactions that are conducted via the insertion of alkenes and alkynes to the arylpalladium intermediates and alkenylpalladium are known as the Heck reactions [1]. In 2010, the Nobel Prize in Chemistry was awarded to Richard F. Heck, Ei-ichi Negishi and Akira Suzuki for their work on palladium-catalysed cross couplings in organic synthesis.

The Heck reaction is the most powerful and widely used chemical reaction used to form substituted alkenes with organic moieties bearing a suitable leaving group such as halide, triflate or diazonium. Several intermediates in pharmaceutical and fine chemicals are synthesized using this method. For example, the synthesis of guaianolide-type sesquiterpene lactone derivatives as active compounds inhibiting resistant acute leukemic cells, and the synthesis of cyclotene as an important material in wafer level chip packaging for smartphone screen manufacturing [2, 3].

Palladium (Pd) behaves as a catalyst in the Heck reaction and it can be present in several forms which are homogeneous and heterogeneous. Pd catalysts commonly form in these two types: Pd(0) bonded to phosphine ligand i.e., $Pd(PPh_3)$, or as a palladium salt such as $PdCl_2$ or $Pd(OAc)_2$ [1]. Generally, the palladium catalyst contains a phosphine ligand, thus it can facilitate organic transformation efficiently. However, the phosphine ligand has a few disadvantages including high toxicity, costly and air sensitivity [4]. Some phosphine-free ligands such as Schiff base ligands, carbocyclic, thiols and N-heterocyclic carbene (NHC) have been tested and developed.

Among the phosphine-free ligands developed, NHC ligands show good quality for the palladium catalyst as it has stronger σ-donor properties and stability to air, moisture, and heat when compared to phosphine ligands [5]. This study, therefore, focuses on the synthesis and characterization of two palladium catalysts (Pd-NHC-H and Pd-NHC-Cl) and their catalytic activities in the Heck reaction.

2. Experimental

2.1. Reagents and Instruments

The chemicals and solvents used were palladium chloride, benzimidazole, benzyl bromide, 4-chlorobenzyl bromide, tetrahydrofuran (THF), methanol, dimethyl sulfoxide (DMSO), 1-bromo-4-nitrobenzene, styrene, N,N-dimethylacetamide (DMA), dimethyl sulfoxide (DMSO), dimethylformamide (DMF), ethyl acetate, ether, dichloromethane, sodium carbonate (Na_2CO_3), sodium hydroxide (NaOH) and magnesium sulphate ($MgSO_4$). All of the above were purchased at analytical grade and used without further purification.

The characterization of the synthesized complexes was conducted using instrumentation methods such as Fourier Transform Infrared (FT-IR), ^1H and ^{13}C Nuclear Magnetic Resonance (NMR) and UV-Vis spectroscopy. The catalytic study for the Heck reaction performance was analyzed using GC-FID.

2.2. Synthesis of Palladium-Benzimidazole Catalyst

The synthesis routes used in the preparation of NHC ligands are shown in figure 1. Benzimidazole (1.1814 g; 10 mmol) was dissolved in 10 mL of THF in a beaker and stirred. Benzyl bromide (3.7629 g; 20 mmol) was then added into the beaker. The mixture was stirred for 24 hours. Then, the expected compound, 1,3-bis(phenylmethyl) benzimidazolium bromide (NHC-H) was separated using vacuum filtration and washed with THF and water. Then it was dried in a vacuum desiccator. The steps were repeated to produce 1,3-bis[(4-chlorophenyl)methyl]-1-benzimidazolium bromide (NHC-Cl) by reacting the Benzimidazole (1.1814 g; 10 mmol) with 4-chlorobenzylbromide (4.1096 g; 20 mmol).

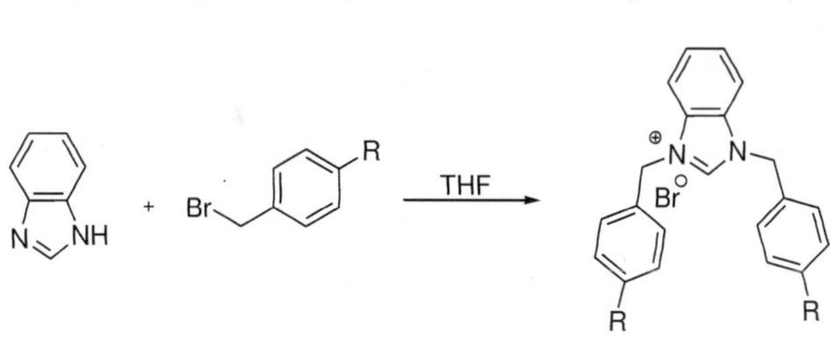

Figure 1. Synthesis route of NHC ligands, where R= H or Cl

The synthesis steps were continued with the stirring of NHC-H ligand (0.2 mmol) in 15 mL dry methanol, and then palladium chloride (0.1 mmol) was added slowly into the solution. The solution was refluxed for 2 hours. After refluxing, 37% NaOH (3.5 mL, 2.0 M) was added dropwise over 30 minutes. Then, it was further stirred for 24 hours at room temperature. The suspension obtained was filtered, washed with diethyl ether and dried in a desiccator. The product obtained was the Pd-NHC-H catalyst.

The steps were repeated using NHC-Cl ligand to obtain the Pd-NHC-Cl catalyst. The synthesis route is shown in figure 2.

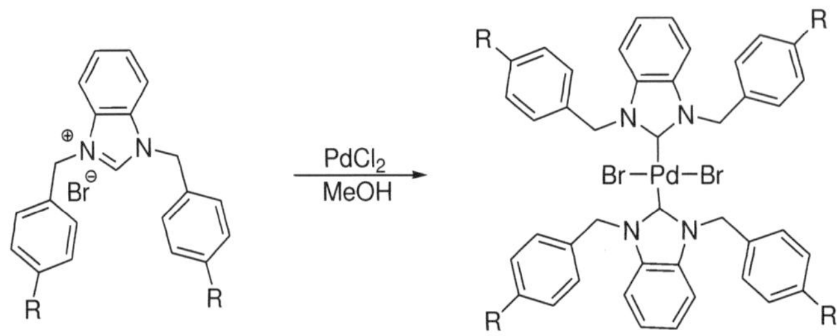

Figure 2. Synthesis route of palladium catalyst, where R= H or Cl

2.3. Preliminary Complexation Study
The UV-Vis titration experiments on both N-heterocyclic carbene ligands with palladium metal were conducted to understand the stoichiometry of the formed complexes of N-heterocyclic carbene ligands and Pd^{2+} cation. N-heterocyclic carbene ligand in DMSO ($2.3x10^{-6}$ mmol) and palladium chloride in DMSO ($1x10^{-4}$ mmol) were prepared. 3 ml of N-heterocyclic carbene ligand was put into a cuvette and the palladium metal was added bit by bit until the ratio of metal added to the N-heterocyclic carbene ligand was 2:1 [6].

2.4. Catalytic Study
Both synthesized palladium catalysts, Pd-NHC-H and Pd-NHC-Cl, were tested for their catalytic activity in a Heck reaction by reacting 1-bromo-4-nitrobenzene with styrene monomer to produce 3-(4-nitrophenyl)-benzene, respectively (figure 3). 1-bromo-4-nitrobenzene (0.20 g;1 mmol), styrene monomer (0.2083 g; 2 mmol), Na_2CO_3 (0.25 g, 2.4 mmol), palladium(II) catalyst (1.0 mmol%) and DMA (10 mL) as solvents were mixed together in a three-neck round bottom flask and refluxed in the presence of nitrogen gas for 1 hour. The mixture was poured into separating funnels with the addition of water (30 mL) and dichloromethane (3 × 10 mL). The mixture was let to sit for 30 minutes, and organic material was extracted and removed. Then, it was dried over $MgSO_4$ and the mixture was filtered in vacuo, leaving the crude product. Finally, the product was evaporated using a rotary evaporator and dried in an oven [7]. The FT-IR spectroscopy was used to characterize the final product. The procedure was repeated with 0.50 mmol% catalyst loading. All reactions were monitored every 15 minutes. The sampling was taken at 15, 30, 45 and 60 minutes and the crude product was analyzed using GC-FID.

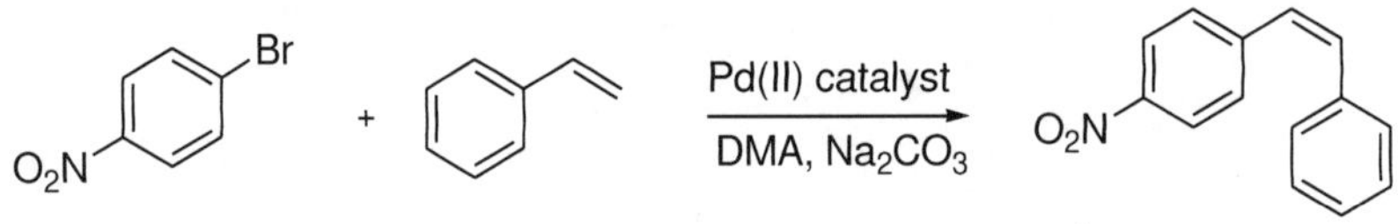

Figure 3. Catalytic reaction of 1-bromo-4-nitrobenzene and styrene, where Pd(II) catalyst = Pd-NHC-H or Pd-NHC-Cl

3. Results and discussion

3.1. Characterisation
NHC ligands and their Palladium complexes (catalyst) were synthesized by the method shown in figure 1 and figure 2. The physical properties of the compounds are listed in table 1. After complexation, the product changed in color and increased in melting temperature when compared with its ligand. This is

due to the presence of the palladium metal attached to the structure of the ligand as proof of the successfulness of the complexation reaction.

Table 1. Physical properties of synthesized compounds

Compound	Colour	Melting point (°C)	% Yield
NHC-H	White	153	78.4
NHC-Cl	Orange	178	90.1
Pd-NHC-H	Brown	167	70.4
Pd-NHC-Cl	Grey	224	82.8

The structure of synthesized compounds was confirmed using FT-IR, ^{1}H-NMR, and ^{13}C-NMR. The FT-IR spectra in figure 4 show the transformation from benzimidazole to palladium catalyst. Observe that in the spectrum of benzimidazole, C=N peaks appeared at 1589 cm^{-1}. However, after introducing benzyl substituent to the benzimidazole nitrogen atom, the C=N peak of NHC ligand (NHC-H) shifted to a lower frequency of 1558 cm^{-1} due to the conjugation of unshared pair electrons on the nitrogen atom that had increased the single bond character [1]. After complexation, the FT-IR spectrum of the synthesized palladium catalyst (Pd-NHC-H) showed a peak of C-N bond at 1658 cm^{-1} to replace the C=N bond. However, the C-N absorption peaks appeared at an unusually high value due to the electronegativity of palladium which developed a polarised Pd-Br bond and led to a relative strengthening of Pd-C and C-N bonds [8]. A similar observation can also be seen in the spectra for the transformation of NHC-Cl to Pd-NHC-Cl where the C=N peak at 1601 cm^{-1} disappeared after the complexation process, followed by the appearance of new C-N peaks at 1672 cm^{-1}.

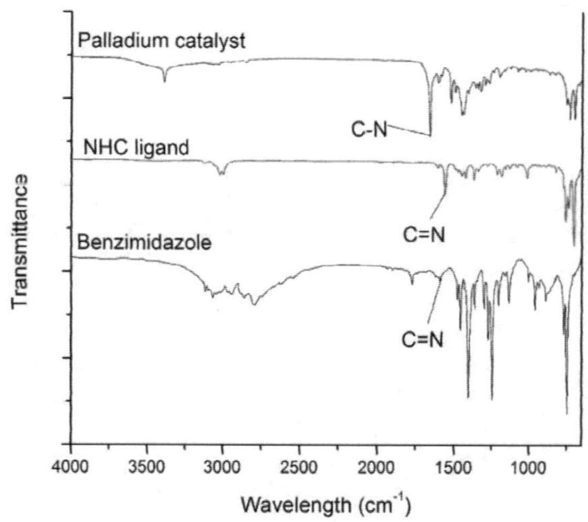

Figure 4. FT-IR spectra of synthesis compounds

In the ^{1}H-NMR spectra (figure 5), a notable change was observed by the disappearance of the chemical shift of the carbene proton in the spectra of Pd-NHC-H and Pd-NHC-Cl at δ 10.32 ppm and 10.20 ppm, which were present in the spectra of the NHC-H and NHC-Cl ligands, respectively. This is upon the deprotonation and coordination of Pd(II) metal cation with the carbene carbon position. The results from the ^{1}H NMR spectra data for synthesized ligands and palladium catalysts are listed in table 2.

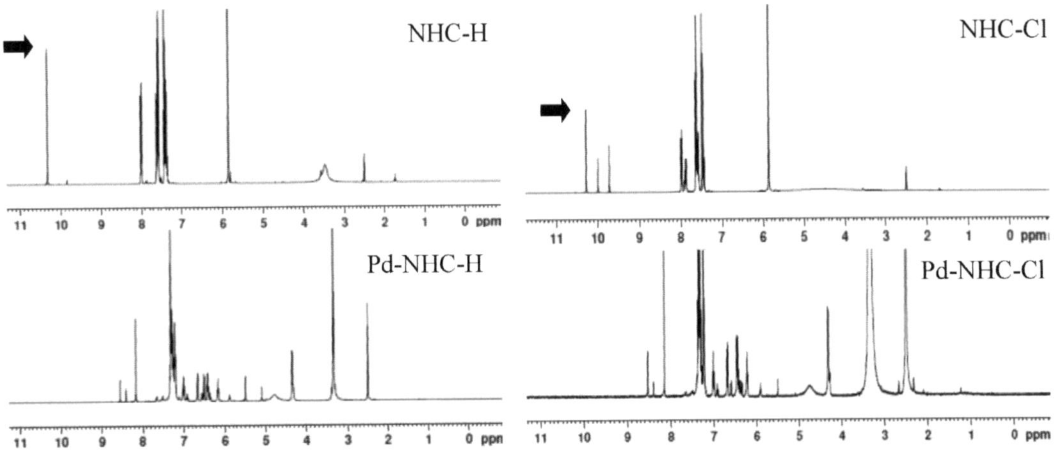

Figure 5. ^{1}H-NMR spectra of synthesis compounds

Table 2. Significant attributions ^{1}H NMR data of synthesis compounds

Attributions	Chemicals Shift, δ (ppm); [multiplicity]			
	NHC-H	NHC-Cl	Pd-NHC-H	Pd-NHC-Cl
N-CH-N (H carbene)	10.32 [s]	10.20 [s]	-	-
CH-Ar	7.99 [m]	8.00 [m]	8.16 [m]	8.10 [m]
CH-Ar	7.61 [m]	7.80 [m]	7.25 [m]	7.20 [m]
CH-Ar	7.39 [m]	7.60 [m]	7.02 [m]	7.00 [m]
Ar-CH_2-N	5.86 [s]	5.90 [s]	5.50 [s]	5.60 [s]

The ^{13}C-NMR spectra in figure 6 show that most of the resonation of the palladium catalyst is almost the same as the resonation of the NHC ligand. Only the carbene carbon chemical shift showed a significant change after the complexation reaction with Pd(II) metal cation. In the Pd-NHC-H spectrum, it was observed that the carbene carbon was resonated at δ 163.72 ppm, while in NHC-H ligand, it resonated at δ 143.19 ppm. Comparable results can also be seen in the NHC-Cl and Pd-NHC-Cl spectra. After complexation, the chemical shift representing the carbene carbon was resonated at a higher chemical shift from δ143.3 ppm to 163.76 ppm. The corresponding signal is due to the attachment of palladium metal at the carbene carbon positions. Other significant chemical shifts of the ^{13}C NMR spectra are summarised in table 3.

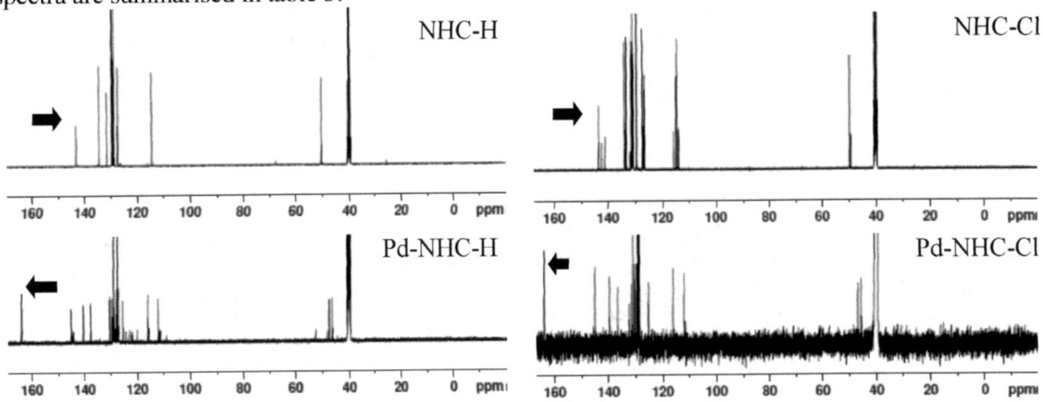

Figure 6. ^{13}C-NMR spectra of synthesis compound

Table 3. Significant attributions ^{13}C NMR data of synthesis compounds

Attributions	Chemicals Shift, δ (ppm)			
	NHC-H	NHC-Cl	Pd-NHC-H	Pd-NHC-Cl
N-CH-N (C carbene)	143.19	143.30	163.72	163.76
C-Cl	-	140.90	-	139.50
CH-Ar	131.51-114.59	130.90-113.80	130.35-115.85	132.30-111.90
Ar-C-N	51.10	49.30	49.8	46.85

3.2. Preliminary Complexation Study

The UV-Vis titration method was used to determine the stoichiometry between the Pd^{2+} metal cation and NHCL (NHC-H or NHC-Cl) in the formation of complexes. The graph plotted in figure 7 is assigned for the absorbance vs. [Pd2+]/[NHCL]. This graph indicates that as the molar fraction of Pd^{2+} metal cation was added into the NHC-H and NHC-Cl, the absorption spectra produced gradually increased as the concentration of Pd^{2+} metal cation increased. However, it is notable that there is a breaking point at 0.5 $[Pd^{2+}]/[NHCL]$ ratio for both graphs which indicates the 1:2 stoichiometry in the formation of complexes [6]. Thus, the suggested general chemical reaction of the complexation is as follows:

$$Pd^{2+} + 2NHCL \rightarrow 2NHCL\text{-}Pd$$

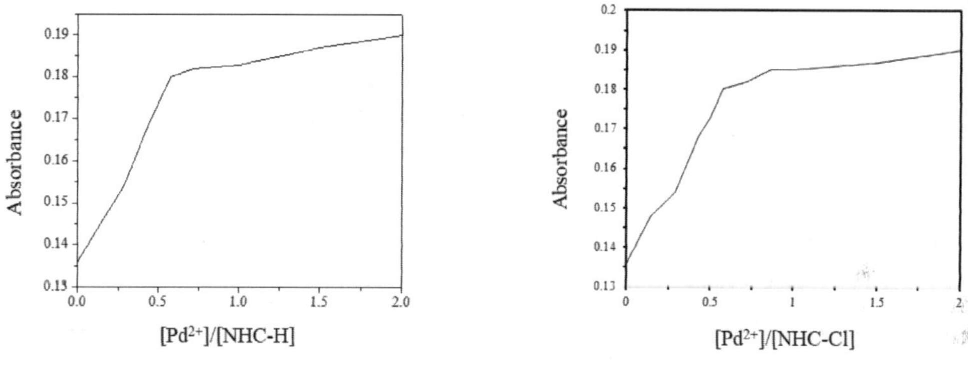

Figure 7. The graph plotted based on absorbance versus $[Pd^{2+}]/[NHC\text{-}H]$ and absorbance versus $[Pd^{2+}]/[NHC\text{-}Cl]$

3.3. Catalytic Activity Study

The reaction between 1-bromo-4-nitrobenzene and styrene (figure 3) was successfully synthesized in the presence of the Pd(II) catalyst to obtain a brown solid 1-nitro-4-strylbenzene with a melting point of 155 °C. The structure of the synthesized compound can be confirmed by the presence of C=C peaks in the FT-IR spectrum at 1641 cm^{-1} which is higher than C=C peaks in the FT-IR spectrum of styrene at 1629 cm^{-1}. It shows the successful bonding between styrene and 1-bromo-4-nitrobenzene. This consequence is due to the transformation of alkene functionality from monosubstituted double bond to disubstituted double bond where the stability of the structure increased, thus increasing the FT-IR stretching vibration [1].

The catalytic performance of Pd-NHC-H and Pd-NHC-Cl were studied at different amounts of catalysts used which are 1.0 and 0.5 mmol%, respectively. The reaction was then monitored at a 15-minute interval until 60 minutes reaction time. Table 4 shows the results obtained from the GC-FID chromatogram analysis. Results show that the conversion rate of the product increased as the catalyst loading and reaction time increases.

From the data, the best catalyst activity condition for the Heck reaction based on TONs is using the palladium catalyst of Pd-NHC-Cl with a catalyst loading of 0.5 mmol% and reaction time of 60 minutes.

The presence of the Cl electronegativity group increases the stability of the Pd catalyst structure, thus leading to higher TONs [8].

Table 4. Catalytic activity at different catalyst loading and reaction time

Entry	Catalyst Loading (mmol %)	Reaction time (minutes)	Conversion of 1-bromo-4-nitrobenzene (%), (TON)	
			Pd-NHC-H	Pd-NHC-Cl
1	0.0	15	4.97 (0)	17 (0)
2		30	-10.38 (0)	3 (0)
3		45	9.16 (0)	20 (0)
4		60	-1.76 (0)	11(0)
5	0.5	15	18.18 (36.36)	9 (19)
6		30	39.00 (78.00)	16 (32)
7		45	41.22 (82.44)	40 (81)
8		60	44.82 (89.64)	62 (124)
9	1.0	15	14.52 (14.52)	26 (26)
10		30	40.14 (40.14)	49 (49)
11		45	44.16 (44.16)	68 (68)
12		60	46.10 (46.10)	78 (78)

4. Conclusions

This study revealed that the synthesized palladium catalysts, Pd-NHC-H and Pd-NHC-Cl, can be used as catalysts in Heck reactions. Pd-NHC-Cl shows good catalytic activity with a catalyst loading of 0.5 mmol% within 60 minutes reaction time in DMA as a solvent and Na_2CO_3 as a base.

Acknowledgments

The authors gratefully acknowledge to the Universiti Teknologi Mara (UiTM), Cawangan Negeri Sembilan, Kampus Kuala Pilah for the facilities and financial support.

References

[1] Tsuji J 2006 *Palladium Reagents and Catalysts: New Perspectives for the 21st Century* (Great Britain-John Wiley & Sons) p 110-112
[2] Ding Y H, Fan H X, Long J, Zhang Q and Chen Y 2013 *Bioorganic & Medicinal Chemistry Letters* **23(22)** 6087
[3] Oberholzer M and Frech C M 2014 *Journal of Visualized Experiments*: JoVE **85** 51444
[4] Yuan D, Chen L, Yuan L, Liao S, Yang M and Zhang Q 2016 *Chemical Engineering Journal* **287** 241
[5] Yılmaz U, Küçükbay H, Deniz S and Şireci N 2013 *Molecules* **18(3)** 2501
[6] Said N R, Rezayi M, Narimani L, Al-Mohammed N N, Abdul Manan N S and Alias Y 2016 *Electrochimica Acta* **197** 10
[7] Arvela R K, Leadbeater N E, Sangi M S, Williams V A, Granados P and Singer R D 2005 *The Journal of Organic Chemistry* **70(1)** 161
[8] Jagtap S 2017 *Catalysts* **7(9)** 267

Determination of Non-Invasive Lung Cancer Biomarker by Quartz Crystal Microbalance Coated with Pegylated Lipopolymer

Nur Athirah Awatif Abdul Rahman[1], Akmal Hadi Ma'Radzi[1,] and Ammar Zakaria[2]

[1] School of Bioprocess Engineering, University Malaysia Perlis, 02600, Arau, Perlis, Malaysia.
[2] Centre of Excellence for Advanced Sensor Technology, University Malaysia Perlis, 02600, Arau, Perlis, Malaysia.

Email: akmalhadi@unimap.edu.my

Abstract. This study presents the application of pegylated lipopolymer to coat a sensor device called quartz crystal microbalance (QCM) for analysis of volatile organic compounds (VOC) on the non-invasive detection of lung cancer biomarkers. The chosen biomarker for this study was propanol, ethyl benzene, hexanal, and decane. QCM with fundamental frequency 30 MHz AT-cut with gold electrode was coated with sensing material composed of lipid, and pegylated lipopolymer with functional group carboxyl and succinyl to increase sensor sensitivity for biomarker detection purposes. The fabricated sensors could detect all presented biomarker with various selectivity and sensitivity responses. Furthermore, the results from biomarker exposure indicated that the sensitivity of sensors coated with succinyl pegylated lipopolymer film was better than non-coated sensor or coated with carboxyl lipopolymer and lipid films to each selected biomarkers.

1. Introduction

Studies in the detection of lung cancer biomarker have been conducted by a research group from the past decades using high advance's technology. One of the promising methods for early detection of lung cancer biomarker involves advances analytical technique. There is exhaled breath condensate (EBC) [1], Fourier transform infrared (FTIR) spectrometry [2], gas chromatography and mass spectrometry [4], solid phase microextraction - gas chromatography (SPME-GC) [3], and electric nose (e-nose) [5]. However, the present methods are using expensive instrumentation, and the analysis data is rather time-consuming since detection of the lung cancer analyzer present in the market nowadays gives non-online result data where it still requires breath to be collected at a different time and separately instrument for analysis. Due to this fact, technology is leaning to a sensor device called quartz crystal microbalance as a replacement for the technology. These have created a great opportunity for scientists to develop a new generation of gas sensors using quartz crystal microbalance device with the help of conductive polymer. QCM can offer a great promising sensor technology based on its low cost, rapid response, portability, non-hazardous label-free, real-time procedure and high sensitivity, which are ideal for detection of online analysis as well as holds a great promise for the next generation of sensors. According to our previous review [6], QCM used as a detector for lung cancer biomarker were few studied by the research group; however, the detection sensitivity is quite low. In order to increase the responsiveness of the sensor, appropriate coating elements based quartz surface can be a great help in determining the sensor response towards interest analyte. It can as well increase the absorption rate to boost the sensor frequency data. In this study, the aim was to detect that specific biomarker for lung cancer found in exhaled breath by using QCM with suitable sensing elements as coated film in order to increase sensor sensitivity towards gas detection. The VOC compound was also chosen based on a previous study [6].

2. Experimental

2.1. Materials

Lipopolymer 1,2-distearoyl-sn-glycero-3-phosphoethanolamine-N-[carboxy (polyethylene glycol) 2000] (Carboxy), 1,2-distearoyl-sn-glycero-3-phosphoethanolamine-N-[succinyl(polyethylene glycol) 2000] (Succinyl) and a lipid, 1,2-distearoyl-sn-glycero-3-phosphoethanolamine (Lipid) were purchased from Avanti Polar Lipids, Alabama for sensor fabrication purposed. Biomarker samples were decane (94.0 % purity), ethyl benzene (99.0 % purity), hexanal (98.0 % purity), and propanol (98.0 % purity) were purchased from Merck, United States. All materials were analytical grade reagents and used as received.

2.2. Sensor Fabrication

Sensors used in the present study were Quartz Crystal Microbalance (QCM) gold electrode with fundamental frequency 30 MHz AT-cut. Before deposition, sensors were washed with acetone and dried under ambient air in a covered petri dish for 24 hours. Sensing material Carboxyl and Succinyl was received in a liquid form from the supplier with concentration of 10 mg/ml of 25 mg lipopolymer and 2.5 ml chloroform. Since acquired ethanolic concentration for this research study was 1 mg/ml, the original lipopolymer solutions were diluted with chloroform to meet the aimed concentration value. Lipid material was received in powder form thus was dissolved in chloroform to produced ethanolic solution with concentration 1 mg/ml. Approximately 2 μL of ethanolic solution was dropped into the center of quartz crystal both side area using micropipette. The coated sensor was then dried under ambient air in a covered petri dish on fume hoods for 48 hours before any gas fabrication process, leaving a deposited thin-film on the surface of the quartz crystal.

2.3. Sampling Setup

Before starting any gas fabrication process, the setup system as depicted in Figure 1 was purged with ambient air pumped using an air pump system and passed through the sensor chamber to establish an initial frequency baseline. Estimated time for stable baseline was 15 minutes. For the fabrication process, liquid VOC was transferred into biomarker sample bottle and inserted into the gas chamber. Ambient air was pumped through the system and was kept at the level of 600 mL/min controlled by the LabView 2012 programme. The gas flowed over the volatile compound for five minutes for vaporization and homogenization. After that, the vaporized gas was passed through the sensor chamber for three minutes allowing interaction with the volatile compound. The frequency reading recorded by the network analyzer was monitored using Multisens installed on a personal computer. The measurement data which is the resonance frequencies of the sensor was measured before (0 second), during (300 seconds) and after (780 seconds) deposition of gas. Each sensor was exposed to each volatile compound for five times.

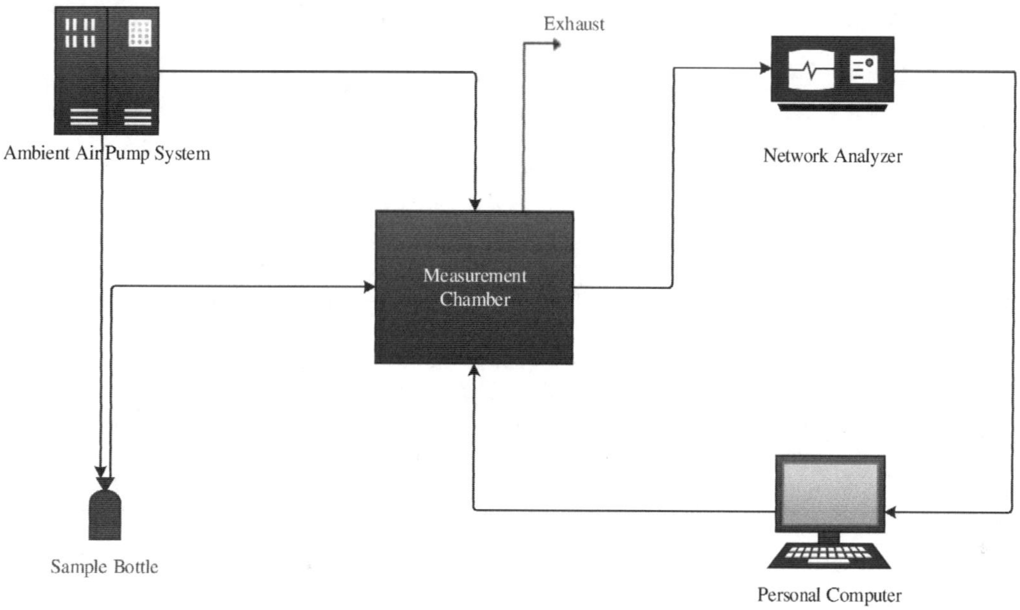

Figure 1. Schematic diagram of measurement system

3. Results and Discussion

3.1. Evaluation of fabricated sensors

Four sensors were used throughout this study with three of them were fabricated with sensing material, and the other one served as blank. The three sensors were coated with lipid (C1), carboxyl (C3) and succinyl (C4) while the non-coated sensor was C2. Pegylated lipopolymer used as sensing material for this study were derived from lipid derivatives and polyethylene compound. Lipid was known by having great ability as chemical sensor and when attach with pegylated compound, it helps enhance the stability and mobility of the lipopolymer [7]. This property is remarkable to be used as film sensing for the coating material. In the exposure experiment, sensors were exposed to each volatile compound five times in order to make sure the repeatability of the sensor is ensured. Sensor response obtained was in negative form and decreasing in frequency value leading to higher volatile compound absorption by the coated sensor, the more sensitive the sensor will be [8]. The frequency value was obtained from the subtraction of initial frequency at 300 s with a value taken at time 390 s. In order to get the limit concentration of VOC that can be detected; each sensor was exposed to biomarker in the concentration range of 0 ppm to 50 ppm. The lowest estimated limits were depicted in Table 1 and illustrated in Figure 2 for a better comparison.

Table 1. Characteristics of the sensors exposed to volatile compounds.

Volatile Compounds	Frequency Shift, Hz				Concentration, PPM			
	C1	C2	C3	C4	C1	C2	C3	C4
Propanol	6	5	9	25	6	20	6	6
Ethyl Benzene	15	2	19	39	2	2	6	8
Hexanal	19	9	35	69	1	2	1	2
Decane	33	9	21	50	1	1	4	4

The lowest propanol concentration detected was 6 PPM by C1 (6 Hz), C3 (9 Hz) and C4 (25 Hz) sensors with the highest frequency shifted was sensor C4 (25 Hz). For ethyl benzene, the lowest concentration can be detected was 2 PPM by sensor C1 (15 Hz) and C2 (2 Hz) however, the decreasing in frequency value was much smaller compared to C3 (19 Hz) and C4 (39 Hz). Hexanal can be detected at much lower concentration value, which is 1 PPM and 2 PPM for sensor C1-C3 (19 Hz - 35 Hz) and C2-C4 (9 Hz – 69 Hz). The highest decreasing value detected by C4 during exposure to hexanal was 69 Hz. Sensor C1 (33 Hz) and C2 (9 Hz) showed the lowest detection limit for decane, which was 1 PPM while sensor C3 (21 Hz) and C4 (50 Hz) were 4 PPM. The highest decreasing value showed by each sensor meaning that the sensor showed high-sensitivity potential towards the volatile compounds. Among four sensors used for this study, sensor C4 coated with pegylated lipopolymer with succinyl as the functional group was better in the detection of any volatile compounds, although the concentration limit that can be detected by C4 is not the lowest among studied sensors but still the highest among the rest sensors. This shows that sensor coated with succinyl material can be absorbed analyte better than lipopolymer with carboxyl functional group and lipid. Figure 3 illustrated the detection of each analyte over each sensor at their lowest detection limit.

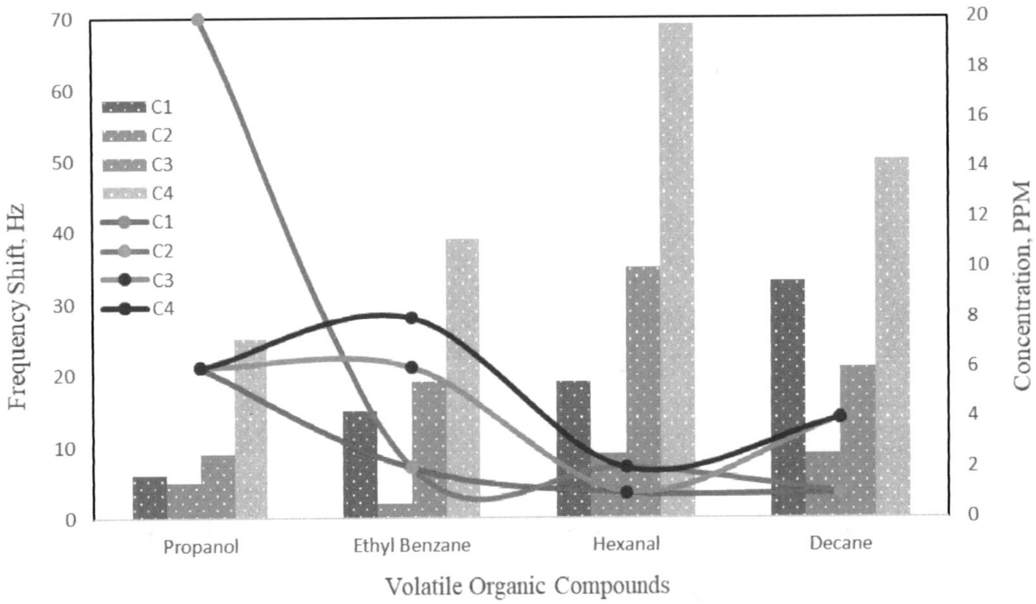

Figure 2. The representative frequency shift of coated sensor in the detection of volatile organic compounds.

Each sensor used in this study demonstrates the ability to detect organic compounds, and each has different detection limits. Although some compounds have the same detection limit, however, the frequency shift value between them varies. This clearly shows which sensors are more sensitive to selected biomarkers. To discuss which sensors provide better shift in frequency value, comparisons are made between C1 to C4 where the highest transitions have been reported by C4 while the lowest by C2. C4 coated with succinyl tends to detect each compound used in this experiment better than other sensors. This is due to succinyl composition, which has more recognition sites that can interact with biomarkers better and increase the adsorption rate of the thin film. Since more recognition sites are available in C4 to form bonding with the organic compound compared to C1, C2 and C3, it indirectly improves sensors' sensitivity. Given that propanol and hexane is polar compounds, they have more tendencies to share their electron more than ethyl benzene and decane do as the two are non-polar compound. Figure 3 illustrated the detection of each analyte over each sensor at their lowest detection limit.

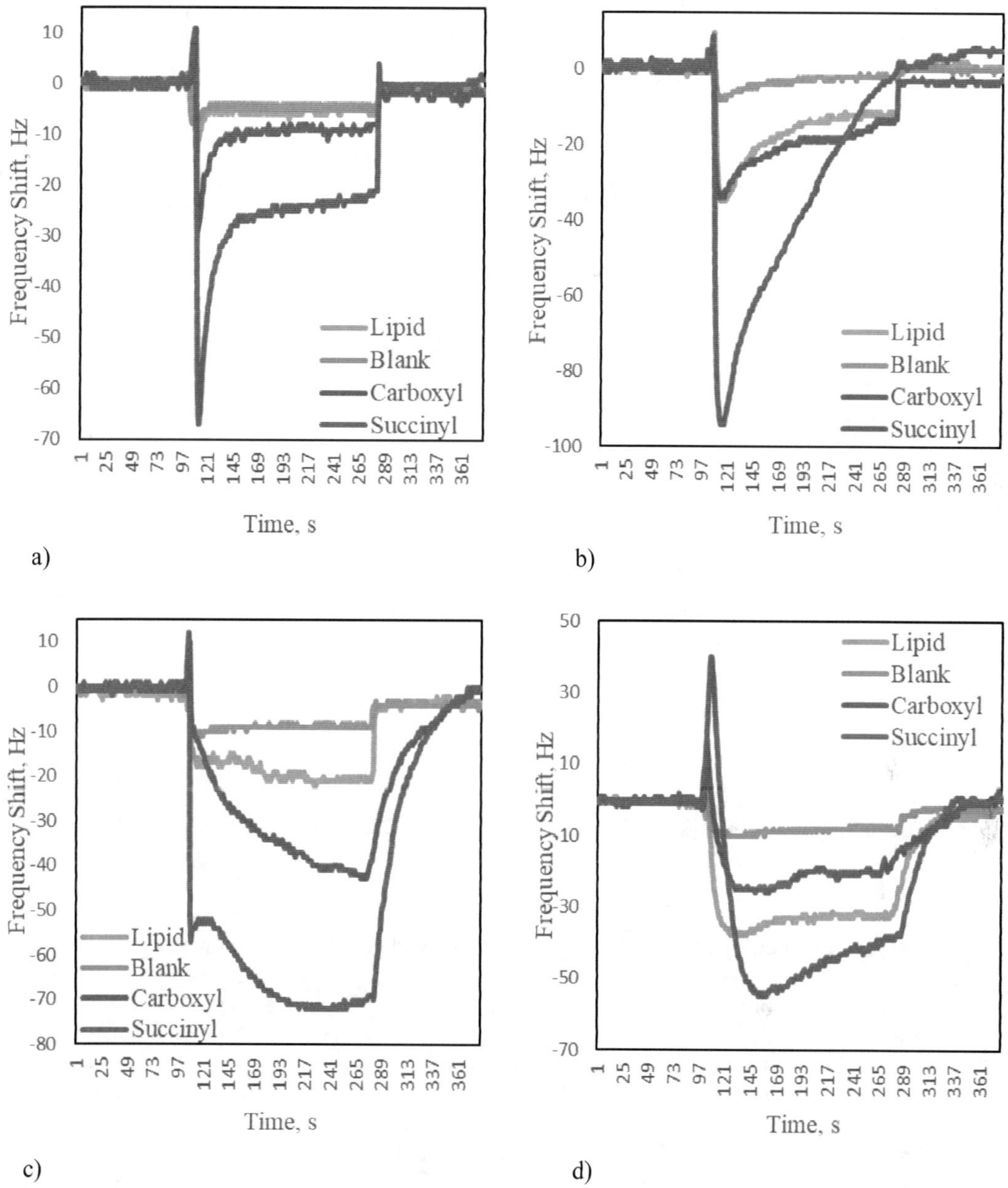

Figure 3. Decrement in frequency reading by each sensor during exposure to (a) propanol, (b) ethyl benzene, (c) hexanal, and (d) decane within 380 s (time range 200 s to 480 s).

3.2. Evaluation of sensor responses

Obtained sensor responses were evaluated in terms of sensing film ability to detect which biomarker better. Since the concept of applying coating film to the surface of QCM is to increase its sense ability, the greater decreasing of frequency reading, the more sensitive the sensor will be. This was said the coating material can form interaction with the exposed analyte by absorbing more analyte thus increasing its masses and indicate in negative response on frequency value. Figure 4 illustrated comparison between the lowest detection of biomarkers by sensor coated with lipid (C1). Each biomarker presented throughout this study was varied in terms of the functional group, for example, propanol is an alcohol group; ethyl benzene is benzene, hexanal are alkyl aldehyde and decane is from an alkane group. Selected biomarker also possess a variety of numbers of density value with the lowest among them is decane and the highest one is ethyl benzene; however, when exposed to each biomarker, sensor coated with lipid films was better in detecting the decane compounds than ethyl benzene, although the lowest detection was propanol. This could be due to the presence of single bonds between carbon atoms in decane, which is easier to break down and interact with lipid compound than another biomarker.

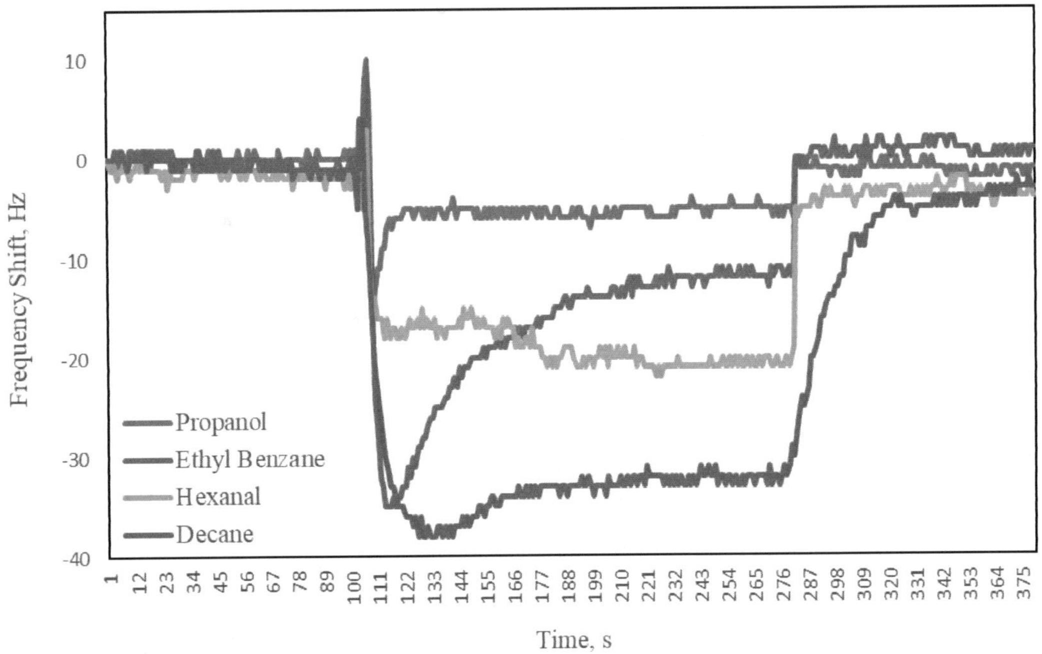

Figure 4. Frequency pattern for biomarkers exposed on sensor coated with lipid.

The detection of biomarker using non-coated sensor was the lowest among coated sensors. The frequency change for each analyte was within 2 to 9 Hz only compared to the lipid sensor which was 6 to 33 Hz. Figure 5 will illustrate detection of the biomarker at lowest concentration value. Unlike C1, the lowest detection biomarker by C2 was ethyl benzene, and the highest detection was hexanal. Since no sensing material was used to coat the sensor surface, the detection of analyte was believed with the touch of the biomarker with the surface of gold-quartz without any absorption process occurred. From here we can see that blank QCM has more tendencies to detect compounds at lowest density better compared to the one with high density where decane (0.73 g/ml), hexanal (0.81 g/ml), propanol (0.835 g/ml) and ethyl benzene (0.87 g/ml) in decreasing order.

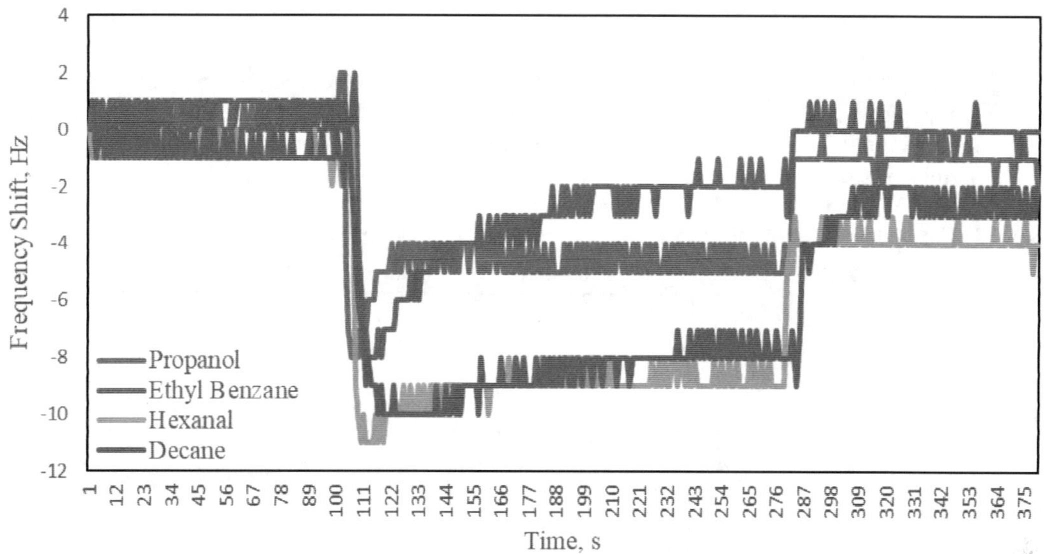

Figure 5. Frequency pattern for biomarkers exposed on the non-coated sensor.

Sensor coated with lipopolymer with a functional group carboxyl showed greater detection towards hexanal followed by decane, ethyl benzene, and propanol. The detection of hexanal by C3 seems to increase proportionally to the time and with decane, it is taking quite some time before it can reach the initial baseline after the sensors were stopped from biomarker exposure. Compared to propanol and ethyl benzene, their sensor reading will reach its initial baseline at a shorter time. This could be due to the properties of the carboxyl as the functional group for lipopolymer film, where it has the tendency to increase its bond stability when interacting with both alkane and alkyl aldehyde compounds. The sensor responses were shown in Figure 6.

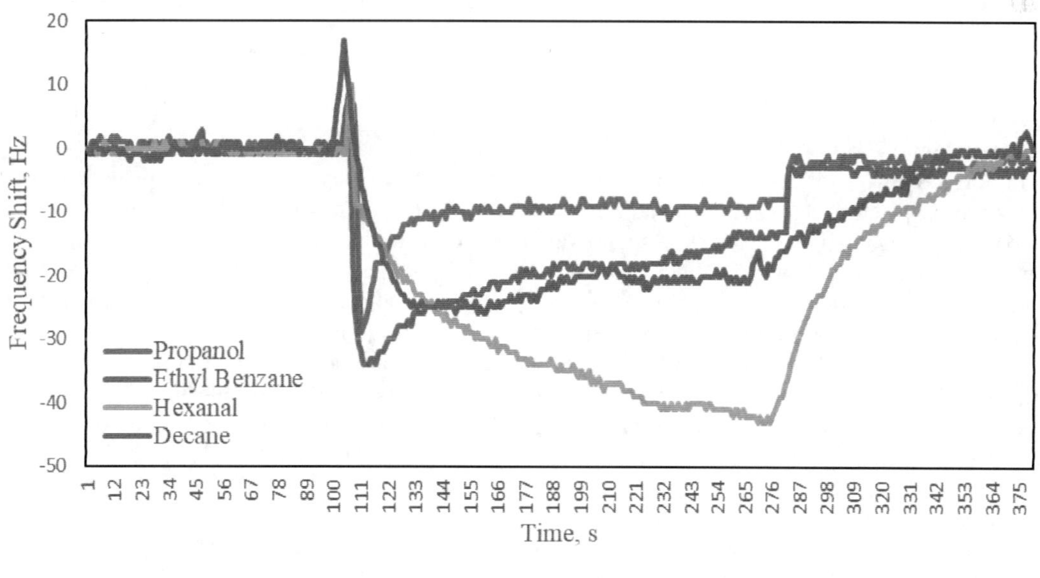

Figure 6. Frequency pattern for biomarkers exposed on sensor coated with carboxyl.

Inspecting the response by sensor coated with succinyl lipopolymer in Figure 7, one can observe that detection towards biomarker was the highest compared to sensor C1, C2, and C3. Although at early stage detection towards ethyl benzene was the higher than the rest of biomarker, however, the frequency reading value kept increasing proportionally with time until it reached the initial baseline after 273 s. Detection on decane also showed the same pattern with ethyl benzene, but, there was only a slight increment each second for comparison. As referred to the method, reading data for frequency value was taken at 390 s, which is 190 s on Figure 7, the lowest biomarker detected by C4 was propanol. The highest order for biomarker detection from hexanal, decane to ethyl benzene.

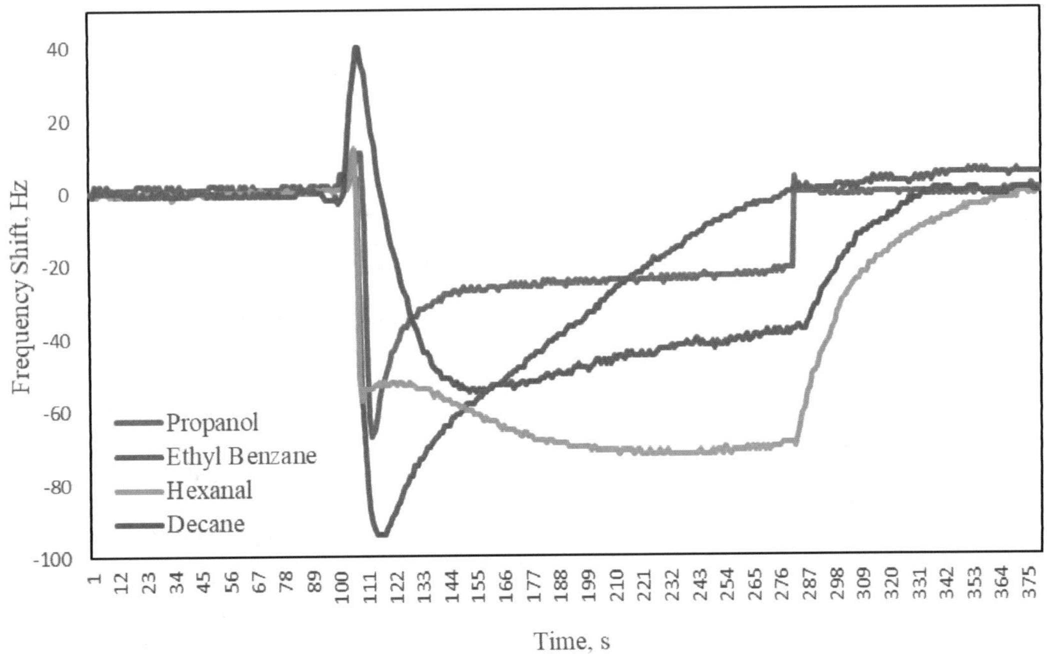

Figure 7. Frequency pattern for biomarkers exposed on sensor coated with succinyl.

4. Conclusion

In the present paper, the application of non-invasive lung cancer biomarker detection was developed using fabricated quartz crystal microbalance with pegylated lipopolymer sensing material. Based on the evaluation of sensor response towards selective biomarker (decane, hexanal, ethyl benzene, and propanol) at ppm level, sensor coated with succinyl-lipopolymer showed the best selectivity and sensitivity than the remaining sensor used. Sensor coated with carboxyl-lipopolymer comes in second in the detection of biomarker followed by lipid and non-coated sensor. This indicates that using lipopolymer as a sensing material for QCM can be expected to be used in the detection of a biomarker for lung cancer. Through further research, it is advisable to study the interaction of the functional group present in lipopolymer material with sensor sensitivity and groups of biomarker exposure.

5. Acknowledgments

This work was financially supported by the Ministry of Higher Education Malaysia under the Fundamental Research Grant Scheme (FRGS/1/2015/TK05/UNIMAP/02/8).

References

[1] Li M, Yang D, Brock G, Knipp RJ, Bousamra M, Nantz MH, and Fu XA 2015 Breath carbonyl compounds as biomarkers of lung cancer *Lung Cancer* 90(1) 92–97

[2] Dai X, Liu X, Liu L, Zhu B, and Fang Z 2015 A novel image-guided FT-IR sensor using chalcogenide glass optical fibers for the detection of combustion gases *Sensors and Actuators B: Chemical* 220 414–419

[3] Curran K, Underhill M, Gibson LT, and Strlic M 2016 The development of a SPME-GC/MS method for the analysis of VOC emissions from historic plastic and rubber materials *Microchemical Journal* 124 909–918

[4] Kim KH, Jahan SA, and Kabir E 2012 A review of breath analysis for diagnosis of human health *TrAC Trends in Analytical Chemistry* 33 1–8

[5] Delgado-Rodríguez M, Ruiz-Montoya M, Giraldez I, López R, Madejón E, and Díaz MJ 2012 Use of electronic nose and GC-MS in detection and monitoring some VOC. *Atmospheric Environment* 51 278–285

[6] Nur Athirah Awatif Abdul Rahman, Ammar Zakaria, Nora Jullok, and Akmal Hadi Ma'Radzi 2017 Applications of Sensing Material on Quartz Crystal Microbalance in Detection of Volatile Organic Compounds : A Review *Global Journal of Engineering and Technology Review* 2(2) 10-22

[7] Ulrich JK, and Michael T 1985 Lipid membrane technology for chemical and biosensor development *TrAC Trends in Analytical Chemistry* 4(4)

[8] Wyszynski B, Kim D, and Nakamoto T. 2013 Stabilization of coating for QCM odor sensors with liquid GC materials supported by lipopolymers and lipids *Sensors and Actuators B: Chemical* 179 81–86

IOP Conf. Series: Materials Science and Engineering **458** (2018) 012021 doi:10.1088/1757-899X/458/1/012021

Production of bio-oil and bio-char from different biomass wastes

***M Saleem[1] and N Baig[1]**

[1] Institute of Chemical Engineering and Technology, University of the Punjab, Lahore 54000, Pakistan

*Corresponding Email: msaleem.icet@pu.edu.pk

Abstract. Environmental pollution is rising due to the excess use of fossil fuels causing severe global warming and acid rains. High market price and the limited amount of fossil fuels have caused energy crises worldwide especially in the underdeveloped countries. The huge amount of biomass organic wastes is being widely used as an alternative feedstock as a potential source for second generation biorefining process. In this research four different biomass wastes are used to investigate the yield of bio-oil and biochar using a conventional pyrolysis method. Rice husk, bagasse, sawdust (Dalbergia sisso Roxb.), and sawdust (Populus caspica Bornm.) are operated in a fixed bed reactor. The rationale of this study is to evaluate the effect of temperature on the yield of bio-oil and biochar at three different temperatures (300 °C, 400 °C, and 500 °C) with heating rate of 3.69 °C/min. The particle size of each biomass waste used was 0.498 mm. Sulfuric acid was used as a catalyst to enhance the rate of reaction. The proximate analysis of biomass wastes was performed in order to evaluate the moisture content, ash content, and volatile matter and fixed carbon. The results showed that with the increase of temperature, the bio-oil yield increases and biochar yield decreases. The highest yield of bio-oil (32.39%) was achieved at 500 °C from bagasse feedstock and the highest yield of biochar (37.50%) was obtained at 300 °C from rice husk feedstock.

1. Introduction

The world is facing huge energy crisis due to high prices and the limited amount of fossil fuels available. New alternatives that are sustainable and environmentally friendly can overcome energy crisis worldwide [1]. Wood wastes and agricultural residues are the significant renewable solid energy resources available which may be converted ito value added products bytheir conversion at high temperature in the absence of oxygen to form bio-oil, biochar, and syngas. Biomass contains cellulose, hemicellulose, lignin, ash, and extractives[2]. Generally, wood wastes and agricultural wastes are different in terms of their species and structural constituents. Wood wastes mainly consist of hardwoods and softwoods. The structural composition of hardwoods waste contains 43% cellulose, 35% hemicellulose, and 22% lignin while softwoods contain about 43% cellulose, 28% hemicellulose, and 29% of lignin on an extractive free basis. The structural composition of agricultural residues such as rice husk, bagasse, and corncob is dependent on the type of biomass species [2].

The most emerging technologies are pyrolysis that converts different kinds of biomass wastes for the synthesis of biofuels and biochemical. Organic end products of wood wastes and agricultural wastes are converted into more valuable products which find many useful applications. [3]. In the presence of the catalyst carbon-carbon bonds in biomass compounds break down. Different hydrocarbons, carbon dioxide, and water are extracted by the process of decarboxylation and dehydration of hydrocarbon compounds via cracking reactions[1]. Condensation of vapors forms bio-oil at temperatures of 400 °C to 500 °C as a significant product than the other pyrolysis products. The color of the oil is dark brown and has a strong acrid smell. Bio-oil is more beneficial as compared to biomass because it can be easy to use, store and transpor. The intensity of energy in bio-oil is five times more than that of the initial raw material[4]. Biochar is alsoa significant product obtained from pyrolysis of biomass, which has more than 50% carbon. Biochar can be used not only as a fuel but also used in soil amendments. İt is quite stable both biologically and chemically due to its porous structure.It remains settled in the soil for 1,000 years[5]. Therefore, from the production of biochar via pyrolysis technique the quantity of carbon dioxide is reduced in the environment. Biochar can be beneficial for adsorption and building processes[6].

Generally, catalysts are used to boost up reaction kinetics by breaking down higher molecular weight compounds into lighter hydrocarbon compounds. Therefore, various catalyst show different end products in various operating conditions[7]. The reaction mechanisms of pyrolysis of biomass feedstock are complicated but can be generally described in three stage mechanism.

$$\text{Biomass feedstock} \longrightarrow \text{Water} + \text{Unreacted solid} \tag{1}$$

$$\text{Unreacted solid} \longrightarrow (\text{Volatile} + \text{Gases})_1 + (\text{Char})_1 \tag{2}$$

$$(\text{Char})_1 \longrightarrow (\text{Volatile} + \text{Gases})_2 + (\text{Char})_2 \tag{3}$$

The 1st step of feedstock degradation, which occurred between 122 °C and 202 °C, can be called pre-hydrolysis. During this stage, some internal modification such as water removal, bond breakage, a free radicals formation, and the carbonyl, and carboxyl groups formation takes place during conversion of biomass into volatiles[8]. The 2nd step of the biomass feedstock degradation correlates to the key pyrolysis technique. Starting with a high rate that ends with the formation of pyrolysis oil, solid, and gas. İn the 3rd step, the char degrades at a very slow rate and carbon-rich char is left behind. At low temperature, cellulose carbonization starts with a gradual degradation and char formation on heating, and fast volatilization proceeds on pyrolysis at higher temperatures. During 1st stage degradation reactions such as depolymerization, oxidation, dehydration, and decarboxylation; the pyrolysis reactions produce nearly 62-72% liquid distillate and 10-18% charred solid [9].

During heating, the hemicelluloses reacts more readily than cellulose. The thermal degradation of hemicellulose begins above 100 °C during heating for 48 h. The hemicelluloses breaks down first, at the temperature of 197-257 °C and cellulose breaks down in the temperature range of 237 to 347 °C, with lignin pyrolyzing at temperatures of 277 to 497 °C [10]. The most significant temperature for the production of the pyrolysis products such as liquid, solid, and gas is between 377 and 527 °C.

Bio-oil has various potential applications for commercial purposes and as well as in domestic for heating purposes. Bio-oil is a complex mixture of oxygenated compounds. Functional groups of bio-oil provide the potential for utilization. Bio-oil is also significant for the production of valuable biochemicals such as furfural and derivatives of furfural. On the other hand, biochar consist of a strong carbon matrix which is known as biochar. At low-temperature pyrolysis, higher amount of biochar is obtained due to the low rates of devolatilization and low conversion of carbon. Biochar also contains hydrogen and inorganic species in two structure: stacked crystalline sheets and randomly ordered amorphous aromatic structure [7]. The potential applications are to stabilize the quality of soil and enhanced vegetative production. Biochar enables the long-term sequestration of carbon in soil due to its inert nature [11]. It is also applicable in soil amendment, co-firing power generating plant, and enhance poultry production.

İn this study different types of biomass wastes such as rice husk, bagasse, sawdust (Dalbergia sisso Roxb.) and sawdust (Populus caspica Bornm.) are pyrolyzed in the presence of catalyst for the production of bio-oil and biochar at various temperatures.

2. Methodology

2.1. Sampling
The samples of raw materials; rice husk, bagasse, sawdust (Dalbergia sisso Roxb. and sawdust (Populus caspica Bornm.); were collected from the field. Air dried samples were ground and then sieved to achieve a particle size of 0.498 mm. The 100g of each sample was impregnated in water solution in the presence of the specific concentration of catalysts before the experiments for the purpose of hydrolysis.

2.2. Experimental Setup
The conventional pyrolysis experiment was conducted in a fixed bed reactor to study the effect of temperature on pyrolysis products such as bio-oil and biochar. The reactor is made up of stainless steel (SS316). The length of the furnace is 80 cm and its inside diameter is 6.3 cm. Nichrome wire is wounded throughout the furnace length as a heating element. The furnace is insulated to facilitate maximum heat utilization. All the biomass wastes samples are pyrolysed in the fixed bed reactor as shown in figure 1.

2.3. Proceducer
In this study, 10 to 20 g of raw material was placed in the stainless-steel sample tube. The tube was hanged in the tubular furnace. After closing the flange and placing measuring instruments in place, nitrogen gas purging was begun. The reactor temperature was set and heating switched on. Cooled water was circulated throughout the condensers when the pump was turned on. For each run continuous flow rate of nitrogen gas was 0.15 lit/min. The vapours were condensed by a two-stage condensation using chilled water. When condensate formation stopped, the heating was continued for 20 minutes to ensure the completion of reaction. The condensate was collected in the conical flasks. The quantity of liquid was more in the first condenser than in the second condenser. The liquid obtained from both flasks was weighed and stored for characterization. Furnace could cool down which took around 1 hour. Then the tube was taken out through the flange and remaining solid was collected from the tube for mass balance. The equations (4), (5), (6), and (7) were used to calculate the conversion and yields of bio-oil, biochar, and syngas using weight of biomass and char on dry basis (db).

$$\text{Conversion (\%)} = \frac{(W_{Biomass,db} - W_{Solid,db})}{W_{Biomass,db}} \times 100 \qquad (4)$$

$$\text{Liquid yield (wt\%)} = \frac{(W_{Liquid,db})}{W_{Biomass,db}} \times 100 \qquad (5)$$

$$\text{Solid yield (wt\%)} = \frac{(W_{Solid,db})}{W_{Biomass,db}} \times 100 \qquad (6)$$

$$\text{Gas yield (wt\%)} = 100\% - \text{liquid yield (wt\%)}\ \ \text{solid yield (wt\%)} \qquad (7)$$

Where $W_{Biomass,db}$ and $W_{Solid,db}$ are the weights of initial biomass and remaining solid, respectively, on the dry basis.

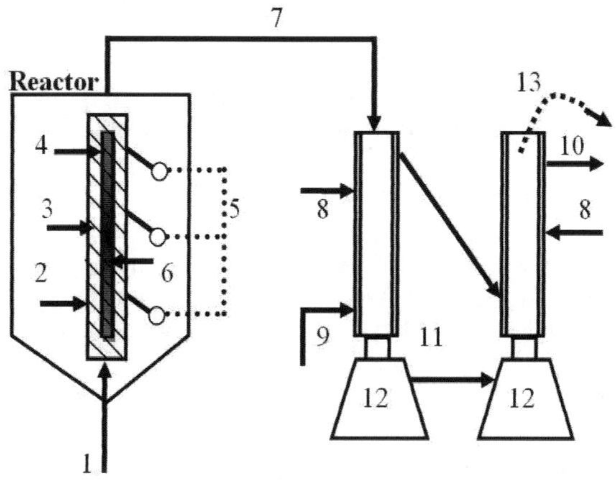

Figure 1. Experimental setup; three-zone tube furnace 1) Fluidizing Gas (Nitrogen) 2) Tube Furnace 3) Heating coil 4) Basket for biomass 5) Thermocouple 6) Raw material (Biomass) 7) Volatiles 8) Condenser 9) Water in 10) Water out 11) Vapors 12) Bio-oil Collector 13) Non-condensable gases

3. Results and Discussion

3.1. Proximate Analysis

Proximate analysis was conducted using four types of feedstock. The volatile matter was high in the sawdust (Populus caspica Bornm.) and low in rice husk which is shown in table 1. High moisture content was 6.80% calculated in rice husk while low moisture content was 1.25% reported for sawdust (Populus caspica Bornm.). High ash content was calculated 17.45% in rice husk. Lower volatile matter in biomass wastes leads to produce a rigid charcoal. The low quality of bio-oil and corrosion in pyrolysis system has been improved due to the low content of moisture.

Table 1. Proximate analysis of selected biomass wastes experimental values

Biomass	Moisture %	Ash%	Volatile Matter %	Fixed Carbon %
Sawdust (Dalbergia sisso Roxb.)	5.25 ± 1.43	1.5 ± 0.5	88.25 ± 2.75	6.0 ± 1.95
Sawdust (Populus caspica Bornm.)	1.25 ± 0.25	1.5 ± 0.28	93.75 ± 1.43	3.0 ± 1.35
Baggage	4.25 ± 0.50	2.0 ± 0.50	89.75 ± 1.32	5.0 ± 0.76
Rice husk	6.80 ± 2.06	17.45 ± 0.69	63.06 ± 1.72	15.85 ± 0.10

3.2. Effect of temperature on products yield

The conversion (%), bio-oil yield (%), Char yield (%) and gas yield(%) for the tested samples at three temperatures (300, 400 and 500°C) are presented in Table 2. The percentage of the biomass sample that is converted to vapours on heating is called conversion.

The conversion varies from 62.50% to 69.44%, 77.09% to 80.28%, 72.91% to 77.09%, and 82.40% to 83.31% for Rice husk, Baggasse, Sawdust (Dalbergia sisso Roxb.) and Sawdust (Populus caspica Bornm.) respectively. One can observe that the conversion increases with temperature for first three samples although to a different extent. The conversion declines with temperature for the fourth sample although variation is not much. The conversion is higher for baggasse (80.28%) and Sawdust (Populus caspica Bornm.) (83.31%). Rice husk undergoes lowest conversion 62.50% and 69.44% at the lowest temperature as well as at the highest temparature respectively. Where as Sawdust (Populus caspica Bornm.) undergoes highest conversion 82.40% and 83.31% at the lowest temperature as well as at the highest temperature respectively.

The bio-oil yield varies from 22.05% to 23.33%, 21.37% to 32.39%, 22.43% to 27.94%, and 21% to 24.83% for Rice husk, Baggasse, Sawdust (Dalbergia sisso Roxb.) and Sawdust (Populus caspica Bornm.) respectively. It is obvious that the yield increases with temperature but to a varying extent. Increase in yield of bio-oil for a temperature change from 300 to 500°C is relatively higher for Rice husk and Baggasse samples as compared to that of the saw dust samples. Baggasse gives highest (32.39%) and lowest (21.37%) bio-oil yields at 500°C and 300°C respectively.

The char yield varies from 37.50% to 30.55%, 22.90% to 19.71%, 27.08% to 22.90%, and 17.60 to 16.68% for Rice husk, Baggasse, Sawdust (Dalbergia sisso Roxb.) and Sawdust (Populus caspica Bornm.) respectively. It is notable that the yield decreases with temperature in all cases but to a varying extent. The maximum variation is depicted by Rice husk samples and the least decrease is shown by the Saw dust (Populus caspica Bornm.) samples.The gas yields are calculated by difference.

This effect of temperature on the yield of bio-oil is mainly due to the maximum primary and secondary degradation of the biomass and char, thus leading to a greater conversion and higher yields at higher temperatures. The yield of biochar is mainly affected by secondary cracking of pyrolysis vapors and decomposition of char at higher temperatures. The effect of temperature is not linear in terms of yields of pyrolysis gases.

Table 2. The conversion and production of bio-oil and biochar yields through pyrolysis of biomass wastes at three temperatures with the catalyst.

Biomass Sample	Temperature (°C)	Conversion (%)	Bio-Oil yield (%)	Char yield (%)	Gas yield (%)
Rice Husk	300	62.50	22.05	37.50	40.45
	400	67.20	21.50	32.79	45.69
	500	69.44	23.33	30.55	46.11
Bagasse	300	77.09	21.37	22.90	55.72
	400	80.46	24.21	19.53	56.25
	500	80.28	32.39	19.71	47.88
Sawdust (Dalbergia sisso Roxb.)	300	72.91	22.43	27.08	50.48
	400	71.75	26.93	28.24	44.81
	500	77.09	27.94	22.90	49.14
Sawdust (Populus caspica Bornm.)	300	82.40	21.60	17.60	60.80
	400	83.17	24.50	16.82	58.67
	500	83.31	24.83	16.68	58.47

Gaseous products from pyrolysis are carbon monoxide, carbon dioxide, hydrogen, methane and lower hydrocarbons such as oxygen and nitrogen. At high temperatures, the large molecules in bio-oil and biochar are depolymerized into smaller molecules, which also contain the gaseous fractions [12].

Conclusion

The yield of pyrolysis products, bio-oil and char converted from different biomass wastes, using conventional pyrolysis process at different temperature (300-500 °C) was investigated in this research. Biomass wastes used in this research were rice husk, bagasse, sawdust (Dalbergia sisso Roxb.) and sawdust (Populus caspica Bornm.). They are operated at 300 °C, 400 °C, and 500 °C temperature in the presence of a catalyst. The catalyst selected for this process was sulfuric acid. The proximate analysis of biomass wastes was performed in order to evaluate the moisture content, ash content, volatile matter and fixed carbon. Yield of bio-oil obtained from bagasse feedstock at the temperature of 500 °C is the highest as compared to others which is 32.39%. On the other hand, yield of biochar produced by rice husk feedstock at 300 °C temperature is highest among others which is 37.50%. The results reveal that the yield of bio-oil increases with increase in temperature and vice versa for the char.

Acknowledgement:

The authors acknowledge the Institute of Chemical Engineering and Technology, University of the Punjab Lahore, Pakistan for financial assistance.

References

[1] Durak H and Aysu T 2015 Effect of pyrolysis temperature and catalyst on production of bio-oil and bio-char from avocado seeds *Res Chem Intermediat* **41** 8067-97
[2] Demirbas A 2007 Effect of temperature on pyrolysis products from biomass *Energ Source Part A* **29** 329-36
[3] Hanif M U, Capareda S C, Kongkasawan J, Iqbal H, Arazo R O and Baig M A 2016 Effects of Pyrolysis Temperature on Product Yields and Energy Recovery from Co-Feeding of Cotton Gin Trash, Cow Manure, and Microalgae: A Simulation Study (vol 11, e0152230, 2016) *Plos One* **11**
[4] Abnisa F, Arami-Niya A, Daud W M A W, Sahu J N and Noor I M 2013 Utilization of oil palm tree residues to produce bio-oil and bio-char via pyrolysis *Energ Convers Manage* **76** 1073-82
[5] Zheng J L 2008 Pyrolysis oil from fast pyrolysis of maize stalk *J Anal Appl Pyrol* **83** 205-12
[6] Chaiwong K, Kiatsiriroat T, Vorayos N and Thararax C 2013 Study of bio-oil and bio-char production from algae by slow pyrolysis *Biomass Bioenerg* **56** 600-6
[7] Jahirul M I, Rasul M G, Chowdhury A A and Ashwath N 2012 Biofuels Production through Biomass Pyrolysis-A Technological Review *Energies* **5** 4952-5001
[8] Shafizadeh F 1982 Introduction to Pyrolysis of Biomass *J Anal Appl Pyrol* **3** 283-305
[9] Fengel D and Wegener G 1984 *Wood Chemistry, Ultrastructure, Reactions*: Berlin: Walter de Gruyter)
[10] Kucuk M H and Demirbas A 1993 Delignification of Ailanthus-Altissima and Spruce-Orientalis with Glycerol or Alkaline Glycerol at Atmospheric-Pressure *Cell Chem Technol* **27** 679-86
[11] Stavi I 2012 The potential use of biochar in reclaiming degraded rangelands *J Environ Plann Man* **55** 657-65
[12] Maschio G, Koufopanos C and Lucchesi A 1992 Pyrolysis, a Promising Route for Biomass Utilization *Bioresource Technol* **42** 219-31

IOP Conf. Series: Materials Science and Engineering **458** (2018) 012022 doi:10.1088/1757-899X/458/1/012022

Development and characterization of semi-refined carrageenan (SRC) films from *Eucheuma cottonii* incorporated with glycerol and α-tocopherol for active food packaging application

K H A Hamid, N A Z M Saupy, N M Zain, S K A Mudalip, S M Shaarani and *N A M Azman

Faculty of Chemical and Natural Resources Engineering, Universiti Malaysia Pahang, Lebuhraya Tun Razak, Pahang 26000, Malaysia

*Corresponding e-mail: ainiazman@ump.edu.my

Abstract. This works focuses on the development of biodegradable active films packaging using all natural compounds with the aim of contributing sustainable film not only by reducing the plastic waste to environment but also as a potential substitute of synthetic preservative in food. Active film packaging was formulated using semi-refined carrageenan (SRC) biopolymer extracted from *Eucheuma cottonii* plasticized with glycerol (G) and incorporated with different concentrations of α-tocopherol (Tp) at 0.1, 0.2, 0.3 and 0.4% v/v as natural antioxidants. Opacity, water solubility and mechanical properties of the resultant film were analysed whereas the migration behaviour of antioxidant films into food simulant during 30 days storages were determined using total phenolic content and DPPH assay. The addition of Tp in SRC+G film decreased the tensile strength but increased the value of elongation at break significantly ($p<0.05$). Opacity and water solubility were improved with increasing concentration of Tp ($p<0.05$). The highest and prolonged release of antioxidant to food simulant was determined by SRC+G+Tp0.4% through 30 day storage with total phenolic content value of 2 mg GAE/ L sample and 88.59% inhibition for DPPH assay. Hence, the new formulation of SRC+G with tocopherol could be an alternative degradable active packaging that might reduce the need of adding antioxidants directly into food products and also prolong the food shelf life.

1. Introduction

The development of biodegradable film from natural sources is becoming important nowadays as it is considered as an alternative to the petrochemical derived plastic that is not safe for the environment. Many researchers have focused on the development of edible and/or biodegradable films made from natural and renewable polymers such as polysaccharides, protein and lipids [1–2] whereby biopolymer from agro-resources showed potentially good film-forming properties [3]. Many natural biopolymers are not only edible and degradable but also possess biocompatibility properties with other functional materials that offer many advantages in food packaging industries.

Carrageenan is a natural polysaccharide polymer obtained from the extraction of red seaweed species from Rhodophyceae family. Previous studies reported that carrageenan is one of the potential biopolymers that has the ability to form a strong gel and provides efficient barrier against gas, lipids and oils [3-4]. The presence of 3,6-anhydrobridges compound in kappa (κ) and iota (ι) carrageenans demonstrating the potential as a film forming ability and the κ-carrageenan having strong properties in forming hard, strong and brittle gels are widely found in *Eucheuma cottonii* species [5]. In industrial process of producing carrageenan, the process involves two distinct types of carrageenan; 1) semi-refined carrageenan and 2) refined carrageenan. Semi-refined carrageenan (SRC) is available at cheaper cost compared to refined carrageenan where it also has excellent binding properties in film-forming solution [6].

Polymer blending with active compound is one of the innovations to obtain new film material containing desired functional properties that offers specific protection for food. Active film packaging is normally derived from natural polymer coated with active compounds such as antioxidant for the purposes of extending food shelf life as well as improving safety and sensorial properties in the food packaging. Various natural antioxidants such as α-tocopherol [7] and phenolic compounds [8] have been used in food films and coating systems. Medina-Jaramillo et al. [9] reported that natural extract from green tea that was blended into starch films demonstrated good film forming properties and exhibited high antioxidant activity in food simulant. Other than that, active film developed from herbs such as lavender, oregano and bearberry leaves successfully showed their antioxidant and antimicrobial properties in food protection by previous studies [10–11].

Vitamin E or known as α-tocopherol is a nontoxic natural antioxidant possessing many pharmacological benefits to human health including anti-cancer and reduces the risk of degenerative diseases such as cardiovascular diseases [12]. Many researches have been conducted to investigate the potential use of the α-tocopherol as antioxidant coating in packaging since the compound is an excellent stabilizer and shows high solubility during polymer processing [13-14]. Furthermore, Shojaee-Aliabadi et al. [4] showed that the addition of active and hydrophobic compounds of α-tocopherol not only gives the protective effect in food, but also improves the mechanical and barrier properties in the development of biodegradable films.

Nevertheless, the addition of plasticizers within the production of plastic film products is not new. Incorporating plasticizer into the active film packaging is able to modify the polymer properties and allow the mass transport of the active agents to the foods. Vieira et al. [3] reported that several properties of the film are improved including increasing flexibility of film and decreasing the glass transition temperature by adding plasticizer. Currently, focus is on natural plasticizers such as glycerol, sorbitol, sunflower oil and linseed oil as substitutes to the synthetic plasticizers due to their low toxicity and migration, and they have been widely used to prepare edible film.

A recent study found that SRC blended with glycerol has good potential to be used as an alternative polysaccharide material for the production of edible biodegradable packaging films with improved mechanical and barrier properties [6]. However, the formulation of SRC with glycerol (G) and α-tocopherol (Tp) as new functional material in active packaging film has not been fully developed. In addition, the mechanical properties, opacity, solubility of film in water and antioxidant release of active packaging film from SRC + G + Tp have been not fully documented. Hence, this research aims to develop and characterize SRC based film with enhanced functional properties by adding glycerol (0.9% v/v) and different concentrations of α-tocopherol (0.1, 0.2, 0.3 and 0.4%, v/v) for potential use as active packaging in food.

2. Materials and Methods

2.1. Materials

Semi-refined carrageenan (SRC) was extracted from *Eucheuma cottonii* seaweed obtained from Sabah, Malaysia. Food grade glycerol, 2,2-diphenyl-1-picrylhydrazyl (DPPH), sodium chloride, folin & ciocalteu's phenol reagent and α-tocopherol were purchased from Sigma Aldrich (England).

2.2. Preparation of semi-refined carrageenan (SRC) from Eucheuma Cottonii seaweed

SRC film was prepared according to the established method of Mustapha et al. [24]. *Eucheuma cottonii* seaweed was cleaned under running tap water to remove debris and dried under sunlight for 8 h. Extraction process was prepared by adding 150 g of dried seaweed into 1.0M KOH solution at 80°C for 2 h and soaked in water for 12 h. Then, the neutralised seaweeds were dried at 50°C for 24 h using a laboratory oven (MEMMERT, EQP004, Schwabach, Germany). Finally, the SRC samples were ground at mesh 0.5 mm using laboratory grinder (Retsch, ZM 200, Haan, Germany) and stored in the dessicator (25°C) for further analyses.

2.3. Preparation of SRC films

SRC based films were prepared according to the method of Farhan et al. [6] with slight modifications. SRC film solution was prepared by dissolving 2 g of SRC in distilled water (2% w/w) under continuous stirring and the solution was heated at 70 °C for 10 min using hot plate magnetic stirrer. After dissolution, plasticizer glycerol was added at 0.9% v/v into the solution. The film-forming solution was heated again and maintained for ± 10 min prior to the addition of antioxidant α-tocopherol at different concentrations (0.1%, 0.2%, 0.3%, and 0.4% v/v) under continuous stirring. Adequate amount of SRC film forming solutions were casted on casting plates and dried using laboratory oven at 40 °C and 30% relative humidity (RH) for 24 hr. The films was then cooled in room temperature and RH for 30hr prior to experimental use.

2.4. Mechanical properties of films

Tensile strength (TS) and elongation at break (EAB) of the film samples were measured according to ASTM D882 method [16] using a Universal Testing Machine (AG-Xplus Series, Shimadzu, Japan). The film samples were uniformly cut (10 cm × 1.5 cm) and clamped between tensile grips with a crosshead speed set at 50 mm/min. Tested film strips were equilibrated at 25 °C and 50% relative humidity (RH) in desiccators for 48 h prior to testing. TS of the films were calculated using the following equation:

$$TS(MPa) = \frac{F_{max}}{\phi} \tag{1}$$

where F_{max} is the maximum load and Φ is the cross-sectional area. EAB of the films was calculated using the following equation:

$$EAB(\%) = (\frac{\Delta_l}{l_0}) \times 100 \tag{2}$$

where Δ_l is the film extension and l_0 is the initial length of the film sample.

2.5. Solubility in water

Solubility of films in water were determined according to the method described by Farhan et al. [6] with slight modifications. The film samples were cut uniformly (2 cm × 2 cm) and dried at 100°C in a laboratory oven for 24 h and weighted to determine their initial dry weight. Each film samples were placed with 30 ml of distilled water and immersed into laboratory water bath (Memmert Waterbath, WNE14, Tokyo, Japan) under constant shaking at 25°C for 24 h. After that, undissolved film samples were filtered using Whatman No. 1 filter paper and the samples were dried at 100°C for 24 h to determine their final dry weight. The solubility of films in water was calculated as a percentage (WS %) using the following equation:

$$WS(\%) = (\frac{W_0 - W_f}{W_0}) \times 100 \tag{3}$$

where W_0 is the initial dry weight of the film and W_f is the final weight of the dried undissolved film.

2.6. Opacity measurement

The opacity of the film samples was performed as described by Shojaee-Aliabadi et al. [4]. The opacity of the films was evaluated using UV-visible spectrophotometer (HITACHI, U-1800, Tokyo, Japan) at an absorbance of 600 nm. The films (3 x 0.3 cm^2) were placed on the plastic cuvette. The opacity of the films was calculated using the following equation:

$$Opacity(OP) = \frac{Abs600}{\gamma} \tag{4}$$

where Abs600 is the value of absorbance at 600 nm and γ is the thickness of film (mm).

2.7. Determination of antioxidant activity

2.7.1. Migration test. A study on the release of the active compounds from the films was carried out by determining the specific migration from polymer into food simulant (90% ethanol) as reported by López De Dicastillo et al. [17]. Film samples (2 x 3 cm^2) were placed in separate 30 mL amber glass vials and added with 10 ml of 90% ethanol. The vials were hermetically sealed with open-top cap and stored at room temperature and protected by light for 30 days storage period. The migration of α-Tocopherol in the simulants was analysed to determine their 1) total phenolic content (TPC) and 2) antioxidant activity.

2.7.2. Determination of TPC and antioxidant activity. Total phenolic content and antioxidant activity was measured using Folin-Ciocalteu method and DPPH scavenging activity respectively and these assays was performed according to previous study [18]. Folin-Ciocalteu method: The final concentration (v/v) for the mixture was prepared with 10 μL samples (90% ethanol of food simulant), 50 μL 10% Folin reagent and 150 μL 7% sodium carbonate solution and diluted with 790 μL MiliQ water. Absorbance at 765 nm was measured using UV-visible spectrophotometer. The results were expressed as mg of Gallic acid equivalents/L sample (mg GAE/ L sample). DPPH assay: A solution of 0.6 mM DPPH dissolved in ethanol appropriate dilutions were made to allow the fall in DPPH concentration to be in the range of 10-90 % (v/v). One aliquot (0.025 ml) of the diluted ethanol solution (90% ethanol of food simulant) was mixed with 0.975 ml of 0.6 mM DPPH. The solution was added to the plastic cuvette and the absorbance was measured at 585 nm using UV-visible spectrophotometer after 4 h. The results were expressed as % inhibition of DPPH radical.

2.8. Statistical Analysis

Statistical analysis was performed using one-way analysis of variance (ANOVA) using SPSS statistical software version 17.0 (SPSS Inc., Chicago, IL, USA). Bonferroni's test was used to determine significant differences at $p < 0.05$ level. Each sample was measured in triplicate and the average standard deviation for each sample was less than 5%.

3. Results and Discussion

3.1. Mechanical properties of SRC based films

Tensile strength (TS) and elongation at break (EAB) assays are to measure the ability of film resistance breakage when used as packaging materials. The value of TS indicated the sustainability of the film when applied with the maximum tensile stress. Table 1 showed the value of TS (MPa) and EAB (%) of SRC based films. SRC-control exhibited lowest TS value compared to SRC+G film and SRC+G film demonstrated the highest TS value compared to all samples with value of 58.20 MPa ($p<0.05$). This was supported by Farhan et al. [6] who revealed the adding glycerol as plasticizer improved the mechanical properties of SRC based film. They pronounced that the increase of TS value of SRC+G film could be attributed to the strong polymer-plasticizer interaction form via hydrogen bonding between SRC matrix chains and glycerol molecules [6]. Meanwhile, the incorporation of α-tocopherol with SRC+G experienced a slight decrease of TS value inversely with the concentration of α-tocopherol ($p<0.05$). These results indicated that the addition of antioxidant α-tocopherol affected the TS value of SRC films might be due to the interaction that modifies the chains between polymeric matrix of SRC with plasticizer and α-tocopherol and formed discontinuities between polymeric structure [1]. The presence of phenolic compound in α-tocopherol weakened the intermolecular interactions between SRC and glycerol resulted to the decrease of TS value. Therefore, the addition of active compounds can decrease the film strength. Martins et al. [7] also proved in their study where the TS value of chitosan based films decreased when the films incorporated with α-tocopherol compared to control film. Although many authors found similar effect on films with Tp incorporation, however the substantial decrease of TS in the active film did not cause much changes and able to use as packaging or coating systems in foods. Biopolymer films with an appropriate concentration plasticizer can be classified as having good mechanical properties under the range of 10–100 MPa of TS and EAB >10% [19]. EAB showed the maximum value of the film allowed to stretch. On the contrary, film with glycerol and addition of α-tocopherol increased the value of EAB % significantly ($p<0.05$) with increase in Tp concentration (Table 1). SRC+G+Tp0.4% film significantly demonstrated the highest value of EAB with 20% value compared to all film produced ($p<0.05$). The results showed that SRC+G with incorporation of Tp (at any concentration) exhibited good flexibility and stretch ability by presenting higher value of EAB % compared to SRC-control film ($p<0.05$). In addition, EAB value of SRC+G film was higher when compared to SRC-control film, and these results convinced that plasticizer was added to polymer to increase the flexibility of film [6]. Theoretically, film with antioxidant addition decreased the TS which would lead to the increase in EAB %, the intermolecular interaction between SRC and glycerol molecules as weakened by α-tocopherol may increase the polymer mobility. Thus, good synergic effect and compatibility SRC polymer with plasticizer and α-tocopherol can increase the mobility of polymeric structure and produce stretchable films. Ma et al. [20] found similar finding on the effect of blending curcumin as antioxidant, where tara gum acted as plasticizer with PVA film on decreasing the TS value and increasing the EAB%. To the best of our knowledge, this is the first report of SRC blending with glycerol and α-tocopherol assessed using TS and EAB%.

3.2. Opacity and solubility of films

The transparency values of all films are presented in Table 1. The lower opacity values indicate the greater transparency of the film, making the film attributes more attractive and clear. The SRC-control film showed highest non-transparency value ($p<0.05$) while adding glycerol reduced the opacity value significantly ($p<0.05$). Similar observation of OP applied to bioplastics made from 2% (w/w) SRC plasticized with glycerol was reported by Farhan et al. [6]. Opacity value increased with the increasing concentration of Tp blended with SRC+G film. These results found that Tp incorporation caused a film reduce in transparency when compared to SRC+G film. The addition of Tp resulted in decreasing of light transmission might be due to the light-scattering induced by the oily droplets of Tp in the film matrix. Consequently, light-scattering event is depending on the particle size of the dispersed phase, where the intensity of the light-scattering is affected by the amount of droplets added in solution [4]. The similar behaviour of films incorporated with oily materials was observed by others literature [4,10]. Film morphology affected the transparency and opacity traits, whereby, the presence of glycerol in films did not permit the SRC polymer to crystalize due to the action of glycerol that interferes with the ice formation by interposing itself within the water hydrogen-bond network. This was achieved by keeping

Table 1. Mechanical properties, opacity and water solubility of SRC based films.

Film Types	Mechanical Properties		Opacity	Water Solubility (%)
	TS (MPa)	EAB (%)		
SRC-control	46.63±2.88^{a}	6.54±0.27^{a}	9.03±0.25^{a}	83.40±2.05^{a}
SRC+G	58.20±2.76^{a}	14.5±0.27^{b}	6.24±0.14^{b}	74.59±1.92^{a}
SRC+G+Tp0.1%	43.88±1.17^{b}	14.7±1.70^{c}	6.88±0.13^{a}	58.61±1.68^{b}
SRC+G+Tp0.2%	42.87±0.24^{c}	16.2±1.84^{d}	8.39±0.47^{a}	52.18±3.62^{c}

SRC+G+Tp0.3%	39.33±0.13^{d}	17.7±1.70^{e}	8.70±0.60^{a}	49.69±2.46^{d}
SRC+G+Tp0.4%	38.23±1.54^{e}	20.0±0.63^{f}	8.05±1.46^{a}	48.65±2.45^{e}

the film material amorphous that permit to light pass through. Furthermore, the crystallinity observed may be attributed by the polysaccharide present in the SRC as a barrier to light, resulting in the increase of the opacity. Thus, incorporation glycerol in the SRC film has a profound impact on light transmittance and transparency of films.

Values are given as mean ± standard deviation. Different letters in the same column indicate significantly different ($p<0.05$) when analysed by Bonferroni's Test. Film solubility is used an indicator to measure film integrity, water resistance and biodegradability of films when used as packaging materials. The solubility of films was improved with the addition of plasticizer and α-tocopherol as shown in Table 1. The result demonstrated that the incorporation of α-tocopherol (at any concentration) in SRC based film significantly decreased ($p<0.05$) the solubility of films in water compared to the SRC-control film. α-tocopherol is hydrophobic in nature due to the present of aliphatic group and lower number of O-H bonds in α-tocopherol makes the film less soluble [21]. Meanwhile, control SRC film dissolved almost completely during the analysis due to the hydrophilic nature of carrageenan. Cerqueira et al. [21] has proven in their study that incorporating oil into the carrageenan lead to significant decrease ($p<0.05$) of the film solubility. This was due to the interaction between the hydroxyl groups of carrageenan chain and plant oil component decreased the availability of hydroxyl groups, thereby reducing the polysaccharides-water interactions and decreasing solubility of film in water [4]. Solubility affects the potential of biodegradable films to protect food products from moisture and contamination in the present of water. However, water solubility may be required in the consumption of food products such as prepackaged products ready for cooking. Therefore, the solubility of films is depending on the application and intended use, whereby the solubility of water also can be considered as indicator for film biodegradability. Thus, adding antioxidant, Tp enhanced the water resistance and integrity of film during contact with water.

3.3. Antioxidant activity of SRC based films incorporated with α-tocopherol

Food simulant of 90% ethanol was used and recommended according to the Commission Regulation (EU) No. 10/2011 that was assigned for food with a lipophilic character [17]. The migration of α-Tocopherol into food simulants was measured by total phenolic content (TPC) and DPPH assay and the release test was performed for 30 days. The idea was to demonstrate the possible practical application of the produced active films in the food packaging field [22]. Overall migration tests with food simulants were carried out and the results are shown in Figure 1 and Figure 2. SRC-control film with or without glycerol showed no polyphenol content whilst all films coated with α-tocopherol showed increased TPC value proportionally to the concentration of α-tocopherol ($p<0.05$). SRC+G+Tp0.4% exhibited the highest release of TPC through 30 day storages compared to all samples with value of 3.5 mg GAE/ L sample ($p<0.05$). Meanwhile, the release test of TPC showed a decline at 20 day storages onwards. These results indicated that the declining behaviour may be due to the longer storage time that affected the degradation polyphenol compounds in α-tocopherol. The degradation of polyphenol was caused by many factors, one of them was the constant release of Tp that began to oxidize the remaining Tp dissolved in 90% ethanol. Similar effect was observed by Jamshidian et al. [23] for which the release of PLA film containing TBHQ in ethanol 95% began to oxidize PLA film with TBHQ throughout storage time.

Figure 2 demonstrated migration study measured using DPPH assays of produced films through 30 days storages. Overall, the films with Tp (at any concentration) displayed increment of percentage inhibition of DPPH radicals ($p<0.05$) whereas the SRC-control and SRC+G displayed no significant difference until 30 days' storages ($p>0.05$). SRC+G+Tp0.4% film experienced highest radical scavenging activity of DPPH with value reaching 80% ($p<0.05$) at 20 days' storages due to the high content of α-tocopherol in the film. Moreover, the release trend of total phenolic contents (Figure 1) was correlated with DPPH scavenging activity to all samples. According to Rafiquzzaman et al. [24], alkaline treated carrageenan exhibited marginally antioxidant activities measured by various assays including DPPH radical while glycerol showed no study on the antioxidant activities towards DPPH radical. Chemical composition of active compounds, the interaction between polymer and active compounds, and the medium conditions may be the factors that influenced the release behaviour of active components from films to food simulant. After 30 day storages, all film still remained intact, and the polymer signals and film traits remained similar as produced observed visually.

In this study, SRC+G+Tp0.4% seems to be reasonably attractive as active film development, indicates the sustainable release kinetics of antioxidant in the food simulants during 30 day storage time. Engagement of α-tocopherol as active compounds in active packaging films was not new, since Sun et al. [14] demonstrated synergic effect α-tocopherol with LDPE film whereas Byun et al. [13] suggested addition of α-tocopherol gives dramatic increase on the antioxidant activity in PLA films development. The migration behaviour of SRC+G+Tp film into food simulant can be considered suitable for application in food packaging field. To our knowledge, no scientific work have been carried out so far on synergic of antioxidant release from α-tocopherol incorporated with SRC-film plasticised with glycerol.

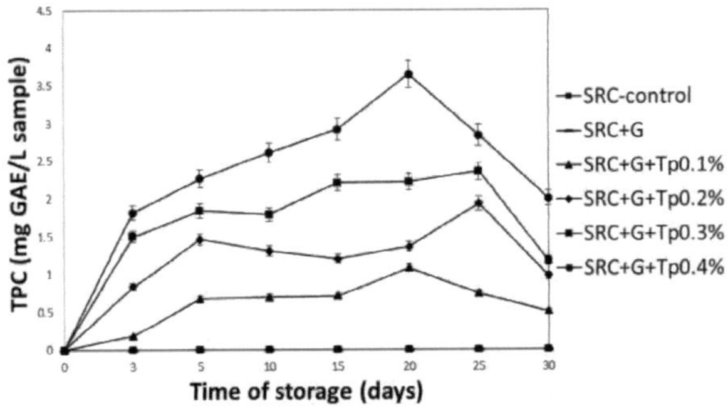

Figure 1. Changes in TPC values (mg GAE/ L sample) of produced film during 30 days storage at 25 ± 1 °C.

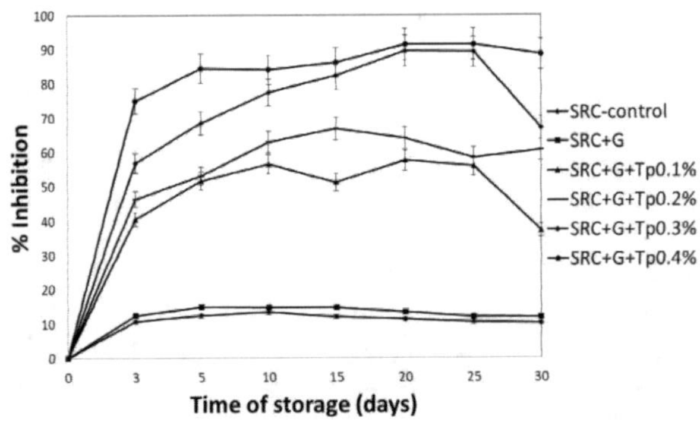

Figure 2. Changes in DPPH values (% inhibition) of produced films during 30 days storage at 4 ± 1 °C.

4. Conclusion

Incorporation of α-tocopherol in SRC plasticized with glycerol reduced slightly the tensile strength but the elongation and stretch ability of the produced film increased significantly. The solubility of film and opacity of the treated film improved significantly. The addition of α-tocopherol into SRC+G+0.4%Tp showed high antioxidant activity release to food simulant through 30 days storages may use in the food packaging application for extending shelf-life and retarding the oxidation process. In conclusion, the formulation of SRC incorporated with α-tocopherol might be an alternative degradable packaging not only that it can prolong shelf life but can also avoid synthetic preservative in contact direct with foods.

Acknowledgments
Authors would like to acknowledge University Malaysia Pahang for the facilities and funding under grant of RDU160381.

References
[1] Shen Z and Kamdem D P 2015 Development and characterization of biodegradable chitosan films containing two essential oils *Int. J. Biol. Macromol.* **74** 289–296
[2] Ma W, Tang C, Yin S W, Yang X Q, Wang Q, Liu F and Wei Z H 2012 Characterization of gelatin-based edible films incorporated with olive oil *Food Res. Int* **49** 572–579
[3] Vieira M G A, Da Silva M A, Dos Santos L O and Beppu M M 2011 Natural-based plasticizers and biopolymer films: A review *Eur. Polym. J.* **47** 254–263
[4] Shojaee-Aliabadi S, Hosseini H, Mohammadifar M A, Mohammadi A, Ghasemlou M, Hosseini S M and Khaksar R 2014 Characterization of κ-carrageenan films incorporated plant essential oils with improved antimicrobial activity *Carbohydr. Polym.* **101** 582–591
[5] Zia K M, Tabasum S, Nasif M, Sultan N, Aslam N, Noreen A and Zuber M 2017 A review on synthesis, properties and applications of natural polymer based carrageenan blends and composites *Int. J. Biol. Macromol.* **96** 282-301
[6] Farhan A and Hani N M 2017 Characterization of edible packaging films based on semi-refined kappa-carrageenan plasticized with glycerol and sorbitol *Food Hydrocoll.* **64** 48–58
[7] Martins J T, Cerqueira M A and Vicente A A 2012 Influence of α-tocopherol on physicochemical properties of chitosan-based films *Food Hydrocoll.* **27** 220–227
[8] Azman, N A M, Skowyra M, Muhammad K, Gallego MG, Almajano M P 2017 Evaluation of the antioxidant activity of Betula pendula leaves extract and its effects on model foods. *Pharm. Biol.* **55** 912-919
[9] Medina-Jaramillo C, Ochoa-Yepes O, Bernal C and Famá L 2017 Active and smart biodegradable packaging based on starch and natural extracts *Carbohydr. Polym.* **176** 187–194
[10] Atarés L and Chiralt A 2016 Essential oils as additives in biodegradable films and coatings for active food packaging *Trends Food Sci. Technol.* **48** 51–62
[11] Martucci J F, Gende L B, Neira L M and Ruseckaite R A 2015 Oregano and lavender essential oils as antioxidant and antimicrobial additives of biogenic gelatin films *Ind. Crops Prod.* **71** 205–213
[12] Jiang Q 2017 Natural Forms of Vitamin E as Effective Agents for Cancer Prevention and Therapy *Adv. Nutr.* **8** 850–867
[13] Byun Y, Kim Y T and Whiteside S 2010 Characterization of an antioxidant polylactic acid (PLA) film prepared with a-tocopherol, BHT and polyethylene glycol using film cast extruder *J. Food Eng.* **100** 239–244
[14] Sun L N, Lu L X, Qiu X L and Tang Y L 2017 Development of low-density polyethylene antioxidant active films containing α-tocopherol loaded with MCM-41(Mobil Composition of Matter No. 41) mesoporous silica *Food Control* **71** 193–199
[15] Normah O and Nazarifah I 2003 Production of semi-refined carrageenan from locally available red seaweed Eucheuma cottonii on a laboratory scale *J. Trop. Agric. Fd. Sc.* **31** 207–213
[16] ASTM D882 2000 Standard test method for tensile properties of thin plastic sheeting method D882-00 (Philadelphia: American Sociaty for Testing and Materials)
[17] López De Dicastillo C, Ares Pernas A, Castro López M D M, López Vilariño J M and González Rodríguez M V 2013 Enhancing the release of the antioxidant tocopherol from polypropylene films by incorporating the natural plasticizers lecithin, olive oil, or sunflower oil *J. Agric. Food Chem.* **61** 11848–11857
[18] Azman N, Segovia F, Martínez-Farré X, Gil E and Almajano M 2014 Screening of Antioxidant Activity of Gentian Lutea Root and Its Application in Oil-in-Water Emulsions *Antioxidants*

3 455–471
[19] Hun J H and Cennadios A 2005 *Innovative in Food Packaging* p 241–262
[20] Ma Q, Ren Y and Wang L 2017 Investigation of antioxidant activity and release kinetics of curcumin from tara gum/ polyvinyl alcohol active film *Food Hydrocoll.* **70** 286–292
[21] Cerqueira M A, Souza B W S, Teixeira J A and Vicente A A 2012 Effect of glycerol and corn oil on physicochemical properties of polysaccharide films - A comparative study *Food Hydrocoll.* **27** 175–184
[22] Yang W, Fortunati E, Dominici F, Giovanale G, Mazzaglia A, Balestra G M, Kenny J M and Puglia D 2016 Effect of cellulose and lignin on disintegration, antimicrobial and antioxidant properties of PLA active films *Int. J. Biol. Macromol.* **89** 360–368
[23] Jamshidian M, Tehrany E A and Desobry S 2012 Release of synthetic phenolic antioxidants from extruded poly lactic acid (PLA) film *Food Control* **28** 445–455
[24] Rafiquzzaman S M, Ahmed R, Lee J M, Noh G, Jo G and Kong I S 2016 Improved methods for isolation of carrageenan from Hypnea musciformis and its antioxidant activity *J. Appl. Phycol.* **28** 1265–1274

Demulsification of Crude Oil in Water (O/W) Emulsions using Graphene Oxide

Nur Hidayati Othman[1*], Ahmad Fadhil Jahari[1,2], Nur Hashimah Alias[1], Husna Hayati Jarni[1], Munawar Zaman Shahruddin[1], Mohd Fazril Irfan, 'Aqilah Dollah[1] and Nor Hadhirah Halim[2]

[1] Department of Oil and Gas Engineering, Faculty of Chemical Engineering, Universiti Teknologi Mara, 40500 Shah Alam, Selangor, Malaysia

[2] Flow Assurance and Formation Damage Group, Petronas Research Sdn. Bhd, Lot 3288 & 3289, Off Jalan Ayer Itam, Kawasan Institusi Bangi, 43000, Kajang, Selangor, Malaysia.

Abstract. Implementation of nanotechnology in oil and gas industries, particularly for enhanced oil recovery (EOR) has recently gained interest. In this work, graphene oxide (GO) has been synthesized and used as nano-demulsifier for crude oil in water (O/W) emulsions. GO was synthesized by using modified Hummers' method and characterized using X-Ray diffractometer (XRD) and Fourier Transform Infrared (FTIR) spectroscopy. Subsequently, the nano-demulsifier's performances were tested to Field D crude oil-water emulsions by varying nano-demulsifier concentrations and contact time. The performances of GO for demulsication of crude oil-in-water emulsions were then determined by measuring the oil contents in water using a UV–Vis spectrophotometer. Demulsification tests indicated that the optimum GO concentration was around 20-40 ppm, while the optimum contact time was 30 min. The residual oil content in the separated water was as low as 100 ppm. This might be due to the amphiphilic properties of GO. It indicates that the GO is a promising demulsifier and by further understanding the intrinsic interaction between GO and crude oil, the performances of GO can be easily tailored in the future.

1. Introduction

Demulsification is a complicated process as oilfield emulsions is a stable liquid-liquid colloidal suspension. However, the demulsification of O/W emulsions can be accelerated by several methods such as thermal, electrical, mechanical and filtration methods. Among those, chemical-based demulsification method is more preferred owing to the simplicity of the process. Seeking for highly efficient, rapid, universal, and low-cost demulsification materials to break the crude oil-water emulsion(O/W) and emulsified oily wastewater at ambient condition has been the goal of the petroleum industry [1].

Field D is one of the offshore field in Malaysia that carried out enhanced oil recovery (EOR) development project by using enhance water alternating gas (EWAG). The is due to the abundance of gas with adequate water and gas injection facilities. An emulsion study has been conducted on Field D where surfactant M was added. However, based on the current analysis, it was found that the level of oil in the water is still higher than the targeted value. Therefore, it is important to ensure that the concentration of oil in water is as low as possible prior discharged to wastewater treatment system at platform to avoid auxiliary process that incurred high cost.

The use of nanotechnology in the petroleum industry has gained enormous interest in the recent years whereby most investigations have shown that nanoparticles are promising for future enhanced oil recovery (EOR). Nanotechnology is now chosen as an alternative method for viscosity reduction (aluminium oxide, copper (II) oxide, iron oxide), interfacial reduction (silica oxide, polyacrylamide micro-gel nano sphere), wettability alteration (tin oxide, silica oxide, alumina coated silica) and enhance sweep efficiency (polymer coated nanoparticles) [2].

Recently, graphene oxide (GO) has find its application in separating oil from oil in water emulsion by behaving as a highly efficient demulsifier [3]. In general, GO comprises of oxygen-containing functional groups including, epoxide, hydroxyl, carbonyl and carboxyl. The presence of these oxygen-containing functional groups provided potential advantages for using GO in numerous applications. The polar oxygen functional groups in GO sheet render it strongly hydrophilic, which gave GO good dispersibility in many solvents, particularly in water. This material is hydrophilic and forms stable suspension in water and other solvents, as GO has many hydrophilic functional groups such as carboxyl and hydroxyl [3]. Due to behaving strongly hydrophilic with these properties they will consequently forms a stable oil in water emulsion.

Demulsification starts with the flocculation of water droplets that tend to clump together and forming aggregates called 'floccs'. These droplets are close enough to each other but may not coalesce as coalescence only takes place when there is a weak emulsifier film that surround the water droplets [1]. Such factors that affect the rate of flocculation; water content in the emulsion (high), temperature of the emulsion (high), viscosity of the oil (low), density difference between oil and water (high), and application of electrostatic field. Subsequently, coalescence takes place in demulsification whereby water droplets start to coalesce together to form a larger drop eventually leads to a decrease presence of water droplets until completion of this process [5].

Such factors that improvised the coalescence process; rate of flocculation (high), absence of mechanically strong films, interfacial tension (high), water cut (high), interfacial viscosity (low), temperature of the emulsion (high), and application of chemical demulsifier [6]. Sedimentation or creaming can also take place during demulsification in which either water droplets settle down in an emulsion or rising of oil droplets in the water phase. Both these two processes result from the density difference between oil and water and may not cause in destabilizing emulsion. Incomplete emulsion droplets may accumulate at the oil/ water interface and likely form an emulsion pad [6]. These emulsion pads are caused by ineffective or insufficient demulsifier, other chemical that nullify demulsifier's effect, low temperature and presence of accumulating solids. Several problems may arise due to the presence of these pads, like reduction in the retention time, increases the BS&W, increase residual oil in water, and equipment upset frequency [7]. Meanwhile, this research aims to synthesis and characterize GO by using modified Hummers' method and study its performances as nano-based demulsifier for demulsification of oil in water emulsion.

2. Methodology

2.1 Chemicals and Materials

Crude oil sample was obtained from Field D. Synthetic brine (produced and sea water) was prepared according to the Field D formulation. The chemicals used for preparing synthetic brine calcium chloride dihydrate ($CaCl_22H_2O$), magnesium chloride hexahydrate ($MgCl_2.6H_2O$), potassium chloride (KCl), barium chloride dihydrate ($BaCl_2.2H_2O$), strontium chloride hexahydrate ($SrCl_2.6H_2O$), sodium bicarbonate ($NaHCO_3$), and sodium chloride (NaCl). The chemicals used for preparing GO was graphite, sodium nitrate ($NaNO_3$), concentrated sulfuric acid (H_2SO_4), potassium permanganate ($KmNO_4$), hydrogen peroxide (H_2O_2, 30% w/w), hydrochloric acid (HCl, 37%). All chemicals were purchased from Merck, R&M Chemicals, Syterm and used without further purification.

2.2 Preparation and Characterization of Graphene Oxide(GO)

Graphene oxide (GO) was synthesized by using modified Hummers' method which involved oxidation and exfoliation of graphite [7]. The gel-like formed GO solution was dried in an oven for 24 hours at 60 °C to form GO nano-demulsifier. The GO nano-demulsifier was then characterized using Fourier transform infrared spectroscopy (FTIR, Perkin Elmer) in the range from 400 cm^{-1} to 4000 cm^{-1} to confirm the presence of oxygen-containing functional group. X-ray diffractometer (XRD, Rigaku) was also used to observe the crystal structure of GO. The emission current and accelerating voltage used were 40 mA and 40 kV. Data was collected using Cu-K radiation in the range of 2θ from 5° to 50°.

2.3 Preparation of Emulsion

Crude oil was retrieved from Field D and kept inside a sampling bottle for further laboratory analysis. The sampling bottle contained with crude oil was dipped inside a water bath at a reservoir condition temperature (54 °C). Brine solution (produced water, PW and sea water, SW) was then prepared according to the Field D formulations. A pre-determined amount of $CaCl.2H_2O$, $MgCl_2.6H_2O$, KCl, $BaCl_2.2H_2O$, $SrCl_2.6H_2O$ and NaCl were added into a beaker and continuously stirred to form a homogeneous brine solution. $NaHCO_3$ was then added into the brine solution prior to lab analysis. An O/W emulsion has been prepared using M surfactant based on the formulations used in pilot study for Field D EWAG. The emulsion test was conducted at pilot condition with 80% water cut by mixing 20 ml of crude oil with 80 ml of brine solution. The brine formulation was based on the the combination of seawater (SW) and produced water (PW) at a ratio of SW:PW = 40:60. SW was first mixed with M surfactant at a ratio of 8 ml of M to 400 ml of SW. Then, the new integrated SW with surfactant was mixed with PW to form the brine solutions, which then mixed with crude oil. All sample tubes were shake for 2 minutes and then placed in water bath for 5 minutes at 54 °C.

2.4 Demulsification Test

The concentrations of GO nano-demulsifier and contact time in O/W emulsion during demulsification tes were varied from 20 to 100 ppm and 30-240 minute, respectively. The concentration of oil in the brine solutions was then determined using a UV–Vis spectrophotometer.

3. Results and Discussions

3.1 Characterization of Graphene Oxide

XRD is used to identify crystalline phases and determine the structural properties of GO [9]. Figure 1 compares XRD spectra for graphite and GO. It can be seen that a sharp and intense peak was observed at 2θ=26.4, which indicates that graphite is a highly crystalline material. Upon oxidation and exfoliation process, the peak at 2θ= 26.4 disappeared and lower angle peak was observed at 2θ=9.58. There is a little broad at bottom of the peak which might indicates to non-uniform strain condition affect by grain size or hardness factors.

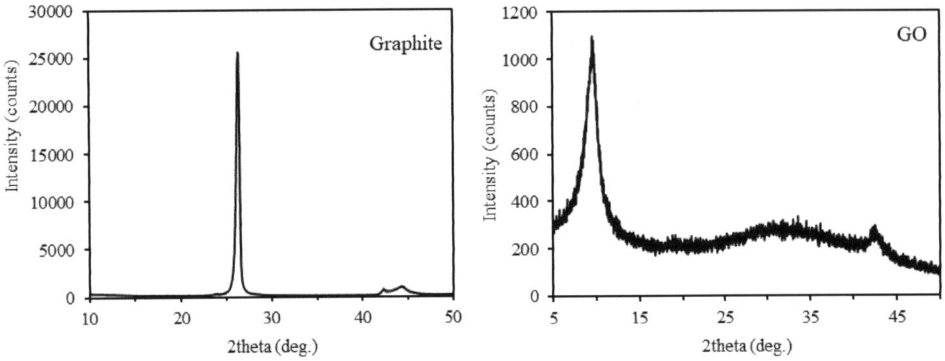

Figure 1. XRD data of graphite and GO

Figure 2 compares the FTIR spectrum for graphite and GO. It can be seen that no functional group is observed for graphite. However, upon the oxidation process of graphite, several new functional groups appeared. The broad spectrum observed at around 3178.3 cm^{-1} is due to O-H stretching vibrations, which might be attributed to adsorbed water on GO. The spectrum at 1716.96 cm^{-1} is known as the stretching vibrations of COOH group, corresponding to carbonyl and carboxyl groups, whereas the spectrum at 1622.56 cm^{-1} is attributed to in-plane vibration (C – O stretching vibration of C – OH). The spectrum at 1043 cm^{-1} is attributed to (C – O – C) stretching vibrations mode. The appearance of these functional groups confirms the formation of GO.

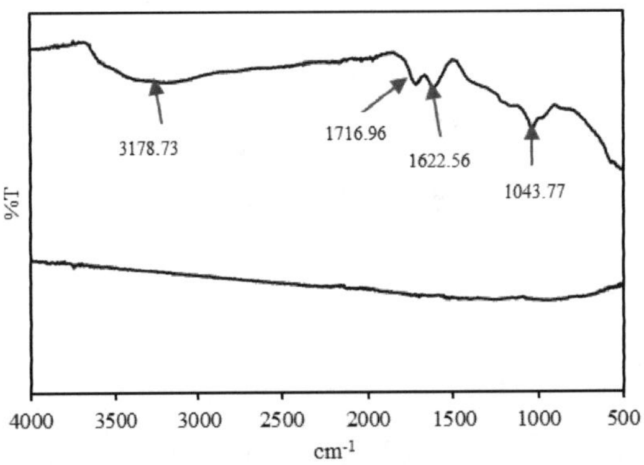

Figure 2. FTIR spectrum for graphite and GO

3.2 Performance Study of Graphene Oxide (GO) demulsifier

Figure 3 shows the demulsification performance of GO nano-demulsifier at various concentrations and contact time. It has been found that GO can effectively break the emulsion even at a very low concentration owing to the nano properties. At 20 ppm of GO concentration and 30 min of contact time, the concentration of O/W was found to be 90.5 ppm. As the contact time of GO nano-demulsifier was increased, the concentration of O/W decrease slightly. This indicates that the demulsification of O/W can be achieved at a shorter time using GO.

Interestingly, it was found that as the concentration of GO nano-demulsifier used increases to 100 ppm, the value of O/W was higher. This indicates that the adsorption of oil was lowered although more GO was used. This might be because as the amount of GO added increased, GO cannot be fully dispersed in the solution. Therefore, the contact area for demulsification to occur decreased.

Oil content in Field D's emulsion was observed to reduce significantly when GO nano-demulsifier was added, where the lowest oil content corresponding to the maximum demulsification efficiency was achieved between 20-40 ppm of GO. When the dosage of GO exceeds the threshold, the oil content in the water was higher. This probably be due to the subsequent distribution of GO dispersion after demulsification. Because GO is more distributed in the water phase, the adsorbed oil on the GO surface causes a slight increase of oil in the separated water samples.

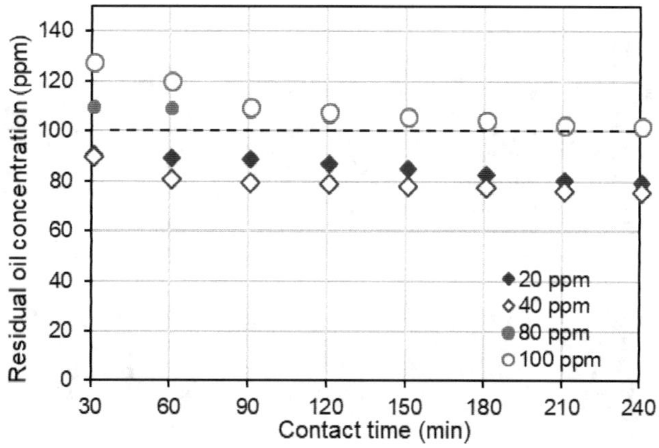

Figure 3. Residual oil content with graphene oxide as nano-demulsifier

3.3. Observation Study of Graphene Oxide

Figure 4 visualizes the demulsification study at selected contact time. It was observed that longer contact time leads to a better separation efficiency. Higher oil adsorption was observed at the beginning, where nearly more than 75% emulsion were separated during the first 30 min. By increasing the contact time to 240 min, better separation of emulsions was observed. As GO concentrations increased, high amount of solid residue was observed to settle at the bottom of separating funnel. These solid residues occurred as some part of the emulsion presence elements might adhere and attracted to GO adsorbent and eventually breaking down the emulsion into two layers.

When GO nano-demulsifier was applied into the emulsion, abundance of irregular oil floccules in the water phase with larger sizes than that in original emulsion was observed. Such findings denote that GO suspension can interact with fine oil drops in emulsion after they contact each other, eventually lead to clear colors of the emulsion just after immediate addition of GO suspension. GO dispersion promote combining of fine oil drops in emulsion to form big oil drops [14]. After completely shaking, oil floccules aggregate quickly to form accumulated floating oil phase on the water surface. The demulsification mechanism involved could be related with the presence of repulsive force that form such stability of the crude oil-water emulsion. The repulsive force normally comprising of molecules likely asphaltenes, resins, and naphthenic acid located at the interface of oil and water. In order to break the stable emulsion, such alternatives implemented should focusing on decreasing the repulsive force and destruct the viscoelastic film situated at oil-water interface [15].

IOP Conf. Series: Materials Science and Engineering **458** (2018) 012023 doi:10.1088/1757-899X/458/1/012023

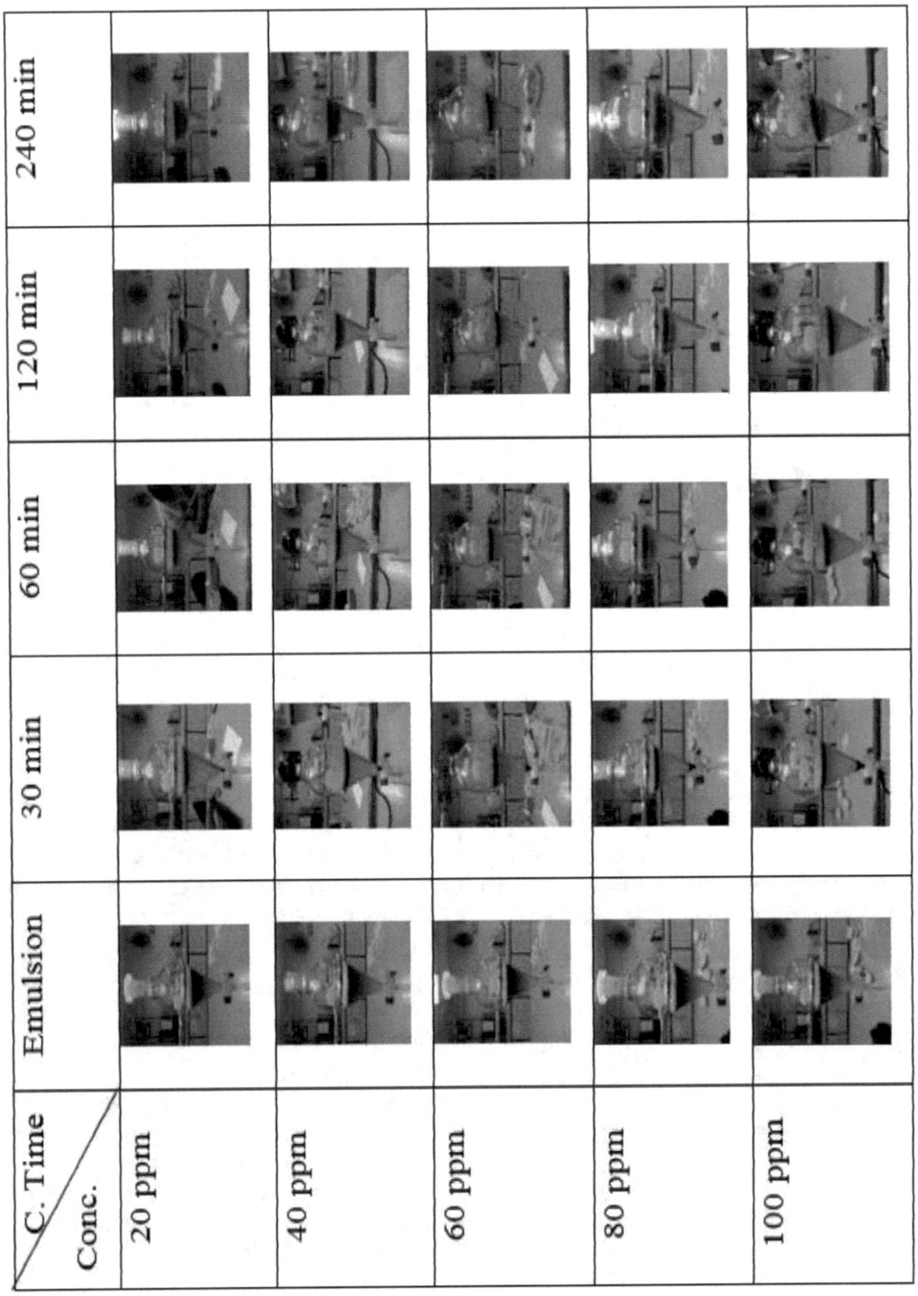

Figure 4. Visualization of O/W demulsification using various GO concentrations and contact times

4. Conclusions

From the results obtained, it has been found that GO nano-demulsifier can effectively break the crude oil in water emulsion. Lower concentration of GO and shorter contact time were preferred due to better demulsification efficiency. The residual oil content in the separated water was found to be as low as 100 ppm that might be due to the amphiphilic properties of GO. By further understanding the intrinsic interaction between GO and crude oil, the performances of GO can be easily tailored in the future to achieve the targeted value of OW content.

Acknowledgments

The authors are grateful to Universiti Teknologi MARA and Ministry of Higher Education (MOHE) for the financial support through FRGS research funding (600-IRMI/FRGS 5/3 (087/2017)).

References

[1] M. Balsamo, A. Erto, and A. Lancia, 'Chemical demulsification of model water-in-oil emulsions with low water content by means of ionic liquids', *Brazilian J. Chem. Eng.*, vol. 34, no. 1, pp. 273–282, 2017.
[2] C. Negin, S. Ali, and Q. Xie, 'Application of nanotechnology for enhancing oil recovery – A review', *Petroleum*, vol. 2, no. 4, pp. 324–333, 2016.
[3] J. Liu, X. Li, W. Jia, Z. Li, Y. Zhao, and S. Ren, 'Demulsification of Crude Oil-in-Water Emulsions Driven by Graphene Oxide Nanosheets', *Energy and Fuels*, vol. 29, no. 7, pp. 4644–4653, 2015.
[4] M. Yimei, 'Research on Demulsification Mechanism of Polymer Surfactant Flooding Produced Fluid', vol. 6, no. 1, pp. 15–21, 2014.
[5] A. A. Vallejo-Cardona *et al.*, 'Demulsification of crude oil-in-water emulsions by means of fungal spores', *PLoS One*, vol. 12, no. 2, pp. 1–17, 2017.
[6] C. Dalmazzone, C. Noik, and L. Komunjer, 'Mechanism of Crude-Oil/Water Interface Destabilization by Silicone Demulsifiers', *SPE J.*, vol. 10, no. 1, pp. 5–8, 2005.
[7] N.H. Othman, N. H., Alias, M. Z. Shahruddin, S.N. C. M. Hussein, and A. Dollah, 'Supported graphene oxide hollow fibre membrane for oily wastewater treatment', *AIP Conference Proceedings* (Vol. 1901, No. 1, p. 020008).
[8] C. Giannini, M. Ladisa, D. Altamura, D. Siliqi, T. Sibillano, and L. De Caro, 'X-ray Diffraction: A Powerful Technique for the Multiple-Length-Scale Structural Analysis of Nanomaterials', *Crystals*, vol. 6, no. 8, p. 87, 2016.

Multiwalled Carbon Nanotubes beaded with Nickel Molybdate for superior electrochemical characteristics

E R Ezeigwe [1,2], K C Ng [3], P S Khiew [2], C W Siong [4], I Kong [5], M T Tan [1,2*]

[1]Department of Electrical and Electronic Engineering, Faculty of Engineering, The University of Nottingham Malaysia Campus.
[2]Center of Nanotechnology and Advanced Materials, Faculty of Engineering, The University of Nottingham Malaysia Campus, Jalan Broga, 43500, Semenyih, Selangor, Malaysia.
[3]R&D Center, MyBig Sdn Bhd, No1. Jalan P4/7, Seksyen 4, Bandar Teknologi Kajang, 43500 Semenyih, Selangor Darul Ehsan, Malaysia.
[4]Low Dimensional Materials Research Center, Department of Physics, Faculty of Science, University of Malaya, 50603 Kuala Lumpur, Malaysia.
[5]Department of Mechanical, Materials and Manufacturing Engineering, The University of Nottingham Malaysia Campus, Jalan Broga, 43500, Semenyih, Selangor, Malaysia.

*Corresponding Author E-mail: Michelle.Tan@nottingham.edu.my

Abstract. $NiMoO_4$ were synthesised on multiwalled carbon nanotubes (MWCNTs) via a simple and environment-friendly solvothermal method. The nanostructures were interconnected forming a sponge-like network providing a large surface area easily accessible to electrons and ion during intercalation-deintercalation process. The MWCNT/$NiMoO_4$ nanocomposite exhibited a specific capacitance of 455 F/g at 4 A/g as an electrode material for an electrochemical capacitor and 66 % of its initial capacitance was retained after 1000 cycle. This study presented a research strategy for developing ternary based electrode material for electrochemical capacitors.

1. Introduction

$NiMoO_4$ as a binary transition metal oxide belonging to the metal-molybdate family has drawn advances in research due to its remarkable electrochemical activity in terms of chemical stability, abundant reserve, low cost and environmental benignity for its use as an electrode material for electrochemical capacitors [1]. Nevertheless, the poor electrical conductivity and cycling stability of metal oxides limits its specific capacitance [2]. To improve this drawback, the integration of carbon materials provide accessible electroactive sites, rapid electron-transport and an attractive synergetic characteristics which can deliver better electrochemical performance [2,3].

The open mesoporous network of carbon nanotubes (CNTs) allows easy ion diffusion to the active surface of the nanocomposite electrodes coupled with the high resiliency of the nanotubes to easily adapt to the volumetric changes during the charge and discharge process. These features of CNTs including its high conductivity, largely accessible surface area and its double layer capacitance improve the overall electrochemical performance of the CNT/metal oxide nanocomposite representing an essential factor for developing the next generation electrochemical capacitors [4]. In this study, we incorporate a facile solvothermal approach for the synthesis of the MWCNT/$NiMoO_4$ ternary nanocomposites. The structural and electrochemical properties of the MWCNT/$NiMoO_4$ metal oxide ternary composite used as electrodes for electrochemical capacitors were analysed comprehensively. Furthermore, different weight ratios of MWCNT to $NiMoO_4$ were prepared and their electrochemical performance evaluated respectively.

2. Experimental

2.1. Synthesis of the multi-walled carbon nanotubes

Using the water assisted catalytic chemical vapor deposition (CCVD) growth method, multi-walled carbon naotubes were produced in quartz tube furnace. Ethylene (40–400 sccm) was mixed hydrogen (30-150scm) before being diluted with argon (100-350scm). Water was added continuously in the chamber by refluxing Ar (5-50scm) using a water bubbler. Fe (1nm) deposited on a silicon substrate and additional Al_2O_3 (10nm) as top layer was used as the catalyst. Before the process, the furnace temperature was set at 750°C and the substrate was used for the growth of the multi-walled carbon nanotubes.

2.2. Synthesis of MWCNTs/$NiMoO_4$ ternary composites

0.29g of MWCNTs was dispersed in a solution of ethanol and water at an apt ratio (2:3) by sonication for 20 minutes. A different mixture of 0.29g of nickel (II) nitrate hexahydrate and with 0.24g of sodium molybdate was prepared in 30 ml of deionized water under stirring. The later solution was then added to the MWCNT solution under stirring for 15 minutes to achieve homogeneity. The solution was transferred into a 50 ml Teflon autoclave and heated to 160°C for 6 hours. The precipitate collected was washed thoroughly and dried overnight. The dried composites were calcined at 400°C for 4 hours. Ratios (1:1, 1:4, 1:8 and 1:16) of MWCNTs to $NiMoO_4$ were prepared and labeled MWCNT-NM, MWCNT-4NM, MWCNT-8NM and MWCNT-16NM respectively

2.3. Characterization

The X-ray diffraction (XRD, Bruker D4) pattern of the nanocomposites are recorded from 10 to 70 ° using Cu-Kα radiation (λ=1.54056 Å) with step size 0.03. The Raman spectra were measured on a Raman spectrometer (Renishaw inVia Raman microscope) with a diode laser excitation source of 532 nm and a CCD detector. The surface characteristics of the nanocomposites were investigated using a Field electron scanning electron microscope (FE-SEM, FEI Quanta-400 FE-SEM) with an EDX analyser and a (TEM, JEOL JEM-2100F) at an operating voltage of 200 kV.

The working electrodes were fabricated from a mixture of 70 wt% MWCNT/$NiMoO_4$ composites, 20 wt% carbon black and 10 wt% polyvinylidene fluoride (PVDF). The electrodes weighting about 1 mg were tested in a two electrode cell using 2 M KOH solution on an Arbin Instrument (BT-2000) and Metrohm Autolab (PGSTAT302F) electrochemical working stations.

3. Results and discussion

The diffraction peaks of MWCNT/$NiMoO_4$ and $NiMoO_4$ (Figure 1) can be indexed to α-phase $NiMoO_4$ with JCPDS card no 086-0361 [5]. The peak at about 26.6° and 44.4° observed in the CNT/$NiMoO_4$ composites can be attributed to the (002) and (101) reflection of carbon respectively. These peaks almost overlaps with the (-112) and (330) peaks of $NiMoO_4$ at a low concentration of MWCNT which makes it challenging to identify.

Energy dispersive X-ray (EDX) analysis depicts the element signals of C, Ni, Mo, and O seen in the spectrum (Figure 2a). The element Si is due to the silicon substrate used during characterization. The Raman spectrum of MWCNT/$NiMoO_4$ composites (Figure 2b) displays the typical carbon peaks located at the 1346 cm^{-1} (D band), 1574 cm^{-1} (G band) and 2688 cm^{-1} (2D band) respectively [6]. The other peaks at 946 , 899, 816, 701, 382 and 245 cm^{-1} respectively can be attributed to the α-phase $NiMoO_4$ as reported in literatures [7].

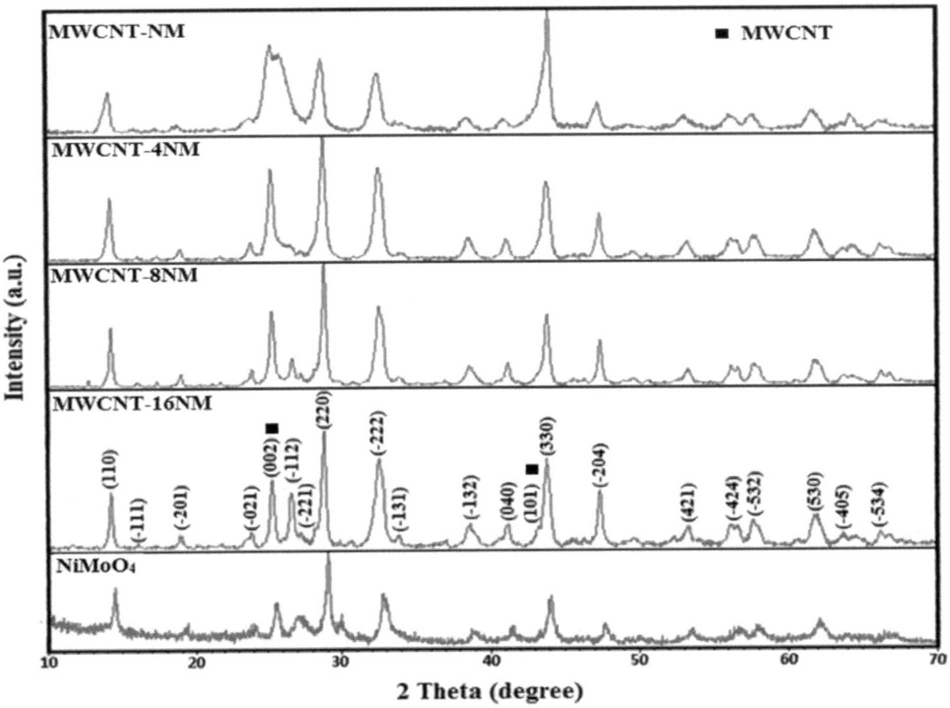

Figure 1: XRD pattern of $NiMoO_4$ and CNT-$NiMoO_4$ composites.

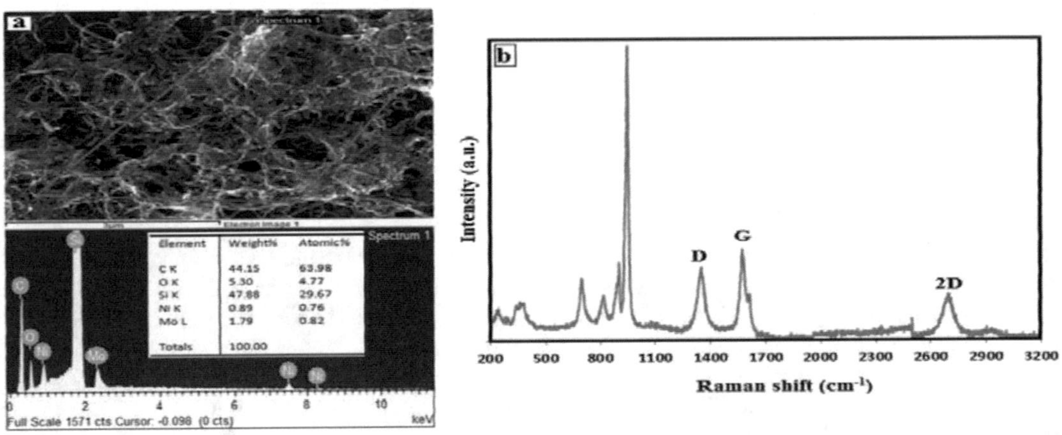

Figure 2: Images of (a) EDX spectrum and (b) Raman spectrum of MWCNT-$NiMoO_4$ composites respectively.

The morphology of the CNTs, the CNT/$NiMoO_4$ composites was investigated by FE-SEM and TEM as shown in figure 3. It can be observed in Figure 3a that the diameter of the MWCNTs is about 30 to 50 nm. Figure 3 (b-e), confirms that the solvothermal synthesis was successfully carried out with the $NiMoO_4$ rods inter connected with MWCNTs forming a sponge-like structure. The concentric layers can be easily in the HRTEM image (figure 3f) and with lattice fringes of 0.34 nm correlating with the (220) reflection of $NiMoO_4$

in the XRD study. Figure 4a depicts the cyclic voltammograms (CV) curves of the MWCNT-$NiMoO_4$ composites and $NiMoO_4$ electrodes at a scan rate of 30 mV/s^{-1}. The redox peaks seen in each curve can be attributed to the reversible faradaic process of Ni^{2+}/Ni^{3+} [8]. Compared with $NiMoO_4$, the nanocomposite electrodes exhibited higher integrated area under the curve implying better charge transfer due to the double layer capacitance and pseudocapacitance resulting from the MWCNT and $NiMoO_4$ respectively. At different scan rates of the MWCNT-4NM electrode (figure 4b), the shape of the curves were not significantly influenced indicating good rate capability and reversibility of the electrode. Charge/discharge analysis were performed on the electrodes at current density of 4 A/g. The MWCNT-4NM electrode displayed the longest discharge time signifying that the electrode having the best charge storage ability and this evident of the specific capacitances calculated from the discharge curves (figure 4c) to be $NiMoO_4$ (166 F/g), MWCNT-NM (280 F/g), MWCNT-4NM (455 F/g), MWCNT-8NM (397 F/g) and MWCNT-16NM (267 F/g) respectively. Notably, the highest capacitance was achieved via the optimal mixture of MWCNT: NiMoO4 = 1:4, in which this combination provided the best synergistic effect between the high electrical conductivity of CNT and the high pseudocapacitance of the metal oxide.

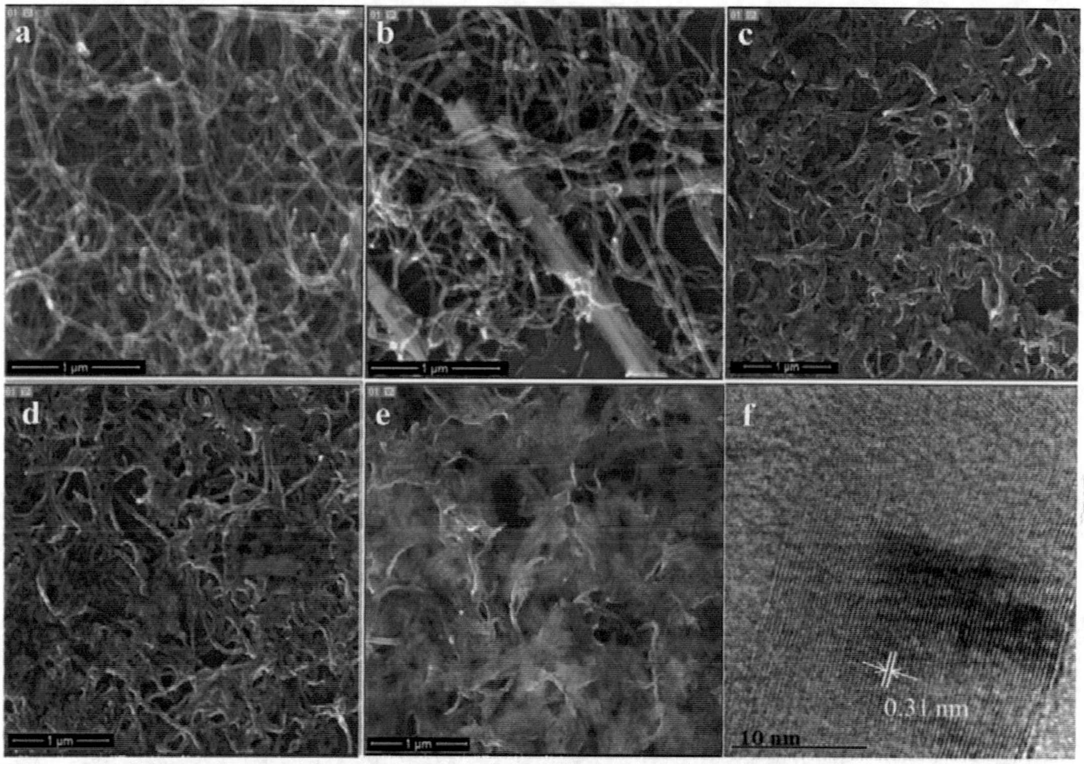

Figure 3: FE-SEM images of (a) MWCNT, (b) MWCNT-NM, (c) MWCNT-4NM, (d) MWCNT-8NM (e) (e) CNT-16NM and (f) HRTEM image of CNT-$NiMoO_4$ nanocomposites.

At different current densities (figure 4d), there were no immense changes in the charge/discharge curves of the MWCNT-4NM electrode implying good rate capability and reversibility as seen in figure 4b. The stability of the MWCNT-4NM electrode was measured at a current density of 3 A/g between 0 – 1 V (figure 4e) and 66 % of its initial capacitance were retained after 2000 cycles. Figure 4f displays the EIS of the composite electrodes; all the curves exhibited a semi-circle at high frequency and a linear component at low-frequency. The semi-circle indicates low charge transfer resistance (R_{ct}); MWCNT-4NM electrode having the smallest (R_{ct}) is in agreement with other electrochemical performances recorded. This performance is mainly due to the large surface area exploited by the $NiMoO_4$ pseudocapacitance and MWCNT double-layer capacitance [9].

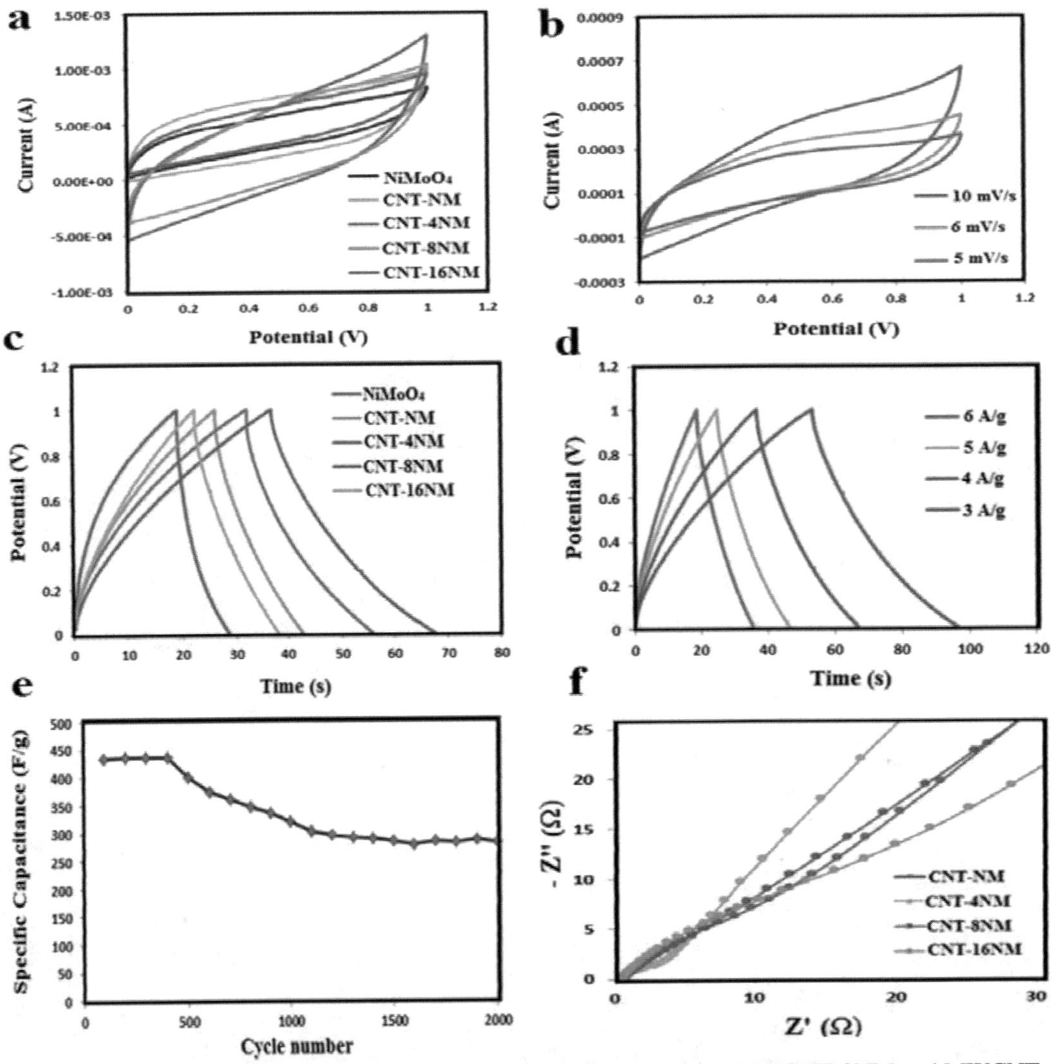

Figure 4: CV curves of (a) $NiMoO_4$, MWCNT-NM, MWCNT-4NM, MWCNT-8NM and MWCNT-16NM at a scan rate of 30 mV/s, (b) MWCNT-4NM electrode at different scan rates ranging from 5-10 mV/s, (c) Charge-discharge curves of $NiMoO_4$, MWCNT-NM, MWCNT-4NM, MWCNT-8NM and MWCNT-16NM at a current density of 4 A/g (d) Charge-discharge curves of MWCNT-4NM at different current densities (e) cycle life of the MWCNT-4NM electrode at a current density of 3 A/g (f) Nyquist plot of the MWCNT-NM, MWCNT-4NM, MWCNT-8NM and MWCNT-16NM composites electrodes.

4. Conclusion

MWCNT-$NiMoO_4$ nanocomposites were successfully synthesised by a facile solvothermal method. The combination of carbon nanotubes and metal oxides complements each other in terms of conductivity and pseudocapacitance. The MWCNT-$NiMoO_4$ composite electrodes exhibited capacitance of 455 F/g at 3 A/g with excellent rate capability. This strategy provides a promising electrode material for electrochemical capacitors.

Acknowledgements
The authors would like to thank the financial support from the Fundamental Research Grant Scheme (Project Codes: FRGS/1/2016/STG02/UNIM/02/1 and FRGS/2/2014/SG06/UNIM/03/1) by Ministry of Higher Education Malaysia and The University of Nottingham.

References
[1] Xiao K, Xia L, Liu G, Wang S, Ding L X, and Wang H 2015 *J. Mater. Chem. A* **3** 6128–35.
[2] Guo D, Luo Y, Yu X, Li Q, and Wang T 2014 *Nano Energy* **8** 174–182.
[3] Liu T, Chai H, Jia D, Su Y, Wang T, and Zhou W 2015 *Electrochim. Acta* **180** 998–1006.
[4] Pan H, Li J, and Feng Y 2010 *Nanoscale Res. Lett.* **5** 654–668.
[5] Jothi P R, Kannan S, and Velayutham G 2015 *J. Power Sources* **277** 350–9.
[6] Xia H, Wang Y, Lin J, and Lu L 2012 *Nanoscale Res. Lett.* **7** 33
[7] Abdel-Dayem H M 2007 *Ind. Eng. Chem. Res.* **46** 2466–72.
[8] Wan H, Jiang J, Ji X, Miao L, Zhang L, Xu K, Chen H, and Ruan Y 2013 *Mater. Lett.* **108** 164–7.
[9] He G, Wang L, Chen H, Sun X, and Wang X 2013 *Mater. Lett.* **98** 164–7.

IOP Conf. Series: Materials Science and Engineering **458** (2018) 012025 doi:10.1088/1757-899X/458/1/012025

Sequence analysis and comparative modelling of nucleocapsid protein from *pseudomonas stutzeri*

S Omar, F Mohd Tap, K Shameli, R Rasit Ali, NW Che Jusoh and N B Ahmad Khairudin*

Chemical Energy Conversion and Application iKohza, Department of Chemical Process Engineering, Malaysia-Japan International Institute of Technology, Universiti Teknologi Malaysia, Jalan Sultan Yahya Petra, Kuala Lumpur 54100, Malaysia

*r-bahiah@utm.my

Abstract. Protein structure can be determined by either experimental method or by computational prediction, which commonly known as homology modelling. The objective of this study is to predict the structure of protein from *Pseudomonas stutzeri* using homology modelling. The suitable template was identified with 80.35% of sequence identity determined by sequence alignment. MODELLER was used to predict the model using the method of satisfaction of spatial restraints. The model was then analysed by computational analysis tools such as Ramachandran`s Plot, Errat, Verify3D Profile evaluation, Secondary Structure Prediction Analysis and Protein Binding Site (ProBiS). The results suggest that the model is reliable and has good stereochemical properties.

1. Introduction

The purpose of homology modelling is to predict and develop three-dimensional structure of unknown protein based on the similar protein structure with an accuracy that could be equivalent to the results achieved from the experiment. Since, the experiment methods using NMR analysis and X-ray diffraction are time consuming, costly and hard to perform, it is more convenient to apply comparative modelling. Besides, the structural genomics prediction can use low resolution structures provided by homology modelling methods to find the target protein structure. According to Protein Data Bank (PDB) [1], the total amount of protein structure that can be accessed is about 125,000 structure in 2016, which year by year, the number of new found of protein structure escalate significantly. It is a big leapt in the world of protein in having computational prediction of 3D structure to determine the proteins function and their significant [2]. By the comparative modelling as alternative method, the protein prediction structure become easier and the known structure rapidly progress compare to before this method been introduced. The objective of this study is to generate and predict a tertiary model of nucleocapsid protein by homology modelling using method of spatial restraints.

2. Methods

The potential template was identified using BLAST [3]. The template was chosen based on the homologous of the known protein structure and their similarities must exceed more than 30% to avoid alignment error. Sequence alignment between the template and the target sequence was performed using UCSF Chimera program [4]. MODELLER program was used to model the structure by using method of spatial restraints [5]. It uses either distances or optimization techniques to satisfy the spatial restrains. The model was analysed based on Ramachandran Plot and protein tertiary structure environment.

3. Results and Discussion

Position-Specific Iterated BLAST (PSI-BLAST) is another method to search for the template. It is an extended program that can search in more wide scope compared to BLAST. Furthermore, it can help to find more distant relative sequences by searching with a custom position-specific scoring matrix (PSSM). Therefore, in this study, PSI-BLAST method is chosen to get a better template for protein *Pseudomonas stutzeri* (R4RRT6). Protein 1EX9 has higher percentage of identities that is 80.35%, compared to 4HS9 which is only 44.44%, it is strongly preferable to use protein 1EX9 as a template to create the 3D homology modelling for the target protein. However, this study employs multiple sequence alignment as shown in figure 1 that involved both template proteins to gain better model.

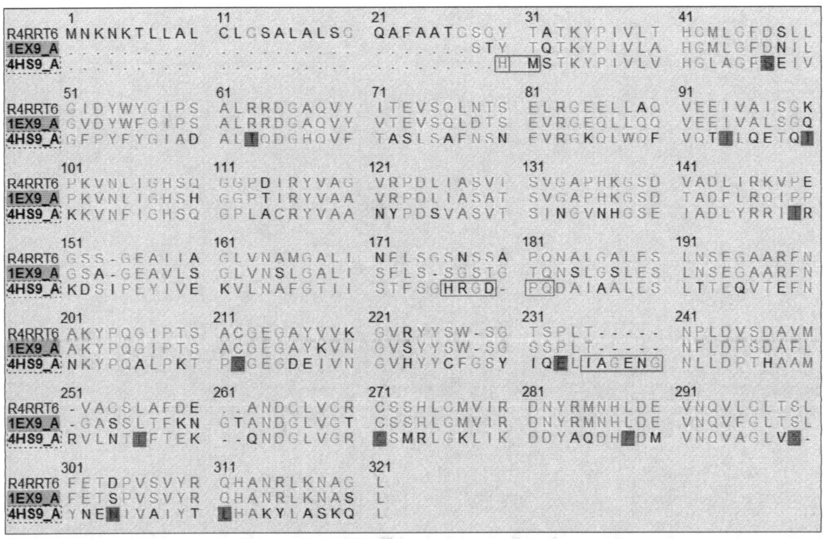

Figure 1. Multiple Sequence Alignment of *Pseudomonas stutzeri* (R4RRT6) with 1EX9 and 4HS9.

For the structure of protein R4RRT6, it was constructed using MODELLER version 9.18 which generated 20 models. The top three models (Model 1, Model 2 and Model 20) that had the lowest energy values were selected for further analyses. Objective function is the free energy inside of the model structure that had been generated. High objective function means that it has higher free energy which cause a bumping effect that happens between molecules in the model and consequently lowered the stability of a model. So, it is more preferable to choose a model with lower energy value for further analysis as it has high stability and no bumping effect.

Figure 2 shows the tertiary structure prediction for protein R4RRT6 based on the template 1EX9. The secondary structure prediction analysis was done using Polyview program and the result is presented in figure 3. In the figure, the red line represents helices, the blue line represents the coils or loops while the green line represents beta strands. It was shown that the model contains eight helices. Table 1 shows the different percentage for alpha-helix, beta-sheet and random coil for each structures of the model. In this table, Model 2 has been analysed that the number of percentage for secondary protein structure which is the α-helix, β-sheet and random coil is about the same with Model 20 which then followed with Model 1.

Table 1. Percentage of Secondary Protein Structure for the Comparison on Model 1, Model 2 and Model 20 of Protein Structure.

Secondary Protein Structure	Model 1	Model 2	Model 20
α-helix	36.27%	38.38%	38.73%
β- sheet	11.27%	14.44%	14.79%
Random Coil	52.46%	47.18%	46.48%

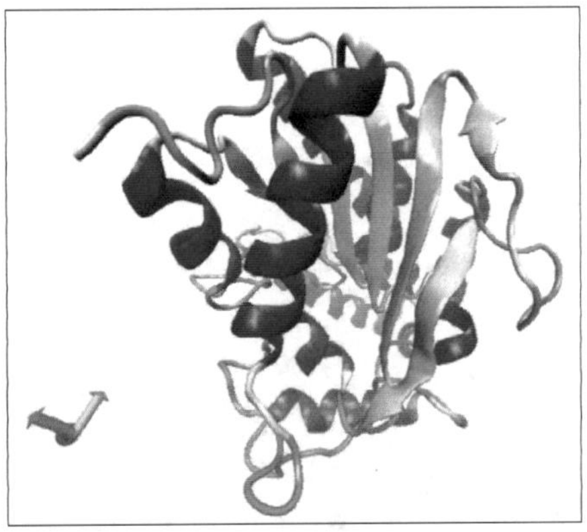

Figure 2. Three Dimensional Structure of Protein for *Pseudomonas stutzeri* (R4RRT6)

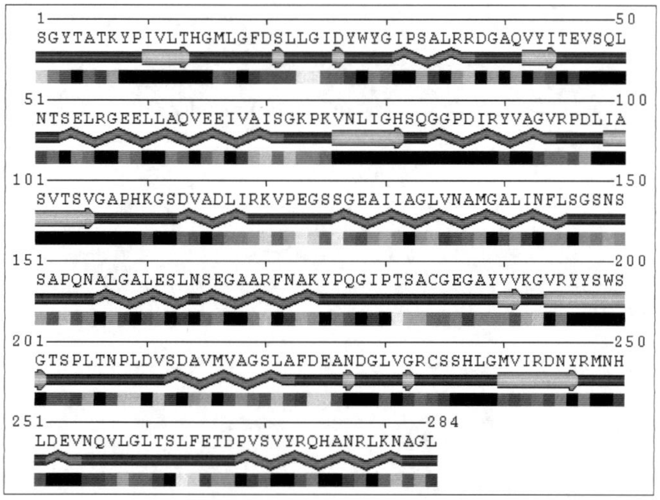

Figure 3. Secondary Structure Prediction

The first analysis on assessing the quality of the model was done using ERRAT [6]. It is beneficial to assess the arrangement of different types of atoms in each protein models. ERRAT is an "overall quality factor" for non-bonded atomic interactions which the higher scores mean higher quality. Normally, the accepted range is more than 50 for a high quality model. In this current case, the Errat score for the model is 94.928. Figure 4 shows Errat plot for Model 2. Since, the overall quality factor is 94.928, it is proven that the tertiary structure of protein R4RRT6 that been generated is a high quality model.

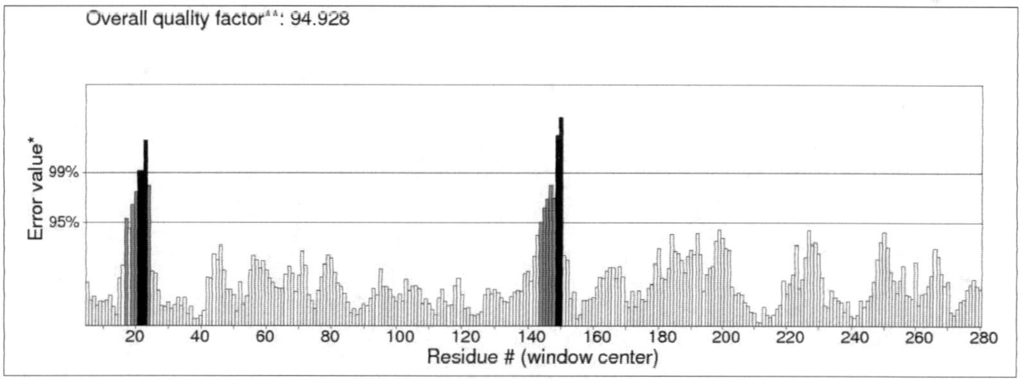

Figure 4. Errat Plot for Model 2

Results from Ramachandran Plot analysis shows that 96.1% of the residues were located in the most favorable core region and 0.0% was in the disallowed region. This result can be considered as significant because of the high percentage of residues in favoured region (>90%). Table 2 shows the summary of the result when comparing the tertiary structure between Model 1, Model 2 and Model 20. Figure 5 shows the Ramachandran Plot for the best model which is Model 2.

IOP Conf. Series: Materials Science and Engineering **458** (2018) 012025 doi:10.1088/1757-899X/458/1/012025

Table 2. Percentage of Amino Acid Residues in Each Region for Ramachandran Plot

Model	Ramachandran Plot (%)			
	Core	Allowed	Generous	Disallowed
1	96.5	2.8	0.7	0.0
2	96.1	3.2	0.7	0.0
20	95.4	3.9	0.7	0.0

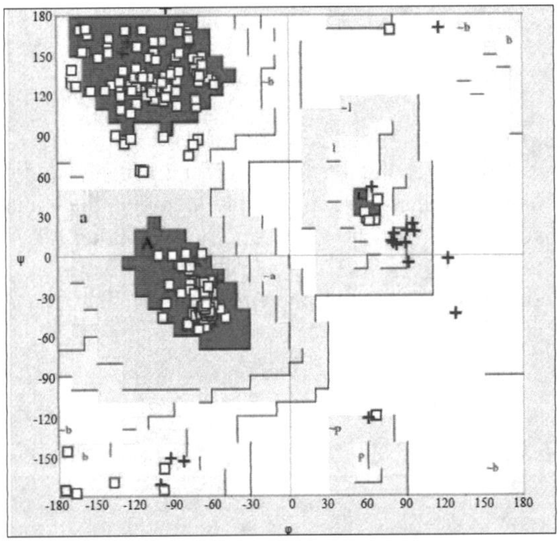

Figure 5. Ramachandran Plot of Predicted Structure of R4RRT6 (model 2)

Figure 6 shows the result of Verify 3D Profile for Model 2 [7]. In the figure, all of the plotted lines were in the allowed range. Therefore, it was indicated that each of the model structures obtained represents a high quality model. The score was represented by vertical axis while horizontal axis represents the protein`s residues [Rosetti et al., 2008]. Verify 3D Profile assessment is used to analyse the compatibility of an atomic model (3D) with their own amino acid sequences. Each residue is assigned a structural class which based on their location and environment (alpha, beta, loop, polar, non-polar and others). Furthermore, in the structural class, a high resolution structure is used to obtain score for each of the 20 amino acids. The scores range from -1 (bad score) to +1 (good score).

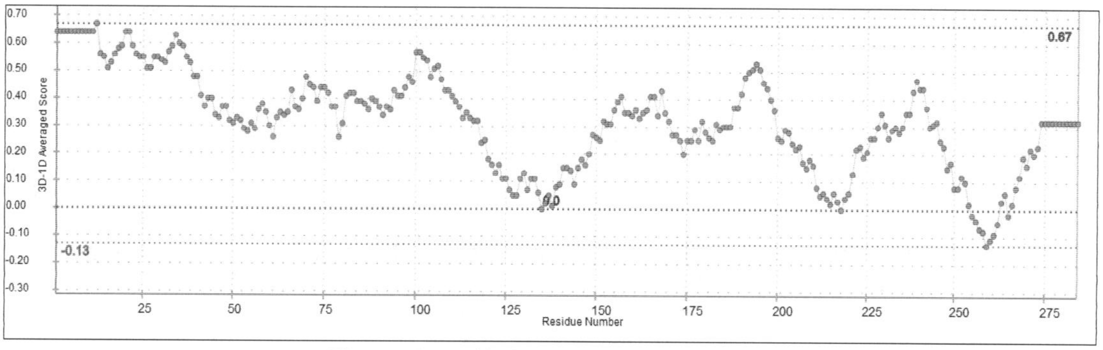

Figure 6. Verify 3D Result for Model 2

The protein binding site analysis was completed using ProBiS which is commonly use to detect structures which have similar binding sites in proteins and local pairwise alignment of crystallographically or NMR determined protein structure from the PDB. Binding site is a location where the protein binds to a ligand. Since only a few residues actually participate in binding the ligand, the other residues usually act as a framework to provide correct conformation and orientation. Figure 7 shows the proposed binding site for the nucleocapsid protein from *Pseudomonas stutzeri* (R4RRT6).

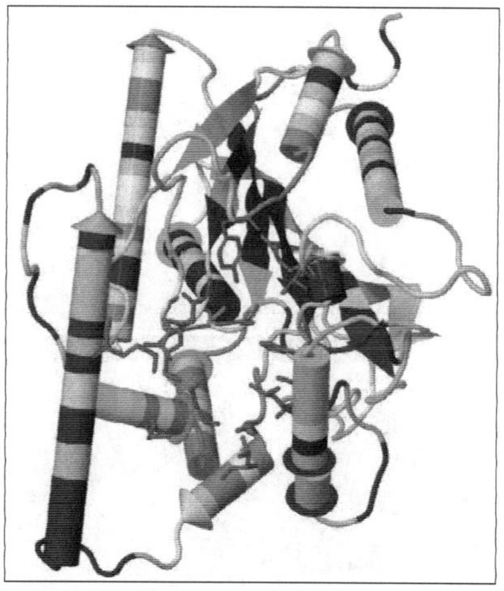

Figure 7. The Predicted Binding Site (Red Region) for Nucleocapsid Protein from *Pseudomonas stutzeri* (R4RRT6)

4. Conclusion

In this study, homology modelling is used to generate 3D structure of protein R4RRT6 by satisfying the spatial restraints of MODELLER program. The suitable template for protein R4RRT6 was selected from the result of multiple sequence alignment which was 1EX9 with sequence identity of 80.35%. Various tools were used to refine and assess the quality of the predicted model.

5. Acknowledgement
The authors would like to thank Universiti Teknologi Malaysia for providing the scholarship and financial support for this project (GUP Vot No: 19H73).

6. References
[1] Berman HM, Westbrook J, Feng Z, Gilliland G, Bhat TN, Weissig H, Shindyalov IN and Bourne PE 2000 *Nuc. Acids Res* **28** 235

[2] Schoenrock A, Samanfar B, Pitre S, Hooshyar M, Jin K, Phillips CA, Wang H, Phanse S, Omidi K and Gui Y 2014 *BMC Bioinformatics* **15** 383.

[3] Altschul SF, Gish W, Miller W, Myers EW and Lipman DJ 1990 *J. Mol. Biol.* **215** 403

[4] Pettersen EF, Goddard TD, Huang CC, Couch GS, Greenblatt DM, Meng EC and Ferrin TE, 2004 *J. Comput. Chem.* **25** 1605

[5] Sali A and Blundell TL1993 J. *Mol. Biol.* **234** 779

[6] MacArthur A 1994 Curr. *Opin. Struct. Biol.* **4** 731

[7] Bowie JU, Lüthy R and Eisenberg D 1991 *Science* **12** 164

IOP Conf. Series: Materials Science and Engineering **458** (2018) 012026 doi:10.1088/1757-899X/458/1/012026

Characterisation and mechanical properties of unsaturated polyester/acrylated epoxidised palm oil polymer blend at different acrylated epoxidised palm oil processing method

N B Thalib, S N H Mustapha[*] and C K Feng

Faculty of Industrial Sciences & Technology, Universiti Malaysia Pahang, Lebuhraya Tun Razak, 26300, Gambang, Kuantan, Pahang, Malaysia.

[*]snhidayah@ump.edu.my

Abstract. Acrylated epoxidized palm oil (AEPO) prepared by the reaction of epoxidized palm oil (EPO) with acrylic acid, triethylamine initiator and hydroquinone inhibitor prepared at various temperature and reaction times. The unsaturated polyester (UPE) resin was blended thoroughly with 10 wt% of the modified palm oil at room temperature, and thermally cured at 100 °C and post-cured at 160°C. The prepared palm oil derivative was characterised by using Fourier Transform Infrared (FTIR) and the test proved the formation of acrylate functional group with characteristic absorbance at 1600-1600 cm^{-1} that indicates C=C functional group of the acrylate. The polymer blend tensile strength of the palm oil acrylate processed at 110°C for 16 hours (M2) showed reduction from 25.56 MP to 13.26 MP (48%), reduction in tensile modulus from 1.47 GPa to 1.18 GPa (19.7%) and the highest impact strength among the four acrylates of polymer blend is 21.97 J/m from palm oil acrylate processed at 105°C for 16 hours (M3). However, polymer blend from M2 acrylate showed increased in elongation at break from 2.805 % to 3.493 % (24.5 %) compared to neat unsaturated polyester and it outweighed reduction in tensile and impact of the polymer blend.

1. Introduction

Polyester is the most versatile synthetic copolymer known nowadays [1]. Polyesters are widely used in industry as plasticizers, laminates, industrial constructions, insulations, molding compounds, tanks, boat building, truck roof and other field [2,3]. The vast applications of polyester are due to the low cost, ease excess and ease combination of polyester with reinforcement materials [2,4]. Among well known polyester is unsaturated polyester resin (UPE). The unsaturated (C=C) functional groups in the polyester chain acting as cross-linking site which can be chemically interacted with other reinforcement additive or crosslinking agent.

Petroleum derived from unsaturated polyester resin has been an interest in industry since last four decades due to its low cost, ease of handling and have good mechanical and thermal properties [5]. The continuous reliance on unsaturated polyester resin (UPE) rises the awareness on continuation of depleting petroleum reserved [4]. Synthetic unsaturated polyester resin (UPE) which synthesized from petroleum source is known with its excellent performance in many applications. One way to overcome the issue on petroleum resources without sacrificing polymer properties is by blending the synthetic polymer with plant oil [4]. Plant oils have become a significant material for both academic and industrial practitioners in intensive research as replacement to synthetic polymers in polymeric materials. The rise of awareness in benefiting vegetable oil is due to its availability, inherent biodegradability and low price [6, 7]. Among research in vegetable oil / thermoset, blending includes epoxy with waste vegetable oil [8], UPE blends with epoxidized methyl linseedate [9], vinyl ester blends with acrylated epoxidized palm oil [10] and unsaturated polyester resin (UPE) blends with epoxidized palm oil [11]. One third of vegetable oil is covered by palm oil [12]. Malaysia is the second largest palm oil producer in the world as has covered 39% of world palm oil production and 44% of world exports [13]. Palm oil is categorized as non-drying oil due to lower unsaturation level in its fatty acid chain. It restricted the use of palm oil as main constituent in synthesizing polymer resin.

In order to produce new unsaturated polyester resin (UPE) resin using palm oil blend, the palm oil fatty acid chain needs to be modified to allow the chemical interaction within the synthetic unsaturated polyester resin. Acrylated epoxidized palm oil, styrene monomer and unsaturated polyester resin (UPE) share the same property with the presence of C=C functional group. The functional group is essential in cross-link the individual entities in the polymer to become one large continuous polymer chain. A continuous polymer chain allows propagation of energy along chain, thus expected to improve the mechanical and thermal properties of the blend polymer. Therefore, a specific method to synthesis acrylated epoxidized palm oil (AEPO) needed to be investigated so that maximum acrylate functional groups will be introduced in the palm oil. The additional rate of acrylic acid and reaction temperature must be controlled during preparation of epoxy acrylate adducts [15]. A study has been conducted on acrylation of epoxidized soy bean oil (AESO) at 110°C and was found that the lowest acid value in the prepared AESO was observed after 36 hours reaction [15]. Fakhari et al. studied on mechanical properties of unsaturated polyester resin/acrylated epoxidized palm oil (UPE/AEPO) blend and conducted acrylation of epoxidized palm oil at 80°C at 16 hours reaction [16]. Later, Salih et al. [17] conducted acrylation of epoxidized palm oil at 110°C at 16 hours reaction and characterized the UV cured acrylated epoxidized palm oil (AEPO) using FTIR and NMR analysis. Current limitation of the studies is continuation to the investigations on the effect of different acrylated epoxidized palm oil (AEPO) processing conditions on properties of acrylated epoxidized palm oil (AEPO) can be done to find the best synthesis methods to get the best and balance of unsaturated polyester resin/ acrylated epoxidized palm oil (UPE/AEPO) blending in term of mechanical and thermal behavior.

This study aimed at developing and characterising bio-based polymer blend of unsaturated polyester resin with acrylated epoxidized palm oil. Acrylated epoxidized palm oil (AEPO) processing method was studied by varying the temperature, processing period and washing methods in order to determine optimum temperature and processing period for ideal use of energy in producing environment friendly polymer without sacrificing the mechanical properties of the material. This works is aimed to demonstrate that this synergistic balance can be achieved. Different process of producing acrylated epoxidized palm oil (AEPO) were conducted and being characterized using FTIR test. The synthesized acrylated epoxidized palm oil (AEPO) was blended in unsaturated polyester resin (UPE) at constant amount, 10 wt% and cured in an oven. The mechanical properties of the polymer were then conducted through standard test. The results provide a benchmark to identify the best processing condition of acrylated epoxidized palm oil (AEPO) in terms of temperature, reaction period and washing techniques to the novel, efficient and multifunctional unsaturated polyester resin/ acrylated epoxidized palm oil (UPE/AEPO) blend can be designed.

2. Materials and method

2.1. Materials

The unsaturated polyester resin (UPE) graded as Reversol P-9539 used as received from Impian Zente Sdn. Bhd. Epoxidised palm oil 031 (EPO 031) received from Budi Oil Sdn. Bhd. Acrylic acid, purchased from Merck, Germany. Triethylamine, supplied by Aldrich, UK. Hydroquinone, purchased from Nacalai Tesque, Japan. Benzoyl peroxide, purchased from Merck, Germany. Diethyl ether, received from R&M Chemicals.

2.2. Preparation of acrylated epoxidized palm oil (AEPO)

Epoxidised palm oil (EPO) was measured and placed in a beaker with a magnetic stirrer. About 1% of triethyl amine and hydroquinone was then added to the beaker and was stirred for 1 hour with a speed of 350 rpm at 40°C. Under continuous stirring, acrylic acid with 1:9 ratio to EPO was added to the beaker dropwise and the temperature of the solution was gradually raised. The slow addition of acrylic acid is to control the reaction progress and to avoid homo-polymerization of the acrylic acid [15]. The mixture was stirred at different period depending on the batches. The resulting AEPO then was washed with diethyl ether, distilled water or by both solvents. The washed AEPO then was dried using rotary evaporator at 95°C until a constant weight of the flask was recorded.

Four batches of AEPO was prepared with each batch had different condition in processing the method. The first batch, M1 AEPO was synthesized at 80°C with 350 rpm stirring for 16 hours before it was washed with diethyl ether and distilled water. The second batch, M2 AEPO was synthesised at 110°C with 350 rpm stirring for 16 hours and washed with distilled water. The third batch, M3 AEPO was synthesised at 105°C with 350 rpm stirring for 16 hours and washed with diethyl ether and distilled water. The fourth batch, M4 AEPO was synthesised at 110°C with 350 rpm stirring for 36 hours and washed with distilled water.

2.3. Preparation of the UPE/AEPO blend

Total volume of 363 ml resin was measured. The initiator, benzoyl peroxide was measured by 1.5 phr and was added to the resin. The mixture was then manually stirred for 20 minutes. For the polymer blend, 10 wt% of the resin was replaced by the prepared AEPO. Four different AEPO were used which is M1 AEPO, M2 AEPO, M3 AEPO and M4 AEPO with densities of 0.98 g/ml, 0.802 g/ml, 1.028 g/ml and 0.836 g/ml, respectively. The AEPO then was blend with UPE and were manually stirred for 20 minutes. The polymer blend was further stirred for 20 minutes after added with benzoyl peroxide. The polymer blend was then poured in aluminium mould and cured at 100°C for 2 hours and post cured at 160°C for another 2 hours.

2.4. FTIR analysis

Fourier Transform Infra-red (FTIR) spectroscopy was used to determine the functional groups in EPO and AEPO and was investigated by FTIR Spectroscopy (Perkin Elmer Spectrum 100) using attenuated total refraction (ATR). The FTIR spectra were recorded in the wavenumber range 700 to 4000 cm-1 with 50 – 100 % recorded transmission.

2.5. Tensile test

The tensile strength, E-modulus, elongation at break and tensile toughness was tested using Universal Testing Machine (UTM) following the procedure of ASTM D-638. The machine (INSTRON 3367 Universal Testing) fitted with 5kN load cell operated at a cross-head speed of 1mm/min. The dimension of the dumbbell shape sample was 115 mm (length) x 25 mm (width) x 5 mm (thick) with a narrow section of 33 mm length and 6 mm width.

2.6. Impact test

The impact test was conducted according to ASTM256 at ambient temperature. The test was conducted using Impact pendulum tester (Zwick). The samples were prepared with dimension of 63.5 mm (length) x 12.7 mm (width) x 4 mm (thick) and notch depth at 0.75 mm. The sample was stand as a vertical cantilever beam and a single swing of 1kN pendulum impacted on the notch face.

3. Results and discussions

3.1. FTIR analysis

The FTIR spectroscopy of the four synthesized AEPO and neat EPO from wavenumber 2000 cm^{-1} to 700 cm^{-1} are shown as in figure 1 and table 1. A smaller range of spectrum wavenumber was highlighted because of the expected presence of spectrums representing functional groups in AEPO at the range of wavenumber. The characteristic of C-O-C (oxirane) functional group stretching was observed at the wavenumber between 840 cm^{-1} – 820 cm^{-1}. Meanwhile the presence of saturated fatty acid backbone was determined from C-H stretching band at C-H in-plane deformation band at 1377 cm^{-1}.

The FTIR spectrums clearly describe the functional groups in chemical structure of AEPO through new IR absorption band in the four AEPO spectrums. During acrylation process, ring opening of epoxy by nucleophile of acrylate will form hydroxyl (-OH) functional group. The fingerprint region for primary alcohol (-OH) is the two bands with strong intensity at range 1075 cm^{-1} to 1000 cm^{-1} and at range 1350 cm^{-1} to 1260 cm^{-1}. Referring to the FTIR spectrums, the M1, M2, M3 and M4 show a stretching at wavenumber 1294 cm^{-1} with 90%, 90%, 83% and 87% transmission (10%, 10%, 17% and 23% absorbance) respectively and strong stretching at 1030 cm^{-1} for M1 only. The spectrums indicate the ring opening of epoxy forming -OH group. The ring opening of epoxy allows introduction of acrylate in the opened epoxy bond. Acrylate functional group can be characterised from C=C stretching at range 1680 cm^{-1} to 1580 cm^{-1}. The presence of alkene functional group in M1, M2, M3 and M4 were supported by C=C stretching vibrations at 1618 cm^{-1} and 1635 cm^{-1}. All the four AEPOs show 95-96% transmission at C=C stretching vibrations. In addition, the four AEPOs show the out of plane C-H deformation through two bands at wavenumber between 980 cm^{-1} – 960 cm^{-1}. It is a useful identification of asymmetrically substituted double bond. M1 and M2 show 90% transmission (10% absorbance), M3 shows 84% transmission (26% absorbance) and M4 shows 91-92% transmission (8-9% absorbance). The high absorbance in M3 indicate the stronger vibration of the C-H deformation due to higher intensity of the out of plane C-H in the AEPO structure formed. The characteristic of acrylate functional group was also observed in the four AEPOs spectrum at wavenumber 1400 cm^{-1}-1420 cm^{-1}, showing $-RCOO^-$ functional group with the strongest vibration observed in M2 AEPO spectrum. This characteristic indicates higher $-RCOO^-$ proportions in M2 which is not removed after washing because the ion trapped in between UPE/AEPO structure. From the AEPO data, it can be concluded that all the four AEPO produced from four different methods had the stated major functional groups in AEPO. However, the recommended processing conditions will be M2 and M3 with temperature at range 105 to 110°C and processing time of 16 hours respectively due to higher absorption and stronger vibrations of the functional groups show in M2 and M3. Table 1 below simplified the major functional groups in EPO and AEPO.

Table 1: Assignment of FTIR spectrum of EPO and AEPO functional groups.

	Functional group	Wavenumber (cm^{-1})	
		EPO	AEPO
A	C-O-C	826	810 – 840
B	C-H (in C=C)	-	980-960
C	-CH-	1377	1377
D	C=C (trans)	-	1618
E	C=C (trans)	-	1635
F	-OH	-	1294 and 1030
G	$-RCOO^-$	-	1400-1420

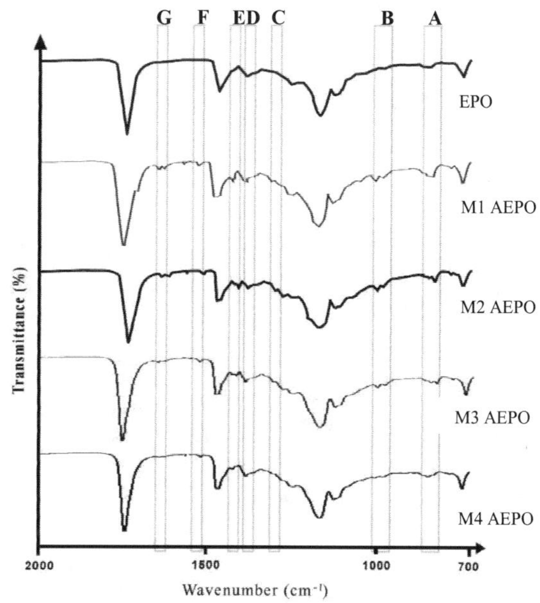

Figure 1: FTIR spectrum of neat UPE, M1 AEPO, M2 AEPO, M3 AEPO and M4 AEPO.

3.2. Mechanical properties

The change in tensile, Young's modulus, strain at break and impact for neat UPE UPE/AEPO blend were varied at M1, M2, M3 and M4 methods (figure 2). From the graph, it is apparent that the addition of 10 wt% AEPO in UPE reduced the tensile strength and tensile modulus of the UPE. The tensile strength of UPE in M2 was reduced from 25.56 MPa to 13.26 MPa (48 %). It is the highest reduction of UPE tensile property. This effect is attributed to the higher molecular weight of M2, no strength and low stiffness effect of AEPO. Therefore, it is expected that higher proportion of AEPO obtained through M2.

Meanwhile the blending of M2 shows the smallest reduction in tensile modulus property, from 1.47 GPa to 1.18 GPa (19.7%). Acrylation reaction is an exothermic reaction [15]. Thus, increase in temperature will favour the formation of acrylates. Therefore, from this experiment, synthesis of AEPO by M4 and M2 (110°C) has introduced more acrylate functional groups at epoxy functional groups in EPO compared to reaction at 80°C (M1) and 105°C (M3). The replacement of UPE by bio-resins reduced fragility of the components, due to the long chain of fatty acid chain [18]. Besides, the lower crosslinking between synthetic polymer and bio-resins also contributed to the observed result [19]. Therefore, in this study, it can be concluded that the lower crosslinking between synthetic polymer and bio-resins also contributed to the observed result [20]. However, the decrease in tensile modulus of M2 is the lowest among the other AEPOs. This suggest to the balance of C=C in M2 to crosslinking with UPE and the crosslink of UPE with styrene compare to the other methods. A similar result was achieved in a work of UPE/AEPO blend and UPE/AESO blend [16, 21].

The elongation at break of neat UPE and UPE blends were as shown in figure 3. From the graph, there was an increase in elongation at break of UPE/M2 AEPO and UPE/M3 AEPO compared to neat UPE. Method 2 showed the highest increment of elongation at break from 2.805 % to 3.493 % (24.5 %). The result suggested that blending of UPE with AEPO had sacrifice the tensile strength but had increased ductility of the UPE. The increased in elongation at break property was due to partly successful cross-linking between stiff UPE and ductile AEPO. This cause the polymer chain to become flexible when elongated. The M2 and M3 processing method with temperature 105°C and 110°C and process time of 16 hours had produced AEPO that readily form flexible UPE.

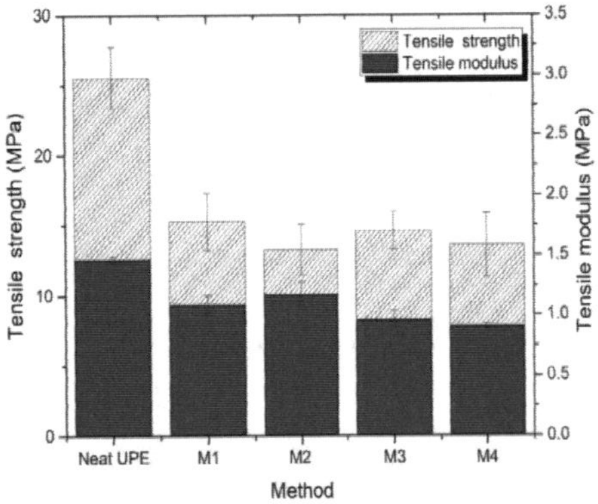

Figure 2: Tensile strength and tensile modulus of neat UPE, M1, M2, M3, and M4 AEPOs with UPE blend.

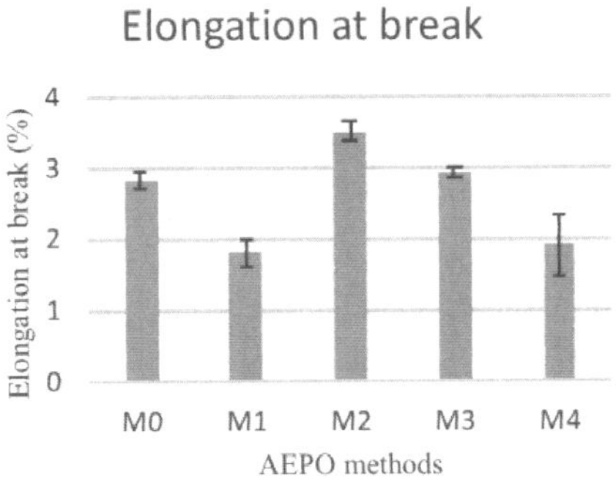

Figure 3: Elongation at break of neat UPE, M1, M2, M3 and M4 AEPOs with UPE blend.

The Izod impact strength of neat UPE and UPE blend at 10 wt% AEPO was shown in Figure 4. As can be seen in the figure, the impact strength of the polymer was decreased in the presence of 10 wt% AEPO in UPE polymer. The highest impact strength among the UPE/AEPO blends is M3 (21.97 J/m), while the lowest is M4 (13.80 J/m). Meanwhile, M1 and M2 (17 J/m) blends showed moderate impact strength. The reduced impact strength in all polymer blends are due to greater cross-linking density in the polymer chain and result to brittle polymers [19]. The cross-linking is between C=C in AEPO with C=C in styrene and UPE to form C-C single bond reduced free C=C in the polymer chain. C-C single bond is weaker than C=C double bond. Thus, the decreased density of C=C in the polymer chain reduced the strength of polymer chain to overcome the supplied impact. Due to high density of C=C in UPE/M1 AEPO and UPE/M2 AEPO blends, 20 minutes polymer blend is not enough to allow all the available C=C to cross-link with UPE. Moderate impact strength of M1 and M2 AEPO show the balance between formation of cross-linking between the polymer blends and the energy absorbed by the remaining C=C that does not involved in the cross-linking.

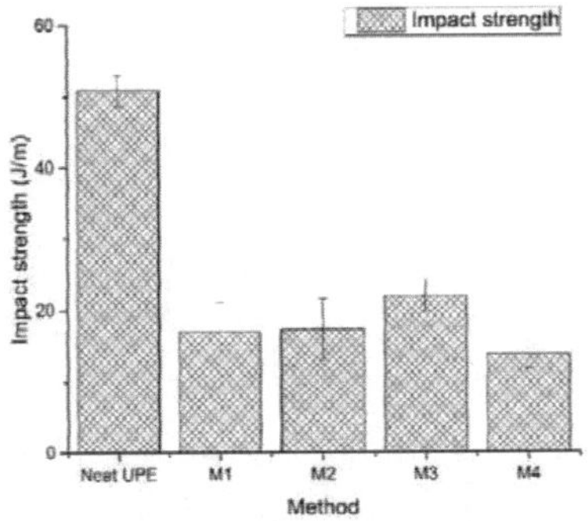

Figure 4: Izod impact strength of neat UPE, M1, M2, M3 and M4 AEPO/UPE blend

4. Conclusion

This work shows that it is possible to blend AEPO in UPE and cure the sample to from thermoset polymer. In this study AEPO was synthesised using four different conditions includes different temperature of 80°C, 100°C and 105°C and different processing period which is 16 hours and 36 hours. The prepared AEPO from the respective methods were labelled as M1, M2, M3, M4 and neat UPE. All the methods showed that AEPO were successfully synthesised as all the samples show functional group absorption of C=C, -OH and -HOC=O in FTIR test. The blending of constant 10 wt% AEPO in UPE had decreased the tensile modulus, and tensile strength. The polymer blend tensile strength of the palm oil acrylate processed at 110°C for 16 hours (M2) showed reduction from 25.56 MP to 13.26 MP (48%), reduction in tensile modulus from 1.47 GPa to 1.18 GPa (19.7%) and the highest impact strength among the four acrylates of polymer blend is 21.97 J/m from palm oil acrylate processed at 105°C for 16 hours (M3). The phenomena were expected due to introduction of stiffer AEPO in the UPE system and failure of acrylates to cross-linking with UPE. Besides, decrease in impact strength is due to decrease of strong C=C in the polymer chain. The presence of AEPO in the polymer chain sacrifice tensile and impact strength but improved elongation at break of UPE blend due to long and flexible chain of AEPO. Polymer blend from M2 acrylate showed increased in elongation at break from 2.805 % to 3.493 % (24.5 %) compared to neat unsaturated polyester. In conclusion, AEPO produced from Method 2 shows balanced mechanical properties when blend with UPE. Therefore, longer processing time is not an essential factor in producing AEPO as the best condition in preparing AEPO can be concluded at 110°C and processed in 16 hours. As the addition of bio-resin in synthetic UPE had reduced UPE mechanical performance, it is suggested that reinforce material is needed to regain and improved the mechanical properties of the polymer blend.

Acknowledgement
The author would like to extend their gratitude to the Universiti Malaysia Pahang for facility support. This work is also supported by the Research University Grant no RDU 170361, which the authors gratefully acknowledge.

References
[1] Dholakiya, B 2012 Unsaturated Polyester Resin for Specialty Applications *Polyester* ed H E-D M Saleh (London: Intech) chapter 7, pp 167-201

[2] Simpson, J M 2010 Synthesis and characterization of unsaturated polyesters for use in multi-vesiculated particles (MVPs) Doctoral dissertation, University of Stellenbosch

[3] Raquez, J M, Deléglise, M, Lacrampe, M F and Krawczak, P 2010 Thermosetting (bio)materials derived from renewable resources: A critical review *Prog. Polym. Sci.* **35** 487-509

[4] Mustapha, S N H, Rahmat, A R and Arsad A 2014 Bio-based thermoset nanocomposite derived from vegetable oil: A short review *Rev. Chem. Eng.* **30** 167–82

[5] Panda, S, Mohanty, P, Behere, D and Kumar, B T 2016 ESOA modified unsaturated polyester hybrid networks: A new perspective *J. Appl. Polym. Sci.* JAPS133(2016)012

[6] Miao, S, Wang, P, Su, Z and Zhang, S 2013 Vegetable-oil-based polymers as future polymeric biomaterials *Acta Biomater.* **10** 1692-704

[7] Salleh, W N F W, Tahir, S M, Mohamed, N S 2017 Synthesis of waste cooking oil-based polyurethane for solid polymer electrolyte *Polym. Bull.* **75** 109-20

[8] Fernandes, F C, Lehane, D, Kirwan, K and Coles, S R 2017 Epoxy resin blends and composites from waste vegetable oil *Eur. Polym. J.* EPJ7706(2017)020

[9] Haq, M, Burgueno, R, Mohanty, A K and Misra, M 2014 Hybrid Bio-Based Composite from UPE/EML Blends, Natural Fibres, and Nanoclay *Macromol. Mater. Eng.* **299** 1306-15

[10] Fakhari, A, Rahmat, A R, Wahit, M U, Alsagayar, Z S and Mustapha, S N H 2014 Influence of

nanoclay on mechanical properties of vinyl ester / acrylated epoxidized palm oil blends *Appl. Mech. Mater.* ***695*** 293-6

[11] Mustapha, S N H, Rahmat, A R, Arsad, A and Yeong, S K 2014 Novel bio-based resins from blends of functionalised palm oil and unsaturated polyester resin *Mater. Res. Innovations* **18** 326-30

[12] Czub, P and Franek, I 2013 Epoxy resins modified with palm oil derivatives-- preparation and properties *Polimery* ***58*** 135-41

[13] Keat, L W 2014 A Study Of The Curing Kinetics Of Polyester Resin From Palm Oil Master's thesis Universiti Malaysia Pahang

[14] Zhang, C, Garrison, T F, Madbouly, S A, Kessler, M R 2017 Recent advances in vegetable oil-based polymers and their composites *J. Prog. Polym. Sci.* JPPS1011(2017)179

[15] Ibrahim, M S, Kandile, N G, Hossam, M S and Moussa, I M 2005 Development of radiation curable surface coationg based on soybean oil. Part I. Preparation and characterization of acrylated oil 2005 *8th Arab Int. Conf. Polym. Sci. Tech.* Cairo-Sharm El-Sheikh Egypt

[16] Fakhari, A, Rahmat, A R, Wahit, M U, Khalili, A and Alsagayar, Z S 2015 Curing and Mechanical Properties of Bio-Based Hybrid Thermoset Resins from Unsaturated Polyester and Acrylated Epoxidized Palm Oil *Adv. Mater. Res.* **1113** 23– 7

[17] Salih, A M, Ahmad, M, Ibrahim, N A, Hj Mohd Dahlan, K Z, Tajau, R, Mahmood, M H and Wan Yunus, W M Z 2015 Synthesis of radiation curable palm oil-based epoxy acrylate: NMR and FTIR spectroscopic investigations *Molecules* **20** 14191– 211

[18] Mosiewicki, M A and Aranguren, M I 2013 A short review on novel biocomposites based on plant oil precursors *Eur. Polym. J.* **49** 1243-56

[19] Das, K, Ray, D, Banerjee, C, Bandyopadhyay, N R, Mohanty, A K and Misra, M 2011 Novel Materials from Unsaturated Polyester Resin/Styrene/Tung Oil Blends with High Impact Strengths and Enhanced Mechanical Properties *J. Appl. Polym. Sci.* **119** 2174-82

[20] Fakhari, A, Rahmat, A R, Wahit, M U and Shoot Kian, Y 2014 Synthesis of new biobased thermoset resin from palm oil *Adv. Mater. Res.* **931** 78-82

[21] Haq, M, Burgueño, R, Mohanty, A K and Misra, M 2008 Hybrid bio-based composites from blends of unsaturated polyester and soybean oil reinforced with nanoclay and natural fibers *Compos. Sci. Technol.* **68** 3344–51

IOP Conf. Series: Materials Science and Engineering **458** (2018) 012027 doi:10.1088/1757-899X/458/1/012027

Feasibility study of kenaf as fluid loss additive in water-based mud

H A Hussein[1,*], A R Ismail[1,2], S I Amran[3] and A M Arshad[1]

[1]Faculty of Chemical and Energy Engineering, Universiti Teknologi Malaysia, 81310 Johor Bahru, Malaysia.
[2]Institute for Oil and Gas, Universiti Teknologi Malaysia, 81310 Johor Bahru, Malaysia.
[3]Faculty of Bioscience and Medical Engineering, Universiti Teknologi Malaysia, 81310 Johor Bahru, Malaysia.

*halami2@live.utm.my

Abstract. Fluid loss is one of the problems encountered when drilling oil and gas wells. It may occur at any depth when the hydrostatic pressure exerted the formation exceeds the formation pressure at porous and permeable zones. Reducing the fluid loss is essential in drilling operation in order to maintain the mud properties. Mud added fluid loss additives will also trigger the changes in other mud properties. The objective of this research is to study the potential of kenaf as fluid loss additive as an alternative to commercial additives available in the market. Kenaf is a fibrous material which derived from plant and environmentally friendly. Laboratory studies were conducted to compare the rheological performance of mud at different concentrations and conditions. The results revealed that the commercial additive (PAC) performed better than kenaf as fluid loss additive. However, better results were achieved when both kenaf and PAC were mixed together at certain proportions. The results of API filtration test showed that the minimum fluid loss achieved was 7 ml. However, HPHT filtration test recorded a very small amount fluid loss.

1. Introduction

Drilling operation is one of the most important stages of hydrocarbon recovery process from the subsurface [1, 2]. Some of the key roles of drilling fluid used in every drilling operation are to bring rock cuttings to surface, cooling the drilling bit, provides hydrostatic pressure to counter formation pressure and to suspend rock cuttings when drilling is stopped [3]. These roles of drilling fluid are unchanged regardless of type of drilling mud used; either water-based mud (WBM), oil-based mud (OBM) or synthetic-based mud (SBM).

Fluid loss is worldwide drilling problem faced by all drilling operators. Fluid loss happens when hydrostatic pressure of the mud is higher than formation pressure inside the borehole, where this is the case for overbalanced drilling. When the formation is porous and permeable, the pressure difference will cause the fluid inside the mud to flow or filtrate into the formation instead of returning to the surface.

It is merely impossible to avoid fluid loss. However, it is acceptable if fluid loss can be minimized to a certain degree, where it varies between drilling operators. Some examples of fluid loss control measures include reducing mud weight during drilling through fluid loss zones, increasing wellbore strength and using air drilling technique. Some techniques might be classified as preventive measure and some are remedial.

Fluid loss into formation means mud invasion has occurred. The product of mud invasion into formation is mud cake. Particles or additives in the mud will be deposited onto wellbore surface at the loss zone. Mud cake is an important mud filtrate product which will help to stop further invasion which causes more fluid loss. The characteristics of materials that are added the mud will determine the two important characteristics of mud cake, which are permeability and thickness. Mud cake need to be less permeable to prevent fluid transfer and thin to avoid other problems due to formation of mud cake such as pipe sticking.

To date, poly-anionic cellulose (PAC) is the most widely used additive in drilling mud to control fluid loss [4]. Since PAC need to be imported and having high price tag, more studies are needed to substitute PAC as fluid loss additive [2, 5].

Due to the high production cost and environmental effect of chemical additives, there is a huge effort by researches to move towards environmental friendly additives. The use of biodegradable materials will also reduce its disposable cost because it can be safely disposed to the environment without the need to be treated. In addition, drilling mud additives are often not available in local industry market.

They need to be imported which leads to higher overall cost of drilling operation. Thus, the need to use local materials is seen as an important aspect to be considered when formulating additive for drilling mud [2]. In short, the potential fluid loss additive need to be biodegradable, locally available, abundance, cheaper and more efficient or comparable to PAC in terms of controlling fluid loss amount.

Kenaf is used in many mud formulations to reduce fluid loss since it is fibrous and biodegradable. There are also patents published that use kenaf as one of fibre sources as fluid loss additive [6, 7]. Furthermore, the most preferred fibrous additive by a researcher came from kenaf [8]. Kenaf is also commonly used as reinforcing agent to avoid fluid loss [9]. Kenaf is also found to be an effective additive to reduce fluid loss in both WBM and OBM [10].

In addition, kenaf is mentioned as one of the specialty products that is used to control spurt losses during drilling at Gulf of Mexico [11]. In the US, there is an additive product called DrillWall™ which is composed mainly from kenaf plant's core. Production company claims that this additive can mix with all types of drilling fluid, non-toxic, can form mud cake faster and can be used in less amount (about 7-10 lbs/bbl) for the same fluid loss results compared to other additive [12].

There are also a few more studies related to thermal degradation of kenaf fibres which suggest that kenaf fibre can be classified as thermally stable up to 300°C [13, 14]. At 250°F, untreated kenaf fibre only loose approximately 5% of its weight, mostly due to moisture evaporation. Furthermore, kenaf fibre also has high thermal stability. Azwa and Yousif (2013) reported that decomposition temperature of untreated kenaf fibre epoxy is 378.64°C [15]. In terms of kenaf plant itself, around 15-20 tons of dry kenaf can be produced by one hectare of kenaf in 120 days only. It also can absorb carbon dioxide (CO_2) more than any other crop plants and also able to remove heavy metals from soil [16]. These are superior advantages of kenaf fibre as fluid loss additive and also kenaf plant benefits to the environment.

The present study is a preliminary analysis of kenaf to see whether it can be used as fluid loss additive in WBM. Kenaf is fibrous and biodegradable. Plastic viscosity of the mud is related to friction between particles and mud, which affected the mud's ability to flow. Yield point and gel strength are important parameters when it comes to bringing cuttings to the surface and suspending them while drilling is ceased.

Fluid loss and mud cake thickness were focused parameters in this study which affect amount of fluid filtrate and will determine probability of occurrences of other problems such as pipe sticking and loss circulation. The effects of the addition of kenaf into WBM in terms of mud rheological properties were studied in this work and their interactions are detailed in this paper.

2. Materials and experimental procedures

2.1. Materials

Raw kenaf were sieved using multiple sieves of different sizes. These nested columns of sieves were placed on sieve shaker as shown in figure 1 and kenaf sample placed on the most top sieve. The shaker was operated for 40 minutes before sieved kenaf from sieves of sizes 45 μm, 75 μm and 106 μm were collected and labelled; ready to be used for experimental procedures. The kenaf of sizes above and below the range were discarded.

Figure 1. Nested columns of sieves arranged on a sieve shaker for sieving process.

2.2. Mud preparation

Water based drilling fluid was prepared based on the composition shown in table 1. The drilling fluid was weighed to achieve the density of 11 lb/gal (1.32 g/cc).

Table 1. Typical composition of water based drilling fluid that contains PAC and kenaf.

Components	Function	Amount
Distilled water	Base fluid	298 ml
Potassium chloride	Shale inhibitor	30 g
Caustic soda	pH control	0.2 g
Flowzan	Viscosifier	0.8 g
Barite	Weighting agent	127 g
PAC	Fluid loss additive	2 g
Kenaf	Fluid loss additive	2 g

Each component was added steadily into the distilled water in the order as shown in table 1, which was placed in a multi-mixer's cup. The mud was mixed using Hamilton beach mixer at the speed of 10,000 rpm. After a component was added into the mud, 5 minutes' interval was allowed before adding the next component to allow homogenous and thorough mixing of the component inside the mud. After all components were added, the mud was mixed for 30 minutes before moving to measuring its rheological properties. For this study, PAC was used as commercial additive.

The rheological properties of mud were measured using viscometer. Apart from plastic viscosity, other properties measured were yield point and gel strength. The fluid loss properties of the mud samples were evaluated by using API filter press and HPHT filter press. The data gathered in these experiments were volume of filtrate volume collected from the tests and thickness of mud cake after test period of 30 minutes. Electronic vernier calliper was used for measuring mud cake thickness. The filtration tests were conducted as per API RP 13B.

3. Results and discussions

3.1. Effect of each additive on rheology

Table 2 summarizes rheological properties of mud samples that contained different additives at different concentration of 2 grams and 4 grams. It also shows typical rheological properties of a mud.

Table 2. Results of rheological properties using different types of additive at 2 and 4 grams.

Additive	Plastic Viscosity (cP)	Yield Point (lb/100ft^2)	Yield Point (Pa)	Gel 10s (lb/100ft^2)	Gel 10s (Pa)	Gel 10m (lb/100ft^2)	Gel 10m (Pa)
Typical values	35-45	30-40	14.4-19.2	7	3.4	12	5.7
2g PAC	30	33	15.8	6	2.9	8	3.8
2g Kenaf	27	6	2.9	4	1.9	4	1.9
4g PAC	60	140	67.1	21	10.1	30	14.4
4g Kenaf	30	12	5.7	5	2.4	5	2.4

Additive	API Fluid loss (ml)	API Mud Cake (/32-inch)	HPHT Fluid Loss (ml)	HPHT Mud Cake (/32-inch)
Typical values	<5	<8	<5	<8
2g PAC	12.8	1.6	16.5	8.3
2g Kenaf	41.5	4.1	21	8.1
4g PAC	4.5	3	6	6.9
4g Kenaf	30.5	4.6	19.5	9.1

Rheological properties achieved using 2 grams of PAC were generally the best in terms of getting the values closest to typical values. This was true for plastic viscosity, yield point and gel strength. For fluid loss and mud cake thickness for both API and HPHT tests, lowest values were recorded using 4 grams of PAC. When kenaf was added, the values were still out of ranges of typical especially on fluid loss amount recorded, both using 2 and 4 grams.

PAC is hydrophilic in nature and soluble in water. PAC and water particles are attracted to each other which causes an increase of plastic viscosity, yield point and gel strength. Higher content of

PAC increased the values of these properties. In addition, when more PAC was added, thicker mud cake build-up happened which blocks water to filtrate and reduced amount of fluid loss. Kenaf is not soluble in water. Kenaf fibre also not absorbing water and does not swell as much as PAC. This explains why addition of kenaf did not improve plastic viscosity, yield point and gel strength.

As observed in table 2, acceptable values were achieved when 2 grams of PAC was added into the mud. However, fluid losses were still way higher than acceptable range. The next experimental procedures were to combine both PAC and kenaf into the same mud sample.

3.2. Effect of mixture of additives on rheology

Table 3 depicts results obtained when both PAC and kenaf were added into the mud. Amount of PAC used was constant at 2 grams for all mud sample, and kenaf amount was increased incrementally at 0.5 grams.

Table 3. Results of rheological properties when both additives were added into the same mud sample.

Additive	Plastic Viscosity	Yield Point		Gel 10s		Gel 10m	
	(cP)	(lb/100ft^2)	(Pa)	(lb/100ft^2)	(Pa)	(lb/100ft^2)	(Pa)
Typical values	35-45	30-40	14.4-19.2	7	3.4	12	5.7
2g PAC	30	33	15.9	6	2.9	8	3.8
2g PAC + 0.5g Kenaf	22	18	8.6	3	1.4	3	1.4
2g PAC + 1.0g Kenaf	20	26	12.5	3	1.4	3	1.4
2g PAC + 1.5g Kenaf	30	28	13.4	4	1.9	4	1.9
2g PAC + 2.0g Kenaf	37	30	14.4	4	1.9	5	2.4

Additive	API Fluid loss (ml)	API Mud Cake (/32-inch)	HPHT Fluid Loss (ml)	HPHT Mud Cake (/32-inch)
Typical values	<5	<8	<5	<8
2g PAC	12.8	1.6	16.5	8.3
2g PAC + 0.5g Kenaf	7.5	2.2	2.0	8.3
2g PAC + 1.0g Kenaf	7.5	2.3	1.0	8.3
2g PAC + 1.5g Kenaf	7.2	2.3	0.5	8.4
2g PAC + 2.0g Kenaf	7.0	2.4	0.0	8.8

Based on table 3, rheological properties achieved using 2 grams of PAC were generally the best in terms of getting the values closest to typical values. This was true for plastic viscosity, yield point and gel strength. For fluid loss and mud cake thickness for both API and HPHT tests, the data are illustrated in figures 2 and 3 respectively.

From Figure 3, higher fluid loss recorded via HPHT test compared to API test without presence of kenaf, but thickness of mud cake was considerable. Mud cake was getting slightly thicker with presence of only 0.5 gram of kenaf while fluid loss was drastically reduced. The fluid loss was reduced for about 85%, from 16 ml to 3 ml. Reduction of fluid loss continued to occur at higher kenaf concentrations but in a much slower rate. Thickest mud cake recorded at 2 grams kenaf concentration which is 8.8/32 inch which is an increase of only around 7% compared to the thickness without presence of kenaf. At HPHT condition, some amount of PAC and kenaf would get degraded and transformed into char and gas. Cellulose char was in a smaller size than its original form. This suggests rapid decrease in fluid loss once kenaf was introduced. More of these smaller particles will get deposited, trapped and accumulated as mud cake which helps in forming denser mud cake thus reducing fluid filtrate though it. Even some portion of the cellulose undergone degradation due to high temperature, there are still some that are in their original form. This explains why mud cake thickness continued to increase with higher kenaf concentration.

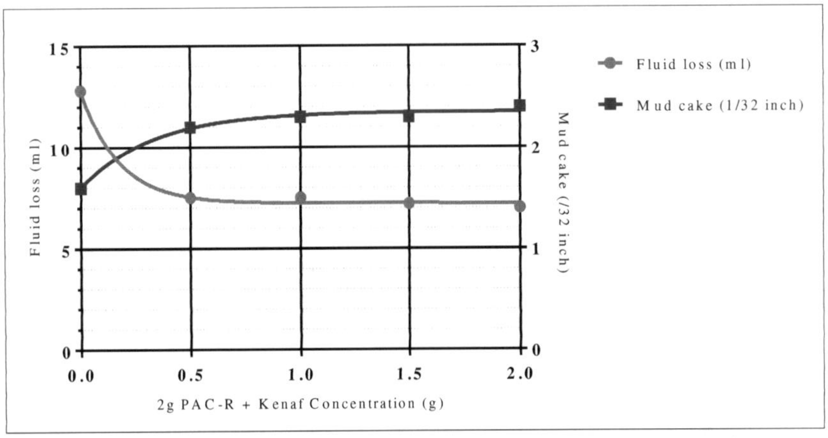

Figure 2. Fluid loss and mud cake thickness obtained using API filter press.

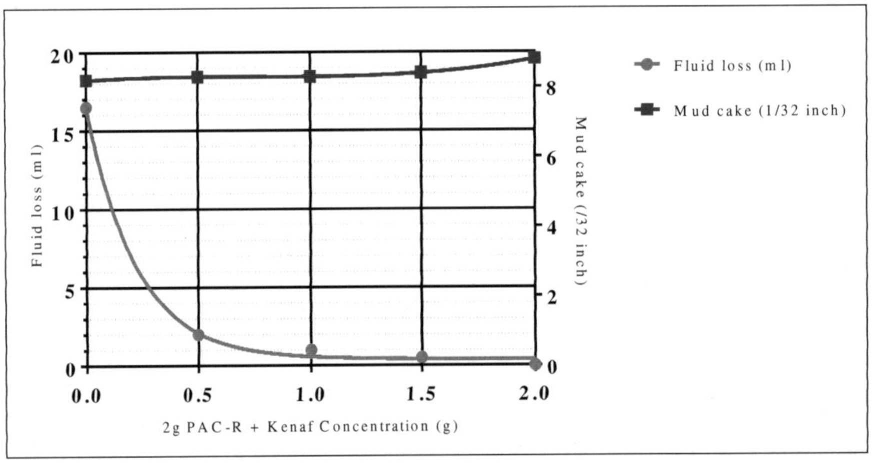

Figure 3. Fluid loss and mud cake thickness obtained using HPHT filtration test.

4. Conclusions

The experimental results show that the commercial fluid loss additives (PAC) performed better than kenaf as fluid loss additive when they were used separately. However, mixing the both the kenaf and PAC at certain proportions produced better results. The API filtration test showed that the minimum fluid loss achieved was 7 ml. However, at HPHT filtration test, a very minimal amount fluid loss was obtained which mean that it is nearly prevents the fluid loss into the formation.

Nomenclatures

μm	–	micrometer
Pa	–	pascal
cP	–	centipoise

Acknowledgments

The authors wish to thank Universiti Teknologi Malaysia for funding this project under Research University Grant vote Q.J130000.2546.17H77. The authors also appreciate Scomi Oiltools Sdn. Bhd. for providing the chemicals and additives to run the experiments.

References

[1] Skalle P 2011 *Drilling fluid engineering* Pål Skalle Ventus Publishing ApS

[2] Olatunde A O et al. 2012 Improvement of Rheological Properties of Drilling Fluid Using Locallly Based Materials *Petroleum & Coal* **54**(1) p. 11

[3] Kosynkin D V et al. 2012 Graphene Oxide as a High-Performance Fluid-Loss-ControlAdditive in Water-Based Drilling Fluids *ACS Applied Materials & Interfaces* **4**(1) p. 222-227.

[4] Fereydouni M et al. 2012 Effect of Polyanionic Cellulose Polymer Nanoparticles onRheological Properties of Drilling Mud *International Journal of Nanoscience and Nanotechnology* **8**(3) p. 171-174.

[5] Abusabah E Elemam, et al. 2015 Impacts of Polyanionic Cellulose Polymer (PAC-LV)on Drilling Fluids Properties SUST *Journal of Engineering and Computer Science (JECS)* **16**(3) p. 30-36.

[6] Heying T L 2003 *Methods for reducing lost circulation in wellbores* (Google Patents)

[7] Kefi S et al. 2012 *Methods for Controlling Lost Circulation in A Subterranean Well andMaterials There for* (Google Patents)

[8] Von Krosigk J R 2004*Composition for oil and gas drilling fluids with solidificationagent, cell transport agent and cellulosic additive* (Google Patents)

[9] Luzardo J et al. 2015 *Alternative Lost Circulation Material for Depleted Reservoirs* (Offshore Technology Conference)

[10] Pilehvari A A and V R Nyshadham 2005 *Effect of Material Type and Size Distributionon Performance of Loss/Seepage Control Material* (Society of Petroleum Engineers)

[11] Watson P A et al. 2005 *Management Issues And Technical Experience in Deepwater andUltra-Deepwater Drilling*(Offshore Technology Conference)

[12] Industrial Hemp Manufacturing 2015 *DrillWallTM; Drilling Fluid Lost CirculationMaterial (LCM).* [cited 2016 17 Nov] (Available from: http://ihempman.com/index.html)

[13] Tian Y S-i K and H Kubota 2007 Thermal Stability of Kenaf Fiber and KenafFiber/Low-density Polyethylene Composite *Journal of Materials Life Society* **19**(4) p. 192-194.

[14] Tajvidi M and A Takemura 2010 Thermal degradation of natural fiber-reinforcedpolypropylene composites *Journal of Thermoplastic Composite Materials* **23**(3) p. 281-298.

[15] Azwa Z N and B F Yousif 2013 Characteristics of kenaf fibre/epoxy compositessubjected to thermal degradation *Polymer Degradation and Stability* **98**(12) p. 2752-2759.

[16] Industries K G 2015 *About Kenaf.* (Available from:http://www.kenafibers.com/kenaf.html)

IOP Conf. Series: Materials Science and Engineering **458** (2018) 012028 doi:10.1088/1757-899X/458/1/012028

Preparation and characterization of polyimide membranes in iron wastewater removal by membrane distillation

S S Lee[1*], K C Chong[1], S O Lai[1], C H Koo[1], W J Lau[2], T W Ngu[1] and C H De[1]**

[1] Lee Kong Chian Faculty of Engineering and Science, Universiti Tunku Abdul Rahman, Jalan Sungai Long, Bandar Sungai Long, 43300 Kajang, Malaysia
[2] Advanced Membrane Technology Research Centre (AMTEC), Universiti Teknologi Malaysia, 81310 Skudai, Johor, Malaysia

*lsshin@utar.edu.my, **chongkc@utar.edu.my

Abstract. Polyimide (PI) polymer is a polymeric material with superior thermal, chemical and mechanical properties allowing it to be possesses the advantages of commercially viable materials to be applied in membrane distillation (MD) process. The PI hollow fiber membranes were fabricated by phase inversion method using N-Methyl-2-pyrrolidone (NMP) as solvent. The effect of feed inlet temperature and polydimethysiloxane (PDMS) coating concentration on the permeate flux performance were studied. The characteristics test conducted suggest that the membrane possesses dense morphology with tear-drop structure. The results from energy dispersive X-ray (EDX) depicted PDMS solution was successfully coated on the membrane surface. In addition, direct contact membrane distillation experiment in the iron (III) oxide rejection shows that the feed inlet temperature had a positive effect on the permeate flux performance attributed to the higher transmembrane vapor pressure while the rejection rate was found to be consistent at 95.10 ± 0.15%. Besides, the rejection rate of iron (III) oxide for PI hollow fiber membrane coated with PDMS increases from 95.13 to 97.58% with the increase of PDMS coating concentration. This is due to the enhanced surface layer nonetheless further optimization on the concentration is required to improve the permeate flux performance.

1. Introduction

Fresh water supply has been an alarming issue in the 21st century as reported by United Nation [1]. The accelerating industrialization, urbanization as well as population growth increases the need of reliable water supply. Ironically, the rapid growth destroys the balance between the demand and supply of water [2]. As a result, the quality of water is plummeting in some places around the globe. Yang et al. reported that there are more than a billion peoples in this world are unable to access to clean drinking water supply [3]. The water stress and shortage issue are going to upsurge for the years to come. Separation technologies are greatly demanded to remove hazardous inside the water so as to meet the requirements for municipal and industrial purposes. Currently, the commercial membrane separation technologies in wastewater treatment are predominantly pressure driven process such as reverse osmosis (RO), ultra-filtration (UF), nano-filtration (NF) [4].

Membrane distillation (MD) is the emerging technology which is less energy extensive due to thermal driven nature. The concept MD was first introduced in the late 1960s and subsequently being patented in between year 1963 to 1975 by Bodell and Findley, respectively [5]. Despite its early emergence, membrane distillation was not favoured at that time due to the technical incompetency in achieving comparable performance in terms of vapour permeability and membrane characteristics to pressure driven process [6]. The advancement of polymer industry in early 80s created a huge market for membrane technology in water separation reignite the interest on MD.

The commonly used polymeric materials in MD membrane fabrication are polyethylene (PE), poly(vinylidenefluoride) (PVDF) and polytetrafluoroethylene (PTFE) attributed to their enhanced permeability and hydrophobicity. Polyimide (PI) is the recently interested arising polymeric materials in membrane technology owning to its superior thermal stability, mechanical strength and chemical resistance [7]. The characteristics owned by PI materials allowed it to be effectively applied in MD due to the thermal driven nature of MD in water separation. Lately, several researchers focus performance of the membrane coated with a layer of coating materials such as aerogel (CX) and/or polydimethysiloxane (PDMS) attributed to its ability to improve the surface defect and permeate flux enhancement in both water and gas separation [8, 9].

In the wastewater rejection research, recent studies found that MD is promising in wastewater rejection with superior rejection rate. Dao et al. reported the rejection rate of more than 99% in MD for aluminium and arsenic wastewater removal [10]. Additionally, Qtaishat et al. focus on the saline waste water removal by DCMD using hydrophilic/hydrophobic polyetherimide membrane [11]. Lately, Zuo et al. investigate the performance of vacuum membrane distillation by using dual-layer hydrophobic/hydrophilic membrane with membrane surface coating for permeate flux enhancement [12].

In this work, PI hollow fiber membranes were fabricated by dry-jet wet phase inversion spinning method to evaluate the performance of PI membrane on the iron (III) oxide synthesized wastewater rejection. Subsequently, the membranes were coated by different concentration of PDMS solution to investigate the effect of the surface coating on the permeate flux and rejection rate of iron (III) oxide using direct contact membrane distillation system.

2. Materials and method

2.1. Membrane dope solution preparation

In this study, PI hollow fibre membranes were fabricated with 12 wt% of polyimide and 88 wt% of N-methyl-2-prrrolidone (NMP). PI was purchase in flake form from Alfa Aesar, USA and NMP was acquired from Sigma Aldrich, USA. Polydimethylsiloxane (PDMS) used as coating materials that applied on the membrane surface and hexane used as solvent for coating solution preparation were both obtained from Sigma Aldrich, USA. The iron (III) oxide used to fabricate the laboratory synthesized wastewater was obtained from Sigma Aldrich, USA.

Prior to the dope solution preparation, PI flakes was dried in vacuum oven at 70°C for 24 h to remove the moisture content. Then, the dope solution was prepared by dissolving PI flakes in the NMP solution and stirred thoroughly to produce a homogeneous solution. Then, the solution was degassed to remove potential air bubbles produced during mixing and stirring. The presence of bubble will reduce the quality of membrane formed.

The PI hollow fibre membranes were fabricated using dry-jet wet phase inversion spinning method as depicted in figure 1 where the spinning parameter used in this study was tabulated in table 1. The dope solution was firstly poured into the storage tank before spinning. The dope solution was forced through the spinneret by gear pump while the bore fluid flows through the annulus to solidify the membrane. The nascent hollow fibre membranes enter the coagulation bath where the solvent exchange occurs. The hollow fibre membranes were collected by winding drum and later the membranes were stored in pure water for two days in order to completely removing the residue solvent. Lastly, the membrane was dried at room temperature for two days before usage.

PDMS coating solution was used in this study to improve the membrane surface structure as depicted by Zulhairun *et al* [13]. The concentrations of the coating solution were at 3, 5 and 10 wt% respectively. The PDMS coating solution was prepared by mixing the PDMS solution with hexane solvent at predetermined volume to acquire the desired solution concentration. The solution was stirred by mechanical stirrer for 2 h in order to obtained homogenous solution.

Dip coating method was engaged in this work where the PI hollow fibre membranes were immersed in conical flask for ten minutes and this procedure was repeated five times to attain a good coating surface. Lastly, the membrane was dried in vacuum oven at 70°C for 24 h before usage. The pristine PI hollow fibre membrane used in this study was labelled as PI whereas the PI hollow fibre membrane with PDMS coating concentration of 3, 5 and 10 wt% were marked as PI-3, PI-5 and PI-10, respectively, for ease of reference.

Table 1. Spinning parameter of hollow fibre membrane.

Spinning parameter	Value
Bore fluid type	Distilled water
Bore fluid temperature (°C)	25
Bore fluid flow rate (mL/min)	1.0
External coagulant type	Tap water
External coagulant temperature (°C)	25
Spinneret dimension OD/ID (mm/mm)	0.6/0.3
Air gap distance (cm)	30
Room relative humidity (%)	55 ± 5

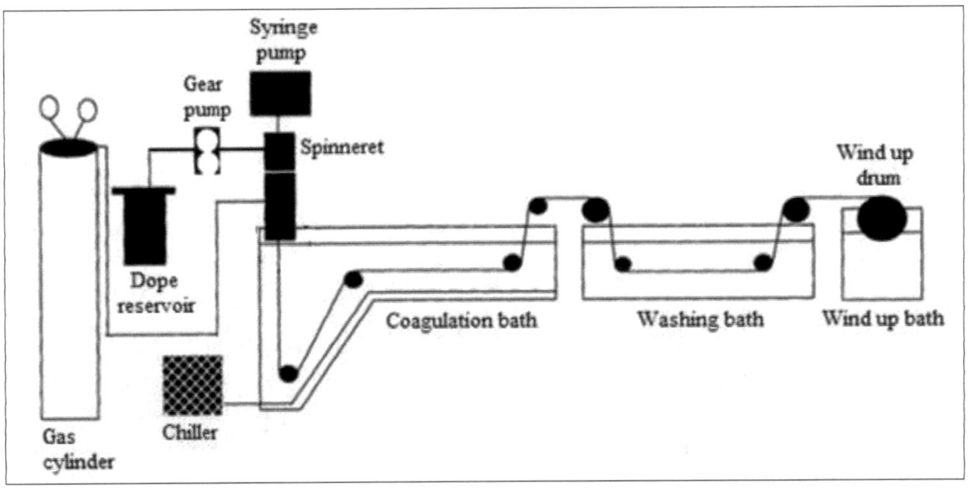

Figure 1. Schematic diagram of dry-jet wet phase inversion spinning method [14].

2.2. Membrane characterization

The membrane morphology was observed with scanning electron microscope (SEM) (Hitachi, Japan). Prior to the characterization, pre-treatment was undergone where the membranes were cryogenically crack using liquid nitrogen. The membrane samples were then fixed to the sample holder using carbon tape and subsequently sputtered with gold by a sputter coater (Emitech, United Kingdom).

The membrane porosity in this work was measured using gravitational method. Dry weight of the membrane was first measured by electronic weight balance (AND, Japan). Later, the membrane was wetted by being immersed into 2-butanol to fully fill up the membrane pores. The membrane was then dried at room temperature to eliminate excess alcohol on the surface of the membrane before the wet weight was measured. The membrane porosity can be calculated by the following equation.

$$Porosity, \in = \frac{\frac{(W_w - W_d)}{\rho_a}}{\frac{(W_w - W_d)}{\rho_a} + \frac{W_d}{\rho_b}} \tag{2.1}$$

where W_w is the weight of wet membrane (g), W_d is the weight of dry membrane (g), ρ_a is the density of PI polymer (g/cm^3) and ρ_b is the density of 2-butanol (g/cm^3).

In this work, liquid entry pressure was determined by the method similar to Khayet and Matsuura [4]. Five hollow fiber membranes with an average length of 23 cm were attached to a tubular module by applying epoxy adhesive. Distilled water was flow into the lumen of the membrane with the aid of compressed nitrogen gas. The pressure of the gas was increased in the stepwise rate of 0.5 bar until first distilled water droplet was seen, i.e. the liquid entry pressure of the membrane.

2.3. Membrane distillation experiment

In this work, direct contact membrane distillation (DCMD) was utilized to remove iron (III) oxide synthesized wastewater where the experiment setup was schematically shown in figure 2 which is similar to the work reported by Chong *et al* [15]. Ten hollow fiber membranes with average length of 23 cm were inserted into the cylindrical tubular membrane module. Both ends of the module were sealed up with epoxy glue. The feed and permeate were in cross-flow pattern in the membrane module. The feed solution flew along the lumen side of the membrane while the permeate solution flew along the shell side. Two booster pumps were used to re-circulate the feed and permeate solution.

The laboratory synthesized feed solution was similar the wastewater from mining industry (1.0 wt% Fe_2O_3). The solution was heated to the temperature ranging from 40 to 55°C on an electrical heating plate (LMS, Japan). Concurrently, the permeate solution was cooled to 18°C by electrical chiller (Eyela, Japan) before being pumped into the shell side of the membranes. Water rotameter (Blue White, USA) was used to measure the flow rate of feed and permeate. Meanwhile, the flow rates were regulated by needle valve (Swagelok, USA). The quality of permeate was monitored by an electronic conductivity meter (4520 by Jenway, United Kingdom). The weight of the permeate obtained was measured using an electronic weight balance (AND, Japan) with data logger (AND, Japan).

The permeate flux of the membranes, J, was determined using equation (2.2) [16],

$$J = \frac{\Delta W}{A \Delta t} \tag{2.2}$$

where J is the permeate flux (kg/m^2.hr), ΔW is the difference between the initial and final permeate weight (kg), A is the effective surface area of the membrane (m^2), and Δt is the sampling time (hr). On the other hand, equation (2.3) was used to calculate the rejection of iron (III) oxide, R (%) [16].

$$R = \frac{C_f - C_p}{C_f} \tag{2.3}$$

where C_f and C_p are the iron (III) oxide concentration (ppm) in the feed and permeate solution, respectively.

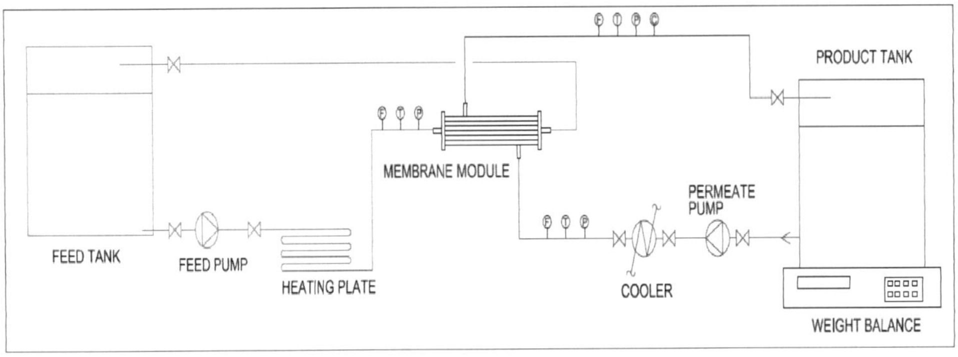

Figure 2. Schematic diagram of DCMD setup [14].

3. Results and discussions

3.1. Membrane morphology

The SEM images of the as-spun hollow fibre membrane morphology were as shown in figure 3. The hollow fibre membrane fabricated with 12 wt% of PI and 88 wt% of NMP illustrated a dense membrane structure extending from the inner to the outer structure. The formation of the dense membrane structure found in the hollow fibre membranes was attributed to the high viscosity of the membrane dope solution (4285 cP) which lead to the delay in formation of membrane structure during phase inversion. It is noteworthy to mention that the membrane illustrated tear-drop like microvoids extended from the middle of the cross section to the inner and outer layer. This phenomenon was due to the used of NMP as strong solvent encourage rapid phase inversion but restricted by the high viscosity of the dope solution which similar to the reported in literature study [5].

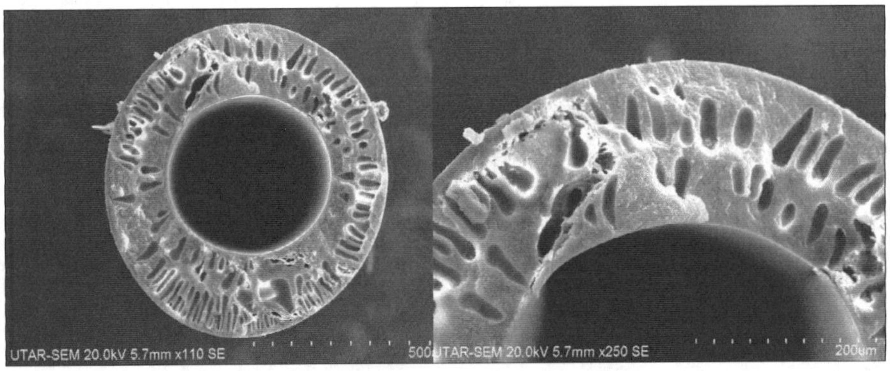

Figure 3. Membrane morphology.

The membrane characteristics of the as-spun PI membrane in this study were tabulated in table 2. The diameter of the hollow fibre membranes was recorded between the range of 390 to 420 μm whereas the thickness of the membranes was found to be 170 to 185 μm. The PDMS coating concentration had a positive effect on the membrane thickness where the higher the membrane coating concentration, the higher the thickness of the membrane attributed to the increase of PDMS content in the coating solution. EDX elemental analysis performed on the membrane surface was tabulated in table 3. The EDX result suggested that the membrane was coated by the PDMS solution attributed to the presence of silicon element which was not presence in the pristine PI membrane. The presence of the silicon trace element on PI-3, PI-5 and PI-10, respectively, was due to the silanol group contained in PDMS where the results was well agreed with the literature study [8].

Table 2. Characteristic of as-spun membrane.

Membrane	Inner diameter (µm)	Thickness (µm)
PI	390 ± 20	170 ± 10
PI-3	410 ± 23	174 ± 16
PI-5	420 ± 18	179 ± 11
PI-10	410 ± 21	185 ± 14

Table 3. EDX elemental analysis of surface coating.

Membrane	Composition (At%)			
	Carbon	Oxygen	Nitrogen	Silicon
PI	72.08	14.35	13.57	Not detected
PI-3	71.86	13.28	12.14	2.72
PI-5	71.39	13.12	11.36	4.13
PI-10	70.67	9.87	10.72	8.74

3.2. Effect of feed temperature in iron (III) oxide removal

Figure 4(a) illustrates the permeate flux performance of as-spun PI hollow fibre membrane on iron (III) oxide synthesized wastewater with respect to different feed inlet temperature where the permeate inlet temperature was set at 18°C, while feed inlet and permeate inlet flow rate were both set at 0.3 l/min. The results illustrated the feed inlet temperature possesses positive effect on the permeate flux performance where the flux increase from 10.20 to 11.65 kg/m^2.hr with the feed inlet temperature increase from 50 to 60°C. This phenomenon was identical to the report by Khayet and Matsuura where the higher the feed temperature, the higher the vapor pressure difference between the feed and the permeate stream induced higher transmembrane pressure which eventually lead to higher permeate flux performance [4]. As shown in figure 4(b), the rejection rate of the iron (III) oxide was found to be consistent at 95.10 ± 0.15% for all the feed inlet temperate indicating the membrane possess good rejection rate.

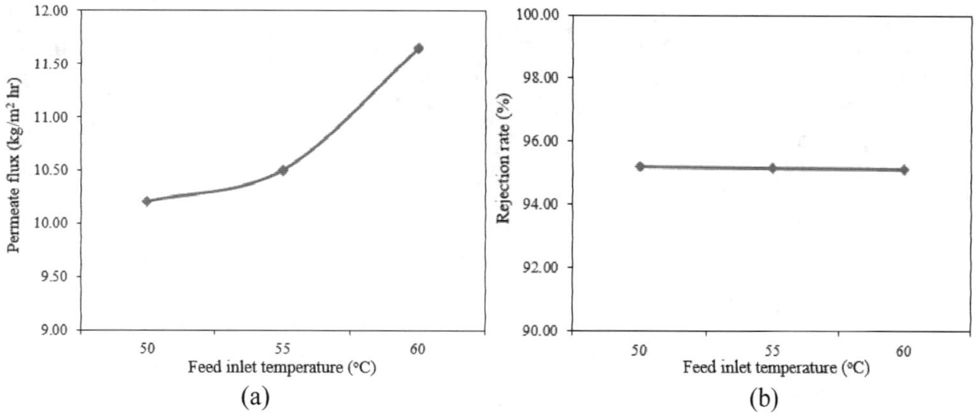

Figure 4. Performance of iron (III) oxide rejection by PI membrane operated at different feed inlet temperature (a) permeate flux, (b) rejection rate.

3.3. Effect of PDMS coating concentration in iron (III) oxide removal

Figure 5(a) shows the permeate flux performance of as-spun PI hollow fiber membrane on iron (III) oxide synthesized wastewater with respect to different PDMS surface coating concentration where the feed inlet temperature was set at 60°C, permeate inlet temperature was set at 18°C, feed inlet and permeate inlet flow rate were both set at 0.3 l/min. The results illustrated the PDMS surface coating concentration possesses negative effect on the permeate flux performance where the flux decrease from 11.65 to 6.56 kg/m^2.hr with the PDMS concentration increase from 0 to 10 wt%. The reduction of the permeate flux across the membrane resulted from the increase of PDMS coating concentration was attributed to the increase of surface layer thickness which lead to the increase of mass transfer resistance and hence reduces the amount of water vapor to diffuse across the membrane. However, as shown in figure 5(b), it is noteworthy to mention that the rejection rate of the iron (III) oxide increases consistently from 95.13 to 97.58% with the increase of PDMS coating concentration showing the additional layer of coating on the surface restricted the iron (III) oxide particle to move across the membrane. The result suggested that further optimization of the PDMS concentration is required in order to obtain a good permeate flux and rejection rate for the iron (III) oxide rejection.

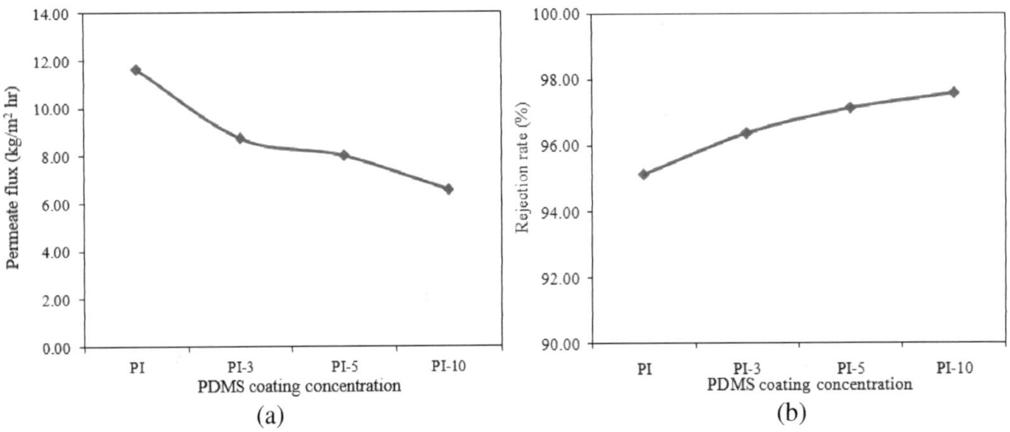

Figure 5. Performance of iron (III) oxide rejection by PI membrane at different PDMS coating concentration (a) permeate flux, (b) rejection rate.

4. Conclusions

In this work, PI membrane was successfully fabricated by dry-jet wet phase inversion spinning method in laboratory scale with the PDMS coating concentration at 3, 5 and 10 wt%, respectively. The hollow fibre membranes were characterized in terms of morphology, thickness and elementally analysed on the coating quality. The membrane morphology obtained from this work was generally dense membrane with tear-drop like structure due to rapid phase inversion of the solvent in highly viscous dope solution. The direct contact membrane distillation experimental study showing the permeate flux performance increase with the increment of feed inlet temperature which was induced by the increase of transmembrane vapor pressure. On the other hand, the PI hollow fibre membrane with high PDMS coating concentration illustrated a drop in permeate flux performance but improved rejection rate of iron (III) oxide. The findings from this study suggested that the PI hollow fibre membrane coated with PDMS possesses the potential to be used in the iron (III) oxide wastewater rejection subjected to the further optimization of the concentration in our future works.

Acknowledgement

The authors are grateful to the financial support from Universiti Tunku Abdul Rahman under the UTARRF grant number IPSR/RMC/UTARRF/2017-C1/L08.

References

[1] Satulu V, Mitu B, Altynov V A, Lizunov N E, Kravets L and Dinescu G 2017 Synthesis and characterization of porous composite membranes with hydrophilic/hydrophobic sides *Thin Solid Films* **630** 92–99

[2] González D, Fane A G and Wang R 2017 Membrane distillation: Perspectives for sustainable and improved desalination *Renewable and Sustainable Energy Reviews* **80** 238–259

[3] Yang X, Fane A G and Wang R 2014 *Membrane distillation: Now and future Desalination* John Wiley & Sons, Inc. pp 373–424

[4] Khayet M and Matsuura T 2011 *Introduction to membrane distillation Membrane Distillation* Amsterdam Elsevier pp 1–16

[5] Chong K C, Lai S O, Lee K M, Lau W J, Ismail A F and Ooi B S 2014 Characteristic and performance of polyvinylidene fluoride membranes blended with different polymeric additives in direct contact membrane distillation *Desalination and Water Treatment* 1–9

[6] Hitsov I, Maere T, Sitter K D, Dotremont C and Nopens I 2015 Modelling approaches in membrane distillation: A critical review *Separation and Purification Technology* **142** 48–64

[7] Khayet M 2011 Membranes and theoretical modeling of membrane distillation: A review *Advances in Colloid and Interface Science* **164(1-2)** 56–88

[8] Chong K C, Lai S O, Lau W J, Thiam H S, Ismail A F and Zulhairun A K 2016 Fabrication and characterization of polysulfone membranes coated with polydimethysiloxane for oxygen enrichment *Aerosol Air Quality Research* 1–8

[9] Magueijo A M, Anderson L G, Fletcher A J and Shilton S J 2013 Polysulfone mixed matrix gas separation hollow fibre membranes filled with polymer and carbon xerogels *Chemical Engineering Science* **92** 13–20.

[10] Dao T D, Laborie S and Cabassud C 2016 Direct As(III) removal from brackish groundwater by vacuum membrane distillation: Effect of organic matter and salts on membrane fouling *Separation and Purification Technology* **157** 35–44

[11] Qtaishat M, Rana D, Khayet M and Matsuura T 2009 Preparation and characterization of novel hydrophobic/hydrophilic polyetherimide composite membranes for desalination by direct contact membrane distillation *Journal of Membrane Science* **327(1-2)** 264–273

[12] Zuo J, Chung T-S, O'Brien G S and Kosar W 2017 Hydrophobic/hydrophilic PVDF/Ultem® dual-layer hollow fiber membranes with enhanced mechanical properties for vacuum membrane distillation *Journal of Membrane Science* **523** 103–110

[13] Zulhairun A K, Fachrurrazi Z G, Izwanne M and Ismail A F 2015 Asymmetric hollow fiber membrane coated with polydimethylsiloxane–metal organic framework hybrid layer for gas separation *Separation And Purification Technology* **146** 85–93

[14] Chong K C, Lai S O, Lau W J, Thiam H S, Ismail A F and Roslan R A 2018 Preparation, Characterization and Performance Evaluation of Polysulfone Hollow Fiber Membrane with PEBAX or PDMS Coating for Oxygen Enhancement Process *Polymers* **10** 126

[15] Chong K C, Lai S O, Lau W J and Ooi B S 2014 Preparation, Characterisation and Performance of Polyvinylidene Fluoride Membrane for Sodium Chloride Rejection in Direct Contact Membrane Distillation *Materials Research Innovation* **18** 359 – 363

[16] Hou D Y, Wang J, Qu D, Luan Z K and Ren X J 2007 Fabrication and characterization of hydrophobic PVDF hollow fiber membranes for desalination through direct contact membrane distillation *Separation And Purification Technology* **69** 78–86

Investigation of choline based ionic liquids as an oil spill dispersant

Mansoor Ul Hassan Shah[1], Magaret Sivapragasam[2], Muhammad Moniruzzaman[1,2,*], Md. Mahabubur Rahman Talukder[3] and Suzana Bt Yusup[1]

[1]Department of Chemical Engineering, Universiti Teknologi PETRONAS, 32610 Seri Iskandar, Perak, Malaysia
[2]Centre of Research in Ionic Liquids (CORIL), Universiti Teknologi PETRONAS, 32610 Seri Iskandar, Perak, Malaysia
[3]Institute of Chemical and Engineering Sciences (ICES), Agency for Science, Technology and Research, 627833, 1 Pesek Road, Jurong Island, Singapore

*m.moniruzzaman@utp.edu.my

Abstract. Chemical dispersants are universally used to overcome oil spill damages. However, the toxicity related with chemical dispersants minimizes its usage as oil spill dispersants. Thus, the employment of biodegradable and environmental benign dispersants in marine environment is much needed. In this research, the capability of choline myristate [Cho][Myr] and choline oleate [Cho][Ol] were investigated to be used as oil spill dispersants. Results reported in this study illustrates that by combining these two amphiphilies, an effective dispersion is achieved. A maximum dispersion of 71 % was achieved by combining [Cho][Myr] and [Cho][Ol] with an optimal ratio of 60:40 (w/w). The dispersant to oil ratio (DOR) was 1:10 (v/v) at this optimal combination. Further, by increasing the DOR the effectiveness increases and about 78% dispersion is achieved at 1:1 (v/v). Moreover, the blend of [Cho][Myr] and [Cho][Ol] at 60:40 (w/w) ratio formed a stable emulsion than its single components.

1. Introduction

Several oil spills occurred in the past, for instance, in the year 2010 Deepwater Horizon (DWH) oil spill occurred due to which about 4.9 million barrels of oil was transferred to the sea [1, 2]. Similarly, about 2000 barrel of oil was shifted to the Gulf of Mexico in the year 2000 [3]. In the same way Exxon Valdez oil spill (in year 1989) released about 11 million gallons of crude oil to the Gulf of Alaska[4]. In Malaysia, there remains a regular risk of oil spill and about 121 cases of oil pollution reported in between the year 2009 and 2015[5].

When dispersants are applied, it reduces the interfacial tension between oil and water. The wave action transfers the large oil slick to discreate droplets and the adsorbed surfactant molecules stabilizes each droplet [6]. The stabilized oil droplets are dispersed through the water surface, and with the passage of time most of the oil is degraded by ocean microbes [7].

The dispersant toxicity is one of the basic parameter to be considered before the employment of dispersants to the sea environment [8]. Accelerated environmental awareness and strict rules has made dispersants environmental compatibility a key factor in their applications [9, 10]. A blend of surfactants is used in the recent research for overcoming oil spill damages [11, 12]. Although Corexit is the most employed formulation in this regard, the components and solvent used in its formulation are nonbiodegradable and toxic. [13]. Therefore, various researches have endeavored to determine the oil spill dispersant formulations as a replacement to the conventional toxic dispersants [14, 15].

Choline based ionic liquids has recently attracted plenty of attention of academia and researchers [16]. They are used exclusively in various fields including; biomass pretreatment, carbon dioxide adsorption and as demulsifying agent [16, 17]. In this research, the capability of choline-based biocompatible ionic liquids (ILs) as oil spill dispersants were explored. Two choline based ILs, [Cho][Myr] and [Cho][Ol] (figure 1 A) and B)) were synthesized and combined with different proportion using water as a solvent.

2. Experimental

2.1. Materials

The chemicals employed to synthesize the ionic ILs include; myristic acid (Merck, New Jersey, USA), oleic acid (Sigma, St. Louis, Missouri, USA), and choline hydroxide (46 wt% water, Sigma-Aldrich, St. Louis, Missouri, USA). The chemicals were used as supplied without any further purification. Appropriate amount (about 1%) of salts were dissolved in distilled water to prepare synthetic sea water. Crude oil emulsions were prepared by employing Tapis crude oil (provided by the Petroleum Industry of the Malaysia Mutual Aid Group).

2.2. Synthesis of choline based ILs

A previously reported method was employed for the synthesis of [Cho][Myr] and [Cho][Ol] [18]. In this method, the required acid was imparted to the aqueous choline hydroxide solution with a molar ratio of 1:1 at 25°C. The mixture was continuously stirred, and the reaction mixture was refluxed for 24 h at 80°C. The synthesized ILs were then dried in rotary evaporator (BUCHI, Switzerland) under vacuum (80°C, 2 h). Figure 1 (A and B) representing the molecular structures of the synthesized ILs.

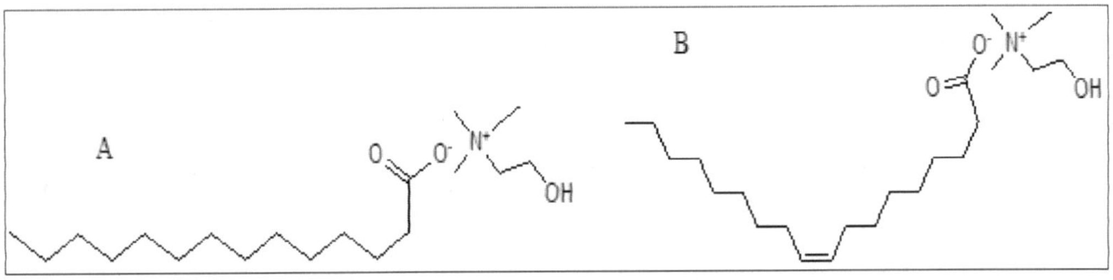

Figure 1. Structure of choline based ILs A) Choline myristate B) Choline oleate.

2.3. Measurement of interfacial tension

The crude oil and synthetic sea water interfacial tension (IFT) was measured using the pendant drop method (OCA20, Data Physics). In this method, the synthetic sea water was placed in a glass tube followed by injection of dispersant crude oil mixture. The IFT was evaluated by drop shape analysis. The IFT of the pure components as well as of the combined system were investigated by mixing appropriate volume of the stock solutions of [Cho][Myr] and [Cho][Ol].

2.4. Baffled Flask Test (BFT)

The effectiveness of dispersant was determine using U.S.A Environmental Protection Agency (EPA) revised protocol [19]. In the Baffled Flask Test (BFT), 120 mL of synthetic sea water was introduced in the baffled flask. Then 100 μL of crude oil was gently dispensed onto the surface of synthetic sea water. On the top of crude oil, 4 μL dispersant was poured carefully so that the dispersant should not touch the water first. This gives dispersant-to-oil ratio (DOR) of 1:25. Different DOR was obtained by varying the dispersant volume while keeping the oil volume constant. The solution was then mixed by placing the flask in an orbital shaker (at 200 rpm) for 10 min. Allowing a settling time of 10 min, 30 mL sample from the center of the flask was then taken and the dispersed oil in the aqueous sample was extracted with dichloromethane (DCM). Absorbance readings at 340, 370 and 400 nm was recorded by utilizing a spectrophotometric measurement. The dispersant effectiveness was calculated using equation

$$Eff(\%) = \frac{Total\ Oil\ Dispersed}{\rho_{oil} V_{oil}} \times 100 \quad (1)$$

ρ_{oil} = Density of Tapis crude oil
V_{oil} = Volume of Tapis crude oil added to the test flask

2.5 Emulsions preparation
Crude oil emulsions were prepared by adding 1 mL of crude oil to 10 mL of synthetic sea water. Appropriate amount of dispersant was added, and the sample was then placed on a vortex mixer for 3 min. After emulsion preparation, vials were placed on the benchtop for visual observation at room temperature (25°C). Images were taken over a period of time to evaluate the stability.

3. Result and discussion

3.1. Interfacial tension
The IFT between sea water and crude oil were evaluated and the DOR was kept constant at 1:10 (v/v) for all measurements. The IFT of pure and combined system of [Cho][Myr] and [Cho][Ol] are presented in figure 2. It is expected that both [Cho][Myr] and [Cho][Ol] pack closely at oil water interface due to favorable interactions between both the tails of [Cho][Myr] and [Cho][Ol]. As it can be seen that [Cho][Ol] tail has one cis unsaturation and their existence ensures that the tail remain flexible and liquid like at room temperature. Therefore, the flexible, bent tail of [Cho][Ol] may aid in the packing of these molecules close to each other and results in low IFT compared to its pure components. This lower IFT of the mixture as compared to its pure components is due to the synergistic interaction between the molecules and this synergistic interaction was maximum at 60/40 (w/w) ratio of [Cho][Myr] and [Cho][Ol] which resulted in least IFT (2.18 mN/m). Athas *et al.* [20] noticed a similar trend where the IFT of the mixture (Lecithin and Tween 80) was less than individual components IFT, showing the synergism between the components. IFT results presented in this research work illustrates the synergistic behavior of the two components ([Cho][Myr] and [Cho][Ol]).

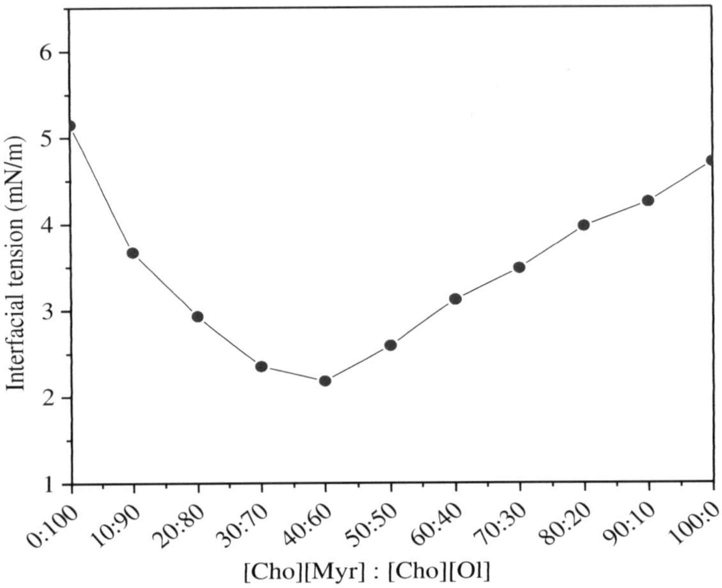

Figure 2. Interfacial tension of [Cho][Myr] and [Cho][Ol] at different ratio (w/w) of two components at 25 °C.

3.2. Effectiveness of dispersant

The dispersants effectiveness was determined using BFT (as mentioned in section 2.4) and the results are presented in figure 3 as a function of [Cho][Myr] and [Cho][Ol] (w/w) ratio. It was noticed that the individual components exhibited less effectiveness as compared to the combined system. The most effective dispersion was achieved at 60:40 ratio of [Cho][Myr]: [Cho][Ol]. At this optimal ratio 71% effectiveness was observed. The IFT measurement (figure 2) between sea water and crude oil (DOR 1:10 (v/v)) revealed that low IFT is achieved at this ratio. The sharp decrease in IFT explains the corresponding high effectiveness at this optimal ratio. Further, for [Cho][Myr]:[Cho][Ol] ratios other than 60:40, results in high IFT, which accounted for the decline in dispersant effectiveness.

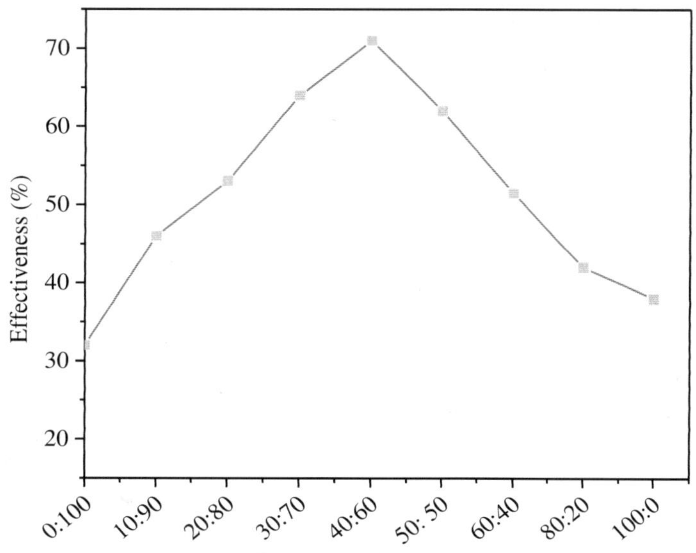

Figure 3. Effectives of dispersant at different [Cho][Ol] and [Cho][Myr] ratios. The DOR was kept constant at 1:10 (v/v) for all combinations.

3.3. Dispersant oil ratio

The effectiveness of dispersant over a range of DOR are presented in figure 4. The dispersant consists of 60:40 [Cho][Myr]: [Cho][Ol] (w/w), the most effective [Cho][Myr]: [Cho][Ol] ratio (figure 3). The DOR used in this work include; 1:1, 1:10, 1:25, 1:50 and 1:100, and it was observed that the effectiveness increased by increasing the DOR. The highest effectiveness of about 78 % was achieved at 1:1 DOR. This high effectiveness at high DOR is due to the fact that small oil droplets formed at high DOR, which are more stable compared to the larger oil droplets, hence resulting in high effectiveness. The results presented in this work are in line with the previous study conducted by Nyankson *et al.* [21], observed high effectiveness at high DOR.

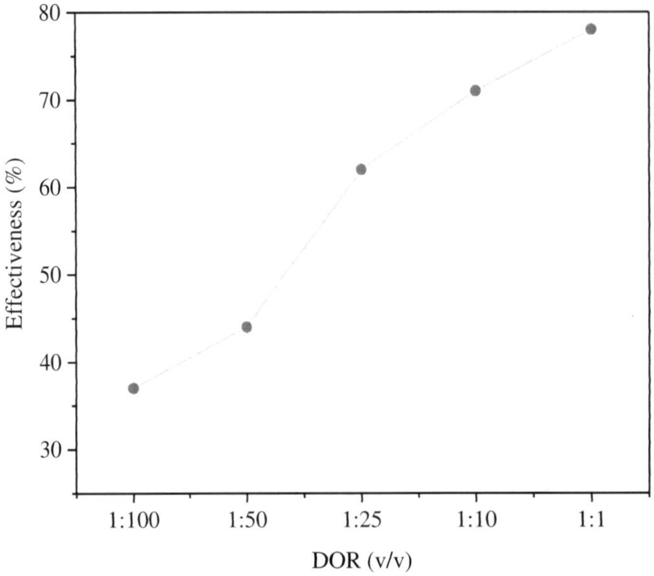

Figure 4. Effectives of dispersant at different DOR. The composition of the components are 60:40 ([Cho][Myr]:[Cho][Ol]) for all DOR.

3.4. Crude oil and water emulsions

Crude oil sea water emulsions were prepared by maintaining the oil water ratio at 1:10 (w/w) and presented in figure 5. A dispersant oil ratio of 1:1 (w/w) was kept constant for all experiments. Results presented in figure 5 are for three cases: [Cho][Myr] alone, [Cho][Ol] alone and a 60:40 ratio of [Cho][Myr] and [Cho][Ol]. For the first two cases initially, a homogeneous brown coloured emulsion is formed (at t = 0 min) after vortex mixing. However, after 30 min of vortex mixing the oil droplets rises to the top by forming a distinct layer of oil above the aqueous phase, indicating that the emulsions in both cases are unstable. In contrast, the emulsion prepared with 60:40 ratio of [Cho][Myr] and [Cho][Ol] behaved differently. In this case, the emulsion formed at (t = 0 min) and after 30 min are almost identical and there is no indication of oil separation from aqueous phase. Therefore, we can infer from visual observations that blend of [Cho][Myr] and [Cho][Ol] formed a much stable emulsion and acted as an efficient dispersant than either its stand alone components.

Figure 5. Emulsification formed between crude oil and sea water (a) 100% [Cho][Ol] (b) 100% [Cho][Myr] (c) 60% [Cho][Myr] + 40% [Cho][Ol].

4. Conclusion

In this study the choline based ILs were employed to combat oil spill issues. The most effective [Cho][Myr] and [Cho][Ol] mixture ratio is 60:40 (w/w). A stable emulsion was observed at this ratio compared to the individual components. At this optimal ratio, the dispersant effectiveness was 71 % having DOR 1:10 (v/v). For ratio [Cho][Myr]:[Cho][Ol] higher/lower than this, effectiveness is decreased by high interfacial tension. Furthermore, by increasing the DOR ratios the effectiveness increases, and 78 % effectiveness is achieved at 1:1 (v/v). Moreover, instead of toxic organic solvents, water was used as a solvent which will ultimately make it an environmentally benign dispersant.

Acknowledgements

The authors would like to gratitude, centre of research in ionic liquids (CORIL), Universiti Teknologi PETRONAS, Malaysia for providing research facilities and funding (0153AA-E35). M.U.H Shah gratefully acknowledge the financial assistance through Graduate Research Assistance (GRA) scheme by the centre of research in ionic liquids (CORIL), Universiti Teknologi PETRONAS.

References

[1] Kostka J E, Prakash O, Overholt W A, Green S J, Freyer G, Canion A, Delgardio J, Norton N, Hazen T C and Huettel M 2011 Hydrocarbon-degrading bacteria and the bacterial community response in Gulf of Mexico beach sands impacted by the Deepwater Horizon oil spill *App. and env. micro.* **77** 7962-74

[2] Gong Y, Zhao X, Cai Z, O'reilly S, Hao X and Zhao D 2014 A review of oil, dispersed oil and sediment interactions in the aquatic environment: influence on the fate, transport and remediation of oil spills *Marine pol. bul.* **79** 16-33

[3] Stoermer S, Butler G and Henry C 2001 *Application of dispersants to mitigate oil spills in the Gulf of Mexico: The Poseidon Pipeline spill case study* International Oil Spill Conference: American Petroleum Institute 1227-9

[4] Carson R T, Mitchell R C, Hanemann M, Kopp R J, Presser S and Ruud P A 2003 Contingent valuation and lost passive use: damages from the Exxon Valdez oil spill *Env. and res. eco.* **25** 257-86

[5] Hamid A G 2017 Filling the legal vacuum in combatting maritime crimes: a law reform strategy for Malaysia *Int. J. of Public Law and Policy* **6** 1-20

[6] Board O S 2005 *Oil spill dispersants: efficacy and effects. Committee on Understanding Oil Spills: Efficacy and Effects* National Research Council National Academies Press, Washington

[7] Varadaraj R, Robbins M, Bock J, Pace S and MacDonald D 1995 *Dispersion and biodegradation of oil spills on water* International Oil Spill Conference: American Petroleum Institute 101-6

[8] Almeda R, Hyatt C and Buskey E J 2014 Toxicity of dispersant Corexit 9500A and crude oil to marine microzooplankton *Eco. and env. safety* **106** 76-85

[9] Maier R and Soberon-Chavez G 2000 Pseudomonas aeruginosa rhamnolipids: biosynthesis and potential applications *App. Micro. and Biotech.* **54** 625-33

[10] Shah M U H, Sivapragasam M, Moniruzzaman M, Talukder M M R, Yusup S B and Goto M 2017 Production of sophorolipids by Starmerella bombicola yeast using new hydrophobic substrates *Bio. Eng. J.* **127** 60-7

[11] Wang A, Li Y, Yang X, Bao M and Cheng H 2017 The enhanced stability and biodegradation of dispersed crude oil droplets by Xanthan Gum as an additive of chemical dispersant *Marine pol. bul.* **118** 275-80

[12] Michel J, Adams E, Addassi Y, Copeland T, Greeley M, James B, Mcgee B, Mitchelmore C, Onishi Y and Payne J 2005 *Oil spill dispersants: efficacy and effects* The National Academy of Sciences: Washington, DC

[13] Nyankson E, DeCuir M J and Gupta R B 2015 Soybean lecithin as a dispersant for crude oil spills *ACS Sus. Chem. & Eng.* **3** 920-31

[14] Owoseni O, Nyankson E, Zhang Y, Adams S J, He J, McPherson G L, Bose A, Gupta R B and John V T 2014 Release of surfactant cargo from interfacially-active halloysite clay nanotubes for oil spill remediation *Langmuir* **30** 13533-41

[15] Nyankson E, Olasehinde O, John V T and Gupta R B 2015 Surfactant-loaded halloysite clay nanotube dispersants for crude oil spill remediation *Ind.l & Eng. Che. Res.* **54** 9328-41

[16] Gadilohar B L and Shankarling G S 2017 Choline based ionic liquids and their applications in organic transformation *J. of Mol. Liq.* **227** 234-61

[17] Aburto J, Márquez D M, Navarro J C and Martínez-Palou R 2014 Amphiphilic Choline Carboxylates as Demulsifiers of Water-in-Crude Oil Emulsions *Ten. Surfac. Deter.* **51** 313-7

[18] Petkovic M, Ferguson J L, Gunaratne H N, Ferreira R, Leitao M C, Seddon K R, Rebelo L P N and Pereira C S 2010 Novel biocompatible cholinium-based ionic liquids—toxicity and biodegradability *Green Chem.* **12** 643-9

[19] Sorial G A, Venosa A D, Koran K M, Holder E and King D W 2004 Oil spill dispersant effectiveness protocol. II: Performance of revised protocol *J. of Env. Eng.* **130** 1085-93

[20] Athas J C, Jun K, McCafferty C, Owoseni O, John V T and Raghavan S R 2014 An effective dispersant for oil spills based on food-grade amphiphiles *Langmuir* **30** 9285-94

[21] Nyankson E, Demir M, Gonen M and Gupta R B 2016 Interfacially active hydroxylated soybean lecithin dispersant for crude oil spill remediation *ACS Sus. Chem. & Eng.* **4** 2056-67

IOP Conf. Series: Materials Science and Engineering **458** (2018) 012030 doi:10.1088/1757-899X/458/1/012030

Single stage extraction and recovery of hexavalent chromium using blended TOA-TOMAC in palm oil-based diluent via supported liquid membrane process

N Othman[1, 2,*], R N R Sulaiman[1] and S A Ning[1]
[1] Department of Chemical Engineering, Faculty of Chemical and Energy Engineering, Universiti Teknologi Malaysia, 81310, UTM, Johor Bahru, Johor, Malaysia
[2] Centre of Lipids Engineering and Applied Research (CLEAR), Ibnu Sina Institute for Scientific and Industrial Research, Universiti Teknologi Malaysia, 81310, UTM, Johor Bahru, Johor, Malaysia

*norasikin@cheme.utm.my

Abstract. Diluent is an important components in the supported liquid membrane (SLM) process. The conventional diluents used in SLM are usually flammable, volatile, and toxic. To promote a sustainable development, the palm oil was incorporated in the SLM for the removal and recovery of chromium. SLM is a three-phase system with an organic phase containing the carrier in diluent which immobilized in a membrane support and is set in between the simulated chromium and sodium hydroxide solutions that act as a feed and stripping phases respectively. Both solutions were pumped into the membrane support cell in a recycled operation for about 5 hours. To monitor the changes of the chromium ion concentration for both phases, the chromium ion concentration in the feed and stripping phases as a function of time was analysed using an atomic absorption spectrometry (AAS). Several parameters namely type of stripping agent, diluent composition and carrier concentration were investigated. Results showed that about 75 and 73% of chromium was extracted and recovered, respectively at the best conditions of using sodium hydroxide, mixture of palm oil-kerosene (50:50) and TOA-TOMAC (0.20-0.20M). Thus, palm oil is regarded as feasible as a greener diluent in SLM process for chromium ion extraction.

1. Introduction

Chromium is widely used in various types of industry owing to its good features of magnetic properties, hardness and anti-corrosion. It is also known as alloying material for steel, which is commonly used for surface coating and refractory material. Other applications of chromium are in the preservation of wood, leather tanning, synthetic manufacturing, industrial catalysts, and colour pigments for paints [1-2]. The most common oxidation state of chromium that exists in aqueous is hexavalent (Cr (VI)) and trivalent chromium (Cr (III)). However, due to the high solubility and bioavailability, hexavalent chromium is more toxic compared with the trivalent chromium due to its ability to form various anionic species in aqueous solution, which are harmful, carcinogenic and mutagenic compounds towards environment [3]. World Health Organization (WHO) and USEPA have declared chromium as one of the most toxic metal for the environment [4]. Thus, removal of chromium from the industrial effluents have become a necessary task in ensuring environmental safety.

Previously, a few conventional techniques have been employed in order to remove chromium from industrial waste and wastewater such as precipitation [5], solvent extraction [6], reverse osmosis [7], and electrodialysis [8]. Every conventional method has its own advantages and limitations. For instance, precipitation process achieves high percentage of heavy metal removal with an addition of other chemicals. However, this process generates the sludge that leads to an extra disposal cost. Meanwhile, solvent extraction involves high consumption of chemical whilst reverse osmosis is a simple operation with effective removal of contaminants but requires expensive monitoring system and high energy consumption. On the other hand, electrodialysis implies high capital and operating cost due to fouling and scaling of the membrane. However, these methods only remove without recovering the chromium from wastewater. Hence, in view of the drawbacks of the conventional methods, the liquid membrane (LM) technology has attracted much interest and appears to be an advanced technique due to its great potential in the field of separation process for various organic compounds and metallic ions [9]. The main advantages of LM technology is the removal and recovery in one single step. Therefore, the targeted solute can be simultaneously removed and recovered.

Principally, LM technology is composed of three main configurations or designs namely bulk liquid membrane (BLM), emulsion liquid membrane (ELM) and supported liquid membrane (SLM). In LM system, there are three main phases involved including feed, membrane and stripping phases. Theoritically, the mechanism of LM occurs as the targeted solute is transported from feed to stripping phases across the LM phase that acts as the barrier [10]. Mostly, the diffusion of the targeted solute is facilitated by the carrier in LM phase. In BLM, a relatively thick layer of immiscible LM contained carrier in organic diluent separate the bulk feed and stripping phases [11]. LM phase contains carrier as well as surfactant in organic diluent to form emulsion known as ELM [12]. Meanwhile, LM phase contains carrier in organic diluent and is placed in a microporous polymer membrane support known as SLM[13].

Among these configurations, SLM is one of the simplest, efficient and easy to be scaled up.The main components in SLM system are carrier, stripping agent and diluent. Carrier is chosen based on the selectivity of the targeted solute ion present in the feed phase and can either be acidic, basic and solvating. Acidic carriers such as 2-hydroxy-5-nonylacetophenone oxime (LIX841),5,8-diethyl-7-hydroxy-dodecan-6-oxime(LIX63), phosphorus derivative (di-2-ethylhexyl phosphoric acid (D2EHPA) and di-2,4,4-trimethylpentyl phosphinic acid (Cyanex 272) are generally used for extraction based on the cationic exchange mechanism [14-17]. Meanwhile, basic carriers such as trioctylamine (TOA),tridodecylamine (TDA), tri-n-octylmethylammonium chloride (TOMAC) and etc undergo metal ion extraction through anion exchange reaction [18-20]. Subsequently, the solvating or neutral carriers are basic in nature, hence extracting either neutral metal complexes or acids through the solvate formation [21-22]. To date, an application of mixed carriers is gaining attention to improve the metal ion extraction efficiency via synergistic effect. Previously, Sulaiman and Othman [23] prove that about 83% of nickel ions were successfully extracted via the mixture system of 0.08 M LIX63 and 0.02 M D2EHPA with the maximum synergistic enhancement factor, R_{max} of 29.56. This is also supported by Singh *et al.* [24] who reported similar observation as more than 95% of uranium (VI) were recovered in 360 minutes of SLM process using a binary mixture of 0.60 M PC88A and 0.15 M Cyanex 923 in dodecane. Subsequently, good diluent should provide good characteristics such as high solubility, low flash point [25]. So far, the petroleum based diluents have been employed for metal ion extraction in liquid membrane technology namely kerosene, n-heptane and toulene and etc [26]. Nevertheless, this type of diluents is toxic, non-renewable, non-biodegradable, flammable and volatile in nature. An introduction to the green organic diluent such as vegetable oils attracts the attention of several reseachers as a way to promote greener process in the future [27]. Acording to Jusoh *et al.* [28], the combination of 30/70 kerosene to palm oil also leads to high separation of succinic acid from the fermentation broth. Another observation reported by Chakrabarty *et al.* [29] claimed that in about 95% of mercury were extracted using coconut oil as diluent through the SLM process. Besides, Othman *et al.* [30] also found that almost 100% of chromium were extracted and recovered using the diluent containing kerosene-palm oil mixture with the ratio of 3:7.On the other hand, stripping agent also plays a crucial role as a binder for the targeted solute at the membane-stripping interface. Commonly, metallic ions extracted by basic carriers seems suitable to be stripped using neutral or alkaline solutions and vice versa in order to create the chemical potential between both membrane and stripping phase [31].

In this present investigation, removal and recovery of chromium (VI) using SLM was studied. A liquid membrane formulation placed in the membrane support composed of a blended carrier of TOA and TOMAC in the mixture of kerosene and palm oil as organic diluent. Several parameters affecting the extaction and recovery performance of chromium (VI) such as the type of stripping agents, the composition of diluent and the concentration of carrier were examined.

2. Experimental

2.1. Reagent and materials

Potassium dichromate, ($K_2Cr_2O_7$), supplied by Sigma Aldrich in powdered form, was used as a source of hexavalent chromium Cr (VI) in the simulated wastewater. Trioctylamine (TOA) and Tri-n-octylmethylammonium chloride (TOMAC) as carriers were procured from Fluka. Kerosene and cooking palm oil as diluents were obtained from Merck and Mart, respectively. Meanwhile, sodium hydroxide and sulfuric acid as stripping agents were purchased from Merck. Besides, polyvinylidenfluoride (PVDF) ordered from Millipore was used as a membrane support for organic solution with an average effective pore diameter of 0.22μm, an average thickness of 125μm and porosity of 75%. All these materials were of analytical reagent grade and used directly as received from the manufacturer without further purification.

In fact, tertiary amines (TOA) and quaternary ammonium salts (TOMAC) are the most widely used ionic carriers in chromium ion extraction using SLM process owing to the high coordination ability and stability of the complex strength [32-33]. Besides, in terms of the carrier-chromium interaction perspective, more highly structured basic mixed carrier (TOA-TOMAC) is favored due to the high degree of solvation of carriers with the chromium ion in the membrane, hence enhancing the stability of the carrier-chromium complexation [34].

2.2. SLM set up and extraction

SLM rig set up is composed of membrane cell, feed and strip vessel, double head peristaltic pump, flowmeter and tubing. There are three main phase involve in SLM process which are feed, membrane and stripping phases. The feed phase contains simulated chromium (VI) solution, which was prepared by mixing an appropriate quantity of potassium dichromate in deionized water with measured pH of 2. Subsequently, organic liquid membrane was prepared by dissolving the corresponding volume of carriers in the organic diluent to obtain carrier solutions of different concentrations, ranging from 0.05 to 0.35M. The PVDF membrane support (10cm x 3.5 cm) was impregnated with the organic solution for 24 hours before leaving it to drip for a few seconds followed by placing it in the membrane cell as shown in figure 1. Meanwhile the stripping solution (sodium hydroxide) of desired concentration was prepared by dissolving appropriate weight of pellets in deionized water. About 300 mL of feed and strip solutions were added into feed and strip vessel, respectively. These solutions were pumped into the membrane cell with a recycled operation. Both aqueous feed and stripping solutions were magnetically stirred to avoid concentration polarization conditions at the membrane interfaces and in the bulk of the solutions. Ten mL sample of each feed and stripping solutions were periodically taken every 30 minutes for 5 hours to determine the changes in the chromium ion concentration using Atomic Absorption Spectrometry (AAS) with a wavelength of λ=540 nm. The new membrane support was used for each experiment. The experiment was carried out at ambient temperature (25 ± 1°C) with standard deviations of chromium ion concentrations less than ± 5%.

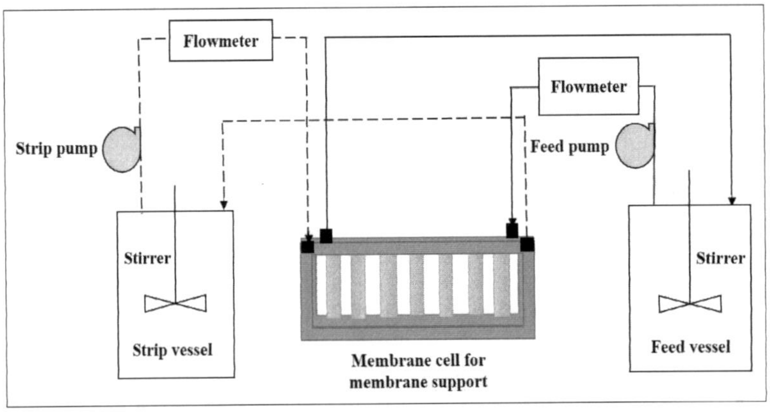

Figure 1. Schematic diagram of SLM process.

2.3 Data analysis

2.3.1. Extraction and recovery performance. The extraction and recovery performance of chromium ion through SLM process are determined using equations (1) and (2), respectively:

$$\text{Removal (\%)} = \frac{[Cr]_{f_o} - [Cr]_{f_t}}{[Cr]_{f_o}} \times 100 \quad (1)$$

$$\text{Recovery (\%)} = \frac{[Cr]_s}{[Cr]_{f_o}} \times 100 \quad (2)$$

Where $[Cr]_{f_o}$ represents the initial chromium in the feed phase; $[Cr]_{f_t}$ indicates the final concentration of chromium in the feed phase and $[Cr]_s$ denotes the concentration of chromium in the stripping phase.

2.3.2. Determination of permeability value. Membrane permeability is defined as the ability of a membrane to allow the desired solute for passing through. Permeability of the chromium ions extracted from feed to stripping phases is determined using equation (3) [35]:

$$ln\frac{c}{c_i} = -p\left(\frac{A}{V}\right)t \quad (3)$$

where c_i is the initial concentration of chromium ions in the feed phase, c is the concentration of chromium ions at a given time, p is the permeability value (cms^{-1}), A is the effective area of the membrane (cm^2), V is the volume of aqueous feed phase (cm^3) and t is the time.

2.3.3. *Determination of liquid membrane loss.* Liquid membrane loss, Δm is used to evaluate an instability of SLM extraction of chromium ion and can be determined using equation (4) [36]:

$$\Delta m\ (gcm^{-2}) = \frac{(m_1 - m_2) - (m_1 - m_0)}{A} \quad (4)$$

Where m_0, m_1 and m_2 are the weights of the dry, wet, and used membrane support, and A is the effective area. A weight of dry membrane is a weight membrane before used while weight wet membrane designates the weight membrane after impregnation with organic liquid membrane. Then, the weight of used membrane refers to the weight of membrane after extraction process.

3. Results and discussion

3.1. Transport mechanism of chromium (VI) ion extraction in SLM

Apparently, chromium ion is able to exist in aqueous solution as $HCrO_4^-$, CrO_4^{2-}, $HCr_2O_7^-$ and $Cr_2O_7^{2-}$ depending on the pH value of the solution and the total concentration of chromium. As the pH is lower than 0.5 or too acidic, the chromic acid ($H_2Cr_2O_7$) is predominant. Meanwhile, $Cr_2O_7^{2-}$ appears to be the main anion in an acidic aqueous phase. However, $Cr_2O_7^{2-}$ is converted into $HCrO_4^-$, in acidic aqueous solution as the concentration of chromium lies in the range lower than (1.26 – 1.74) x 10^{-2} mol/L. Beyond this critical concentration values, Cr(VI) is generally found as CrO_4^{2-} [37]. As the lowest concentration used lies in the range lower than the concentration mentioned above, $HCrO_4^-$ is the prevailing anion that existed in the chromium solution as a feed phase in this present work. The mechanism for simultaneous extraction and recovery of chromium ion using SLM process is illustrated in figure 2. In this study, both blended TOA-TOMAC and NaOH act as carrier and stripping agent, respectively.

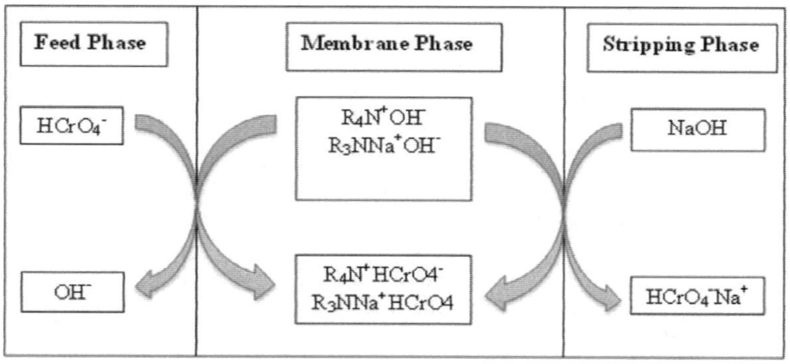

Figure 2. Transport mechanism of chromium ion extraction via SLM process.

In SLM transport process, the diffusion of chromium ions takes place via the following steps:

a) Both basic carriers, TOA (R_3N) and TOMAC ($R_4N^+Cl^-$) in the membrane phase were protonated by the stripping agent, NaOH at the membrane-stripping interface as shown in equation (5):

$$[R_3N]_{org} + [R_4N^+Cl^-]_{org} + 2[Na^+OH^-]_{aq} \leftrightarrow [R_3NNa^+OH^-]_{org} + [R_4N^+OH^-]_{org} + [Na^+]_{aq} + [Cl^-]_{aq} \quad (5)$$

b) During the extraction, the protonated carrier molecules in the membrane phase react chemically with the hydrochromate ions at the feed-membrane interface, hence forming chromium-carrier complexes as represented in equation (6):

$$[R_3NNa^+OH^-]_{org} + [R_4N^+OH^-]_{org} + 2[HCrO_4^-]_{aq} \leftrightarrow [R_3NNa^+HCrO_4^-]_{org} + [R_4N^+HCrO_4^-]_{org} + 2[OH^-]_{aq} \quad (6)$$

c) Subsequently, these chromium-carrier complexes of $R_3NNa^+HCrO_4^-$ and $R_4N^+HCrO_4^-$ complexes diffuse across the membrane phase from feed-membrane interface to the membrane-stripping interface reversibly. At the membrane-stripping interface, the stripping reaction with NaOH take place as represented in equations (7) and (8):

$$[R_3NNa^+HCrO_4^-]_{org} + [Na^+OH^-]_{aq} \leftrightarrow [R_3NNa^+OH^-]_{aq} + [Na^+HCrO_4^-]_{org} \quad (7)$$

$$[R_4N^+HCrO_4^-]_{org} + [Na^+OH^-]_{aq} \leftrightarrow [R_4N^+OH^-]_{aq} + [Na^+HCrO_4^-]_{org} \quad (8)$$

3.2. *Effect of stripping agent type*

To elucidate the effect of stripping agent type, two types of stripping agent namely sodium hydroxide (NaOH) and sulphuric acid (H_2SO_4) were employed for simultaneous extraction and recovery of chromium ion in SLM process. The extraction and recovery performance of chromium ion as a function of different types of stripping agent after 5 hours of experiment are tabulated in table 1. It can be clearly seen that NaOH showed a better performance with 75 and 73% of extraction and recovery percentage, respectively. Conversely, H_2SO_4 only provided up to 42 and 10% of extraction and recovery percentage, respectively. As both carriers used are the basic type, the alkaline and neutral stripping agent are preferable in creating the chemical potential between the membrane and stripping phases [31]. Fundamentally, both ionization energy and electron affinity become higher as we go up a column of a periodic table. Since oxygen atom (-O-) is above sulphur atom (-S-) in the periodic table, it has higher ionization energy as well as electron affinity. Therefore, it provides high tendency to rapidly break the ionic bonding with sodium ion, hence enhancing the stripping efficiency. In addition, the lowered stripping efficiency for sulfuric acid solution is probably due to the competition between sulfate and chromium ions which working in favor of forming a more stable carrier–chromium complex of larger solvation degree, allowing chromium to stay in the organic phase [34]. As a result, NaOH is employed as stripping agent throughout this study.

Table 1. Effect of stripping agent types towards extraction and recovery efficiency of chromium

Stripping agent	Extraction (%)	Recovery (%)
Sodium hydroxide (NaOH)	75	73
Sulphuric acid (H_2SO_4)	42	10

3.3. *Effect of diluent composition*

The feasibility of palm oil as a substitute diluent for SLM process is evaluated by studying the effect of various diluents towards the extraction and recovery performance of chromium ion as shown in figure 3. Meanwhile, the variation of the permeability values, viscosity and liquid membrane loss with respect to the different diluent composition are tabulated in table 2. According to figure 3, both liquid membrane containing 100% kerosene and mixtures of palm oil and kerosene (50:50) achieved high percentages of chromium ion extraction (73%). Likewise, the recovery efficiency also showed that the liquid membrane with 100% kerosene provided the highest recovery (81%) followed with the one containing the mixtures of palm oil and kerosene (73%) within 5 hours of experiment. Besides, the changes in the permeability values from 6.3 to 4.2 x10^{-4} cms^{-1} upon mixing with the 50% of palm oil also was observed. The low permeation of chromium ion across the membrane phase is caused by the increment of the viscosity liquid membrane from 32 to 41 cP. Theoretically, liquid membrane with high viscosity tends to retard the permeation of chromium ion across the membrane phase, thereby inhibiting a substantial amount of chromium ion being transported to the stripping phase. This is in line with Kumar *et al.* [37] who claimed that high viscosity solvent is able to create high resistance as well as hindering the mass transfer of solute ion in the membrane phase.

On the other hand, the liquid membrane containing 100% palm oil provided the lowest extraction (22%) and recovery (8%) performance. This is probably due to the highest viscosity of 72cP which thereby hindering the permeation of chromium ion across the membrane phase. The low permeation of chromium ion into the membrane phase is in accordance with Chakrabarty *et al.* [29] who indicated that the accumulation of higher amount of oil phase on the membrane surface as a consequence of high viscosity solvent seems to block the chromium ion from reacting with the carrier at the feed-membrane interface. Hence longer time is needed for chromium ion to pass through the membrane phase. Besides, Chang *et al.* [38] also found that the non-polarity nature of vegetable palm oil tends to interact weakly with polar compounds such as chromium resulting in poor solubility of solute ion in the organic membrane phase. As referring to the liquid membrane loss analysis, liquid membrane containing palm oil suffered higher loss compared to the mixture of palm oil and kerosene as well as fully kerosene. This means the liquid membrane containing fully palm oil seems unstable and easily loss the liquid membrane into the aqueous phase.

Thus, it can be inferred that the sequence of diluent composition which significantly enhanced the extraction and recovery efficiency is as in the following order: 100% kerosene> mixture of kerosene and palm oil (50:50)> 100% palm oil. However, as a way to promote green technology, mixture of kerosene and palm oil (50:50) was employed for the next investigation.

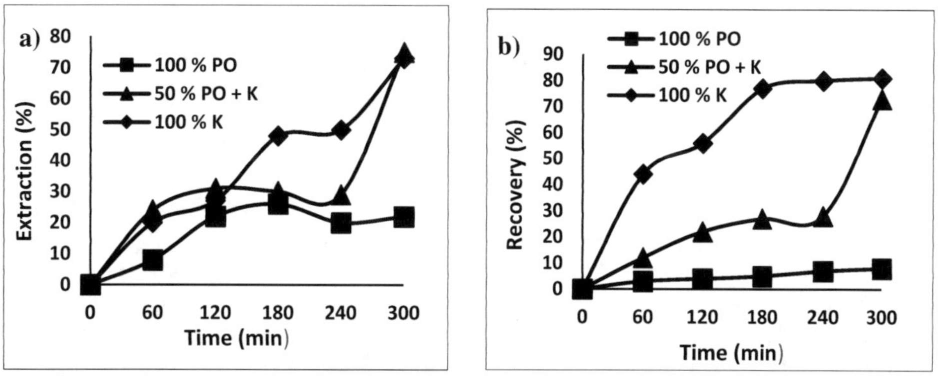

Figure 3. (a) Effect of diluent composition on chromium extraction and (b) Effect of diluent composition on chromium recovery where PO=palm oil and K=kerosene.

Table 2. Variation of permeability and liquid membrane loss with respect to the effect of diluent composition.

Diluent composition	Permeability, ρ $x10^{-4}$ (cms^{-1})	Liquid membrane loss (gcm^{-2}) $x10^{-3}$	Viscosity (cP)
100% palm oil	2.1	4.7	72
50% palm oil + 50% kerosene	4.2	1.1	41
100% kerosene	6.3	2.2	32

3.4. Effect of carrier concentration

The carrier has a profound significance in SLM process by acting as a shuttle in transporting metal ion from feed to the stripping phases through membrane phase. Theoretically, the rate of metal ion permeation increases with a rise in carrier concentration [36]. Therefore, in order to understand the effect of carrier concentrations towards extraction and recovery performance of chromium ion, an equimolar concentrations of blended TOA-TOMAC were varied from 0.05 to 0.35M as shown in figure 4. In the meantime, the variations of the permeability values, viscosity, and liquid membrane loss with respect to the different carrier concentration are tabulated in table 3. According to figure 4(a), it can be noted that the extraction percentage of chromium ion within 5 hours of extraction time showed a steady increase from 21 to 75% upon enhancing the blended TOA-TOMAC concentration from 0.05 to 0.20M, respectively. Obviously, low permeation of chromium ion in the membrane phase (2.1 x 10^{-4} cms^{-1}) is triggered by low carrier concentration of 0.05M. Apparently, low carrier concentration is inadequate to assist the mass transfer of chromium ion transportation in the membrane phase. This seems to be consistent with Altin et al. [39] who found that the low quantity of carrier molecules in the membrane phase cause less transportation of targeted solute into the stripping phase. However, an increment of carrier concentration up to 0.20M actually provides high number of carrier molecules available for the formation of chromium-carrier complex at the feed-membrane interface. This behaviour led to higher permeation rate of chromium ion which was almost doubled (4.2 x 10^{-4} cms^{-1}) as well as improving the extraction efficiency. Beyond 0.20M, a sudden fall of extraction efficiency (26%)was observed. The reduction of permeation rate to 3.1 x 10^{-4} cms^{-1} can be adversely affected by the viscosity effect. It was proven by an increment trend of viscosity from 36 to 53cP as the carrier concentration ws increased from 0.05 to 0.35M, respectively. The viscous membrane phase seems to inhibit the diffusion of chromium ion across the membrane phase, which thereby leading to an extraction inefficiency. Similar result was reported by Rehman et al. [40] who indicated that according to Stokes–Einstein in equation (10), diffusity is inversely proportional to viscosity. Thus, lower viscosity has higher solute ion diffusion through membrane phase and vice versa.

$$D=\frac{KT}{6n\pi r} \quad (10)$$

where T is the absolute temperature, K is the Boltzmann constant, r is the ionic radius of solute and n is the viscosity of the organic phase in cP.

On the other hand, figure 4(b) illustrates the recovery performance of chromium ion as a function of different carrier concentrations. Notably, recovery efficiency showed similar trend with extraction which implies a simultaneous extraction and stripping reaction during SLM process. At low concentration of 0.05M, the recovery efficiency provided a slow and steady increase up to 15% in 5 hours of experiment. A possible explanation for this situation is the insufficient number of carrier molecules present in the membrane phase, thus retards the chromium ions transportation into the stripping phase. This behaviour was also experienced by Sulaiman and Othman [3] who believed that the low carrier concentration tends to reduce the mass transfer of chromium ions, hence accumulating these complex in the feed-membrane interface without being transported into the stripping phase.

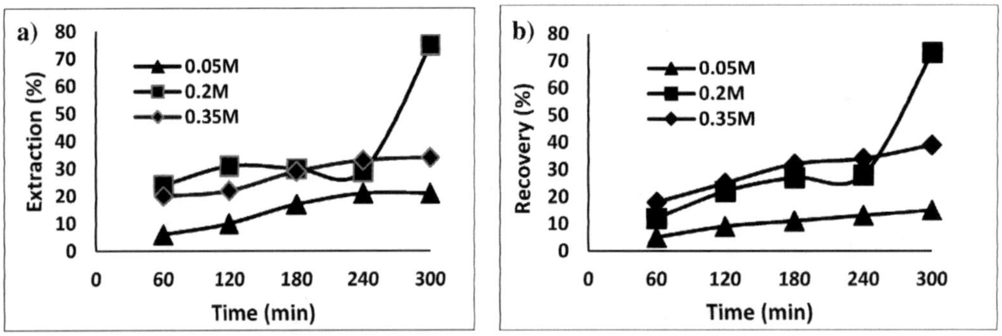

Figure 4. (a) Effect of carrier concentration on chromium extraction and (b) Effect of carrier concentration on chromium recovery where PO=palm oil and K=kerosene.

Surprisingly, further increment of carrier concentration up to 0.20M, the recovery percentage significantly increased up to 73% in 5 hours of experiment. It is due to the high number of carrier molecules which facilitates the permeation of chromium complexes in the membrane phase. Basically, the basic carriers of TOA and TOMAC containing amine group has high tendency in forming hydrogen bonding with water molecules in the aqueous phase, thus is able to create water channels through the membrane pores. This behaviour promotes direct channeling between the feed and stripping phases. For a longer experiment, the liquid membranes in the pores slowly oozes out into the aqueous phase, thus affecting the recovery efficiency.This is in accordance with Zhang et al. [35] who claimed that the lifetime of SLM process not only relies on the porosity but also on the ability of the membranes to keep the pores free from water channels.

Meanwhile, further increment of carrier concentration to 0.35 M, resulted to steady increment of the recovery percentage up to 29% in 5 hours of extraction time. The low recovery might be due to the high viscosity membrane phase that is crowded with high number of chromium-carrier complexes without being stripped. It can also be noted that the liquid membrane loss seemed to increase in about twice (2.1×10^{-3} g/cm^2). According to Huidong et al.[41], the liquid membrane is not completely insoluble in an aqueous solution and a certain degree of solubility exists at the aqueus-membrane interface. An increase in carrier concentration, increases the solubility of carrier at the aqueus-membrane interface. As a result, this phenomena reduce the interfacial tension among both phases and the aqueous solution is able to wash away the liquid membrane. Thus the best condition was obtained at 0.20 M of TOA-TOMAC with 75 and 73% of extraction and recovery, respectively

Table 3. Variation of permeability, viscosity, and liquid membrane loss as a function of the effect of carrier concentration.

[TOA-TOMAC](M)	Permeability, ρ x 10^{-4} (cms^{-1})	Viscosity (cP)	Liquid membrane loss (gcm^{-2}) x 10^{-3}
0.05 -0.05	2.1	36	1.9
0.20 -0.20	4.2	41	1.1
0.35 -0.35	3.1	53	2.1

Conclusion

Through this work, the SLM process is shown to be a promising method due to its capabilities in extracting the hazardous hexavalent chromium complex from aqueous solution. Besides, the feasibility of palm oil as a substitute green diluent in SLM process has been proven. This green process has high potential to be applied in industrial level for the treatment of wastewater containing heavy metal ions.

Acknowledgments

The authors would like to acknowledge the Ministry of Higher Education (MOHE) (FRGS: R.J130000.7846.4F949) and Universiti Teknologi Malaysia (UTM) for the financial support of this research. Besides, Raja Norimie Raja Sulaiman also would like to express her sincere gratitude to the Universiti Teknologi Malaysia for the sponsorship of UTM Zamalah.

References

[1] Rajewski J and Religa P 2016 *J. Mol. Liq* ***218*** 309

[2] Tapiero Y, Sánchez J and Rivas B L 2017 *Chi. J. Chem. Eng* **25** 938

[3] Sulaiman R N R and Othman N 2017 *Malays. J. Anal. Sci* **21** 416

[4] W.H. Organization 2008 *Guidelines for drinking-water quality: Incorporating first and 2nd Addenda. Recommendations, (third edition)* Geneva, Swiss: World Health Organization Vol. 1, p 668

[5] Sowmya P T, Mahadevraju G K, Ramesh A and Sreenivas V 2013 *Int. J. Innov. Res. Develop* **2** 65

[6] Mane C P, Mahamuni S V, Kolekar S S, Han S H and Anuse M A 2016 *Arabian J. Chem* **9** *S1420*

[7] Cimen A 2015 *Russian Journal of Physical Chemistry* A **89** 1238

[8] Sadyrbaeva T Z 2016 *Chem. Eng. Process. Process Intensif* **99** 183

[9] Sulaiman R N R, Othman N and Amin N A S 2013 *Jurnal Teknologi* **65** 33

[10] Sulaiman R N R, Othman N and Amin N A S 2016 *Desalin. Water treat* **57** 3339

[11] Han A, Zhang H, Sun J, Chuah G K and Jainicke S 2017 *J. Water. Process Eng* **18** 63

[12] Goyal R K, Jayakumar N S and Hashim M A 2011 *Desalination* **278** 50

[13] Venkateswaran P and Palanivelu K 2005 *Hydrometallurgy* **78** 107

[14] Ochromowicz K and Apostoluk W 2010. *Sep.Purif.Technol* **73** 112

[15] Hosseini T, Mostoufi N, Daneshpayeh M and Rashchi F 2011 *Hydrometallurgy* **105** 277

[16] Batchu N K, Jeon S H and Lee M S 2011 *J. Ind. Eng. Chem* **26** 286

[17] Andrade F and Elizalde M P 2005 *Solvent Extr. Ion Exch* **23** 85

[18] Rajasimman M and Sangeetha R 2009 *J. Hazard. Mater* **168** 291

[19] Nayl A A and Aly H F 2015 *Nonferrous Met. Soc. China* **25** 4183

[20] Rahul K G, Jayakumar N S and Hashim M A 2011 *J. Hazard. Mater* **195** 383

[21] Kumbasar R A 2009 *J. Hazard. Mater* **167** 1141

[22] Kumbasar R A 2010 *J. Hazard. Mater* **178** 875

[23] Sulaiman R N R and Othman N 2017 *J. Hazard. Mater* **340** 77

[24] Singh S K, Misra S K, Tripathi S C and Singh D K 2010 *Desalination* **250** 19

[25] Rajasimman M and Sangeetha R 2009 *J. Hazard. Mater* **168** 291

[26] Alguacil F J, Coedo A G and Dorado M T 2000 *Hydrometallurgy* **57** 51

[27] Bjorkegren S, Fassihi K R, Martinelli A, Jayakumar N S and Hashim M A 2015 *Membranes* **5**168

[28] Jusoh N, Othman N and Nasruddin N A 2016 *Malays. J. Anal. Sci* **20** 436

[29] Chakrabarty K, Saha P and Ghoshal A K 2010 *J. Membr. Sci* **350** 395

[30] Othman N, Noah N F M, Poh K W and Yi O Z 2016 *Procedia Engineering* **148** 765

[31] Yıldız Y, Manzak A and Tutkun O 2015 *Desalin. Water treat* **53** 1246

[32] Cezary A K and Wladyslaw W 2005 *J. Membr. Sci* **266** 143

[33] Supriyo K M and Prabirkumar S 2018 *Chem. Eng. Res. Des* **132** 564

[34] Aynur S 2015 *Sep. Sci. Technol* **50** 1010

[35] Zhang B, Gozzelino G and Baldi G 2001 *Colloids Surf* A **193** 61

[36] Eyupoglu V, Surucu A and Kunduracioglu A 2015 *Pol. J. Chem. Tech* **17(2)** 34

[37] Kumar A, Manna M S, Ghoshal A K and Saha P 2016 *J. Environ. Chem. Eng* **4** 943

[38] Chang S H, Teng T and Ismail N 2010 *J. Hazard. Mater* **181** 868

[39] Altin S, Yildirim Y and Altin A 2010 *Hydrometallurgy* **103** 144

[40] Rehman S, Akhtar G, Chaudry M A, Ali K and Ullah N 2012 *J. Membr. Sci* **389** 287

[41] Huidong Z, Biyu W, Yanxiang W and Qilong R 2009 *Chi. J. Chem. Eng* **17(5)** 750

Carboxymethyl cellulose from cavendish banana peel as a drag reduction agent

H Kaur[1*], A Japper-Jaafar[1], S Yusup[2]

[1]Department of Mechanical Engineering, Universiti Teknologi PETRONAS, 32610 Seri Iskandar, Perak, Malaysia.
[2]Department of Chemical Engineering, Universiti Teknologi PETRONAS, 32610 Seri Iskandar, Perak, Malaysia.

*harvingurchran@gmail.com, harvin_10663@utp.edu.my

Abstract. Reducing drag is reducing frictional pressure loss under turbulent flow conditions and can be achieved with addition of additives. These additives such as small amount of polymer can dramatically affect the turbulent structures of the flowing fluid and increase the flow rate. Synthetic polymers which are widely employed harm our environment when used excessively. A biopolymer is then soughed for such purpose which is produced by living organisms and they are fairly shear stable and biodegradable. A biopolymer, derived from the Cavendish banana peel, was synthesized in this study. The polymer was converted to carboxymethylcellulose by etherification process. The drag reduction was assessed in this study using a rheometer where a reduced produced torque was perceived as a reduction of drag. The biopolymers produced were then rheologically characterized where the viscoelastic effects produced by these biopolymers were utilized to explain the drag reduction phenomena. The findings shows that the biopolymer, synthesized at the right conditions, is capable of drag reduction up to 40% demonstrating the potential of the biopolymer from the green Cavendish banana peel at drag reduction when dissolved in water. The biopolymer synthesized at reaction temperature of 50°C and NaOH concentration of 40% works best at drag reduction.

1. Introduction

Drag reduction is observed as a reduced frictional pressure loss under turbulent flow conditions which will then substantially increase the flow rate of the fluid. Practical applications include water flooding systems, firefighting, agricultural spraying, and pipeline transport and drainage systems. In cases of heavy rain, for instance, existing drainage systems may not be able to cope with the amount of water introduced in the system and is further worsened with the accumulation of mud, leaves and other debris. Drag reduction agents can be introduced to increase the flow rate of water, reducing the water accumulation in the system and subsequently lessen the possibility of heavy flooding.

Widely studied polymers are usually employed to reduce drag. In a vessel, when a small amount of polymer is added, it can substantially reduce drag that the fluid has created. The important aspect of the polymers are that they possess a high molecular weight in terms of content of cellulose, chain flexibility and long linear chain structure. Most polymers are derived from petroleum and based on the chemistry of carbon, although some polymers have non-carbon based compounds.

An organic biopolymer produced by living organisms, such as polysaccharide, is a macromolecule. Cellulose is the most common organic compound and biopolymer on Earth where about 34% of all plant matter is cellulose. It is synthesized in vivo through polymerization of glucose units and association of chains by the actions of syntheses. Functionalized cellulose is strategic in the developments and applications of new biomaterials where it could be converted by etherification [1], [2]. Carboxymethylcellulose (CMC) is a copolymer of two units: B-D-glucose and B-D-glucopyranose 2-O-(carboxymethyl)-monosodium salt, not randomly distributed along the macromolecule, which are linked via B-1, 4-glycosidic bonds. It is an anionic polyelectrolyte prepared by reaction of sodium chloroacetate with alkali cellulose. CMC is non-toxic and soluble in water when the degree of substitution is higher than 0.5 [3]. It has a thickening effect, film-forming ability and excellent behavior as protecting colloid and adhesive [9] as well as shear stable. Most of the CMC applications are due to its rheological properties. The ability of CMC to function as a thickener or flow-control agent depends largely on its degree of substitution, the molecular weight, and also on the relative stiffness of the cellulosic backbone. The number of substituted hydroxyl groups per monomeric units is known as the degree of substitution (DS). [4] reported that the DS of CMC depends upon the sodium hydroxide concentration, reaction temperature and reaction time. The solubility of CMC tends to increase with increasing DS [5].

Based on research conducted, between corn cobs, sugarcane bagasse, wheat straw, banana peel and durian husk, the highest two biopolymers that contains the most cellulose are the banana peel and durian husk, the only limitation with durian husk is that the durian fruit is seasonal and not produced all year round. Cellulosic polysaccharides can be isolated from banana. Banana peels which sums up to 40% of the total weight of fresh bananas are underutilized [6]. By synthesizing the banana peel and converting it into Carboxymethylcellulose, the banana peel can be utilized. The potential applications of banana peel depend on its chemical composition and the important functional components of banana peel are the Resistant Starch (RS) and dietary fiber [7]. To produce a high amount of Carboxymethylcellulose, a banana peel should have a high content of fiber and carbohydrates with low levels of lignin [8]. [9] performed a study to compare the compositions of ripe and unripe Banana peel flours from the Cavendish (Musa acuminate L. cv. Cavendishii) and Dream (Musa acuminata colla AAA cv. Berangan) bananas and found that the green Cavendish banana peel has the highest carbohydrate level and crude fiber. The moisture content of the ripe banana peel ranges from 9.21% to 11.20% and was higher than that for unripe Banana peel [9]. This difference could be due to differences in varieties of bananas and processing parameters used. The crude fat contents of the peels were moderate (4.81%- 5.96%), and the carbohydrate contents of the peels ranged from 54.01% to 60.81%). The higher carbohydrate content of green banana peels could be related to its relatively low moisture content [9]. The high carbohydrate content is expected, as banana is known to contain sugars, starch and dietary fibers [9]. From this study, it can be concluded that the Cavendish green banana peel has a high carbohydrate level and crude fiber. The work presented in this paper focuses on the synthesis of modified cellulose from green Cavendish banana peel. The degree of substitution (DS) is varied during the synthesis by varying the sodium hydroxide concentration, reaction temperature and reaction time.

2. Methodology

2.1. Materials

The Cavendish green banana peel was utilized in this study. The chemicals used in the preparation and analysis of the carboxymethylcellulose (CMC) from the banana peel were purchased from Avantis Sdn. Bhd., Malaysia with 99% purity. They were of Analytical Reagent (AR) grade or equivalent as the AR grade had a higher purity than the standard laboratory reagent (SLR). The chemicals were sodium hydroxide, isopropanol, 90% ethanol, absolute methanol and sodium monochloroacetate. The water utilized as the solvent in the drag reduction assessment was the tap water obtained from the Universiti Teknologi PETRONAS (UTP) laboratory.

2.2. *Synthesizing of CMC*

The method of synthesizing the Carboxymethylcellulose (CMC) was adopted from the method employed by the Centre of Excellence for Polysaccharides Research, Friedrich Schiller University of Jena, Germany [10]. This method was chosen as it utilized the least amount of chemicals and relied mostly on the organic raw material. In this study, a total of 27 samples were synthesized under different parameter combinations.

2.3. *Extraction of Cellulose*

Banana peel was rinsed with tap water and sun-dried for 4 days. The dried peels were cut into small pieces below 20 mm using a Cole Parmer mortar grinder. It was then cooked with NaOH at 100°C for 3 hours on a magnetic stirrer at a constant speed of 200 rpm. The black slurry obtained was then filtered and washed with water for three times. The residue was dried in an oven at 55°C for 24 hours which resulted in pulps. The pulp was subsequently washed with water, filtered and dried again at 55°C for another 24 hours. The dried pulps were grounded into powder with an average size of 1 mm and stored in polyethylene bags to maintain minimal moisture content.

2.4. *Synthesis of Carboxymethylcellulose from Banana peel*

About 15.0g of cellulose powder from the steps above, 50ml of various concentrations of NaOH (20, 40 and 60%) and 450ml of isopropanol were mixed under an overhead stirrer in the beaker for 30 minutes at 200 rpm. The carboxymethylation reaction was initiated by adding 18.0g of monochloroacetic acid. The sample mixture was then covered with aluminium foil to avoid evaporation and heated up to the desired reaction temperature (50°C, 55°C and 60°C) in an oven for a variation of reaction time (60 minutes, 150 minutes and 240 minutes). The mixture was separated into the liquid phase and solid phase using a sieve. The liquid phase was removed and the solid phase was suspended in 100ml of methanol (70%v/v), neutralized with glacial acetic acid and filtered.

The final product was rinsed five times by suspending it in 300ml of ethanol (70%v/v) for 10 minutes to remove any undesirable by-products, and washed with 300ml of absolute methanol before being filtered for the next cycle of rinsing. The residue from filtration was dried at 55°C in the oven for 24 hours to obtain the final product: the Carboxymethylcellulose.

2.5. *Degree of Substitution (DS)*

The degree of substitution (DS) of CMC is the average number of hydroxyl group in cellulose structure which is substituted by carboxymethyl and sodium carboxymethyl group at C—2, C—3 and C—6. The DS of the CMC sample was determined using a standard ASTM method (D1439-94) known as the back titration method. The DS value is determined via:

$$A = \frac{BC-DE}{F} \quad (1)$$

$$Degree\ of\ Substitution = \frac{(0.162)xA}{1-(0.058xA)} \quad (2)$$

where A is the Mili-equivalents of consumed acid per gram of specimen, B is the Millimeters of added sodium hydroxide, C is the Normal sodium hydroxide, D is the Millimeters of consumed hydrochloric acid, E is the Normal hydrochloric acid and F is the Specimen gram used.

2.6. *Drag reduction assessment*

A standard sample preparation procedure was adopted where a 100 ppm of banana peel CMC was mixed with UTP tap water. An overhead stirrer was used with a constant low speed around 200 rpm for a duration of 3 hours to ensure the biopolymer was well mixed with water. Once the mixing was completed, the polymer solutions appeared to be visibly homogeneous and the beaker was sealed to avoid water loss by evaporation which caused an increase of concentration in the solution. Solutions were then left for at least 8 hours to ensure complete de-aeration.

To load the solution onto the controlled stress rheometer, a pipette is utilized. Once the sample was placed into the double concentric cylinder geometry on the rheometer, the solution was left to rest for 2 minutes to ensure that the molecules were sufficiently relaxed after the high shear applied during loading. A method adopted from [11] was utilized for the drag reduction assessment. A strain sweep step was performed from 0 rad/s to 300 rad/s (the maximum attainable limit by the rheometer) and the torque was monitored under the imposed angular velocity. The temperatures for all the assessments were maintained constant at 25°C with the gap within the geometry set to 2000 micron. All measurements were repeated at least three times to confirm repeatability and only a representative data sample was presented. The repeatability was found to be within 10%.

The drag reduction ability was assessed under a controlled stress rheometer by measuring the difference of the flow curves (torque vs. speed) of a treated solution to untreated solution. Any reduction in torque was believed to be contributed by the drag reduction mechanism and the drag reduction percentage can be calculated via

$$DR\% = \left[\frac{T_u - T_t}{T_u}\right] x100 \tag{3}$$

where;

T_u: Torque obtained with the untreated fluid, N.m

T_t: Torque obtained with the treated fluid (addition of CMC), N.m

A double concentric cylinder is the geometry utilized for the drag reduction assessment where the inner cylinder will rotate to induce shear. The Reynolds number equation corresponding to this geometry, Re, is given by,

$$R_e = \frac{\Omega \rho R_m (\Delta R)}{\mu} \tag{4}$$

where Ω is the angular velocity (rad/s), ρ is the fluid density (kg/m^3), R_m is the moving cylinder radius of the rheometer (m), ΔR is the difference between the outer and inner cylinder radius (m) and μ is the fluid viscosity (Pa.s).

3. Result and Discussion

Twenty-seven samples of the green Cavendish banana peel CMC are synthesized with different parameter combinations. However, only selected representative samples are discussed in this paper for simplicity. Similar observations could be deduced from other samples.

3.1. Polymer characterization

The biopolymer that is characterized has been synthesized with a reaction time of 150 minutes, reaction temperature of 50°C and NaOH concentration of 40%. This sample is chosen at random for the characterization.

The X-ray diffraction (XRD) results indicate that the highest peak is the cellulose with d values of 7.33741, 5.38695 and 3.89250. The second highest component is the polysaccharide with d values of 4.55430 and 5.33168. Other different types of starch are also present in the sample but not as significant as the cellulose and polysaccharide. The surface morphology as scanned via Field Emission Scanning Electron Microscopy (FESEM) is rough with the fibers twisted indicating that the bonds between the molecules are strong and stable. The strong hydrogen bonds in the structure are associated with the structure of molecular cellulose. Surface area of the CMC is quantified utilizing a pore size analyser. The surface area of the banana peel CMC is sufficiently large (0.0558 ± 0.0699 m²/g) enabling it to interact with other molecules to dissolve easily in solvents such as water. The Pore volumes and pore sizes of CMC will show that the size is sufficiently large to facilitate the adsorption and desorption process where 10 Å is recommended to be sufficient for the process to progress [12].

The average DS value of CMC calculated from Equation (3) and (4) is 0.62, above the recommended limit of 0.5 [3], and is regarded to be sufficient for solubility. Table 2 summarizes the spectrum of elements by weight % and atomic %. The element most found in the biopolymer is carbon (C) and

IOP Conf. Series: Materials Science and Engineering **458** (2018) 012031 doi:10.1088/1757-899X/458/1/012031

oxygen (O) which is 49.60 and 43.23 by weight %, respectively indicating that the cellulose occupies 92.83% of the biopolymer which is the most important element needed to synthesize a good drag reduction agent [13].

3.2. Drag Reduction Assessment

The centrifugal force due to the rotation of the inner cylinder of a double concentric cylinder leads to the destabilization of the flow field. The point at which the streamlines ceases to be circular for a Newtonian liquid and at which the flow field presents Taylor instabilities can be found by [14] and it should occur at a critical Reynolds Number defined by:

$$R_e = \frac{\Omega \rho R_m (\Delta R)}{\mu} > \left(\frac{R_m}{\Delta R}\right)^{0.5} \qquad (5)$$

where $R_m = R_3$ and $\Delta R = R_4 - R_3$. Prior to any further measurements of drag reduction, fresh tap water sample is tested under the rheometer as a reference.

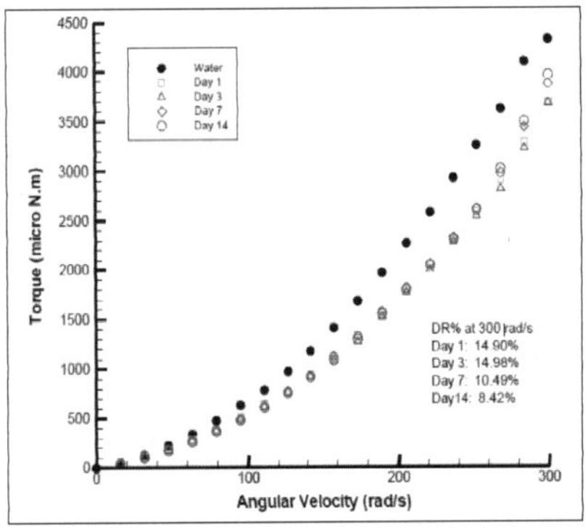

Figure 1. Torque profile for water and 100 ppm of biopolymer solution synthesized at 50°C.

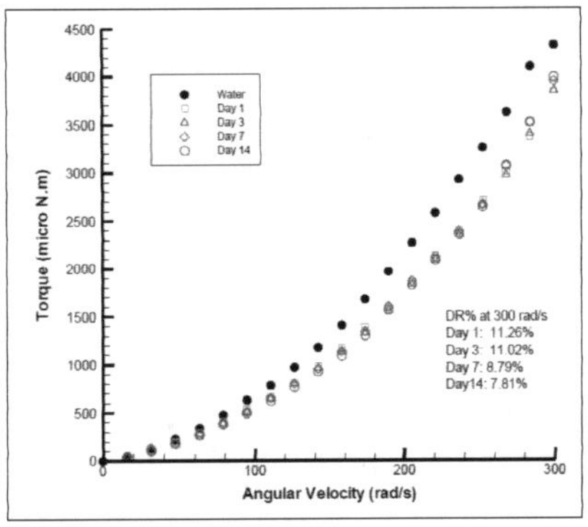

Figure 2. Torque profile for water and 100 ppm of biopolymer solution synthesized at 55°C.

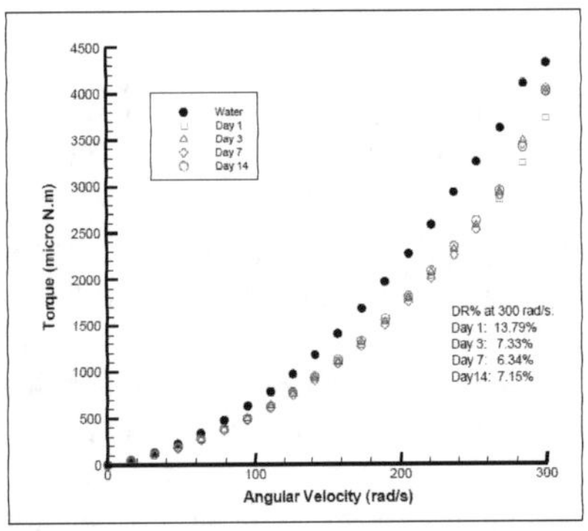

Figure 3. Torque profile for water and 100 ppm of biopolymer solution synthesized at 60°C.

3.3. Effect of Reaction Temperature

The 100ppm biopolymers produced at different reaction temperatures of 50°C, 55°C and 60°C for 240 minutes using 40% NaOH are mixed with UTP tap water. All the samples are left overnight after mixing for de-aeration and subsequently tested using the rheometer on Day 1, Day 3, Day 7 and Day 14. The results are summarized in figures 1, 2 and 3.

The graphs show that the drag reduces when the biopolymer is mixed with water. Sample 1 gives the best drag reduction. This shows that banana peel synthesized at 50°C with constant reaction time and concentration of NaOH yield higher Carboxymethylcellulose as indicated by the highest drag reduction performance followed by reaction at the temperature of 55°C and finally at 60°C. Temperature of 50°C is hence the optimum temperature to convert the cellulose to CMC above which the CMC structure is thought to degrade under such high temperatures. Over the duration of 14 days, the drag reduction percentage reduced by as much as 6.64%, calculated at 300 rad/s, indicating that the sample degrades over time.

3.4. Effect of Reaction Time

With the reaction temperature kept constant at 50°C and concentration of sodium hydroxide remained at 20% during synthesis, the reaction time is varied at 60, 150 and 240 min. All the samples are mixed at 100 ppm and left overnight for de-aeration and subsequently tested on Day 1, Day 3, Day 7 and Day 14. Figures 4, 5 and 6 summarize the findings.

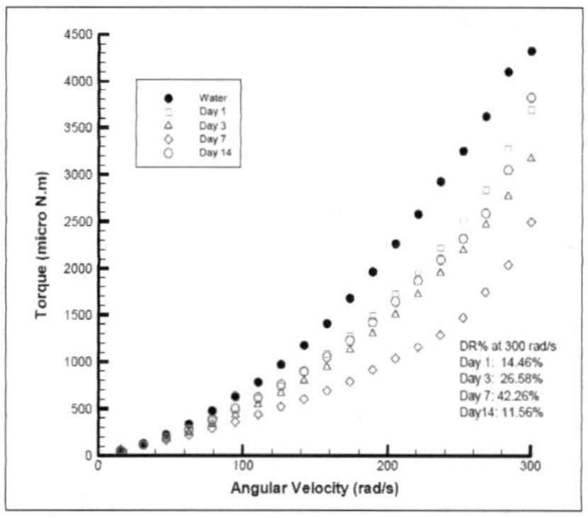

Figure 4. Torque profile for water and 100 ppm of biopolymer solution synthesized at a reaction duration of 60 min.

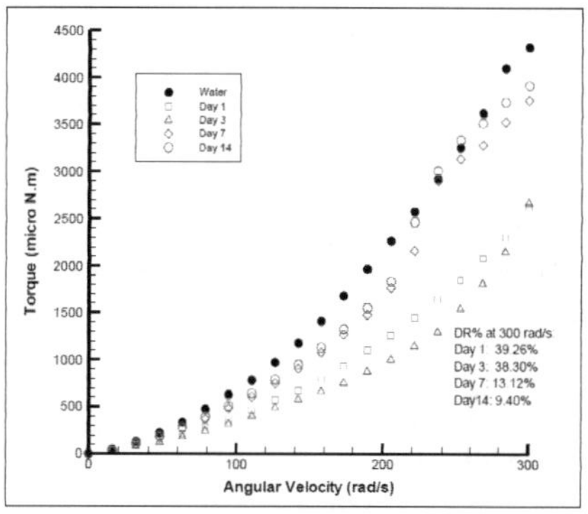

Figure 5. Torque profile for water and 100 ppm of biopolymer solution synthesized at a reaction duration of 150 min.

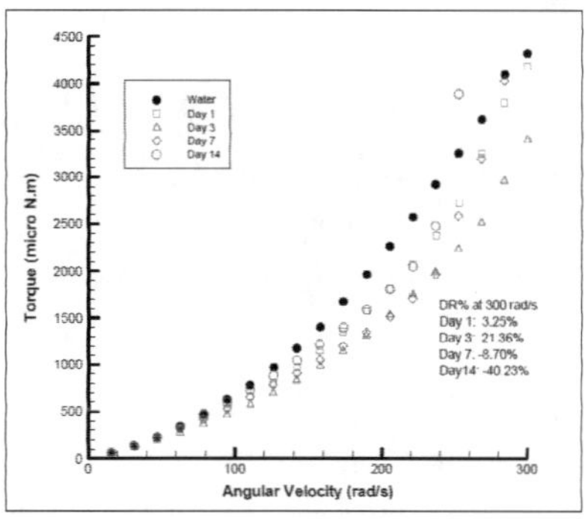

Figure 6. Torque profile for water and 100 ppm of biopolymer solution synthesized at a reaction duration of 240 min.

It is observed that a reaction time of 60 minutes produced CMC with progressively good drag reduction capability up until day 7 beyond which the drag reduction ability reduces. However, when the sample is subjected to reaction times of 150 and 240 minutes the CMC produced exhibit reduction in drag reduction capability after day 3. Such performance is thought to be due to the prolonged reaction time the sample is exposed to during synthesis. The reaction time will affect the swell-ability of the cellulose.

IOP Conf. Series: Materials Science and Engineering **458** (2018) 012031 doi:10.1088/1757-899X/458/1/012031

Though swell-ability provides a better environment for carboxymethylation reaction even higher reaction time may significantly cause degradation of the cellulose chain and the CMC structure. The degree of substitution (DS) of the Carboxymethyl group increases with the increase of reaction time.

3.5. Effect of NaOH Concentration

The reaction time is maintained at 150 minutes with the reaction temperature set at 55°C. The concentrations of NaOH used for the synthesis are 20w/w%, 40w/w% and 60w/w% respectively. All the samples are mixed at 100ppm and left overnight for de-aeration and subsequently tested on Day 1, Day 3, Day 7 and Day 14. Figures 7, 8 and 9 summarize the findings.

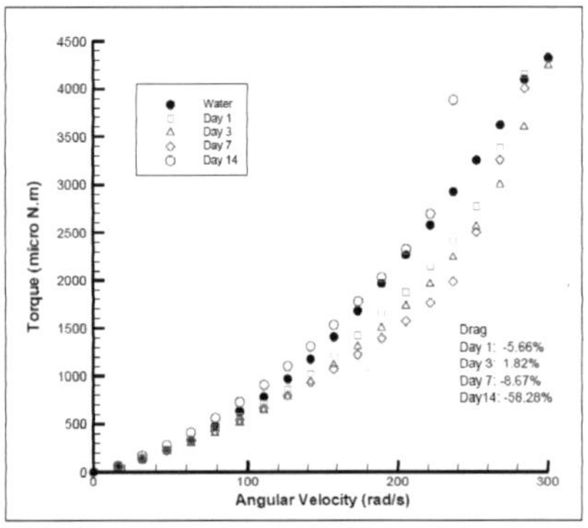

Figure 7. Torque profile for water and 100 ppm of biopolymer solution synthesized with 20% w/w NaOH concentration.

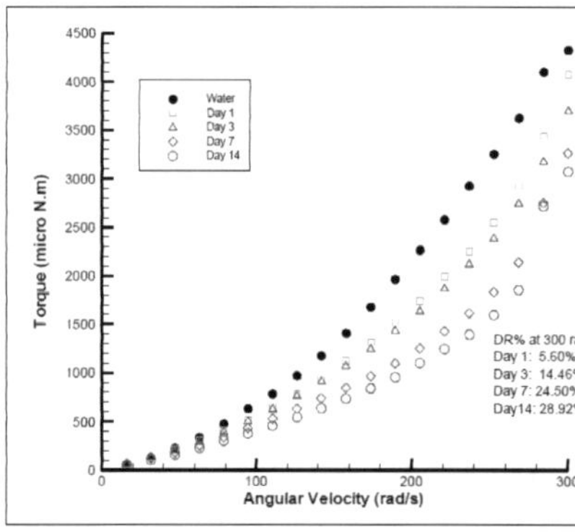

Figure 8. Torque profile for water and 100 ppm of biopolymer solution synthesized with 40% w/w NaOH concentration.

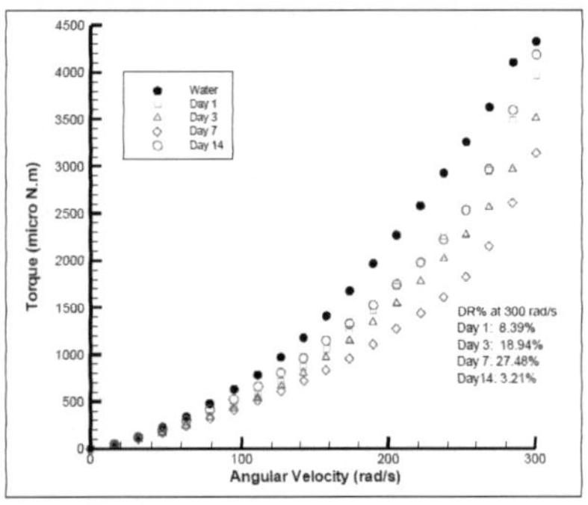

Figure 9. Torque profile for water and 100 ppm of biopolymer solution synthesized with 60% w/w NaOH concentration.

When determining the production of CMC, the concentration of NaOH plays a significant role. The proportion of NaOH must be optimum to obtain a high value of degree of substitution. The increase of NaOH concentrations affects the percentage yield of CMC, which is indirectly linked to the drag reduction performance. The reduced yield is due to the side reaction which takes place and results in the formation of sodium glycolate from the monochloroacetic acid. A side reaction is a carboxymethylation process which converts the monochloroacetate to sodium glycolate. The monochloroacetic acid molecules react with the NaOH leading to the destruction of cellulose chains or CMC polymer [15], [1]. From the figures, the yield of CMC is highest at 40% of NaOH where the drag reduction is progressively increasing from day 1. This finding could be explained by considering the carboxymethylation procedure described above, where two competitive reactions took place simultaneously; the first reaction to produce the CMC and the second reaction resulting in a by- product. This result shows that the first reaction prevails over the second reaction up to a NaOH concentration of 40%, beyond which the sodium glycolate is formed instead. 20% NaOH concentration is considered insufficient for the first reaction to be completed. Polymer degradation could also take place at high concentrations of NaOH. Similar observations are documented in maize starch [16], corn starch [17], sago waste [18], potato starch [19] and pigeon pea starch [20]. The nature and composition of organic solvents used as reaction media plays an important part in the structural transformation of cellulose during alkalization. The nature of biopolymer affects the biodegradability as the isopropyl alcohol having dielectric constant is a bad solvent for sodium hydroxide and it favors higher concentration of NaOH in the vicinity of cellulose, giving rise to substantial change of polymorphism.

4. Conclusion

For the first time, Carboxymethylcellulose (CMC) is produced from the Cavendish banana peel using the Carboxymethylation process. Twenty-seven samples are synthesized to cover all the different parameters. Material characterization is performed on the banana peel CMC via X-ray Diffraction (XRD), Field Emission Scanning Electron Microscopy (FESEM) and Pore Size Analyzer (BET). The characterization analysis shows that the cellulose is the strongest component in the sample occupying 92.83% of the biopolymer. The Double concentric cylinder (DCC) geometry in the stress-controlled rheometer is utilized in this study to assess the drag reduction. The drag reduction performance is thought to be linked to the amount and quality of the CMC produced under various parameters. Varying the reaction temperature during synthesis, the biopolymer synthesized at the lowest chosen reaction temperature of 50°C works best at drag reduction. The drag reduction performance starts to decrease when it is synthesized at higher temperatures due to degradation of CMC. Another synthesis parameter varied is the reaction time. The experimental results imply that prolonged reaction time results in the degradation of CMC structure and hence decreases the drag reduction capability. When the NaOH concentration is varied, the solution with an optimum NaOH concentration of 40% utilized during synthesis gave the best drag reduction capability. During synthesis, two reactions are anticipated to have taken place in producing the CMC; the first reaction involves cellulose hydroxyl with monochloroacetic acid in the presence of NaOH to produce CMC and the second reaction involves the conversion of monochloroacetate to sodium glycolate as a by-product. This study implies that sodium glycolate might have formed at higher NaOH concentrations. The findings presented in this paper prove that green Cavendish banana peel CMC has a great potential at drag reduction when synthesized at the right conditions.

References

[1] French A D, Bertoniere N R, Brown R M, Chanzy H, Gray D, Hattori K and Glasser W 2003 "Cellulose", *Kirk-Othmer Encyclopedia of Chemical Technology, 4th Ed.* **10**, pp. 375-376.

[2] Bono A, Ying P H, Yan F Y, Muei C L and Sarbatly R 2009 Synthesis and characterization of carboxymethyl cellulose from palm kernel cake *Adv. in Nat. and App. Sci* **21**, no. 2. pp. 5-11.

[3] Rowell R M, Han J S and Rowell J S 2000 Characterization and factors effecting fiber properties *Nat. Pol. and Agro. Based Comp.* **8**, no.1, pp. 28-34.

[4] Waring MJ and Parsons D 2001 Physio-chemical characterisation of carboxymethylated spun cellulose fibres *Biomaterials,* **15**, no. 4, pp. 903-912.

[5] Girard M, Turgeon S L and Paquin P 2002 Emulsifying properties of whey protein-carboxymethylcellulose complexes *J. of Food Sci.***8**, no.1, pp. 113-119.

[6] Tchobanoglous G, Theisen H and Vigil S 1993 *Integrated solid waste management: Engineering principles and management issues* New York McGraw-Hill

[7] Emaga T H, Andrianaivo R H, Wathelet B and Tchan 2007 Effects of the stage of maturation and varieties on the chemical composition of banana and plantain peels *Food Che.* **5**, no. 1 pp. 590-600.

[8] Emaga T H, Robert C, Ronkart S N and Wathelet B 2008 Dietary fiber components and pectin

chemical features of peels during ripening in banana and plantain varieties *Bio. Tech.,* **4**, no.8 pp. 4346–4354.

[9] Ramli S, Ismail N, Alkarkhi AFM and Easa A M 2015 The use of principal component and cluster analysis to differentiate banana peel flours based on their starch and dietary fibre components *Trop Life Sci Res*, **18**, vol. 21, pp. 91-100.

[10] Heinze T 2005 Carboxymethyl ethers of cellulose and starch-a review *Chem. Plant Raw Mater*, **22**, vol.4, pp. 13-29.

[11] Henaut I, Darbouret M, Palermo T, Glenat P and Hurtevent C 2009 Experimental Methodology to Evaluate DRA: Effect of Water Content and Waxes on Their Efficiency *Society of Petroleum Engineers, Texas,* **4**, no. 5, pp.14-19.

[12] Biswal D R and Singh R P 2014 Characterisation of carboxymethyl cellulose and polyacrylamide graft copolymer *Carbohydrate Polymers*, **2,** no. 1, pp. 379-387.

[13] Inaba H, Haruki N and Horibe A 2000 Flow drag and heat transfer reduction of flowing water containing fibrous material in a straight pipe *Int. J. of Ther. Sci.* **6,** no. 4, pp. 18-29.

[14] Chhabra R P and Richardson J F 1999 *Non-Newtonian Flow in the Process Industries* Butterworth-Heinemann, Oxford.

[15] Rachtanapun P, Simasatitkul P, Chaiwan W and Watthanaworasakun Y 2012 Effect of sodium hydroxide concentration on properties of carboxymethyl rice starch *Int. Food Res. J.* **19**, no. 3, pp. 923-931.

[16] El-Sheikh M A 2010 Carboxymethylation of maize starch at mild conditions *Carb. Pol.* **7**, no.1 pp. 875-881.

[17] Bhattacharya D, Singhal R S and Kulkarni P R 1995 A comparative account of conditions for the synthesis of sodium carboxymethyl starch from corm and amaranth starch *Carb. Pol.* **7**, no.1, pp. 247-253.

[18] Pushpamalar V, Langford S J, Ahmad M and Lim Y Y 2006 Optimization of reaction conditions for preparing carboxymethylcellulose from sago waste *Carb. Pol.,* **7**, no.1, pp. 312-218.

[19] Tijsen C J, Scherpenkate H J, Stamhuis E J and Beenackers A A C M 1999 Optimisation of the process conditions for the modification of starch *Chem. Eng. Sci*, **5**, no. 3, pp. 2765-2772.

[20] Lawal O S, Lechner M D and Kulicke W M 2008 The synthesis conditions, characterizations and thermal degradation studies of an etherified starch from an unconventional source *Pol. Deg. and Sta.*, **6**, vol. 4, pp. 1520-1528.

IOP Conf. Series: Materials Science and Engineering **458** (2018) 012032 doi:10.1088/1757-899X/458/1/012032

The use of factorial design in screening of factors influencing hexavalent chromium extraction by continuous green emulsion liquid membrane

N F M Noah[1], N Othman [1,2 *] and N Jusoh[1]
[1]Department of Chemical Engineering, Faculty of Chemical and Energy Engineering, Universiti Teknologi Malaysia, 81310 UTM Johor Bahru, Johor, Malaysia.
[2]Centre of Lipids Engineering & Applied Research (CLEAR), Ibnu Sina Institute for Scientific and Industrial Research (Ibnu Sina ISIR), Universiti Teknologi Malaysia, 81310 UTM Johor Bahru, Johor, Malaysia.

*norasikin@cheme.utm.my

Abstract. Recently, increasing concerns on hexavalent chromium as an environmental pollutant can be observed due to its build-up to toxic levels in the environment, resulting from various industrial and agricultural activities such as electroplating, stainless steel production, leather tanning and textile manufacturing. This work screens the significant factors that largely affect the efficiency of hexavalent chromium (Cr(VI)) removal from electroplating wastewater in continuous green emulsion liquid membrane using fractional factorial design. There are six factors investigated for affecting the removal of chromium which are residence time (t), TOMAC concentration (M), stripping agent concentration (M), rotational speed (rpm), treat ratio, and modifier concentration (w/v). The response variable of the chromium removal was identified using the two-level fractional factorial (two level) design and the results were analyzed statistically. Five factors were identified to have significant influence on chromium removal which are retention time, rotational speed, treat ratio, modifier concentration and carrier concentration. Regression models for chromium extraction were developed and the adequacy of the model was examined. The results of this study indicate that fractional factorial design is capable to predict the significant factors affecting hexavalent chromium extraction by using continuous green emulsion liquid membrane.

1. Introduction

Chromium is a transition metal with a number of industrial uses due to its toughness and resistance to heat and corrosion. It is widely used for manufacturing stainless steel such as chromate manufacturing, chrome plating, ferrochrome production, and stainless steel welding. Nevertheless, these applications increase the concentrations of chromium that is known to be a toxic and mutagenic substance in wastewater. Chromium exposures in both occupational and environmental settings have the ability to trigger human carcinogen. Several restrictions have been imposed on the use of chromium compounds in many parts of the world due to these health and environmental issues.

Previously, various techniques have been investigated in order to choose the best technique to remove the economically hazardous metals of hexavalent chromium from manufacturing wastewater such as chemical precipitation [1], reverse osmosis[2], adsorption[3], ion exchange[4], solvent extraction processes [5], and so forth. On top of that, emulsion liquid membrane (ELM) is regarded as one of the potential and promising technologies for the extraction of chromium ions from industrial wastewater. This is due to the pronounced advantages of ELM which are low capital and operating costs, low energy and reagent consumption, high concentration factors, and high mass fluxes [6]. Even though the two substantial problems of the ELMs which are membrane fouling and extractant leakage are still in need of a solution, the research proposed in this paper was proceeded in order to investigate the commercial applicability of chromium extraction in large scale industrial application using continuous green emulsion liquid membrane (CGELM).

In CGELM, various factors such as residence time (t), TOMAC concentration (M), stripping agent concentration (M), rotational speed (rpm), treat ratio, and modifier concentration (w/v) affect the removal of chromium. Generally, the optimum condition of these factors was preceded by screening experiments that were carried out by one-factor at a time (OFAT) method where the effect of one factor is assessed by varying the value of only one factor while all other independent variables remain constant [7][8][9]. However, this traditional approach was reported to be very costly and time consuming due to many kinds of high materials expense and large number of experiments [10]. Furthermore, this method can only be used to study one factor at a time thus cannot capture the interactions between the factors[11].

Experimental designs, such as the full factorial designs are useful for studying systems with a small number of factors as these experimental designs provide information concerning interactions between factors [12]. A frequently stated advantage of fractional factorial (FF) designs over OFAT designs is their high relative efficiency. However, as the number of factors become larger, the total number of experimental runs required often becomes impractical, if not impossible[13]. In order to save on costly runs, experimenters usually perform only a fractional factorial design, which can overcome this problem as it usually requires fewer experimental running which is just a partial number of a full factorial design. Thus, its relatively inexpensive and efficient way to improve a process has become its major attractiveness. A fractional factorial design is a modified standard factorial that permits gaining information on main effects and low-order interactions without having to run the full design. It consists of confounding the factorial into blocks, and run only one of them [10]. This method has been used in various fields such as in studying the role of eight parameters in both stability and thermal conductivity of carbon nanotubes/water nanofluids [14], in investigating the main competitive factors in the adsorption of arsenic(V) onto reclaimed iron-oxide coated sands [15] as well as in identifying the key factors influencing the extraction efficiency of Cu(II) from aqueous solutions using soybean oil-based organic solvents[16].

Nevertheless, so far there is no work reported on the usage of fractional factorial design in screening the effect of several controllable factors on the extraction efficiency of chromium from real industrial effluent in continuous green emulsion liquid membrane in vegetable oil-based organic solvents which is more renewable, environmental-friendly and economic than the petroleum-based ones. In this work, six parameters were screened in order to evaluate their effects on a response using a two-level fractional factorial design. Furthermore, analysis of variance (ANOVA) and predicted chromium removal from the model at 95% confidence interval against the experimental result is used to examine the adequacy of the developed empirical relationship regression model.

2. Materials, equipment and methods

2.1 Materials

Diluents used were palm oil (Buruh) which is refined cooking oils obtained from Lam Soon Edible Oils. Extractant tricaprylmethylammonium Chloride (Aliquat 336/TOMAC) was obtained from Fluka. Sorbitan monooleate (Span 80) as a nonionic surfactant and 1-octanol were used as surfactant and

IOP Conf. Series: Materials Science and Engineering **458** (2018) 012032 doi:10.1088/1757-899X/458/1/012032

modifier, respectively were purchased from Fluka. Thiourea and Sulphuric acid (H_2SO_4) solution were used as stripping agents and were purchased from Merck. Real rinse electroplating process wastewater sample was obtained from the Electroplating Company at Masai, Johor. All chemicals and reagents used were of analytical reagent grade.

2.2 Preparation of Water in Oil (W/O) Emulsion

The primary W/O emulsion was produced by mixing equal volume of (50mL) organic membrane solution (Span 80, 1-octanol and extractant with diluent) with stripping liquid solution (Thiourea acidic) using the Heidolph Silent Crusher-M Homogenizer. The homogenizer speed was set to be 7000 rpm in order to produce a stable emulsion with 1 minute of emulsification time. Then, the white milky emulsion was ready for extraction process. The emulsion needs to be freshly prepared before each step of the experiment.

2.3 CGELM Chromium Extraction

2.3.1 Rig set up for CGELM. Figure 1 shows the schematic diagram of the single-stage, continuous-flow mixer settler apparatus used in this work. The rig consists of 1.5 L extraction vessel with 4 baffled, which is about 12cm in diameter and 15cm height. There are two peristaltic pumps (Cole-Parmer, Masterflex Models) equipped with quick-load heads used to pump the emulsion and feed solutions through two inlet lines mounted on the upper wall of the vessel. Digital drives were used to control the pump speed variations within 1%. The pumps were calibrated before running the experiment. An outlet port mounted at the extractor bottom was controlled for the liquid level to be constant along the experiment (1000 mL). An outlet port mounted on the separator bottom wall was used to take the samples.

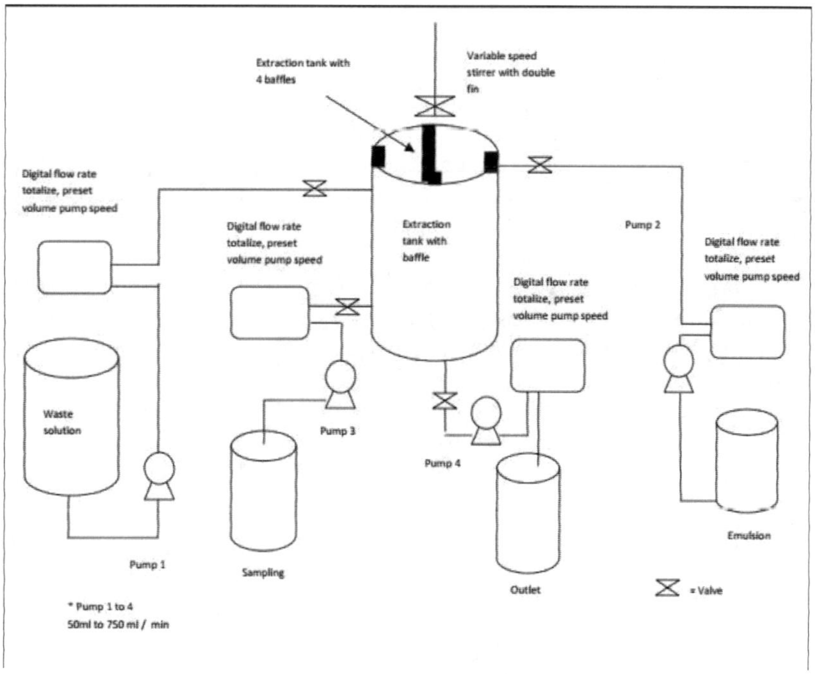

Figure 1. Schematic diagram of continuous ELM extraction system

2.3.2 Chromium removal in CGELM. A series of 100 mL W/O emulsion of optimized condition was prepared using 120 mL beaker and continuously supplied into the emulsion vessel. The emulsion need to be freshly prepared before each step of experiment. Then, the prepared emulsion and rinsed electroplating solution were combined continuously in the extractor using a digital mixer system (IKA RW 20 Digital Dual Range Mixers) at agitation speed of 350 rpm for 3 minutes retention time. Pump controller 1 and pump controller 2 were adjusted to control the flow rate of feed phase and emulsion solution, respectively into the extraction vessel to maintain the treat ratio. The experiment started with low flow rate of 20ml/min and 180ml/min for emulsion and feed phase, respectively.

Along the process, the sample was taken out every five minutes of extraction process. Sample taken from the screening process undergoes phase separation using separating funnel within 10 minutes. The aqueous phase is the treated chromium solution that was analysed using AAS to determine the efficiency of extraction as well as the removal percentage of chromium in the liquid solution. Steady state was indicated as the chromium concentrations did not change over a period of two or three samples. After achieving the steady state concentration, the pumps and mixer were stopped to separate the emulsion and the aqueous phase. The separation of emulsion was conducted in a separating funnel within 15 minutes, where the three-phase dispersion was settled into the emulsion and the external phase. The treated chromium solution (external feed phase) at the bottom of the separating funnel was taken for concentration measurement. Then, the internal of the liquid membrane was separated to recover the chromium ions. The emulsion was demulsified using high voltage electrostatic coalescer.

The efficiency of chromium removal from the feed and the percentage of swelling/breakage occurred during the process were calculated using Equations 1 and 5, respectively.

$$\%\ chromium\ removal = \frac{c_i - c_f}{c_i} x\ 100 \tag{1}$$

where C_i is the initial concentration of chromium in feed phase and C_f is the final concentration of chromium in feed phase after ELM process.

2.4 Design of experiment

2.4.1 Two-level fractional factorial design. A 2^{6-3} fractional factorial design with resolution III was applied to screen the effects in chromium removal with variations in residence time (t), TOMAC concentration (M), stripping agent concentration (M), rotational speed (rpm), treat ratio, and modifier concentration (w/v). Resolution III was chosen due to the expensive runs and plentiful factors. The variables of interest and their real values at the specified levels in the design are exhibited in table 1. The delimitation of experimental region for each factor was determined from the preliminary experiments. Each independent variable was set to a high (+1) and a low (−1) level. Other factors such ratio aqueous to organic (1:1), Span 80 concentration (3%), initial feed concentration (40 mg/L) and operating temperature (25°C) were fixed. In the design matrix, the effects of six variables on the removal efficiency of chromium were evaluated. A total of 8 experimental runs were randomized for statistical purposes. The statistical significance of each individual factor and their combinations at 5% significance level were evaluated using the Statistica 7 software.

Table 1. The high and low levels of the different factors applied in 2^{6-3} fractional factorial design

Factors	Symbols	Units	Levels	
			High (+1)	Low (-1)
Residence time	X_1	min	5	1
TOMAC concentration	X_2	M	0.04	0.004
Stripping agent concentration	X_3	M	1.0	0.1
Rotational speed	X_4	Rpm	450	150
Treat ratio	X_5	Emulsion:Feed	0.25	0.1
Modifier concentration	X_6	% (w/v)	5	1

2.4.2 Regression analysis. A multiple regression analysis was performed on a regression model which corresponds to the following first-order response function:

$$y = \beta_o + \sum_{i=1}^{k} \beta_i x_i + \sum\sum_{j<i} \beta_{ij} x_i x_j + \varepsilon \quad (2)$$

Where β_o, β_i and β_{ij} are the regression coefficients for the intercept, linear and interaction coefficients, respectively, y is the dependent variable or the response, x_i and x_j are the independent variables in coded units, and ε is the error term. Estimation of the regression coefficients that caused the model to best fit a set of collected response variable data was carried out by the least squares method [10]. All analysis was performed using the Statistica 7 software.

2.4.3. Model adequacy checking. The adequacy of regression models obtained in fitting the observed data was examined by the analysis of variance (ANOVA), coefficient of determination (R^2) and the absolute average deviation (AAD). ANOVA evaluates the significance of regression by determining whether there is a relationship between the response variable and a subset of the regression variables via the Fisher's statistical test (F-test). R^2, which has a value from 0 to 1, measures the global fit of a model and is calculated by equation (3):

$$R^2 = \frac{SST - SSE}{SST} = 1 - \left(\frac{SSE}{SST}\right) \quad (3)$$

where SST and SSE are the total sum of squares and error sum of squares, respectively. However, adding a variable to a model will always increase its R^2 value regardless of whether the additional variable is statistically significant or not [10]. Therefore, Pilkington *et al* [17] suggested the analysis of average absolute deviation (or mean absolute deviation) (AAD) as a method to describe the magnitude of deviation present in equation (4):

$$AAD = \left[\frac{\sum_{i=1}^{N}\left(\frac{|y_{i,exp} - y_{i,pre}|}{y_{i,exp}}\right)}{N}\right] x100 \quad (4)$$

where $y_{i,exp}$ and $y_{i,pre}$ are the experimental and predicted responses, respectively, and N is the total number of experimental points. Evaluation of the R^2 and AAD values together would give a better judgment about the adequacy of the model.

3. Results and discussions

3.1 Screening of factors affecting chromium removal

The design matrix used in the 2^{6-3} fractional factorial design to screen the factors affecting the removal percentage is shown in table 2. In this study, a total of 8 runs of experiment were conducted under consistent conditions in one block of measurements and the experimental sequence (standard order) was randomized so that the effects of uncontrollable factors are minimized. The results showed that the removal efficiency was found to be in the range of 81 to 100%. This indicates that continuous green emulsion liquid membrane is applicable to remove chromium from electroplating wastewater.

Next, all important factors that significantly affect the chromium removal was evaluated based on a half-normal probability plot of standardized effects, t-value and p-value, a Pareto chart, main effects and interaction plots at 5% significance level using the Statistica 7 software.

Table 2. Design Matrix for 2^{6-3} fractional factorial design and chromium removal performance

Std Order	Run Order	Blocks	Variables						% Removal
			X_1	X_2	X_3	X_4	X_5	X_6	
4	1	1	5	0.040	0.1	450	0.10	1	97.84
2	2	1	5	0.004	0.1	150	0.10	5	91.05
7	3	1	1	0.040	1.0	150	0.10	5	81.16
1	4	1	1	0.004	0.1	450	0.25	5	100.00
3	5	1	1	0.040	0.1	150	0.25	1	94.05
5	6	1	1	0.004	1.0	450	0.10	1	94.63
8	7	1	5	0.040	1.0	450	0.25	5	100.00
6	8	1	5	0.004	1.0	150	0.25	1	100.00

X_1: t(min), X_2: [TOMAC] (M), X_3: [Tu Acidic], X_4: Rotational speed (rpm), X_5: treat ratio (Emulsion:Feed), and X_6: [Octanol] (% w/v); All variables are in uncoded units.

3.1.1 Half-Normal probability plot of standardized effects. The half normal probability plot of the effects in figure 2 shows the absolute values of the estimated effects from the largest effect to the smallest effect. Generally, the estimated effects of an unimportant variable will typically manifest themselves as being near zero and tend to be on or close to a near-zero line in the plot, while the estimated effects of an important factor will typically manifest themselves by being off the line and well-displaced from zero. Hence, the larger the important effects, the further they are from the zero line in the plot.

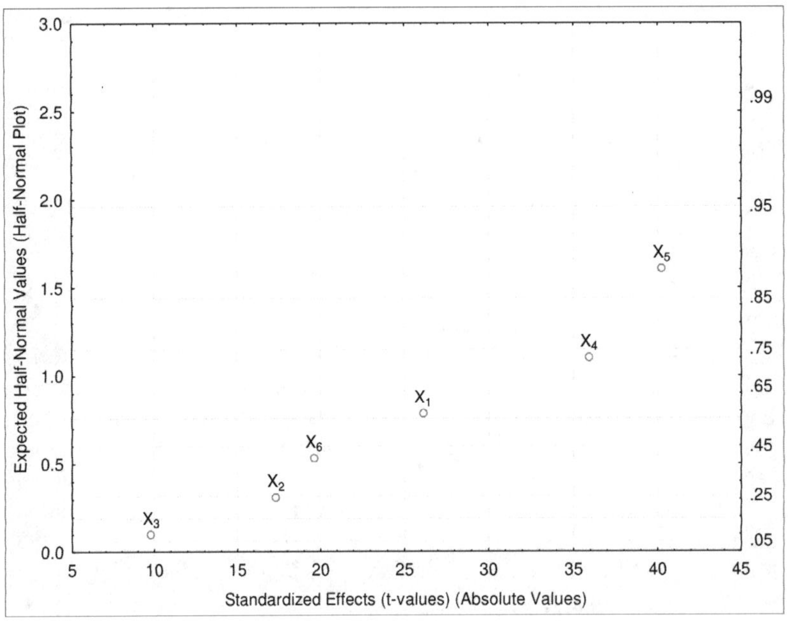

Figure 2. Half-normal probability plot of standardized effects on chromium removal

From this analysis, it could be observed that the largest effect on chromium recovery that is well-displaced in the plot is the X_5 (treat ratio) followed by X_4 (rotational speed), X_1 (residence times), X_6 (modifier concentration) and X_2 (extractant concentration). The sequence of the significant main effects with respect to decreasing influence on chromium removal is found to be $X_5 > X_4 > X_1 > X_6 > X_2$. The analysis indicates the results as consistent to the CGELM system principles and is explained below.

According to the plot, treat ratio is the most significance effect on the chromium removal efficiency due to treat ratio controlled the interfacial mass transfer across an ELM. Treat ratio is the volume ratio of emulsion to the aqueous external phase solution (volume feed solution). Basically, its increment will increase the volume of emulsion as a whole. Therefore, the surface area for mass transfer is increased owing to the formation of a larger number of emulsion globules. As a result, a higher degree of chromium removal is obtained. This finding is supported by Sulaiman *et al.* [18] and Noah *et al.* [6], who observed that high treat ratio significantly increased the emulsion phase hold up in the feed phase, which then increased the removal capacity of the solutes.

Meanwhile, the second and third most significant process variables were rotational speed and retention time, respectively. Generally, in order for the extraction to be effective in the continuous extractor, fluid circulated by the impeller (rotational speed) must be swept into the entire vessel in a reasonable time (residence time) [7][8]. This implies that the effective mixing of fluids in the extractor depend to that the rotational speed and retention time of the operation system.

On the other hand, modifier concentration gives a significance effect in stabilizing the ELM resulting to an increment in the removal efficiency. Lastly, as the removal of chromium in CGELM system is a facilitated transport mechanism, an extractant is needed to promote solute transportation through the membrane. The TOMAC was reacted with chromium to form a TOMAC-chromium complex and diffused through the liquid membrane. This means that the more carrier compound existed in the liquid membrane phase, the more chromium was removed within the retention time. However, increasing the amount of extractant will increase the viscosity of the membrane phase, which limits the removal rate. This finding is in line with our previous study which change in TOMAC concentration will change the rate of chromium transport as well as viscosity of the liquid membrane significantly [19].

It should be noted that the aliased of the main factors with interaction of the 2 factor and higher is negligible due to the use of resolution III in the experimental design. Moreover, the 3-way interactions are usually not significant [10]. Therefore, it can be assured that X_5, X_4, X_1, X_6 and X_2 are the subset of important factors. On the contrary, the linear individual terms (X_3) is found to be insignificant or less significant. This indicates that thiourea acidic concentration in the internal phase did not affect chromium removal in the CGELM system.

3.1.2 Pareto chart. Another quick way to screen the significant factors is by analysing the pareto chart of the estimated effects. In order to validate the results obtained from the half-normal probability plot of effects in figure 2, a Pareto chart was generated and tabulated as shown in figure 3. The p-value serves as a tool for checking the significance of each coefficient and can be used to indicate the strength of interaction between variables. Low values of $p < 0.05$ indicate high significance of the corresponding coefficients. Pareto chart analyses revealed that there are 5 factors that are statistically significant, and clearly have much larger effect than the other factor. The 5 factors are X_1 which is the retention time, X_2 which is the extractant concentration, X_4 which is the rotational speed, X_5 which is the treat ratio, and X_6 which is the modifier concentration. Compared to figure 2, the sequence of the significant main and interaction effects with respect to decreasing influence on chromium removal is parallel with that obtained from the half-normal probability plot of standardized effects, that is $X_5 > X_4 > X_1 > X_6 > X_2$.

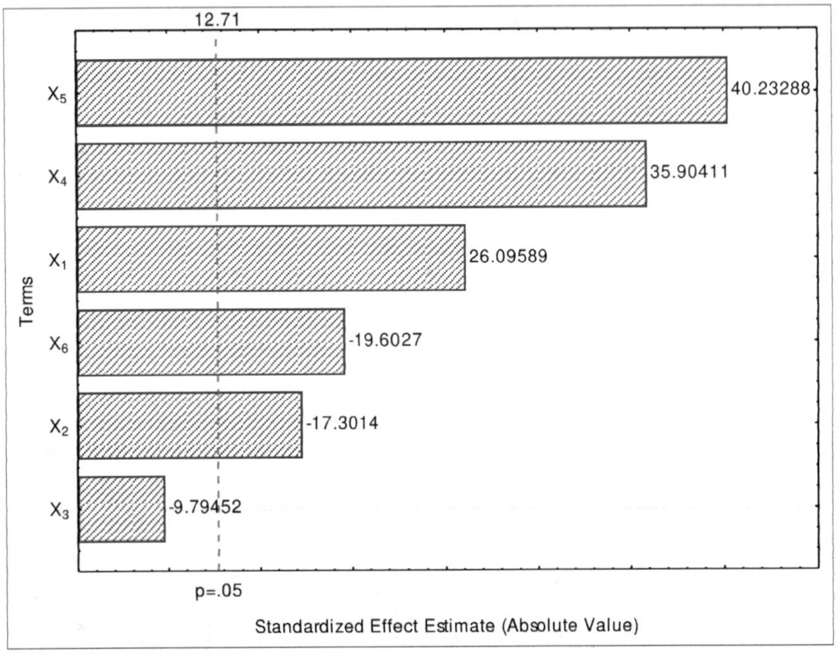

Figure 3. Pareto chart of each parameter coefficient for chromium extraction

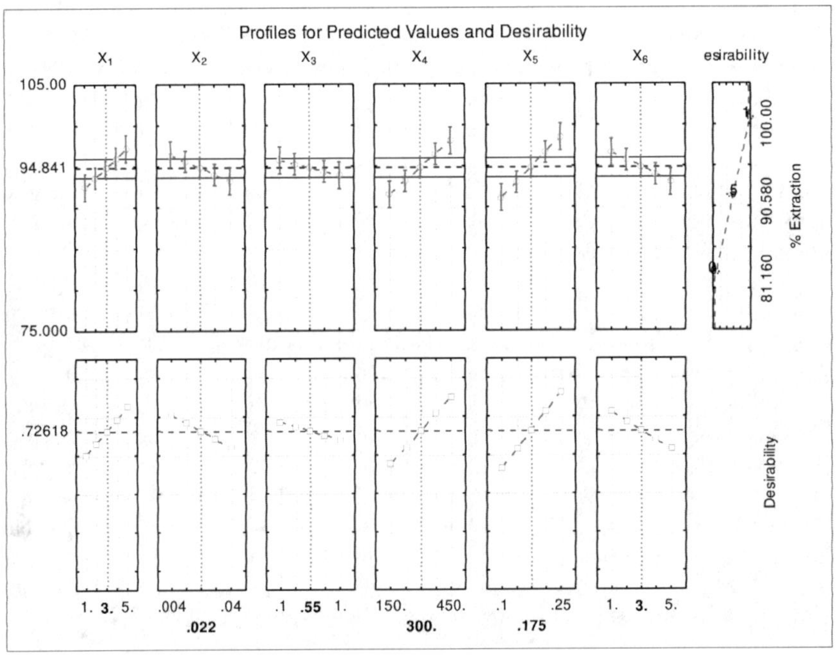

Figure 4. Prediction profiler and desirability plot in fractional factorial design

3.1.3 Main effects and interactions plots. The standardised main effects of the main factors were also observed and shown in figure 4. Based on the specified goals for each response, the value of each factor is obtained as $X_1=3$, $X_2=0.22$, $X_3=0.55$, $X_4=300$, $X_5=0.175$, $X_6=3$, in the screening design. According to the graph, the main effects of X_1, X_2, X_4, X_5, and X_6 obtained are shown as the steep lines indicating their significant effects toward chromium removal. Meanwhile, the nearly flat effect lines attained for the main effects of X_3 reveals their unimportant effects on chromium removal. As the main factor X_3 has no significant effects on the chromium removal, the value of 0.5M of thiourea acidic can be used as fixed conditions in future study.

3.2 Regression Model and Analysis of Variance (ANOVA) for Response

Table 3 shows the estimated effects and regression coefficients (Coef) together with the corresponding standard deviation (SD_{coef}), t-value (T) and probability (P) values for the main effects terms attained from the linear least square method using the Statistica 7 software. Generally, greater magnitude of T, F and lower value of P value indicate that the corresponding coefficient terms are more significant. The linear regression model of the response (chromium removal) with six independent variables (X_1, X_2, X_3, X_4, X_5 and X_6) was given with their first order model in coded units as illustrated in equation (5):

$$Y = 94.84 + 2.38X_1 - -1.57X_2 - 0.89X_3 + 3.27X_4 + 3.67X_5 - -1.78X_6 \quad (5)$$

where each regression coefficient was computed according to the linear least square method and the results produced are as tabulated in table 3. In equation (5), a positive relationship with residence time (X_1), rotational speed (X_4) and treat ratio (X_5) can be seen, whereas, a negative relationship with extractant concentration (X_2) and modifier concentration (X_6) can be observed for chromium removal. This is in agreement with work conducted by [20] stated that the positive sign in the equation represents synergistic effects and the negative sign denotes antagonistic effects on the dependent variable, denoted as Y. This suggests that longer retention time, higher rotational speed and treat ratio will enhance chromium removal. Meanwhile, an increase in the amount of extractant and modifier concentration tends to reduce the chromium removal efficiency. In addition, the size of regression coefficients in equation (5) denotes the degree of significance of each independent variable. The order of decreasing significance with respect to the influence on chromium removal, is $X_5 > X_4 > X_1 > X_6 > X_2$. This is in good agreement with the findings attained earlier.

Table 3. Estimated effects and coefficients of the regression models for chromium removal

Term	Effect	Coef	SD_{coef}	T	P
Constant	94.84125	94.84125	0.091250	1039.356	0.000613
X_1	4.76250	2.38125	0.091250	26.096	0.024383
X_2	-3.15750	-1.57875	0.091250	-17.301	0.036755
X_3	-1.78750	-0.89375	0.091250	-9.795	0.064773
X_4	6.55250	3.27625	0.091250	35.904	0.017727
X_5	7.34250	3.67125	0.091250	40.233	0.015820
X_6	-3.57750	-1.78875	0.091250	-19.603	0.032448

3.3 Model adequacy checking

The adequacy or goodness of fitted model of the regression model for chromium removal is checked by analysis of variance (ANOVA) using Fisher F-test at 5% significance level. Table 4 shows the analysis of variance (ANOVA) model for the percentage of chromium removal at high confidence level (95%) to obtain a good prediction model. According to table 4, the F-value calculated for response is 713.31 which is higher than the F-value tabulated ($F_{0.05,9,5}$ = 233.99) indicating that the model contributes significantly to the response and rejects the null hypothesis at the 0.05 level of significance. Null hypothesis will comply if all variables do not give significant outcome with the response. This is in agreement with the result obtained by Ooi *et al.* [21] which stated that the calculated F-value must be greater than the tabulated F-value in order to ensure the model to accurately predict the experimental result.

On the other hand, the value of R^2 in the model obtained is 0.9998 (99.98%) which is closer to 1 denoting that only 0.02% of the total varians is not explained by the regressors in the model. Therefore, it can be concluded that the predicted values calculated from Equation 5 are in good agreement with the experimental values. This is strongly supported by Wan Omar and Saidina Amin [22] who reported that if R^2 value is more than 0.75 it is considered as acceptable value, however, R^2 value of more than 0.8 is much better.

As there is only small differences between the R-squared (R^2 = 0.9998) and the adjusted R-squared (Adjusted R^2 = 0.9984) values which is 0.14%, this implies that the model is in good reliability to produce good estimation of the response [10]. This finding is in agreement with Nam *et al.* [19] who observed that smaller gap between R^2 and the adjusted R^2 is desirable for the judgement of the goodness of the model. This can be explained by the fact that the addition of input (independent) variables which are significant to a dependent variable will increase the value of adjusted R^2, whereas if non-significant variables being added into the model, the value of adjusted R^2 will decrease, while the R^2 value will continuously increase with the addition of more variables, regardless of the input variables' significance to the model.

The predicted chromium removal from the model at 95% confidence interval against the experimental result using the Statistica 8.0 is plotted in figure 5 with its AAD of 0.00041%, which was computed from equation (5). The high R^2 and small AAD values signify the model as able to give a reasonably good estimate of response (chromium removal) for the system in the range studied. Meanwhile, the observed experimental values are evenly distributed close to the straight line, lending support to that the model was accurately predict the experimental result. Hence, the regression model (equation (5)) can be used as a predictive tool to obtain chromium removal over the entire uncertainty range of significant effects studied.

Table 4. Analysis of variance (ANOVA) for first order model chromium removal by CGELM

	SS	DF	Mean Square	F-value	F-tabulated (α=0.05)
Regression	291.03	6	48.51	713.31	233.99
Residual	0.06	1	0.067		
Total SS	291.09	7			

ANOVA; Var.:% Extraction; R-sqr=.99977; Adj:.9984 (Spreadsheet6) 2**(6-3) design; MS Residual=.0666125 DV: % Extraction.

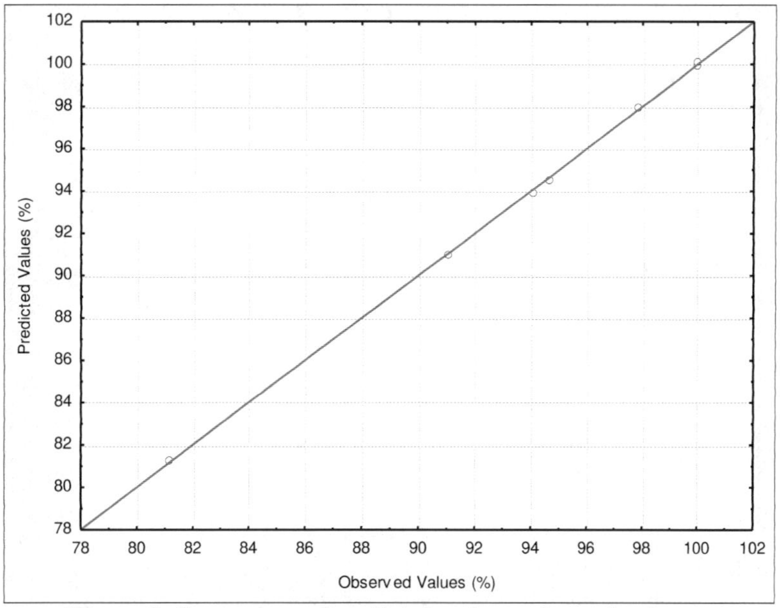

Figure 5. Comparison of predicted and experimental value on chromium removal

3.4 Hexavalent chromium removal using CGELM

One of the important parts is whether the screened condition gives high efficiency of chromium removal using the continuous green emulsion liquid membrane system. Based on the specified goals for each response, the value of each factor is obtained as X_1 =3, X_2 =0.22, X_3 =0.55, X_4 =300, X_5 =0.175, X_6 =3, in the screening design with 95% chromium removal. In order to confirm the prediction, further experiment was carried out at the screening condition for the chromium removal and the result shows that 99% of chromium was recovered with less only 4% deviation with predicted values as tabulated in table 5. This result indicates that continuous green emulsion liquid membrane is applicable to remove chromium from electroplating wastewater.

Table 5. Best Stability Conditions for Chromium Extraction by ELM

Factors	Symbols	Screened value	Chromium extraction (%)		Error (%)
			Observed value	Predicted value	
Residence time	X_1	3	99	95	4
TOMAC concentration	X_2	0.22			
Thiourea acidic	X_3	0.55			
Rotational speed	X_4	300			
Treat ratio	X_5	0.175			
1-octanol	X_6	3			

4. Conclusion

A systematic experimental design approach has been applied to screen six factors which is residence time (t), TOMAC concentration (M), stripping agent concentration (M), rotational speed (rpm), treat ratio, and modifier concentration (% w/v) that affects the chromium removal in continuous green emulsion liquid membrane using a 1/8 fraction of 2^{6-3} full factorial design. The results reveal that the degree of significance of each independent variable in the order of decreasing significance with respect to the influence on chromium removal is $X_5 > X_4 > X_1 > X_6 > X_2$. A regression model for chromium removal was developed and its R^2 (99.98%), R^2 (adj) (99.84%) and AAD (0.00041%) values were determined. The high R^2 and small AAD values signify that the model obtained is able to give a reasonably good estimate of chromium removal for the system in the range studied.

Reference

[1] F Fu, L Xie, B Tang, Q Wang and S Jiang 2012 Application of a novel strategy-Advanced Fenton-chemical precipitation to the treatment of strong stability chelated heavy metal containing wastewater *Chem. Eng. J.* **189-190** 283–287.

[2] A Tripathi and A K Dwivedi 2012 Studies on recovery of chromium from tannery wastewater by Reverse Osmosis *J. Ind. Pollut. Control.* **28** 29–34.

[3] K S Padmavathy, G Madhu and P V Haseena 2016 A study on Effects of pH, Adsorbent Dosage, Time, Initial Concentration and Adsorption Isotherm Study for the Removal of Hexavalent Chromium (Cr (VI)) from Wastewater by Magnetite Nanoparticles *Procedia Technol.* **24** 585–594.

[4] A S Dharnaik and P K Ghosh 2014 Hexavalent chromium [Cr(VI)] removal by the electrochemical ion-exchange process *Environ. Technol.* **35** 2272–2279.

[5] M Kul and K O Oskay 2015 Separation and recovery of valuable metals from real mix electroplating wastewater by solvent extraction *Hydrometallurgy* **155** 153–160.

[6] N F M Noah, N Othman and N Jusoh 2016 Highly selective transport of palladium from electroplating wastewater using emulsion liquid membrane process *J. Taiwan Inst. Chem. Eng.* **64** 134–141.

[7] A Bhowal, G Bhattacharyya, B Inturu and S Datta 2012 Continuous removal of hexavalent chromium by emulsion liquid membrane in a modified spray column S*ep. Purif. Technol.* **99** 69–76.

[8] S C Lee 2000 Continuous extraction of penicillin G by emulsion liquid membranes with optimal surfactant compositions *Chem. Eng. J.* **79** 61–67.

[9] S C Lee, K H Lee, G H Hyun and W K Lee 1997 Continuous extraction of penicillin G by an emulsion liquid membrane in a countercurrent extraction column *J. Memb. Sci.* **124** 43–51.

[10] R H Myers, D C Montgomery and C Anderson-Cook 2009 Response Surface Methodology: Process and Product Optimization Using Designed Experiments *Wiley Ser. Probab. Stat.* **704**.

[11] M A Bezerra, R E Santelli, E P Oliveira, L S Villar and L A Escaleira 2008 Response surface methodology (RSM) as a tool for optimization in analytical chemistry *Talanta.* **76** 965–977.

[12] M R Hubble 2012 Engineering statistics, Uma Ética Para Quantos XXXIII 81–87.

[13] T Lundstedt, E Seifert, L Abramo, B Thelin, Å Nyström, J Pettersen and R Bergman 1998 Experimental design and optimization *Chemom. Intell. Lab. Syst.* **42** 3–40.

[14] M E Meibodi, M Vafaie-Sefti, A M Rashidi, A Amrollahi, M Tabasi and H S Kalal 2010 The role of different parameters on the stability and thermal conductivity of carbon nanotube/water nanofluids, *Int. Commun. Heat Mass Transf.* **37** 319–323.

[15] J C Hsu, C J Lin, C H Liao and S T Chen 2008 Evaluation of the multiple-ion competition in adsorption of As(V) onto reclaimed iron-oxide coated sands by fractional factorial design, *Chemosphere.* **72** 1049–1055.

[16] S H Chang, T T Teng and N Ismail 2011 Screening of factors influencing Cu(II) extraction by soybean oil-based organic solvents using fractional factorial design *J. Environ. Manage.* **92** 2580–2585.

[17] J L Pilkington, C Preston and R L Gomes 2014 Comparison of response surface methodology (RSM) and artificial neural networks (ANN) towards efficient extraction of artemisinin from Artemisia annua *Ind. Crops Prod.* **58** 15–24.

[18] R N Raja Sulaiman, N Othman and N A Saidina Amin 2016 Recovery of ionized nanosilver by emulsion liquid membrane process and parameters optimization using response surface methodology *Desalin. Water Treat.* **57** 3339–3349.

[19] N Othman, N F M Noah, K W Poh and O Z Yi 2016 High Performance of Chromium Recovery from Aqueous Waste Solution Using Mixture of Palm-oil in Emulsion Liquid Membrane, in: Procedia *Eng.* **148** 765–773.

[20] N Sathyamoorthy, D Magharla, P Chintamaneni and S Vankayalu 2017 Using Box Behnken design *J. Basic Appl. Sci.* **6** 362–373.

[21] Z Y Ooi, N Othman and N F Mohamed Noah 2016 Response surface optimization of kraft lignin recovery from pulping wastewater through emulsion liquid membrane process *Desalin. Water Treat.* **57** 7823–7832.

[22] W N N Wan Omar and N A Saidina Amin 2011 Optimization of heterogeneous biodiesel production from waste cooking palm oil via response surface methodology *Biomass Bioenergy.* **35** 1329–1338.

Acknowledgments

The authors would like to acknowledge the Ministry of Higher Education (MOHE) for the financial support under Research grant: Vote 4F949 and Universiti Teknologi Malaysia to make this research possible.

Comparison of Total Inorganic Carbon (TIC) Analysis with Titration Method Analysis in Determining Carbon Dioxide (CO_2) Content of Amine Solvent

Chan Zhe Phak, Syafiqa Bt M Saleh, Muhammad Afiq Muzammil

PETRONAS Research Sdn. Bhd., Kawasan Institusi Bangi, 43000 Kajang, Selangor, Malaysia.

chanzhephak@petronas.com.my

Abstract: Total Inorganic Carbon (TIC) is an analysis method to determine carbon content in a sample. In this paper, applicability of TIC method to determine carbon dioxide (CO_2) content in amine solvent is explored in detailed. Both commercial and self-blended amine solutions, consisting of MDEA and PZ, were used in this study and bubbled with CO_2. TIC method was calibrated for both mid-level mode and high level mode. Difference in results were found when comparing TIC against the standard titration method; this is largely due to error in titration method during the methanol pH adjustment step. TIC method is considered to be better in terms of limit of detection, limit of quantification, recovery percentage, and susceptibility to error.

1. Introduction

In the present scenario, anthropogenic global warming is acknowledged as one of the environmental issues. Fossil fuels are considered as the leading source of energy consumed in the world where electricity, heat generation and transportation depend heavily on fossil fuel and this is responsible for two-third of the worldwide CO_2 emission [2]. As stated by C. Nwaoha et al. (2009), among various technologies for capturing CO_2, CO_2 capture by chemical absorption through the use of amine-based solvent is acknowledged as the most proven and commercially available technology [3]. Conventional industrial methods to reduce acid gas employ gas absorption device, such as scrubber and spray tower, but these current method have several disadvantages such as difficulty of obtaining accurate estimation of the gas-liquid mass transfer area. As an alternative to the current technologies, non-dispersive gas-liquid contact via a micro porous membrane is approached. According to Mansourizadeh and Ismail (2009), fluids can contact on opposite side of the membrane and gas-liquid interface is formed if suitable membrane configuration such as hollow fibre is used [1].

To select a suitable liquid absorbent, some criteria have to be taken into consideration, and one of the main criteria is its reactivity. Highly reactive solvent can lead to higher absorption rate and reduction of liquid mass transfer resistance [1]. Amine solvent which contains aqueous MDEA and PZ is the focus in this paper. The absorption of CO_2 by amine-based solvent can analogously be represented by an acid-base reaction, where the amine solvent acts as the base while CO_2 is the acid, to form salt like amine carbamate ($AmineCOO^-$) and bicarbonate (HCO_3^-) [3]. According to Xu Zhang, the reaction of CO_2 and PZ can be regarded as the rapid pseudo-first-order reaction in parallel with that of CO_2 with MDEA [4]. Below is the several reaction steps that can occur through absorption process [2, 3]:

$$2H_2O \leftrightharpoons H_3O^+ + OH^- \quad (1)$$
$$CO_2 + 2H_2O \leftrightharpoons H_3O^+ + HCO_3^- \quad (2)$$
$$HCO_3^- + H_2O \leftrightharpoons H_3O^+ + CO_3^{2-} \quad (3)$$
$$MDEAH^+ + H_2O \leftrightharpoons MDEA + H_3O^+ \quad (4)$$
$$MDEA + CO_2 + H_2O \leftrightharpoons MDEAH^+ + HCO_3^- \quad (5)$$
$$PZH_2^{2+} + H_2O \leftrightharpoons PZH^+ + H_3O^+ \quad (6)$$

$$PZH^+ + H_2O \leftrightharpoons PZ + H_3O^+ \tag{7}$$
$$PZ + CO_2 + H_2O \leftrightharpoons PZCOO^- + H_3O^+ \tag{8}$$
$$PZCOO^- + CO_2 + H_2O \leftrightharpoons H^+PZCOO^- + HCO_3^- \tag{9}$$
$$H^+PZCOO^- + H_2O \leftrightharpoons PZCOO^- + H_3O^+ \tag{10}$$
$$PZCOO^- + CO_2 + H_2O \leftrightharpoons PZ(COO^-)_2 + H_3O^+ \tag{11}$$

Analytical methods used in determining CO_2 content and CO_2 loading in this paper are titration and TIC. Amine content is determined by titrating solvent with hydrochloric acid (HCl) while for the CO_2 content and loading, potassium hydroxide (KOH) is used. From the data of amine content and CO_2 content, the CO_2 loading of the solvent can be determined. TIC analytical method determines the content of CO_2 in the sample by acidifying it with phosphoric acid (H_3PO_4) in the reaction chamber of the analyser in order to convert inorganic carbon compounds (carbonate and bicarbonate) into CO_2 which is then measured by solid state non-dispersive (SSNDIR) detector. Comparing with titration method, TIC method produces less chemical waste and consumes less chemicals, making its operating cost cheaper. Besides that, its throughput is much higher than titration method through auto sampler which can free up the technicians' time for other tasks and analyses. The objective of this study is to compare titration analysis method with TIC analysis method in detailed.

2. Methodology

2.1. Total Inorganic Carbon (TIC) Calibration

TIC calibration was done by using 1000 ppm sodium carbonate (Na_2CO_3) solution (stock solution) which was bought from Sigma-Aldrich. 8 standard solutions were prepared by diluting 1000 ppm Na_2CO_3 solution with distilled water to 0 ppm (distilled water), 1 ppm, 5 ppm, 10 ppm, 50 ppm, 100 ppm and 500 ppm and 1000 ppm in 50 mL volumetric flask. The volume of 1000 ppm Na_2CO_3 solution to be diluted was calculated by using the following equation:

$$V_{stock} = \frac{M_{standard} \times V_{standard}}{M_{stock}} \tag{12}$$

The prepared standards were then transferred into the vial bottles until the volume of the standards were about three-quarter of the bottles. TIC method was set to two different modes; high level mode and mid-level mode. Standard of 0 ppm, 1 ppm, 10 ppm, 50 ppm and 100 ppm were used for mid-level mode calibration. Meanwhile, standard of 0 ppm, 10 ppm, 100 ppm, 500 ppm and 1000 ppm were used for high level mode calibration. 1.3 M H_3PO_4 acid was used as the reagent to perform the analysis in OI Analytical Aurora 1030C TOC Analyser (Aurora) with the sample volume of 2 mL and 0.5 mL for high level and mid-level mode respectively. Sample injection was done by the auto sampler and the same sample volume was used during calibration procedure.

2.2. Sample Preparation

Amine solvent was prepared by blending MDEA (purity > 98%), PZ (purity > 99%) and distilled water in a 1 L bottle. The mass of each component was measured using electronic balance (FR-300 MK II). The solvent was prepared with 5 different concentration of amine (30wt%, 35wt%, 40wt%, 45wt% and 50wt %) with mass ratio between PZ and MDEA was 0.183. The blended amine solvent was bubbled with CO_2 gas with different time intervals (0s, 2s, 4s, 6s, 8s, 10s, 60s, 120s, 240s, 360s and 480s) at atmospheric pressure. Commercial amine solvent that had been used in membrane contactor test rig was regenerated by using rotary evaporator at the temperature of 105°C for different time intervals (0 min, 15 min, 30 min, 45 min and 60 min).

2.3. Reagent Preparation

Three reagents were prepared for titration analysis and TIC analysis. HCl acid and KOH were used in titration while H_3PO_4 acid was used for TIC reagent.

2.3.1 KOH Preparation and Standardization

66 g of KOH pellet was weighed using electronic balance (FR-300 MK II). Distilled water was added in to completely dissolve the pellet. The KOH solution was then diluted in a 2 L volumetric flask. 1 g

of potassium hydrogen phthalate (KHP) was dissolved in about 40 mL of distilled water. About 6 drops of phenolphthalein were added into the KHP solution. The KHP solution was then titrated with the prepared KOH solution until the colour just turned to pink. The following equation was used to calculate the normality of the KOH solution:

$$N_{KOH} = \frac{W_{KHP}}{204.22} \times \frac{1000}{V_{KOH}} \quad (13)$$

2.3.2 HCl Preparation and Standardization

84 mL of concentrated HCl (37%) was diluted in 2 L volumetric flask with distilled water. The diluted HCl then was titrated with 25 mL of KOH (6 drops of phenolphthalein was added) until the colour turned to colourless. The following equation is the HCl normality calculation:

$$N_{HCl} = \frac{V_{KOH} \times N_{KOH}}{V_{HCl}} \quad (14)$$

2.3.3 H_3PO_4 Preparation and Standardization

69 mL of H_3PO_4 was diluted with distilled water in a 1 L volumetric flask. The diluted H_3PO_4 then was titrated with 25 mL of KOH (6 drops of phenolphthalein was added) until the colour turned to colourless. The following equation is the HCl normality calculation:

$$N_{H_3PO_4} = \frac{1}{3} \times \frac{V_{KOH} \times N_{KOH}}{V_{H_3PO_4}} \quad (15)$$

2.4. Titration

2.4.1 Amine Content Determination

4.5 g of sample (amine solvent) was diluted to 100 mL volume in a beaker with distilled water. The mixture then was stirred at 400 rpm by using magnetic stirrer (WiseStir). The mixture was titrated with the standardized HCl acid until the pH dropped to 4.5 (measured by pH meter probe). The following equation was used to calculate amine content in the sample:

$$Amine\ content, wt\% = \frac{V_{HCl} \times N_{HCl} \times A}{W_{sample}} \quad (16)$$

$$A = \frac{r\,(MW_{MDEA} - MW_{PZ}) + MW_{PZ}}{10\,(2-r)} \quad (17)$$

Where r is the mole fraction of MDEA in a mixture of MDEA and PZ only.

2.4.2 CO_2 Content and CO_2 Loading Determination

120 mL of methanol (CH_3OH) was titrated with the standardized KOH until its pH rose up to 11.2. 20 g of the sample was then added and stirred at 400 rpm with magnetic stirrer while titrating it with the KOH until its pH rose up to 11.2. The equations below were used to calculate CO_2 content and CO_2 loading of the sample: (Note: Regenerated amine is the 300 mL of lean solution that was refluxed for 24 hours by using electromantle)

$$CO_2\ content, wt\% = \left(\frac{V_{KOH}}{W_{sample}} - \frac{V_{KOH,regenerated\ amine}}{W_{regenerated\ amine}}\right) \times N_{KOH} \times 4.4 \quad (18)$$

$$CO_2\ loading, mol/mol = \frac{CO_2\ content, wt\% \times B}{Amine\ content, wt\%} \quad (19)$$

$$B = \frac{MW_{Amine}}{MW_{CO_2}\,(1+y)} \quad (20)$$

Where y is the mole fraction of PZ in a mixture of MDEA and PZ only.

2.5. Total Inorganic Carbon (TIC) Sample Preparation and Determination

1 mL of the sample was diluted 100 times in a 100 mL volumetric flask. The diluted sample was transferred into two TIC vials until three-quarter full. The two TIC vials were then analysed with two different modes of TIC respectively; high mode and mid-level mode, as shown in Table 1 below.

Table 1. Parameter setting for different modes of TIC

Mode	Sample Volume, mL	Acid Volume, mL	Reaction Time, min	System Pressure, psi
High level	0.2	2.0	3.0	25
Mid-level	2.0	0.5	2.5	20

3. Result and Discussion

3.1 Titration

As CO_2 is dissolved in the solvent, CO_2 initially reacts with H_2O to produce carbonic acid (H_2CO_3), which subsequently dissociates to HCO_3^-[6]. Referring to Equation (1), the presence of H_3O^+ would reduce the pH value of the sample. In determining the CO_2 content through titration, OH^- is needed to neutralize H_3O^+ and according to Chung et al. (2010), OH^- is also needed to convert free dissolved CO_2 into non-volatile ionic species [7]. Meanwhile, to determine the amine content in solvent, the pH of the solvent needs to be reduced by adding acid in order to protonate the amine. According to S.A. Freeman, acid is used to protonate PZ by adding two protons to the molecule, forming diprotonated PZ molecule (PZH_2^{2+}) [5]. Stated by Jessica Narku-Tetteh et al. (2017), the protonation of amine in the aqueous solution was performed by titrating the amine with HCl until methyl orange endpoint was achieved with the measurement of pH after every addition of 0.5 mL of HCl [8]. The following equations describe the chemistry behind titration method:

Amine content determination

$$MDEA + H_3O^+ \rightleftharpoons MDEAH^+ + H_2O \quad (21)$$

$$PZ + 2H_3O^+ \rightleftharpoons PZH_2^{2+} + H_2O \quad (22)$$

CO_2 content determination

$$OH^- + H_3O^+ \rightleftharpoons 2H_2O \quad (23)$$

$$OH^- + CO_2 \rightleftharpoons HCO_3^- \quad (24)$$

Amine content was obtained by titrating it until its pH value dropped to 4.5 (specified in method for commercial amine solvent), but it turned out that equivalence point for the blended amine was less than 4.5. Similar observation for equivalence point was made when the commercial amine solvent was titrated. If titration is stopped at pH 4.5 as specified in the method, it will result in a bias in amine content determination. The accuracy in determining the amine content is important since it directly affects the result for CO_2 loading of the solvent, as shown in Equation (18). Therefore, it is suggested that the solvent needs to be titrated to the equivalence point in amine content titration.

Referring to Figure 1 and Figure 2, the pH value at the equivalence point of the blended solvent are approximately 3.4 and 3.2, respectively (determined from the graph). Meanwhile, the pH value at the equivalence point of the commercial amine solvent is 3.2, regardless of the concentration of CO_2 in the solvent.

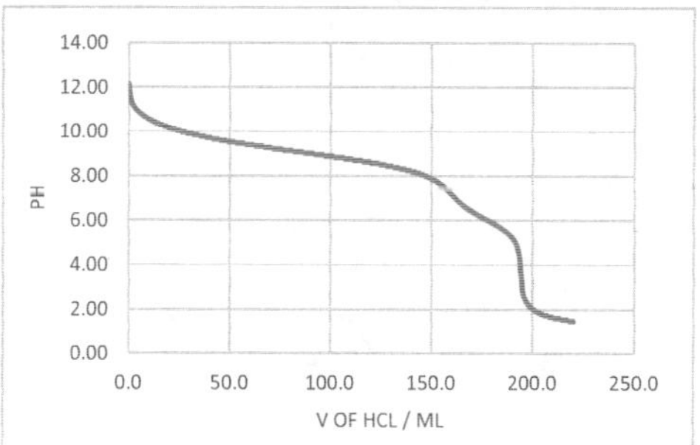

Figure 1. Equivalence point determination graph of 50 wt. % for the blended amine solvent.

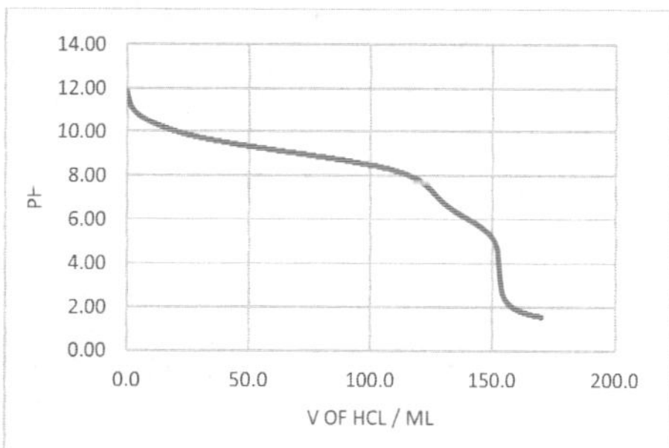

Figure 2. Equivalence point determination graph of 30 wt. % for the blended amine solvent.

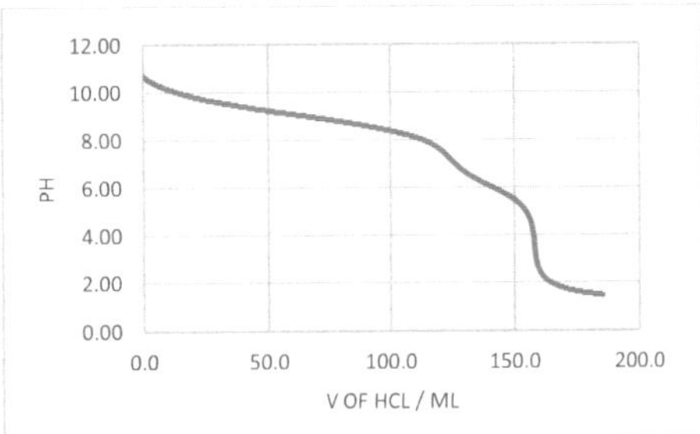

Figure 3. Equivalence point determination graph of the commercial amine solvent at 0.02 wt. % CO_2.

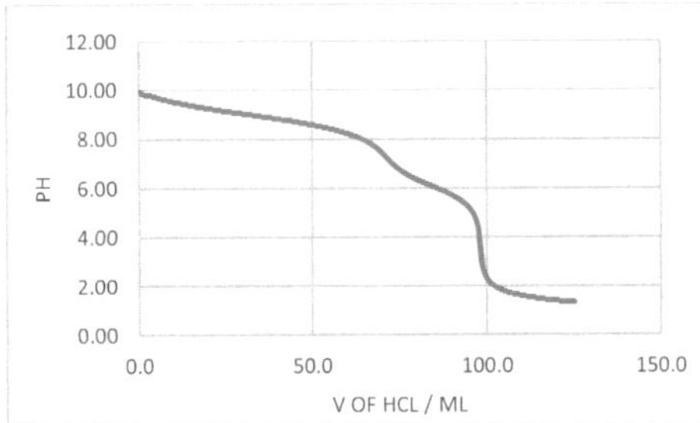

Figure 4. Equivalence point determination graph of the commercial amine solvent at 3 wt. % CO_2.

In determination of CO_2 content, pH of methanol was first adjusted to 11.2 using KOH before addition of sample. This was found as a source of error because it was difficult to adjust the pH of methanol to 11.2 exactly. Table 2 shows a few examples of titration where the pH of methanol ranges from 11.24 to 11.40, instead of 11.2. These has resulted in error as the titration end-point is set at pH 11.2. The results will be more accurate if the titrations are allowed to proceed until the pH of methanol after KOH adjustment.

Table 2. Initial pH of methanol after KOH adjustment

Sample	pH of methanol after KOH adjustment	CO_2 content, wt%	
		Titration to pH of methanol	Titration to pH of 11.2
30wt% amine	11.24	0.48	0.48
30wt% amine	11.30	0.50	0.48
30wt% amine	11.30	0.50	0.48
30wt% amine	11.35	0.46	0.45
30wt% amine	11.30	0.53	0.52
Commercial	11.30	0.72	0.71
Amine Solvent	11.40	0.74	0.72
T-test result		$T_{stat}(4.317) > T_{critical}(2.447)$	

As shown in Table 3, Limit of Detection (LOD) and Limit of Quantification (LOQ) were established from amine solvent (30wt%) with low concentration of CO_2. LOD is defined as three times the standard deviation meanwhile LOQ is ten times the standard deviation.

Table 3. Measurement of low concentration of CO_2 in amine solvent through titration

Titration to pH of 11.20	
	0.05 wt%
	0.07 wt%
	0.05 wt%
	0.03 wt%
	0.04 wt%
	0.02 wt%
Average	0.04 wt%
Standard Deviation	0.02 wt%
Limit of Detection (LOD)	0.05 wt%
Limit of Quantification (LOQ)	0.18 wt%

3.2 Total Inorganic Carbon (TIC)

In TIC, CO_2 content is determined by acidifying the sample in order to convert aqueous CO_2, carbamate and bicarbonate species into gaseous CO_2 [5]. CO_2 then will be carried in a nitrogen gas stream to be analysed by the detector.

$$H_3O^+ + HCO_3^- \rightleftharpoons CO_2 + 2H_2O \qquad (25)$$

As being stated in the methodology part, the equipment needs to be calibrated with Na_2CO_3 standard. The area under the curve, formed from the signal detected by the SSNDIR, is matched with the respective standard concentration. Table 4 show the calibration data for the high level and mid-level mode respectively by using Na_2CO_3 solutions as standards. Figure 5 show the calibration curve, plotted from the calibration data in Table 4. Linearity of both calibration curves are good as the R^2 (coefficient of correlation) values are both higher than 0.999.

Table 4. Calibration data for high level and mid-level mode

Calibration data for high level mode		Calibration data for mid-level mode	
Carbon Concentration, ppm	Average (Avg) Area Under The Curve	Carbon Concentration, ppm	Average (Avg) Area Under The Curve
0.000	164	0.000	2270
10.000	12230	1.000	12343
100.000	110741	10.000	103617
500.000	538201	50.000	455942
1000.000	1041923	100.000	905914

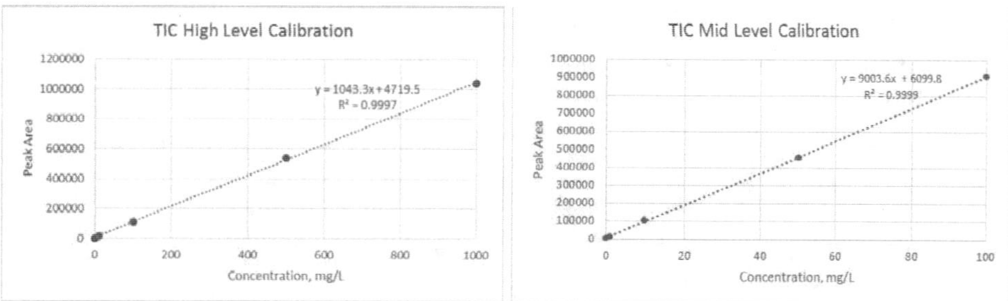

Figure 5. Calibration curve for TIC high level and mid-level mode.

The sample volume injected into the chamber are 0.2 mL and 2 mL for high level and mid-level mode respectively. The difference in the sample volume injected in the chamber affect the amount of CO_2 released, as bigger sample volume will release higher amount of CO_2. This is illustrated in Table 4 where the average area for mid-level mode is higher than high level mode for 0 ppm, 10 ppm and 100 ppm. Limit of detection (LOD) and limit of quantification (LOQ) were established from blank analysis, as shown in Table 5 below. LOD is expressed as three standard deviations of the blank and LOQ is ten standard deviations of the blank. The LOD and LOQ values for both high level mode and mid-level mode are sufficiently low, well below the sample range of 2ppm to 350ppm (carbon content).

Table 5. Measurement of blank by TIC at high and mid-level mode

Carbon Concentration	High Level	Mid-level
	-4.39 ppm	1.08 ppm
	-4.29 ppm	1.19 ppm
	-4.28 ppm	1.23 ppm
	-4.29 ppm	1.25 ppm
	-4.27 ppm	1.22 ppm
	-4.27 ppm	1.22 ppm
Average	-4.30 ppm (-0.00158 wt%)	1.20 ppm (0.00044 wt%)
Standard Deviation	0.05 ppm (0.00002 wt%)	0.06 ppm (0.00002 wt%)
Limit of Detection (LOD)	0.14 ppm (0.00005 wt%)	0.18 ppm (0.00007 wt%)
Limit of Quantification (LOQ)	0.46 ppm (0.00017 wt%)	0.60 ppm (0.00022 wt%)

The original intention of having mid-level mode is to lower down the LOQ. Results from Table 5 shows that similar LOQ can be achieved with high level mode. Therefore, mid-level mode is not required. Besides, Table 6 shows that there is no significant difference in results between high level mode and mid-level mode.

Table 6. Comparison between high level mode and mid-level mode of 45 wt% self-blended amine

CO2 Bubbling Time, s	CO_2 content, wt%		CO_2 loading, mol/mol	
	High Level Mode	Mid-level Mode	High Level Mode	Mid-level Mode
2	0.4305	0.8873	0.0205	0.0422
4	0.5125	0.5041	0.0244	0.0240
6	0.6617	0.6949	0.0315	0.0331
8	0.8551	0.9166	0.0409	0.0438
10	0.9380	0.9972	0.0445	0.0473
T-test result	$T_{stat}(1.417) < T_{critical}(2.776)$		$T_{stat}(1.413) < T_{critical}(2.776)$	

3.3 Comparison of Titration and Total Inorganic Carbon (TIC) Analysis

The data for the blended amine solvent is tabulated in Table 7, where comparison is made between TIC method at high level mode and titration method. It is observed that the difference between titration results and TIC results for most samples are above 5%. The significant difference in the results is very much likely due to the fact that titration method is subjected to error in methanol pH adjustment step. Due to this reason, TIC method is considered more accurate in comparison to titration method.

Table 7. Data comparison for blended solvent

Initial Amine Conc., wt.%	CO_2 Bubbling Time, s	Amine, wt.%	CO_2, wt.%		CO_2 loading, mol/mol	
			Titration	TIC	Titration	TIC
30.00	10	29.48	1.1472	1.2091	0.0827	0.0872
	60	28.50	3.4670	4.5247	0.2587	0.3376
	120	28.27	4.1244	5.0523	0.3103	0.3801
	240	27.71	5.4347	7.3498	0.4170	0.5640
	360	27.68	6.5159	7.7182	0.5005	0.5929
	480	27.56	6.1995	8.4774	0.4784	0.6542
40.00	60	38.59	3.7499	3.8051	0.2066	0.2097
	120	37.96	5.4783	5.8692	0.3068	0.3287
	240	37.28	7.3431	8.3391	0.4188	0.4756
	360	36.98	8.4903	8.8870	0.4882	0.5110
	480	36.89	8.3952	8.3295	0.4838	0.4801
45.00	2	44.71	0.3885	0.4305	0.0185	0.0205
	4	44.65	0.4847	0.5125	0.0231	0.0244
	6	44.60	0.6418	0.6617	0.0306	0.0315
	8	44.49	0.8261	0.8551	0.0395	0.0409
	10	44.84	0.8974	0.9380	0.0426	0.0445
	60	43.66	3.4942	3.5057	0.1702	0.1707
	120	42.80	5.3222	5.0957	0.2644	0.2532
	240	42.10	7.0289	7.2115	0.3551	0.3643
	360	41.63	7.9250	7.8397	0.4048	0.4004
	480	41.60	8.5047	8.7715	0.4348	0.4484
50.00	2	49.36	0.4279	0.4932	0.0184	0.0212
	4	49.24	0.5485	0.6515	0.0237	0.0281
	6	49.30	0.6944	0.7617	0.0300	0.0329
	8	49.43	0.7590	0.7828	0.0327	0.0337
	10	49.34	0.8708	0.8971	0.0375	0.0387
	60	48.01	3.3173	3.8132	0.1469	0.1689
	120	47.35	4.8674	4.9855	0.2186	0.2239
	240	46.41	6.6617	6.9454	0.3052	0.3182
	360	46.01	7.6994	8.3453	0.3559	0.3857
	480	45.79	7.9213	8.6877	0.3678	0.4034
T-test result			$T_{stat}(3.699) > T_{critical}(2.042)$		$T_{stat}(3.252) > T_{critical}(2.024)$	

Table 8. Data comparison for the commercial amine solvent

Sample	Amine / wt.%	CO_2 / wt.%		CO_2 loading / mol/mol	
		Titration	TIC	Titration	TIC
A	47.54	0.7010	0.7928	0.0305	0.0345
B	47.47	0.6797	0.7786	0.0296	0.0339
C	47.34	0.6518	0.7891	0.0285	0.0345
D	38.74	3.6101	3.7103	0.1927	0.1980
E	40.96	4.4251	5.0123	0.2234	0.2530
F	37.60	3.5278	4.0307	0.1940	0.2217
G	41.41	3.9014	3.9253	0.1948	0.1960
T-test result:		T_{stat} (2.583) > $T_{critical}$ (2.447)		T_{stat} (2.449) > $T_{critical}$ (2.447)	

The data for the commercial amine solvent is tabulated in Table 8. Similar observation can be made, where there is a significant difference between TIC results and titration results

4. Conclusion

Solvent analysis is very important to monitor the performance of solvent based acid gas removal technology. In this work, the applicability of TIC method to determine CO_2 content and CO_2 loading in amine solutions was studied. There is no need to use mid-level mode when analysing sample with low CO_2 content, as there is no significant difference between high level mode and mid-level mode in terms of LOD, LOQ, and results. In titration method, the error during the methanol pH adjustment step contributed to the difference in results with TIC method. TIC analysis is chosen as the preferred method due to lower LOD, lower LOQ, and less susceptible to error. In addition, TIC method does not involve volatile organic solvent which reduces cost and makes the analysis more environmental friendly. Besides, titration is less accurate compared to TIC, TIC is also seen to produce less chemical waste. Due to that, TIC analysis method is studied as a substitute to titration method, where TIC is seen as more convenient and requires less cost.

References

[1] Mansourizadeh A and Ismail A F 2009 Hollow fiber gas-liquid membrane contactors for acid gas capture: A review *J. Haz. Mat.* **171** 38-53

[2] Khan A, Halder GN. and Saha A K 2017 Experimental investigation on efficient carbon dioxide capture using piperazine (PZ) activated aqueous methyldiethanolamine (MDEA) solution in a packed column *Int. J. Greenhouse Gas Control* **64** 163-73

[3] Nwaoha C, Odoh K, Ikpatt E, Orji R and Idem R 2017 Process simulation, parametric sensitivity analysis and ANFIS modelling of CO_2 capture from natural gas using aqueous MDEA-PZ blend solution *J. Env. Chem. Eng.* **5** 5588-98

[4] Zhang X, Zhang C F, Qin S J and Zheng Z S 2001 A kinetics study on the absorption of carbon dioxide into a mixed aqueous solution of methyldiethanolamine and piperazine *Ind. Eng. Chem. Res.* **40** 3785-91

[5] Freeman S A, Dugas R, Wagener D H V, Nguyen T and Rochelle G T 2009 Carbon dioxide capture with concentrated, aqueous piperazine *Int. J. Greenhouse Gas Control* **4** 119-24

[6] Ghalib L, Ali B S, Ashri W M, Mazari S and Saeed I M 2016 Modeling the effect of piperazine on CO_2 loading in MDEA/PZ mixture *Fluid Phase Equilibria* **434** 233-43

[7] Chung P Y, Soriano A N, Leron R B and Li M H 2010 Equilibrium solubility of carbon dioxide in the amine solvent system of (triethanolamine + piperazine + water) *J. Chem. Thermodynamics* **42** 802-07

[8] Narku-Tetteh J, Muchan P, Saiwan C, Supap T and Idem R 2017 Effect of side chain structure and number of hydroxyl groups of primary, secondary and tertiary amines on their post-combustion CO_2 capture performance *Energy Procedia* **114** 1811-27

[9] Samanta A and Bandyopadhyay S S 2007 Kinetics and modelling of carbon dioxide absorption into aqueous solutions of piperazine *Chem. Eng. Sci.* **62** 7312-19

IOP Conf. Series: Materials Science and Engineering **458** (2018) 012034 doi:10.1088/1757-899X/458/1/012034

A new recycling process for recycled plastic production from glass fiber reinforced plastics

K Ikenaga*, H Nagamitsu, T Kikukawa and K Kusakabe

Department of Nanoscience, Sojo University, 4-22-1 Ikeda, Nishi-ku, Kumamoto 860-0082, Japan.

*ikenaga@nano.sojo-u.ac.jp

Abstract. The decomposition of glass fiber reinforced plastics (GFRP) with ethylene glycol monoallyl ether (EGMA) and benzyl alcohol (BnOH) by non-catalytic transesterification under high pressure, microwave irradiation was investigated. The decomposition product of GFRP with EGMA (EGMA-P) has a high reactivity for the cross-linking reaction to unsaturated polyester (UP) compared to the product of GFRP with BnOH (Bn-P) because of the presence of many allyl groups derived from EGMA. Therefore, the hardened material (EGMA-HM) with high thermal stability was prepared from EGMA-P, unsaturated polyester, and curing agent. Finally, the recycled GFRP with high bending breaking point was produced from EGMA-P.

1. Introduction

GFRP is a composite made of polymer reinforced with glass fiber. Unsaturated polyester is generally used as thermosetting resin composites and styrene resin behaves as a binder for the cross-linking polymerization. GFRP composites are widely used in many industries including windmill, automobile, shipbuilding, construction and bathtub as a hard material. Nowadays, annual production volume of GFRP in Japan is about 250,000 ton and the annual generation volume of GFRP waste is about 200,000 ton. Most GFRP wastes (ca. 95 %) end up in a landfill. The remaining GFRP wastes (ca. 5 %) become an alternative fuel in the cement industry. The mechanical recycling process of waste GFRP was difficult because of high hardness property of GFRP [1].

To realize the sustainable society, chemical recycling of GFRP has been attracted much attention from many researchers in academia and industry. The decomposition method of GFRP in supercritical or subcritical fluid without catalyst required high-temperature and high-pressure operation [2-4]. The reaction pressure was able to be reduced in supercritical methanolysis of GFRP with N, N-dimethyl-4-aminopyridine catalyst [5]. Furthermore, solvothermal decomposition of GFRP was demonstrated under ordinary pressure in the presence of tripotassium phosphate as an alkali catalyst for a prolonged time (12 hr.) [6]. The reproduction of the decomposed resin was performed through the transesterification or dehydration esterification of oxygen-containing groups in the resin with strong acid catalyst. The reproduced polyesters became black colored materials in almost all cases.

The desired full conversion of GFRP to a recycle polymer was not accomplished until now despite many efforts. We proposed the convenient solvolysis process of GFRP without catalyst under high pressure and microwave heating [7]. In this study, more developed solvolysis process of GFRP with

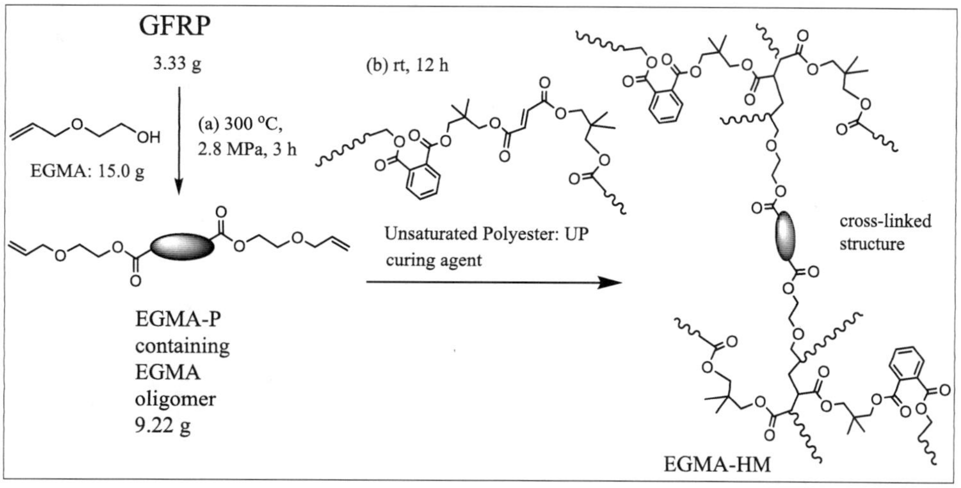

Figure 1. Postulated reaction schemes: (a) decomposition of GFRP resin by EGMA-solvolysis and (b) cross-linking reaction of allyl groups of EGMA-P with UP.

ethylene glycol monoallyl ether (EGMA) under high pressure and microwave heating was investigated. After the solvolysis, the recycle polymer was obtained through the cross-linking reaction of decomposition products having allyl group with unsaturated polyester (UP) as shown in figure 1. Finally, the recycled GFRP was produced from EGMA-P.

2. Experiments

2.1 Decomposition of GFRP

The piece of GFRP used in this study consists of 65 % network resin and 35 % glass fiber from the combustion test. GFRP 3.33 g and EGMA 15.0 g, were added in a 10 mL sealed tube equipped with a stir bar. Microwave was irradiated to the tube attached in the microwave apparatus (Initiator+8 with single mode microwave irradiation, Biotage AB). The reaction temperature of 300°C and the reaction pressure of 3.0 MPa were achieved by the maximum irradiation (2.45 GHz, 300 W) of the microwave apparatus for 3 h. After the microwave irradiation, the products were thermally extracted from the reaction mixture using chloroform (90 mL). The chloroform solution was separated by vacuum filtration from the glass fibers and the unreacted resin of GFRP. Chloroform was evaporated from the solution and then the remaining high-boiling solvent was distilled under reduced pressure to obtain the product (EGMA-P) as a tarry residue (Figure 2). The decomposition ratio of the GFRP was calculated from the weight reduction of GFRP as follows.

$$Decomposition\ ratio\ of\ resin = \frac{reduction\ amount\ of\ GFRP}{initial\ amount\ of\ GFRP} x100 \qquad (1)$$

GFRP was decomposed with benzyl alcohol (BnOH) for comparison to give the product (Bn-P) under the same condition with EGMA experiment. The products were identified by IR, NMR, and GPC analysis.

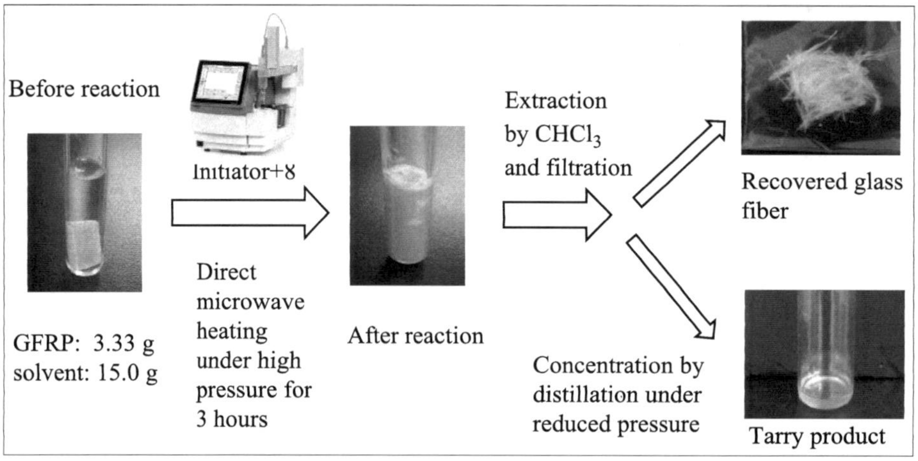

Figure 2. Sequential procedure of decomposition of the GFRP resin.

2.2 Preparation of hardened material (HM) by curing

The EGMA-hardened material (EGMA-HM) was produced by the curing reaction of EGMA-P (5.0 g) with UP (15 g), and curing agent (methyl ethyl ketone peroxid; 0.30 g) at room temperature according to the standard preparation method. The BnOH-hardened material (Bn-HM) was prepared from Bn-P with UP and curing agent. The styrene-hardened material (St-HM) as a standard substance was formed with pure styrene with UP. The thermogravimetric analysis of hardened materials was performed by TG-DTA apparatus (30-500°C, 1.0°C / 1.0 min., Thermo plus TG8120, Rigaku Co.).

2.3 Preparation of GFRP

The hand lay-up method was used for the preparation of GFRP. The weight composition ratio of recycle polymer, styrene, UP, curing agent and glass fiber was fixed at 3:2:15:0.3:10, respectively according to the standard preparation of GFRP including 65 % resin and 35 % glass fiber. GFRPs prepared from EPMA-P and Bn-P was called as GFRP-E and GFRP-B, respectively. Standard GFRP was synthesized from styrene (5.0 g), UP (15 g), the curing agent (0.30 g), and fresh glass fiber cloth (10 g). According to JIS K 7017, the bending breaking points of GFRPs were determined from three-point point bending test of GERPs with AG-Xplus (Shimadzu Co.).

3. Results and discussion

GFRP resin was easily non-catalyzed decomposed by transesterification with EGMA and BnOH under high pressure, microwave heating. The both products (EGMA-P and Bn-P) were tarry brownish oils. Table 1 presents the decomposition ratio of GFRP evaluated from the weight reduction of GFRP. These decompositions were carried out using direct microwave heating apparatus with high pressure (at 300°C, 3.0 MPa, Initiator + 8 and Biotage AB). The decomposition ratio in EGMA (69.3%) was almost the same as that in BnOH (70.7 %) for 3-hour reaction. After the filtration, the recovered solid was fully identified as glass fiber. This means the complete esterification of GFRP was achieved in EGMA and BnOH regardless of aliphatic and aromatic alcohol. Figure 3 shows the NMR spectra of EGMA-P and separately-synthesized EGMA oligomer. The molecular weight distribution of EGMA-P is 1200-17000. The NMR analysis of EGMA-P indicated the existences of allyl groups of EGMA-P suggesting the occurrence of the cross-linking reaction to UP.

Table 1. Decomposition of GFRP resin by direct microwave heating with high pressure

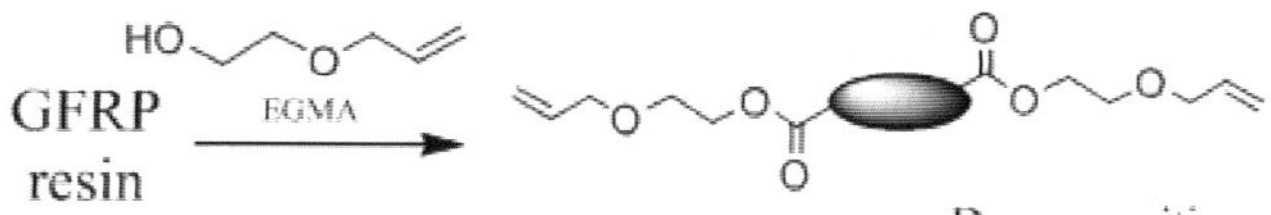

Entry	GFRP / g	Solvent	Time / h	Product	Decomposition ratio of resin / %
1	3.11	BnOH	3	Bz-P	70.7
2	3.33	EGMA	2	EGMA-P	67.2
3	3.40	EGMA	3	EGMA-P	69.3

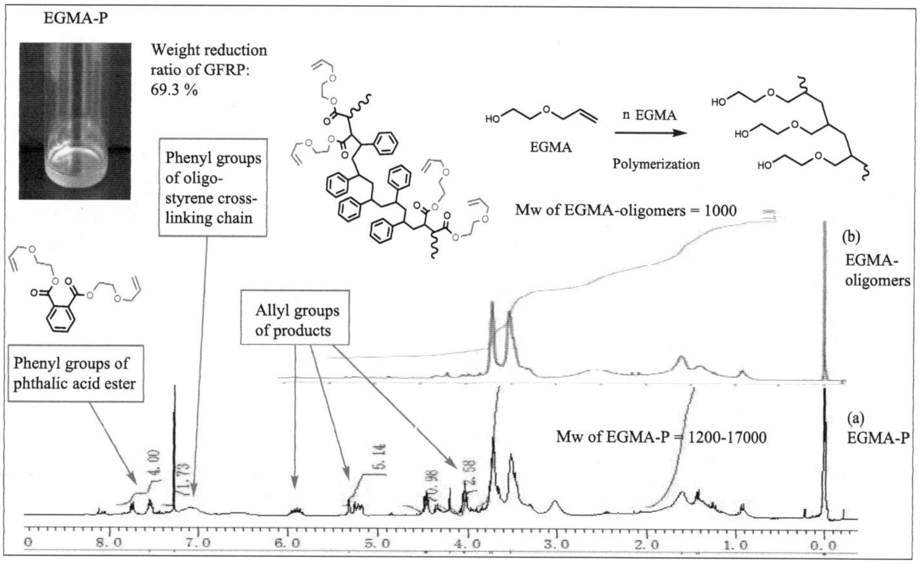

Figure 3. The NMR spectra of: (a) EGMA-P and (b) EGMA-oligomers.

EGMA-P containing EGMA oligomers reacted with UP by using curing agent at room temperature overnight to give the EGMA-HM with a sufficient hardness. Figure 4 shows the TGA curve of EGMA-HM, Bn-HM, and St-HM. The total weight loss of Bn-HM and St-HM reached almost 100% at 500 °C. Meanwhile the weight loss of EGMA-HM at the temperature of more than 300 °C was low indicating the highest heat resistance among them. These results strongly suggested that EGMA-P, which was decomposed under high pressure, microwave irradiation, has the sufficient cross-linking reactivity to UP. The bending breaking point of GFRP-E was the same as that of standard GFRP prepared without recycle polymer. Meanwhile the bending breaking point of GFRP-B was about 14 % lower than that of GFRP-E due to the insufficient cross-linking of UP with Bn-P having no double bond. EGMA-P was found to behave alternative linker instead of styrene for the preparation of GFRP.

The complete recycling process of waste GFRP was achieved and EGMA-P with the cross-linking reactivity was prepared by the transesterification of EGMA [8]. Finally, the regenerated GFRP prepared from EGMA-P indicates sufficient hardness.

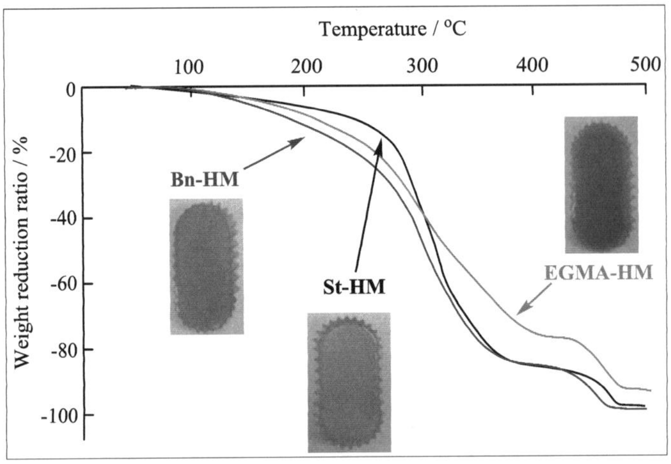

Figure 4. Thermogravimetric analysis curves of HMs.

Conclusion

The decomposition of GFRP with EGMA and BnOH readily was performed by non-catalytic transesterification under high-pressure microwave irradiation. The EGMA-P with many allyl groups had a high reactivity for the cross-linking reaction to UP compared to the product of GFRP with Bn-P. The EGMA-HM had a high thermal stability. The recycled GFRP had a high bending breaking point by using EGMA-P. The complete recycling process of waste GFRP by using EGMA solvolysis could be achieved in future.

Acknowledgements

HORIKO Manufacturing Co., Ltd. (Japan) is cordially acknowledged for the technical support.

References

[1] Stuaib N A and Mativenga D T 2006 *J Clean. Prod.* **120** pp.198-206
[2] Sugeta T, Nagaoka S, Otake K and Sako T 2001 *Kobunshi Ronbunshu* **58** pp. 557-563.
[3] Nakagawa T 2008 *J. Network Polym.* **29** pp. 158-165.
[4] Goto M, Sasaki M and Hirose T 2006 *J. Mater. Sci.* **41** pp.1509-1515.
[5] Kamimura A, Akinari Y, Watanabe T, Yamada K and Tomonaga F 2010 *J. Mater. Cycles Waste Manag.* **12** pp. 93-97.
[6] Fukuzawa H, Shibata K and Izawa H 2002 *Proc. of the 13th conf. of the Jap. soc. of waste man. exp.* **I** pp. 428.
[7] Ikenaga K 2015 *Jidousyagijyutsu* **69** pp. 57-62.
[8] Ikenaga K, Horie S and Futagoishi O JP, 2015-247597, A.

IOP Conf. Series: Materials Science and Engineering **458** (2018) 012035 doi:10.1088/1757-899X/458/1/012035

Effects of wire diameter, yarns size and wire configuration to wire cloth electrode produced from textile technology for dielectrophoresis application

Fadhila Nadzri[1], Zurina Zainal Abidin[1,*], Jamil Mohd Salleh[2]

[1]Department of Chemical and Environmental Engineering, Faculty of Engineering, Universiti Putra Malaysia, 43400 Serdang, Selangor
[2]Department of Applied Science, Universiti Teknologi Mara, 40450 Shah Alam, Selangor

[*]zurina@upm.edu.my

Abstract. Dielectrophoresis (DEP) is one of an alternative way for cell separation. It has mainly been limited to processing small volumes due to constrain in fabrication of microelectrode over large surface areas. This work focuses on optimizing the wire cloth electrode fabricated from the use of textile technology. The plain-weave wire cloth consists of stainless steel wires as the microelectrode arrays and polyester yarn as the insulator. The study involved producing wire cloth electrode with variations of wire diameters, yarn sizes and wire configuration (single or double twisted). Simulation using FEMLAB software was conducted to study the electric field generated from the wire cloth. The performance of the finished wire cloth electrode was then tested by conducting a small-scale separation experiment. It was found that larger diameter wire electrode and smaller yarn size resulted to higher electric field strength. In regards to wire configuration, double twisted wire led to higher electric field intensity and better nonuniformity with 1.25% and 9.29% higher cell separation yield recorded than single wire for 100μm and 71μm, respectively.

1. Introductions

Biochemical and bioprocessing has gained an overwhelming recognition due to need for rapid and robust technologies. More and more industries nowadays were moving towards developing process based on biochemical and biotechnology because of their inherent advantages [1]. However, when dealing with bioprocessing, the usual constraint is the downstream processing or the bioseparation which is the recovery and purification of the desired product from the upstream process whereby the percentage of recovery is low. These kinds of trends and challenges in modern bioseparation have led to various discoveries and novel approaches in the techniques used.

Dielectrophoresis (DEP) is one of an alternative way for bioseparation, particularly cell separation. DEP is a motion of particles caused by the interaction of cell induced dipole in non-uniform AC electric fields [2]. It has certain advantages over other methods mostly because of its high resolution or selectivity. Additionally, by DEP, cell viability can be maintained and the process can be conducted under sterile condition. Previous works regarding DEP involved separating live and dead cells or monocytes [3,4], manipulating cells using travelling electric fields [5,6], and application in the water [7] and wastewater treatment [8]. In nanotechnology, numerous works have been done regarding viruses [9,10]. Alternatively, research on the use of higher throughput have increased with 3-

dimensional insulator [7] or electrode separator [3,11,12], wire cloth electrode separator [13], filter chip separator [14] and integrated microfluidic chip [15].

DEP has been limited to processing small throughput due to restriction in fabrication of microelectrode for large surface areas which mostly used lab-on-a-chip device and printed circuit board. Hence, the preparation of microelectrode using textile technology is implemented to overcome the throughput issue. While previous study [13] proved to be effective for fungi species such as yeast (*Saccharomyces cerevisiae*), this study targeted smaller microorganism using bacteria *Escherichia coli* with improvement of the microelectrode gap which involved wire diameter, yarn size and wire configuration. Electric field generated from the wire cloth system is simulated using FEMLAB and wire cloth performance is investigated following its production.

2. Materials and Methods

The studies were grouped into the production of wire cloth, the simulation of the electric field and the separation experiment.

2.1. Production of wire cloth

A simple plain-weave pattern is chosen for the wire cloth design in which stainless steel wire (Ormiston Wire Ltd, UK) with diameter of 71μm and 100μm was used as weft whilst the warp consisted of the textured multifilament polyester yarns (Hualon Recron Pte Ltd, Nilai, Malaysia) with yarn density of 58 decitex and 83 decitex (respective diameter of 10μm and 16μm). The maximum weft density of the weaving machine was 93 pick per inch (PPI). In general, weaving process was done in stages started with warping, followed by threading and weaving.

2.1.1. Warping. The yarns were taken from yarn cones assembled in parallel lines by rewinding process to give a warp. The warp density of 30 ends per inch for a total of 4 inches was used. The prepared warp was then transferred to a weaver beam of a weaving machine (Swiftec Mfg., Kajang, Malaysia).

2.1.2. Threading. The threading stage involved positioning the warped yarn into series of slot of the weaving machine (shafts, heddles and reed) according to the pattern required. The weaving machine has 4 shafts with mounted heddles on each shaft. For the plain weave pattern, the first yarn of the prepared warp was hooked in the eye of the first heddle on shaft 4. Then the subsequent yarns were threaded through all the first heddles on shaft 3, shaft 2 and finally shaft 1. This sequence of threading made one complete cycle. Once a cycle was finished, the same procedures were repeated onto the next set of heddles on all shafts following its order until all the yarn warps were threaded. After the heddles, the warp ends were inserted into each dent in the reed (30 dents/ inch) to regulate the yarn spaces.

2.1.3. Weaving. Each weaving cycle involves shedding, picking and beating. During shedding process, the warp threads which were under tensioned by the heddles were separated into 2 layers to form a tunnel with half of the heddle shafts (shaft 2 & 4) to be at the top layer and the other half (shaft 1 & 3) to be at the bottom layer. Then, in the picking process, the stainless steel wire, acting as the weft part were individually inserted into the tunnel prior to the beating in which the reed is drove forward to tighten the newly inserted weft. Simultaneously, this action caused the heddle frames on shaft 1 and 3 to be raised up while the heddles frames on shaft 2 and 4 to be lowered, locking the weft with warp and forming a new tunnel alternating the initial configurations. These steps were repeated until the required length of wire cloth was obtained. For this study, variation of yarn size (58 and 83 decitex), wire diameter (71μm and 100μm) and wire configuration (single and double twisted) were used.

2.2. Simulation study on the electric field pattern
A study of the electric field strength and pattern generated by electrodes of the wire cloth was performed using FEMLAB 3.0 (COMSOL™) using the Electromagnetic module [13,16]. The electric field was simulated with parameters that were wire diameter, yarn size and wire configuration. The yarn was drawn to wrap the wire alternately above and below mimicking the plain weave pattern. A rectangle shape was drawn to represent the water that encapsulated the wire cloth area.

2.3. Separation experiment
After the wire cloth production, its effectivity was investigated using the separation experiment in which the data acquired includes the visual of the separation as well as the percentage of the separation achieved. Prior to the experiment, all variation of wire cloth produced were trimmed and tested for any connectivity issue and proceeded for a small chamber construction using microscope glass slides.

2.3.1. Cell preparation. Escherichia coli (Biotechnology Lab, Universiti Putra Malaysia) was cultured overnight in nutrient broth (Oxoid, UK) at 37°C in a 150 rpm incubator shaker (Stuart, Fisher Scientific, UK). The cell was then harvested, centrifuged at 5000 rpm (Heraeus, Thermo Scientific, USA) and washed four times with distilled water. Early concentration of cells suspended in distilled water was measured through 600nm wavelength of a UV spectrophotometer (Thermo Electron Corporation, UK).

2.3.2. Experimental setup. The flow diagram for the experiment is shown schematically in figure 1. The DEP chamber was placed under a camera-equipped microscope (Leica Qwin, Germany) to observe and capture visuals of dielectrophoresis of bacteria cells through the software provided. The aforementioned *E. coli* cells suspension was placed in a syringe and the syringe was positioned at the syringe pump. Once the chamber was filled with distilled water, the cells was pumped into the chamber and immediately 1 MHz frequency and 20 Vpk-pk voltage was applied to the system using the frequency generator (TG120, Thurlby Thandar Instruments, UK). The visuals of the DEP was captured through the software of the microscope camera and afterwhile, the experiment was stopped, the DEP sample was processed and the yield was analyzed using spectrophotometer.

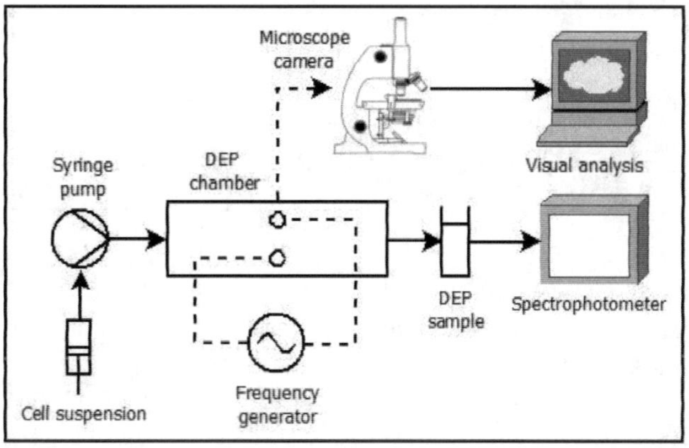

Figure 1. The schematic diagram of the separation experiment as explained in section 2.3.2.

3. Results and Discussion
The discussion covers all the said effects which are yarn size and wire diameter and configuration towards the wire cloth product, the simulation study as well as to the separation experiment.

IOP Conf. Series: Materials Science and Engineering **458** (2018) 012035 doi:10.1088/1757-899X/458/1/012035

3.1. Wire cloth product

The objective for the fabrication of wire cloth was to determine the best combination of settings needed to produce an optimized wire cloth for dielectrophoresis application. The key factor was to obtain the smallest gap between electrodes for a more efficient separation.

3.1.1. Effect of wire diameter. The distance between the electrodes was optimized using 71 μm and 100 μm stainless steel wires. As can be seen in Figure 2, the electrode wires are the black bold lines in the horizontal direction, whilst the multifilament polyester yarns are perpendicular to it. The increment of the wire diameter caused the gap between the wires decreased, hence stronger dielectrophoretic force experienced by the cells [9,17]. This is due to the weft setting fixed to 93 PPI (picks per inch) in which the machine beats exactly 93 wires for every inch produced resulted either smaller or larger gap based on the compression of weft which in this case the electrode wires. Table 1 summarized the average gap calculated from all wire cloth produced.

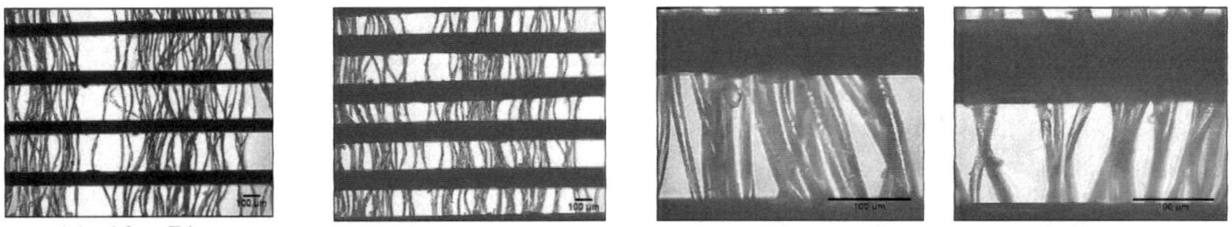

(a) 10x, 71μm (b) 10x, 100μm (a) 40x, 71μm (b) 40x, 100μm

Figure 2. The wire cloth under the light microscope with magnification of 10x and 40x. The electrode diameter used was single 71μm (a) and 100μm (b) with similar yarn size.

Table 1. Summary of gap between electrodes for each wire variables for both yarn sizes

Wire Diameter (μm) & configuration	Average Real Gap (μm)	
	58 dtx	83 dtx
71 single	146	198
71 double	94	104
71 double (twisted)	126	139
100 single	119	150
100 double	39	70
100 double (twisted)	69	122

3.1.2. Effect of wire configuration.

Next, the effect of wire configuration to the wire cloth was studied. The wire diameter used were both 71 μm and 100 μm with three comparable wire configurations; single wire, double wire and double wire twisted. Figure 3 shows the images of the electrodes for double configuration with significant change of wire gap from previous single configuration. The 100 μm double wire obtained smaller electrode gap compared to 71 μm double wire because double configuration makes the diameter value doubled. This leads to the aforementioned compression of the weft that resulted to the small electrode gap value. Similarly, the usage of double twisted wire for the 100μm wires (figure 4b) still gave smaller gap than the 71μm double twisted (figure 4a). Numerically, for a 93PPI of weft setting, 14% gap difference was recorded from both wire diameters (table 1).

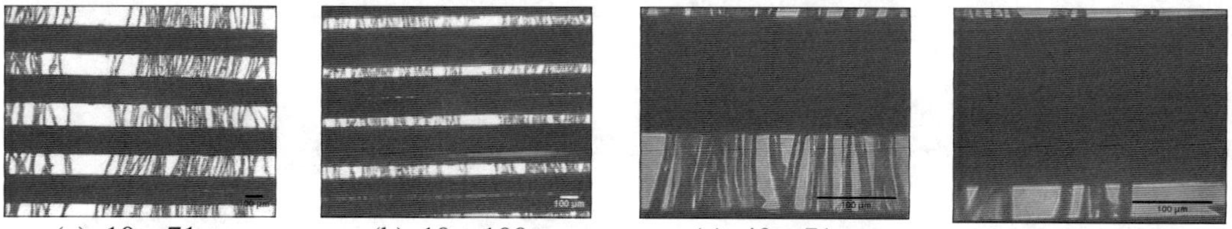

(a) 10x, 71μm (b) 10x, 100μm (a) 40x, 71μm (b) 40x, 100μm

Figure 3. The double non-twisted wire cloth under the light microscope with magnification of 10x and 40x. The electrode diameter used was single 71μm (a) and 100μm (b) with similar yarn size.

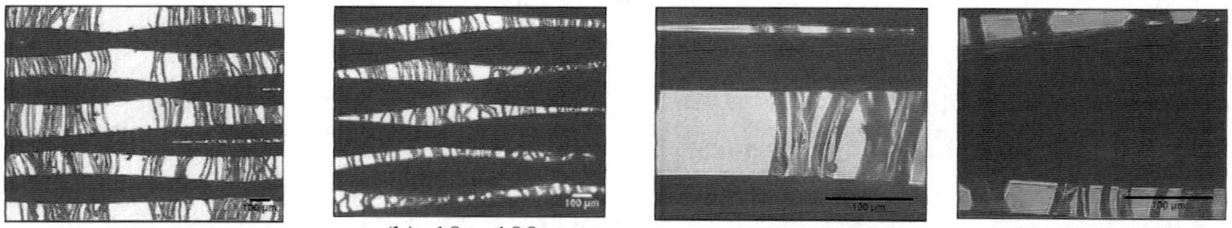

(a) 10x, 71μm (b) 10x, 100μm (a) 40x, 71μm (b) 40x, 100μm

Figure 4. The double twisted wire cloth under the light microscope with magnification of 10x and 40x. The electrode diameter used was single 71μm (a) and 100μm (b) with similar yarn size.

3.1.3. Effect of yarn size.

Polyester yarns comprised of very fine and slightly elastic filaments, very flexible for weaving stainless steel wire and helped reducing the distance between the wires. Two polyester yarn size of 58 dtx and 83 dtx were used in order to compare the electrode gap achieved. Referring to table 1, although different yarn sizes were used, the pattern of its corresponding real gap values was not quite faraway with the smallest electrode gap again produced by 100μm double wire. Other than the gap, the significance of using smaller yarn size was to promote better electric field intensity around the system during DEP process which is discussed next.

3.2. Electric field simulation of the wire cloth electrode

In discussing dielectrophoresis theoretically, the main factors affecting its process were the electric field strength and electric field non-uniformity. Again, the electric field generated with variation of wire diameter, wire configuration and yarn size were modelled with the terminal wires alternately connected to the electric potential (20 Vpk-pk) and ground [13].

3.2.1. Effect of different wire diameter.

As shown in figure 5, the blue circle represents cross-section of the alternate charged and ground wire electrodes while the zigzag shape represents the polyester yarn with the red area indicates high electric field strength. Figure 5(a) simulated the 71μm wire cloth system while the 100 μm wire was represented by figure 5(b). Maximum values of electric field generated were 1.340×10^6 V^2 m^{-3} and 1.341×10^6 V^2 m^{-3} respectively showed that the diameter influence the electric field strength in the wire cloth system which was the smaller the gap, the higher the electric field achieved.

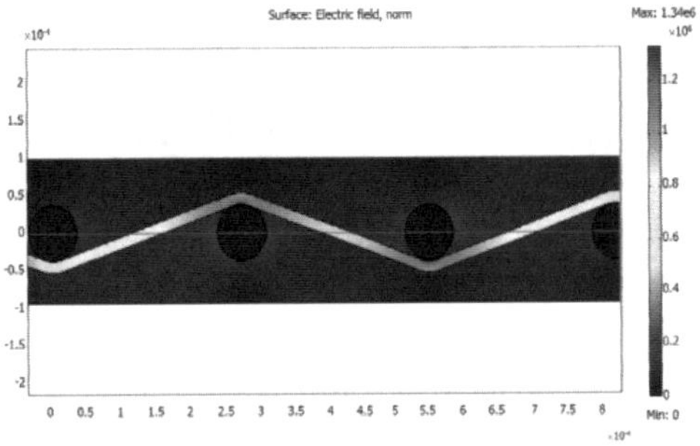

(a)

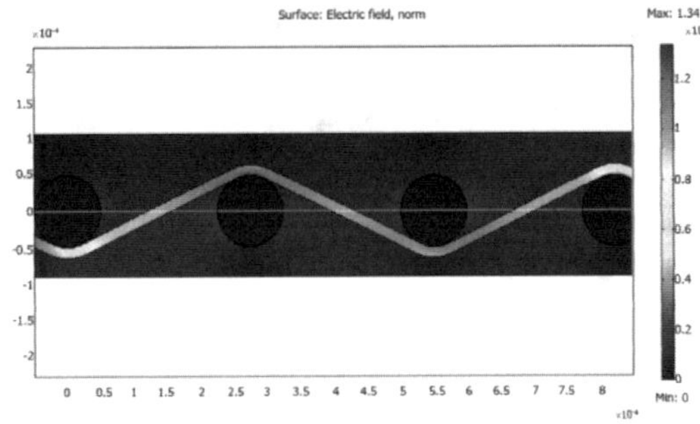

(b)

Figure 5. The simulation of single 71µm(a) and 100µm(b) wire cloth system. Maximum value of electric field obtained were 1.340×10^6 V^2 m^{-3} and 1.341×10^6 V^2 m^{-3} respectively.

3.2.2. Effect of different wire configuration. ∇E_{rms}^2 defines the average local non-uniform electric field strength and gradient, and is dependent on the applied voltage as well as the electrode geometry and size. Hence, some alteration to the wire arrangement might modify the electrode geometry for instance the introduction of the twisted wire configuration. From the simulation done, the highest value of electric field obtained was 2.042×10^6 V^{2}m^{-3} through 100µm double twisted wire. Table 2 summarizes the maximum electric field attained for all the simulated wire cloth.

Table 2. Comparison of the electric field with respect to wire setting and yarn diameter

Wire Setting	Yarn Diameter (µm)	Max Electric Field (V^{2}m^{-3})
71 µm single	16	1.340×10^6
100 µm single	16	1.341×10^6
	10	1.591×10^6
100 µm twisted	10	2.042×10^6

The twisted wire configuration furthermore produced diverse range of non-uniformity due to its different geometry. The contour option was selected during simulation yielded patterns as seen in figure 6. For the single wire, the contour generated repetitive define patterns while the twisted wire induced more coverage with inconsistent patterns. This was beneficial to dielectrophoresis as the cells required non-uniformities to achieve a good degree of separation [18,19].

3.2.3. Effect of warp size (polyester yarn).

Next, the simulation done using a smaller yarn size of 58 dtx resulted to a higher value of electric field produced compared to 83 dtx yarn size with values of 1.341×10^6 V^{2}m^{-3} and 1.591×10^6 V^{2}m^{-3} each (table 2). This was due to the fact that the yarns has its own permittivity value (=3.0) and this property was responsible for distorting the electric potential and electric field of a DEP system. As the yarn also act as an insulator, its size determined the intensity of the electric field distortion experienced by the wire cloth, hence the higher electric field value for smaller yarn. Research [14,17] proved that a higher permittivity of insulator resulted to a lower electric field and vice versa.

3.3. Separation experiment
After theoretical simulation of the wire cloth behavior, they were brought to test for separation experiment for further understanding. A small DEP chamber was prepared using glass slides to

evaluate the performance of the wire cloth for *E. coli* collection. The observation consisted of both microscopic views of the DEP as well as quantitative results derived from the number of cells collected. Figure 7 shows a successful DEP with pearl chain formation (as indicated by the red circles) of bacteria cells seen at the edge of the electrodes; the formerly neutral-charged cell experienced permittivity complex relative to the medium (which was water), hence resulted to the cells movement towards the high field region of electrode that was known as the positive dielectrophoresis (positive DEP) [4].

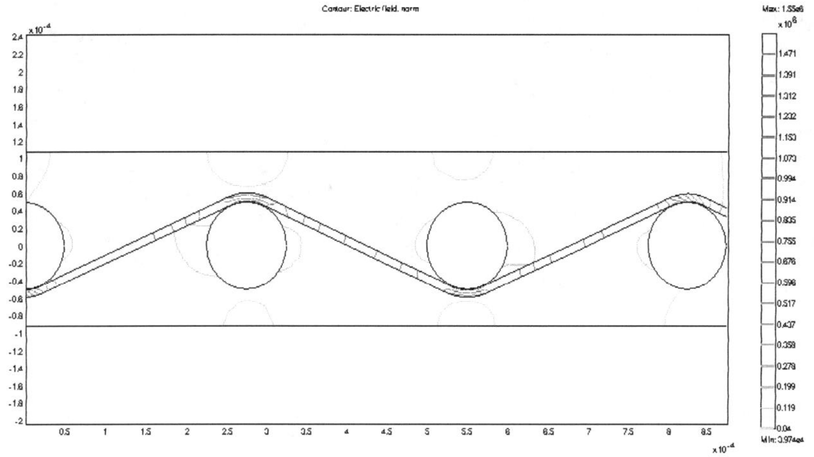

(a)

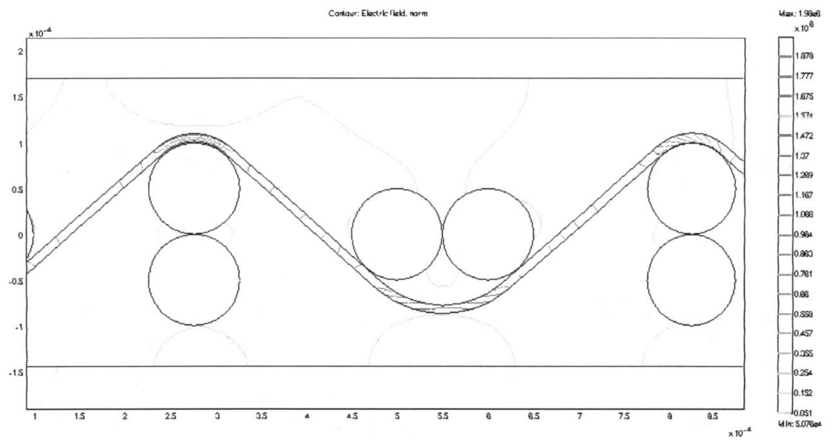

(b)

Figure 6. The contours represents the non-uniformity of electric field simulation of (a) 100μm single wire and (b) 100μm double twisted wire

In terms of the wire cloth efficiency, the yield was calculated [20] by percentage of cells experiencing positive DEP during the separation experiment illustrated as in Figure 8. The percentage cells collected was compared through the three parameters which were the wire diameters, yarn size and wire configurations. Figure 8(a) proved that larger wire diameter led to smaller electrode gap that resulted a stronger DEP force experience by the cells, hence the higher yield. This led to higher accumulation of cells at the electrodes that form pearl chain.

Similarly, stronger DEP force in the smaller yarns system cause a higher cell collection (figure 8(b)). Lastly, the wire configuration reflected a different pattern compared to the simulation study done in the previous section. While double twisted wires offered the most yield (Figure 8(c)), the double wire (non-twisted) gave lowest yield due to the electrode heating that possibly caused the cells to be non-viable and not responsive to the DEP force [21]. The double twisted wire features on the other hand was good in heat dissipation.

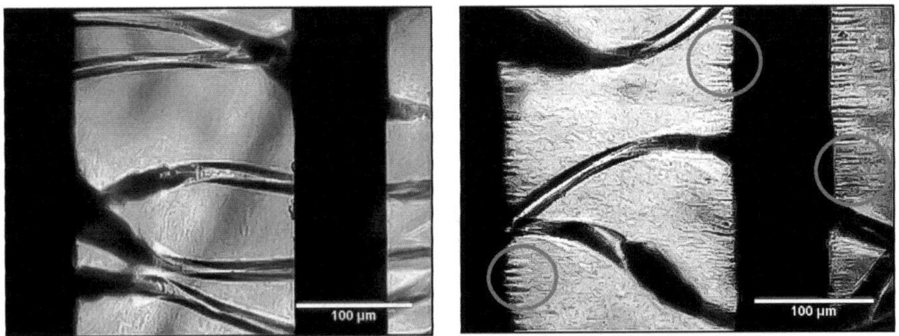

(a) (b)

Figure 7. The 40x magnification of wire cloth for single 71µm electrode during DEP with (a) The image of the system after the introduction of cell suspension prior to DEP process, and (b) The positive DEP of E. *coli* after voltage of 20Vpk-pk was applied.

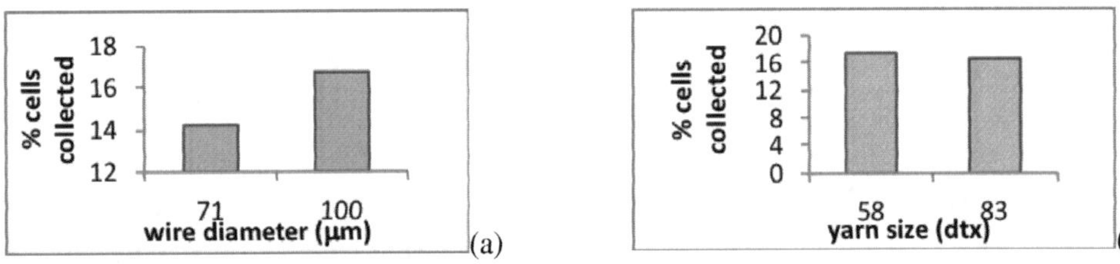

(a) (b)

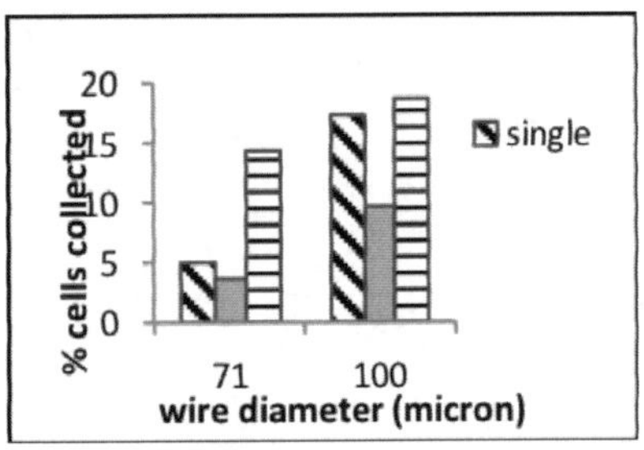

(c)

Figure 8. The percentage cells (yields) collected in comparison with regards to (a) wire diameter, (b) yarn size and (c) wire configuration.

4. Conclusions

There were variable stages of determining a better DEP process in a higher throughput system in which this study covered from the wire cloth electrode production, the wire cloth simulation study towards the DEP experiment. The electrode diameter and configuration as well as yarn size maintained the important relationship for a more high-throughput friendly microelectrode in which smaller physical gap was desired. In the simulation study of the electric field intensity, the maximum electric field achieved was 2.042×10^{6} V^2m^{-3} through double 100µm twisted wire. We have established that smaller electrode gap from greater wire diameter led to a higher electric field strength and double twisted configuration yields better electrode non-uniformity. Polyester yarn on the other hand served as a double-edge sword of holding the steel wire in the desired manner in wire cloth production as well as to distort the electric field, promoting more non-uniformities in which smaller yarn size was better as it lessens the insulative element of the system generating greater electric field intensity. From the experimental results it can be concluded that a larger wire diameter and smaller yarn size produced smaller electrode gap that led to higher electric field strength which caused larger DEP force for better cell separation. The twisted configuration on the other hand showed most efficient separation of all three wire configurations due to its non-uniformity of the electric field yields that triggered a better DEP force. The double wire configuration did better in reducing electrode gap but resulted to electrode heating that caused possible cell death. Thus, the use of twisted wire geometry by further enhancing the non-uniformity coverage for the cell separation without compromising the heating effects.

Acknowledgements

The project and student has been funded by MOSTI through Escience Fund (01-03-04-SF0841).

5. References

[1] Doble M, Kruthiventi A and Gaikar V 2018 *Biotransformations and Bioprocesses* M. Doble, A.K. Kruthiventi, V.G. Gaikar. 1-10

[2] Pohl H A and Crane J S 1971 Dielectrophoresis of Cells *Biophys. J.* **11** 711–27

[3] Yildizhan Y, Erdem N, Islam M, Martinez-Duarte R and Elitas M 2017 Dielectrophoretic Separation of Live and Dead Monocytes Using 3D Carbon-Electrodes *Sensors (Basel)* **17** 2691

[4] Markx G H, Talary M S and Pethig R 1994 Separation of viable and non-viable yeast using dielectrophoresis *J Biotechnol* **32** 29–37

[5] Pethig M S T, J P H B, J A T and R 1996 Electromanipulation and separation of cells using travelling electric fields *J. Phys. D. Appl. Phys.* **29** 2198

[6] Cheng I F, Froude V E, Zhu Y, Chang H C and Chang H C 2009 A continuous high-throughput bioparticle sorter based on 3D traveling-wave dielectrophoresis *Lab. Chip.* **22** 3165-3312

[7] Nakidde D, Zellner P, Alemi M M, Shake T, Hosseini Y, Riquelme M V, Pruden A and Agah M 2015 Three dimensional passivated-electrode insulator-based dielectrophoresis *Biomicrofluidics* **9** 14125

[8] Abidin Z, Fakhru'l-Razi A, York Chow W and Hafifudin N 2013 Construction of ASMC by Dielectrophoresis Using Wirecloth Electrode for the Treatment of Wastewater *Dev. Sustain. Chem. Bioprocess. Technol.* 65-71

[9] P. H M 2002 Strategies for dielectrophoretic separation in laboratory-on-a-chip systems *Electrophoresis* **23** 2569–82

[10] Hughes M P, Morgan H and Rixon F J 2001 Dielectrophoretic manipulation and characterization of herpes simplex virus-1 capsids *Eur. Biophys. J.* **30** 268–72

[11] Fatoyinbo H O, Kamchis D, Whattingham R, Ogin S L and Hughes M P 2005 A high-throughput 3-D composite dielectrophoretic separator *IEEE Trans. Biomed. Eng.* **52** 1347–9

[12] Faraghat S A, Hoettges K F, Steinbach M K, van der Veen D R, Brackenbury W J, Henslee E A, Labeed F H and Hughes M P 2017 High-throughput, low-loss, low-cost, and label-free cell separation using electrophysiology-activated cell enrichment *Proc. Natl. Acad. Sci.* **114** 4591 LP-4596

[13] Abidin Z Z, Downes L and Markx G H 2007 Novel electrode structures for large scale dielectrophoretic separations based on textile technology *J. Biotechnol.* **130** 183–7

[14] Ciprian I, Guolin X, Celeste L F, Lam O P and H. T F E 2007 A 3-D dielectrophoretic filter chip *Electrophoresis* **28** 1107–14

[15] Li S, Li M, Bougot-Robin K, Cao W, Yeung Yeung Chau I, Li W and Wen W 2013 High-throughput particle manipulation by hydrodynamic, electrokinetic, and dielectrophoretic effects in an integrated microfluidic chip *Biomicrofluidics* **7** 24106

[16] Z. A Z, Les D and H. M G 2006 Large scale dielectrophoretic construction of biofilms using textile technology *Biotechnol. Bioeng.* **96** 1222–5

[17] Abidin Z Z, Abdullah A G L, Yunus Z, Markx G H and Science A 2007 *Wire Cloth Electrodes : A Study of Electric Field for Dielectrophoretic Separation of Cells.* **4** 205-212

[18] Wakizaka Y, Hakoda M and Shiragami N 2004 *Effect of electrode geometry on dielectrophoretic separation of cells.* **20** 13-19

[19] Green N A M Y and N G 2008 Continuous separation of submicron particles using Angled electrodes *J. Phys. Conf. Ser.* **142** 12068

[20] Pethig R R 2017 *Dielectrophoresis: Theory, Methodology and Biological Applications* Wiley. 40

[21] Donato S S, Chu V, Prazeres D M F and Conde J P 2013 Metabolic viability of Escherichia coli trapped by dielectrophoresis in microfluidics *Electrophoresis,* **34**(4) 575-582

IOP Conf. Series: Materials Science and Engineering **458** (2018) 012036 doi:10.1088/1757-899X/458/1/012036

Preparation of activated carbon from oil palm empty fruit bunch by physical activation for treatment of landfill leachate

Y M N S Ismail, N Ngadi*, M H Hassim, M J Kamaruddin, A Johari and M A A Aziz[1]

Faculty of Chemical and Energy Engineering, Universiti Teknologi Malaysia, 81310 Skudai, Johor, Malaysia

*norzita@cheme.utm.my

Abstract. The present work demonstrates the preparation of oil palm empty fruit bunch derived activated carbon (EFBAC) by air in a closed furnace for the adsorptive removal of COD, ammoniacal nitrogen and colour from landfill leachate. The removal of COD, ammoniacal nitrogen and colour were investigated with EFBAC activated at different activation temperatures and activation time. Yield of EFBAC at different activation temperatures and time were also examined. The experimental results manifested that increasing activation temperature increased COD removal but decreased colour removal. It has no significant removal on ammoniacal nitrogen. Increasing activation time increased COD and colour removal but has no significant removal on ammoniacal nitrogen. Furthermore, yield of EFBAC decreased with increasing activation temperature and time. The finding illustrated the applicability of EFBAC as a promising solution for pollution from landfill leachate due to its economic feasibility for real application.

1. Introduction

Municipal solid wastes have become a major issue globally. Increasing amount of municipal solid wastes occur from many factors such as growing of population, productivity, consumption habits, lifestyles, and resource use and also development of technologies [1]. Along with municipal solid wastes, landfill leachate is the real problem to the groundwater and also water course such as rivers and lakes.

Landfill leachate is known as a contaminated liquid effluent generated from rain water percolating through deposited municipal solid wastes. Typically, stabilized landfill leachate constitutes a high strength of ammoniacal nitrogen (>400 mg/L), moderate strength of chemical oxygen demand (COD) (5000 – 20000 mg/L) and also low BOD/COD ratio (< 0.1) [2]. Landfill leachate imposes negative impacts to humans as well as the environment as it may seep into groundwater and flow to the water course. Ammoniacal nitrogen is the main toxicant to the living organism. High concentration of untreated ammoniacal nitrogen can promote algae growth and decrease dissolved oxygen level through algae bloom (eutrophication). It is also toxic to aquatic lives [3].

There are various types of treatment that can be adopted to treat landfill leachate such as physical, chemical, or biological, such as adsorption, membrane filtration, chemical precipitation, air stripping, and reverse osmosis. Biological treatments are preferred if the landfill leachate is young due to their reliability, and cost-effectiveness [4]. However, most landfill leachate is considered stabilized leachate, therefore, biological treatment is not favourable. Adsorption is the most common treatment for stabilized landfill leachate. The process is simple but the drawbacks include (i) the need of frequent regeneration, and (ii) high cost of production of adsorbent [2]. Hence, researchers all around the globe seek for renewable sources of activated carbon such as banana frond [5–6], coffee ground [7], durian peel [8], palm shell [9], tamarind fruit seed [10], sugarcane bagasse [11–13], orange peel [14], and rice husk [15]. However, no study has yet been found using palm oil empty fruit bunch (EFB) as precursor for the treatment of landfill leachate.

Malaysia produced 19.92 million tonnes of crude palm oil in 2017 and it is estimated that approximately 19.92 million tonnes of EFB were generated as their wastes [16]. The abundance of EFB attract researcher to explore the usage of EFB as a precursor of activated carbon [17–20]. Previously, EFB derived activated carbon focussed on removal of phenol and its derivatives, dyes and metals such as boron and iron.

The objective of this study is to examine the effect of preparation condition of EFB-based activated carbon by physical activation in the limited air environment for the treatment of the post-treatment landfill leachate. The effect of activation temperature and time were evaluated in terms of removal of COD, ammoniacal nitrogen, and colour.

2. Material and Methods

2.1. Landfill leachate

The leachate was collected from a municipal solid waste sanitary landfill, Tanjung Langsat Landfill Site, located at Pasir Gudang, Johor, Malaysia. The coordinates are 1°28'22"N 103°59'36"E. The samples were taken after the leachate undergoes anaerobic and aerobic treatment, before discharged into the nearby water course. The leachate samples were then stored in refrigerator at temperature of 4°C for preservation purpose. The characteristics of the leachate are presented in table 1. COD of leachate was measured according to ASTM D1252 (Test Method B—Micro COD by Sealed Digestion and Spectrometry). For ammoniacal nitrogen and colour, it was measured by using Hanna Multiparameter Photometer (HI 83099).

Table 1. Characteristic of leachate samples.

Parameter	Unit	Average	Min	Max
COD	mg/L	974	504	1279
BOD	mg/L	77.1	63	102
Suspended Solid	mg/L	213	180	280
Ammoniacal Nitrogen	mg/L	129	117	160
Colour	PCU	4042	2960	7450

2.2. Oil palm EFB

EFB was collected from Alaf Palm Oil Mill located in Kulai, Johor, Malaysia.

2.3. Preparation of EFB based activated carbon

EFB was washed and dried at 105°C for 24 hours. Then, it was ground into particles sized 1.0 – 2.0 mm. The precursor then underwent carbonization in a closed furnace at 450°C for 2 hours before activated at various temperatures (600 – 900°C) and various activation times (15 minutes – 4 hours) as in table 2. The activated carbon (EFBAC) produced was cooled down to room temperature overnight. The yield of the products were calculated. EFBAC underwent batch adsorption study by putting in 0.2 g of adsorbent in 30 mL of leachate and was shaken in isothermal shaker for 2 hours at 200 rpm. The removal efficiency of all parameters was calculated as follows,

$$Percentage\ removal = \frac{C_o - C_e}{C_o} \ x\ 100\% \quad (1)$$

Where,
C_o = Initial concentration of parameters, mg/L
C_e = Final concentration of parameters, mg/L

Yield percentage of samples were measured by using equation as follows,

$$Yield = \frac{m_f}{m_i} \ x\ 100\% \quad (2)$$

Where,
m_f = mass of activated carbon (after activation), g
m_i = mass of precursor (before activation), g

Table 2. Preparation of EFBAC

Sample	Activation Temperature (°C)	Activation Time (min)
1	600	15
2	700	15
3	800	15
4	900	15
5	800	30
6	800	45
7	800	60
8	800	90
9	800	120
10	800	180
11	800	240

2.4. Characterization of EFBAC
EFB and EFBAC (Sample 11) were characterized by their textural properties by nitrogen adsorption using Micromeritics 3 Flex Surface Characterization analyzer to obtain surface area, pore volume, and pore size.

3. Results and Discussion

3.1. Characterization of EFBAC
Surface area, pore volume, and pore size are important characteristics of activated carbon as it can explain the removal performance of adsorbate. Nitrogen adsorption is a standard procedure for determination of porosity of the carbonaceous adsorbents. Detailed characteristics of the porosity of EFB and EFBAC are listed in table 3. From the data, it is evident that BET surface area, Langmuir area, and total pore volume of EFB were 1.1163 m^2/g, 1.7836 m^2/g, and 0.0006 cm^3/g, for EFBAC 694.86 m^2/g, 803.62 m^2/g, and 0.3170 cm^3/g respectively. It is noted that micropore area and micropore volume for EFBAC were 559.31 m^2/g, and 0.2208 cm^3/g, respectively. However, there was no significant difference on average pore size between EFB and EFBAC.

Increased surface area and pore volume were due to increasing amount of micropores. The EFBAC developed in this study was basically microporous since 69.6% of total pore volume was micropore volume. It was also noted that 80.5% of total surface area of EFBAC was micropore area. This phenomenon happens because at high temperature, low molecular weight volatiles are released from the matrix structure, and generated a fragile structure, resulting in increasing pore size [20]. Insignificant difference on pore size between EFB and EFBAC may be due to the volatilization of low molecular weight volatiles which used the same route or pores that made the pores not widened but deepened in EFBAC. It was proven by data from this study that the pore size is almost similar but the pore volume increases as the EFB is activated into EFBAC.

Table 3. Textural properties of EFB and EFBAC.

Parameter	Unit	EFB	EFBAC
BET surface area	m^2/g	1.1163	694.86
Langmuir surface area	m^2/g	1.7836	803.62
Micropore Area	m^2/g	N.A[a]	559.31
External surface area	m^2/g	1.7260	135.54
Total pore volume	cm^3/g	0.0006	0.3170
Micropore volume	cm^3/g	N.A[a]	0.2208
Average pore size	nm	1.896	1.837

[a]The micropore area and volume are below the detection limit of the instrument.

3.2. Yield of activated carbon from different activation condition

Yield, in activated carbon preparation, is the percentage final weight of activated carbon produced after activation, from its initial weight of raw material on a dry basis. Apart from removal performance, yield of the activated carbon is a relatively important information for the industry since it can predict the amount of activated carbon generated after activation. Figure 1 represents the yield percentage of EFBAC at different activation temperatures and also with reference to carbonized EFB. Increasing activation temperature reduces the yield of the EFBAC. This happens as increasing temperature will increase volatile releases, resulting in decreasing yield [21].

The effect of activation time on EFBAC yield is displayed in figure 2. It is apparent that the yield percentage decreases with increasing activation time. This phenomenon is due to more volatiles which might be released with longer activation time [21]. It is also suggested that the weight loss is due to the development of pore structure at longer activation time [22].

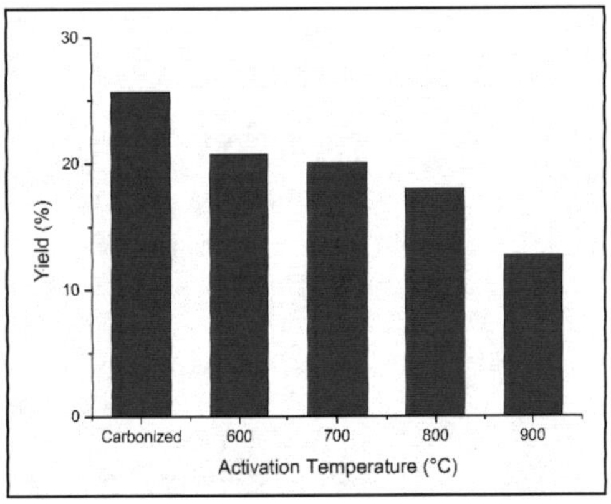

Figure 1. Yield percentage vs activation temperature.

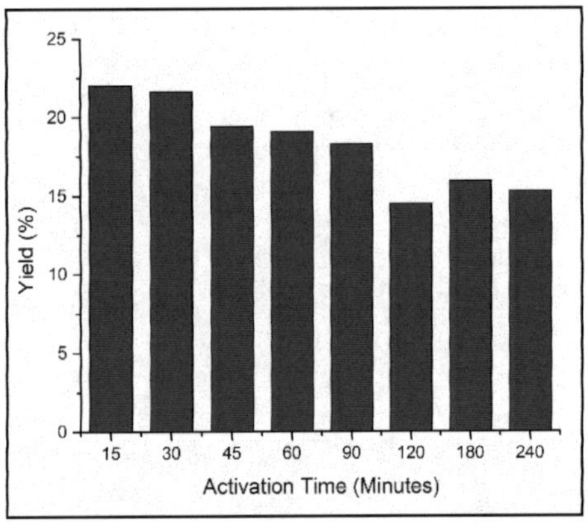

Figure 2. Yield percentage vs activation time.

Table 4 lists a comparison of yield of activated carbon derived from different precursors under different activation conditions. In can be concluded that the activated carbon prepared in this work showed significant yield compared to other activated carbon by physical activation with 23.08% yield. This finding can reduce the cost of producing activated carbon.

Table 4. Comparative evaluation of yield of various activated carbon by physical activation.

Precursor	Activator	Yield (%)	Reference
EFB	Air	23.1	Present study
Date stones	Steam	28.4	[23]
Almond shell	CO_2	25.0	[24]
Sugarcane bagasse	Steam	23.0	[25]

3.3. Removal performances of activated carbon from different activation condition

The removal of COD, ammoniacal nitrogen, and colour by EFBAC synthesized at different temperatures with reference to EFB and also carbonized EFB were depicted as figure 3. The removal of COD ranged from 23% to 81%. The removal of ammoniacal nitrogen ranged from 16% to 32% and removal of colour ranged from 55% to 74%. It was observed that higher removal of COD by using EFBAC was activated at higher activation temperature, and the highest COD removal was by using EFBAC activated at 900°C (81%). For ammoniacal nitrogen, there was no significant effect of activation temperature towards removal of ammoniacal nitrogen. For colour, the removal was decreasing with increasing activation temperature.

Using EFBAC activated at high temperature gives high COD removal due to its highly developed surface area. According to Aishah [20], high activation temperature produces high surface area of activated carbon. High surface area of activated carbon gives more active sites to adsorb organics that contributes to COD. Apart from surface area, pore size also contributes to adsorption of COD from landfill leachate. It was reported in previous study that activated carbon derived from EFB has average pore diameter ranging from 1.2 to 16.4 nm [21, 26–28]. This pore size is suitable for adsorption of humic substances such as humic and fulvic acid since they are the main components that make landfill leachate highly non-biodegradable [29]. It was also reported that aquatic humic substances have particle sizes that vary from 0.9 to 6.6 nm in diameter [30]. This finding is supported by textural properties of EFBAC depicted in table 3.

In contrast, EFBAC activated at high temperature gives lower colour removal from landfill leachate which might be due to decreasing total pore volume. Closure of pores can occur at high temperature due to shrinkage of activated carbon [20]. Currently, there are no study yet to investigate correlation of activation temperature and pore volume on physically activated EFB. However, Tan *et al.* (2012) found that increasing carbonization temperature can increase pore volume but after 200°C, pore volume decreases [31]. Although high surface area of activated carbon is achieved by using high activation temperature, the high temperature also ruptures pores that are already developed making the remaining pores to have less volume but higher in number. On the other hand, suspended solids in leachate may contribute to colour reading. The most plausible explanation is that suspended solid particle size in leachate is larger than pores developed in EFBAC which makes the removal to decrease with increasing activation temperature. This finding is in contrast with Aziz *et al.* (2007) where they found that colour in landfill leachate was mainly contributed by organic matters [32].

However, ammoniacal nitrogen removal was not significantly affected by activation temperature and time. This might be due to EFBAC has no functional group since it is physically activated. This is in agreement with Aziz *et al.* (2004) who concluded that the adsorption capacity of activated carbon is determined not only by its total surface area, but also by its internal porous structure and the presence of functional groups on the pore surface [33].

The optimum activation temperature of EFBAC was found at 800°C, where at activation temperature higher than 800°C has no significant increment of COD and ammoniacal nitrogen removal, and also decreased colour removal.

Removal of COD, ammoniacal nitrogen, and colour by using EFBAC activated at different activation times are shown in figure 4. It is observed that increasing activation time can increase the removal of COD and colour in leachate with maximum removal of 72.3% and 68.2%, respectively. However, there was no significant removal in ammoniacal nitrogen using EFBAC activated at different activation times (16% to 20% removal).

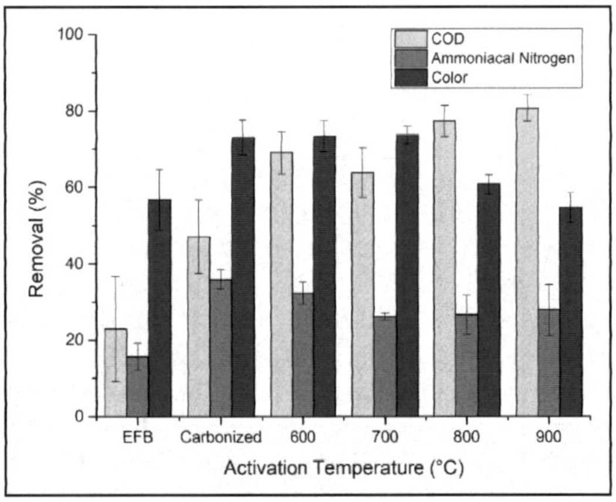

Figure 3. Removal percentage of COD, ammoniacal nitrogen and colour vs activation temperature.

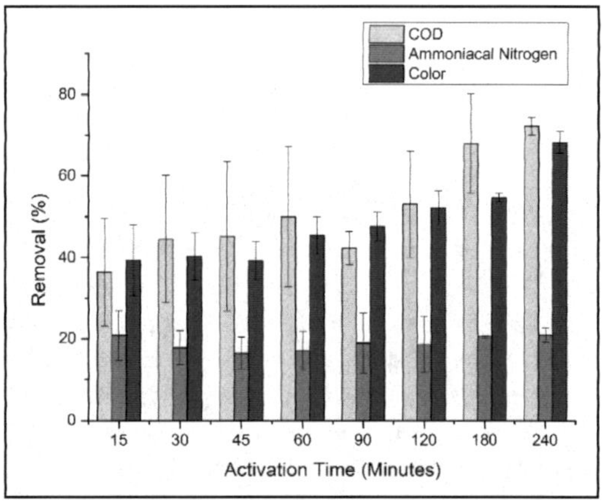

Figure 4. Removal percentage of COD, ammoniacal nitrogen and colour vs activation time.

Increasing removal of COD and colour by increasing activation time might be due to increase in surface area. This may be due to the activator (air) which was penetrated deeper inside particle making the development of porosity higher at long activation time compared to the porosity of particles at short activation time [20]. High surface area gives more sites for adsorption to occur. However, insignificant amount of ammoniacal nitrogen removal with respect of activation time may be due to desorption of the small size of ammonia particle in leachate relative to pore size in EFBAC.

4. Conclusion

The potential of EFB based activated carbon by air was examined. The adsorption of COD, ammoniacal nitrogen, and colour by using EFBAC has been demonstrated. The experimental data show that high activation temperature and high activation time produce activated carbon that can adsorb significant amount of COD, ammoniacal nitrogen, and colour from landfill leachate. It also shows that significant yield percentage obtained by using physical activation hints its economic feasibility for real application.

Acknowledgement

The authors would like to extend their sincerest gratitude to Alaf Palm Oil Plantation, Kota Tinggi, for supplying oil palm EFB sample to be utilized for this study. The authors also acknowledge the financial support by Universiti Teknologi Malaysia under university grant No. Q.J130000.2546.17H65 and Q.J130000.2646.14J29.

References

[1] Foo K Y and Hameed B H 2009 An overview of landfill leachate treatment via activated carbon adsorption process *J. Hazard. Mater.* **171** 54–60
[2] Kurniawan T A, Lo W H and Chan G Y 2006 Physico-chemical treatments for removal of

recalcitrant contaminants from landfill leachate *J. Hazard. Mater.* **129** 80–100
[3] Zhu G, Peng Y, Li B, Guo J, Yang Q and Wang S 2008 Biological removal of nitrogen from wastewater *Rev. Environ. Contam. Toxicol.* **192** 159–95
[4] Halim A A, Sidi S F A and Hanafiah M M 2017 Ammonia removal using organic acid modified activated carbon from landfill leachate *Environ. Ecosyst. Sci.* **1** 28–30
[5] Foo K Y, Lee L K and Hameed B H 2013 Preparation of banana frond activated carbon by microwave induced activation for the removal of boron and total iron from landfill leachate *Chem. Eng. J.* **223** 604–10
[6] Ghani Z A, Yusoff M S, Zaman N Q, Zamri M F M A and Andas J 2017 Optimization of preparation conditions for activated carbon from banana pseudo-stem using response surface methodology on removal of color and COD from landfill leachate *Waste Manag.* **62** 177–87
[7] Ching S L, Yusoff M S, Aziz H A and Umar M 2011 Influence of impregnation ratio on coffee ground activated carbon as landfill leachate adsorbent for removal of total iron and orthophosphate *Desalination* **279** 225–34
[8] Kamaruddin M A, Yusoff M S and Ahmad M A 2011 Optimization of durian peel based activated carbon preparation conditions for ammoniacal nitrogen removal from semi-aerobic landfill leachate *J. Sci. Ind. Res.* **70** 554–60
[9] Lim Y N, Shaaban M G and Yin C Y 2009 Treatment of landfill leachate using palm shell-activated carbon column: Axial dispersion modeling and treatment profile *Chem. Eng. J.* **146** 86–9
[10] Foo K Y, Lee L K and Hameed B H 2013 Preparation of tamarind fruit seed activated carbon by microwave heating for the adsorptive treatment of landfill leachate: A laboratory column evaluation *Bioresour. Technol.* **133** 599–605
[11] Foo K Y, Lee L K and Hameed B H 2013 Preparation of activated carbon from sugarcane bagasse by microwave assisted activation for the remediation of semi-aerobic landfill leachate *Bioresour. Technol.* **134** 166–72
[12] Azmi N B, Bashir M J K, Sethupathi S, Wei L J and Aun N C 2015 Stabilized landfill leachate treatment by sugarcane bagasse derived activated carbon for removal of color, COD and NH3-N - Optimization of preparation conditions by RSM *J. Environ. Chem. Eng.* **3** 1287–94
[13] Azmi N B, Bashir M J K, Sethupathi S and Ng C A 2014 Anaerobic stabilized landfill leachate treatment using chemically activated sugarcane bagasse activated carbon: kinetic and equilibrium study *Desalin. Water Treat.* **57** 3916–27
[14] Xie Z, Guan W, Ji F, Song Z and Zhao Y 2014 Production of biologically activated carbon from orange peel and landfill leachate subsequent treatment technology *J. Chem.*
[15] Halim A A, Abidin N N Z, Awang N, Ithnin A, Othman M S and Wahab M I 2011 Ammonia and COD removal from synthetic leachate using rice husk composite adsorbent *J. Urban Environ. Eng.* **5** 24–31
[16] MPOB 2018 *Overview of the Malaysian Oil Palm Industry*
[17] Md Arshad S H, Ngadi N, Aziz A A, Amin N S, Jusoh M and Wong S 2016 Preparation of activated carbon from empty fruit bunch for hydrogen storage *J. Energy Storage* **8** 257–61
[18] Osman N, Shamsuddin N and Uemura Y 2016 Activated Carbon of Oil Palm Empty Fruit Bunch (EFB); Core and Shaggy *Procedia Eng.* **148** 758–64
[19] Wirasnita R, Hadibarata T, Yusoff A R M and Mat Lazim Z 2015 Preparation and characterization of activated carbon from oil palm empty fruit bunch wastes using zinc chloride *J. Teknol.* **74** 77–81
[20] Aishah S, Kadir S A, Matali S, Mohamad N F, Hidayu N and Rani A 2014 Preparation of activated carbon from oil palm empty fruit bunch (EFB) by steam activation using response surface methodology *Int. J. Mater. Sci. Appl.* **3(5)** 159-163
[21] Alam M Z, Muyibi S A, Kamaldin N and Md. Alam A M and N K 2008 Production of activated carbon from oil palm empty fruit bunches for removal of zinc. *Dep. Biotechnol. Eng. Fac. Eng. Int. Islam. Univ. Malaysia* 1–11

[22] Liu Q-S, Zheng T, Wang P and Guo L 2010 Preparation and characterization of activated carbon from bamboo by microwave-induced phosphoric acid activation *Ind. Crops Prod.* **31** 233–8
[23] Bouchelta C, Medjram M S, Bertrand O and Bellat J P 2008 Preparation and characterization of activated carbon from date stones by physical activation with steam *J. Anal. Appl. Pyrolysis* **82** 70–7
[24] Toles C A, Marshall W E, Wartelle L H and McAloon A 2000 Steam-or carbon dioxide-activated carbons from almond shells: physical, chemical and adsorptive properties and estimated cost of production *Bioresour. Technol.* **75** 197–203
[25] Qureshi K, Bhatti I, Kazi R and Ansari A K 2008 Physical and chemical analysis of activated carbon prepared from sugarcane bagasse and use for sugar decolorisation *Int. J. Chem. Biomol. Eng.* **1** 145–9
[26] Hameed B H, Tan I A W and Ahmad A L 2009 Preparation of oil palm empty fruit bunch-based activated carbon for removal of 2,4,6-trichlorophenol: Optimization using response surface methodology *J. Hazard. Mater.* **164** 1316–24
[27] Lee T, Zubir Z A, Jamil F M, Matsumoto A and Yeoh F Y 2014 Combustion and pyrolysis of activated carbon fibre from oil palm empty fruit bunch fibre assisted through chemical activation with acid treatment *J. Anal. Appl. Pyrolysis* **110** 408–18
[28] Hidayu A R, Mohamad N F, Matali S and Sharifah A S A K A K 2013 Characterization of activated carbon prepared from oil palm empty fruit bunch using BET and FT-IR techniques *Procedia Eng.* **68** 379–84
[29] Kjeldsen P, Barlaz M A, Rooker A P, Baun A, Ledin A and Christensen T H 2002 Present and long-term composition of MSW landfill leachate: A review *Crit. Rev. Environ. Sci. Technol.* **32** 297–336
[30] Thurman E M, Wershaw R L, Malcolm R L and Pinckney D J 1982 Molecular size of aquatic humic substances *Org. Geochem.* **4** 27–35
[31] Tan W C, Othman R, Matsumoto A and Yeoh F Y 2012 The effect of carbonisation temperatures on nanoporous characteristics of activated carbon fibre (ACF) derived from oil palm empty fruit bunch (EFB) fibre *Journal of Thermal Analysis and Calorimetry* **108** 1025–31
[32] Aziz H A, Alias S, Adlan M N, Faridah, Asaari A H H and Zahari M S 2007 Colour removal from landfill leachate by coagulation and flocculation processes *Bioresour. Technol.* **98** 218–20
[33] Aziz H A, Adlan M N, Zahari M S M and Alias S 2004 Removal of ammoniacal nitrogen (N-NH_3) from municipal solid waste leachate by using activated carbon and limestone *Waste Manag. Res.* **22** 371–5

Depolymerization of polycarbonate by methanol under pressurized microwave irradiation

K Ikenaga*, K Higuchi, S Kohri and K Kusakabe

Department of Nanoscience, Sojo University, 4-22-1 Ikeda, Nishi-ku, Kumamoto 860-0082, Japan.

*ikenaga@nano.sojo-u.ac.jp

Abstract. Depolymerization of polycarbonate (PC) waste by methanolysis without catalyst under pressurized microwave irradiation was proposed in order to get bisphenol A (BPA). The depolymerized products were identified by NMR and FTIR spectroscopy. A maximum BPA yield of 94% was achieved at a reaction temperature and pressure of 190°C and 3 MPa for 3 hours. On the other hand, dimethyl carbonate could not be detected by FTIR analysis. In order to clarify the reaction mechanism, benzyl alcohol was used instead of methanol under the similar condition. Dibenzyl ether was found to be directly produced by the simultaneous reactions of transesterification and decarbonation during the microwave irradiation. The proton ratio of dibenzyl carbonate:dibenzyl ether = 11.2:88.8 was determined from the ^{1}H-NMR chemical shifts of the reaction mixture. Therefore, dimethyl carbonate in the case of the methanolysis was converted to dimethyl ether, which vaporized at atmospheric pressure after the reaction under pressurized microwave irradiation. The proposed methanolysis method of waste PC can achieved the high efficient recovery of BPA without catalyst use and avoid the need for a further separation process of dimethyl carbonate.

1. Introduction

Polycarbonate (PC) is one of the important engineering plastics having excellent properties of impact resistance, transparency, heat resistance and workability. Commercial PCs have been produced mainly through a polycondensation of 2,2-bis(4-hydroxyphenyl) propane (bisphenol-A, BPA) with carbonyl chloride, dimethyl carbonate (DMC) or diphenyl carbonate. A major application of PC is the data storage products such as CD, DVD and Blue-ray Disc. Extended production of PC results in huge amount of PC waste, which should be recycled for the fossil fuel resources conservation and carbon dioxide emission control. However, the main treatments of PC waste are the landfill and the incineration disposal. Chemical depolymerization methods of waste PC to recover starting materials, such as BPA and carbonate, have attracted attention over the years. Alkali-catalyzed methanolysis of PC in a mixed solvent of methanol and toluene was performed and gave free BPA (96 %) in a solid form and DMC (100 %) in solution [1]. Chemical recycling of PC by the thermal decomposition with steam at 500°C was investigated. The addition of $Ca(OH)_2$ catalyst in the chemical recycling of PC with steam at 500°C led to the phenol yield of 70 mol% [2]. The solubility of PC was improved by using ionic liquid for the methanolysis depolymerization of PC [3]. To maintain the yield of product is the main advantage in this recycle use of ionic liquid. In the method for the depolymerization of PC in supercritical ethanol, PC was completely degraded into diethyl carbonate and BPA [4]. The microwave irradiation for the depolymerization of PC could accelerated the reaction rate [5, 6]. In our previous research, efficient recycling processes were achieved by using pressurized microwave irradiation for the hydrolytic depolymerization of poly(ethylene terephthalate) [7] and the alcoholysis of polyester resin in glass fiber reinforced plastics [8] due to the enhancement of compatibility between waste plastic and solvent in addition to the rapid heating.

In this study, non-catalyzed depolymerization of PC in methanol was carried out under the condition of pressurized microwave irradiation at 190°C and 3.0 MPa. In order to clarify the reaction mechanism, depolymerization of PC in benzyl alcohol was examined under the similar condition to the methanolysis. The formation of the ether compounds by decarbonation reaction of the carbonate products was investigated.

2. Experiments

2.1 Materials and Reagents

Pure PC pellets (L-1225LM, diameter = *ca.* 3 mm, thickness = *ca.* 3 mm) were supplied from the Teijin Chemical Ltd., Japan. Commercially available analytical reagent grade methanol (MeOH), benzyl alcohol (BnOH), 2,2-bis(-hydroxyphenyl) propane (BPA), dibenzyl carbonate (DBnC) and dibenzyl ether (DBnE) were used without further purification. MeOH and BnOH were used as solvents, and BPA, DBnC, and DBnE were used as standard reagents.

2.2 Depolymerization of PC with methanol and water

The pellets of PC (0.257 g, 1.01 mmol) was added to MeOH (4.02 g, 1.25 mmol) for methanolysis or to water (5.22 g, 290 mmol) for hydrolysis in a 10 mL sealed tube equipped with a stir bar. Microwave was irradiated to the tube attached in the microwave apparatus (Initiator+8 with single mode microwave irradiation, Biotage AB). Microwave irradiation (2.45 GHz, 300 W) was applied for 3 h at the reaction temperature of 190°C, and reaction pressure of 3.0 MPa. After the microwave irradiation, the reaction mixture was transferred from reaction tube to Kugelrohr flask by using MeOH (30 mL). MeOH was distilled from the solution by using Kugelrohr apparatus (Sigma-Aldrich Chemie Gmbh). MeOH containing low boiling components was trapped the tube in chilled bath (40°C, 110 Pa). BPA as a white solid product was remained in Kugelrohr flask. The BPA yield was calculated as the actually produced amount of BPA and the maximum amount of BPA obtained theoretically as the following.

$$BPA\ yield\ (\%) = \frac{produced\ amount\ of\ BPA}{maximum\ amount\ of\ BPA} x100 \qquad (1)$$

2.3 Depolymerization of PC with benzyl alcohol

The pellets of PC (0.256 g, 1.01 mmol) and BnOH (4.01 g, 125 mmol), were added in a 10 mL sealed tube equipped with a stir bar. The subsequent reaction procedures were almost similar to the methanolysis described in experimental section 2.2. The reaction temperature and pressure were changed to 250°C and 0.3 MPa, respectively. The unreacted BnOH and the solvent mixture of DBnC and DBnE containing trace BnOH were distillated from the reaction solution by Kugelrohr distillation apparatus at 80°C / 140 Pa and at 130°C / 120 Pa, respectively. Trace BnOH was completely recovered from the distillated solvent mixture by column chromatography (silica-gel: Wakosil C-200, eluent (50 mL); chloroform to hexane weight ratio of 3:1) Finally the eluent was evaporated under reduced pressure to obtain the mixture of DBnC and DBnE.

2.4 Products characterization

The chemical structure of BPA, DBnC, and DBnE were confirmed with FTIR spectrometer (Spectrum 100, PerkinElmer Inc.), FT-NMR spectroscopy (JNM-EX270, JEOL Ltd.). The proton ratio of DBnC:DBnE was determined from the chemical shifts of methylene moiety of benzyl group in ^{1}H-NMR spectra.

3. Results and discussion

3.1 Depolymerization of PC with H_2O and MeOH

As indicated in Entry 1 of table 1, hydrolysis of PC at the microwave irradiation condition of 190 °C at 3.0 MPa achieved at the maximum output of the microwave apparatus for 3 hours gave little BPA (BPA yield = 6 %) and left the unreacted PC due to the low affinity of PC to water. Hydrolysis of PC under microwave irradiation required acid or basic catalyst [5, 6]. Meanwhile, methanolysis of PC under the same reaction condition readily progressed to give BPA in a higher yield (94 %) as indicated in Entry 4. As shown in figure 1, the BPA yield became the maximum value at the reaction time of 3 hours and then gradually decreased due to the decomposition of BPA. On the other hand, DMC produced from methanolysis could not detected by IR analysis of the reaction mixture. Methyl group on DMC can barely detected by the NMR analysis. DMC and the related compounds might disappear unexpectedly from the reaction mixture during the depolymerization process.

Table 1. Depolymerization of polycarbonate under pressurized microwave irradiation[a]

Entry	PC /mmol	R-OH /mmol	Time /hour	Yield of BPA /%
1	1.06	H_2O 290	3.0	6
2	1.06	MeOH 127	12.0	90
3	1.06	MeOH 127	6.0	92
4	1.01	MeOH 127	3.0	94
5	1.01	MeOH 127	1.0	20

[a]The reaction temperature and pressure were fixed at 190°C and 3.0 MPa, respectively.

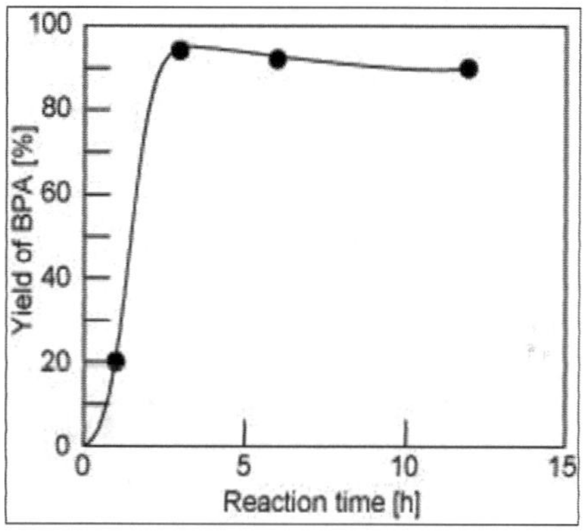

Figure 1. Yield of BPA

3.2 Depolymerization of PC with BnOH

In order to understand the disappearance of DMC from the reaction mixture, the alcoholysis of PC with benzyl alcohol (BnOH) was investigated. After the reaction at 250°C and at 0.3 MPa for 3 hours, BPA yield was as high as 100 %. The evaporation of the eluent in column chromatography operation gave the clear, colorless oil residue (0.254 g) which might be the carbonate compound with high boiling point. Figure 2 shows the ^{13}C-NMR spectra of the oily residue, dibenzyl carbonate (DBnC) and dibenzyl ether (DBnE). The chemical shifts of CH_2 of benzyl group (A = 69.88 ppm), quaternary carbon of phenyl group (B = 135.31 ppm), and carbonyl carbon of carbonate (C = 155.23 ppm) were observed in the standard samples of DBnC in Figure 2(b). The chemical shifts of CH_2 of benzyl group (A = 72.12 ppm) and quaternary carbon of phenyl group (E = 138.29 ppm) were observed in the standard samples of DBnE in Figure 2(C). The ^{13}C-NMR chemical shifts of the oily residue (Figure 2(a)) was compared with chemical shifts of figure 2 (b) and (c). Consequently, the oily residue was found to be composed of DBnC and DBnE. As a result of ^{1}H-NMR analysis of benzyl alcohol solution recovered from the reaction mixture of Entry 3, the proton ratio of DBnC:DBnE = 11.2:88.8 was determined from the chemical shifts of methylene moiety of benzyl group.

IOP Conf. Series: Materials Science and Engineering **458** (2018) 012037 doi:10.1088/1757-899X/458/1/012037

The main reaction products of alkali-catalyzed methanolysis were reported to be BPA and DMC [1]. However, BPA and dimethyl ether might be produced predominantly in the non-catalytic methanolysis under pressurized microwave irradiation as shown in figure 3. Therefore, dimethyl ether with the boiling point of –24°C can easily escape at the depressure procedure of the microwave apparatus.

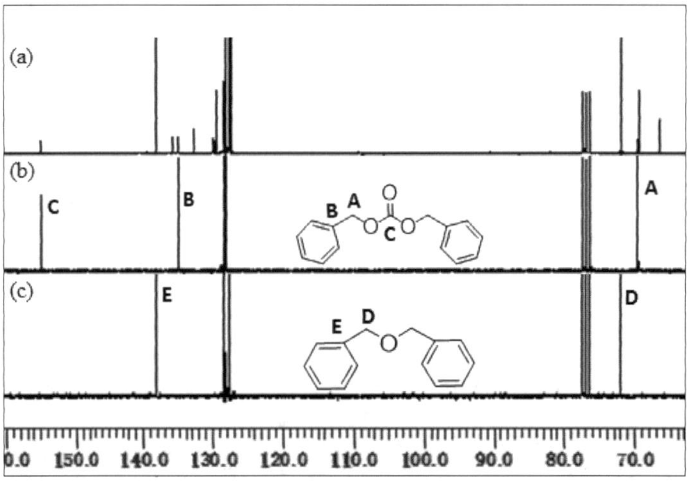

Figure 2. ^{13}C-NMR spectra of (a) the mixture removed BnOH by Silicagel column chromatograpy with $CHCl_3$-C_6H_8 eluent in Entry 4, (b) the standard sample of DBnC, and (c) the standard sample of DBnE.

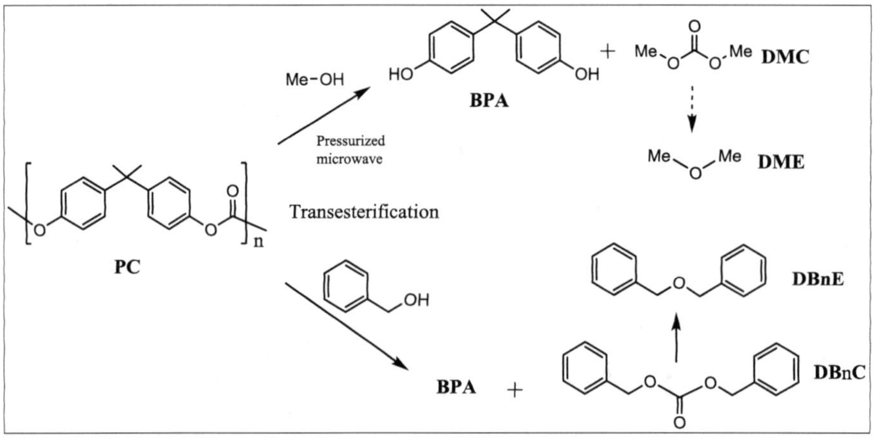

Figure 3. Depolymerization of polycarbonate by methanol and benzyl alcohol under pressurized microwave irradiation

Figure 4 shows the postulated production mechanism of dialkyl ether in the non-catalyzed decomposition of PC under pressureized microwave irradiation. The pathway A indicates the step-wise production of dialkyl ether by way of decarbonation of dialkyl carbonate. The pathway B indicates the simultaneous reaction of transesterification. To determine the production pathway of DBnE, the decomposition of PC with MeOH in the presence of DBnC was examined. The decarbonation of DBnC could not proceed during the methanolysis. This result strongly supports the formation of DBnE by the simultaneous reaction pathway B.

The proposed methanolysis method of waste PC under pressurized microwave irradiation can achieved the high efficient recovery of BPA without catalyst use and avoid the need for a further separation process of dimethyl carbonate.

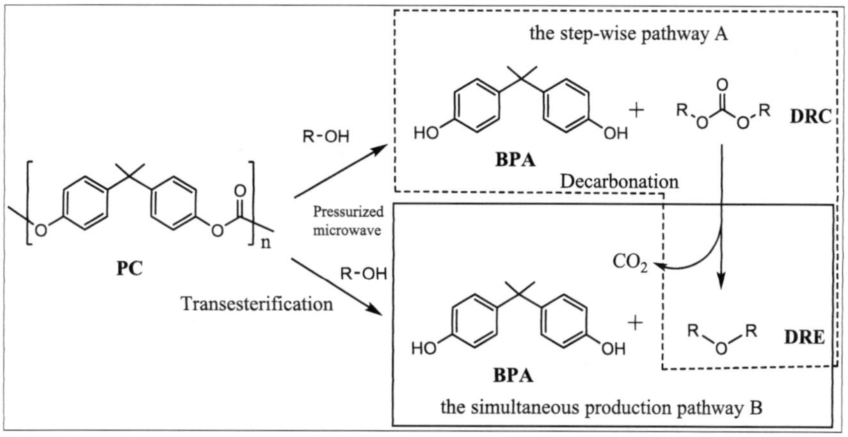

Figure 4. Postulated production mechanism of dialkyl ether in the non-catalyzed decomposition of PC under pressurized microwave

4. Conclusion

Chemical recycling of polycarbonate waste by non-catalyzed methanolysis under pressurized microwave irradiation was studied in this paper. The maximum yield of BPA of 94% was achieved at reaction temperature and pressure of 190°C and 3.0 MPa for 3 hours. On the other hand, trace amount of dimethyl carbonate was detected by NMR. From the result of depolymerization of PC with BnOH to confirm the reaction pathway to DBnE, BPA and DBnE were found to be directly formed by way of the transesterification. Less reactivity for the decarbonation of DBnC during the methanolysis strongly suggests the direct formation of DBnE in the transesterification of PC with alcohol. The depolymerization process can be considered as a high efficiency, convenient process from the viewpoint of non-catalyst use and the easy separation of BPA from dimethyl ether.

References

[1] Hu L-C, Oku A and Yamada E 1998 *Polymer,* **39** 3841–3845.

[2] Yoshioka T, Sugawara K, Mizoguchi T and Okuwaki A 2005 *Chem. Lett.* **34** 282-283.

[3] Song X, Liu F, Li L, Yang X, Yu S and Ge X 2013 *J. Hazard. Mater.* **244-245**, 204-8.

[4] Jie H, Ke H, Zhou Q, Chen L, Wu Y and Zhu Z 2006 *Polym. Degrad. Stabil.* **91** 44-50.

[5] Tsintzou G P, Antonakou E V and Achilias D S 2012 *J. Hazard. Mater.* **241-242**, 137-145.

[6] Deirram N, Rahmat AR, Rashidi S, Ghanbari T and Moradour M 2012 *Sci. Res. Essays* **7** 3399-3405.

[7] Ikenaga K, Inoue T and Kusakabe K 2016 *Procedia Eng.* **148** 314-318.

[8] Ikenaga K, Nagamitsu T, Kikukawa K and Kusakabe K 2018 *Proceedings of ICPEAM2018* inpress.

Estimation of safe environmental concentrations of ionic liquids towards bacteria by chemical toxicity distribution method

Muhammad Ishaq Khan, Dzulkarnain Zaini*, Azmi Mohd Shariff, Muhammad Athar and Muhammad Yasir Shamim

Centre for Advanced Process Safety (CAPS), Chemical Engineering Department, Universiti Teknologi Petronas, 32610 Bandar Seri Iskandar, Perak Darul Ridzuan, Malaysia.

*dzulkarnain.zaini@utp.edu.my

Abstract. Ionic Liquids (ILs) are considered as the potential replacement of industrial organic solvents because of their salient physical and chemical properties, for example, ionic nature, low vapor pressure, and good solubility properties. These properties make ILs feasible for potential application. Industrialization of ILs has a potential to adversely affect and pose risk to the ecosystem because of their toxic nature. However, the conventional research on the ILs is mainly carried out by performing acute the toxicity assessment of individual species for individual ILs which is termed as effect assessment. Effect assessment for individual ILs is insufficient for the ecotoxicological risk assessment of ILs. Subsequently, effect assessment of single ILs on one species is not sufficient to evaluate risk. The main objective of the current research was to estimate the safe environmental concentrations (SECs) of ILs towards three bacterial strains *V fischeri*, *E coli* and *S vacuolatus* by Chemical Toxicity Distribution (CTD) method. CTDs provided the SECs for a group Imidazolium NTf2 ILs toward three bacteria. *E coli* was the most sensitive species amongst three species when exposed to the selected group of ILs with 2.976 x 10^{-4} mmol/L as SEC.

1. Introduction

Ionic liquids (ILs) are novel chemicals that gained a lot of interest in the industrial sector as well as in academic research because of their role as potential industrial and environmentally friendly solvents. ILs are salts at room temperature and usually have melting point below 100°C [1; 2]. ILs are the chemical composed of cations and anions. The cationic part is organic and usually anionic part is inorganic. Cationic part of ILs is considered as the most responsible part for the physical and chemical properties of the compound itself. The anionic part consists of inorganic anions. Most of the research on ILs have been published in the last two decades which shows that there is an extensive focus of academia and industries has shifted towards ILs in last recent years. Initially, ILs were thought to be green replacements for volatile organic compounds [3]. The literature reported that ILs are useful in enhancing the capability of chemical processes with no negative impact on the atmosphere compared to volatile organic solvents because of their very low or negligible vapor pressure. Due to negligible vapor pressure, ILs do not significantly contribute to air pollution [4].

The increasing use of ILs especially in the industrial sector which ultimately leads to increase their discharge into an environment. The possible hazards of ILs may be due to accidental spills, leaching of landfill sites or via the industrial discharge which heavily affects the aquatic and terrestrial life [5].

Therefore, the toxicological studies are conducted for individual ILs for individual species called effect assessment. Effect assessment for individual ILs is insufficient for the risk assessment because different types of metabolism and different routes of exposure and other factors affect environment differently. Subsequently, effect assessment of single ILs on one species is not sufficient to evaluate risk.

Potential hazards of ILs demand the assessment of the risk caused by the toxicity [6]. However, because of unavailability of exposure data of ILs, it is impossible to get the frequency or probability of the exposure concentrations. An extensive research is carried out to assess the acute toxicity of ILs at the laboratory level. Acute toxicity experiments are conducted to assess the toxicity of individual ILs towards individual test species.

2. Methodology

In ecotoxicology, there are two types of assessments needed to be done i.e. the effect and exposure assessment. Effect assessment can be done by acute toxicity testing and the effect concentration either can be taken from real exposure data. However, data on ILs exposure is not available yet [7]. In current work, we developed a methodology to model the safe environmental concentrations from effect data (toxicity). There are four steps involved in the estimation of SECs from toxicity data of ILs. These steps are described in a methodological framework which is presented in Figure1.

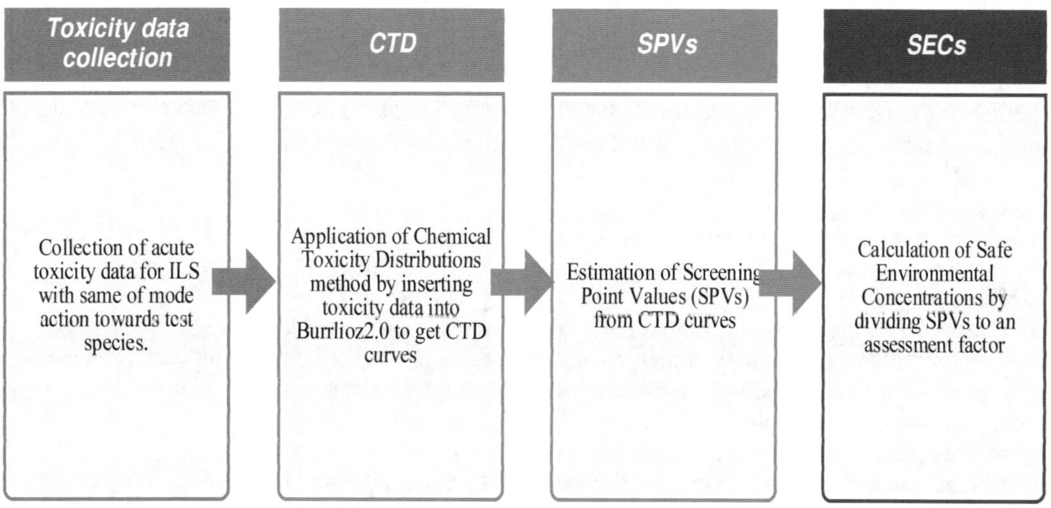

Figure 1. Methodological Framework

2.1. Toxicity data collection

The solubility of imidazolium ILs in water is very high[8] which makes these ILs highly toxic towards aquatic species including. We have collected toxicity data of imidazolium ILs with NTf_2 anions towards three bacteria *E coli*, *V fischeri* and *S vacuolatus* from literature. Toxicity data of [EMIM]NTf_2 [9; 10], [BMIM]NTf2[9; 11-13], [HMIM]NTf2[9; 11; 14] and [OMIM]NTf_2 is tabulated in Table 1. The trend of increasing toxicity with enlargement of the alkyl chain is observed in the data tabulated above. It is noticeable that same cation and the same anion show different toxicities when alkyl chain length is varied.

Table 1. Toxicity of Imidazolium NTf_2 ILS towards different Bacterial species

Ionic Liquids	Ec50(mmol/L)		
	E coli	*V fischeris*	*S vacuolatus*
[EMIM]NTf_2	-	0.844[9]	0.10[10]
[BMIM]NTf_2	2.0[11]	0.339[9]	0.057[12] [13]
[HMIM]NTf_2	1.0 [14]	0.051[9]	0.0011[13]
[OMIM]NT_2	0.7[14]	0.016[15]	-

2.2. Chemical toxicity Distributions

Chemical Toxicity Distribution method was used by William et al. [16] to assess the effect of a group of same chemicals with the same mode of action. In the current research, we used the CTD method to assess the effect of a group of ILs to single species to find safe environmental concentration (SECs). For each species, the EC50 of ILs with the same mode of action are used to calculate CTDs according to the methods outlined by Williams et al. [9]. The EC50 data for all ILs towards target species were ranked. The ranks were converted to % ranks (j) using the Weibull formula [17]

$$j = 1000 * (i)/(n + 1) \tag{1}$$

where i is the rank and n is the total number of compounds. The CTDs curves were plotted using log EC50 against the probit of the % rank, calculated as =NORMINV((j),5,1).

3. Results and Discussions

Toxicity data (EC50 concentrations) of selected ILs towards three bacteria *E coli*, *V fischeri* and *S vacuolatus* were selected. EC50 of the Imidazolium ILs with the same mode of action for each species was statistically analyzed by CTD method. The resulting CTD curves provided the concentrations called screening point values. Toxicity data taken from literature was arranged according to increasing alkyl chain length. The selected data was good enough as the toxicity of the imidazolium ILs to three of the bacterial strain followed the accepted trends of increasing alkyl chain length. "Probit of % effect" against the Log EC50 was plotted in Figures 2-4. Input data was taken from the CTDs spreadsheet of ILs towards algae.

A graph was plotted from CTD spreadsheets of *E coli* and presented in Figure 2. The equation of graph was used to calculate the concentrations on 1st and 5th percentiles. These concentrations were Termed as Screening point values (SPV). These SPVs are then divided by an assessment factor of 1000 to convert these concentrations to SECs[7].

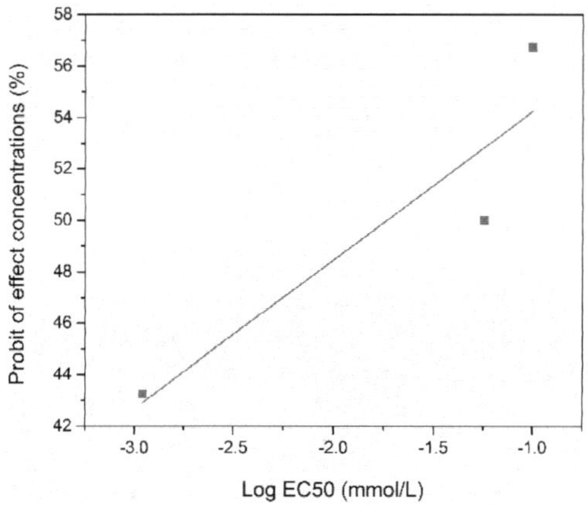

Figure 2. CTD curve of Imidazolium NTf2 ILs toward *E coli*

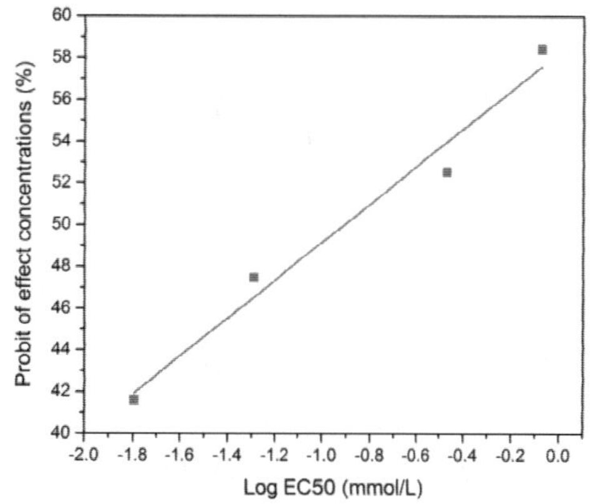

Figure 3. CTD curve of Imidazolium NTf2 ILs toward *V fischeris*

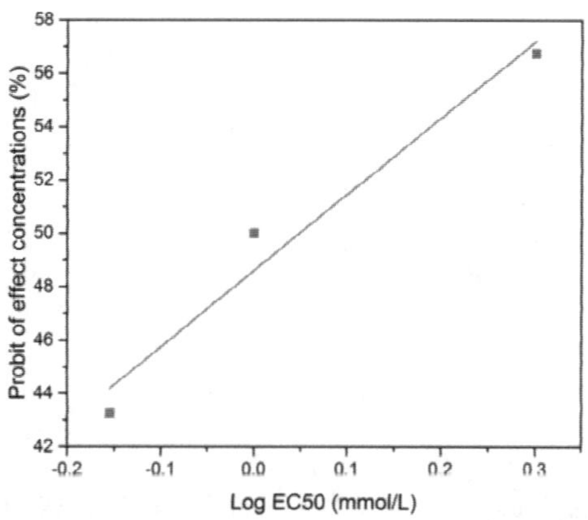

Figure 4. CTD curve of Imidazolium NTf2 ILs toward *S vacuolatus*

The Figure 3 and Figure 4 were plotted from CTD spreadsheets of *V fischeris* and *S vacuolatus* respectively. In each figure, the probit of % ranking was plotted against the log of toxicity values to form an equation which was used to calculate percentiles. The concentration at the 5th percentile was taken as Screening Point values. SPVs were then divided by an assessment factor of 1000 to convert these concentrations to SECs. The results are presented in Table 2.

Table 2: SECs of Imidazolium NTf_2 ILs Based on CTDs

Organism	Centile	SPV(mmol/L)	SECs (mmol/L)
E coli	5th	0.297674309	2.976 x 10^{-4}
V fischeris	5th	0.001915	1.915 x 10^{-6}
S vacuolatus	5th	2.68487 x 10^{-5}	2.68487 x 10^{-8}

The 5th percentile was calculated from the equation of the above graph and taken as screening point values. SPVs were divided by an assessment factor of 1000 to get Safe Environmental Concentration (SEC). SEC for *E coli* was 2.976 x 10^{-4} mmol/L. Similarly, the SECs of imidazolium ILs towards *V fischeris* and *S vacuolatus* were 1.915 x 10^{-6} mmol/L and 2.68487 x 10^{-8} mmol/L. SECs indicated that *E coli* are the most sensitive species amongst three species when exposed to the selected group of ILs. 2.976 x 10^{-4} mmol/L is the concentration which could be considered as a safe concentration for aquatic ecosystem containing *E coli*, *V fischeris*, and *S vacuolatus.*

4. Conclusion

Chemical Toxicity Distribution method is used to assess the toxicological effect the ILs which do not have any major industrial accident data so for because of lack of industrial applications. The outcome of the CTD method is the safe environmental Concentrations. SECs are the concentrations at which the 95 % species are protected from the hazards of toxic chemicals. In current work, *E coli* (having 0.0002967 mmol/L as SEC) was proved to be more sensitive species amongst *V fischeris* and *S vacuolatus*. It means that even a small amount of imidazolium NTf2 ILs can affect 5 % of the *E coli* in an aquatic ecosystem containing *E coli, V fischeris* and *S vacuolatus*. SECs could be used as the standard value to compare with real environmental concentrations to quantify the risk. If the ratio of measured environmental concentration to SECs is less than 1, the environment will be considered as safe. If this ratio exceeds 1, there is a need to control ECs. Risk assessment of ILs requires leakage data and CTDs is the best alternative techniques because of the unavailability of risk assessment data of ILs. In future, a huge number of data may be used to effectively assess the risk posed by ILs to different organisms of ecosystems. Furthermore, new modeling techniques on the toxicity of ILs can be adopted to reduce the impact of a group of ILs towards different compartment s of the ecosystems.

Acknowledgment
The authors would like to gratitude Universiti Teknologi PETRONAS, Malaysia for providing research facilities and YUTP funding (0153AA-E19) to make this research feasible.

References

[1] Araque, J.C., Hettige, J.J., and Margulis, C.J., Modern room temperature ionic liquids, a simple guide to understanding their structure and how it may relate to dynamics, The Journal of Physical Chemistry B 119 2015, 12727-12740.

[2] Moniruzzaman, M. and Goto, M., Ionic liquids: future solvents and reagents for pharmaceuticals, *J. Chem. Eng. Jpn.* **44** 2011, 370-381.

[3] Gutowski, K.E., Broker, G.A., Willauer, H.D., Huddleston, J.G., Swatloski, R.P., Holbrey, J.D., and Rogers, R.D., Controlling the aqueous miscibility of ionic liquids: aqueous biphasic systems of water-miscible ionic liquids and water-structuring salts for recycle, metathesis, and separations, *J. Am. Chem. Soc.* **125** 2003, 6632-6633.

[4] Bier, M. and Dietrich, S., Vapour pressure of ionic liquids, *Mol. Phys.* **108** 2010, 211-214.

[5] Ventura, S.P.M., e Silva, F.A., Gonçalves, A.M.M., Pereira, J.L., Gonçalves, F., and Coutinho, J.A.P., Ecotoxicity analysis of cholinium-based ionic liquids to Vibrio fischeri marine bacteria, *Ecotoxicol. Environ. Saf.* **102** 2014, 48-54.

[6] Bubalo, M.C., Radošević, K., Redovniković, I.R., Halambek, J., and Srček, V.G., A brief overview of the potential environmental hazards of ionic liquids, *Ecotoxicol. Environ. Saf.* **99** 2014, 1-12.

[7] Khan, M.I., Zaini, D., Shariff, A.M., and Moniruzzaman, M., Framework for Ecotoxicological Risk Assessment of Ionic Liquids, *Proc. Engg.* **148** 2016, 1141-1148.

[8] Freire, M.G., Carvalho, P.J., Gardas, R.L., Marrucho, I.M., Santos, L.M., and Coutinho, J.A., Mutual solubilities of water and the [C n mim][Tf2N] hydrophobic ionic liquids, *J. Phys. Chem. B* **112** 2008, 1604-1610.

[9] Ventura, S.P., Gonçalves, A.M., Sintra, T., Pereira, J.L., Gonçalves, F., and Coutinho, J.A., Designing ionic liquids: the chemical structure role in the toxicity, *Ecotoxicology* **22** 2013, 1-12.

[10] Steudte, S., Stepnowski, P., Cho, C.-W., Thöming, J., and Stolte, S., (Eco) toxicity of fluoro-organic and cyano-based ionic liquid anions, Chem. Commun. 48 2012, 9382-9384.

[11] Łuczak, J., Jungnickel, C., Łącka, I., Stolte, S., and Hupka, J., Antimicrobial and surface activity of 1-alkyl-3-methylimidazolium derivatives, *Green Chemistry* **12** 2010, 593-601.

[12] Matzke, M., Stolte, S., Thiele, K., Juffernholz, T., Arning, J., Ranke, J., Welz-Biermann, U., and Jastorff, B., The influence of anion species on the toxicity of 1-alkyl-3-methylimidazolium ionic liquids observed in an (eco)toxicological test battery, *Green Chemistry* **9** 2007, 1198-1207.

[13] Izadiyan, P., Fatemi, M., and Izadiyan, M., Elicitation of the most important structural properties of ionic liquids affecting ecotoxicity in limnic green algae; a QSAR approach, *Ecotoxicol. Environ. Saf.* **87** 2013, 42-48.

[14] Deng, Y., *Physico-chemical properties and environmental impact of ionic liquids*, PhD Thesis, Université Blaise Pascal-Clermont-Ferrand II, 2011.

[15] Mutalib, M.I.A. and Ghanem, O.B., Ecotoxicity of Ionic Liquids Towards Vibrio fischeri: Experimental and QSAR Studies, in: *Progress and Developments in Ionic Liquids,* S. Handy, ed., InTech, 2017, pp. 429-449.

[16] Williams, E.S., Berninger, J.P., and Brooks, B.W., Application of chemical toxicity distributions to ecotoxicology data requirements under REACH, *Environ. Toxicol. Chem.* **30** 2011, 1943-1954.

[17] Dobbins, L.L., Usenko, S., Brain, R.A., and Brooks, B.W., Probabilistic ecological hazard assessment of parabens using Daphnia magna and Pimephales promelas, *Environ. Toxicol. Chem.* **28** 2009, 2744-2753.

Utilizing Lipid-extracted Microalgae Residue for Removal of Methylene Blue from Aqueous Solution

Muhammad Rais Rosli[1], Man Kee Lam[1,2,*], Wan Nadiah Amalina Kadir[1,2], Yoshimitsu Uemura[1,2] and Siew Hoong Shuit[3]

[1]Chemical Engineering Department, Universiti Teknologi PETRONAS, Seri Iskandar 32610, Perak, Malaysia.
[2]Centre for Biofuel and Biochemical Research, Institute of Self-sustainable Building, Universiti Teknologi PETRONAS, Seri Iskandar 32610, Perak, Malaysia.
[3]Department of Chemical Engineering Technology, Faculty of Engineering Technology, Universiti Malaysia Perlis, Padang Besar, Malaysia.

*lam.mankee@utp.edu.my

Abstract. Recently, microalgae biomass have been highlighted as the most feasible feedstock for biodiesel production, particularly due to their fast growth rate and they are able to synthesize high content of lipid within their cells. However, after extraction of lipid from the microalgae biomass, the remaining microalgae residues are usually discarded as waste. Thus, the present paper is aimed to recycle the lipid-extracted microalgae residue as adsorbent to remove methylene blue (MB) from aqueous solution. It was found that the microalgae residue could remove 74 % of the MB with biomass dosage of 6.25 g/L, stirring speed of 200 rpm at initial MB concentration of 2 mg/L. The adsorption of MB on microalgae residue fitted well with the pseudo-second order kinetic model and the equilibrium adsorption data indicated that the adsorption behavior followed the Langmuir isotherm.

1. Introduction

Adsorption is a separation process where adsorbate like soluble organics, dyes, pesticides, etc., is concentrated from a bulk liquid or vapour attached on to the surface of adsorbent (porous solid) by either physical or chemical attraction. It is commonly used in environmental engineering practices for colour removal of wastewater caused by dyes or other substances but the effectiveness and the sustainability in terms of economic is depending on the type of adsorbent [1]. In the past years, there are a lot of studies on biosorption (adsorption using biomass products) and its potential as a process for treatment on metal or dye bearing in the wastewater. Biosorption can be defined as the passive uptake of organic or inorganic species from solution by either microbial biomass, such as bacteria, yeast and algae, or waste from industry such as pinewood, rice husk, etc. [1, 2].

Various adsorbent can be used to remove the colour or dye from water, such as activated carbon, organic substances like clay, sludge and zeolites, biosorbents like fungi and algae, solid waste from agriculture like leaves, seeds and fibres, industrial solid wastes and many more type of adsorbents [3]. Studies have proven that adsorption using activated carbon is the best technique for water treatment especially when dealing with dye that contains toxic pollutant as activated carbon has high sorption capacity. However, the activated carbon is usually expensive and may not economically feasible to be used in large scale [2].

On the other hand, the lipid extracted microalgae residue can be potentially recycled as an absorbent instead of discarded after the lipid extraction process. Microalgae biomass are potential precursors for biodiesel production by extracting their lipid and convert the lipid through transesterification reaction. However, after the lipid is extracted from the microalgae biomass, the remaining microalgae residue is usually discarded to the environment. However, the microalgae residue has the potential to be recycled as an absorbent due to the pore structure created in the biomass during lipid extraction process.

Thus, in the present paper, microalgae residue will be recycled and used as alternative adsorbent for the dye removal. The experiments were conducted under lab-scale and removal efficiency of MB by adsorption onto lipid-extracted microalgae residue were determined, as well as the associated adsorption isotherm and kinetic.

2. Methodology

2.1. Preparation of Dye Solution
MB (Sigma-Aldrich) powder was used as the basic dye in the present work. The MB solution was prepared in the concentrated stock solution, which was 1 L of 200 mg/L solution. From this stock solution, the sample solutions were prepared later through dilution process.

2.2. Microalgae Lipid-extracted Residue
The dried microalgae biomass, *Chlorella vulgaris*, was supplied by Centre for Biofuel and Biochemical Research, UTP. 70 g of the dried microalgae biomass was weighed and mixed with 400 ml of methanol and 200 ml of chloroform. Then, the microalgae and the solvent solution was stirred for 5 hours by using magnetic stirrer before it was filtered using filter funnel and filter paper. The microalgae residue on the filter paper was taken and dried by both air dry and oven before used in the subsequent adsorption process. The microalgae before extraction and after extraction were characterized to determine their morphology and functional groups by using Scanning Electron Microscopy (SEM) and Fourier-transform Infrared Spectroscopy (FT-IR), respectively.

2.3. Effect of Methylene Blue Concentration
Four concentration of MB was produced; 2, 3, 4 and 5 mg/L. The solution then was put inside the stirrer and stirred for 10 minutes. Then, 10 ml of solution from each of the solution was taken and read by UV-VIS spectrophotometer, as the initial reading. The stirring was proceeded and 1 g of adsorbent (lipid-extracted microalgae residue), was inserted into each of the concentration of the solution. The 10 ml sample collection was repeated in 20 minutes interval until the system reached equilibrium.

2.4. Effect of Biomass Dosage
Five solutions of MB solution were prepared with equal concentration based on the optimized value from Section 2.4. The initial reading was taken by taking 10 ml of sample using 10 ml cell square and absorbance was determined using UV-VIS spectrophotometer. The process was proceeded with addition of 0.5 g, 1 g, 1.5 g, 2 g and 2.5 g of microalgae residue into the solution, respectively. The stirring process continued and 10 ml sample collection was repeated in 20 minutes interval until the system reached equilibrium.

2.5. Effect of Stirring Speed
Four solutions of MB solution were prepared with equal concentration based on the optimized value from Section 2.4. The initial reading was taken by taking 10 ml of sample using 10 ml cell square and absorbance was determined using UV-VIS spectrophotometer. Then, 2.5 g of microalgae residue was added into each of the solution and the stirring proceeded with different stirring speed for each solution; 50, 100, 150 and 200 rpm respectively. The 10 ml sample collection was repeated in 20 minutes interval until the system reached equilibrium.

2.6. *Removal Efficiency*
The removal efficiency of MB was calculated by using equation (1) as follow:

$$\boldsymbol{Removal\ efficiency} = \left(\mathbf{1} - \frac{C_t}{C_0}\right) \times \mathbf{100}\ \% \quad (1)$$

where, C_0 is initial concentration and C_t is concentration at time t.

2.7. *Characterization of Lipid-extracted Microalgae Residue*
The morphology and functional groups of pure microalgae biomass and lipid-extracted microalgae residue were carried by using Scanning Electron Microscopy (SEM, Hitachi TM3030) and Fourier Transform Infrared Red (FT-IR, Thermo Scientifi Nicole iS10) at Centre for Biofuel and Biochemical Research, UTP.

2.8. *Adsorption Isotherm Model*
The adsorption isotherms, Freundlich and Langmuir, were used in the present work (table 1). The original and linearized equation for each Freundlich and Langmuir isotherm are shown in equation (2) and (3), where q_e is the amount of dye adsorbed at equilibrium time (mg/g), C_e is the equilibrium concentration of dye in solution (mg/L), q_m is the maximum adsorption capacity (mg/g), K_a is the isotherm constants for Langmuir (L/mg), K_f is the capacity of the adsorbent constant for Freundlich and n is the intensity of adsorption constant for Freundlich [4-7].

Table 1. The equation model and linearized model for Freundlich and Langmuir isotherm

Isotherm Models	Equation model	Linearized model	
Langmuir	$q_e = \frac{q_m K_a C_e}{1 + K_a C_e}$	$\frac{1}{q_e} = \frac{1}{K_a q_m}\frac{1}{C_e} + \frac{1}{q_m}$	(2)
Freundlich	$q_e = K_f C_e^{\frac{1}{n}}$	$lnq_e = lnK_f + \frac{1}{n}(lnC_e)$	(3)

2.9. *Adsorption Kinetic Model*
The adsorption kinetic was determined based on the result from Section 2.4, and was evaluated based on the highest R^2 value on the plotted graph depending on their equation (table 2). For the present work, only two types of kinetic was analyzed and compared, which were kinetic pseudo-first order and pseudo-second order, as shown in equation (4) and (5), where, q_t is the concentration of adsorbate on adsorbent at time t, q_e is the equilibrium concentration of adsorbate on adsorbent at time t, K_1 is the kinetic constant for pseudo-first order and K_2 is the kinetic constant for pseudo-second order [8, 9].

Table 2. The equation and linearized model for Pseudo-first order and Pseudo-second order kinetic.

Kinetic Models	Equation model	Linearized model	
Pseudo-first order	$\frac{dq_t}{dt} = K_1(q_e - q_t)$	$ln\left(1 - \frac{q_t}{q_e}\right) = lnq_e - K_1 t$	(4)
Pseudo-second order	$\frac{dq_t}{dt} = K_2(q_e - q_t)^2$	$\frac{t}{q_t} = \frac{1}{K_2 q_e^2} + \left(\frac{1}{q_e}\right)t$	(5)

3. Results and Discussions

3.1. SEM Analysis

Samples of microalgae before and after lipid extraction were characterized to describe their morphology by using SEM. Figure 1 shows the result of SEM for microalgae before and after lipid extraction, respectively. From the figure, the sample of microalgae biomass after lipid extraction was ruptured and scatted. This is probably due to the extraction process itself as the lipid was extracted out from the cell body, as mention by Lee, *et al.* [10], that the cells were disrupted in extraction process.

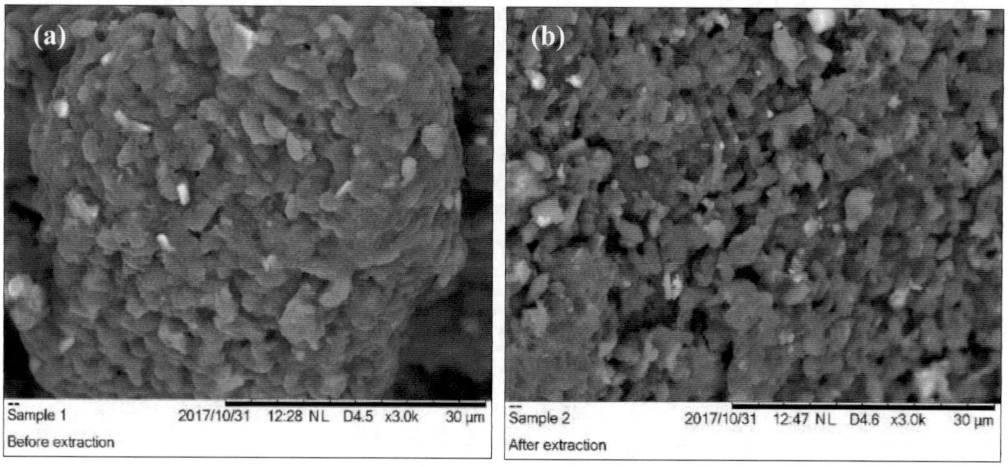

Figure 1. The SEM results for microalgae sample (a) before lipid extraction (b) after lipid extraction.

3.2. FT-IR Analysis

The FT-IR result shown in figure 2 compared microalgae functional groups before and after lipid extraction process. As can be seen, the trend of both graphs is similar while the peaks intensity has only minor differences (table 3).

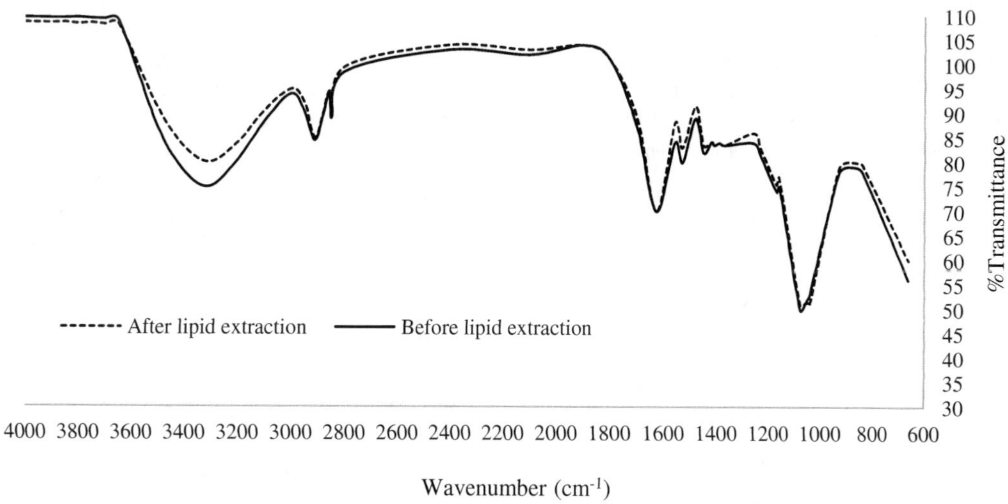

Figure 2. FT-IR result for microalgae sample before and after extraction.

The result indicated that there were no significant changes in functional groups occurred during the lipid extraction process. Even though there are slightly different in intensity of peaks, the values are still remained in the same range.

Table 3. Functional groups of microalgae samples

Peak value	Functional group	Frequency range
Before: 3304.82 After: 3287.54	Alcohol (O-H)	3200-3600
Before: 2919.51 After: 2919.33	Alkane (C-H)	2850-3000
Before: 2850.81 After: 2850.62	Carboxylic Acid (O-H)	2500-3300
Before: 1627.09 After: 1625.09	Alkene (C=C)	1620-1680
Before: 1538.43 After: 1537.76	Nitro (N-O)	1515-1560
Before: 1409.49 After: 1455.19	Alkane (-C-H)	1350-1480
Before: 1076.48 After: 1088.94	Amine (C-N)	1080-1360

3.3. Effect of MB Concentration

Figure 3 shows the MB removal efficiency (in percentage) for all four samples. From the figure, the highest MB removal efficiency was 73.6 % at 2 mg/L MB concentration, followed by samples 3 mg/L (69.5%), 4 mg/L (66.7%) and 5 mg/L (62%), respectively. From this finding, it can be concluded that the lower concentration of methylene blue solution resulted the highest removal efficiency for this type of adsorbent. Bulut and Aydın [11] also attained similar trend of results, which they indicated that the high driving force for mass transfer was one of the causes of increment in loading capacity of adsorbent in relation to dye ions. Other than that, the adsorption capacity, q, also can be calculated from the concentration by using Equation 6, where q_t is the adsorption capacity at time t (mg/g), C_0 is initial concentration (mg/L), C_t is concentration at time t (mg/L), V is volume of solution (L) and W is weight of adsorbent (g) [12]. Figure 4 shows the adsorption capacity for all four different concentrations, in which the higher concentration of MB producing higher adsorption capacity of MB adsorbed per unit mass of microalgae residue. The increment of the initial concentration of MB solution might increases the driving force of the concentration gradient, thus attained the results as shown in figure 4.

$$q_t = \frac{C_0 - C_t}{W} V \tag{6}$$

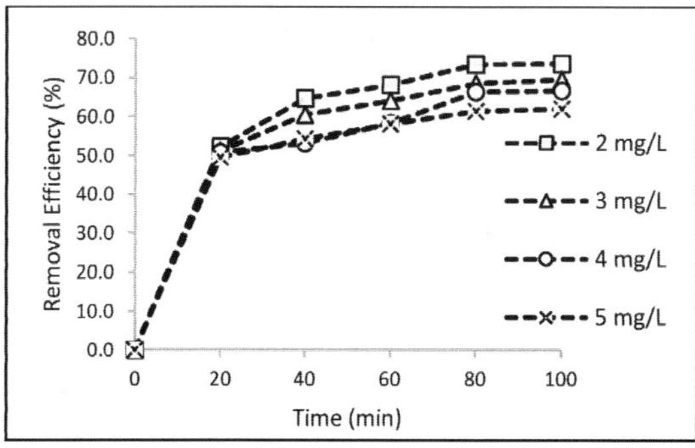

Figure 3. Removal efficiency of MB solution by microalgae residues

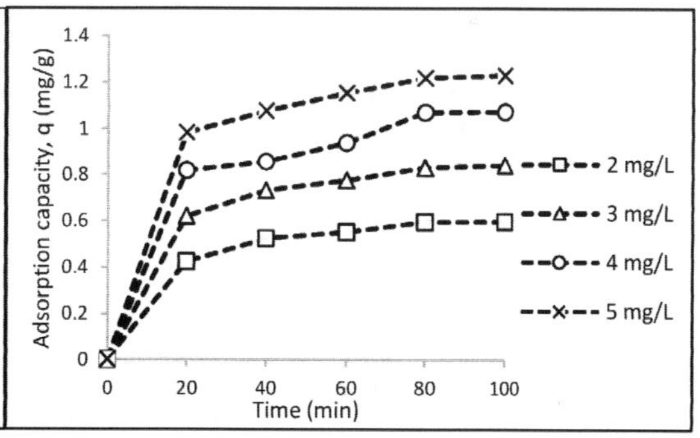

Figure 4. Adsorption capacity, q at different time.

3.4. Adsorption Isotherm

The results attained in Section 3.3 were fitted in the linearized isotherm equations. Table 4 shows that Langmuir and Freundlich isotherms are acceptable as both can fit the data which resulted to the R^2 value more than 0.99. However, Langmuir isotherm is chosen as it gives the higher R^2 value as compared to the Freundlich isotherm. According to Dural, *et al.* [13], the Langmuir isotherm indicates the physical monolayer adsorption, in which the adsorbed molecules do not interact as the uptake only occurred on the homogenous surface [14, 15]. In addition, the q_m attained in the present research was significantly higher as compared to banana peel (0.124 mg/g) [16] and orange peel (1.744 mg/g) [17]. The comparison also indicates the potential of microalgae residue to be utilized as alternative adsorbent and can be further transformed to activated carbon to attain higher adsorption capacity.

Table 4. Langmuir and Freundlich constants for MB adsorption by microalgae residue

Isotherm	Langmuir			Freundlich		
Parameter	q_m (mg/g)	K_a (mg/L)	R^2	K_F ((mg/g)(L/mg)$^{1/n}$)	n	R^2
Value	2.151	0.7158	0.9983	0.874	1.708	0.9926

3.5. Adsorption Kinetic

The data obtained in Section 3.3 were fitted into the linearized kinetic equations and table 5 summarizes the information related to the adsorption kinetic model. The second-order kinetic model was likely to be chosen as the R^2 values for all the concentration were the highest and constant compared to the first-order kinetic model. Beside the R^2 value, the Sum of Error Square (SSE) also can be used to determine the order of kinetic model [12]. The lower of SSE value means less error occurred and thus, determine the best fit of the adsorption kinetic model. Since second-order kinetic model also has the lowest SSE values (except at concentration at 3 mg/L), therefore the second-order kinetic is the best fit for adsorption of MB by using microalgae residue as adsorbent. Besides, the second-order kinetic also indicates the chemisorption is the rate-determining step that controls the adsorption process. This is further proved in the FT-IR analysis that the presence of multiple functional groups resulted to the binding of MB on the surface of the lipid-extracted microalgae residue instead of only depending on the van der Waals force.

Table 5. Adsorption kinetic model constants for methylene blue adsorption by microalgae residue.

Conc. (mg/L)	First-order kinetic model					Second-order kinetic model			
	$q_e$exp (mg/g)	$q_e$cal (mg/g)	K_1 (1/min)	R^2	SSE (%)	$q_e$cal (mg/g)	K_2 (1/Min)	R^2	SSE (%)
2	0.597	0.730	0.0626	0.9044	0.059	0.669	0.1302	0.9989	0.032
3	0.841	0.776	0.0492	0.9593	0.029	0.930	0.1012	0.9992	0.040
4	1.073	1.313	0.0580	0.8146	0.107	1.208	0.0625	0.9882	0.060
5	1.230	1.074	0.0519	0.9525	0.070	1.330	0.0914	0.9989	0.045

3.6. Effect of Biomass Dosage

The effect of dosage of biomass was studied by manipulated the biomass dosage, which were 1.25 g/L, 2.50 g/L, 3.75 g/L, 5.00 g/L and 6.25 g/L, respectively. The initial concentration of methylene blue solution was set to be constant at 5 mg/L with constant stirring speed, at 150 rpm. The result was obtained and the removal efficiency was calculated as shown in figure 5. From the figure, the higher biomass dosage could result the higher MB removal efficiency. By comparison, when the biomass dosage was 1.25 g/L, the final efficiency was 59% compared to the biomass dosage of 6.25 g/L which can achieved efficiency as high as 70.3%. This is probably due to more surface area are available for the adsorption process to occur [11].

3.7. Effect of Stirring Speed

The effect of stirring speed was studied on the efficiency in removal of MB. Four stirring speed were set and the initial concentration of methylene blue solution and biomass dosage was kept constant at 5 mg/L and 6.25 g/L, respectively. Figure 6 shows the result of the removal efficiency for each stirring speed. From the figure, it can be observed that the higher stirring speed could result higher MB removal efficiency. The MB removal efficiency at 50 rpm was about 63%, however, as the speed increased to 100 rpm, 150 rpm and 200 rpm, the removal efficiency increased to 66.5%, 70.3% and 74.1%, respectively. This is probably because the stirring speed may influence the distribution of the adsorbate to be adsorbed. Nevertheless, the results obtained differ with Doğan, *et al.* [15], as according to that study, the stirring speed did not give any different to the removal efficiency.

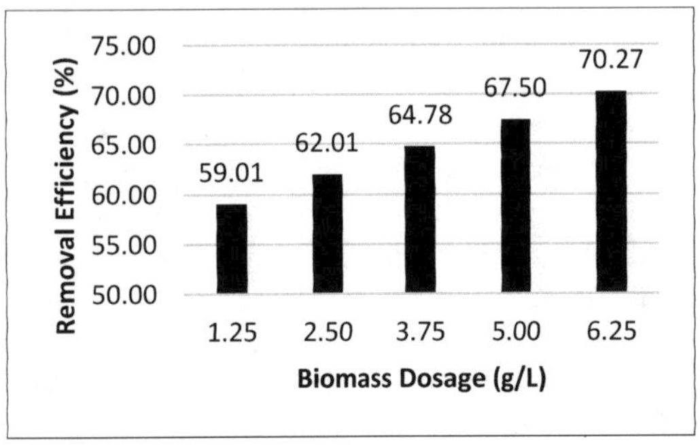

Figure 5. Removal efficiency of different biomass dosage at 100th minute.

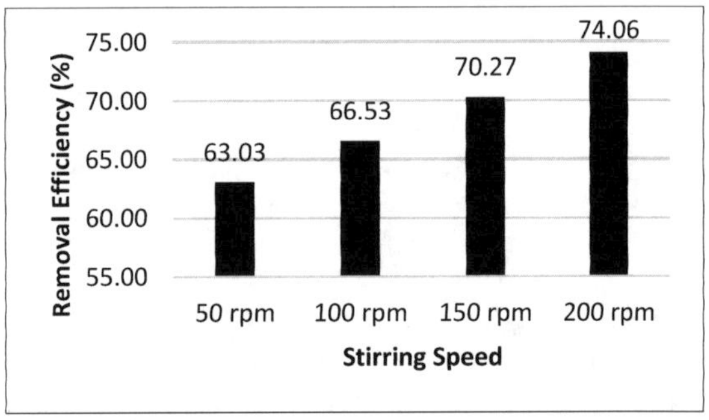

Figure 6. Removal efficiency of different stirring speed at 100th minute.

4. Conclusion

In the present work, it was proven that lipid-extracted microalgae residue could be used as alternative adsorbent to adsorb MB. As for the initial concentration of MB solution, the lower concentration resulted the higher removal efficiency, while for the adsorbent dosage, as the adsorbent dosage increased, the removal efficiency increased. The higher stirring speed also resulted the higher removal efficiency. The highest MB efficiency attained was 74% with the biomass dosage of 6.25 g/L, stirring speed of 200 rpm. The adsorption isotherm determined in the present work was Langmuir isotherm whereas the adsorption kinetic model would be following the second-order kinetic model due to its high R^2 and SSE% values.

Acknowledgment
The authors would like to acknowledge the financial support from Universiti Teknologi PETRONAS (YUTP-FRG with cost center 0153AA-H46). Technical support from Centre for Biofuel and Biochemical Research (CBBR) of UTP is highly appreciated.

References
[1] Russo M E, Di Natale F, Prigione V, Tigini V, Marzocchella A and Varese G C 2010 *Chem. Eng. J.* **162** 537.

[2] Kousha M, Daneshvar E, Sohrabi M S, Jokar M and Bhatnagar A 2012 *Chem. Eng. J.* **192** 67.

[3] Daneshvar E, Vazirzadeh A, Niazi A, Sillanpää M, and Bhatnagar A 2017 *Chem. Eng. J.* **307** 446.

[4] Anastopoulos I and Kyzas G Z 2016 *J. Mol. Liq.* **218** 174.

[5] Rangabhashiyam S, Anu N, Giri Nandagopal M S and Selvaraju N 2014 *J. Environ. Chem. Eng.* **2** 398.

[6] Aksu Z and Tezer S 2005 *Process Biochem.* **40** 1347.

[7] Yagub M T, Sen T K, Afroze S and Ang H M 2014 *Adv. Colloid Interface Sci.* **209** 172.

[8] Dotto G L, Cadaval T R S and Pinto L A A 2012 *Process Biochem.* **47** 1335.

[9] Marzbali M H, Mir A A, Pazoki M, Pourjamshidian R and Tabeshnia M 2017 *J. Environ. Chem. Eng.* **5** 1946.

[10] Lee J Y, Yoo C, Jun S Y, Ahn C Y and Oh H M 2010 *Bioresour. Technol.,* **101** S75.

[11] Bulut Y and Aydın H 2006 *Desalin.* **194** 259.

[12] Lam M K and Zakaria R 2008 *Int. Conf. on Environ. (Malaysia)* (Universiti Sains Malaysia) p 1-12.

[13] Dural M U, Cavas L, Papageorgiou S K and Katsaros F K 2011 *Chem. Eng. J.* **168** 77.

[14] Altenor S, Carene B, Emmanuel E, Lambert J, Ehrhardt J J, and Gaspard S 2009 *J. Hazard. Mater.* **165** 1029.

[15] Doğan M, Özdemir Y and Alkan M 2007 *Dyes Pigm.* **75** 701.

[16] Gautam S and Khan S H 2016 *Int. J. Sci. Environ. Technol.* **5** 3230.

[17] AbdurRahman F B, Akter M and Abedin M Z 2013 *Int. J. Sci. Technol. Res.* **2** 47.

IOP Conf. Series: Materials Science and Engineering **458** (2018) 012040 doi:10.1088/1757-899X/458/1/012040

An improved characteristics length in strouhal number for internal flow induced acoustics in corrugated pipe

Yeong Jin King[1,2,*], M Salman Leong[1], Kok Yung Chong[2] and Jer Vui Lee[2]

[1]Institute of Noise and Vibration, Universiti Teknologi Malaysia, Johor, Malaysia
[2]Universiti Tunku Abdul Rahman, Selangor, Malaysia

*yeongjin.king@gmail.com

Abstract. Corrugated pipes often experience self-sustained oscillations which leads to high intensity noise, known as "whistling" during critical flow conditions. In order to determine the peak whistling Strouhal number (Sr_{p-w}) and whistling amplitude, a two dimensional (2D) numerical simulation method using Fluent software is proposed in this paper. The flow pattern of the fluid passing through the corrugated pipe and the effect of pipe geometries on the Sr_{p-w} and whistling amplitude are investigated. A ratio of $\frac{l}{w+r_{up}}$ is found to be a good indicator of the consistency of Sr_{p-w} where the l is the plateau length, w is the width of the corrugation, and r_{up} is the upstream radius of the corrugation. Furthermore, optimal characteristic length is also assessed among three different characteristic lengths including $w + r_{up}$, w and $\frac{l^2}{w+r_{up}}$. The $\frac{l^2}{w+r_{up}}$ is found to be the optimal as the characteristic length. This finding is significant as it can allow us to avoid critical flow condition to minimize and prevent the whistling phenomenon more accurately and effectively by using the proposed equations.

1. Introduction

There has been significance amount of study being done on the vortex induced vibration of pipe as this phenomenon caused the fatigue damage to the pipe [6] compare to the internal flow-induced acoustics of the corrugated pipe. Corrugated pipes are flexible pipes that are specially designed with corrugations to facilitate fluid flow between sea installations and surface equipment on drilling platforms. The main purpose of corrugations is to provide local stiffness and rigidity to prevent collapse of the pipe during bending while maintaining global flexibility. Nevertheless, corrugated pipes often experience self-sustained oscillations that leads to high intensity noise, known as "whistling" or "singing" during critical flow conditions [8], [9].

A special feature that can be observed in the flow of a corrugated pipe is known as shear layer which divides the high-speed flow region from the low speed flow region next to the pipe wall. This is shown in figure 1[1]. The formation of shear layer is normally noticed in a turbulent flow where the viscous sublayer adjacent to the wall does not have sufficient kinetic energy to travel against an adverse pressure gradient. When this occurs, there will be a back flow along the pipe wall in the opposite direction to the main flow. Eventually, a separation of the boundary layer (shear layer) from the pipe wall is resulted at a cavity where there is abrupt pipe widening. This phenomenon also can be observed on the pipe with multiple side branches [7], [10].

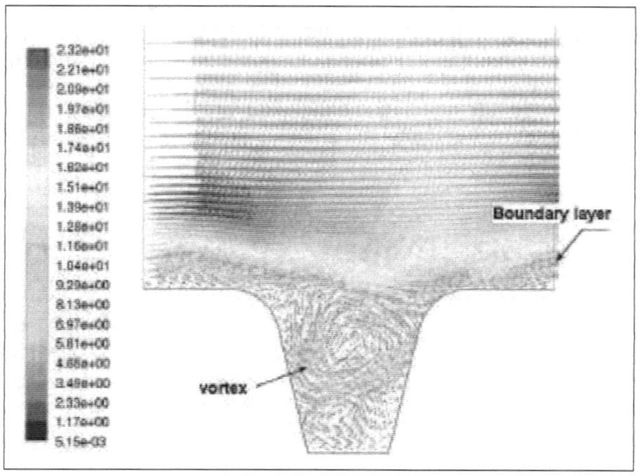

Figure 1. Shear Layer and Vortex in the Corrugation

According to NakiboĞLu et al. [2], self-sustained oscillations is caused by interaction of shear layer instability with longitudinal local acoustic standing waves as defined in a feedback loop as shown in figure 2 [1]. The shear layers that are produced at the upstream of cavities are the source of sound due to its unsteadiness. The acoustic sound waves in turn impose acoustic pressure which then triggers the rolling up of the shear layer into vortices. These vortices detach periodically to form vortex shedding. According to Popescu et al. [1], a lock-in phenomena will occur where there is a resonance between longitudinal standing acoustic waves and vortex shedding when the frequency of vortex shedding coincides with the acoustic natural frequency of the pipe. The minimum fluid velocity for a lock-in to occur is known as onset velocity.

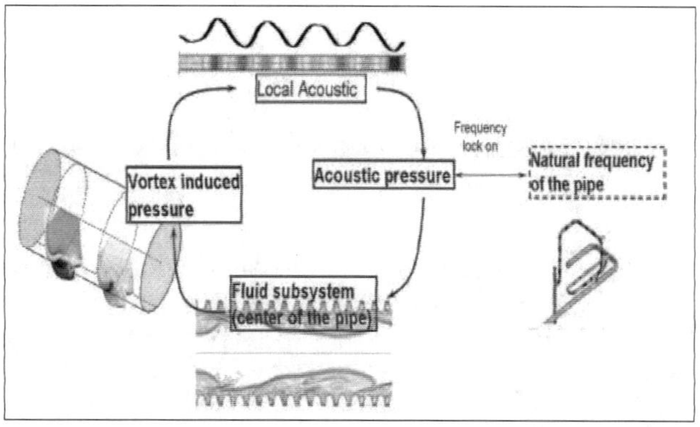

Figure 2. Feedback Loop between Hydrodynamic Subsystem and Acoustic Subsystem

As shown in figure 3[3], the pipe's inner diameter (D_p) is changing along the pipe length (L_p) in both systems periodically. In corrugated pipe, each cavity is in slit shape. Its width and depth are denoted as W and H respectively. The radiuses of cavities are named as r_{up} and r_{down} for upstream and downstream edges, respectively. Besides, the wave length of the variation of pipe diameter is known as pitch (Pt). Last but not least, the plateau (l) is referring to the part with fixed diameter between two consecutive cavities.

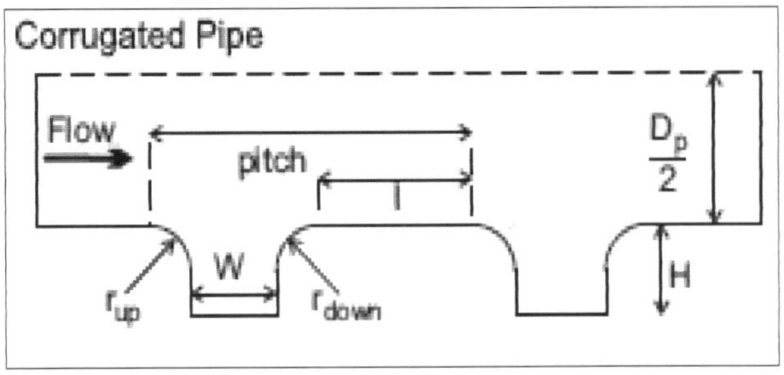

Figure 3. Geometric Characteristics of Corrugated Pipe and Multiple Side Branch System

An important dimensionless parameter that is normally used to characterise the oscillating flow is known as Strouhal number, *Sr*.

$$Sr = {fL_c}/{U} \quad (1)$$

where
f = oscillation frequency, Hz
L_c = characteristic length, m
U = average flow velocity in the pipe, m/s

According to NakiboĞLu et al. [2], as the average flow velocity increases, the frequency of whistling shows only slight increase in a particular resonant mode. As a result, the Strouhal number actually decreases at each mode and hence a range of Strouhal numbers will be noticed. The greatest value of Strouhal number in a specific mode is known as critical Strouhal number (Sr_{cr}). After that, the rise of average flow velocity increases the sound amplitude until it reaches a maximum value. The Strouhal number at peak sound amplitude at a particular acoustic mode is known as peak whistling Strouhal number (Sr_{p-w}). Nakiboğlu et al. [3] explained that Sr_{p-w} number is obtained through a plotting of global gradient of consecutive modes.

In this paper, the method used is first discussed. The effect of pipe geometries including cavity height (*H*), downstream edge radius (r_{down}), upstream edge radius (r_{up}), cavity width (w) and plateau (l) on the Sr_{p-w} and whistling amplitude are then investigated. In Section 3, the optimal characteristic length is identified among $w + r_{up}$, w and $\frac{l^2}{w+r_{up}}$. In the last section, conclusions are drawn on the findings from the simulation work.

2. Numerical Methodology

2.1 Method Used and Benchmarking

2.1.1 Method Used. Before a simulation is conducted by using Fluent, it is necessary to sketch and mesh the 2D model of single cavity corrugate pipe using Gambit 2.3 modelling software. The 2D model is then saved and imported into Fluent for simulation work. Large Eddy Simulation (LES) turbulence model is used as it is very suitable for aero-acoustic simulations. It resolves all eddies with scales larger than grid scale but eliminates small scales of the solution. This is good enough as large eddies possess most energy and interact more strongly with the flow. In the following section, all the results are obtained by using an air velocity of 13.61m/s unless otherwise specified.

There are two important considerations in order to read off the Sr_{p-w} from the graph obtained from Fluent. Firstly, when the lock-on frequency is changed, a system will go through a minimum of energy according Popescu et al. [1]. Therefore, the peak should only be read off after the first global minimum amplitude. Secondly, the Sr_{p-w} should only be referred to the first global peak after the first global minimum amplitude.

In the following sections, the characteristic length used to conduct the simulation is the summation of cavity width, w and upstream edge radius, r_{up}, $(w + r_{up})$. Nevertheless, 3 different types of characteristic lengths including $w + r_{up}$, w and $\frac{l^2}{w+r_{up}}$ are used in section 3 to determine the optimal characteristic length. In order to prevent confusion, Sr_{p-w} are represented in $Sr_{w+r_{up}}$, Sr_w and $Sr_{\frac{l^2}{w+r_{up}}}$ respectively, according to the characteristic length used to calculate the Strouhal number.

2.1.2 Benchmarking. In order to ensure the simulation results obtained from the Fluent software are reliable, several bench markings have been made by comparing the simulation results with the experimental work of Nakiboğlu et al. [3]. The simulation results are shown in the figure 4, 5 and 6 as sound pressure level vs Strouhal Number for sample Geo 2, Geo 4, and Geo 5 respectively and further compile as shown in table 1 with the comparison of experimental results of Sr_{p-w} by Nakiboğlu et al. [3] and the simulation results of Sr_{p-w}.

Table 1. Comparison of Experimental Results and Simulation Results of Sr_{p-w}.

Sample	W (mm)	H (mm)	r_{up} (mm)	r_{down} (mm)	l (mm)	Experimental Sr_{p-w}	Simulation Sr_{p-w}
Geo 2	4	4	2	2	4	0.36	0.34
Geo 4	4	4	2	2	0	0.38	0.35
Geo 5	4	4	2	2	8	0.40	0.40

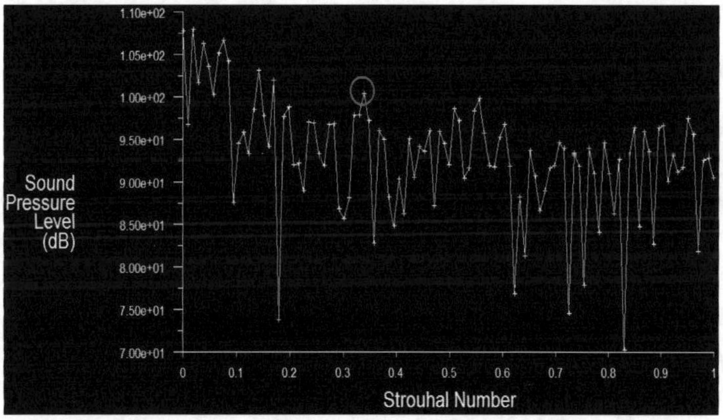

Figure 4. Sound Pressure Level as a Function of Strouhal Number of Geo 2 Corrugated Pipe

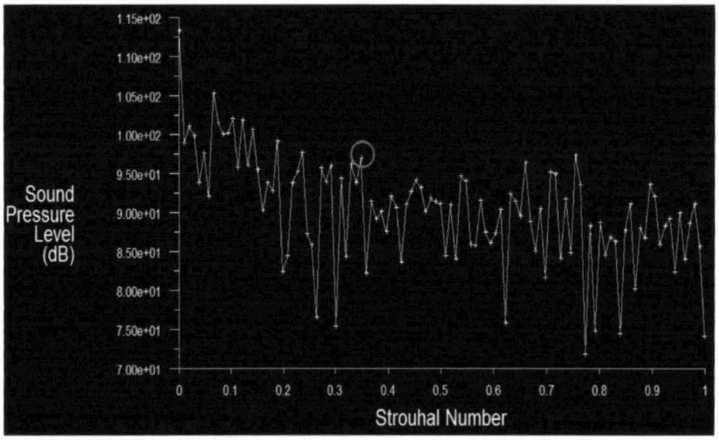

Figure 5. Sound Pressure Level as a Function of Strouhal Number of Geo 4 Corrugated Pipe

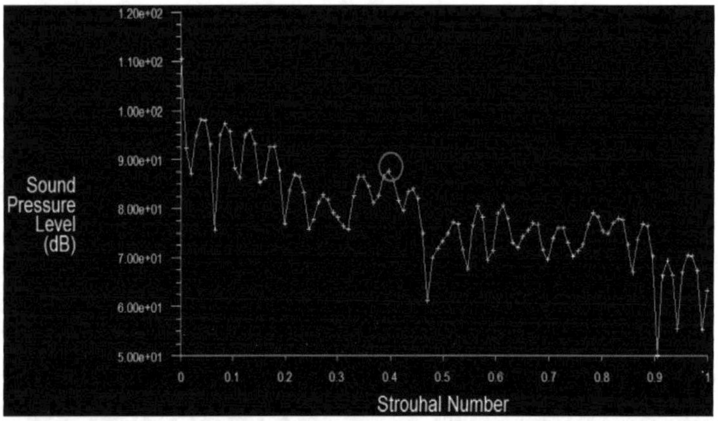

Figure 6. Sound Pressure Level as a Function of Strouhal Number of Geo 5 Corrugated Pipe

2.2 Effect of Different Pipe Geometries on $Sr_{p\text{-}w}$ and Whistling Amplitude

2.2.1 Effect of Cavity Depth, H. From the results obtained from simulation, it is found that the change of cavity depth, H from 4mm to 10mm has only minor effect on the $Sr_{w+r_{up}}$, which is in the range of 0.30 to 0.34. In addition, it has also insignificant effect on the whistling amplitude which only varies between 100dB to 102.5dB.

Table 2. $\frac{l}{w+r_{up}}$ ratio, $Sr_{w+r_{up}}$ whistling amplitude, at Different Cavity Depths.

w (mm)	H (mm)	r_{up} (mm)	r_{down} (mm)	l (mm)	Pitch (mm)	$\frac{l}{w+r_{up}}$	dB	$Sr_{w+r_{up}}$
4	4	2	2	4	12	2/3	100.0	0.34
4	6	2	2	4	12	2/3	102.5	0.34
4	8	2	2	4	12	2/3	102.5	0.30
4	10	2	2	4	12	2/3	102.5	0.30

This can be inferred from the vortex velocity profile in which all the four types of geometries have similar clockwise vortex velocity magnitude around 1.5m/s. Besides, these four geometries also showed approximately same thickness of boundary layer.

2.2.2 Effect of Downstream Edge Radius, r_{down}. The change of downstream edge radius, r_{down} from 2mm to 10mm has varied the $Sr_{w+r_{up}}$ within the range of 0.11 to 0.24, which is very minor. On the other hand, the change of r_{down} has also insignificant effect on the whistling amplitude. The largest difference of whistling amplitude is only 5.7dB. The results is agreed with the work of Nakiboğlu et al. [3] who explained that this is because the shear layer thickness has become less concentrated as the downstream edge is approached. This can also be explained by the small changes of clockwise vortex velocity magnitude from 1.43m/s to 1.52m/s as r_{down} is varied.

Table 3. $\frac{l}{w+r_{up}}$ ratio, $Sr_{w+r_{up}}$ whistling amplitude, at Different Downstream Edge Radius.

w (mm)	H (mm)	r_{up} (mm)	r_{down} (mm)	l (mm)	Pitch (mm)	$\frac{l}{w+r_{up}}$	dB	$Sr_{w+r_{up}}$
4	4	2	2	2	10	1/3	103.8	0.15
4	4	2	4	2	12	1/3	101.3	0.20
4	6	2	6	2	14	1/3	107.0	0.24
4	8	2	8	2	16	1/3	102.0	0.16
4	10	2	10	2	18	1/3	102.0	0.11

2.2.3 Effect of Upstream Edge Radius, r_{up}. From table 4, it is observed that as the upstream edge radius, r_{up} is increased from 2mm to 8mm, the $Sr_{w+r_{up}}$ is increased from 0.15 to 0.62. When r_{up} is further increased to 10mm, the $Sr_{w+r_{up}}$ is dropped to 0.42. Nothing can be concluded so far in terms of the effect of r_{up} to $Sr_{w+r_{up}}$. On the other hand, it is observed from the velocity profile that the boundary layer becomes thinner and the clockwise vortex becomes weaker as the r_{up} is increased. This can therefore explain why the whistling amplitude is decreased as the r_{up} is increased until a saturation is observed at r_{up} of 8mm. There is a reduction of 11.3dB when r_{up} is increased from 2mm to 8mm. The results are also in agreement with [3] who discovered that rounding of r_{up} is able to reduce the initial sound absorption and increase the whistling amplitude by a factor of 3 to 5.

Table 4. $\frac{l}{w+r_{up}}$ ratio, whistling amplitude, $Sr_{w+r_{up}}$ at Different Upstream Edge Radius.

w (mm)	H (mm)	r_{up} (mm)	r_{down} (mm)	l (mm)	Pitch (mm)	$\frac{l}{w+r_{up}}$	dB	$Sr_{w+r_{up}}$
4	4	2	2	2	10	1/3	103.8	0.15
4	4	4	2	2	12	1/4	97.0	0.37
4	6	6	2	2	14	1/5	95.0	0.57
4	8	8	2	2	16	1/6	92.5	0.62
4	10	10	2	2	18	1/7	93.0	0.42

2.2.4 Effect of Cavity Width, w. When the width of the cavity, W is changed, a similar observation is noted as to that of the change of r_{up}. There is no a clear relationship between w and $Sr_{w+r_{up}}$ because the $Sr_{w+r_{up}}$ fluctuates as the w is changed. On the other hand, although thicker boundary layer and

IOP Conf. Series: Materials Science and Engineering **458** (2018) 012040 doi:10.1088/1757-899X/458/1/012040

stronger clockwise vortex are observed, the whistling amplitude is reduced significantly as the w is increased. As seen in table 5, the whistling amplitude is reduced from 107.5dB to 83dB when w is increased from 2mm to 12mm. A closer investigation to the vortex profile reveals that a secondary counter clockwise vortex is formed at the upstream of the primary clockwise vortex when w is 6mm and 8mm. Besides, a tertiary clockwise vortex is also formed at the upstream of the secondary counterclockwise vortex when W is 10mm and 12mm. It is inferred that these secondary and tertiary vortices are contributing to the reduction of whistling amplitude through a "cancelling" effect to the primary vortex.

Table 5. $\frac{l}{w+r_{up}}$ ratio, whistling amplitude, $Sr_{w+r_{up}}$ at Different Cavity Widths.

w (mm)	H (mm)	r_{up} (mm)	r_{down} (mm)	l (mm)	Pitch (mm)	$\frac{l}{w+r_{up}}$	dB	$Sr_{w+r_{up}}$
2	4	2	2	2	8	1/2	107.5	0.30
4	4	2	2	2	10	1/3	103.8	0.15
6	4	2	2	2	12	1/4	92.5	0.31
8	4	2	2	2	14	1/5	89.0	0.58
10	4	2	2	2	16	1/6	84.0	0.57
12	4	2	2	2	18	1/7	83.0	0.44

2.2.5 Effect of Plateau, l. From table 6, it is found that the whistling amplitude does not show an explicit trend as plateau, l is changed. According to Nakiboğlu and Hirschberg [4], there is a shift of peaks as l is varied. In other words, a secondary peak will form and grow at certain plateau length to replace the first peak which diminishes in the process until it disappears. Furthermore, an important finding is observed from table 6. It is noted that the $Sr_{w+r_{up}}$ fluctuates when l is varied from 0mm to 6mm. Nevertheless, the $Sr_{w+r_{up}}$ becomes more constant as I is changed from 6mm to 10mm.

The value of 6mm is exactly the same as the summation of w and r_{up}, $(w + r_{up})$. Therefore, it is inferred that there is a relationship between the l and $(w + r_{up})$. In order to investigate this relationship, the ratio of $\frac{l}{w+r_{up}}$ is computed for every geometry from table 2 to table 5 and tabulated in figure 7.

Table 6. $\frac{l}{w+r_{up}}$ ratio, whistling amplitude, $Sr_{w+r_{up}}$ at Different Plateaus.

w (mm)	H (mm)	r_{up} (mm)	r_{down} (mm)	l (mm)	Pitch (mm)	$\frac{l}{w+r_{up}}$	dB	$Sr_{w+r_{up}}$
4	4	2	2	0	8	0	97.0	0.35
4	4	2	2	2	10	1/3	103.8	0.15
4	4	2	2	3	11	1/2	72.0	0.28
4	4	2	2	4	12	2/3	100.0	0.34
4	4	2	2	5	13	5/6	99.0	0.27
4	4	2	2	6	14	1	83.0	0.39
4	4	2	2	7	15	7/6	97.5	0.39
4	4	2	2	8	16	4/3	87.5	0.40
4	4	2	2	9	17	3/2	96.3	0.40
4	4	2	2	10	18	5/3	92.0	0.43

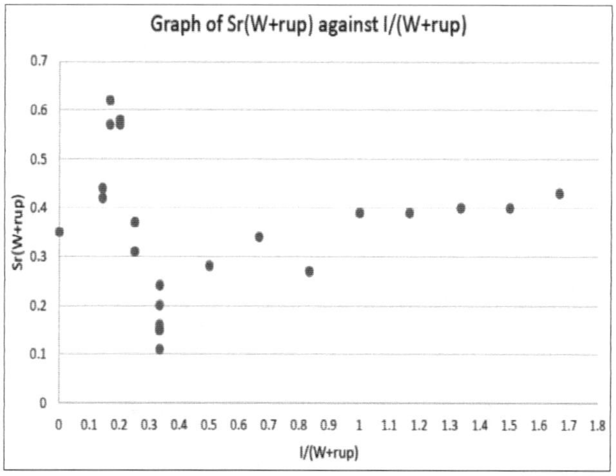

Figure 7. Graph of $Sr_{w+r_{up}}$ against $\frac{l}{w+r_{up}}$

From figure 7, it is observed that when $l < w + r_{up}$, the value of $Sr_{w+r_{up}}$ is fluctuating. $Sr_{w+r_{up}}$ is at a maximum value of 0.62 when $\frac{l}{w+r_{up}}$ is 0.1667 and at a minimum value of 0.11 to 0.24 when $\frac{l}{w+r_{up}}$ is 0.3333. There is a smaller fluctuation for$0.3333 < \frac{l}{w+r_{up}} < 1$.

Nevertheless, when $l \geq w + r_{up}$ or $\frac{l}{w+r_{up}} \geq 1$, the value of $Sr_{w+r_{up}}$ is far more constant in the range of 0.39 to 0.43. A conclusion that can be made at this point is that when $l < w + r_{up}$, w and r_{up} have dominant effect over l in affecting the flow over corrugated pipe, therefore value of $Sr_{w+r_{up}}$ is fluctuating. However, when $l \geq w + r_{up}$, l is dominant over w and r_{up}. The change of $Sr_{w+r_{up}}$ in this range is insignificant, which means that l does not contribute much to the change of $Sr_{w+r_{up}}$ as compared to w and r_{up} as discussed earlier.

Although the ratio of $\frac{l}{w+r_{up}}$ is a good indicator of the consistency of $Sr_{w+r_{up}}$, it cannot be used to infer the value of $Sr_{w+r_{up}}$. In other words, even though two corrugated geometries have the same ratio of $\frac{l}{w+r_{up}}$, they do not necessary have the same value of $Sr_{w+r_{up}}$. This can be seen by comparing table 6 and 7. Although some of the geometries in two tables have the same ratio of $\frac{l}{w+r_{up}}$, their $Sr_{w+r_{up}}$ are significantly different.

Table 7. $\frac{l}{w+r_{up}}$ ratio, whistling amplitude, $Sr_{w+r_{up}}$ at Different Geometries.

w (mm)	H (mm)	r_{up} (mm)	r_{down} (mm)	l (mm)	Pitch (mm)	$\frac{l}{w+r_{up}}$	dB	$Sr_{w+r_{up}}$
11	6	6	2	10	42	10/17	88.8	0.67
11	4	4	2	10	27	2/3	98.8	0.28
6	4	4	2	10	22	1	97.0	0.21
4.5	4	3	2	10	19.5	4/3	104.0	0.22
4.5	4	1.5	2	10	18	5/3	94.0	0.17

It should be noted that table 7 also shows a similar phenomenon as mentioned earlier, where a consistency of value of $Sr_{w+r_{up}}$ is observed when $l \geq w + r_{up}$ or $\frac{l}{w+r_{up}} \geq 1$.

3. Optimal characteristic Length

Numerous works have been done previously to determine the optimal characteristic length and the sum of the width of cavity and the upstream edge radius, $w + r_{up}$, which is also known as modified gap width, is found to be the best representation of characteristic length. Since the ratio of $\frac{l}{w+r_{up}}$ is found to be a good indicator of the consistency of $Sr_{w+r_{up}}$, it is therefore possible to use it as a characteristic length in calculating the Strouhal number provided it can give a lesser variation of Sr_{p-w} than the use of $w + r_{up}$ as characteristic length.

Nevertheless, $\frac{l}{w+r_{up}}$ is only a ratio but not a length. It cannot be used as a characteristic length unless it has a unit of meter. In order to make it a characteristic length with unit of meter, $\frac{l^2}{w+r_{up}}$ is proposed in this project. This parameter is verified to be a good indicator of the consistency of $Sr_{w+r_{up}}$ as a similar pattern in figure 7 is also observed in figure 8, in which the $Sr_{w+r_{up}}$ is plotted against $\frac{l^2}{w+r_{up}}$.

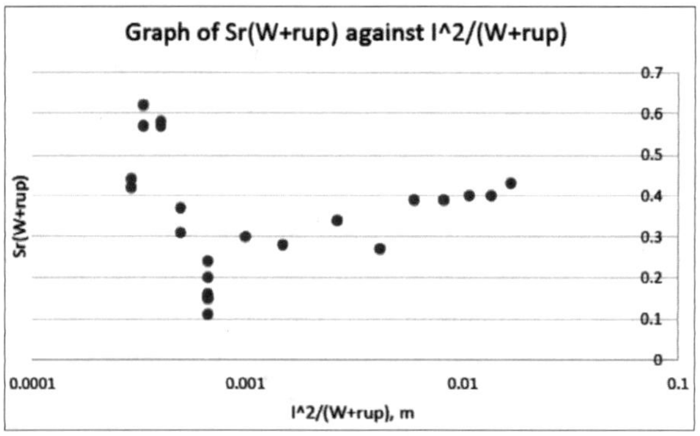

Figure 8. Graph of $Sr_{w+r_{up}}$ against $\frac{l^2}{w+r_{up}}$

Furthermore, w is also used as a characteristic length in this paper and the results of Sr_{p-w} obtained by using air velocity from 11.61m/s to 21.61m/s for three different types of characteristic length are tabulated in table 8.

Table 8. $Sr_{w+r_{up}}$, Sr_w and $Sr_{\frac{l^2}{w+r_{up}}}$ at Different Geometries.

w (mm)	H (mm)	r_{up} (mm)	r_{down} (mm)	l (mm)	Pitch (mm)	Velocity (m/s)	$Sr_{w+r_{up}}$	Sr_w	$Sr_{\frac{l^2}{w+r_{up}}}$
4	4	2	2	4	12	11.61	0.38	0.29	0.10
4	4	2	2	4	12	13.61	0.34	0.25	0.05
4	4	2	2	4	12	15.61	0.30	0.20	0.13
4	4	2	2	4	12	17.61	0.30	0.09	0.06
4	4	2	2	4	12	19.61	0.43	0.17	0.08
4	4	2	2	4	12	21.61	0.30	0.08	0.05

From table 8, it is observed that the use of $w + r_{up}$ as the characteristic length results in variation of Strouhal number of $0.3 \leq Sr_{w+r_{up}} \leq 0.43$. The use of w as the characteristic length has led to larger variation of Strouhal number in the range of $0.09 \leq Sr_w \leq 0.29$. The smallest variation in Strouhal number is observed when $\frac{l^2}{w+r_{up}}$ is used as the characteristic length, which is limited to $0.05 \leq Sr_{\frac{l^2}{w+r_{up}}} \leq 0.13$.

It is therefore concluded that $\frac{l^2}{w+r_{up}}$ is the optimal choice of characteristic length as the use of $\frac{l^2}{w+r_{up}}$ in the calculation of Strouhal number led to smallest variation. This finding is significant as it can allow a better design of flow condition to minimize and avoid the whistling phenomenon. The equation 1 can be rearranged to obtain the following formula:

$$U_{cr} = \frac{f_n L_c}{Sr_{p-w}} \quad (2)$$

where
f_n = acoustic natural frequency of the pipe, Hz
L_c = characteristic length, m
U_{cr} = critical flow velocity in the pipe, m/s
$Sr_{p\text{-}w}$ = Strouhal number

This formula can allow us to predict the critical flow condition in which whistling phenomenon will occur and therefore avoid it. The value of Sr_{p-w} can be obtained from the data obtained from simulations when a particular value of characteristic length $\frac{l^2}{w+r_{up}}$ is selected. The acoustic natural frequency of the pipe, f_n can then be calculated by using Cumming's formula as proposed in Kristiansen and Wiik [5] as follow:

$$f_n = 0.89\frac{nc}{2L\prime} \quad (3)$$

where
n = mode number (1, 2, 3, ….)
c = speed of sound, m/s
L'= effective pipe length, m

When the critical flow velocity, U_{cr} has been determined, it can be easily avoided by manipulating the fluid flow velocity. It is also possible to select or design the parameters of the corrugated pipe if the flow velocity cannot be manipulated.

For instance, given a 2m long corrugated pipe and speed of sound of 340.29m/s, the second mode of acoustic natural frequency is calculated to be 151.43Hz by using equation 3. If the is found to be 0.1 from the simulation for the pipe with $\frac{l^2}{w+r_{up}}$ of 0.005, the U_{cr} can then be determined to be around 7.57m/s from equation 2. In order to avoid whistling phenomenon, the flow condition should be much larger or smaller than this value. It is important to note that other modes of resonance frequency should also be calculated and avoided accordingly. With the proposed method, the whistling phenomenon can be overcome effectively.

In the cases where flow velocity is constant, equation 2 can also be used. For instance, if a 2m corrugated pipe with l = 8mm, w = 10mm and r_{up} = 2.8mm with second mode of acoustic natural frequency of 151.43Hz (calculated using equation 3) is selected for a flow condition of 8m/s, its $Sr_{\frac{l^2}{w+r_{up}}}$ can be determined with equation 2 to be around 0.0946. This value is then compared with simulation method by using $\frac{l^2}{w+r_{up}}$ of 0.005. If the values of $Sr_{\frac{l^2}{w+r_{up}}}$ is different between calculated value and simulation results, this particular design of corrugated pipe is suitable to be used for the flow condition. However, a different pipe should be re-selected shall the values of $Sr_{\frac{l^2}{w+r_{up}}}$ is similar or same between calculated value and simulation results. With the proposed method, the whistling phenomenon can be overcome effectively.

It should also be noted that the simulation method proposed in this project is able to contribute effectively in the research of flow induced acoustics in corrugated pipe. This is because the conduction of simulation is much simpler than running an experiment. Different models of corrugated pipe can be easily drawn with few minutes and simulation results can be obtained easily with high accuracy. Hence, time can be saved and the progress of research can be accelerated.

Furthermore, the use of simulation method is cheaper than the use of experimental method. This is because simulation method can provide a wide range of flow conditions and information without the need of costly experimental setup provided the initial conditions and boundary conditions are specified correctly. This can also reduce waste of material since different configurations of corrugated pipe are not required.

Last but not least, large amount of space is not required by using simulation method since no equipment are needed. This can save the cost indirectly while not compromising the accuracy of the results.

4. Conclusion

In this paper, the effects of various corrugated pipe geometries on the Sr_{p-w} and whistling amplitude are simulated using Fluent software. The simulation method is first compared with the experimental results from Nakiboğlu et al. [3] to ensure the reliability of the results obtained. The 2D numerical simulation method is found to be accurate in determining the Sr_{p-w} and whistling amplitude.

The change of cavity depth, Hand downstream edge radius, r_{down} are found to have insignificant effect on the Sr_{p-w} and whistling amplitude. It is found that the change of cavity depth, H from 4mm to 10mm only changes $Sr_{w+r_{up}}$ from 0.30 to 0.34. In addition, it has also insignificant effect on the whistling amplitude which only varies between 100dB to 102.5dB. The change of downstream edge radius, r_{down} from 2mm to 10mm has varied the $Sr_{w+r_{up}}$ within the range of 0.11 to 0.24, which is very minor. On the other hand, the change of r_{down} does not affect the whistling amplitude greatly. The largest difference of whistling amplitude is only 5.7dB.

However, the increase of upstream edge radius, r_{up} and cavity width, w lead to a fluctuation of Sr_{p-w} and a reduction of whistling amplitude. Whistling amplitude is dropped from 103.8dB to 93dB when r_{up} is increased from 2mm to 10mm. Besides, whistling amplitude is reduced from 107.5dB to 83dB when W is increased from 2mm to 12mm.

The strongest finding is that when $l < w + r_{up}$, the value of $Sr_{w+r_{up}}$ is fluctuating. $Sr_{w+r_{up}}$ is at a maximum value of 0.62 when $\frac{l}{w+r_{up}}$ is 0.1667 and at a minimum value of 0.11 to 0.24 when $\frac{l}{w+r_{up}}$ is 0.3333. There is a smaller fluctuation for$0.3333 < \frac{l}{w+r_{up}} < 1$.

Nevertheless, when $l \geq w + r_{up}$ or$\frac{l}{w+r_{up}} \geq 1$, the value of $Sr_{w+r_{up}}$ is far more constant in the range of 0.39 to 0.43. It can be concluded that when $l < w + r_{up}$, w and r_{up} have dominant effect over l in affecting the flow over corrugated pipe, therefore value of $Sr_{w+r_{up}}$ is fluctuating. However, when $l \geq w + r_{up}$, l is dominant over w and r_{up}. The change of $Sr_{w+r_{up}}$ in this range is insignificant, which means that l does not contribute much to the change of $Sr_{w+r_{up}}$ as compared to w and r_{up} as discussed earlier.

Since the ratio of $\frac{l}{w+r_{up}}$ is found to be a good indicator of the consistency of $Sr_{w+r_{up}}$, it is therefore possible to use it as a characteristic length in calculating the Strouhal number. Nevertheless, $\frac{l}{w+r_{up}}$ is only a ratio but not a length. It cannot be used as a characteristic length unless it has a unit of meter. In order to make it a characteristic length with unit of meter, $\frac{l^2}{w+r_{up}}$ is proposed in this project.

Lastly, it is found that the use of $w + r_{up}$ as the characteristic length results in variation of Strouhal number of $0.30 \leq Sr_{w+r_{up}} \leq 0.43$. The use of W as the characteristic length has led to larger variation of Strouhal number in the range of $0.09 \leq Sr_w \leq 0.29$. The smallest variation in Strouhal number is observed when $\frac{l^2}{w+r_{up}}$ is used as the characteristic length, which is limited to$0.05 \leq Sr_{\frac{l^2}{w+r_{up}}} \leq 0.13$. It is therefore concluded that $\frac{l^2}{w+r_{up}}$ is the optimal choice of characteristic length.

This finding is significant as it can allow us to avoid critical flow condition to minimize and prevent the whistling phenomenon more accurately and effectively by using equation 2 and 3. It is also possible to design the pipe parameters if the flow condition cannot be manipulated in certain conditions. With the proposed method, the whistling phenomenon can be overcome effectively.

Lastly, the simulation method proposed in this project is able to contribute effectively in the research of flow induced acoustics in corrugated pipe. This is because the simulation method is able to save cost, time and space compare to experimental method while not compromising the accuracy of the results obtained. In other words, large amount of information can be obtained by using simulation methods which are especially useful for the researchers in understanding the science and arts behind this topic of interest. There have been multiple proposals on the mitigation of the flow-induced acoustics in corrugated pipe through the smoothing the internal surface of the corrugated pipe or by introducing liquid droplet into the flow [11]. However, these method still not feasible for commercial practice.

References

[1] Popescu M, Johansen S T and Shyy W 2011 Flow-Induced Acoustics in Corrugated Pipes *Comm. in Comp. Phy.* **10** 120-139

[2] Nakiboğlu G, Belfroid S P C, Golliard J and Hirschberg A 2011 On the whistling of corrugated pipes: effect of pipe length and flow profile *Journal of Fluid Mechanics* **672** 78-108

[3] Nakiboğlu G, Belfroid S P C, Willems J F H and Hirschberg A 2010 Whistling behavior of periodic systems: Corrugated pipes and multiple side branch system *Int. J. of Mech. Sci.* **52** 1458-1470

[4] Nakiboğlu G. and Hirschberg A 2012 Aeroacoustic power generated by multiple compact axisymmetric cavities: effect of hydrodynamic interference on the sound production *Physics of Fluids* **24** 067101

[5] Kristiansen U R and Wiik G A 2007 Experiments on sound generation in corrugated pipes with flow *The Journal of the Acoustical Society of America* **121** 1337–1344

[6] Gao Y, Zong Z and Sun L 2011 Numerical prediction of fatigue damage in steel catenary riser due to vortex-induced vibration *Journal of Hydrodynamics* **23** 154-163

[7] Martinez-Lera P, Golliard J and Schram C 2010 Numerical study of flow-induced pulsations in pipe systems with closed branches *16th AIAA/CEAS Aeroacoustics Conference* (Stockholm, Sweden)

[8] Nakiboğlu G, Manders H B M and Hirschberg A 2012 Aeroacoustic power generated by a compact axisymmetric cavity: prediction of self-sustained oscillation and influence of the depth *Journal of Fluid Mechamics* **703** 163-191

[9] Nakiboğlu G, Rudenko O and Hirschberg A 2012 Aeroacoustics of the swinging corrugated tube: Voice of the Dragon *Journal of the Acoust. Soc. of America* **131** 749-65

[10] Tonon D, Landry B J T, Belfroid S P C, Willems J F H, Hofmans G C J and Hirschberg A 2010 Whistling of a pipe system with multiple side branches: Comparison with corrugated pipes *Journal of Sound and Vibration* **329** 1007-1024

[11] Andries V, Eckeveld, Jerry W and Christian P 2017 Flow-induced noise in corrugated pipes - why does liquid reduce whistling? *23rd AIAA/CEAS Aeroacoustics Conference* (Denver, Colorado)

IOP Conf. Series: Materials Science and Engineering **458** (2018) 012041 doi:10.1088/1757-899X/458/1/012041

Development of low-cost and user-friendly sustainable portable particulate sensor

M A Fathihah[1], M P Khairunnisa[1], M Rashid[1], J NorRuwaida[1], M Dewika[2], Y Ito[3] and I Wuled Lenggoro[3,4]

[1] Air Resources Research Laboratory, Malaysia Japan International Institute of Technology, 54100 Universiti Teknologi Malaysia Kuala Lumpur, Malaysia
[2] Centre of American Education, Sunway University 47500 Subang Jaya, Selangor, Malaysia
[3] Graduate School of Bio-Application and System Engineering Tokyo University of Agricultural and Technology, 2-24-16 Nakacho, Koganei, Tokyo 184-8588, Japan
[4]Department of Chemical Engineering, Tokyo University of Agriculture and Technology, 2-24-16 Nakacho, Koganei, Tokyo 184-8588, Japan

E-mail: khairunnisa.kl@utm.my

Abstract. Particulate Matter (PM) has various effects on the quality of air, health, and environment. A newly developed on-line PM_{10} (particulate matter less than 10 µm) measurement using Arduino as the microcontroller which has multi-connections was able to provide real-time measurement. Shinyei dust sensor (PPD42NS) is used to measure the concentration of the particulate matter at ambient air along with temperature and humidity sensor (DHT22). The aims of the study are to build a portable and user-friendly device for the monitoring of PM_{10} as a low-cost and sustainable particulate sensor. The device was tested for five days to determine the particulate concentration indoor and outdoor at one site in Dungun, Terengganu. The results showed that the average of the particulate concentration at the indoor and outdoor for five days were 329 and 1587 µg/m^3 respectively. The custom-made device can be treated as a low-cost measuring device for particulate matter in ambient air which will be further refined and verified.

1. Introduction

The quality of air can be determined by the concentration of particulate matter (PM) in the atmosphere. According to United Environmental Protection Agency (USEPA), the particulate matter is the mixture of all solid and liquid particles suspended in air, which are hazardous [1]. There has two categorizes of PM that considered by USEPA which are $PM_{2.5}$ (with a diameter range of 2.5 - 0.1 µm) and PM_{10} (with a diameter range of 10 - 2.5 µm).

The microscopic solid and liquid droplets are so small that they can be inhaled and cause serious health problems to human health. The detrimental effects of respirable suspended particulate (RSP) on human health and the relationship between air pollutants and health has been widely studied. The human lung can regularly be exposed to a variety of particles including those from cigarette smoke, ceramic industry [2], burning of biomass, dust storms, mining or processing of coal [3] and mineral oxides, agricultural work, wildfires, environmental tobacco smoke, emissions from traffic, and gas and wood stoves [4].

Not only that, particulate matter can cause atmospheric visibility impairment. It gives an impact on global climate change [5]. Haze is one of the major concern when it comes to global warming. This is because of the large amount of particulate that suspended in air exert a significant cooling effect on the earth's climate. $PM_{2.5}$ is the major contributors to this problem. Hence, air monitoring is necessary in order to resolve these issues.

The advanced instruments such as (scanning mobility particle spectrometer (SMPS), condensation particle counter (CPC) and electric low pressure impactor (ELPI) are categorized in air monitoring research. These instruments helped the researchers to obtain the concentration of the particulate by the light scattering principle on-site in real time. The majority of these sensors work on the light scattering principle to determine the concentration of the particulate matter [6].

Advanced instruments are precise, suitable to monitor various properties of PM, and can be also used for mobile monitoring. However, these instruments cannot be applied in large monitoring networks mainly due to their high cost [7]. Currently, the low-cost sensors can measure the number of PM features and suitable for monitoring. This concern has triggered the idea that the development of the low cost and portable sensor is necessary for monitoring the air quality. Therefore, the most recent solution targeting micron-sized particulates are presented in this study.

2. Experimental methods and materials

2.1. Design and specification

2.1.1. Arduino Platform. The device is using open-source hardware platform, Arduino [8] as illustrated in Figure 1 (a) as the source to read the measurement data of the particulate concentration as well as humidity level from the sensor. According to Ali et al., the Arduino platform provides various development circuit boards, some of which utilize Atmel's low-power CMOS 8-bit microcontrollers based on AVR enhanced RISC architecture [9]. These microcontrollers are capable of computing approximately 300,000 lines of program code per second, which is more than sufficient for most of the input and output applications required for the time scales of typical building data collection (e.g., seconds or minutes) [10]. Arduino provides an integrated development environment (IDE) that is capable of running on all major operating systems and has support for a simplified C/C++ programming language.

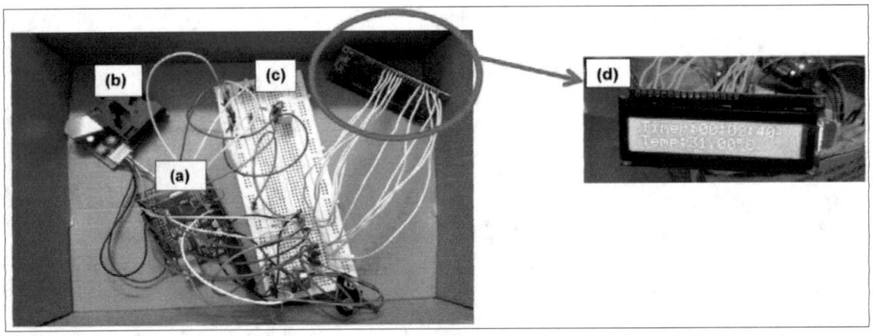

Figure 1. Hardware setup of monitoring environment (a) Arduino (b) Shinyei PPD42NS (c) Humidity and Temperature DHT22 and (d) 16x2 LCD

2.1.2. Shinyei PPD42NS. The particulate sensor unit, Shinyei PPD42NS [11] as shown in Figure 1 (b) is built based on the light scattering method. This sensor detects airborne particulate continuously after turning on. The original detection method is similar to the particle counter that can be obtained the pulse output that corresponds to concentration per unit volume [11]. The low-cost optical sensors are counting the amount of time that particles are detected by the light receptor sensor [12]. The detectable size is more than 1.0 µm. The schematic diagram of this hardware is depicted in the Figure 2 (a) and (b). Updraft airflow is generated by heater resistor. The infrared light beam from light emitter (LED) is focused with a lens to the sensing point at the center. airborne particulate inflow into the sensor box and pass through the sensing point [13]. Light receptor (Photo IC) receives scattered light through the lens and transforms it into a pulse signal. The size of particle will be detected by the signals resulting from the detection of scattered light are passed through filtering and amplification circuitry that are externally visible on the PPD42NS, resulting in 0 – 5 V pulses of approximately 10 – 100 m in length [11].

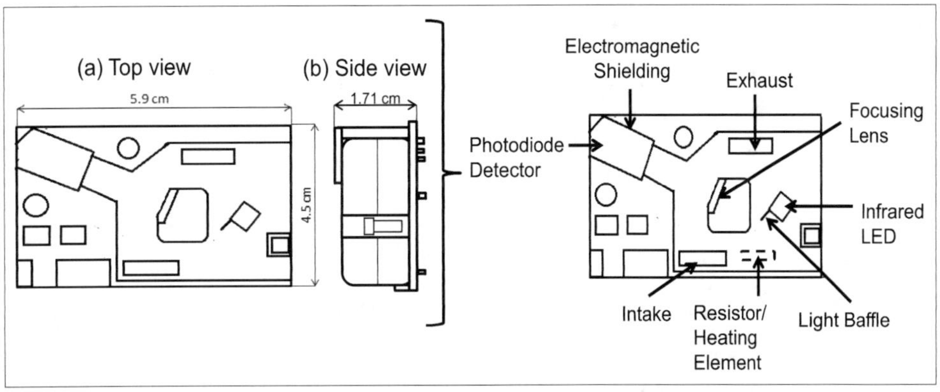

Figure 2. The schematic diagram of Shinyei PPD42NS (a) front view and (b) side view

2.1.3. Humidity and Temperature Sensor. The temperature and humidity sensor (DHT22) was added to the Arduino platform as visualized in Figure 1 (c) to measure the temperature and the relative humidity of the ambient air. Apart from that, this sensor was used to analyse the effects of temperature and humidity on the performance PPD42NS sensor by ensuring the instruments remained undisturbed, upright and unexposed to extreme conditions. DHT22 consists of two electrodes with moisture substrate within them. As the conductivity of the substrate or resistance changes, the humidity will change. For temperature sensor, DHT22 provide the "Negative Temperature Coefficient (NTC)" sensor or known as thermistor that made by sintering of semi-conductive material such as ceramics or polymers which gave the bigger changes in resistance with small changes in temperature [14].

2.2. Study location

The investigation involved analyzing the particulate concentration as well as the relative humidity and temperature. The site was selected in Dungun area as depicted in Figure 3, which is located in the east of Malaysia. The device is exposed to indoor and outdoor for 1 hour in the morning for five days respectively. Dungun is nearer to the South China Sea, made it a suitable place to consider as having the high wind movement. The device was attempted at the end of the January (29th January 2018) until early of the February (9th February 2018).

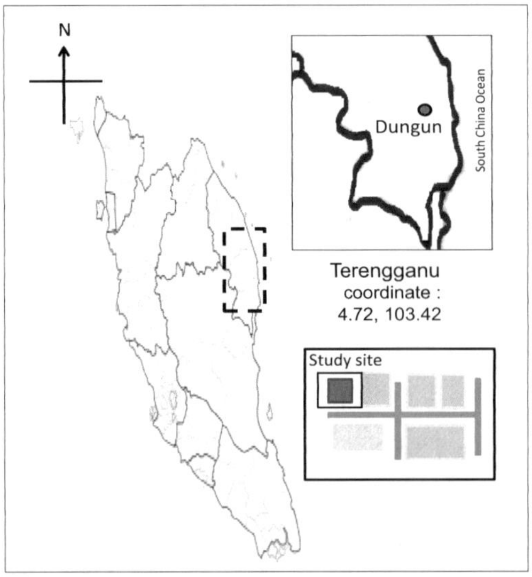

Figure 3. Site map for particulate monitoring in Dungun, Terengganu

Shinyei PPD42NS sensor can measure the PM_{10} by the photodiode inside it. While passing through the sensor, particulates scatter light and the intensity of the light received by the phototransistor is directly correlated with the concentration of particulates. The indoor location was done in the house at a fixed height (61 cm) to ensure the region does not change for the investigation. For outdoor, the height is set at 44 cm outside the house due to some limitations. The data logging in COM6 port of Arduino was saved into the Excel document for further analysis.

3. Results & Discussion

3.1. Device Operating

The device is operating by using Arduino as the microcontroller to activate the sensor. The device is able to monitor the mass concentration of the particulates as shown in Figure 4 and Figure 5 for each study location. It also compatible with the feature of the humidity and temperature sensor to measure the temperature and the relative humidity of the study site as illustrated in Figure 7. The liquid crystal display (LCD) is attached to the device in order to display the particulates concentration with the humidity and temperature (Figure 1 (d)). The device was easy to operate, easy to implement and user-friendly which is considered as a green technology that can contribute towards the sustainable environment.

IOP Conf. Series: Materials Science and Engineering **458** (2018) 012041 doi:10.1088/1757-899X/458/1/012041

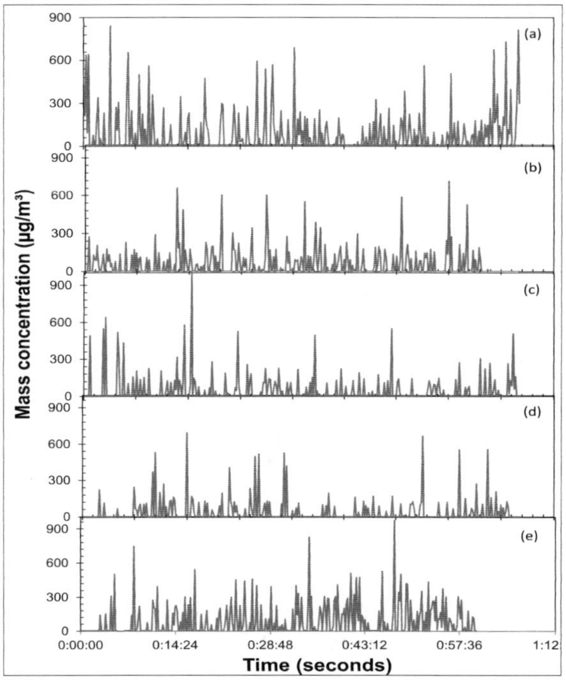

Figure 4. The mass concentration of the PM_{10} for indoor location in five days at 7am-8am (a) 29th January 2018, (b) 30th January 2018, (c) 31st January 2018, (d) 1st February 2018 and (e) 2nd February 2018

3.2. PM concentrations versus time

The study found that the average concentration of PM_{10} value depicted lower value when tested indoor (329 µg/m) as visualized in Figure 4 as compared to outdoor (1587 µg/m3) as shown in Figure 5 in the morning at consecutive five days. The particulate concentrations for PM_{10} were taken for one hour in the morning (7 am-8 am) for total five days at two different locations, respectively. Figure 4 (a),(b),(c),(d),(e) visualized the PM_{10} concentration sensed by the dust sensor. The mass concentration of PM_{10} for five days were in ranged below than 900 µg/m^3. The concentration was quite high because the sensor was quite senstive. The study was highlighted that the low-cost optical sensors are counting the amount of time that particles are detected by the light receptor sensor instead of counting the individual particles like the high-cost sensors. As a result, the data obtained was quite noisy as compared to expensive sensors which lead to quality of near-real-time data is reduced [12]. The high PM_{10} concentration in the morning also affected by the particulate that came from the outdoor because the door keep opened and closed within this period as mentioned by Diapouli *et al.*, [15].

Figure 5 (a), (b), (c), (d), (e) depicted the particulate concentration for the outdoor area. The mass concentration of PM_{10} for five days were in ranged below than 6000 µm. The concentration PM_{10} was higher in the morning because the collection data was taken on a weekday which was the peak hour for people went to work as the emission from the exhaust car might be influenced by high in PM_{10} concentration [16]. This phenomenon also was discussed by Lawrence *et al* about the particulate from exhaust vehicle might be contributed to the high concentration of PM_{10} as the sensor was located in front of the car [17].

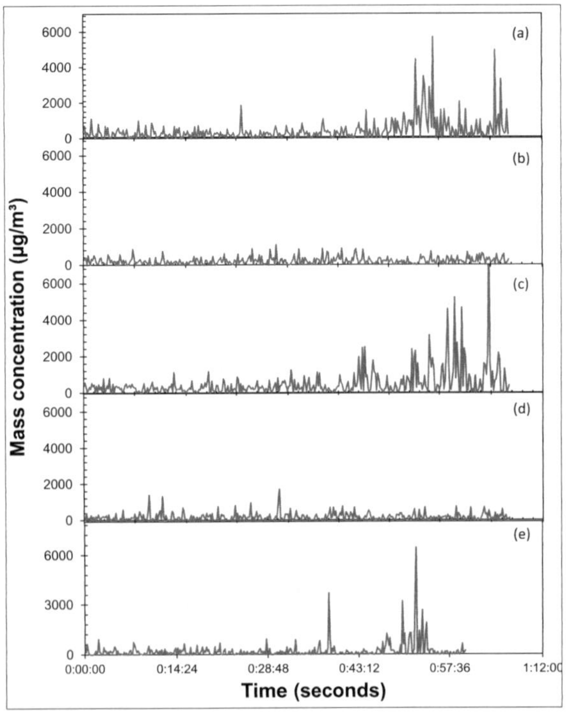

Figure 5. The mass concentration of the PM_{10} for outdoor location in five days at 7am-8am on (a) 4th February 2018, (b) 5th February 2018, (c) 6th February 2018, (d) 8th February 2018 and (e) 9th February 2018

The back-trajectories analysis was done using Meteorological Data Explorer (METEX) [18] developed by Centre for Global Environment Research (CGER) in Japan to calculate the air trajectories within one week of the outdoor experiment. Figure 6 visualized the particulate trajectories for one week in Dungun site. It showed that the particulate originated came from the east region which was contributed to the high concentration of the PM_{10} within that week. Malaysia experiences northeast monsoon in November to March which dominated to meteorological conditions such as wind and rainfall [19]. There were event of thick fog due to cold weather and turbulent sea coming to Dungun site during that period of time [20]. Fog was a liquid-particles aerosol formed by condensation with sizes ranging from submicrometer to about 200 µm, one of the sources that contributed to the high concentration of PM_{10} [21].

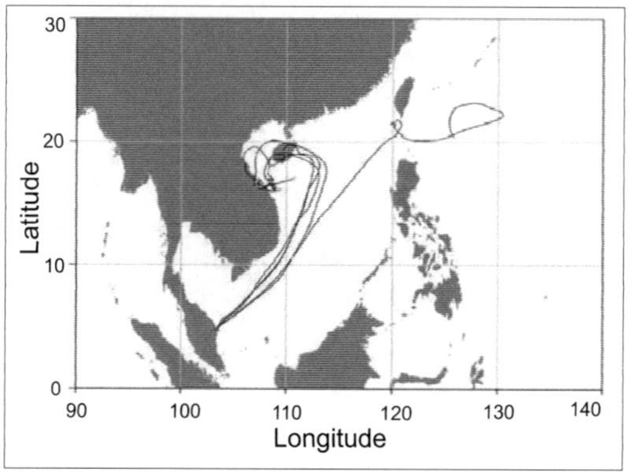

Figure 6. Back-trajectories analysis for five days from 4th February 2018 to 9th February 2018 at 7 am (100 m).

3.3. Hourly temperature and relative humidity

The device was located in the living room for an indoor while, for outdoor was in front of the house's door. The intake of temperature and humidity level was done parallel to the reading of the particulate concentration. Figure 7 depicted the mean daily temperatures trended from approximately 20 to 30 °C while the relative humidity indoor and outdoor ranged between 60-80 % levels. The humidity and temperature level was influenced by the warm and humid air in Malaysia [19]. As the temperature decrease, the air could not hold more water vapour as compared when the temperature is rise, the humidity is increase. This phenomena was explained by Nguyen *et al.*, which was the differences in thermal preferences, levels of temperature control, socioeconomic status, and the health states of the homes' occupants were likely contribute to this variability [22]. The result is compared with the humidity and temperature index showed from the Meteorology of Malaysia to observe the variation between the theoretical and experimental result with diurnal variations of approximately ± 5 °C and ± 1 % respectively.

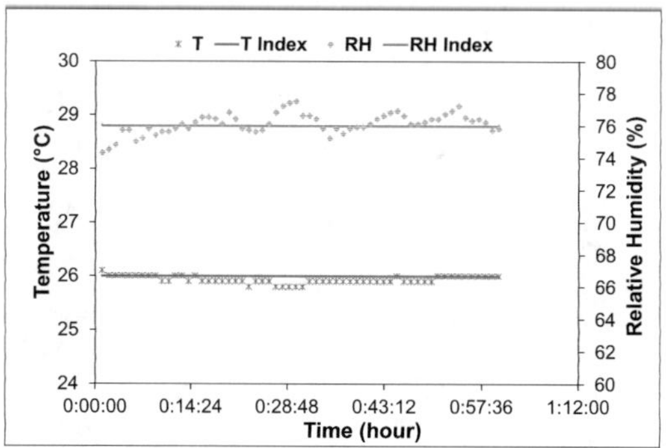

Figure 7. The relative humidity and the temperature outdoor at 7 am – 8 am outdoor (4th February 2018) in Dungun, Terengganu.

4. Conclusion

The development of a low cost and custom-made particulate sensor has been tested and reported in this paper. The device has the ability to measure the mass concentration of PM_{10} along with the humidity and temperature at selected sampling sites. It could be implemented as a simple and user-friendly device for monitoring daily air quality which is easy to operate in a sustainable manner.

References

[1] USEPA. *Particulate Matter (PM) Basics*. 2018 [cited 2017 11 November]; Available from: https://www.epa.gov/pm-pollution/particulate-matter-pm-basics#PM.

[2] Sintorini M M 2018 The impact of total suspended particulate concentration on workers' health at ceramic industry *IOP Conference Series: Earth and Environmental Science* **106** 012034.

[3] Petra P, Jan H and Jan B 2016 Impact of Mining Activities on the Air Quality in The Village Nearby a Coal Strip Mine *IOP Conference Series: Earth and Environmental Science* **44** 032021.

[4] Ghio A J, Soukup J M and Dailey L A 2016 Air pollution particles and iron homeostasis *Biochimica et Biophysica Acta (BBA)* **1860** 2816-2825.

[5] Tai A P K, Mickley L J and Jacob D J 2010 Correlations between fine particulate matter (PM2.5) and meteorological variables in the United States: Implications for the sensitivity of PM2.5 to climate change *Atmos. Env.* **44** 3976-3984.

[6] Chiang C 2017 Design of a High-Sensitivity Ambient Particulate Matter 2.5 Particle Detector for Personal Exposure Monitoring Devices *IEEE Sensors Journal* **18** 165-169.

[7] Gozzi F, Della G Ventura and Marcelli A 2016 Mobile monitoring of particulate matter: State of art and perspectives *Atmos. Pol. Res.* **7** 228-234.

[8] Arduino. *Arduino*. 2018 [cited 2017 12 October]; Available from: https://www.arduino.cc.

[9] Ali A S, Zanzinger Z, Debose D and Stephens B 2016 Open Source Building Science Sensors (OSBSS): A low-cost Arduino-based platform for long-term indoor environmental data collection. *Building and Environment* **100** 114-126.

[10] Ferdoush S and Li X 2014 Wireless Sensor Network System Design Using Raspberry Pi and Arduino for *Env. Mon. App. Pro. Comp. Sci.* **34** 103-110.

[11] Shinyei. *PPD42NJ Particle Sensor Unit Detects airborne particles | Shinyei Technology*. 2017 [cited 2017 October 20]; Available from: http://www.shinyei.co.jp/stc/eng/optical/main_ppd42.html.

[12] The World Air Quality Index. *The Shinyei experiment*. 2017 [cited 2018 10 February]; Available from: http://aqicn.org/sensor/shinyei/.

[13] Liu D, Zhang Q, Jiang J and Chen D R 2017 Performance calibration of low-cost and portable particular matter (PM) sensors *Journal of Aerosol Science* **112** 1-10.

[14] Bogdan M 2016 How to Use the DHT22 Sensor for Measuring Temperature and Humidity with the Arduino Board *ACTA Universitatis Cibiniensis* **68** 22.

[15] Diapouli E, Chaloulakou A and Spyrellis N 2008 Indoor and Outdoor PM Concentrations at a Residential Environment Athens Area *Global Nest Journal* **10** 201-208.

[16] Giechaskiel B, Maricq M, Ntziachristos L, Dardiotis C, Wang X, Axmann H, Bergmann A and Schindler W 2014 Review of motor vehicle particulate emissions sampling and measurement: From smoke and filter mass to particle number *J. of Aer. Sci.* **67** 48-86.

[17] Lawrence S, Sokhi R and Ravindra K 2016 Quantification of vehicle fleet PM10 particulate matter emission factors from exhaust and non-exhaust sources using tunnel measurement techniques *Env. Pol.* **210** 419-428.

[18] Research Centre for Global Environment. *CGER METEX.* Global Environmental Database 2018 [cited 2018 10 April]; Available from: http://db.cger.nies.go.jp/metex/index.html.

[19] Satari S Z, Zubairi Y Z, Hassan S F and Hussin A G 2015 Some statistical characteristic of malaysian wind direction recorded at maximum wind speed: 1999-2008 *Sains Malaysiana* **44** 1521-1530.

[20] Bernama, *MetMalaysia: Cold weather phenomenon in East Coast could recur - Nation | The Star Online*, in *The Star*. 2018, @staronline: Kuala Lumpur.

[21] Hinds W C 1999 *Aerosol Technology: Properties, Behavior, and Measurement of Airborne Particles, 2nd Edition* 504

[22] Nguyen J L, Schwartz J and Dockery D W 2014 The relationship between indoor and outdoor temperature, apparent temperature, relative humidity, and absolute humidity *Indoor air* **24** 103-112.

Adsorption-photocatalysis of organic dyes using empty fruit bunch activated carbon-metal oxide photocatalyst

Wei Wen Loo, Yean Ling Pang*, Kam Huei Wong, Steven Lim and Shee Keat Mah

Department of Chemical Engineering, Lee Kong Chian Faculty of Engineering and Science, Universiti Tunku Abdul Rahman, 43000 Selangor, Malaysia.
*pangyl@utar.edu.my

Abstract. This study reports on the adsorption-photocatalysis degradation of organic dyes by utilizing zinc oxide (ZnO) photocatalyst supported on empty fruit bunch (EFB) derived activated carbon (AC) under visible light irradiation. XRD diffraction patterns of AC/ZnO sample indicated the presence of AC and ZnO. SEM analysis illustrated that ZnO particles were spherical-shaped and agglomerated, AC exhibited porous structure while AC/ZnO had agglomerated particles along the porous structure. EDX results showed that AC/ZnO sample consists of C, O and Zn elements with minimal Si and K impurities. The optimum operating conditions for adsorption-photocatalytic degradation to achieve 100 % in 50 min were initial Congo red concentration of 10 mg/L, AC/ZnO catalyst loading of 1.25 g/L, solution pH of 3 and solution temperature of 55 °C. The determination of oxygen demand analysis (64.29 % removal) indicated that only partial Congo red was being degraded due to the generation of intermediate products which were more difficult to be oxidized compared to their parent organic dye compounds. Kinetic studies which were also carried out at different solution temperatures (35, 40, 45, 50 and 55 °C) suggested that the degradation process of Congo red followed pseudo-second order model with an estimated activation energy of 86.27 kJ/mol.

1. Introduction

The coloured wastewater generated from textile industry affects the public aesthesis of water quality, as small amount of dyes that present in water will induce the recognition of colour due to the high visibility of dyes. It has been reported that about 15 % of organic dyes were discharged into the environment during synthesis and processing of textile fabric [1]. The presence of dyes in drinking water is a health hazard as it causes eye and skin irritation, haemolytic anemia, nausea, vomiting, and abdominal pain [2, 3]. Generally, dye can be classify into two types: natural and synthetic dyes [4]. Natural dyes are extracted from animal or plant sources. Synthetic dyes are organic compounds that synthesized artificially such as basic dyes, acid dyes, direct dyes and mordant dyes. It is estimated that approximately 10,000 different types of textile dyes with an annual production rate of over 7×10^5 metric tonnes [5, 6]. Azo dyes are the largest group (about 60–70%) of widely used synthetic dyes.

Dye removal techniques can be categorized into three methods: physical, chemical and biological method [3]. Most of the common removal techniques such as biodegradation, adsorption and flocculation are effective but time consuming and not favorable due to the high operational costs and sludge accumulation [1]. For instance, biological methods are known to be the most economical option to remove dye as compared to physical and chemical techniques. Despite of that, biological methods are restricted by cause of technical constraints such as the removal of azo dye that was not effective.

Moreover, it generates toxic aromatic amine (a human carcinogen) under anaerobic conditions [7]. Thus, photocatalytic systems based on semiconductors have been widely utilized for the degradation of organic pollutants due to the advantages of low operating costs, the possibility of eliminating complex compounds, no additional chemicals and the ability to be conducted at ambient temperature and pressure [8]. Zinc oxide (ZnO) is one of the most promising photocatalyst showing desirable properties such as high catalytic activity, low toxicity, low cost, ease in preparation, high chemical, thermal and mechanical stability [9]. Concerning the extensive research conducted by using photocatalytic degradation process in the presence of zinc oxide (ZnO) as catalyst under ultraviolet (UV) light as primary light source, the results for dye degradation were not satisfying due to the low surface area and high band gap energy of 3.37 eV. From the literature, it was found that activated carbon with large surface area, high microporous structure and energy band gap in the range (3.0–3.6 eV) is effectively for the removal of textile effluents [10]. Hence, activated carbon (AC) which has high surface to volume ratio is beneficial for adsorption followed by photocatalytic process. In spite of that, limited study focuses on the use of biomass such as agricultural wastes to derive the activated carbon. In furtherance of new idea and different statistics, this research was carried out to degrade the organic dyes by using photocatalysis method and empty fruit bunch was being utilized as the source of activated carbon to couple with ZnO as photocatalyst.

2. Experimental

2.1. Materials

Congo red (40%), methyl orange (85 %), sodium hydroxide (NaOH, 97 %) and zinc oxide (ZnO) were obtained from R & M Chemicals. Hydrochloric acid (HCl, 37 %), potassium hydroxide (KOH, 85%) and n-Hexane were obtained from Merck while zinc nitrate ($Zn(NO_3)_2$, 98 %), reactive blue (35 %) and rhodamine B (95%) were obtained from Sigma-Aldrich. All the chemical reagents were used as received. Distilled water was used throughout the study.

2.2. Photocatalytic degradation of congo red and analysis of liquid samples

Typically, 1.0 g/L of AC-ZnO photocatalyst was put into 100 mL of 10 mg/L congo red aqueous solution in a beaker. The solution was stirred using magnetic stirrer throughout the experiment to ensure good adsorption between congo red molecules and the photocatalyst surface. Liquid samples were withdrawn at certain time intervals. The dye solution was separated from the photocatalyst using syringe filter 0.45 µm and the residual concentration of congo red was determined using a UV–vis spectrophotometer (Jenway 6320D). The measurement of maximum absorbance was performed at 498 nm. Chemical oxygen demand (COD) determination was also conducted using a Colorimetric Determination Method 80000 (DRB 200).

3. Results and discussion

3.1. SEM-EDX analyses of AC, ZnO and AC-ZnO

The surface morphology of three photocatalysts, ZnO, AC and AC-ZnO were studied using SEM analysis as shown in figure 1 (a), (b) and (c), respectively. SEM image of ZnO shows pure ZnO particles was in spherical shape and agglomerated. Meanwhile, the morphology of AC exhibited fibre-like and porous in nature with plenty of grain boundaries. The pores were indicating the available of catalytic sites for adsorption and photodegradation to take place. Whilst, for AC-ZnO, the particles were randomly agglomerated around the surface of AC. The incorporation of ZnO in the carbon matrix partially blocked the porosity of the AC surface. Yang, Liao and Mao [11] reported that ZnO did not enter the inner microporosity of the AC during incorporation process but remained on the outer surface.

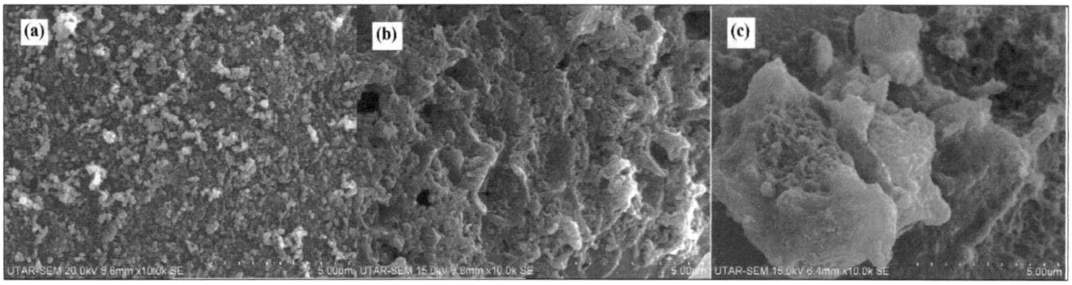

Figure 1. SEM images of (a) ZnO, (b) AC and (c) AC-ZnO.

EDX analyses of ZnO, AC and AC-ZnO photocatalysts showed that ZnO consisted of only two peaks, which were Zn (66.56 at %) and O (33.44 at %). This indicating that there were no additional impurities and it was pure ZnO. However, C (64.29 at %), O (34.68 at %) and K (1.03 at %) were discovered in EDX spectrum of AC sample. Theoretically, only C and O peaks appear on AC spectrum. The appearance of K peak might due to the remaining from KOH which was utilized to activate the carbon atoms. Despite of that, AC sample comprises mainly of carbon and oxygen which confirmed the chemical composition of activated carbon. Similar result was observed for AC-ZnO spectrum where there was occurrence of other slight peaks. The appearance of minority peak such as Si (2.49 at %) and K (3.07 at %) apart from the main peaks of C (45.30 at %), O (39.31 at %) and Zn (9.82 at %) found in this study. The high percentage of C, O and Zn implied that ZnO was successfully incorporated into AC. The existence of slight Si elemental peak might due to the contaminants on EDX machine sample's plate.

3.2. XRD study

The crystalline structure of ZnO, AC and AC-ZnO photocatalysts were analysed by XRD. The results were presented in figure 2 (a), (b) and (c), respectively. The diffraction peaks of ZnO occurred at 2θ = 31.63 °, 34.29 °, 36.12 °, 56.47 °, 62.73 ° and 67.82 ° with the highest peak occurred at 36.12 °. This indicated that there was no impurity appeared in ZnO as well as other phases of ZnO, suggesting the high purity of ZnO photocatalyst [2]. The XRD pattern of AC-ZnO consists of more peaks as compared to ZnO and AC with 2θ = 21.51 °, 23.46 °, 24.72 °, 25.70 °, 32.67 °, 33.56 °, 35.60 ° and 59.02 °. There were two stronger peaks formed at 23.46 ° and 32.67 ° where the peak at 23.46 ° indicated the presence of AC while the peak at 32.67 ° represented the presence of ZnO. This clearly due to the incorporation of ZnO onto the surface of AC. Meanwhile, the diffraction peak of AC occurred at 2θ = 12.05 °, 23.98 °, 24.21 °, 30.08 °, 31.27 °, 31.78 °, 34.10 °, 37.85 °, 39.22 ° and 44.46 °. The overall XRD pattern of AC were weak. The formation of sharper peaks at 24.21 ° and 30.08 ° can be explained by the accumulation of impurities from the preparation of AC such as potassium which complied with the results from SEM-EDX. The crystallite sizes of ZnO, AC and AC-ZnO photocatalysts were calculated by using Scherrer's equation. The calculated crytallite size for ZnO and AC photocatalysts were 41.75 nm and 43.06 nm, respectively. The smaller crystallite size of AC-ZnO, 8.02 nm could be explained the observation of broad and weak peak obtained in XRD patterns.

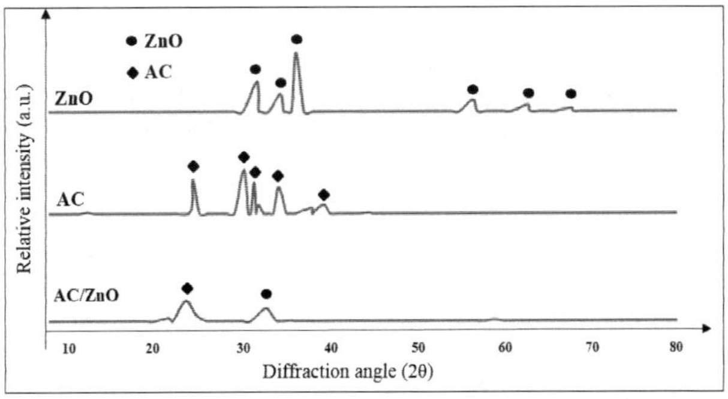

Figure 2. XRD patterns for (a) ZnO, (b) AC and (c) AC-ZnO.

3.3. FTIR study

The functional groups that appeared in ZnO, AC and AC-ZnO photocatalyst were identified using FTIR analysis. Figure 3 illustrates the FTIR spectra of these catalysts in the frequency range of 600 – 4000 cm^{-1}. There was a slight peak intensity for both ZnO and AC-ZnO in the range of 500 – 600 cm^{-1}. This corresponded to the Zn-O stretching modes and were indicative of the formation of ZnO [3]. AC showed an intense peak at 700 – 900 cm^{-1} indicated the bending of C=O-C [2]. A sharp peak was formed for AC and AC-ZnO samples within the range of 950 -1050 cm^{-1} which was due to C-H bending vibrations [12]. Stretching vibration of aromatic ring C=C appeared within the range of 1050 – 1300 cm^{-1} by the reason of incomplete carbonisation of carbon material [13]. The sharp, intense peak for AC in the range of 1300 – 1500 cm^{-1} was due to the O-C-O stretching mode. The peak spotted at around 1600 cm^{-1} was assigned to the stretching mode of C=O in the oxalate [3, 12]. For the band of AC appeared at around 1800 cm^{-1} might be attributed to symmetric and asymmetric stretching vibrations of CH_2 bonds [12]. A broad absorption peak appeared at 3300 – 3600 cm^{-1} was assigned to O-H stretching vibrations [12].

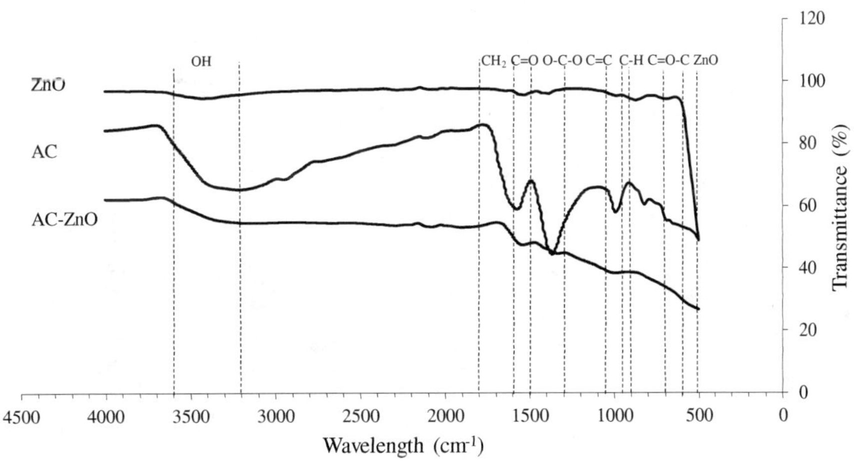

Figure 3. FTIR spectra of (a) ZnO, (b) AC and (c) AC-ZnO.

3.4. Effect of parameter studies

Figure 4 (a) shows the effect of various congo red concentration (10, 20, 30, 40 and 50 mg/L) on photocatalytic degradation of congo red at different time intervals. The results showed that the degradation efficiency decreased with the increased of congo red concentration. It was observed that the degradation efficiency decreased from 88.44% to 1.97% which was almost negligible when the congo red concentration increased from 10 to 50 mg/L. This phenomenon can be explained by the additional of the dye molecules which cause competing effect on the adsorption and reaction of O_2 or OH^- as the source of reactive radicals [5]. In other words, the number of reactants increased with the increase of dye molecules while the available active site of catalyst remains unchanged causing insufficient site for adsorption and photocatalytic reaction of dye molecules [14]. Apart from that, it can also be explained by the path length of photon that entered to the dye solution was decreased due to the large amount of dye molecules. This indicating more light source will be absorbed by the dye molecules instead of the catalyst [2].

Figure 4 (b) depicts the effect of various solution pH (pH 3, 5, 7, 9 and 11) on photocatalytic degradation of congo red at different time intervals. Results show that lower solution pH influenced the degradation efficiency positively. The degradation efficiency increased from 77.30% to 97.75% when the solution pH decreased from pH 11 to pH 3. This phenomenon can be explained by the structure of congo red molecules that contained sulfonic groups which can be ionized easily in acidic medium. Therefore, congo red became soluble anion and prone to be adsorbed and reacted on the positive surface charge of AC-ZnO photocatalyst. Apart from the reason stated above, congo red anions can be directly oxidized by oxygen radicals under light radiation and resulted in the high degradation efficiency. On the other hand, the decrease in degradation efficiency in alkaline solution could be explained by the basis of amphoteric behaviours of AC-ZnO photocatalyst. The surface of AC-ZnO became negatively charged when there was high concentration of hydroxide ions, causing electrostatic repulsion between dye molecules and catalyst led to the decreased in adsorption and reaction process [12]. Whereas, congo red behave as an anionic dye, which is negatively-charged. This enabled the dye molecules easier to be adsorbed and react on the positive AC-ZnO surface by electrostatic interaction [14].

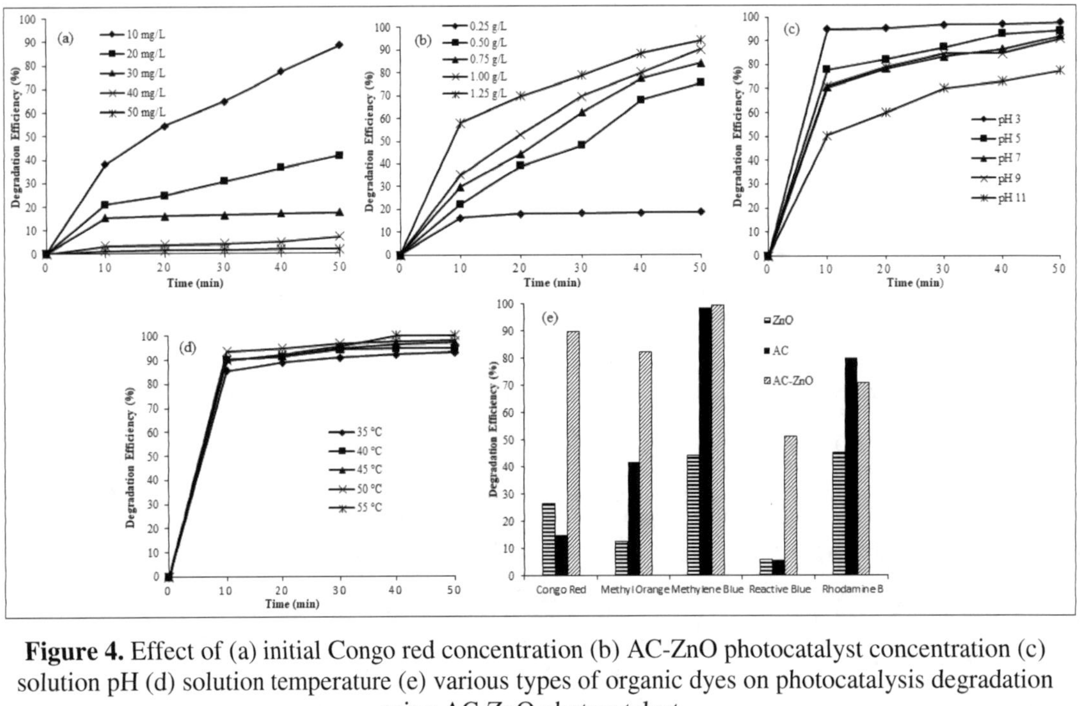

Figure 4. Effect of (a) initial Congo red concentration (b) AC-ZnO photocatalyst concentration (c) solution pH (d) solution temperature (e) various types of organic dyes on photocatalysis degradation using AC-ZnO photocatalyst.

Figure 4 (c) depicts the effect of various AC-ZnO concentration (0.25, 0.50, 0.75, 1.00 and 1.25 g/L) on photocatalytic degradation of congo red at different time intervals. Results showed that the degradation of congo red was positively influenced by the AC-ZnO photocatalyst concentration where the degradation efficiency increased with increased in the concentration of AC-ZnO photocatalyst. It was found that the degradation efficiency of congo red increased from 18.48% to 94.02% when the concentration of AC-ZnO increased from 0.25 to 1.25 g/L. The reason of this observation was due to the fact that the increased in AC-ZnO amount will directly increase the surface area and availability of adsorption site for the dye adsorption followed by degradation to occur. This phenomenon can also be explained by the degradation of congo red was favoured by increasing AC-ZnO loading as increased in active sites led to the enhancement of production and reaction of ·OH radicals [2]. These observation trend is similar to those reported in literature for the adsorption of Rhodamine B by mesoporous carbon material [12].

The effect of solution temperature (35, 40, 45, 50 and 55 °C) on the photocatalytic degradation of congo red was experimented and shown in figure 4 (d). It showed a positive impact when the solution temperature was raising from 35 °C to 55 °C, degradation efficiency increased from 92.88% to 100%. The significant degradation efficiency was clearly observed in the first 10 min, which averagely jot a result as high as 90% degradation. It was worth being noted that a 100% of degradation efficiency was achieved at the end of the experiment when the temperature was set to 55 °C. The increased in degradation efficiency of congo red with rising in solution temperature was owing to the fact that higher solution temperature increased the activation energy of reaction and therefore, more collision occurred between molecules caused the increased in reaction rate. The mass transfer between photocatalyst molecules and dye molecules increased at the same time [6]. Furthermore, this phenomenon might be attributed to the increased diffusion of dye molecules towards the external boundary layer of catalyst. Elevated solution temperature enhanced the diffusion of dye molecules into the pores of photocatalyst [12].

The effect of several types of organic dyes such as Congo Red, Methyl Orange, Methylene Blue, Reactive Blue and Rhodamine B for adsorption and photodegradation processes were investigated by utilizing three types of photocatalysts included ZnO, AC and AC-ZnO. Based on the observation as indicated in figure 4 (e), methylene blue clearly possesses the best degradation efficiency under all three types of catalyst, while reactive blue recorded the lowest degradation efficiency among all types of organic dyes. As for methylene blue, high degradation efficiency was due to its nature as a cationic dye where electrostatic attraction can occur easily between dyes and Ac-ZnO and reacted with photo-generated electrons [15]. Besides, the contribution of adsorption of AC and AC-ZnO on methylene blue were found as high as 76.81 % and 79.58 %, indicated photocatalytic reaction did not contribute much for the degradation efficiency. Rhodamine B is a xanthene dye, which was harder to be degraded due to its superior light resistance as well as ultrahigh solubility in water [16]. Meanwhile, low degradation efficiency of reactive blue might be attributed to the presence of anionic sulfonate groups which lower the chances of adsorption of reactive blue to take place [1]. It was found that adsorption of congo red by ZnO achieved 18.16 %, AC recorded 9.69 % while AC-ZnO achieved 27.39 %. Congo red and methyl orange were anionic dye. A negatively-charged chromophore was produced when ionized in water. This allowed them to be able to adsorb easier by the positive charge photocatalyst surface and thus, recorded better degradation efficiency. AC-ZnO photocatalyst achieved the highest catalytic activity was mainly due to the availability of large amount adsorption sites for reaction as compared to the other two photocatalysts. In addition, the degradation efficiency was influenced by the structures of organic dyes itself [17]. From the results shown, congo red, methyl orange and methylene blue had relatively higher degradation efficiency which was above 80%. Photocatalysis process was preceded by the adsorption of substrate molecules on the surface of catalyst to ensure the effective contact with electrons or holes to occur.

COD analysis was conducted and it was found that 54.76% of COD had been removed after 30 min while 64.29 % of the COD had been removed after 50 min of reaction. The partial removal of COD might be due to various intermediates products that produced during the photodegradation of congo red. These intermediate products that were hard to be oxidized as compared to parent compounds. It can also be explained by the nature of congo red where asymmetric oxidative cleavage led to the formation of mono azo intermediates initially which can be further oxidatively and reductively cleaved into simple metabolites. Thus, it was anticipated that higher percentage of COD removal could only be achieved by proceeding the experiment for a longer period.

3.5. Possible mechanism

The mechanism on AC-ZnO to degrade dye molecules is illustrated in figure 5. When UV light source strikes on the surface of AC-ZnO, the electrons are excited from valence band (VB) into conduction band (CB), leaving the holes behind [14]. The electrons (e^-) and holes (h^+) pairs were available for oxidation and reduction to take place and then lead to the formation of ·OH radicals, which was the most crucial radicals for the photocatalytic degradation to occur. Besides, the oxygen molecules adsorbed on ZnO then formed superoxide anion radicals, $\cdot O_2$ because of their interaction with electrons in the conduction band of ZnO. Meanwhile, the hydroxyl groups on surface produce highly reactive hydroxyl radicals, ·OH by reacting with the holes in the VB of ZnO. The dye molecules are finally degraded by the reaction with both $\cdot O_2$ and ·OH radicals [15]. The presence of AC can enhance the adsorption capacity of Congo red and improve the separation efficiency of electrons and holes [11]. In other words, AC-ZnO was an effective photocatalyst for photocatalysis process. The reactions that take place are indicated by equations (1) – (5).

$$ZnO + \text{light source} \rightarrow h^+ + e^- \quad (1)$$

$$h^+ + H_2O \rightarrow OH^\cdot + h^+ \quad (2)$$

$$h^+ + Dye \rightarrow \text{oxidation products} \quad (3)$$

$$e^- + Dye \rightarrow \text{reduction products} \quad (4)$$

$$OH^\cdot + Dye \rightarrow \text{degradation products} \quad (5)$$

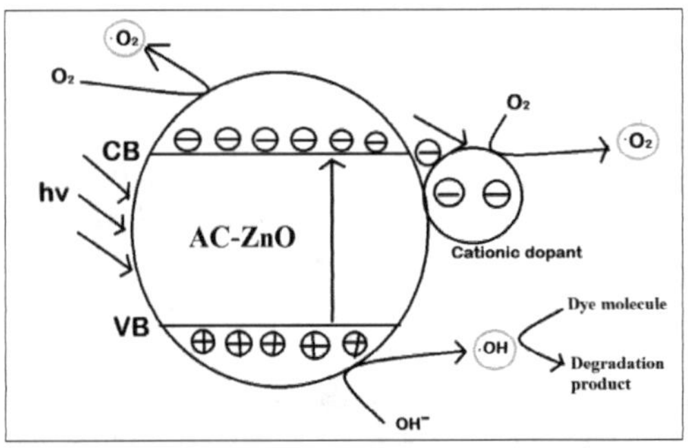

Figure 5. Schematic diagram of photodegradation mechanism of dye molecules.

3.6. Kinetic Study

Kinetic studies were performed on the basis of the rate of disappearance of congo red in order to determine the order of the reaction for photocatalytic degradation of the dye. In figure 6 (a), the plot of $1/C_t$ vs t shows a linear relationship, from which rate constant, k_{app} can be determined from the slope of the plot. This indicated that the adsorption and reaction of congo red at different solution temperatures was recorded as pseudo second-order kinetics. Therefore, the slope of the straight lines should be equal to the apparent second-order rate constant (k_{app}). It was also observed that the pseudo second-order rate constant increased with increasing solution temperature where the k_{app} increased from 0.0887 to 0.3251 L/mg·min when the solution temperature increased from 35 °C to 50 °C. This was due to the accelerated production of OH radicals when the solution temperature increased and therefore, contributed to the gradual increase of rate constant [6].

Based on figure 6 (b), the plot was a considerably good linear graph with R^2 value of 0.9695. The Arrhenius relationship can be implied by equation (6).

$$\ln k_{app} = -10381\,(1/T) + 31.149, \quad R^2 = 0.9695 \quad (6)$$

From equation (6), E_a and A_0 can be determined by the slope (E_a/R) and intercepts (ln A_0) of the linear plot, respectively. E_a and A_0 for degradation of congo red was 86.27 kJ/mol and 3.37x10^{13} L/mg·min, respectively. Tekin and Derya [18] reported that that the E_a for photodegradation of congo red was 56.1 kJ/mol. Their result showed lower E_a value as compared to the photocatalytic degradation of congo red in current study, which was 86.27 kJ/mol. This implied that higher solution temperature was required to overcome the E_a. Higher E_a value indicated that the reaction was mainly influenced by solution temperature as compared to the concentration and life of reactive free radicals.

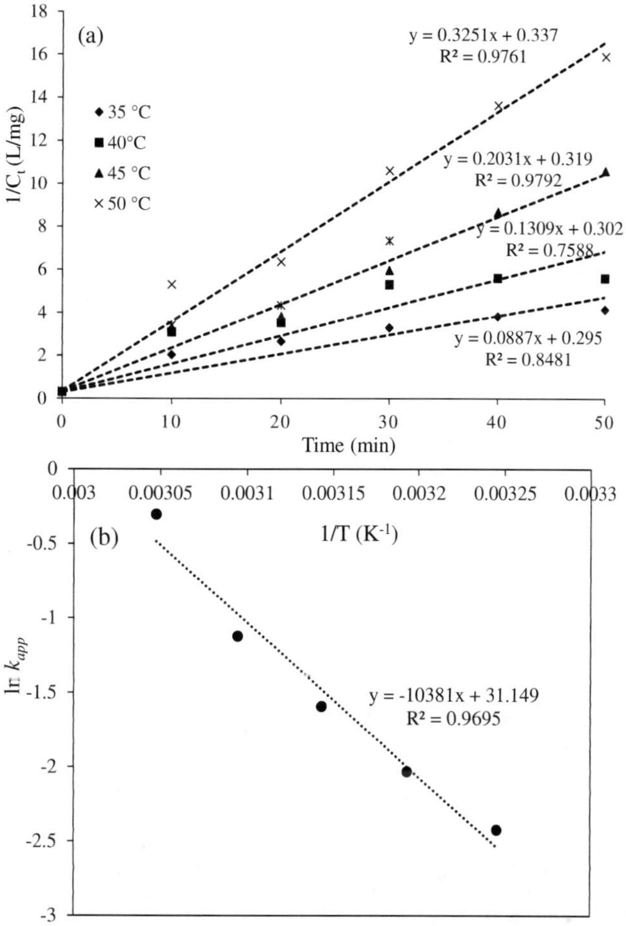

Figure 6. (a) Pseudo second-order reaction kinetics plot for photocatalytic degradation of congo red (b) Arrhenius plot of k_{app} against $1/T$.

4. Conclusions

ZnO, AC and AC-ZnO photocatalyst were successfully synthesised and characterised. AC was prepared by calcination in a furnace at temperature of 450 °C for 2 hrs while ZnO was synthesised by using zinc precursor. These three types of photocatalysts were then characterised by XRD, SEM-EDX and FTIR analyses. XRD diffraction patterns of AC-ZnO sample indicated the presence of AC and ZnO. SEM results showed that ZnO particles were in spherical shape and agglomerated, AC exhibited porous structure while AC-ZnO had agglomerated particles along the porous structure. Whilst, EDX results illustrate that ZnO contained only Zn and O elements, AC contained elements of C and O together with K impurities which resulted from the preparation of AC. Meanwhile, AC-ZnO sample shows the presence of Zn, O, C, Si and K elements impurities. The optimum conditions for degradation efficiency of congo red were obtained at an initial dye concentration of 10 mg/L, initial dosage of AC-ZnO photocatalyst at 1.25 g/L, solution pH of 3 and solution temperature of 55 °C for 50 min to achieve 100 % degradation efficiency. However, only partial COD (64.29 %) was being removed. This was due to the generation of intermediate products which were more difficult to be oxidized as compared to parent organic dye compounds. On the other hand, the reaction kinetics of photocatalytic degradation of congo red at different solution temperature (35, 40, 45, 50 and 55 °C) were followed pseudo second-order kinetics. The E_a and A_o for photocatalytic degradation of congo red were 86.27 kJ/mol and 3.37×10^{13} L/mg•min, respectively.

Acknowledgments
This study was financially supported by Universiti Tunku Abdul Rahman Research Grant through the Centre for Photonics and Advanced Materials Research and Centre for Environment and Green Technology (CEGT), UTAR (IPSR/RMC/UTARRF/2018-C1/P01). The financial support provided by Fundamental Research Grant Scheme by the Ministry of Education (MOE) Malaysia is also gratefully acknowledged.

References
[1] Kakarndee S and Nanan S 2018 *J. Environ. Chem. Eng.* **6** 74-94
[2] Ong C B, Mohammad A W, Rohani R, Ba-Abbad M M and Hairom N H H 2016 *Process Saf. Environ. Prot.* **104** 549-557
[3] Intarasuwan K, Amornpitoksuk P, Suwanboon S and Graidist P 2017 *Sep. Purif. Technol.* **177** 304-312
[4] Lipskikh O I, Korotkova E I, Khristunova Y P, Barek J and Kratochvil B 2018 *Electrochim. Acta* **260** 974-985
[5] Yang C, Yu J, Li Q and Yu Y 2017 *Mater. Res. Bull.* **87** 72-83
[6] Zangeneh H, Zinatizadeh A A L, Habibi M, Akia M and Hasnain Isa M 2015 *J. Ind. Eng. Chem.* **26** 1-36
[7] Ma C, Wang F, Zhang C, Yu Z, Wei J, Yang Z, Li Y, Li Z, Zhu M, Shen L and Zeng G 2017 *Chemosphere* **168** 80-90
[8] Mohamed A, El-Sayed R, Osman T A, Toprak M S, Muhammed M and Uheida A 2016 *Environ. Res.* **145** 18-25
[9] Qi K, Cheng B, Yu J and Ho W 2017 *J. Alloys Compd.* **727** 792-820
[10] Soliman A M, Elsuccary S A A, Ali I M and Ayesh A I 2017 *J. Water Process Eng.* **17** 245-255
[11] Yang Q, Liao Y and Mao L 2012 *Chin. J. Chem. Eng.* **20** 572-576
[12] Saini J, Garg V K, Gupta R K and Kataria N 2017 *J. Environ. Chem. Eng.* **5** 884-892
[13] Mo X, López D E, Suwannakarn K, Liu Y, Lotero E, Goodwin Jr J G and Lu C 2008 *J. Catal.* **254** 332-338
[14] Abdelwahab N A and Helaly F M 2017 *J. Ind. Eng. Chem.* **50** 162-171
[15] Wang S, Zhou Y, Han S, Wang N, Yin W, Yin X, Gao B, Wang X and Wang J 2018 *Chemosphere* **197** 20-25
[16] Lin K-Y A and Lin J-T 2017 *Chemosphere* **182** 54-64
[17] Rao A N S and Venkatarangaiah V T 2014 *Port. Electrochim. Acta* **32** 213-231
[18] Tekin D 2014 *Prog. React. Kinet. Mech.* **39** 249-261

IOP Conf. Series: Materials Science and Engineering **458** (2018) 012043 doi:10.1088/1757-899X/458/1/012043

Removal of anionic surfactant sodium dodecylbenzenesulfonate from water using fly ash adsorbent

Ahmer Ali Siyal[1], **M Rashid Shamsuddin**[1,2,*], **Nurul Ekmi Rabat**[1], **Muhammad Zulfiqar**[1], **Muhammad Ayoub** [1,2] **and Khairun Azizi Azizli**[1]

[1]Department of Chemical Engineering, Universiti Teknologi PETRONAS, 32610 Bandar Seri Iskandar, Perak, Malaysia.
[2]Center for Biofuel and Biochemical Research (CBBR), Institute for Sustainable Living, Universiti Teknologi PETRONAS, 32610, Seri Iskandar, Perak, Malaysia

*mrashids@utp.edu.my

Abstract. Surfactants are organic pollutants that are environmentally hazardous for aquatic and human life. A variety of methods are used for the removal of surfactants but adsorption is the most suitable method due to high efficiency, simple operation and economical. This paper describes the adsorption of an anionic surfactant sodium dodecylbenzenesulfonate (SDBS) using fly ash as adsorbent. The adsorbent dosage of 10 g/L and contact time of 5 hours were optimum for the adsorption of SDBS. The maximum adsorption capacity of 6.83 mg/g was obtained. The adsorption capacity of fly ash for anionic surfactant is higher as compared to the adsorption capacity of granite sand for anionic surfactant and fly ash for anionic dyes (reactive red 23 and reactive blue 171). Fly ash possesses capability to be used for the removal of anionic surfactants.

1. Introduction

Surfactants are the organic pollutants detected in wastewaters. Surfactants are used in a large number of applications for different purposes such as emulsion stabilization, foaming, wetting, detergency, mineral separations, pharmaceutical formulations and other purposes [1]. Surfactants can be classified into four classes based on charge on the head group; anionic, cationic, non-ionic and amphoteric [2, 3]. Surfactants are toxic to terrestrial and aquatic life as they cause health hazards (dermatitis and eye irritation). Surfactants also remain in plants for a longer period of time [4]. The high annual consumption of surfactants in domestic and industrial sectors is posing an increasing threat on the environment [5, 6]. Surfactants must be removed from wastewater before discharging into water resources. A variety of physical, chemical, biological and membrane techniques are used for the removal of surfactants [7-9]. Adsorption is the most suitable technique for the removal of surfactants due to its high efficiency, simple operation and economic feasibility [10]. Activated carbon is the widely used adsorbent material for the removal of surfactants due to its high efficiency. However, it is more expensive than other adsorbents which limits its application on large-scale [11]. Coal fly ash containing some quantity of carbon in the form of unburned carbon can be a good adsorbent for the removal of surfactants.

Coal fly ash (FA) is the solid residue remained after combustion process of coal in thermal power stations. Coal fly ash flies with flue gases and collected by electrostatic precipitators or bag filters. The composition of fly ash depends on combustion and type of coal [12, 13]. The global annual production of coal ash is around 600 million tonnes with 500 million tons of it is fly ash [14, 15]. Fly ash is considered as an environmental threat and its increased production demands safe disposal or utilization in useful applications. The disposal of fly ash in landfills or lagoons is expensive, putting pressure on the industries for finding new applications of fly ash. Coal fly ash is mainly used in concrete production, soil amendment and as filler in polymers [16], however, the bulk quantity remains underutilized. Converting fly ash into adsorbent material could add value proposition on what traditionally has been considered as waste. This in turn could save the land from direct disposal, reduce the overall cost of adsorption and create monetary value for fly ash. This paper describes the adsorption of an anionic surfactant SDBS using fly ash. The characterization of fly ash and the effect of adsorbent dosages and contact time on the adsorption of fly ash are discussed.

2. Experimental section

2.1 Raw materials

Fly ash was obtained from a local coal thermal power plant. The analytical grade (AR) anionic surfactant (SDBS) in powder, HCl solution (purity-97 %) and NaOH in pellets form were purchased from R & M Chemicals, Malaysia.

2.2 Adsorption of anionic surfactant

SDBS surfactant solution was prepared at 100 mg/L initial concentration and pH 2. The pH of surfactant solution was adjusted using 0.1M HCl and 0.1M NaOH solutions. For adsorption experiment, 200 mL of surfactant solution was transferred into a beaker and mixed with fly ash adsorbent at 5, 10, 25, and 50 g/L dosage concentrations. The mixture was stirred on a water bath shaker at 200 rpm and 25 °C. Three 3 ml samples were taken at 0, 5, 15, 30, 60, 120, 180, 240, 300 and 1440 minutes and filtered using a syringe filter (Whatman 0.22 µm) prior to UV analysis. The Shimadzu UV Spectrophotometer 1800 at maximum absorbance wavelength of 223 nm was used for determining the concentration of SDBS surfactant. The adsorption capacity and removal efficiency (%) were calculated using following equations (1) and (2).

$$Q_t = \frac{V_s\,(C_i - C_f)}{m} \tag{1}$$

$$Removal\ efficiency\ (\%) = \frac{(C_i - C_f)}{C_i} \times 100 \tag{2}$$

Where,
Q_t = surfactant adsorbed (mg of surfactant/g adsorbent) at a given time
V_s = Volume of sample solution (L)
C_i = Initial concentration of surfactants (mg/L)
C_f = final concentration of surfactants (mg/L)
m = weight of dry adsorbent (g)

3. Result and Discussion

3.1 Characterization of Adsorbent

The particle size analysis results of fly ash are shown in table 1. It can be observed that the higher portion of fly ash particles are in the size range of 0.95 to 30.20 µm. The lower particle size is more beneficial for adsorption as compared to higher particle size due to higher surface area. The surface area of the adsorbent plays a very important role in the adsorption due to greater contact of the adsorbate molecules with the adsorbent surface. Fly ash possess surface area of 0.85 m^2/g [17]. Although the surface area of fly ash is small compared to activated carbon (600-2000 m^2/g) and other materials but other factors such as its porosity and pore size could enhance the adsorption. The detailed particle size analysis can be seen elsewhere [17].

The microstructural analysis of fly ash is shown in figure 1. Fly ash contains cenospheres, magnetic spheres and quartz. Fly ash is irregular in shape with particles of different size and shape. The major portion of particles are smaller in size which is consistent with particle size analysis results. The smaller particles will be more beneficial for adsorption of surfactant providing more active adsorption sites for the adsorption of anionic surfactant. Some particles are attached to each other forming agglomerates and producing pores in agglomerates. Pores are desirable in adsorption for the same reason mentioned earlier. The detailed microstructural analysis can be seen elsewhere [17].

Table 1. Particle size analysis of fly ash [17]

S.No	Size fraction (μm)	Vol %
1	<0.95	17.7
2	0.95 – 30.200	70.86
3	30.200 -120.226	25.72
4	>120.226	11.68

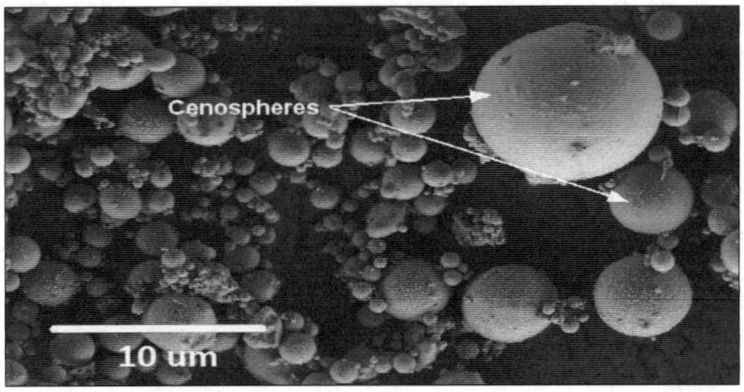

Figure 1. FESEM image of fly ash [17]

3.2 Adsorption of SDBS surfactant

3.2.1 Effect of adsorbent dosage.

The results of the effect of adsorbent dosage are shown in figure 2. The increase of dosage of fly ash from 5 g/L to 10 g/L increased the adsorption capacity of fly ash from 1.63 mg/g to 6.83 mg/g due to increase of active adsorption sites. The increase of dosage of fly ash increased the active adsorption sites on the surface of fly ash which adsorbed higher quantity of surfactant causing increase of the adsorption capacity of fly ash. Further increase in the dosage of fly ash from 10 g/L to 50 g/L deceased the adsorption capacity due to availability of excess number of active adsorption sites. The excess number of active adsorption sites on the surface of fly ash decreased adsorption capacity due to distribution of surfactants on higher quantity of fly ash. The increase of dosage of fly ash from 5 g/L to 25 g/L increased the removal efficiency of anionic surfactant from 8.15 % to 93.35 % and then decreased with further increase of adsorbent dosage to 50 g/L due to excess quantity of fly ash. The adsorbent dosage of 10 g/L is optimum for the removal of anionic surfactant SDBS.

3.2.2 Effect of contact time.

The adsorption of anionic surfactant SDBS on fly ash increases with the increase of adsorption time. The changes in adsorption capacity and removal efficiency were rapid during early period of adsorption and then become slow with the passage of time as shown in figure 3. The adsorption capacity and removal efficiency became constant after 5 hours (300 minutes). Similar adsorption capacity and removal percentage were obtained at 24 hours. It shows that equilibrium has achieved within the first 5 hours of adsorption and further increase in adsorption time does not affect the adsorption capacity and removal efficiency. The small particles size and porous structure of fly ash supports the adsorption of anionic surfactants in 5 hours. The contact time of 5 hours is the optimum for adsorption of anionic surfactant SDBS using fly ash adsorbent.

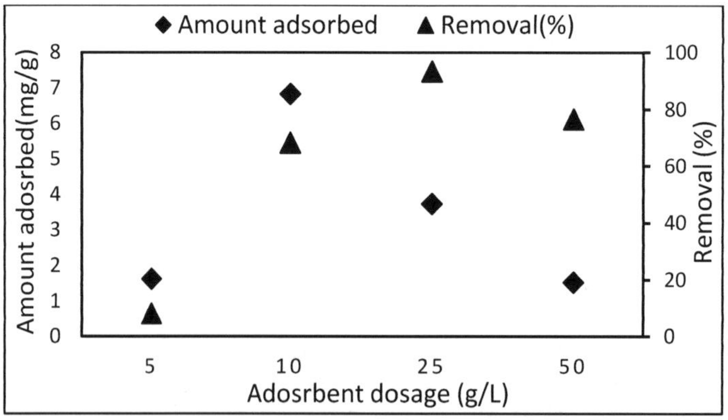

Figure 2. Effect of adsorbent dosage on the adsorption of SDBS (pH-2, SDBS concentration-100 mg/L)

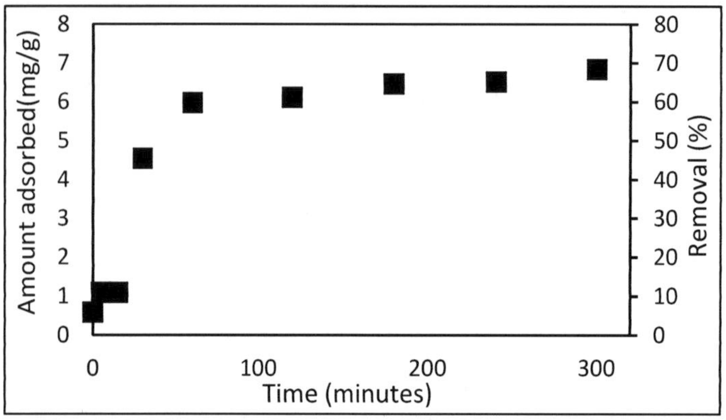

Figure 3. Effect of contact time on the adsorption of SDBS surfactant (pH-2, SDBS concentration-100 mg/L) and adsorbent dosage-10 g/L)

The optimum adsorption capacity of 6.83 mg/g was obtained at adsorbent dosage of 10 g/L, surfactant concentration of 100 mg/L and pH of 2 at room temperature at a contact time of 5 hours. This adsorption capacity is higher than the adsorption capacity of another anionic surfactant sodium dodecyl sulfate (SDS) on fly ash (2.95 mg/g at adsorbent dosage of 25 g/L) [10]. The adsorption capacity of fly ash for removal of an anionic surfactant SDBS is higher as compared to the adsorption capacity of crushed Berea Sandstone [18] and granite sand [19] for anionic surfactant and fly ash for anionic dyes (reactive red 23 and reactive blue 171) [20]. The adsorption of anionic surfactant on fly ash occurs ithrough the mechanism of Van der Waals' forces of attraction. It is obvious that electrostatic interactions are not involved due to anionic nature of adsorbent [21] and adsorbate so Vand er Waals' forces of attraction is the possible mechanism of adsorption of anionic surfactants on fly ash. The detailed mechanism needs further study using kinetics and isotherm models. The adsorption results suggest that fly ash possesses the capability for the removal of anionic surfactant from wastewater. The use of fly ash for the removal of anionic surfactants is economical due to its abundance and low cost.

4. Conclusions

Fly ash possess surface area of 0.85 m^2/g and porous structure. The adsorbent dosage of 10 g/L and contact time of 5 hours were optimum for the adsorption of anionic surfactant SDBS. The maximum adsorption capacity of 6.83 mg/g was obtained. The adsorption capacity of fly ash for anionic surfactant is higher as compared to adsorption capacity of granite sand for anionic surfactant which shows that it can be used for the removal of anionic surfactants. Fly ash possesses the capability to remove anionic surfactants.

References

[1] M I Bautista Toledo, J Rivera Utrilla, J D Méndez Díaz, M Sánchez Polo, and F Carrasco Marín 2014 Removal of the surfactant sodium dodecylbenzenesulfonate from water by processes based on adsorption/bioadsorption and biodegradation *J. of coll. and interf. sci.* **418** 113-119
[2] Z Gönder, I Vergili, Y Kaya and H Barlas 2010 Adsorption of cationic and anionic surfactants onto organic polymer resin Lewatit VPOC 1064 MD PH *Env. geochem. and health* **32** 267-273
[3] S Gupta, A Pal, P K Ghosh and M Bandyopadhyay 2003 Performance of Waste Activated Carbon as a Low-Cost Adsorbent for the Removal of Anionic Surfactant from Aquatic Environment *J. of Env. Sci. and Health* **38** 381-397
[4] T Cserháti, E Forgács and G Oros 2002 Biological activity and environmental impact of anionic surfactants *Env. int.* **28** 337-348
[5] C Edser 2008 Status of global surfactant markets *Focus on Surfact.* 1-2
[6] J Beltrán Heredia, J Sánchez Martín and M Barrado Moreno 2012 Long-chain anionic surfactants in aqueous solution. Removal by Moringa oleifera coagulant *Chem. Eng. J.* **180** 128-136
[7] D Qin, Z Liu, H Bai, D D Sun and X Song 2016 A new nano-engineered hierarchical membrane for concurrent removal of surfactant and oil from oil-in-water nanoemulsion *Scient. reports* **6**, 24365
[8] M Lechuga, A Fernández Arteaga, M Fernández Serrano, E Jurado, A Burgos and F Ríos 2014 Combined use of ozonation and biodegradation of anionic and non-ionic surfactants *J. of Surfact. and Deterg.* **17** 363-370
[9] M Goyal, R Dhawan and M Bhagat 2008 Adsorption of dimethyl sulfide vapors by activated carbons *Coll. and Surf.s A: Physicoche. and Eng. Aspects* **322** 164-169
[10] A Zanoletti et al 2017 Embodied energy as key parameter for sustainable materials selection: The case of reusing coal fly ash for removing anionic surfactants *J. of Cleaner Pro.* **141** 230-236
[11] R Shamsuddin 2013 Protein Intercalated Bentonite for Bio-composites Doctor of Philosophy (PhD) Doctoral, University of Waikato, Hamilton, New Zealand
[12] R A Kruger 1997 Fly ash beneficiation in South Africa: creating new opportunities in the market-place *Fuel* **76** 777-779
[13] E M van der Merwe, L C Prinsloo, C L Mathebula, H C Swart, E Coetsee and F J Doucet 2014 Surface and bulk characterization of an ultrafine South African coal fly ash with reference to polymer applications *Applied Surface Science* **317** 73-83
[14] M Ahmaruzzaman 2010 A review on the utilization of fly ash *Progress in Energy and Combustion Science* **36** 327-363
[15] R C Joshi and R Lohita 1997 *Fly ash in concrete: production, properties and uses*. CRC Press
[16] Z T Yao *et al* 2015 A comprehensive review on the applications of coal fly ash *Earth Sci. Rev.* **141** 105-121
[17] A A Siyal, K A Azizli, L Ismail, Z Man and M I Khan 2016 Suitability of Malaysian Fly Ash for Geopolymer Synthesis *Adv. Mat. Res.* **1133** 201-205
[18] M R Azam, I M Tan, L Ismail, M Mushtaq, M Nadeem and M Sagir 2013 Static adsorption of anionic surfactant onto crushed Berea sandstone *J. of Petrol. Explo. and Prod. Tech.* **3** 195-201
[19] M N Khan and U Zareen 2006 Sand sorption process for the removal of sodium dodecyl sulfate (anionic surfactant) from water *J. of Hazard. Mat.* **133** 269-275

[20] D Sun, X Zhang, Y Wu and X Liu 2010 Adsorption of anionic dyes from aqueous solution on fly ash, *J.of Hazard. Mat.* **181** 335-342
[21] S Wang, Y Boyjoo, A Choueib and Z H Zhu 2005 Removal of dyes from aqueous solution using fly ash and red mud *Water Research* **39** 129-138

Nanocarbon-based enhanced squeeze treatment for improved scale management

A Muhsan[1*], U Ishtiaq[1], A Rozali[1], N Mohamed[2] and T Albarody[3]

[1]Petroleum Engineering Department, Universiti Teknologi PETRONAS
[2]Fundamental and Applied Sciences Department, Universiti Teknologi PETRONAS
[3]Mechanical Engineering Department, Universiti Teknologi PETRONAS

*Corresponding Author: ali.samer@utp.edu.my

Abstract. Oil and gas industry is facing dilemma regarding scale management particularly for offshore production activities in ultra-deep waters under high pressure-high temperature (HPHT) conditions. Scale inhibition squeeze treatment widely employed technique for scale control. However, at HPHT conditions, routine conventional chemistries for scale mitigation are ineffective and unviable. This paper presents promising advantages of employing nanotechnology to enhance conventional scale squeeze treatment. Experimental studies were carried out to examine the potential benefits of using Graphene Oxide (GO) and Carbon nanotubes to increase the adsorption and retention of ETDA scale inhibitor onto reservoir rock in a process coined as 'Nano-carbon Enhanced Squeeze treatment (NCEST)'. This process involves treating the reservoir rock in the near wellbore region with nanomaterials that rendered increased active surface area for scale inhibitor adsorption. Stronger bonding was witnessed between EDTA and reservoir rock which accounts for better retention of EDTA. Test analysis of coreflood experiments were conducted by the aid of FESEM, EDX and UV-visible. Application of the proposed NCEST model was observed to significantly increase adsorption of EDTA on rock core sample treated with GO to achieve maximum scale inhibitor adsorptive capacity of 180mg/g as compared to 51mg/g when treated without nanomaterial.

Keywords: Scale Inhibitor; Squeeze Treatment; Graphene Oxide; Carbon Nanotubes; Organosilane

1. Introduction

The formation of mineral scale associated with production of hydrocarbon has been a major strain in oilfield operations. Scale mass is formed by precipitation due to chemical reactions as ions that are dissolved in produced water react with each other to form an insoluble solid [1]. Relative to the nature of the scale and formation fluid composition, scale deposition can take place within a reservoir which causes formation damage or in the production system where blockage can cause severe operational problems. Solid scale deposition in the producing reservoir matrix can lead to irrevocable and permanent formation damage, eventually cutting-off production by building-up in the near wellbore region by plugging reservoir pore throats and perforations of the production tubing. If not remediated, continuous scale deposition can narrow-down or even completely choke production tubing and flow lines as shown in figure 1. Pipeline plugging due to scale deposition can result in drastic hype in operational pressure drop thereby reducing well productivity which is economically unwarranted and pose acute safety hazards. Scale formation also gives impetus to corrosion phenomenon. The most common type of scale found in wells is caused by the formation of sulfate and carbonate scales of Calcium and Strontium [2, 3]. Brine is the primary cause of calcite and sulfate scales formation.

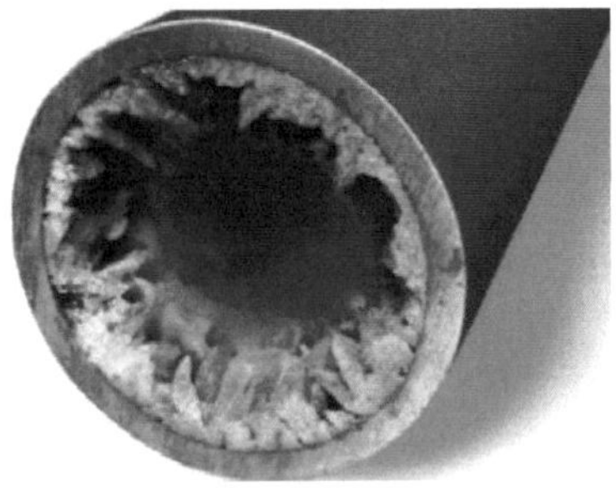

Figure 1. Pipeline with constricted diameter due to scale deposition [1]

Scale removal is a common well-intervention operation with a wide range of mechanical, chemical and scale inhibitor treatment options. Scale inhibitor squeeze treatment is the most frequently used and effective scale prevention and mitigation technique [1, 4, 5]. In this technique, about 10% of the commercial inhibitor chemical is pumped downhole (squeezed) at high injection pressures in order to partially adsorb on the reservoir rock followed by 12-24 hours well shut-off for its retention [5]. The adsorbed and retained chemicals on the reservoir rock act as storage. During production, flow of reservoir formation water desorbs and dissolves this stored scale chemical which flows into the production and distribution system to mitigate scale formation process.

Scale inhibitor adsorption and subsequent retention onto the producing formation rock is of paramount importance which determines the effectiveness (success) of squeeze treatment. Squeeze treatment lifetime can be defined as the volume of water produced by a well before the retained inhibitor concentration dwindles below the critical threshold concentration value referred as minimum inhibitor concentration (MIC) or minimum effective dose (MED) [4]. Inhibitor concentration below MIC/MED makes it incapable to prevent scale formation phenomenon.

This conventional squeeze treatment is exorbitant due to high chemical cost and pumping requirements for its injection into the reservoir. Well has to be shut for about 12-24 hours [4] for implementation of this treatment which causes production loss and consequently translates into lesser revenue generation. Significant quantity of scale inhibitor chemical is lost into cracks of reservoir matrix due to high pressure injection and low adsorption. The efficiency of squeeze treatment needs to be improved as it is essential for petroleum industry by enhancing adsorption and long retention of scale inhibitor chemical onto the reservoir rock surface.

1.1. Nanotechnology application for scale inhibition

In light of recent interest in nanotechnology application in oil and gas industry, research efforts are being oriented towards exploring the role of nanomaterials to improve and enhance conventional squeeze treatment process widely applied for scale control. By analyzing the adsorption of conventional squeeze treatment, it can be deduced that lack of availability of suitable active surface for scale inhibitor adsorption is one of the primary reasons for ineffective squeeze treatment with short squeeze lifetimes [4, 5]. Furthermore, weak interactive bonding between the scale inhibitor and the reservoir rock surface is also mainly responsible for premature inhibitor flush-out when water starts to leach out of the reservoir matrix at certain flow rates [4-6].

In order to address the shortcomings of conventional squeeze treatment, injection of highly dispersed GO and CNTs nanomaterials into the reservoir is suggested and investigated as illustrated in figure 2.

A proactive dispersion strategy is utilized to achieve high dispersion of the stated nanomaterials. It is hypothesized that highly dispersed nanomaterial may provide high active surface area for EDTA scale inhibitor adsorption and retention thereby mitigating the negative outcomes of squeeze treatment by enhancing its lifetime.

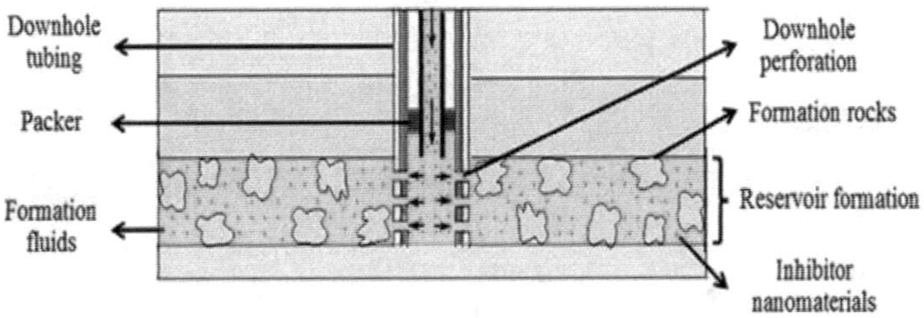

Figure 2. Dispersed nanomaterial injection in downhole region [6]

1.2. MWCNTs and GO nanomaterial

Multiwall carbon nanotubes (MWCNTs) have an elastic modulus of about 1TPa and tensile strength of 100GPa [7], which is over exponentially higher than any industrial fiber. They have high thermal stability and conductivity of 3500 $Wm^{-1} K^{-1}$ at room temperature [8], which can conveniently sustain harsh reservoir conditions. MWCNTs have features of dispersion, deposition and functionalization. They are emerging as a multi-dynamic coating material providing high active surface area for various applications [9]. GO has easy dispersibility in water, organic solvents and different matrixes due to the presence of oxygen functionalities [10, 11]. They can be mixed with ceramic or polymer matrixes to tailor and induce superior mechanical and thermal attributes. They are strong material which can be transformed into paper thin coating films [12]. This structure of CNTs and GO as shown in figure 3 which needs to be explored and hence utilized for wide range of applications.

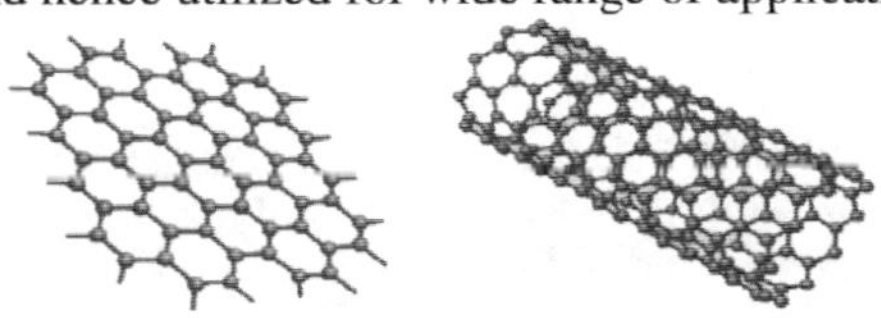

Figure 3. Structure of Graphene (left) and Carbon Nanotubes (right) structure

2. NCEST methodology

The Nano-Carbon Enhanced Squeeze Treatment (NCEST) is a proposed methodology that focuses on enhancing the adsorption of scale inhibitors onto the rock surface to enhance the squeeze treatment lifetimes. This is done by modifying the chemistry of the rock surface using ethyl cellulose as a binder agent. NCEST methodology can be explained in the following steps:

i. Deposition of a binder on a surface of the core sample.
ii. Coating nanomaterials to the surface of the core sample to allow adherence between the GO and CNTs nanomaterials and ethyl cellulose by chemical interaction by the aid of core flooding.
iii. Coating certain concentration of scale inhibitor in the modified core by the aid of core flooding.
iv. Adsorbed scale inhibitor is gradually leached-out from the core during water production.

This methodology is divided into three experiments namely experiment A, B and C as illustrated in figure 4. Experiment A tests out the feasibility of employing nanomaterials to enhance scale inhibitor squeeze treatment by conducting adsorption test at static condition. Whereas experiment B and C test feasibility of employing nanomaterial in dynamic conditions by the aid of core flood testing.

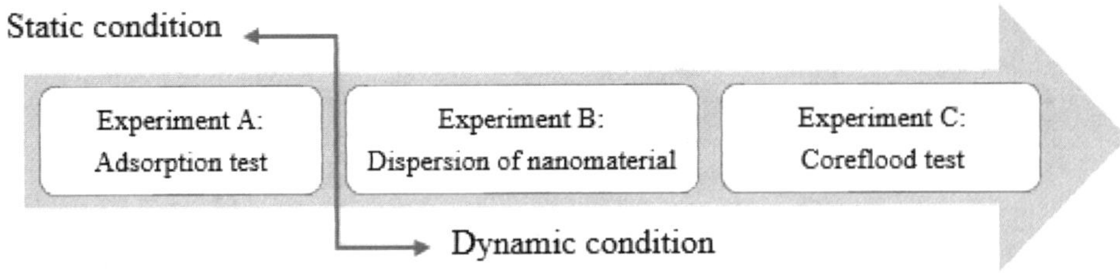

Figure 4. Experimental division of NCEST methodology

3. Materials and methods

Multi-walled carbon nanotubes (MWCNT) were supplied in powder form by Carbon Nano-material Technology Co. Ltd with ~15 nm diameter and ~5 μm length. Reduced Graphene Oxide was procured from Sigma-Aldrich in the form of aqueous dispersed solution stabilized with poly (sodium 4-styrenesulfonate). Calcium chloride dihydrate ($CaCl_2.2H_2O$) and ethylene-diamine tetraacetic acid (EDTA) were used as provided from laboratory. Ethyl cellulose and Sodium Dodecyl Sulphate (SDS) surfactant in powder form were supplied by Merck. Probe Sonicator (model: Q-sonica 500) was used to mechanically disperse nanofluid solutions. Ultra-visible (UV-vis) spectrophotometer (model: Agilent Technologies – Cary 60) was used to characterize concentrations of the prepared solutions. Coreflood tests were performed using Vinci-Technologies Bench Top Permeameter system (BPS-350). The porosity and permeability of Berea sand stone core plug was determined by using Vinci-Technologies Poroperm Coval 30 equipment. Scanning Electron Microscope (SEM) (model: Hitachi H-7500) was used for micro-imaging analysis. PVDF Syringe Filter with pore size of 0.45 μm supplied from Ricco Labstore was used to filter the prepared solutions.

3.1. Experiment A: Adsorption test

The first part of this NCEST methodology involves examination of adsorption of scale inhibitor onto the nanomaterials. It is the most fundamental part in this study as it gives indication of improvement in squeeze lifetimes. EDTA scale inhibitor was employed in this treatment. Adsorption tests were conducted both in static and dynamic conditions. For static conditions, the experiment was carried out by simply stirring the nanomaterials with EDTA for 24 hours in ambient temperature and pressure.

3.1.1. Experimental procedure. 300 ml of 0.05 M EDTA in distilled water (DW) was prepared. 4 ml sample of EDTA in DW solution was taken to be used as initial concentration. 300 ml of brine solution was prepared by diluting 1147.8 mg of calcium chloride dihydrate ($CaCl_2.2H_2O$) salt in 300 ml DW. 0.05 M EDTA was prepared in the brine solution. 4 ml sample of EDTA in brine solution was taken as initial concentration. 50 ml of 0.05 M EDTA in water/brine solution was then poured into 6 beakers. Desired amount of nanomaterial (refer table 1) was added into each of the beaker and magnetically stirred using magnetic stirrer. After 1 hour, 4 ml of the solution was filtered using 0.45 μm syringe filter for UV-Vis measurement. The solution was then left to be stirred for another 23 hours. After 24 hours, 3 ml of the solution is taken and filtered using syringe filter for UV-Vis measurement.

Table 1. Adsorption test preparation

Sample	Nanomaterials	Weight
A	CNTs	100 mg
B	GO	10 ml
C	CNTs + GO	50 mg CNTs + 5 ml GO

3.2. Experiment B: Dispersion of nanomaterials
Nanoparticles have the tendency of being conveniently agglomerate due to van der Waals forces of attraction which may plug the porous medium. Dispersing the nanomaterial is of critical importance in order to avoid plugging of the wellbore during nanofluid injection. Several dispersing agents such as surfactants have been identified as being principally good at dispersing carbon-based nanomaterials such as GO and CNTs [13], which include N-methyl-2-pyrrolidone (NMP), Sodium Dodecyl Sulphate (SDS), Dimethyl Sulfoxide (DMSO), N,N-dimethylformamide (DMF) etc. Highly dispersed CNTs and GO enables to formulate stabilized nanofluid that can offer high active surface area to chemically bind scale inhibitor chemical for enhanced squeeze treatment along with superior rheological characteristics for its injection into the reservoir.

In order to attain high dispersed of CNT/GO nanofluid, 1000 ml of 2% Sodium Dodecyl Sulphate (SDS) surfactant was prepared by diluting it in distilled water (DW). Three beakers each were filled with 200 ml of the SDS solution. Desired amount of nanomaterial (refer Table 2) was added into the SDS solution followed by putting the beaker in an ultrasonic bath for 2 hours followed by sonication for 5 min for each prepared solution. The solutions were left static under the influence of gravity for 24 hours for any unfunctionalized and undispersed CNTs/GO to precipitate and settle at the bottom of the beakers. Only the upper solution suspension (top half) was used in the next experiments to ensure that the nanomaterials used are well dispersed and eviscerate the possibility of agglomerated nanoparticles.

Table 2. Dispersion test preparation

Beaker	Nanomaterials	Weight
A	CNTs	40 mg
B	GO	4 ml
C	CNTs + GO	20 mg CNTs + 2 ml GO

3.3. Experiment C: Coreflood testing
Coreflood testing was performed to simulate NCEST methodology (refer 2.0) at dynamic conditions by using Vinci-Technologies Bench Top Permeameter system (BPS-350). The CNTs/GO nanofluid was injected into the Berea sandstone core plugs. The length, diameter and weight of the core plugs were measured using Vernier caliper and weighing balance, while their porosity, permeability and pore volume were calculated by Poroperm test using Poroperm Coval 30 equipment.

10 PV of brine solution was prepared by adding 3.812 g/l sodium chloride (NaCl) and 3.826 g/l calcium chloride dihydrate ($CaCl_2.2H_2O$) powder in distilled water. 10 PV of 1775 ppm EDTA and DW solution was prepared by adding EDTA salt into distilled water and stirred using magnetic stirrer. The coreflood test was performed at room temperature at the flow rate of 1 ml/min. 10 PV of cellulose was injected into the core followed by shutting it for 1 hour. 10 PV of dispersed nanomaterial solutions were injected into the core shutting it for 24 hours. Then, the core was rinsed with 10 PV of brine solution. 10 PV of EDTA in DW solution was injected into the core and the effluent samples were taken for UV-Vis measurement. The core was again shut for further 24 hours. Post-flushing was performed using DW again UV-Vis measurement of the effluent was performed. Lastly, SEM analysis was conducted on a miniscule piece of the core plug.

3.3.1. Experimental procedure: Baseline core flooding. 10 PV of EDTA and DW solution was injected into the core and effluent sample was taken for UV-Vis measurement. The core was shut for 24 hours to enable the EDTA to adsorb on the rock. The core was afterwards rinsed with 10 PV of brine solution. Background solution (DW) is pumped into the core and effluent samples were taken for UV-Vis measurement. Core sample was analyzed for FESEM micro-imaging.

4. Results and discussion

4.1. Stage 1: Adsorption Test

UV-Vis spectrophotometry is a type of adsorption spectroscopy which operates in ultraviolet-visible region. The concentration of a particular solution is determined by degree of incident light absorbance by the solution which is translated into a concentration value by using Beer-Lambert law. This law states that the absorbance of a solution is directly proportional to the concentration of the absorbing species in the solution and the path length [14]. Concentration of an unknown solution can be calculated by preparing a standard solution with pre-set known concentrations. The standard can range from the lowest to maximum concentrations the sample can achieve. In this experiment, the initial concentration of EDTA in DW/Brine was set at 0.05M [3]. After the adsorption experiment, its concentration is expected to decrease. Hence, the concentrations of standard solutions that were prepared were EDTA in DW/Brine of 0.01M, 0.02M, 0.03M, 0.04M, 0.05M and 0.06M.

Flocculated CNTs are not active in the UV-vis region [15, 16]. Only individual singled-out CNTs absorb in this region. In order to characterize the dispersion and concentration of CNTs in dispersant solution, UV-vis spectroscopy absorbance values were recorded at 200 nm illustrated by UV-vis spectra as shown in figure 5. After the standard solutions were prepared and tested using UV-vis equipment, plot of absorbance vs concentration was drawn as shown in figure 6. The absorbance value must be taken at the same wavelength of 200nm as a control. It is assumed that at this wavelength, the Beer-Lambert law is well-obeyed by CNTs and absorbance is minimally affected by the ambient conditions of CNTs. Plotted graph will validate the assumption of Beer-Lambert law by looking for a straight-line relationship of the data.

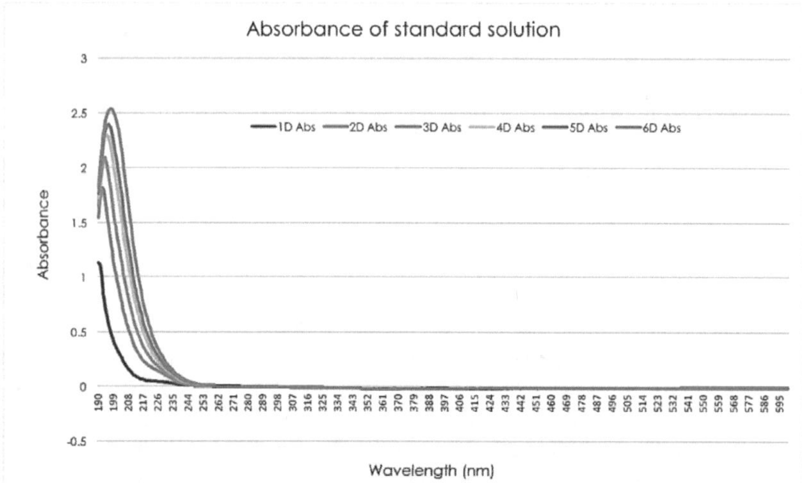

Figure 5. Absorbance spectra of UV-Vis measurement

The line plot shown in figure 6 can be used to determine the concentration of an unknown sample from its absorbance value by interpolation. This is done by drawing a horizontal line from the y-axis at samples' absorbance value until it intersects the best fit line. Then, a vertical line is drawn from that point to the x-axis. The x-value at this point is the concentration of the sample. The second method that can be used to determine the concentration is by simply calculating the value using equation of the straight line of the graph as shown in equation (1).

$$y = 42.451x + 0.0694 \quad (1)$$

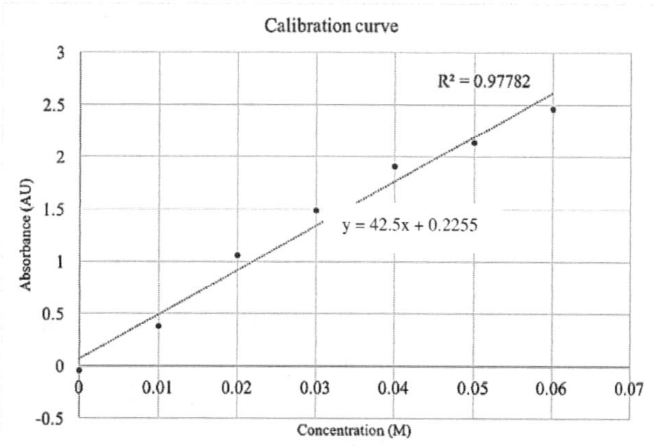

Figure 6. Plot of standard solution

Based on the samples' absorbance values measured by UV-Vis spectrophotometry testing, bar charts of concentration of EDTA in DW/brine against types of nanomaterials were constructed and shown in figure 7 and figure 8.

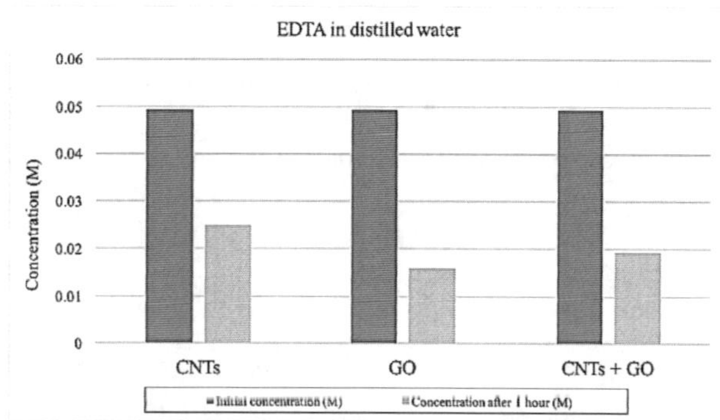

Figure 7. Concentration of EDTA in DW before and after 1 hour of stirring

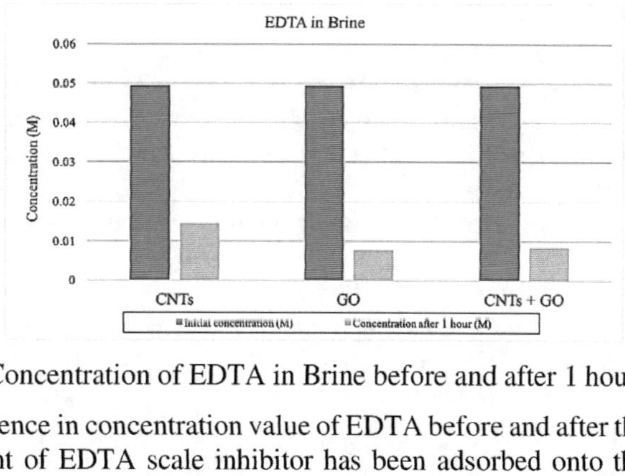

Figure 8. Concentration of EDTA in Brine before and after 1 hour of stirring

The significant difference in concentration value of EDTA before and after the stirring process shows that a significant amount of EDTA scale inhibitor has been adsorbed onto the nanomaterials. Quick adsorption of EDTA on the nanomaterials can also be observed from this experiment as only 1 hour is needed for the EDTA to be effectively adsorbed. From the initial and final concentrations of EDTA in DW/brine for different nanomaterials, the concentration reduction of EDTA in varying nanomaterials was calculated to clearly differentiate which nanomaterials managed to reduce the highest concentration of EDTA over an hour. Apart from that, the adsorption capacities (q) of the nanomaterials were also calculated using equation (2) as below. Its results are tabulated in table 3.

$$q = \frac{(C_i - C_f) \times V}{W} \quad (2)$$

q = Adsorption capacity (mg/g); C_i = Initial concentration (mg/l); C_f = Final concentration (mg/l); V = Volume of adsorbent used (l); W = Weight of the adsorbent (g)

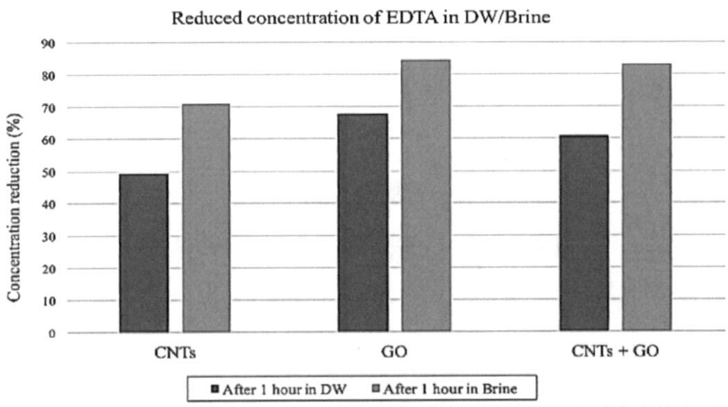

Figure 9. Concentration reduction (%) of EDTA in different nanomaterials and background solution

Table 3. Adsorption capacity of different nanomaterials in varying base solution

Adsorption capacity q (mg/g)	CNTs	GO	CNTs + GO
DW	432.9	594.4	532.8
Brine	621.8	739.4	728.3

Figure 9 and table 3 shows that the best performing nanomaterial to be Graphene Oxide (GO). This can be clearly seen as GO gives the highest reduction in EDTA concentration compared to CNTs and CNTs + GO in both distilled water and brine as the background solutions. This may be due to the structural properties of the GO which are analogous to graphene. Since graphene has a more sheet-like structure when compared with CNTs which are cylindrical nanostructure, graphene particles have lower tendency to agglomerate with each other. On the other hand, CNTs are more prone to agglomeration due to the presence of van der Waals force acting between them. Agglomeration of the CNT nanoparticles reduces its efficiency as an adsorbent as less active surface area is available for adsorption of scale inhibitor. When compared with CNTs, GO provides higher activated surface area rendering more potential to adsorb higher amounts of scale inhibitor required for effective scale squeeze treatments.

Apart from the different nanomaterials, type of background solution used also plays a significant role in enhancing the adsorption of scale inhibitor onto the nanomaterials. In this experiment, two types of background solutions were used which are distilled water and brine. Brine is a solution of calcium chloride dihydrate ($CaCl_2.2H_2O$). The purpose of using brine is to investigate the adsorption ability of EDTA scale inhibitor in the presence of calcium ions (Ca^{2+}). From results shown in figure 9, it was observed that the adsorption of EDTA in brine as the background solution is consistently higher than in distilled water. This indicates that the presence of calcium ions plays an important role to bond/adsorb scale inhibitor with the nanomaterial. The calcium ions in the brine solution calcium-functionalize the nanomaterial which conveniently bonds with the negatively charged scale inhibitor molecules as shown in figure 10.

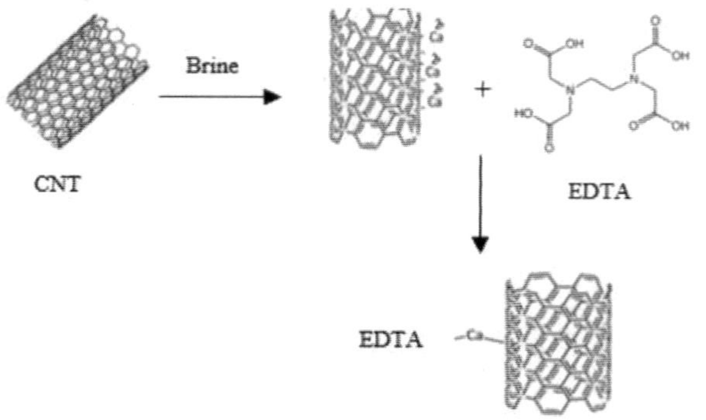

Figure 10. Schematic of EDTA adsorption on CNT in brine solution in the presence of calcium ions

From static adsorption test (Experiment A), it can be concluded that GO is the better nanomaterial to be used as an adsorbent to enhance scale inhibitor adsorption ability. GO was screened and used in the next stage of NCEST methodology which is the dynamic adsorption test.

4.2. Stage 2: Coreflood test

After necessary injection, flooding processes were performed at the flow rate of 1 ml/min. The effluent samples from the coreflood equipment were collected and sent for UV-Vis spectrophotometry measurement. To measure the effectiveness of using nanomaterials to enhance the adsorption of scale inhibitor into the formation rock, the experiment was conducted with two core samples; one with nanomaterial injected and the other without nanomaterial injected (baseline). The results of these two core samples were compared and further analysis were carried out. To analyze the concentration of samples before and after the coreflood run, the process goes through the same flow scheme as in Experiment A. The absorbance value at 200 nm wavelength was taken and interpolated in the calibration curve to reveal its concentration value. After, a graph of concentration against core samples was drawn.

From the results illustrated in figure 11, there is a marked difference in SI concentrations for core sample treated with GO nanomaterial compared to the core sample treated without GO, showing that more adsorption takes place on the core sample that was treated with nanomaterials. This is due to the coating and penetrative potential of GO that provides high surface area platform with active sites for the adsorption of EDTA scale inhibitor on the substrate. Low value of final concentration of EDTA also implies that there is a good retention of EDTA on the formation. The adsorption capacity of EDTA for both cases are calculated and illustrated in figure 12 and table 4.

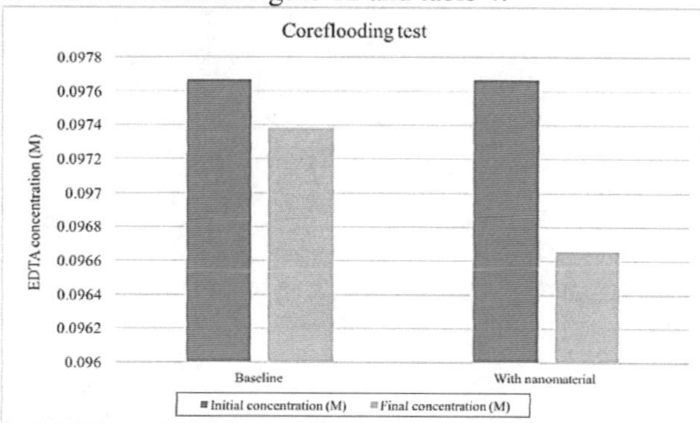

Figure 11. Concentration of EDTA before and after shut in for 24 hours

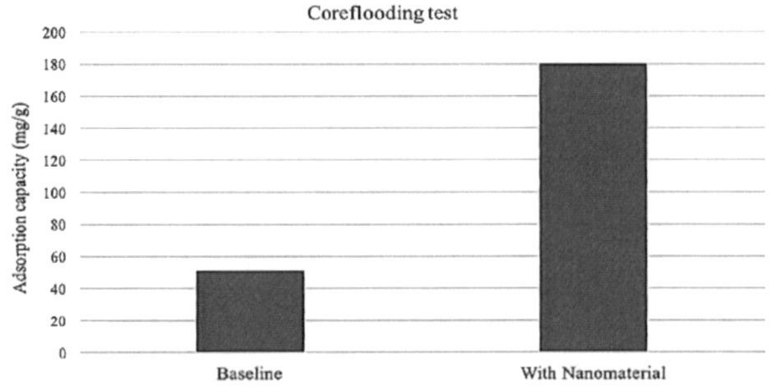

Figure 12. Adsorption capacity of EDTA on core samples

Table 4. Adsorption capacity of EDTA on core samples

Core sample	Adsorption capacity (mg/g)
Baseline	51
With nanomaterials	180

In this dynamic adsorption testing, GO is used as the nanomaterial rendering maximum adsorptive capacity of 180mg/g for EDTA. While in static adsorption testing (Experiment A), the adsorption capacity of EDTA on GO is approximately ~590 mg/g. This is because in dynamic adsorption test, not all of the surface area of GO is in contact with the EDTA. Coreflood test can best simulate the scale squeeze treatment process yielding pragmatic adsorptive capacity estimate value which provides a clear indication of the degree of effectiveness of the treatment.

4.3. Field Emission Scanning Electron Microscopy (FESEM) analysis

After coreflood testing, a small fraction of the core samples was tested for FESEM analysis. This analysis was done to examine the topographical and elemental information of nanoparticles and EDTA on the rock surface at a high magnification. Images in figure 13 depicts the graphene structure on the rock surface, which is clear evidence of nanomaterials attachment on the rock surface. In addition, this FESEM imaging tool has added feature of Energy Dispersive X-ray (EDX) system that analyzes the elemental content on the rock. Results from EDX can be pre-programmed to show the atomic percentage of desired elements as shown in table 5.

Table 5. EDX elemental analysis of core samples

	CORE SAMPLES			
Weight (%)	Baseline (initial)	Baseline (after)	With GO (initial)	With GO (after)
Carbon, C	-3.05	10.83	6.14	49.23
Nitrogen, N	7.78	8.78	2.52	11.3
Oxygen, O	61.3	50.65	63.04	32.59

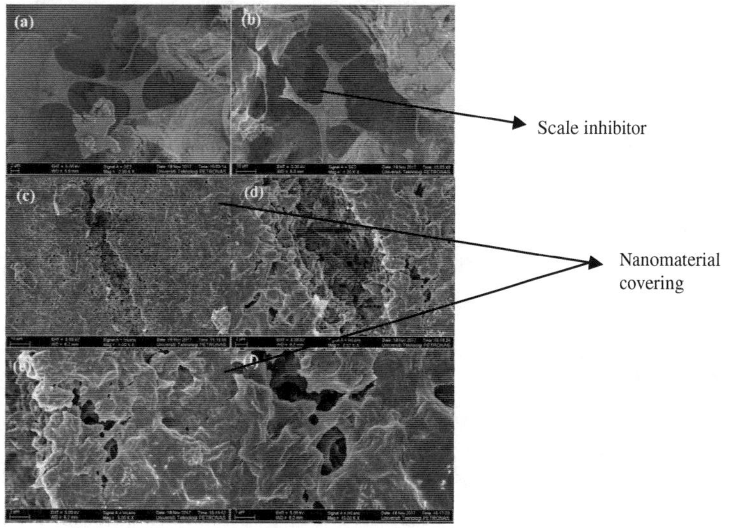

Figure 13. FESEM images of scale inhibitor and graphene particles on rock surface

From the chemical formula of EDTA (C10H16N2O8), the elements that make up this acid are Carbon, Nitrogen and Oxygen. Given results from the EDX analysis shows that the weight percentage of these three elements exist abundantly after the injection of EDTA on both baseline and with nanomaterials core samples. This indicates that the EDTA scale inhibitor has adsorbed onto the rock formation. However, the increment is higher in core sample that was treated with GO nanomaterial. This further supports the result from coreflood testing, which manifests that the adsorption of EDTA onto rock formation treated with nanomaterials to be higher compared with sample without any nanomaterial adsorbance.

5. Conclusion

The main objective of developing new methodology of optimized nanomaterial namely graphene oxide (GO) and Carbon Nanotubes (CNTs) in expanding scale inhibitor squeeze lifetime was achieved successfully. Based on the outcome from this study, nanomaterials have proven to be effective as an agent to increase and facilitate the adsorption of scale inhibitor onto the rock surface. In this study, the authors managed to investigate the performance of two types of nanomaterials. It was found that GO gives the best performance in enhancing squeeze treatment lifetime due to its structural properties compared CNTs. It was also proven that treating the rock surface with nanomaterials can significantly increase the adsorption rate of EDTA scale inhibitor on the formation rock. Adsorption capacity of EDTA on core sample treated with GO was 180mg/g while the core sample with no nanomaterial treatment was to be only at 51 mg/g.

The results show a promising outcome in optimizing nano-based materials to facilitate better adsorption of scale inhibitor. Further development of this research can include:

- Investigating optimum dispersant concentrations to achieve highest dispersions for different nanomaterials.
- Studying the desorption dynamics of scale inhibitor from nanomaterials to ensure that the scale inhibitor can effectively desorbed back into the produced water providing scale inhibitor chemical concentrations to be above the threshold value.
- Probing rock impairment in terms of reduction of porosity and permeability by injecting rock pretreatment binder agent and nanomaterials.
- Applying NCEST methodology on different types of carbon-based nanomaterials or scale inhibitor chemical as different scale inhibitor may have varying compatibility with nanomaterials.

6. Recommendation

Although the experimented concept was encouraging, there are still several areas in the methodology that needs to be extensively studied and refined. This can be done by performing more experimental work that can simulate environments that are close to the real reservoir conditions. Rock impairment was a primary apprehension and concern. The impact of nanomaterial type, injection velocity and salinity on the permeability and porosity of reservoir rock needs to be further investigated in depth.

References

[1] Crabtree M, Eslinger D, Fletcher P, Miller M, Johnson A and King G 1999 Fighting Scale Removal and Prevention *Oilfield Rev.* **11** pp 30-45
[2] Olajire A A 2015 A review of oilfield scale management technology for oil and gas production *Journal of Petroleum Science and Engineering* **135** pp 723-37
[3] Moghadasi J, Jamialahmadi M, Müller-Steinhagen, H and Sharif A 2007 Scale Deposits in Porous Media and Their Removal by EDTA Injection. *Proc. of 7th International Conf. on Heat Exchanger Fouling and Cleaning – Challenges and Opportunities*
[4] Ghorbani N, Wilson M C, Kapur N, Fleming N and Neville A 2014 CarbonNanotubes: A New Methodology for Enhanced Squeeze Lifetime CNTs. *SPE International Oilfield Scale Conference and Exhibition*
[5] Ghorbani, N, Wilson M, Kapur N, Fleming N, Tjomsland T and Neville A 2017 Adsorption of polyphosphinocarboxylic acid scale inhibitor on carbon nanotubes: A prospective method for enhanced oilfield scale prevention *Journal of Petroleum Science and Engineering* **150** pp 305-11
[6] Shen D, Zhang P, Kan A T, Fu G, Alsaiari H A and Tomson M B 2008 Control Placement of Scale Inhibitors In The Formation With Stable Ca-DTPMP Nanoparticle Suspension And Its Transport In Porous Medium.*SPE International Oilfield Scale Conf.*
[7] De Volder M F, Tawfick S H, Baughman R H and Hart A J 2013 Carbon nanotubes: present and future commercial applications *Science* **339** pp 535-39
[8] Pop E, Mann D, Wang Q, Goodson K and Dai H 2006 Thermal conductance of an individual single-wall carbon nanotube above room temperature *Nano letters* **6** pp 96-100
[9] Kaempgen M, Duesberg GS and Roth S Transparent carbon nanotube coatings 2005 *Applied Surface Science* **15** 425-29
[10] Zhao J, Liu L and Li F 2015 Graphene oxide: physics and applications 57-150 (Amsterdam, The Netherlands: Springer)
[11] Perrozzi F, Prezioso S and Ottaviano L 2014 Graphene oxide: from fundamentals to applications *Journal of Physics: Condensed Matter* **27**
[12] An S J, Zhu Y, Lee S H, Stoller M D, Emilsson T, Park S, Velamakanni A, An J and Ruoff R S 2010 Thin film fabrication and simultaneous anodic reduction of deposited graphene oxide platelets by electrophoretic deposition *The Journal of Physical Chemistry Letters* **1** pp 1259-63
[13] Rastogi R, Kaushal R, Tripathi S K, Sharma A L, Kaur I and Bharadwaj L M 2008 Comparative study of carbon nanotube dispersion using surfactants. *Journal of colloid and interface science* **328(2)** 421-428
[14] Mehta A, 2012 Ultraviolet-Visible Spectroscopy-Derivation of Beer-Lambert Law *Analytical chemistry*
[15] Yu J, Grossiord N, Koning C E and Loos J 2007 Controlling the dispersion of multi-wall carbon nanotubes in aqueous surfactant solution *Carbon* **45** (3) pp 618-23
[16] Lauret J S, Voisin C, Cassabois G, Delalande C, Roussignol P, Jost O and Capes L 2003 Ultrafast carrier dynamics in single-wall carbon nanotubes *Physical review letters* **90**

IOP Conf. Series: Materials Science and Engineering **458** (2018) 012045 doi:10.1088/1757-899X/458/1/012045

Integrated constraints optimization for surface and sub-surface towards CAPEX free maximizing production

S A Zainal[1], W R Daud[2], Z A Putra[3] and N Nor[3]

[1]PETRONAS Group Technical Solution, Project Delivery and Technology, PETRONAS, Kuala Lumpur 50050, Malaysia
[2]Chemical Engineering Program, Universiti Kebangsaam Malaysia, Bangi 43000, Selangor, Malaysia
[3]Chemical Engineering Department, Universiti Teknologi PETRONAS, Seri Iskandar 32610, Perak , Malaysia

Abstract. Integrated process operations and optimization are highly demanded in petroleum industries due to increasing costs. One of the advances is monitoring the integrated process by using single modeling and optimizing platform. This paper presents iCON, PETRONAS owned process simulation software, to predict the optimum pressure set point for separator train and choke valves' opening, to maximize production in an existing facility. The source code developed was used to generate pressure and flow equation for each well to obtain material compositions for iCON model. Starting from current topside operating separator pressure, multiple case studies were run in iCON to produce production profiles at different operating pressures. The optimum pressure will be iteratively analyzed against the process constraint to locate the global optimum topside separator pressure set point. A single surface and sub-surface modeling platform ensures seamless data transfer, thermodynamic stability and efficient optimization iteration. The findings from this research are considered reaping the additional gas and oil production just by changing surface facilities separators set points and individual choke valve opening without violating process constraints.

1. Introduction

Fossil fuel continue to be widely used worldwide as primary energy resources, this is due to new resources of sustainable energy have yet to be developed. These conditions have motivated researches in the area of oil and gas production to explore new approaches to the application of optimization techniques to maximize oil and gas production rates and to minimize production cost.

Conventionally a standalone topside process model can only achieve production increment of up to 1%. In order to achieve higher production increment, an integrated topside process and subsurface model is required [1]. This modeling technique requires the integration of different modeling platforms for reservoir, wells, headers, topside facility, interconnecting pipelines between platforms and re-injection wells back to the reservoir. The main challenge is to develop a seamless interface communication between the different modeling platforms.

An integrated surface and sub-surface model and optimization contribute an important role during production because an integrated model can produce comprehensive and realistic operational recommendations. An integrated field model combined with optimization presents many link-ability challenges in terms efficient algorithms to eliminate thermodynamic instability when couple the models with optimization and sufficient hardware capacity to run the complex models.

Standalone topside optimization approach focuses on equipment performance which includes separators and compressors efficiencies as well as stable operation which can be achieved by proper process control strategy. By executing this approach, current performance of the processing facilities typically be benchmarked against the design capacity. The outcome of the work includes new operating separator pressure to ensure efficient gas and oil separation, new compressor speed or control configuration to meet gas export as well as gas lift/injection requirements, and new and/or improved control scheme with better tuning parameters for stable operation with the optimum production set points. By having all the recommendations implemented, the processing platform is able to achieve process stability with minimal tripped and unplanned shutdown.

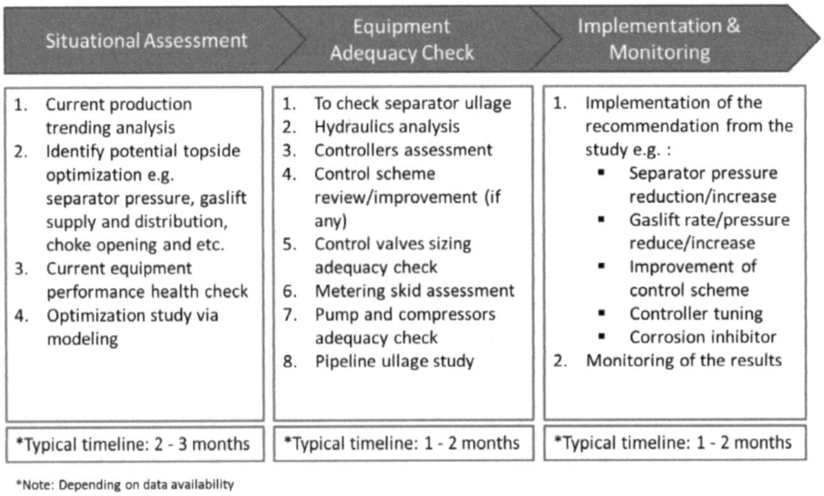

Figure 1. Typical methodology for standalone topside optimization study.

Standalone topside optimization focuses on process stability which reduces the potential of production loss due to tripped or plant shutdown but it does not indicates the impact from topside operations to subsurface and vice versa. Unstable production in multiphase production systems and pipelines can cause serious operational problems for downstream receiving production facilities. Hence, integrated surface-subsurface optimization is the solutions to achieve global production optimization.

Several researchers have conducted research on various integration works Bailey et al [2] and Cullick et al [3] discussed complex petroleum field projects applying uncertainty analysis but the surface process facility was not considered. Integrated ECLIPSE and HYSYS simulators to model integrated field operation in a deepwater oil field was also conducted. These simulators were coupled by using Automation and Parallel Virtual Machine and apply a genetic algorithm for the optimization [4]. The use of integrated optimization in a daily operations setting of the LNG value chain was studied by [5]. To reduce computational time they chose simple models for all system components. A sizable gain could be identified by integrating all models into one model platform as opposed to many modeling platforms. Integrated operation and optimization to represent the value chain from reservoir to export terminal was also studied by Rahmawati et al [6] using realistic assumptions and parameter values. An integrated case study from reservoir to the surface processes and a simplified economic model to maximize field economics was also studied by Rahmawati et al [7].

In this paper an integrated first principle compositional thermodynamic model from well to surface processes is developed in one single platform iCON. The optimization objective function is to locate a CAPEX free operation settings by manipulating top site operating pressure and individual choke valve opening by satisfying surface facilities constraints. The model will create the operating profile to satisfy the maximum production envelope. Top site equipment capacities check are considered as global optimization constraints.

2. Modeling Strategy

An integrated field and optimization models play an important role during production optimization due to its ability to recommend comprehensive operational set points. Zainal and Hussein [1] confirmed that in order to achieve higher production increment, an integrated top site processes and sub-surface model is required. This modeling technique integrates the models using different modeling and thermodynamic platforms for reservoir, wells, headers, top site processes, interconnecting pipelines and reinjection well back to reservoir and/or wells.

Litvak et al [8] developed an integrated compositional model of the reservoir and surface facilities. This model using multiple modeling platforms and only valid to daily optimization for small fields due to unstable thermodynamic translation among the multiple platforms. A practical approach to use Vertical Lift Performance (VLP) curves generated by PROSPER software, to predict wells performance and production at different top site pressure set points, is used. High and medium headers complex network hydraulic and pressure balance analysis typically model in PIPESIM will be modeled within iCON steady state environment as a single modeling platform to avoid thermodynamic translation complications.

In this paper, the significant of using single integrated optimization modeling platform is elaborated and reviewed the effect of well lineup to the optimization interim result. Previous works on surface and subsurface are compared and discussed to prove the significant of this work. Iterative method to locate CAPEX free optimization point by manipulating well lineup, top site pressure, individual choke valve opening are discussed. Finally, CAPEX free optimization envelope is developed to cater for all possible manipulating variables changes.

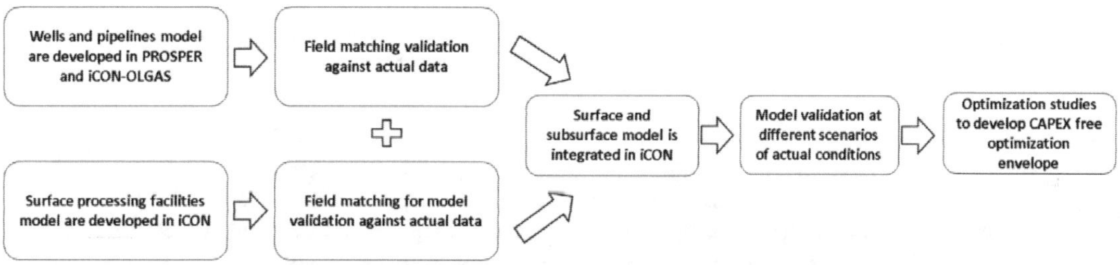

Figure 2. Methodology for Integrated Optimization study.

Figure 2 explains the overall approach for surface sub-surface integrated optimization where the subsurface and topside facilities are modelled in iCON environment. Once the model has been validated with actual data e.g. well test data, pipeline data, topside operating conditions and gas, oil, and water production rate, then the model is ready to be optimized with process constraints.

This research aims to develop an integrated surface and subsurface model from individual well, pipelines and surface process facilities in single modeling platform namely iCON. The integrated model is used to generate CAPEX free optimization region towards maximizing oil or gas production at different well line up scenarios. Reservoir model is not considered in this study as the fact that they operate in different time scale.

The vertical lift well pressure-flow curves, intersecting points between Inflow Performance Curves and Out Flow Performance Curves, data use nodal analysis generated from PROSPER which consists of relationship of Tubing Header Pressure (THP), Water Cut, Gas Oil Ratio (GOR) and gas lift rate with well production rate (gas, oil and water) as illustrated in Figure 3. The subsurface optimization variables consist of gas lift rate, water cut and GOR while topside optimization variables are individual well choke valve opening and individual header pressure which translated to separators pressure. Multivariable regression tool will be coded within iCON software to obtain the correlations between THP, gas lift rate, water cut, GOR to the individual well production.

The source code developed will then be used to generate flow equation for each flowing well to get the final composition to iCON model. Starting from current topside operating separator pressure, multiple case studies will be run in iCON to produce production profiles at different operating pressures and individual choke valve opening to locate maximum oil/gas production. The optimum pressure will be iteratively analyzed against the process constraints to locate the global optimum point at topside maximum production.

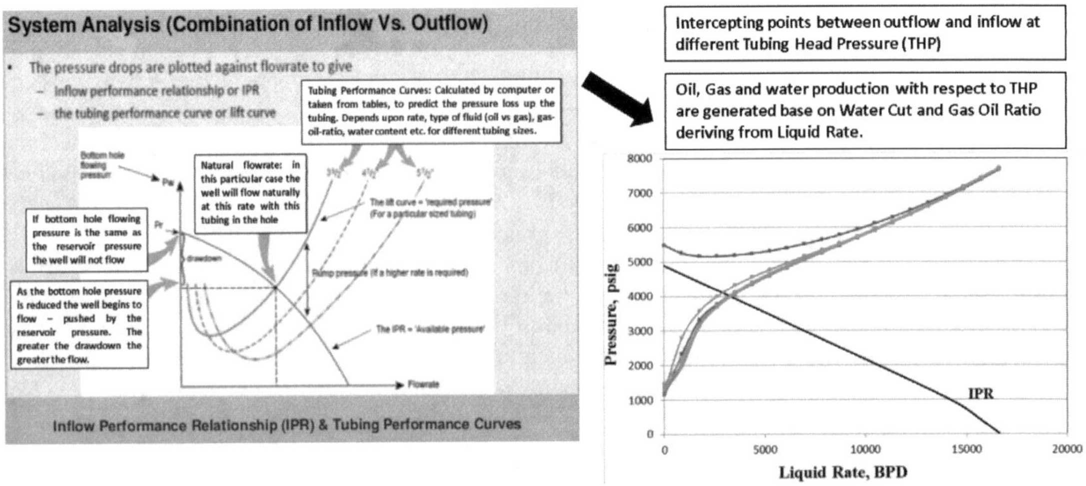

Figure 3. Individual Well Vertical Lift Performance (VLP) Curves.

All the individual well models will be combined by production platform and the production rates for each platform is calculated and subsequently connected to respective production separator model in iCON. The topside optimization is conducted by manipulating the individual header pressure and choke valve opening to each well to determine the impact of those two parameters to gas and oil production from each platform. Figure 4 illustrates the process flow diagram of well models and processing facilities model in iCON. The data use in the well models are the sensitivity analysis from PROSPER and the data used for processing facilities are taken from the installed equipment datasheet and current operation and production data.

Typically, the topside process is controlled using a distributed control system, while the wells are controlled manually by the operators through the choke valve stroke positions. Although this separate control system may not be optimal with respect to maximizing the full processing capacity of the surface facilities, its high operational reliability due to simple design complexity is an advantage. In addition to the natural variation in the flow rates and the multi-phase flow in each well, its dynamics are affected by the choke valve position and the gas lift rate. Also, there is the additional complexity introduced by using multiple production wells with individual process dynamics behavior and compositions. Add the dynamics of the topside process to this, and it is obvious that it may be difficult for a human operator to control the production chokes in an optimal manner.

Furthermore, the integrated model of topside and wells becomes more important when the process disturbances become larger (e.g. slugging wells). With unstable wells present, the platform operators tend to back off from the process capacity constraints to avoid process shut downs and flaring of gas in the event of excessive slugs entering the topside process. This in turn yield lower production rates which lead to a loss of revenue. With integrated well and topside model, it will also be able to operate closer to the process constraints with process disturbances such as slugging wells due to well pressure variations [9].

The production rates are set at the inlet of the topside process by use of the stroke position of the production choke valves for each well. There is no automatic feedback from the process to where the production rates are set. The feedback is handled manually by the operators and the performance will naturally vary depending on the operator's action. Especially, when one or more of the wells show natural variation in the feed rates, the feedback is slow and not optimal with respect to process requirements [10].

The process constraints (which limit the throughput) are typically located in the processing unit and the operational objective is to utilize the available capacity (constraint control) to the maximum. Since the production rate is set at the wells, introducing automatic feedback control from the constraints to the feed stream to the process is assumed to have great economic potentials in term of plant efficiency. However, the number of feed conditions (wells, water cuts, GOR, WGR) and different pressure setting options make automatic pressure flow relationship difficult to implement.

Large fields where oil production is in decline (due to high water cut wells) and the identification of an optimized production strategy will be importance to maximizing the return on investment. It is necessary for an integrated well and topside model to be developed to cater for more representative in maximizing oil and gas production [11]. Recent studies have presented multiple model platforms the required suitable and accurate time stamp for optimization to work. Look up tables were also used to determine multiphase flow from each reservoir without considering surface facilities installed capacity as constraint variables. Potential revenue predicted from recent studies on integrated model was based on single point optimization without giving optimization validity envelope which is difficult to implement.

3. Modeling Results

There are two models were developed in this work namely iCON steady state and iCON dynamic. iCON steady state is focused to optimizing the separator train pressure and individual choke valve opening to maximize production. Dynamic model is used to check the controllability show stoppers due to additional production arising from the new optimized set points coming from steady state model. The comprehensive Surface and Sub Surface models were developed in iCON to represent the interaction between multiple well behaviors and surface process facilities to generate optimized operating envelope at different active/inactive well configurations.

Figure 4 shows the surface and sub-surface simulation model in iCON. Due to the space constraint, only small parts (e.g. well connection in Fig. 4(a), and low-pressure system of a surface facility in Fig. 4(b)) of the overall simulation network is shown. Separator train pressures are optimized in surface model while individual choke valve opening is optimized in sub-surface model. In this work iCON optimizer is used to locate the optimization point and eventually the optimization envelope.

The integrated model is tested and compared against the operating envelope to ensure model robustness prior to production optimization. Optimization objective function (OF) is setup to maximize the oil and gas production. Manipulative variables (MV) include individual header pressure and individual choke valve opening at each well. Existing equipment design are used as the constraints such as capacities of the pipes and valves.

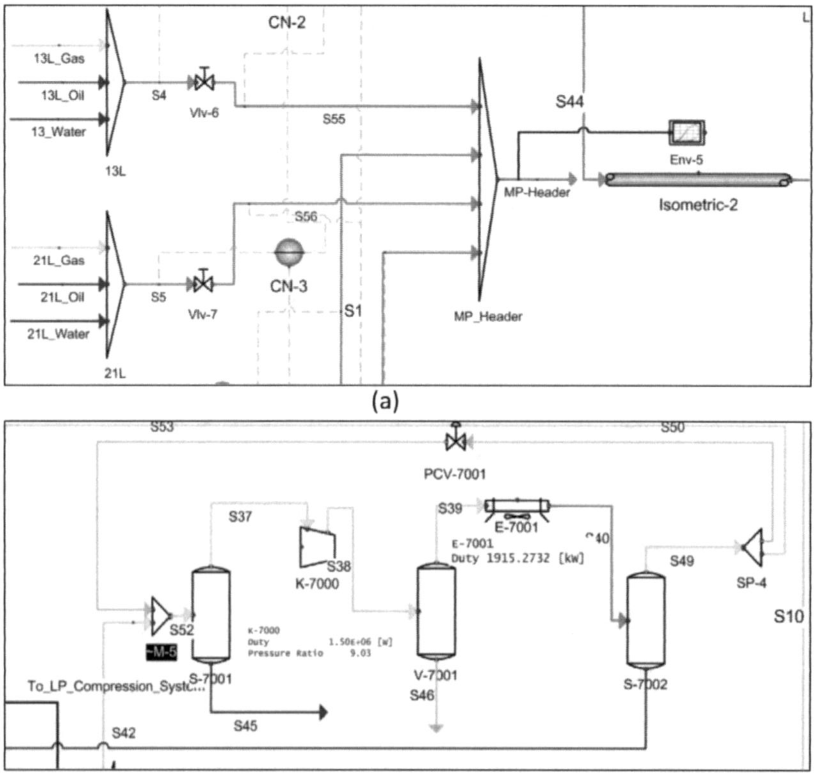

Figure 4. Typical flow diagram of (a) well model connected to a Medium Pressure (MP) header, and (b) some part of surface processing facilities.

Once the constrained optimization point is located, the optimization procedures are repeated at different well line up scenarios. Finally, the optimization validity envelope is constructed to represent CAPEX free optimization region. Table 1 shows the changes of manipulated variables during the optimization run.

Table 1. Optimized manipulated variables versus running period.

Optimization Period	Var 1	Var 2	Var 3	Var 4	Var 5	Var 6	Var 7	Production (bbl/d)	Delta
Initial Value	270	40	50%	50%	50%	50%	800	5669	-
After 3 days	214.5	61.29	51.1%	53.6%	51.1%	51.1%	700	5794	125
After 5 days	212.6	64.26	50.3%	54.1%	49.7%	53%	705.3	5809	15
After 6 days	212.6	64.26	50.3%	54.1%	49.7%	56.2%	705.3	5822	4
After 7 days	212.6	67.47	50.3%	54.1%	49.7%	56.2%	705.3	5826	4
After 8 days	214	68.64	51.1%	55.1%	47.3%	57.2%	707.8	5830	4

Where:
Var 1 : Medium Header Pressure (psig)
Var 2 : Low Header Pressure (psig)
Var 3 - 6: Medium Pressure Well Valve Opening (%)
Var 7 : High Header Pressure (psig)

Production increase:

= (125+15+4+4+4)/5669 = 152/5669 x 100%

= 2.7 % (equivalent to 152 bbl/d of oil production increase)

Figure 5 and 6 are typical iCON Optimizer setup after 3 and 8 days of running respectively. Each day of running results were tabulated in table 1 and plotted as in figure 7. From figure 7, production increase has plateaued after 6 days of running period. Oil production increment is accumulated after it reaches plateau state and is considered as one point in CAPEX free optimization envelope. The same procedures are repeated at different well lineup scenarios to ensure all possible well configurations are covered. Well lineup configuration coverage is crucial to safeguard a realistic optimization set points are applied during the implementation period.

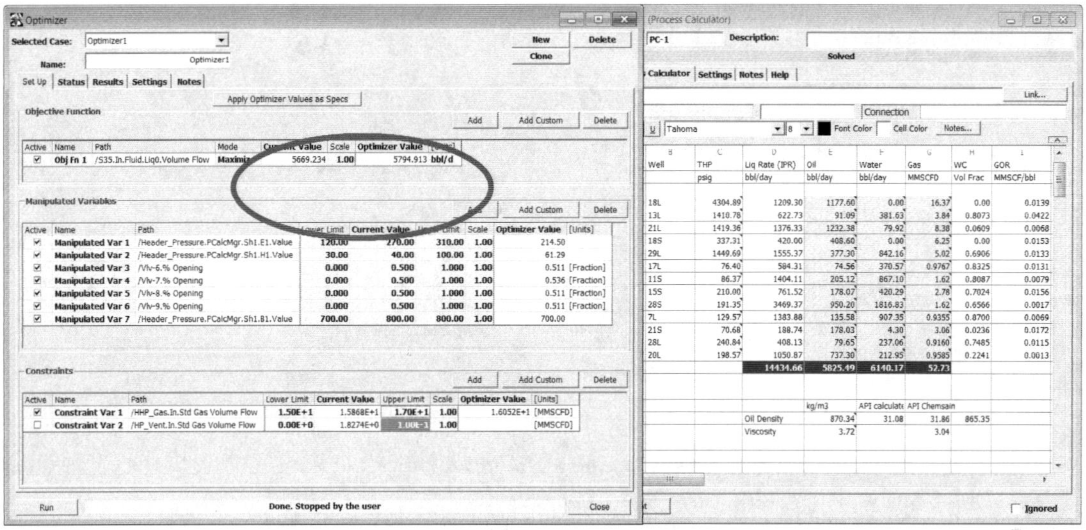

Figure 5. Optimization results after 3 days running.

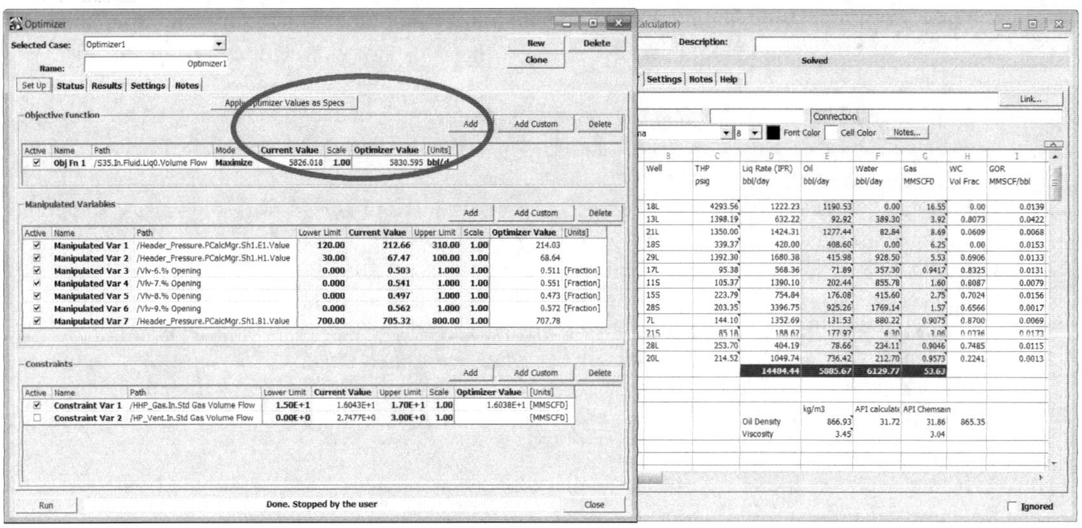

Figure 6. Optimization results after 8 days running.

4. Conclusion

The constrained optimization technique applied above represent realistic optimization set points for implementation within the optimization envelope. Convergence time varies from 5 hours to 6 days to reach plateau increment of oil production depending on top site mode size, well lineup configurations and initial value of steady state point. CAPEX free optimization envelope generated is specific to predetermine well characteristic coming from PROSPER results and only valid as long as the well properties are maintained.

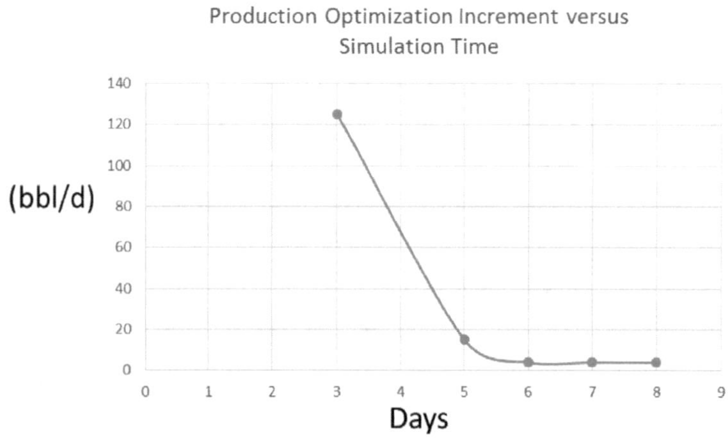

Figure 7. Production increase versus running period.

References

[1] Abidin S A and Hussein N 2014 Integrated Optimisation for Surface and Sub Surface Towards Maximising Production *Offshore Tech. Conf.*

[2] Bailey W J, Couët B and Wilkinson D 2005 Framework for Field Optimization to Maximize Asset Value *Society of Petroleum Engineers* pp 7–21

[3] Cullick A S, Heath D, Narayanan K, April J and Kelly J 2003 Optimizing Multiple-Field Scheduling and Production Strategy with Reduced Risk *Society of Petroleum Engineers*

[4] Nazarian B 2002 Integrated Field Modeling thesis Norwegian University of Science and Technology

[5] Foss B A and Halvorsen I J 2009 Dynamic Optimization of the LNG Value Chain, in: H.E. Alfadala, G.V.R. Reklaitis, M.M. El-Halwagi (Eds.) *Proc. 1st Annu. Gas Process. Symp.*, Elsevier Amsterdam pp 10–18

[6] Rahmawati S D, Whitson C H, Foss B A 2010 A. Kuntadi, Multi-Field Asset Integrated Optimization Benchmark, *Society of Petroleum Engineers*

[7] Rahmawati S D, Whitson C H, Foss B and Kuntadi A 2012 Integrated field operation and optimization *J. Pet. Sci. Eng.* **81** pp 161–170

[8] Litvak M L, Hutchins L A, Skinner R C, Darlow B L, Wood R C, Kuest L J and Bay P 2002 E Field Production Optimization System Based on Integrated Reservoir and Facility Simulation, *Society of Petroleum Engineers*

[9] S Rastoin, Z Schmidt and D R Doty 1997 A Review of Multiphase Flow Through Chokes, *J. Energy Resour. Technol.* **119** 1–10

[10] Offshore 61 (2001) http://www.offshore-mag.com/articles/print/volume-61/issue-8.html (accessed January 26, 2018)

[11] Economides M, Kappos L Petroleum Pipeline Network Optimization Encycl. Life Support Syst. EOLSS. EXERGY, ENERGY SYSTEM ANALYSIS AND OPTIMIZATION

IOP Conf. Series: Materials Science and Engineering **458** (2018) 012046 doi:10.1088/1757-899X/458/1/012046

Interfacial behaviours during biodiesel synthesis in alkaline methanol phase

K Kusakabe1*, H Miyazaki1, M Sakuragi[1], N Ikeda[2], G Guan[3] and Y Uemura[4]

[1] Department of Nanoscience, Sojo University, 4-22-1 Ikeda, Nishi-ku, Kumamoto 860-0082, Japan
[2] Department of Environmental Science, Fukuoka Women's University, 1-1-1 Kasumigaoka, Higashi-ku, Fukuoka 813-8529, Japan
[3] Institute of Renewable Energy, Hirosaki University, 2-1-3, Matsubara, Aomori 030-0813, Japan
[4] Center for Biofuel and Biochemical Research, Universiti Teknologi PETRONAS, Bandar Seri Iskandar, Tronoh, Perak 31750, Malaysia

kusakabe@nano.sojo-u.ac.jp

Abstract. Dynamic phenomena at liquid-liquid interface during transesterification of corn oil was investigated by using a pendant drop method. The drop volume of corn oil gradually decreased as time proceeds because of the dissolution of the products such as fatty acid methyl ester and glycerol in methanol. The drop volume of corn oil with 10 mol% oleic acid was almost constant during first 240 s of the reaction and then decreased. The consumption of KOH during saponification of oleic acid at the interface ceased the transesterification of corn oil. During the saponification, the apparent interfacial tension was initially increased due to the increase in hydrophilicity by transfer of water produced at the interface to methanol phase, and then, decreased due to the dissolution of fatty acid methyl ester formed by the transesterification in both oil and methanol phases.

1. Introduction

Due to the depletion of the fossil fuels and the increasing environmental problem, there is a great demand for alternative sources of fossil fuel. Biodiesel produced from biomass resources is considered as a clean renewable fuel and the best candidate for a diesel fuel substitution. Biodiesel can be used in any compression ignition engines without any modification because it has similar properties to that of diesel produced from crude oil [1-3]. It was reported that biodiesel can reduce net carbon dioxide emissions by 78% on a lifecycle basis when compared to conventional diesel fuel [4]. Biodiesel known as fatty acid methyl ester (FAME) is usually produced from triglyceride (oils and fats) by transesterification with methanol in the presence of an acid or a base catalyst [5-7]. Basic homogeneous catalysts such as NaOH and KOH to improve biodiesel yields are used in conventional processes for production of biodiesel [8,9]. Homogeneous acid catalysts such as H_2SO_4 are also able to catalyze this reaction, but they are seldom used because they are less active and more corrosive. When the oils contains significant amounts of free fatty acids (FFAs) and water content, FFAs react with the alkaline catalyst to form soaps. Usually, esterification of FFA is carried out by acid-catalyzed processes to reduce FFA before alkali-catalyzed transesterification [10,11].

Alkali-catalyzed transesterification is a two-phase liquid reaction system because methanol and oil are mutually-immiscible and therefore the overall reaction rate is controlled by both the interfacial area and the diffusion of chemical species. The biodiesel production is generally carried out at the temperature ranging from 333 to 373 K for several hours in a stirred batch reactor [1-3]. Meanwhile, addition of co-solvent such as dimethyl ether (DME) and tetrahydrofuran (THF) to the liquid-liquid phase for homogenization could greatly promote the transesterification reaction rate [12]. In the homogeneous reaction, oil conversion of 100 and 97% was reached after a 1-min reaction at a KOH concentration of 0.5 wt% even at room temperature for DME and THF additions, respectively. In our previous studies [13,14], transesterification of sunflower oil (SFO) and waste cooking oil (WCO) in a microtube reactor was carried out and the flow patterns were simultaneously observed. We found that the flow patterns when using WCO, changed from a slug flow at the inlet region to a parallel flow at the middle region and then to a homogeneous flow at the outlet region as the reaction progressed at 333 K. These changes of flow pattern are caused by the formation of FAME which acts as a mutual solvent for the reactants. When SFO was used in a microtube reactor, fine droplets composed of the produced glycerol and methanol were dispersed and circulated in the oil segments. Similar phenomena was also observed by other researchers [15,16]. Thus, the reaction properties are clearly influenced by interfacial behaviors between segregated phases.

The pendant drop methods are widely applied for surface tension measurements [17-20]. In order to investigate the interfacial phenomena between oil and methanol phases, a pendant drop method was applied to clear the interfacial behaviors during the transesterification of SFO in the methanol containing 4.5 wt% of KOH at 298 K [14]. If the reaction occurs between the two liquids, accompanying with the progress of the reaction, the interfacial tension should be changed with the reaction time. In this case, it should be difficult to decide the interfacial tension accurately, but we can use it to do some rough estimates, and infer the reaction mechanisms. In addition, this method would be an easy and convenient tools to evaluate the interfacial effect of surfactants [21] and ionic liquid [22,23] on the improvement of the transesterification reaction.

In this study, the interfacial tension between oil and methanol phase was determined by the pendant drop method in more details. In order to examine the surface activity of various compounds formed during the transesterification, oleic acid, monoolein, methyl oleate and glycerol were mixed with corn oil. The influences of these compounds on the interfacial tension were investigated. The changes in the volume and the apparent interfacial tension of a single oil drop were measured during the transesterification.

2. Materials and methods

2.1 Chemicals

Corn oil, dehydrated methanol, potassium hydroxide, glycerol, acetic acid, and methyl oleate (oleic acid methyl ester) were obtained from Wako Pure Chemical Ind. Ltd., Japan. Monoolein (Glycerol monooleate; >40.0%) was obtained from Tokyo Chemical Industry Co., Ltd., Japan. Oleic acid (>99%) was obtained from Sigma-Aldrich. The acid and saponification values of the oil were determined using standard titration methods [24]. The molar amount of triglyceride per gram is equal to one third of the molar amounts of fatty acids per gram which can be calculated by the difference between two values. Accordingly, the molecular weight of the oil was estimated from the reciprocal of the molar amount of triglyceride per gram. Water content in the oil was determined using a Karl-Fischer moisture titrator (MKC-610, Kyoto Electronic Manufacturing Co. Ltd.). The viscosity was determined with a torsion-balanced, oscillation-type viscometer (VM-1G, CBC Materials Co., Ltd.). The density was determined using a pycnometer.

2.2 Phase diagrams

The bimodal curves in the triangular phase diagrams for the systems of methyl oleate/methanol/corn oil and oleic acid/methanol/corn oil were determined from turbidmetric analysis using the titration method. The measurement was carried out in a constant temperature reservoir at 298 K. Corn oil was mixed with oleic acid or methyl oleate to make homogeneous solution. Methanol was fed to the homogeneous mixture through a microtube with a syringe pump and mixed. The point when the mixture transferred from transparent (homogeneous) to turbid (heterogeneous) was considered to be the saturation point of the methanol. The amount of added methanol was determined with an analytical balance. The tie lines were determined for the methyl oleate/methanol/corn oil system at the weight ratio of methyl oleate/methanol/corn oil = 10/45/45, 30/35/35 and 50/25/25. The mixture was shaken for about 2 h to allow intimate contact between the phases and separated into an oil layer and a methanol layer by sedimentation. Tie lines were obtained separately by analysing the samples taken from each layer.

2.3 Pendant drop method

The pendant drop method is one of the most accurate methods to measure interfacial tension. In this research, axisymmetric drop shape analysis-profile (ADSA-P) was used for interfacial tension measurement [25-28]. The interfacial tension during the transesterification at 298K was determined using the pendant drop apparatus as shown in Fig.1. Methanol or a 4.5 wt% methanol solution of KOH was used to fill a rectangular quartz cell (10×10×45 mm). The temperature was controlled at exactly 298 K. Using an injection stainless needle with an outer diameter of 0.7 mm, the oil was slowly introduced into the methanol phase and a single drop was formed on the tip of the needle. The change in the shape of the oil drop was observed using a CCD camera with the reaction time noted. Pictures were taken automatically every 30 s. In ADSA-P, the experimental profile of a drop and density difference between the oil and the methanol phase were entered to the numerical program. In the numerical program, a series of Laplacian curves with known interfacial tension are fitted to the experimental profiles. The best fit identifies the true interfacial tension. Reliability of the apparatus was checked by measuring the interfacial tension in water-toluene system. The volume of a drop, V, was calculated by numerical integration of the profile.

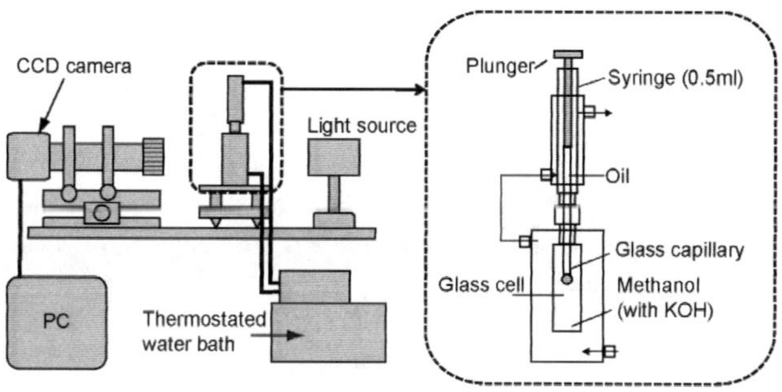

Figure 1. Experimental set-up of pendant drop technique.

3. Results and discussion

The fatty acid composition of corn oil was found to be palmitic (13.6%), stearic (1.9%), oleic (30.8%), linoleic (53.2%) and linolenic (0.5%) by gas chromatography analysis. Density and viscosity of corn oil at 298 K were 0.914 kg m^{-3} and 50.0 mPa s, respectively. Acid value and saponification value of corn oil were 0.0 and 193 mg-KOH g^{-1}. The molecular weight determined from the acid and saponification value was 870.

Figures 2 (a) and (b) present the triangular phase diagram determined from the turbidmetric analysis. The experimental results showed that the area of miscibility regions was large when oleic acid was added into the mixture of methanol and corn oil compared to methyl oleate.

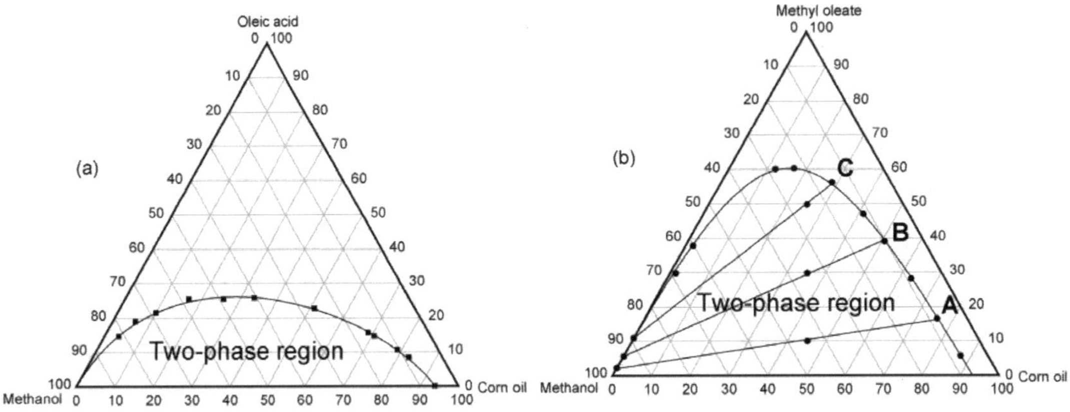

Figure 2. Phase diagram of ternary systems at 25°C. (a); oleic acid/methanol/corn oil system, (b); methyl oleate/methanol/corn oil system, solid lines are tie lines. The composition scales are in weight fraction

In basic transesterification, the fatty acid methyl ester (FAME) yield is inhibited by free fatty acid and/or water. Free fatty acids in oil caused saponification (equation (1)) which reduces the FAME yield by consuming the base catalyst and leads to the difficulty of separation of biodiesel products. Water also reduces the FAME yield by the hydrolysis of FAME in the presence of base catalyst as equation (2).

$$RCOOH + KOH \rightarrow RCOO\text{-}K^{+} + H_2O \quad (1)$$
$$RCOOCH_3 + H_2O \rightarrow RCOOH + CH_3OH \quad (2)$$

The volume ratio and the apparent interfacial tension of the drop for corn oil and corn oil with oleic acid content of 10 mol% were measured in methanol without KOH catalyst. As shown in figure 3 (a), both drop of corn oil and corn oil with oleic acid initially expanded due to the diffusion of methanol into the oil phases. In the presence of oleic acid in corn oil, the drop size after the initial expansion decreased due to the diffusion of oleic acid into the methanol phase. Figure 4 shows the volume change of the drop for corn oils with 10 mol% monoolein, 10 mol% glycerol and 10 mol % methyl oleate. Glycerol and methyl oleate was able to dissolve in methanol and therefore the drop volume ratio was decreased. In the case of monoolein chosen as a model of intermediate product, the volume change behaviour was

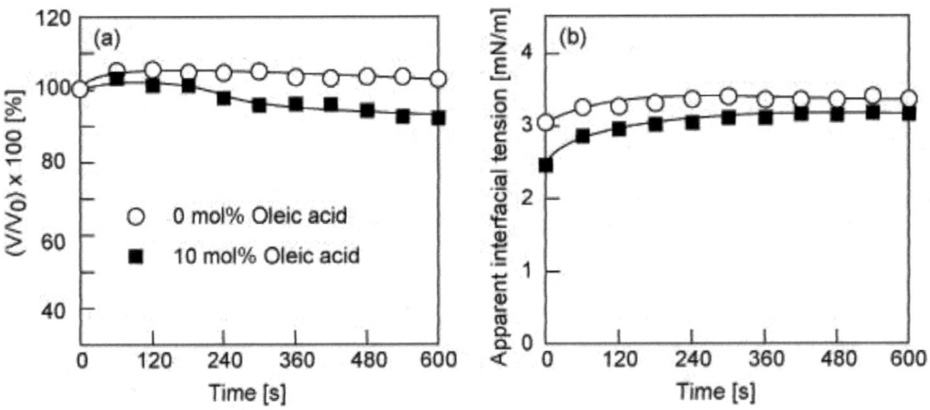

Figure 3. Interfacial behaviours of corn oil (○) and corn oil with 10 mol% oleic acid (■) at 25°C in methanol solution. (a); volume ratio of drop, (b); apparent interfacial tension; without KOH catalyst.

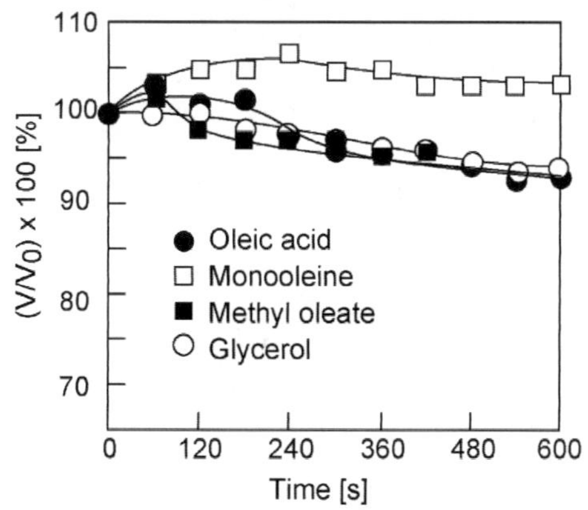

Figure 4. Effect of various compounds on change in volume ratio of drop without KOH catalyst.

similar to that of pure corn oil, indicating that monoolein did not diffuse into methanol phase. As shown in figure 3(b), the apparent interfacial tension of corn oil with oleic acid content was lower than that of corn oil at the initial stage and gradually approached to the value of pure corn oil with the emission of oleic acid.

The volume ratio and the apparent interfacial tension of the drop in methanol with KOH content of 4.5 wt% were measured for evaluating the progress of transesterification. In the initial stage of the measurement, methanol including methoxide ion, which was formed by the reaction of methanol and KOH, could penetrate into the oil phase and then, the contact between corn oil and methoxide ion started the transesterification. As shown in figure 5 (a), the drop volume ratio of corn oil gradually decreased as time proceeds because of the dissolution of the reaction products such as FAME and glycerol. The reaction rate was increased with the increase in specific surface area due to the shrinkage of the drop along the reaction progress. The drop volume ratio of corn oil with oleic acid content of 10 mol% was almost constant for 240 s and then decreased. This means that the consumption of KOH during saponification of oleic acid at the interface ceased the transesterification of corn oil.

The apparent interfacial tension for corn oil was monotonically decreased as time proceeds as shown in figure 5(b). Meanwhile, the apparent interfacial tension of corn oil with oleic acid increased and reached the maximum value. Water exhibits a very high surface tension when compared to other liquids. The increase in the apparent interfacial tension might be due to the presence of water, which was produced at the interface by saponification of oleic acid. The maximum interfacial tension reached 6.3 mN m^{-1} at the oleic acid content of 10 mol%.

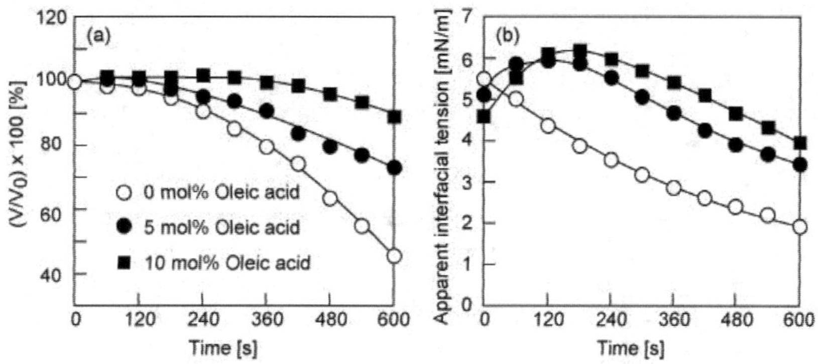

Figure 5. Interfacial behaviors of corn oil (○), corn oil with 5 mol% oleic acid (●) and corn oil with 10 mol% oleic acid (■) at 25°C in methanol solution with 4.5 wt% KOH. (a); volume ratio of drop, (b); apparent interfacial tension.

Monoolein and oleic acid act as surfactant in aqueous solution. The surface activities of monoolein, oleic acid, methyl oleate and glycerol in methanol solution were evaluated from the change of the interfacial tension by their concentration as shown in figure 6. These compounds exhibited no surface activity in methanol solution and thereby did not lead to the remarkable decrease in the interfacial tension during the transesterification. This also means that these compounds are distributed in the drop due to no accumulation on the interface.

When the oil drop was suspended in methanol with KOH content of 4.5 wt%, FAME arose as a consequence of transesterification on the interface and easily transferred to methanol phase. Accordingly, the existence of high concentration FAME on both oil and methanol would reduce the interfacial tension. The interfacial tensions of the oil drops with different compositions at the points of A, B and C in figure 2(b) were determined. As the transesterification progressed, the composition would head in the direction of A, B and C. Figure 7 reveals that the methyl oleate concentrated on the interface by transesterification could reduce the interfacial tension significantly.

The behaviours of glycerol were significantly different from those of FAME. Guan et al. [14] observed the flow behaviour in microtube reactor during the transesterification and found that fine droplets composed of glycerol and methanol were dispersed and circulated in the oil segments. Figure 8 shows the drop of corn oil observed at 900 s after the beginning of the reaction. Fine droplets were observed and moved inside the single drop of corn oil, similar to the results reported by Guan et al. [14]. These phenomena suggested that the transesterification occurred inside the oil phase located adjacent to the interface. In addition, a fine streamline composed of glycerol appeared immediately beneath the oil drop because of the high density of glycerol.

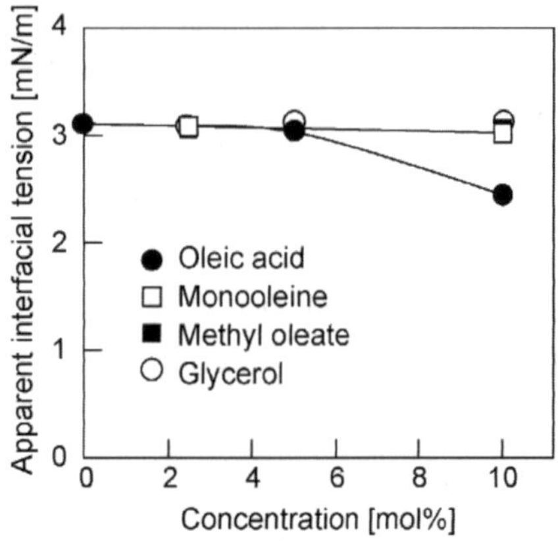

Figure 6. Effect of various compounds on apparent interfacial tension.

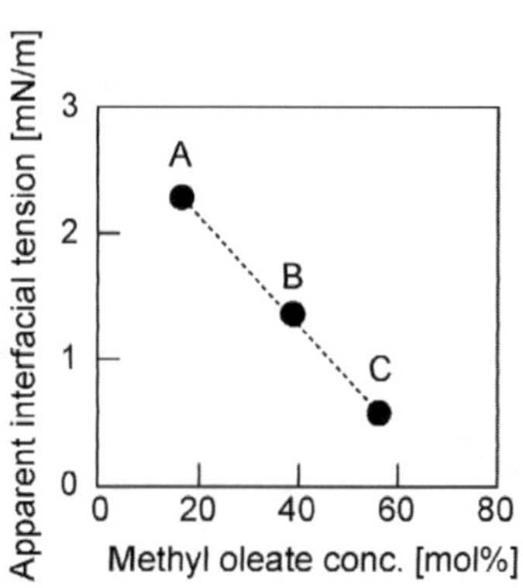

Figure 7. Relationship between apparent interfacial tension and methyl oleate concentration

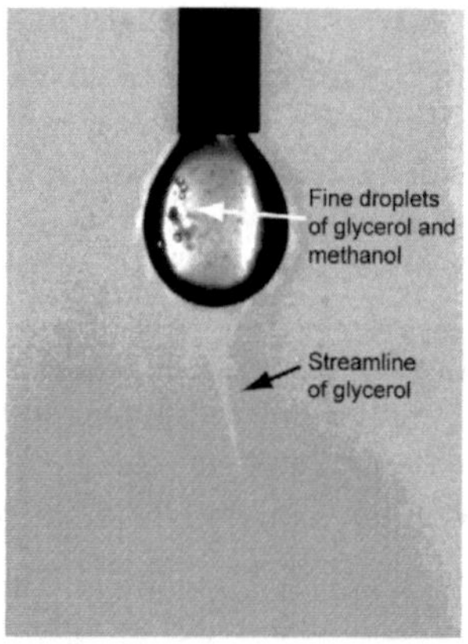

Figure 8. A corn oil drop in methanol solution with 4.5 wt% KOH at 25°C at 900 s.

The effect of water content on the interfacial behaviour was investigated by using corn oil containing 5 mol% oleic acid with and without water. Figure 9(b) indicated that apparent interfacial tensions were almost the same at the initial stage. In addition, the apparent interfacial tension of the water-saturated sample remained almost unchanged with time. Water in oil contributed the reverse reaction of saponification (equation (1)). Therefore, the saponification continued at a slow rate and depressed the transesterification until at least 600 s. Consequently, the volume ratio of the drop was also unchanged as shown in figure 9(a).

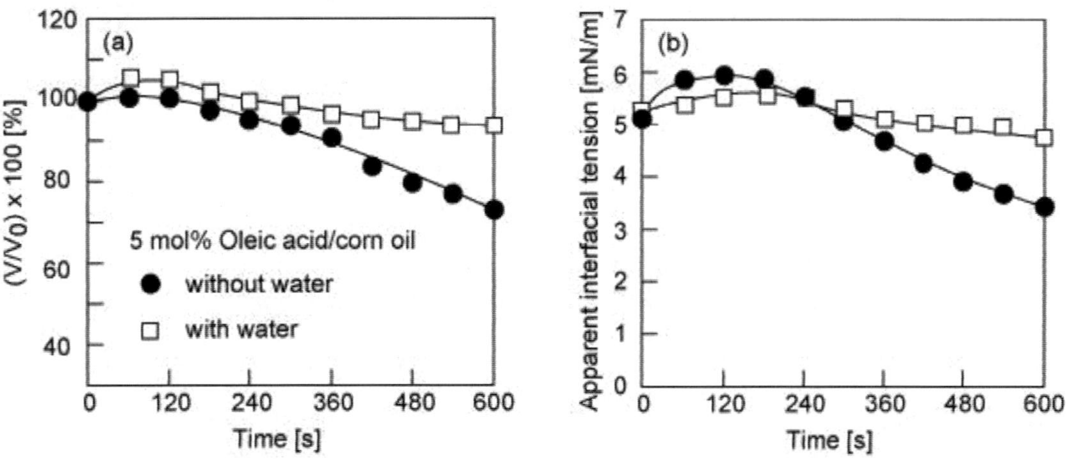

Figure 9. Interfacial behaviours of corn oil with oleic acid (water content 0.074 wt%; □) and with oleic acid + water (water content 0.162wt%; •) at 25°C in methanol solution with 4.5 wt% KOH. (a); volume ratio of drop, (b); apparent interfacial tension.

4. Conclusions

Interfacial behaviours during biodiesel synthesis significantly influenced the reaction rate of mass transfer controlled process in the two phase liquid-liquid system. The pendant drop method was applied to measure the volume and the apparent interfacial tension of the oil drop during the transesterification in methanol phase containing KOH catalyst. The drop volume of corn oil gradually decreased as time proceeds because of the dissolution of FAME and glycerol into methanol. The volume ratio of corn oil with 10 mol% oleic acid was almost constant for 240 s and then decreased. The consumption of KOH produced during saponification of oleic acid ceased the transesterification. During the saponification, the apparent interfacial tension was initially increased due to the increase in hydrophilicity of methanol by transfer of water produced at the interface. FAME formed by the transesterification could dissolve both oil and methanol phases and therefore reduced the apparent interfacial tension.

Acknowledgements

This work was supported by a JSPS Grant-in-Aid for Scientific Research (B) (21360390 and 25289281), Japan.

References

[1] Wan S, Truong-Trieu V M T, Ward T, Whalem J K and Altosaar I 2017 *Biofuel Bioprod Bior* **11** pp 749-764
[2] Demirbas A, Bafail A, Ahmad W and Sheikh M 2016 *Energy Exploration & Exploitation* **34** pp 290-318
[3] Guan G and Kusakabe K 2012 *Journal of the Japan Petroleum Institute* **55** pp 171-181
[4] Tyson K S 2001 *Biodiesel Handling and Use Guidelines* (NREL, Golden Co.)
[5] Lee H V, Juan J C, Taufiq-Yap Y H, Kong P S and Rahman N A 2015 *Journal of Renewable and Sustainable Energy* **7** pp 1-46
[6] Gnanaprakasam A, Sivakumar V M, Surendhar A, Thirumarimurugan M and Kannadasan T 2013 *Journal of Energy* **4** pp 1-11
[7] Van Gerpen J H 2005 *Fuel Processing Technology* **86** pp1097-1107
[8] Freedman B, Butterfield R O and Pryde E H. 1986 *Journal of the American Oil Chemists' Society* **63** pp 1375-80

[9] Klofutar B, Golob J, Likozar B, Klofutar C, Zagar E and Poljansek I 2010 *Bioresource Technology* **101** pp 3333-44
[10] Cankci M and Van Gerpen J 1999 *Transactions of American Society of Agricultural Engineers* **42** pp 1203-10
[11] Son S M, Kimura H and Kusakabe K 2011 *Bioresource Technology* **102** pp 2130-32
[12] Guan G, Kusakabe K, Sakurai N and Moriyama K 2007 *Chemistry Letters* **36** pp 1408-9
[13] Guan G, Kusakabe K, Moriyama K and Sakurai N 2009 *Industrial & Engineering Chemistry Research* **48** pp 1357-63
[14] Guan G, Techima M, Sato C, Son SM, Irfan MF and Kusakabe K 2010 *AIChE J* **56** pp 1383-90
[15] Sun J, Ju JX, Ji L, Zhang L X and Xu N P 2008 *Industrial & Engineering Chemistry Research* **47** pp 1398-1403
[16] Tanawannapong Y, Kaewchada A and Jaree A 2013 *Journal of Industrial & Engineering Chemistry* **19** pp 37-41
[17] Saad SMI, Policova Z and Neumann AW 2011 *Colloids & Surfaces A* **384** pp 442-52
[18] Saad SMI, Policova Z, Acosta EJ and Neumann AW 2010 *Langmuir* **26** pp 14004-13
[19] Alvarez NJ, Walker LM and Anna SL 2009 *Journal of Colloid and Interface Science* **333** pp 557-62
[20] Hoorfar M, Kurz MA and Neumann AW 2005 *Colloids Surface* A **260** pp 277-85
[21] Salam KA, Velasquez-Orta SB and Harvey AP 2016 *Biofuel Research Journal* **3** pp 366-71
[22] Ullah Z, Bustam MA, Man Z, Khan AS, Muhammad N, Sarwono A, Farooq M, Ullah R and Mengal AN 2017 *ChemistrySelect* **2** pp 8583-95
[23] Wahidin S, Idris A and Shaleh SRM 2016 *Bioresouce Technol*ogy **206** pp 150-4
[24] Van Gerpen J, Shanks B, Pruszko R, Clements D and Knothe G 2004 Biodiesel analytical methods. Colorada: National Renewable Energy Laboratory
[25] Motomur K, Matubayasi N, Aratono M and Matuura R 1978 *Journal of Colloid and Interface Science* **64** pp 356-61
[26] Ambwani DS and Fort Jr. T 1979 Prog Coll Pol Sci S 11 (New York, Plenum) pp 93-119
[27] Rotenberg Y, Boruvka L and Neumann AW 1983 *Journal of Colloid and Interface Science* **93** pp 169-183
[28] Lahooti S, Del Rio O I, Cheng P, Neumann A W and Spelt K 1996 *Applied Surface Thermodynamics* (Marcel Dekker, New York) p 441

IOP Conf. Series: Materials Science and Engineering **458** (2018) 012047 doi:10.1088/1757-899X/458/1/012047

Distillation column inherent safety index at preliminary design stage

Azmi Mohd Shariff[1,2,*], Muhammad Athar[1,2], Azizul Buang[1,2], Muhammad Ishaq Khan[1,2], Heri Hermansyah[3]

[1] Center for Advanced Process Safety (CAPS), Universiti Teknologi Petronas, 32610 Bandar Seri Iskandar, Perak Darul Ridzuan, Malaysia.
[2] Chemical Engineering Department, Universiti Teknologi Petronas, 32610 Bandar Seri Iskandar, Perak Darul Ridzuan, Malaysia.
[3] Industrial Bioprocess Engineering Laboratory, Department of Chemical Engineering, Faculty of Engineering, Universitas Indonesia, Kampus UI Depok, Depok 16424, Indonesia.

* azmish@utp.edu.my

Abstract. Distillation columns are widely used in many chemical process industries, especially in oil and gas processing. Accidents in the distillation columns have resulted in enormous loss of human lives and assets. Inadequate design of distillation columns is one of the vital reason for the failure, which can be enhanced through inherent safety concept. Therefore, this paper introduces a new indexing method for designing inherently safer distillation columns right from the preliminary design stage. Distillation Column Inherent Safety Index (DCISI) is based on three sub-indices; chemical index, process index and distillation index. These sub-indices are estimated through individual scores by considering various parameters in each category. For the unacceptable score, modification of process conditions is used to enhance safety level of the design. DCISI is estimated for all combination of conditions and lowest DCISI value indicates the inherently safer design of the distillation column.

1. Introduction

With the progressive development of the process industries, designers are focusing on safety as the most priority aspect, because of hazardous chemicals processing. Therefore, a number of guidelines and procedures have been developed and practiced over the last few decades in order to increase the safety level, as highlighted in Figure 1 [1].

In process industries, distillation columns are frequently used, because of cheap separation agent cost, to perform the separation of liquid components. Analysis of previous accidents has revealed that failure of the distillation column can be catastrophic. For example, accidents of Ojima Town, Gunma and Pascagoula, Mississippi have caused huge structural and human loss inside the facility and in the vicinity of the plant [2,3]. There are several reasons for failure for any process equipment which have been investigated by many authors. Among these reasons, design inadequacy has been identified as the most critical for any process equipment [4-6].

Most of the traditional safety studies are carried out once the process design is complete. At this stage, the risk is controlled rather than minimization or mitigation using passive, active and procedural strategies of process safety [7]. At this stage, any modification in the process can cause extra investment in comparison to the design adjustments at the preliminary design stage [8].

Inherent safety concept proposed by Kletz can fulfill the objective of design modifications during early design phases. It emphasizes the reduction and elimination of the root cause of hazards associated with materials and operation. The key principles for inherent safety involve four major elements; as demonstrated in figure 2 [9].

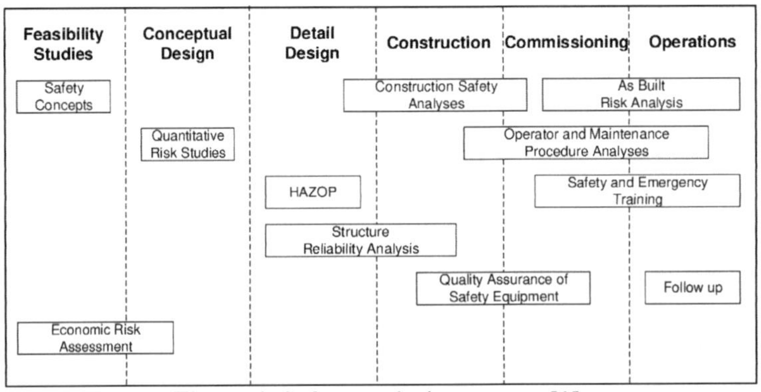

Figure 1. Safety analysis program [1].

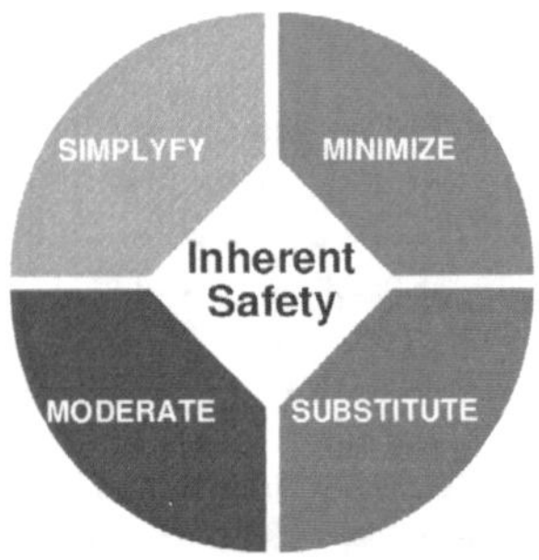

Figure 2. Inherent safety principles [9].

To evaluate numerous design options for a process, quantification of inherent safeness is the biggest challenge. To achieve this target, various inherent safety methods have been proposed. Majority of the methods use indexing technique for inherently safer process route selection [10]. These include inherent safety index (ISI), integrated inherent safety index (I2SI) and process route index (PRI) [10-13]. However, none of the technique is applicable to assess inherent safety level of the distillation column as the nature of individual equipment cannot be taken into account. Consequently, this paper intends to propose a new indexing method for the inherent safety assessment of the distillation column during the preliminary design stage. In the distillation column inherent safety index (DCISI), different design options for the distillation column are proposed and prioritized to identify the inherently safer design for the distillation column. This index would help to identify the safer and cost-effective design for the distillation column at the early design phase.

2. Methodology

A systematic framework to design an inherently safer distillation column at the preliminary design stage is provided in Figure 3. There are various factors, which have a major contribution in distillation column design and are divided into three categories namely chemical, process, and distillation, outlined in Table 1.

Table 1. Parameters for DCISI

Category	Parameters
Chemical	Auto-ignition temperature, flammability, explosiveness
Process	Operating pressure, temperature
Distillation	Reflux ratio, relative volatility

The higher the score for a parameter, the more hazardous is the category. Initially, the factors for distillation column design are analyzed and then merged to provide the value of the distillation column inherent safety index (DCISI). The scoring values for these factors are explained in Table 2. Finally, the ranking of DCISI values identifies the inherently safer design.

Table 2. Score for DCISI

Score	Description
1	Recommended
2	Sound engineering practice
3	Probably unsafe
4	Minor accident
5	Major accident

In next subsections, the contributing factors for safe distillation column design are described. The scoring of all the parameters is provided in table 3.

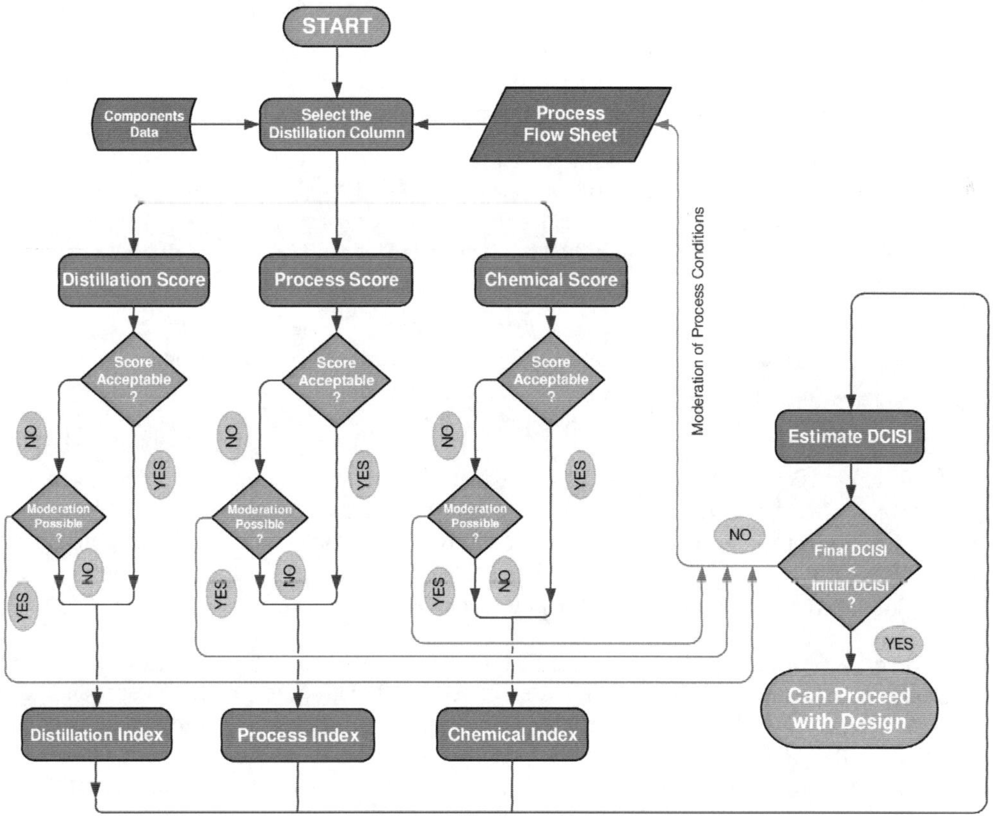

Figure 3. Framework for the distillation column inherent safety index (DCISI).

Chemical Score (CS)
The overall chemical score (CS) is a combination of autoignition temperature score, flammability score, and explosiveness score expressed:

$$CS = CS_{AIT} + CS_{FM} + CS_{EX} \quad (1)$$

Auto Ignition Temperature Score (CS_{AIT}). Auto-ignition temperature (AIT) of the fluid is a key factor to identify the safer scenario for a distillation column. Temperature greater than the AIT enhances the chances of fire and explosion and vice versa. The score in table 3 is based on the comparison of distillation column temperature and autoignition temperature.

Flammability Score (CS_{FM}). Flammability is the capability of any chemical to ignite, which can cause major fire or explosion in the process industry. The score for flammability is described in table 3 which is adapted and modified from ISI [11].

Explosiveness Score (CS_{EX}). Explosiveness is the chemical mixture composition which can convert to the explosion and can be estimated by the difference between lower flammability limit (LFL) and upper flammability limit (UFL) of a mixture. Like flammability, the explosiveness score is adapted from ISI and described in table 3 [11].

Process Score (PS)
The operating conditions of the distillation column are the major elements to define the process score. This score can be estimated by:

$$PS = PS_{P} + PS_{T} \quad (2)$$

Pressure Score (PS_P). The pressure of the distillation column is one of the critical factors to define the safety level of the column. The leak chances are increased due to greater forces created by high pressure. The score for pressure is adapted from ISI method and furnished in table 3 [11].

Temperature Score (PS_T). Temperature is an indication of heat energy in any process. The higher the heat energy, the more are the chances for the process to get out of control. Also, the higher heat energy decreases the vessel material strength, especially in case of thermal shocks. The temperature scoring values are adapted from ISI method and outlined in table 3 [11].

Distillation Score (DS)
The distillation specific characteristics also contribute to deciding the safety level of the distillation column. The distillation score is a function of reflux ratio and relative volatility and can be estimated as follows:

$$DS = DS_{R} + DS_{\alpha} \quad (3)$$

Reflux Ratio Score (DS_R). Reflux ratio is defined as the ratio between the liquid returning back to the distillation column to the distillate collected. Although, increasing the reflux ratio enhances the distillate quality [14, 15], however, the increasing flow rates would increase the inventory in the column and duty requirements of distillation column. Therefore, the smaller the reflux ratio the safer is the column, and the relevant scoring is described in Table 3.

Relative Volatility Score (RS_α). Relative volatility is the comparison of vapor pressures in a liquid chemical mixture. For a multicomponent mixture, the relative volatility is measured between light key comment and heavy key component [14,15]. The higher the relative volatility, the easier is the separation and safe. However, for lower relative volatility, the separation becomes difficult, and distillation column is operating under risky conditions. The scoring values are explained in table 3.

Evaluation of Scores

The score for each category is obtained from equations (1) - (3), and score value describes the safety level of the distillation column. The safety criteria are defined as if any category score for distillation column design option is more than the half of the maximum category score, the design is considered as unsafe. However, if the score is less than the half of maximum category score, the design is safe. For example, the maximum score for the process category is 10, and for a design option, if the process score is > 5, the design is unsafe.

2.1. Estimation of distillation column inherent safety index (DCISI)

The score values are converted to the indexing value for each category and can be estimated by:

$$CI = \frac{CS}{15} \tag{4}$$

$$PI = \frac{PS}{10} \tag{5}$$

$$DI = \frac{DS}{10} \tag{6}$$

The overall value of DCISI is obtained by the summation of the chemical index, process index, and distillation index, as follows:

$$DCISI = CI + PI + DI \tag{7}$$

Distillation column design option with lower DCISI value is preferable in terms of safety in comparison and vice versa.

Table 3. Parameters scoring description for DCISI

Score	Chemical		Process			Distillation	
	Auto-ignition temperature	Flammability	Explosiveness (%)	Operating Pressure (bar)	Temperature (°C)	Reflux Ratio	Relative volatility
1	Column temperature < Auto ignition temperature	Non-flammable	Non explosive	0.5-5	0-70	R ≤1.5	$\alpha > 1.1$
2	-	Combustible (flash point > 55°C)	0-20	6-25	< 0 & 71-150	R ≤ 2.5	-
3	-	Flammable (flash point < 55°C)	21-45	26-50	151-300	R ≤ 3.5	-
4	-	Easily flammable (flash point < 21°C)	46-70	51-200	301-600	R < 5	-
5	Column temperature > Auto ignition temperature	Very flammable (flash point < 0°C & boiling point < 35°C)	71-100	201-1000	> 600	R ≥ 5	$\alpha \leq 1.1$

3. Results and Discussions

The design of distillation column (C-4501) in methanol production process is used to demonstrate the capability of DCISI for the safety assessment of distillation column during the preliminary design stage. Four different design options are considered to design C-4501, which are organized in Table 4.

The simulation of methanol process is available in Figure 4 [16], which has been simulated using the PRSV equations of state due to extended capabilities. Industrially the methanol is produced by synthesis gas in the catalytic environment. Cooling is used for removing the additional heat and the methanol separation from the reaction mixture is achieved by exploiting the thermodynamic characteristics followed by the distillation process. The methanol is obtained as distillate from distillation column, while the vapor separated through phase separation are recycled to increase the process efficiency.

Table 4. Design options for C-4501

Parameter	Design 1	Design 2	Design 3	Design 4
Operating Pressure (bar)	5.00	1.50	1.01	1.01
Operating temperature (°C)	109.2	87.34	78.75	78.05
Reflux Ratio	4	3.5	3	2.5

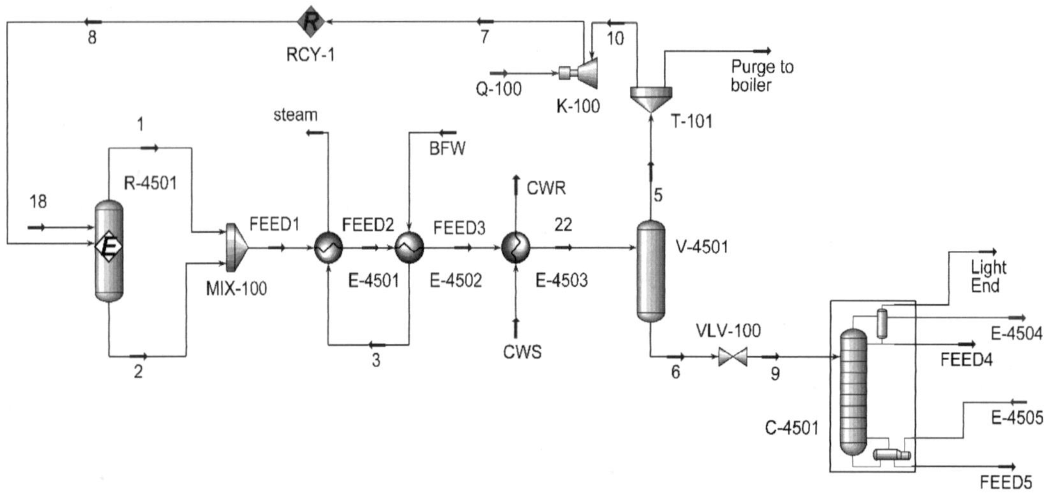

Figure 4. Process simulation of methanol process [16].

The DCISI concept has been applied to the distillation column, C-4501. Initially, process information for the design 1 is extracted from HYSYS and transferred to MS Excel program through VBA coding. The components data is obtained from literature and introduced in the spreadsheet tool using input interface. Scores for the process, chemical and distillation categories are estimated as per equations (1) - (3), presented in table 5. By analyzing the scores, it is recognized that chemical and distillation scores are acceptable, while process score is unacceptable. To minimize these scores, the process conditions for the operation of C-4501 are moderated as design 2, 3 and 4 available in table 4. The calculations of chemical, process and distillation scores are performed one by one for each design option and summarized in Table 5. Although, the moderated process design has not affected the chemical and process scores, however a positive effect on the distillation score is observed. Conclusively, for all modified designs, each category score is in the acceptable range. However, no change in the process and chemical scores are observed, while the distillation score is gradually changing, for the reason of small changes in process conditions, which have created a minute effect on the distillation score. Although, the chemical characteristics are slightly changed, however, the chemical score remains same due to the ranges defined in table 3. For all design options, the DCISI has been estimated using equations (4) to (7), and a slight change in the index value is witnessed. From the score and index values, design 4 is considered as the best design option for the distillation column, C-4501.

Additionally, the effect of changes in the design has been examined in the process to identify the benefits of inherent safety employment in the chemical process design. The safety criteria in this work are the lower value of DCISI, which is achieved through moderation of process conditions and changing the reflux ratio. These moderated conditions have a positive effect on the process efficiency, which is defined in terms of lesser distillation column duty and improved product purity from the distillation column. By moderating the process conditions and reflux ratio, the duty of distillation column is considerably reduced, i.e., 21%. However, the product quality is slightly compromised, around 3%. Conclusively, the design for distillation column can be considered as inherently safer and improved in terms of process characteristics. The effect of moderated design on the performance of the distillation column is depicted in Figure 5.

Table 5. Parameters scores and DCISI for different designs of C-4501

Parameter	Score Value	Design 1	Design 2	Design 3	Design 4
Chemical Aspect					
Auto ignition temperature	1 or 5	1	1	1	1
Flammability	1 - 5	3	3	3	3
Explosiveness	1 - 5	4	4	4	4
Chemical Score		8	8	8	8
Chemical Index		0.533	0.533	0.533	0.533
Process Aspect					
Pressure	1 - 5	1	1	1	1
Temperature	1 or 5	2	2	2	2
Process Score		3	3	3	3
Process Index		0.3	0.3	0.3	0.3
Distillation Aspect					
Reflux Ratio	1 - 5	4	3	3	2
Relative Volatility	1 or 5	1	1	1	1
Distillation Score		5	4	4	3
Distillation Index		0.5	0.4	0.4	0.3
Distillation Column Inherent safety Index (DCISI)		1.333	1.233	1.233	1.133

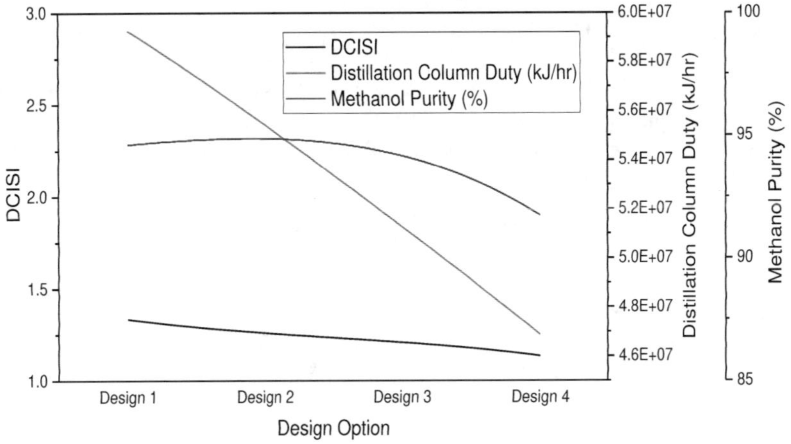

Figure 5. DCISI effect on the performance of C-4501.

For future work, more case studies can be investigated for improvement of the presented indexing method by the inclusion of further significant characteristics of distillation to define the safety level of the column. Furthermore, the type of the internals for columns can be investigated to design an inherently safer distillation from the very early design stages.

4. Conclusion

To achieve the objective of inherently safer distillation column during the early design phase, a new indexing method is proposed in this paper. Distillation column inherently safety index (DCISI) compares different design options of the distillation column focusing the chemical, process and distillation characteristics. The index is estimated by summation of individual indices, which are obtained by converting the score values. DCISI value of all design options are estimated and ranking of DCISI values can identify the inherently safer design. A low DCISI value indicates the design as safe and vice versa. Modification of conditions in distillation column can help to achieve a safer design.

Acknowledgment

The authors would like to gratitude Universiti Teknologi PETRONAS, Malaysia for providing research facilities and funding (0153AB-F07) to make this research feasible.

References

[1] Taylor J R 2003 *Risk Analysis for Process Plant, Pipelines and Transport*, CRC Press
[2] U.S. Chemical Safety and Hazard Investigation Board 2003 Investigation Report, Vapor cloud deflagration and fire at BLSR Operating LTD
[3] Koseki H, Kobayashi M, and Masamitsu T 2017 Explosion and Fire of Hydroxylamine
[4] Jalani J A, Kidam K, Shahlan S S, Kamarden H, Hassan O and Hashim H 2015 An analysis of major accident in the US chemical safety board (CSB) database *J. Tek.* **75** pp 53-60
[5] Kidam K and Hurme M 2012 Design as a contributor to chemical process accidents, *J. Loss Prev. Proc. Indus.* **25** pp 655-666
[6] Kidam K, Sahak H A, Hassim M H, Shahlan S S and Hurme M 2016 Inherently safer design review and their timing during chemical process development and design *J. Loss Prev. Proc. Indus.* **42** pp 47-58
[7] Shariff A M and Zaini D 2010 Toxic release consequence analysis tool (TORCAT) for inherently safer design plant *J. Hazard. Mater.* **182** pp 394-402
[8] Ashford N A and Zwetsloot G 2000 Encouraging inherently safer production in European firms: a report from the field *J. Hazard. Mater.* **78** pp 123-144
[9] CCPS, Center for Chemical Process Safety 2010 *Inherently Safer Chemical Processes: A Life Cycle Approach* (Wiley)
[10] Leong C and Shariff A M 2009 Process route index (PRI) to assess level of explosiveness for inherent safety quantification *J. Loss Prev. Proc. Indus.* **22** pp 216-221
[11] Heikkilä A 1999 *Inherent safety in process plant design: an index-based approach* (VTT Technical Research Centre of Finland)
[12] Khan F and Amyotte P 2004 Integrated inherent safety index (I2SI): a tool for inherent safety evaluation *Process Saf. Prog.* **23** pp 136-148
[13] Khan F and Amyotte 2005 P I2SI: a comprehensive quantitative tool for inherent safety and cost evaluation *J. Loss Prev. Proc. Indus.* **18** pp 310-326
[14] Coulson J, Richardson J, Backhurst J and Harker J 1996 *Coulson & Richardson's Chemical Engineering: Particle technology and separation processes. 5th ed* (Butterworth-Heinemann)
[15] Kister H 1992 *Distillation Design* (McGraw-Hill Education)
[16] Nian C W and You F 2013 Design of Methanol Plant *EURECHA Student Contest Problem Competition ESCAPE-23 Symp.* (Finland, p 25)

IOP Conf. Series: Materials Science and Engineering **458** (2018) 012048 doi:10.1088/1757-899X/458/1/012048

Chemical reactor inherent safety index at preliminary design stage

Muhammad Athar[1,2], Nor Ayuni Binti Zaidi[1], Azmi Mohd Shariff[1,2,*], Azizul Buang[1,2], Muhammad Ishaq Khan[1,2]

[1] Center for Advanced Process Safety (CAPS), Universiti Teknologi Petronas, 32610 Bandar Seri Iskandar, Perak Darul Ridzuan, Malaysia.
[2] Chemical Engineering Department, Universiti Teknologi Petronas, 32610 Bandar Seri Iskandar, Perak Darul Ridzuan, Malaysia.

* azmish@utp.edu.my

Abstract. The design inadequacy of chemical reactor has caused many major accidents in process industries. The absence of safety analysis for chemical reactor especially in the early design stage is one of the reasons for faulty designs. Inherent safety can be used to perform the safety analysis at preliminary design phases. However, the literature is deficient in reporting inherent safety assessment method for chemical reactors. Therefore, this paper introduces a new indexing method for inherently safety assessment of chemical reactors at the preliminary design stage. Chemical Reactors Inherent Safety Index (CRISI) is based on three sub-indices; chemical index, process index and reaction index. These sub-indices are estimated through scores of numerous parameters in each dimension. For the unacceptable score, the process conditions are changed according to the favorable reaction conditions. CRISI is estimated for all combination of process conditions and lowest CRISI value indicates the inherently safer design of the reactor.

1. Introduction

Chemical reactor is absolutely the heart of any process industry as the transformation of raw materials into valuable chemicals is carried out in the reactors. From the literature, it is identified that the reactor design is of critical importance because of the significant energy involvement [1]. The industrial disasters in the last century such as Flixborough and Chernobyl have revealed that the failure of chemical reactors can transform into serious consequences, such as production loss, fatalities, injuries and plant damage. Numerous reasons for Chernobyl accident involve design inadequacy, breaching of safety regulations for the construction followed by inadequate documentation for reactor operation [2]. An analysis of the reactor accidents has revealed that about 85% accidents are caused by runaway reaction. A further study has identified that 60% of runaway reactions occur due to inadequate reactor design, which can be improved through inherently safety concept implementation [3, 4].

The study of past incidents has identified the importance of process safety to operate a process plant. Over the last three decades, many guidelines and methodologies have been implemented to identify, analyze and manage risks from different stages of the process plant, which are demonstrated in figure 1 [5].

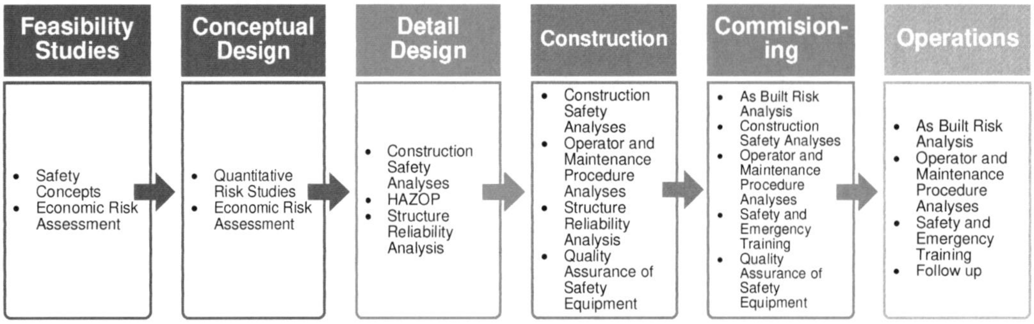

Figure 1. Safety analysis program [5].

The former conventional safety methodologies are used after finalization of process design. Although, these studies identify the consequences and risk of the design, however, at this stage only passive, active and procedural techniques can be implemented to reduce the impact [6]. Further, the application of these strategies to modify the process after design finalization leads to extra investment in comparison to the modifications during the preliminary design stage [7].

The modifications during early design phases can be achieved by inherently safety approach. The aim of inherent safety concept is to reduce and eliminate the root cause of hazards associated with materials and operation. The key principles for inherent safety involve four major elements; as tabulated and described in table 1 [1].

Table 1. Inherently safety elements [1]

Principles	Definition
Minimization	Reduce the quantity of hazardous material
Substitution	Substitute the hazardous material with less hazardous material
Moderation	Operate at less hazardous conditions
Simplification	Eliminates unnecessary design complexity

Quantification of the inherent safety principles has always been a big challenge to evaluate various options for process design. For this reason, various metrics for measurement of inherent safeness have been developed. Majority of the methods use indexing technique for inherently safer process route selection [8]. These include inherent safety index (ISI), integrated inherent safety index (I2SI) and process route index (PRI), process stream index (PSI) [8-12].

However, none of the technique is applicable to assess inherent safety level of the chemical reactor. Therefore, this paper aims to propose a new indexing method for the inherent safety assessment of the chemical reactor during the preliminary design stage. In chemical reactor inherent safety index (CRISI), different design options for the reactor are studied and analyzed to identify the inherently safer design for the reactor. This index would help to identify the more safe and cost-effective design for the reactor at the early design phase.

2. Methodology

There are numerous factors associated with reactor design and are grouped in three dimensions namely chemical, process, and reaction, outlined below in table 2.

Table 2. Parameters for CRISI

Category	Parameters
Chemical	Auto-ignition temperature, flammability, explosiveness
Process	Operating pressure, temperature
Reaction	Reaction parameter, the heat of reaction, yield

A systematic approach based on above-mentioned factors to design an inherently safer chemical reactor at the preliminary design stage is presented in figure 2.

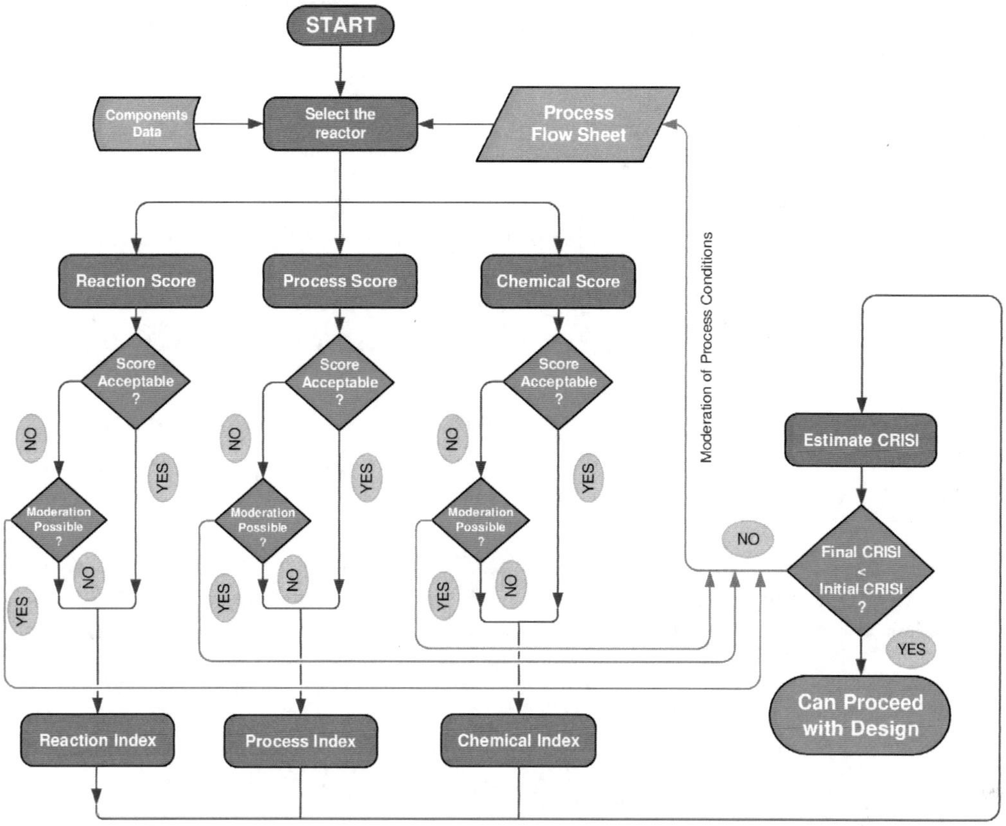

Figure 2. Framework for the chemical reactor inherent safety index (CRISI).

The higher the score for a parameter, the more hazardous is the scenario. At first, these factors for reactor design are analyzed and then combined to define the chemical reactor inherent safety index (CRISI). The scoring values for these factors are explained in table 3. Finally, index values for different design options of chemical reactor relatively ranked.

Table 3. Score for CRISI

Score	Description
1	Recommended
2	Sound engineering practice
3	Probably unsafe
4	Minor accident
5	Major accident

In next subsections, the parameters that contribute to the design and ultimately the safety of chemical reactor are explained. The scoring of all the parameters is provided in table 4.

2.1. Chemical Score (CS)

The overall chemical score (CS) is comprised of the scores for autoignition temperature, flammability, and explosiveness, which can be written as:

$$CS = CS_{AIT} + CS_{FM} + CS_{EX} \quad (1)$$

2.1.1. Auto Ignition Temperature Score (CS_{AIT}). Identifying the auto-ignition temperature of the fluid inside the reactor is vital to identify the safer operating condition for the reactor. Chemical mixture temperature greater than the autoignition temperature favors the fire and explosion scenarios and vice versa. The score in table 4 is based on the comparison of reactor temperature and autoignition temperature.

2.1.2. Flammability Score (CS_{FM}). Flammability indicates the capability of the material to ignite, which can lead to either major fire or explosion scenario. The score for flammability in table 4 is adapted and modified from ISI [9].

2.1.3. Explosiveness Score (CS_{EX}). The calculation of explosiveness is based on the difference between lower flammability limit (LFL) and upper flammability limit (UFL) of a mixture. Like flammability, the explosiveness score is adapted from ISI and can be used to indicate the level of explosiveness in the chemical reactor, available in table 4 [9].

2.2. Process Score (PS)

The operating conditions of the reactor are the contributing factors for the process score. Process Score can be obtained through:

$$PS = PS_P + PS_T \tag{2}$$

2.2.1. Pressure Score (PS_P). The pressure of the reactor is one of the vital factors to define the safety level of the vessel. Operation with a high pressure will increase the chances to leak. The score for pressure is adapted from ISI method and furnished in table 4 [9].

2.2.2. Temperature Score (PS_T). The runaway of a chemical reaction can be controlled if the maximum temperature of synthesis reaction (MTSR) is lower than the maximum allowable temperature (MAT) [13]. The MSTR is a function of adiabatic temperature difference and conversion, given below:

$$MSTR = \Delta T_{ad} \times (1 - \varepsilon) \tag{3}$$

2.3. Reaction Score (RS)

The reaction characteristics also contribute to defining the safety level of the chemical reactor. The reaction score is a function of yield from the reaction, heat involved and reaction parameter and can be written as:

$$RS = RS_{RP} + RS_{HR} + RS_Y \tag{4}$$

2.3.1. Reaction Parameter Score (RS_{RP}). Reaction parameter (B) can be used to estimate the tendency of a runaway reaction in a chemical reactor [14] and can be estimated by using equation (5). Here, E is activation energy, ΔT_{ad} is adiabatic temperature rise, R is gas constant and T is reaction temperature.If, $B \leq 5$, the reactor operation is considered as safe. Meanwhile, if $B > 5$, the reactor is operating under severe conditions and the process temperature needs to be modified [14].

$$B = \frac{E\Delta T_{ad}}{R\,T^2} \tag{5}$$

2.3.2. Heat of Reaction Score (RS_{HR}). The heat of reaction indicates the energy involved in the reaction. If this energy is being released by the reaction, the reaction is exothermic and the scenario is more hazardous. This liberated heat results in a higher temperature in the reactor leading to runaway conditions. The scoring is comprised of heats of the main reaction (HR_{MR}) and the side reaction (HR_{SR}). The same scoring is used for the main and side reactions, which is provided in table 4.

$$HR = HR_{MR} + HR_{SR} \tag{6}$$

2.3.3. Yield Score (RS_Y). Although, a higher yield is required from the reactor, however, the higher yield can be translated as the higher residence time for the mixture inside the reactor. As the residence time is increased, there are chances that the temperature of the reactor becomes high such that the runaway reaction occurs. Reactor with higher yield can have the higher score i.e., operating under severe scenario and vice versa. Scoring of reactor yield is provided in table 4.

2.4. Evaluation of Scores

The score for each category is obtained from equations (1), (2) and (4) and these scores define the safety level of the reactor. The safety criteria are defined as if the category score for the reactor design is more than the half of the maximum category score, the design is unsafe. However, if the score is less than the half of maximum category score, the design is safe. For example, the maximum score for the chemical category is 15, and for a design, if the chemical score is > 8, the design is unsafe.

2.5. Estimation of the chemical reactor inherent safety index (CRISI)

The score values are used to define the indexing value for each category, and can be estimated through:

Chemical Index

$$CI = \frac{CS}{15} \tag{7}$$

Process Index

$$PI = \frac{PS}{10} \tag{8}$$

Reaction Index

$$RI = \frac{RS}{15} \tag{9}$$

Table 4. Scoring description for parameters

Score	Chemical			Process		Reaction		
	Auto-ignition temperature	Flammability	Explosiveness (%)	Operating Pressure (bar)	Temperature (°C)	Reaction Parameter	Heat of Reaction (J/g)	Yield (%)
1	Reactor temperature < Auto ignition temperature	Non-flammable	Non explosive	0.5-5	Maximum allowable temperature > Maximum temperature of synthesis reaction (MTSR)	B ≤ 5	Thermally neutral < 200	0-39
2	-	Combustible (flash point > 55°C)	0-20	6-25	-	-	Mildly exothermic < 600	40-59
3	-	Flammable (flash point < 55°C)	21-45	26-50	-	-	Moderately exothermic < 1200	60-79
4	-	Easily flammable (flash point < 21°C)	46-70	51-200	-	-	Strongly exothermic <3000	80-89
5	Reactor temperature > Auto ignition temperature	Very flammable (flash point < 0°C & boiling point < 35°C)	71-100	201-1000	Maximum allowable temperature < Maximum temperature of synthesis reaction (MTSR)	B > 5	Extremely exothermic >3000	90-100

The overall value of CRISI can be determined by the summation of the chemical index, process index, and reaction index, as follows:

$$CRISI = CI + PI + RI \tag{10}$$

Chemical reactor design with lower CRISI value is preferable in terms of safety in comparison and vice versa.

3. Results and Discussions

The design of methanol synthesis reactor is used to demonstrate the capability of CRISI for chemical reactor safety assessment at the preliminary design stage. Two different design options are considered for the reactor, which is tabulated in table 5.

Table 5. Design options for methanol reaction

Parameter	Design 1	Design 2
Operating Pressure (bar)	72.1	50.0
Operating temperature (°C)	249.3	217.7

The simulation of methanol process is available in figure 3 [15]. Industrially the methanol is manufactured through catalytic reaction of synthesis gas. Extra heat is removed through cooling and the methanol is separated from the mixture using phase separation method. The remaining mixture is recycled to increase the process efficiency.

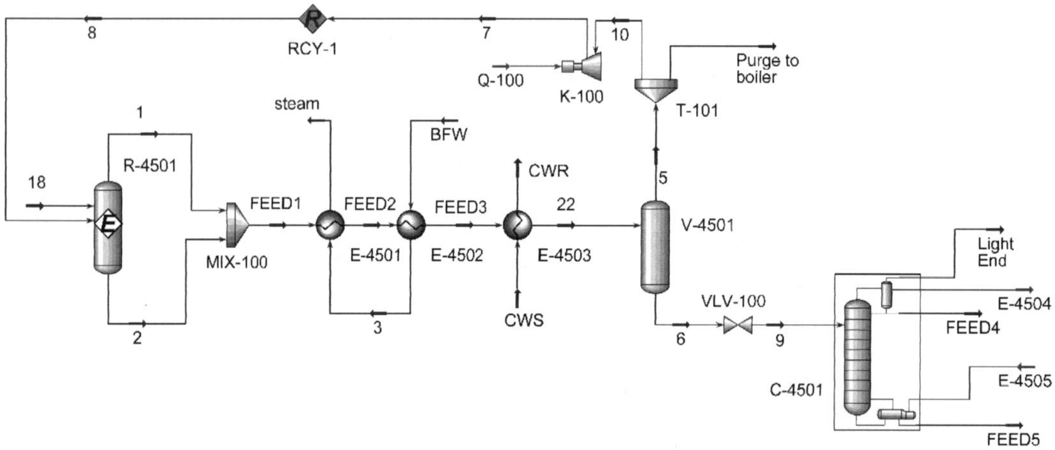

Figure 3. Process simulation of methanol process [15].

CRISI concept has been applied to the methanol synthesis reactor. The process information for the design option 1 is obtained from process simulation software and transferred to MS Excel program through VBA coding. The components data is collected from literature and nested to the spreadsheet tool through user input. The reactor scores for all three categories are estimated as per equations (1), (2) and (4), presented in table 6. By analyzing the scores, it is identified that chemical score is acceptable, while process and reaction scores are unacceptable. To minimize these scores, the process conditions for the reactors are moderated as per the study of favorable conditions for the methanol synthesis reaction and provided as design 2 in table 5. The calculations of chemical, process and reaction scores are performed again, outlined in table 6. Study of these scores has revealed that chemical score is not changed, however, it is acceptable. For process score, it is slightly moderated, and now acceptable. Whereas, the reaction score has not changed. However, the reaction chemistry does not allow to further moderate the process conditions. For the initial and final score values, the CRISI has been estimated using equations (7) to (10), a slight change in the index value is observed. From the score and index values, design 2 is considered as the best design option for the methanol synthesis reactor.

Table 6. Parameters scores and CRISI for methanol synthesis reactor

Parameter	Score Value	Design 1	Design 2
	Chemical Aspect		
Auto ignition temperature	1 or 5	1	1
Flammability	1 - 5	2	2
Explosiveness	1 - 5	4	4
Chemical Score		7	7
Chemical Index		0.467	0.467
	Process Aspect		
Pressure	1 - 5	4	3
MSTR	1 or 5	5	5
Process Score		9	8
Process Index		0.900	0.800
	Reaction Aspect		
Yield	1 - 5	1	1
Heat of Reaction	1 - 5	4	4
Reaction Parameter	1 or 5	5	5
Reaction Score		10	10
Reaction Index		0.667	0.667
Chemical Reactor Inherent safety Index (CRISI)		2.033	1.933

Additionally, the effect of the moderation has been investigated in the process to reveal the advantages of inherent safety concept to design the chemical processes. The safety criteria in this work is the lower value of CRISI, which is achieved through moderation of process conditions. Additionally, the moderated conditions have improved the process efficiency. The process efficiency can be defined as improved with lesser duty requirements and increased product amount to process same amount of feed. For the methanol process, the moderated conditions have affected the reactions in terms of conversion, and an increment of about 53% in the amount of methanol produced is observed. Further, the moderated conditions have reduced the load on all heat exchangers as well as the compressor. For overall process, the duty requirements are improved by 0.7 %. Conclusively, the inherent safety has improved the safety level along with process efficiency, demonstrated in figure 4.

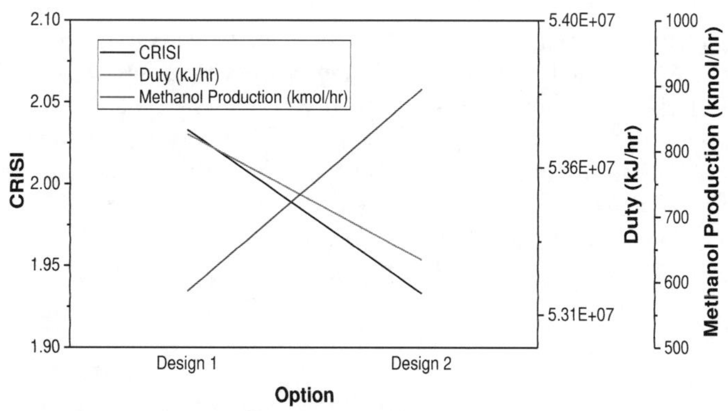

Figure 4. CRISI effect on the methanol process.

For future work, more case studies can be executed for improvement of the suggested methodology by the inclusion of additional influential parameters to define the safety of chemical reactors. Moreover, the type of the chemical reactor can be integrated to design an inherently safer chemical reactor from the very early design stages.

4. Conclusion

A new indexing method to assess the inherent safety level of a chemical reactor in the preliminary design stage is presented in this paper. Chemical reactor inherently safety index (CRISI) evaluates different design options of the chemical reactor by considering the chemical, process and reaction aspects. The score values of each category are converted to index, values and a combined index through relative ranking can identify an inherently safer design for the reactor. High CRISI value identifies the design as unsafe design and inherent safety guide words can be used to modify the design. As a future direction, the type of chemical reactor can be integrated with the aforementioned characteristics.

Acknowledgement

The authors would like to gratitude Universiti Teknologi PETRONAS, Malaysia for providing research facilities and funding (0153AB-F07) to make this research feasible.

References

[1] C. f. C. P. S 2010 CCPS, *Inherently Safer Chemical Processes: A Life Cycle Approach* (Wiley)
[2] Malko M V 2010 The chernobyl reactor: design features and reasons for accident, *Recent Research Activities about the Chernobyl NPP Accident in Belarus, Ukraine and Russia* p 11
[3] Barton J and Nolan P 1989 Incidents in the chemical industry due to thermal runaway chemical reactions *Hazards X: Process Safety in Fine and Speciality Chemical Plants* pp 3-18
[4] Barton K and Rogers R 1997 *Chemical Reaction Hazards*: Elsevier Science
[5] Taylor J 2003 *Risk Analysis for Process Plant, Pipelines and Transport* (CRC Press)
[6] Shariff A M and Zaini D 2010 Toxic release consequence analysis tool (TORCAT) for inherently safer design plant *Journal of Hazardous Materials* **182** pp 394-402
[7] Ashford N A and Zwetsloot G 2000 Encouraging inherently safer production in European firms: a report from the field *Journal of Hazardous Materials* **78** pp 123-144
[8] Leong C T and Shariff A M 2009 Process route index (PRI) to assess level of explosiveness for inherent safety quantification *Journal of Loss Prevention in the Process Industries* **22** pp 216 221
[9] Heikkilä A M 1999 *Inherent safety in process plant design: an index-based approach*: VTT Technical Research Centre of Finland
[10] Khan F I and Amyotte P R 2004 Integrated inherent safety index (I2SI): a tool for inherent safety evaluation *Process Safety Progress* **23** pp 136-148
[11] F. I. Khan and P. R. Amyotte 2005 I2SI: a comprehensive quantitative tool for inherent safety and cost evaluation *Journal of Loss Prevention in the Process Industries* **18** pp 310-326
[12] Shariff A M, Leong C T and Zaini D 2012 Using process stream index (PSI) to assess inherent safety level during preliminary design stage *Safety Science* **50** pp 1098-1103
[13] Bai W, Hao L, Guo Z, Liu Y, Wang R and Wei H 2017 A new criterion to identify safe operating conditions for isoperibolic homogeneous semi-batch reactions *Chemical Engineering Journal* **308** pp 8-17
[14] Grewer T 1994 *Thermal Hazards of Chemical Reactions* (Elsevier)
[15] Nian C and You F 2013 Design of Methanol Plant *ESCAPE-23 Symp.* (Finland, pp 9-12)

IOP Conf. Series: Materials Science and Engineering **458** (2018) 012049 doi:10.1088/1757-899X/458/1/012049

Compliance tracking system for training element of OSHA process safety management

N Azmi[1] **and H Abdul Aziz**[1]

Occupational Safety and Health Program, Faculty of Engineering Technology, Universiti Malaysia Pahang, Lebuhraya Tun Razak, 26300 Gambang, Kuantan, Pahang, Malaysia.

hanidaaziz@ump.edu.my

Abstract. Studies show that many accidents have happened in process industries such as in chemical plants that have taken many peoples' lives. One of the contributing factors of such accidents happening is because of inadequate training provided to employees in particular operational training. One of the established standards highlighting this issue is the training element of OSHA Process Safety Management 29 CFR1910.119(g). Apparently the demand for effective ways to implement Process Safety Management (PSM) is on the rise. Basically the efficiency of standard implementation programme process mostly depends on the systematic framework, which means that all the requirements must be considered, ensuring continuity and completing the implementation process. This paper presents a system to track the performance of PSM training programme based on compliance percentage. Furthermore, the system helps process industries to manage and track information, documentations and required actions regarding the PSM training programme in systematic manner. Besides, it also provides a basis for gap analysis to be carried out for programme improvement. Implementation of this system could serve as a lagging indicator for improvement and to ensure the desired competency of plant personnel.

1. Introduction

Accidents are caused by either unsafe acts or unsafe conditions or both. Many accidents have happened in process industries such, for example, in chemical plants that killed many people. Based on the studies in the USA, inadequate training is one of the causes of accidents happening. The Texas City disaster was an industrial accident that occurred on April 16, 1947, in Port, Texas City. It was the deadliest industrial accident in U.S. history, and one of the largest non-nuclear explosions [1] in which about 15 people were killed and 180 people were injured. This happened because the training department budget was cut in half since 1998 to 2004, so the trainer spent little time on actual training. Similarly, in 2001, there are 785 deaths that are recorded in China while in 2000 about 1092 deaths because of the industrial accidents [2]. Based on the record, the lack of training is one of the main contributing factors as to why accidents take place. It is because the company did not know how to implement the training element, thus accidents were reported to happen year by year [3]. Likewise, on September 27, 2012, a chemical leakage accident occurred at Hube Global Co., Ltd., a chemical products manufacturer located at the 4th National Industrial Complex in Gu-mi City, Gyeongsangbuk Province. In this tragic accident, the hydrogen fluoride was released when the valve of the lorry tank was mistakenly opened by a worker who was injecting the gas into a facility [4]. This human error factor also happened due to the inadequate training of workers.

To overcome or minimize accidents from continuing to occur, every industry should understand and realize that having the trained workers do high risk operating jobs is very important. Occupational Safety and Health Administrative (OSHA) Process Safety Management (PSM) Standard 29 CFR 1910.119(g) is the regulation that is established for process industries. In this regulation, it stated that all employees, including maintenance and contractor employees involved with highly hazardous chemicals, need to fully understand the safety and health hazards of the chemicals and processes they work with so they can protect themselves, their fellow employees, and the citizens of nearby communities. Therefore, there is urgent to have a training system that can ensure company develop, implement and maintain their training program as per PSM requirements.

There are many systems which have been developed to manage trainings such as Training Tracker, Safety Tracker, Pink Staff, Instructional Design System (ISD) and Operational Training Management System (OPTRAMS). The functions of Training Tracker are to schedule training for employees and create training records. Safety Tracker, on the other hand, records accidents to fulfill OSHA logs, Pink Staff provides more than 70 hours of computer-based training and testing [5] while ISD focuses on how to create and deliver highly effective training programs [6]. Of all the systems developed, only OPTRAMS was qualitatively developed based on PSM regulations [7]. In this study, the system that is presented is a quantitatively enhanced OPTRAMS system, whereby the performance of training program is measured based on compliance percentage. The value will become a lagging indicator for improvement.

2. Methodology

2.1 Plan-Do-Check-Act Cycle

In this study, the concept of Plan Do Check Act (PDCA) cycle is adopted to develop the training program performance tracking system as shown in figure 1.

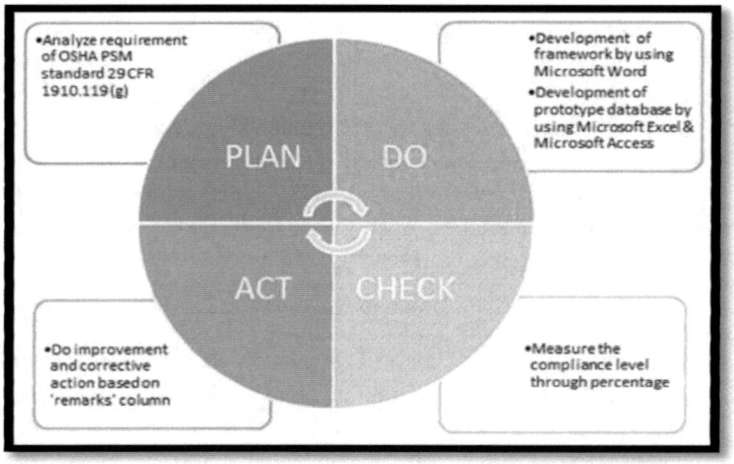

Figure 1. PDCA Cycle adapted from (Chakraborty, 2016) [8].

2.2 Compliance Measurement

In PSM standard, there are four main provisions that need to be fulfilled. They are initial training, in lieu of initial training, refresher training and training documentation. The total requirements of each provision is tabulated in table 1. Each provision can then be assessed by calculating the percentage of the weighted score. The weighted score is assigned using a Likert Scale. Table 2 shows the description of the scores that govern the calculation.

Table 1. Training Element Requirements [9]

PSM Training Provisions	Number of Requirement
Initial training	19
In lieu of initial training	2
Refresh training	4
Training documentation	4
Total	28

Table 2. Description of Answer Options and Scores [10]

Term	Description	Score
NO	The company did not fulfill the requirement, thus there was no documentation	1
YES, but no document	The company has fulfilled the requirements but there was no documentation	2
YES and documented	The company has fulfilled the requirements and documentation was provided to verify the compliance	3

The percentage of the weighted score, maximum score and the percentage of compliance for each phase is calculated based on equations (1), (2) and (3) respectively, as shown below.

$$\frac{Total\ Maximum\ Score\ for\ a\ phase}{Total\ Maximum\ Score\ for\ all\ phases} \times 100\% \quad (1)$$

$$No\ of\ requirement\ in\ a\ phase\ \times 3(\textit{highest score}) \quad (2)$$

$$\frac{Total\ Score\ for\ a\ phase}{84} \mathrm{X}\ 100\% \quad (3)$$

3. Results and Discussion

3.1 Framework of Training Tracking System

The framework of training element in figure 2 was developed based on OSHA PSM requirements. Each of training sub-standards in the framework has its own percentage that needs to be fulfilled. Based on equations (1) - (3), the percentage of weighted score is about 67.9% for initial training, 7.1% for in lieu of initial training, 10.7% for refresher training and 14.3% for documentation. The framework shows that initial training is really important to ensure the competency of the employees prior to any operating tasks.

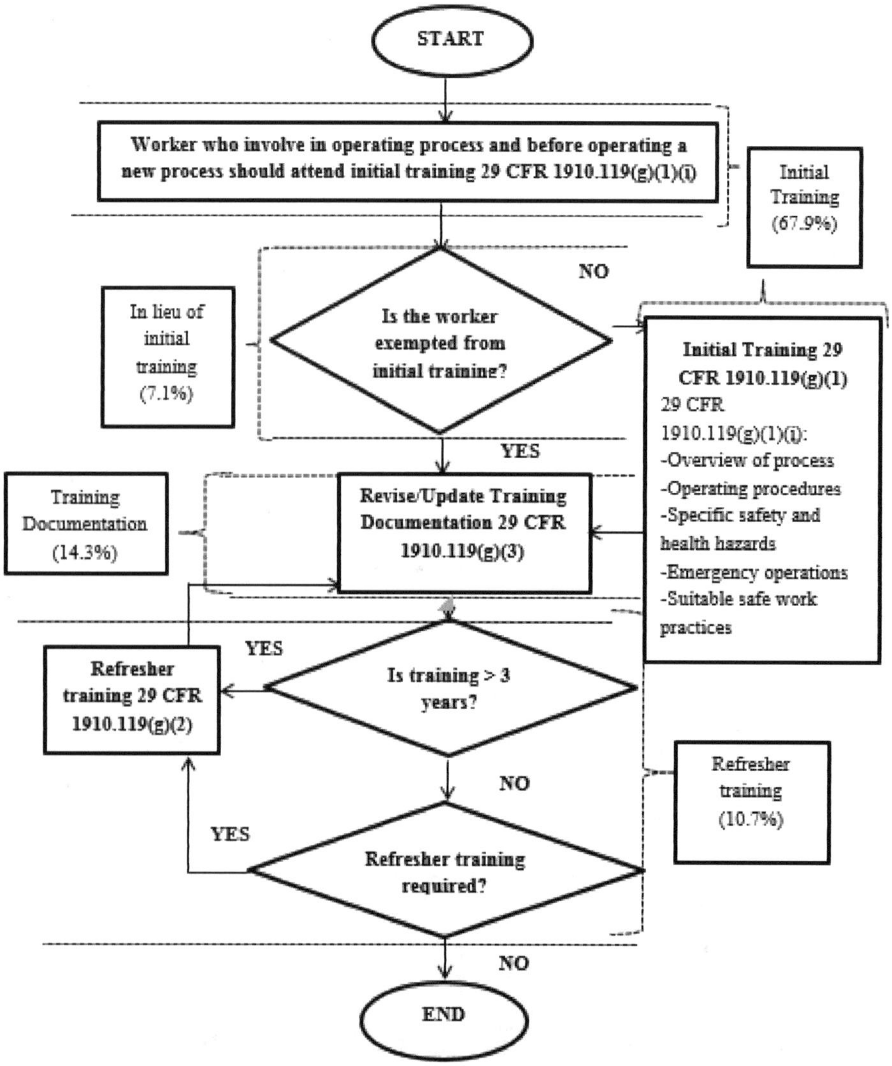

Figure 2. Framework of Compliance Tracking System based on 29 CFR 1910.119 (g).

For the verification of the proposed concept and feasibility study of developed system, a case study was conducted using real plant data at Plant K. In this paper, only a case study on initial training program performance is discussed in detail.

3.1.1 Main Interface

Figure 3 shows the main interface of the Performance Tracking System. All the sub-requirements are listed and linked with the main interface. The 'Percentage (%)' column directly shows the compliance percentage for each sub-standard. The 'Remarks' column is to explain or comments about the incomplete information or condition of the sub-standard. From the comments, the end users can take action and do improvement in order to comply with the standards. The 'Issuance by' column is to allow users to identify the responsible persons when the section needs to be updated or completed.

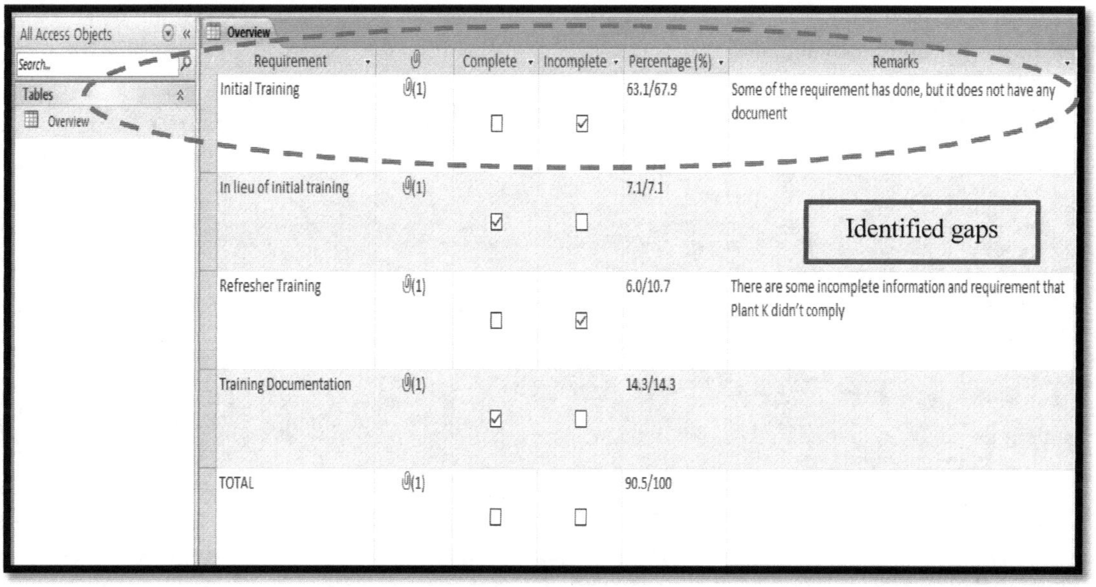

Figure 3. Main Interface.

Based on figure 3, it shows that Plant K has only complied with 'In lieu of initial training' and 'Training Documentation' provisions. According to the 'Remarks' section, there was some incomplete or lack of information identified for 'Initial training' and 'Refresher training' provisions. The 'Percentage' column shows that the compliance percentage of 'initial training' is 63.1/67.9, in lieu of initial training is 7.1/7.1, refresher training is 6.0/10.7 and training documentation is 14.3/14.3. The system shows overall total for the compliance percentage is 90.5%. of PSM compliance as well as the gaps that exist in their training program. The identification of the gaps could serve as a lagging indicator for industries to prioritize their efforts and resources.

3.2 Model for Performance Tracking System with Case Study

3.2.1 Initial Training. The interface of 'Initial training' consists of two parts, which are Part A and Part B. Part A (figure 4) is for initial training requirement while Part B (figure 5 and 6) covers the training program. In this provision, PSM requires new or existing employees who are involved in operating process and a newly assigned process to attend initial training. Figure 4 tabulates that Plant K has fully complied with both of the requirements with the total score of 6.

Figure 5 shows the continuity of interface for Initial Training (Part B). This interface consists of training program that should be conducted for the employees who are involved in the operating process. According to PSM standards, the training should be conducted based on the overview of the process and operating procedures of the process. The training must include emphasis on the steps for each operating phase, operating limits and specific safety and health hazards consideration. Based on the conducted case study (figure 5), Plant K has trained their employees on the consequences of deviation but the document or evidence was not available. Therefore, the score for training of operating limit is 5. Similarly, for 'Related safety and health training', part one of the required document was also not available, bringing the total score to 17 out of 18. Thus the score Plant K obtained for Part B is 47.

In Overall, the score from both parts were automatically calculated and displayed at the 'Total' row. In this case, the total score from both parts shows that the compliance rate of initial training provision in Plant K is 53 out of 57 or 63.1%. In order to ensure that Plant K fully complies with initial training requirement, all the training designed and conducted should be documented to verify the compliance and future reference.

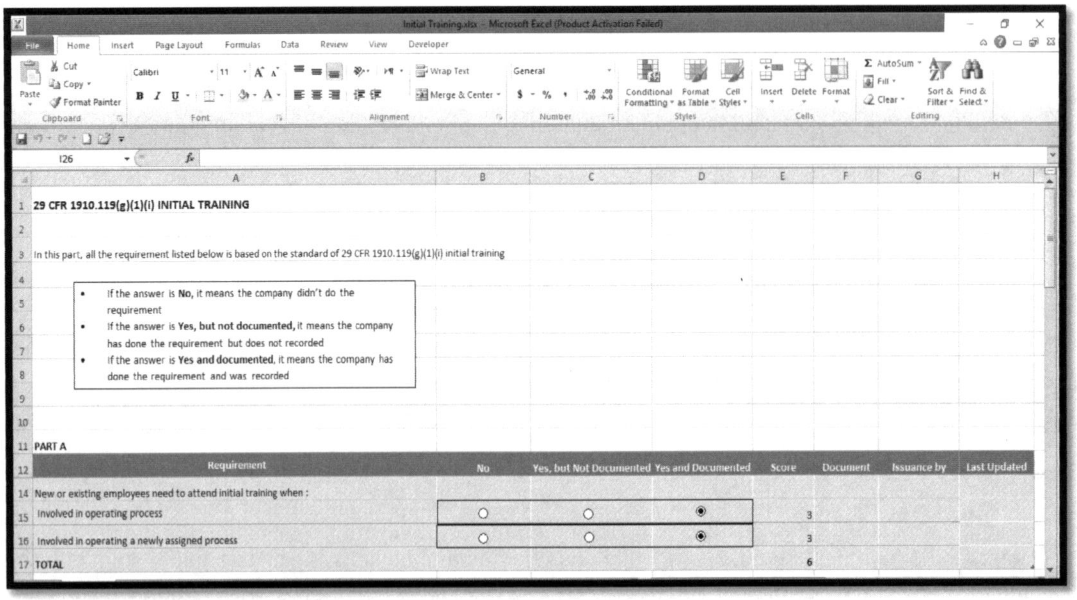

Figure 4. Initial Training Interface (Part A)

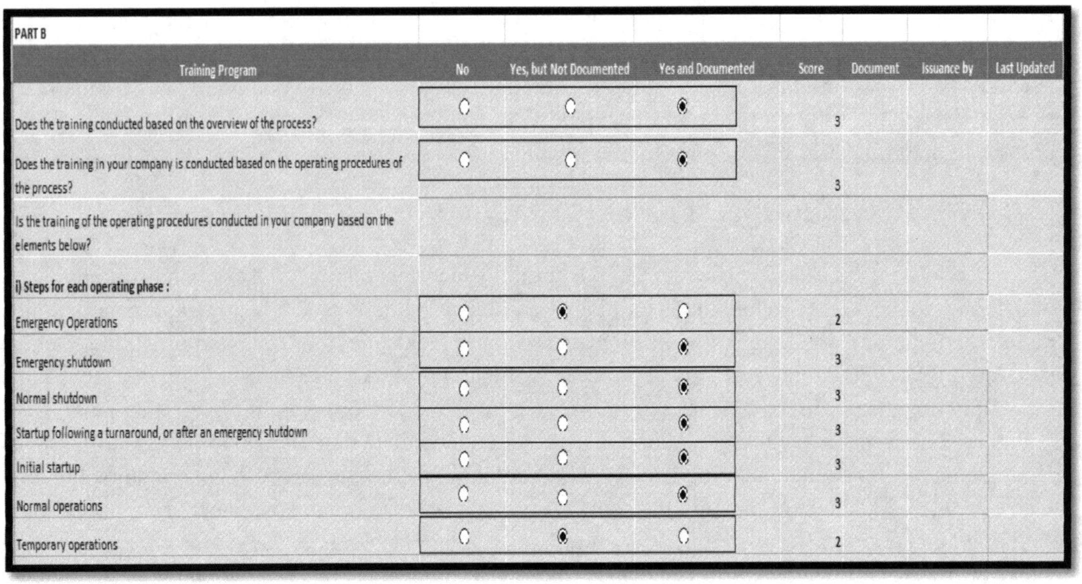

Figure 5. Training Program.

IOP Conf. Series: Materials Science and Engineering **458** (2018) 012049 doi:10.1088/1757-899X/458/1/012049

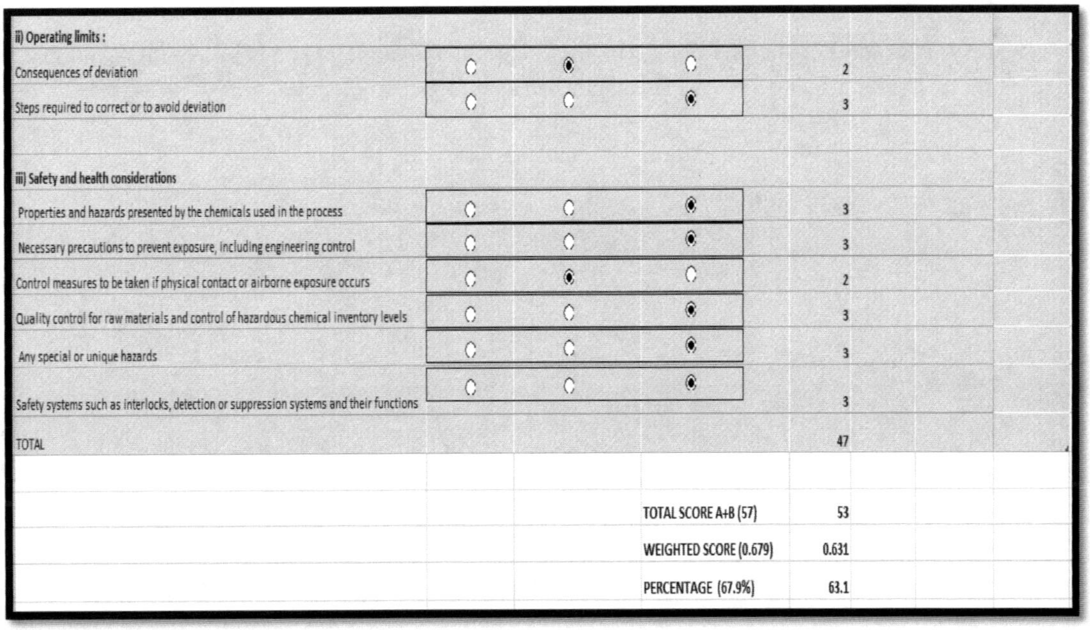

ii) Operating limits :				
Consequences of deviation	○	◉	○	2
Steps required to correct or to avoid deviation	○	○	◉	3
iii) Safety and health considerations				
Properties and hazards presented by the chemicals used in the process	○	○	◉	3
Necessary precautions to prevent exposure, including engineering control	○	○	◉	3
Control measures to be taken if physical contact or airborne exposure occurs	○	◉	○	2
Quality control for raw materials and control of hazardous chemical inventory levels	○	○	◉	3
Any special or unique hazards	○	○	◉	3
Safety systems such as interlocks, detection or suppression systems and their functions	○	○	◉	3
TOTAL				47
			TOTAL SCORE A+B (57)	53
			WEIGHTED SCORE (0.679)	0.631
			PERCENTAGE (67.9%)	63.1

Figure 6. Training Program (continued).

4. Conclusion

This paper introduces a system to track the compliance of PSM training program at the workplace. The indication of performance is shown by the percentage of compliance with the standard requirements. The framework was developed as a guide for employers to manage their training programs in a systematic way. This system prototype is based on the proposed concept that was developed and embedded with the proposed equation and scoring scale for easy implementation and explanation. Implementation of this system could serve as a lagging indicator for improvement in order to ensure the desired competency of plant personnel. Industries need a systematic framework and system to ease and trace their program performance to ensure safe operation, prevent major accidents and also to comply with the training element of PSM standard.

Acknowledgement

The authors would like to thank Universiti Malaysia Pahang for funding this study (RDU1703204) and making this study feasible.

References

[1] Labib A 2014 BP Texas City Disaster. In Learning from Failures 83

[2] Liu T, Zhong M, and Xing J 2005 *J. Saf. Sci.* **43** p 503

[3] Zhao J, Suikkanen J and Wood 2014 *J. Loss Prev. Proc. Ind.* **29** p 170

[4] Lee K, Kwon H M, Cho S, Kim J and Moon I 2016 *J. Loss Prev. Proc. Ind.* **42** p 6

[5] Zaloom V and Ramachandran P 1996 *Proc. of the 19th Int. Conf. on Computers and Industrial Engineering* **31** p 511

[6] Myers P M, Watson B and Watson M 2008 *J. Proc. Saf. Prog.* **102** p 45

[7] Aziz H A, Shariff A M and Rusli R 2014 *J. Adv. Materials Research* **917** p 325

[8] Chakraborty A 2016 *International J. Mech. Eng.* **30**
[9] OSHA. Law and Regulation (29 CFR1910.119) Occupational Safety and Health Administration
[10] Joshi A, Kale S, Chandel S and Pal D 2015 *British J. Appl. Science & Technology* **7** p 396

Dynamic simulation and control of feed conditioning system for CO_2 capture

N A Othman[1 2], H Zabiri[1], L D Tufa[1]

[1] Chemical Engineering Department, Universiti Teknologi PETRONAS, Bandar Seri Iskandar, 32610, Perak, Malaysia
[2] PETRONAS Research Sdn Bhd, Lot 3288 & 3289, Off Jalan Ayer Itam, Kawasan Institusi Bangi, 43000 Kajang, Selangor

nurzatil_16000418@utp.edu.my

Abstract. There are several CO_2 removal technologies currently applied in the commercial and production scale. One of the potential technologies currently being evaluated for offshore CO_2 separation is Centrifugal Fluid Separation technology, specifically Supersonic Gas Separation (SGS). It is potentially better than the conventional solutions such as membrane in terms of CAPEX, hydrocarbon losses, footprint, tonnage and power requirement. However, the stringent requirement of cryogenic temperature and high pressure of its feed requires a robust feed conditioning process plant. There are limited studies on the simulation and control design which are crucial to test the stability and dynamic responses of the feed conditioning plant. This paper investigates the optimal regulatory plant-wide control strategy for the feed conditioning plant subjected by various disturbances in temperature, pressure and feed CO_2 composition.

1. Introduction

The importance of developing a reliable simulation model of technologies has long been recognized by the industry in order to assess the process dynamics, evaluate and optimize equipment design, controllability and operating procedures during design phase, training operators before actual operation as well as sensitivity studies of process disturbance during operation [1]. For a relatively new technology such as SGS technology in CO_2 removal application, prior to being applied at actual field, dynamic simulation of the plant is crucial in order to investigate and assess the variations of feed and process conditions i.e. temperature, pressure and gas compositions as these will impact the CO_2 separation performance inside the separator. Even though the technology has been developed since 1989, the application is more on dehydration and hydrocarbon dew pointing [2]. For CO_2 separation from natural gas, although the proof of concept of this technology has been successfully validated, there is still a lot of development work to be done particularly prior to the field application as there are a lot of uncertainties of feed conditions to be tackled [3].

Supersonic Gas Separation technology for CO_2-NG separation is based on the phase change of the gas mixture component. For Twister, a proprietary-owned SGS technology developed by Twister BV, the CO_2 is condensed by the instantaneous change of the temperature attributed to the expansion of the supersonic flow induced by the Laval nozzle design [4]. For modelling of SGS technology including Twister, most of the data available in literature is on numerical modelling, typically computational fluids dynamic modelling of the supersonic separator. Wen, Cao, Yang and Zhang [5] has developed a supersonic separator to study the swirling effects on the separator performance for natural gas separation. The basis used for the model are continuity equation, momentum equation and energy equation. Fluent CFD software was employed in the numerical study and the experimental result was compared against the CFD simulation which showed good agreement. Prast, Lammers and Betting [6] from Twister BV produced a paper on CFD model of the Twister device, which incorporates nucleation and droplet growth using the Ansys CFX software. The first results of the simulation were compared against experimental values for both component recoveries and wall pressures. It was found that the CFD model gave good indication in predicting the physical behavior of the separator.

While most of the studies done on the SGS technology revolves around the design parameters, there is minimal studies with regards to its feasibility and controllability, particularly for the feed conditioning in CO_2 separation from natural gas.

Dynamic simulation study is essential to evaluate the flexibility of the feed conditioning plant and generate the data required for supersonic separator operating envelope – namely temperature, pressure, gas composition and flow rate. The fluctuations in these key operating parameters, particularly temperature and CO_2 composition will affect the performance of this process, e.g., the changes in these two conditions may reduce the separation efficiency of SGS technology, resulting in higher CO_2 content in sales gas which is undesired. Based on the dynamic simulation and the variations of the feed study results, a process control design would help in achieving the tight control of the supersonic separator's inlet parameters and maintain the stability of the process.

Hence, the main aim of this paper is to develop a dynamic simulation and control strategy of the feed conditioning plant of SGS technology for CO_2 separation plant. The control strategy is designed to meet the specified operating parameters of SGS technology inlet under various disturbances.

2. Methodology

The research methodology employed for this project is summarized in the following flow chart:

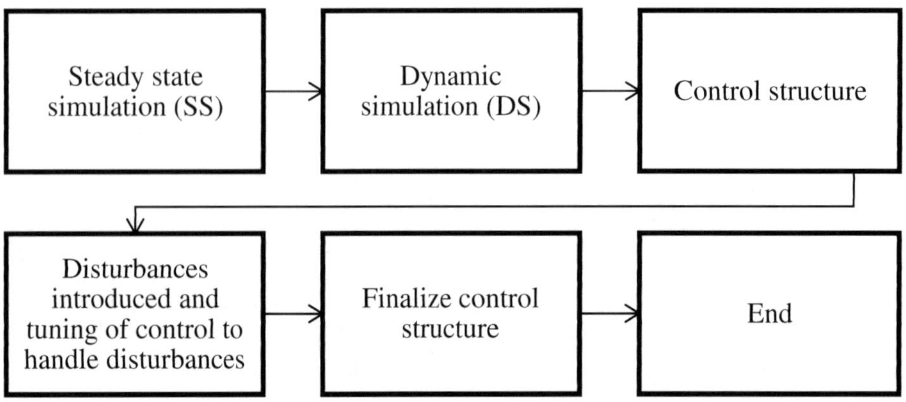

Figure 1. Flow chart of research work

In whole, the research process involves steady state simulation, dynamic simulation and process control design and testing of the feed conditioning plant of CO_2 separation process using supersonic separator. The optimum process parameters of the supersonic separator is identified based on existing data and the dynamic simulation and process control design are conducted based on the steady state simulation work. In this study, process simulator software Aspen HYSYS V.8 was utilized for rigorous steady-state and dynamic simulation of the feed conditioning process. Peng-Robinson equation of state was used to calculate the thermodynamic properties.

IOP Conf. Series: Materials Science and Engineering **458** (2018) 012050 doi:10.1088/1757-899X/458/1/012050

2.1. Feed conditioning plant process descriptions

In the process scheme, the feed (stream 212), a cold stream at cryogenic temperature which consists of 35% CO_2 (mixed gas) goes to the first separator prior to going through the compressor and cold box to knock off most liquid from the stream. This liquid stream is being routed to CO_2 liquid reinjection as it contains high CO_2 content. After the first separator, the feed gas is compressed up to approximately 77 bar and is cooled down to cryogenic temperature, -50°C via a coldbox which uses several process streams (stream 319 and 218) as part of heat integration. The cold feed is fed to the second separator, V-301 to remove remaining liquid phase prior to being fed to SGS technology unit (after stream 300). The mole fraction and other process parameters of feed gas 212 are listed in table 1.

Table 1. Feed gas conditions.

Parameters	Value
Mole fraction of CO_2	0.34
Pressure (bar)	21.30
Temperature (°C)	-56.60
Flow rate (tonne/hr)	773.60

As the steady-state simulation was set up first, some standard modifications are required in order to transition the scheme into dynamic simulation model in Aspen HYSYS.

2.2. Control structure

The feed conditioning plant considered in this study is shown in figure 2:

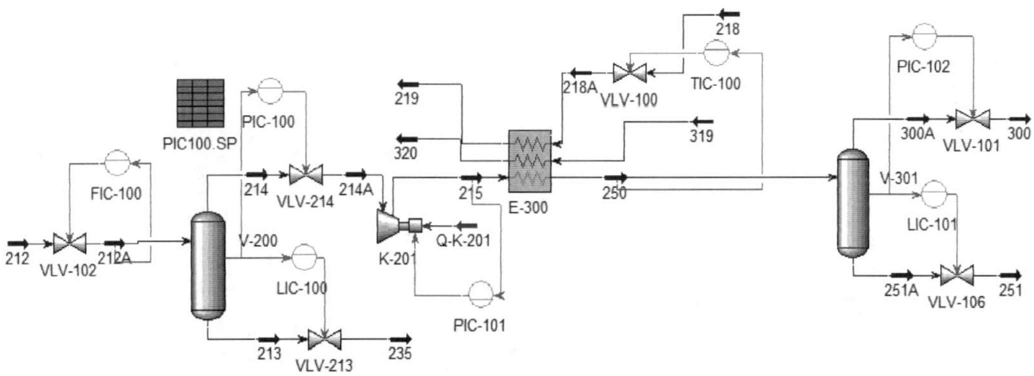

Figure 2. The schematic flow diagram of feed conditioning process of SGS technology for CO2 removal.

As shown in figure 2, there are two liquid levels which need to be controlled to maintain the system, namely the liquid level of first separator, V-200 and second separator, V-301. The main control objectives of this plant are to maintain the CO_2 content of inlet stream to SGS technology i.e. stream 300 within the tolerance limit which is -3 mole% and +1 mole%; and to maintain operating pressure and temperature of stream 300 within the tolerance limit of ±1%. Based on most of the studies conducted on numerical modelling of SGS, the main processes involved in the SGS separator are expansion, condensation of droplets and cyclonic separation of gas and liquid which need a certain design condition i.e. the inlet stream to be at high pressure to ensure sufficient pressure drop and very low temperature due to the phase envelope of CO_2. These parameters are the key process variables to ensure the SGS technology separation efficiency and to meet the CO_2 spec for sales gas.

Based on these explanations, the selected controlled variables (CV) for this system are: CO_2 composition, pressure and temperature of stream 300. Based on earlier explanation, due to the phase envelope, the increase or decrease of these 3 critical process parameters would influence the quality of the product stream of SGS technology. This control structure is designed to cater for disturbances in terms of variation in feed pressure, temperature and CO2 composition (stream 212). As shown in figure 2, there are 7 control valves in the system that can be used as manipulated variables (MV) for this process, including the 2 liquid level controllers while the flow controller FIC-100 is the throughput manipulator. The pressure and temperature of stream 300 are controlled by PIC-101 and TIC-100, while the throughput controller FIC-100 is used to control/manipulate the flow rate in response to disturbances.

3. Simulation results

This section will discuss on the dynamic responses of the plant when disturbances occur at the feed conditioning plant inlet to investigate the dynamic response. The main disturbances are feed gas (stream 212) pressure, temperature and CO_2 composition. The operating pressure, temperature and CO_2 composition of SGS technology inlet, stream 300 are three important parameters which need to be observed when the disturbances are introduced to the system.

3.1 Disturbance of feed gas pressure

The feed gas is varied from +15% to -15% of its original operating pressure, i.e. increase and decrease 15% of 21.3 bar respectively. The red dotted lines in the figures indicate the control objective range for respective process parameters. Dynamic responses of controlled variables after the increase and decrease of feed gas temperature are illustrated in figure 3 below:

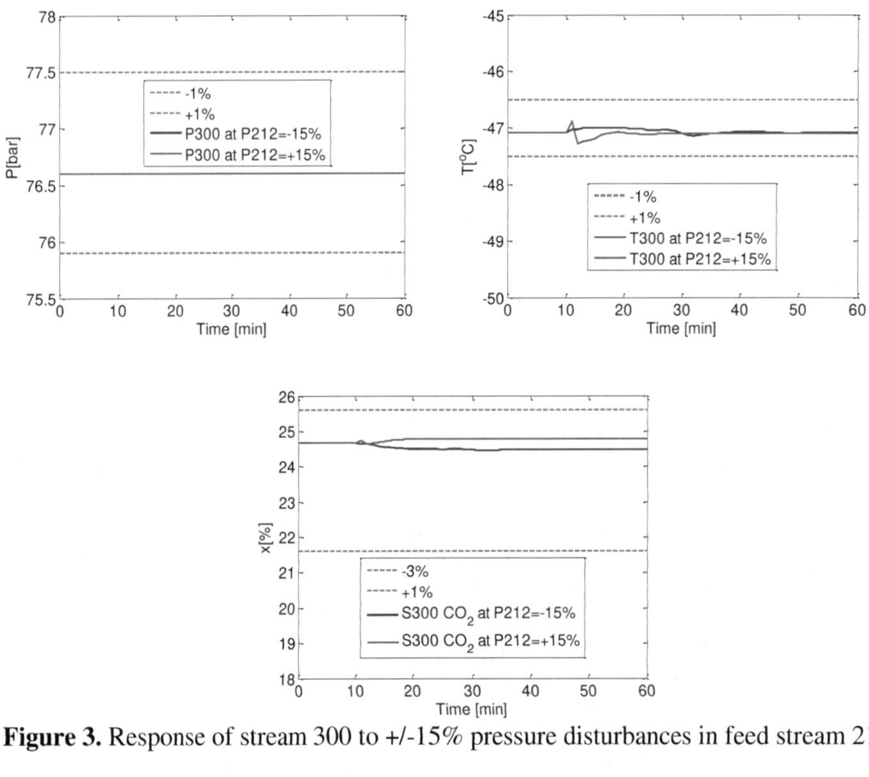

Figure 3. Response of stream 300 to +/-15% pressure disturbances in feed stream 212.

Based on the result, pressure, temperature and CO_2 composition of stream 300 are well within the desired range when disturbance in pressure of feed gas was introduced. The responses are fast, stable and without large overshoot. The controlled variables are minimally affected and can go back to stable state in a short time (short settling time). The control strategy works efficiently in ensuring all the three key variables are well-maintained near or at their desired set-point values.

3.2 Disturbance of feed gas temperature

The feed gas is varied from +15% to -15% of its original operating condition, i.e. increase and decrease 15% of -56.6°C respectively. Dynamic responses of controlled variables after the increase and decrease of feed gas temperature are illustrated in figure 4 below:

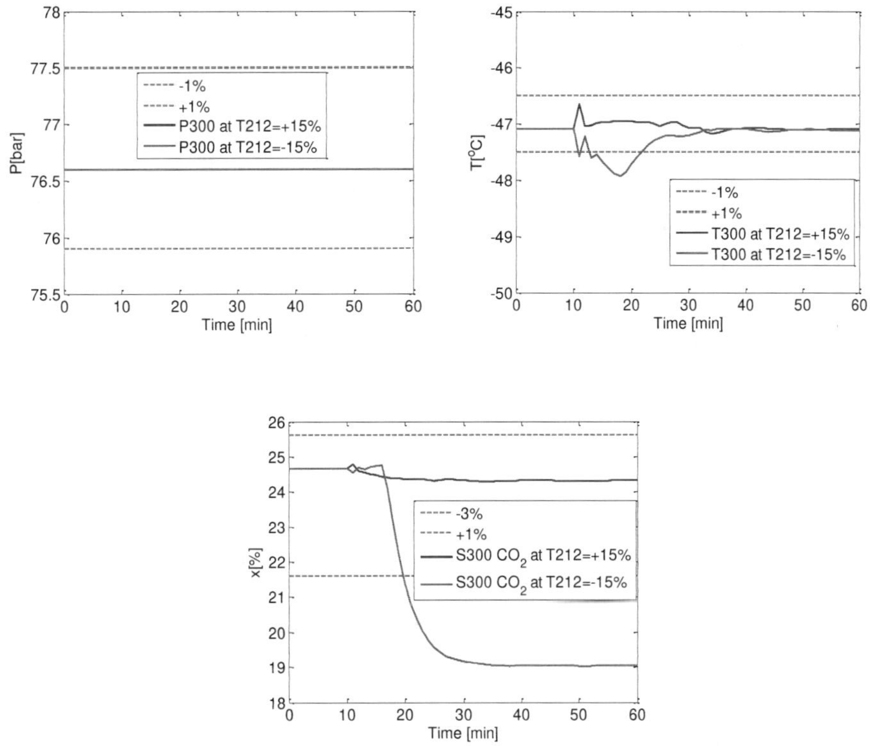

Figure 4. Response of stream 300 to +/-15% temperature disturbances in feed stream 212.

It is observed that pressure and temperature of stream 300 are satisfactorily maintained near the desired steady-state values after some initial transients, however when temperature of feed gas drops too low i.e. at -15% from original condition, CO_2 composition of stream 300 is largely affected. It is due to the lower temperature where more CO_2 condense in liquid form at the first separator hence less CO_2 were left in gas phase, resulting in the lower CO2 content in stream 300 at only 19% compared to original condition, 25%.

However, if the feed temperature drops to only -9.5% the control strategy is able to retain the CO_2 composition within the desired range of -3 mole% and +1 mole% as shown in figure 5 below:

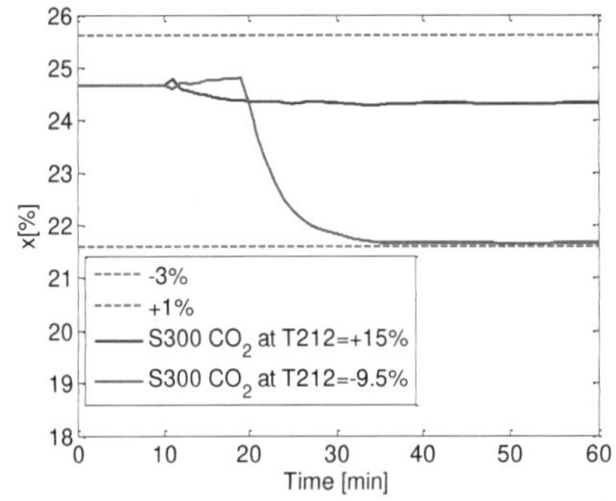

Figure 5. CO_2 composition response of stream 300 to +15% and -9.5% temperature disturbances in feed stream 212.

3.3 Disturbance of feed gas CO_2 composition

The feed gas is varied from +5 mole% to -5 mole% of its original CO_2 composition, i.e. increase and decrease 5 mole% of 0.34 respectively. Dynamic responses of controlled variables after the increase and decrease of feed gas CO_2 composition are illustrated in figure 6 below:

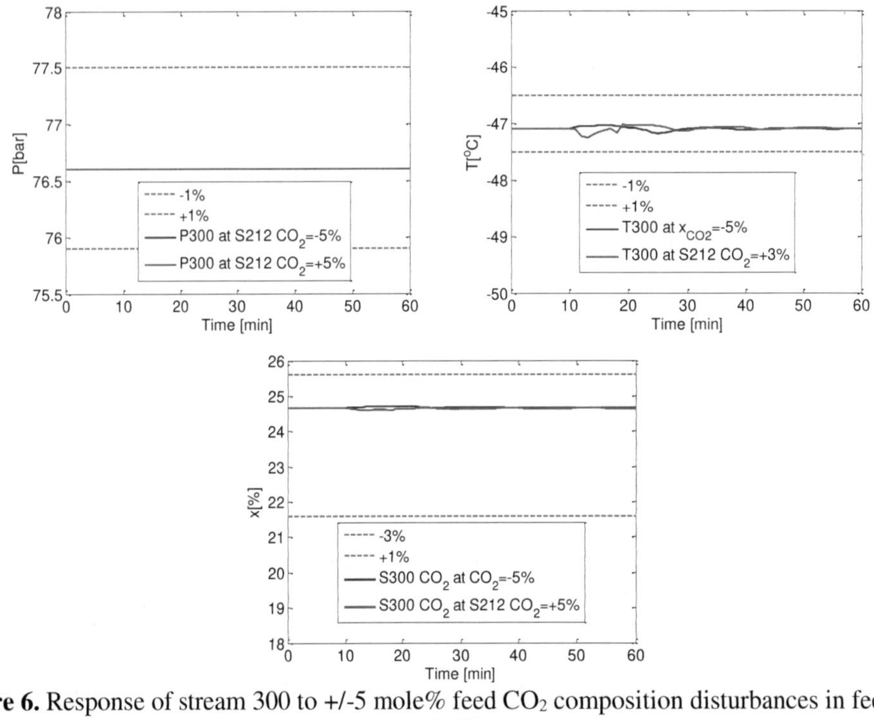

Figure 6. Response of stream 300 to +/-5 mole% feed CO_2 composition disturbances in feed stream 212.

It is observed that variations in feed CO_2 composition have very minimal impact to the system, as all three key variables are hardly perturbed by the changes.

4. Conclusions

In this study, a dynamic simulation and control of feed conditioning process of SGS technology for CO_2 removal was developed to investigate the dynamic behaviours of the plant. The results indicated that the proposed control structure could satisfy the control objectives except for -15% disturbance in feed temperature from its original condition, which results in CO_2 content of stream 300 to be lower than specification. Hence, for future work, the control structure should be improved to address lower temperature disturbances and bring the CO_2 content in the inlet stream to SGS technology to the specified range.

5. References

[1] Seborg D E, Mellichamp D A, Edgar T F and Doyle F J 2010 Process Dynamics and Control
[2] Haghighi M, Hawboldt K A and Abedinzadegan Abdi M 2015 Supersonic gas separators: Review of latest developments *J. Nat. Gas Sci. Eng.* **27** pp 109–21
[3] Samawe R A, Rostani K, Jalil A M, Esa M and Othman N 2014 Concept proofing of supersonic nozzle separator for CO2 separation from natural gas using a flow loop *Offshore Technol. Conf. Asia* Malaysia, pp 2373–76
[4] Schinkelshoek P and Epsom H D 2008 Supersonic gas conditioning - Commercialisation of twister technology *GPA Annu. Conv. Proc.* USA, pp 739–45
[5] Wen C, Cao X, Yang Y and Zhang J 2011 Swirling effects on the performance of supersonic separators for natural gas separation *Chem. Eng. Technol.* **34** pp 1575–80
[6] Prast B, Lammers B and Betting M 2006 CFD for Supersonic Gas Processing *5th Int. Conf. CFD Process Ind.* Australia, pp 1–6

IOP Conf. Series: Materials Science and Engineering **458** (2018) 012051 doi:10.1088/1757-899X/458/1/012051

Routh-Hurwitz tuning method for stable/unstable time-delay MIMO processes

N Mohd[1, 2] and J Nandong[1]

[1] Department of Chemical Engineering, Curtin University, 98009 Miri, Sarawak, Malaysia
[2] Curtin Malaysia Research Institute, Curtin University, 98009 Miri, Sarawak, Malaysia

jobrun.n@curtin.edu.my

Abstract. This paper presents a new method of tuning decentralized Proportional Integral Derivative (PID) control system for an $n \times n$ Multi Input Multi Output (MIMO) processes which can be either open-loop stable or unstable with time-delays. The tuning method has been established via the classical Routh-Hurwitz criteria with the aid of PID stability theorem. This new tuning method requires only 3 common tuning parameters, thus simplifying the conventional multi-loop PID tuning task to obtain the values for *3n* PID parameters. Two numerical examples are presented to demonstrate the effectiveness of the proposed tuning procedure over some of the existing multi-loop PID tuning procedures. The proposed tuning procedure shows a good performance as well as able to simplify the tuning task as compare to the existing tuning method.

1. Introduction

Most real processes in industries are multi-input and multi-output (MIMO) in nature and possess significant time-delays or deadtimes. These criteria lead to an unstable process behaviour. Decentralized (or multi-loop) PID control systems are widely used to control process plants. Controlling a MIMO process is not as simple as controlling a single-input and single-output (SISO) process because of the existence of process interactions or couplings. Process interactions and deadtimes impose an upper bound on the achievable maximum control performance. If a given MIMO process open-loop is unstable, then the unstable poles in the system will impose a lower bound on the allowable minimum control performance. To control an unstable process, the PID controller parameter values must be well tuned within the maximum lower bound and minimum upper bound of the control performance for a stable operation.

For an $n \times n$ MIMO system, there are n PID controllers involved with a total of $3n$ tuning parameters. The decentralized PID tuning methods can be categorized into: (1) detuning, (2) sequential loop closing, (3) iterative or trial-and-error, (4) simultaneous equation solving and (5) independent tuning; see [1]. A significant innovation in the decentralized PID control tuning was proposed in [2], known as the Simultaneous Multi-Loop Multi-Scale Control (SML-MSC) tuning, based on the Multi-Scale Control theory in [3], [4]. The salient feature of the SML-MSC method is that it only requires 2 to 3 scaling parameters in order to tune an arbitrary $n \times n$ multi-loop PID system. However, the technique is restricted to open-loop stable MIMO processes, in which the diagonal processes can be represented by first-order plus deadtime (FOPDT) models. The SML-MSC tuning is easy to apply compared to other tuning methods.

For SISO systems, the closed-loop stability for the standard PID control system has been well studied. Several stability methods have been developed for assessing the closed-loop stability of PID controllers, e.g., the stability analysis based on the Hermite-Biehler theorem [5], the Gain-Phase Margin stability analysis [6], and the classical Routh-Hurwitz criteria [7]. On the contrary, the closed-loop stability analysis that is applicable to the multi-loop PID system remains an open problem.

The present work attempts to address the multi-loop PID control system design via the recently published SISO PID stability theorem, based on the classical Routh-Hurwitz stability criteria [7]. An advantage of the reported PID stability theorem is that, it can be conveniently used to find a stabilizing region of PID parameters both for open-loop stable or unstable processes. Through this PID stability theorem, we can reduce the tuning of multi-loop PID control controllers to just finding the values of 3 common tuning parameters for an arbitrary $n \times n$ MIMO system. Unlike the SML-MSC tuning method in [7], the new simultaneous tuning technique developed in this work is applicable to both open-loop stable and unstable processes.

The main contributions of this work can be summarized as follows:

- Decentralized PID control tuning for a class of stable MIMO systems (Section 4).
- Decentralized PID control tuning for a class of unstable MIMO systems (Section 4).

The rest of this paper is organized as follows. Section 2 presents some preliminaries. Then, Section 3 details the derivations of tuning relations. Next, Section 4 provides two new tuning algorithms followed by Section 5 which demonstrates the applicability of the PID tuning algorithms. Finally, some concluding remarks and future works are highlighted in Section 6.

2. Preliminaries

Consider the ideal form of Proportional-Integral-Derivative (PID) controller

$$G_c(s) = K_c\left(1 + \frac{1}{\tau_I s} + \tau_D s\right) \tag{1}$$

where K_c, τ_I and τ_D represent the controller gain, reset time and derivative time respectively.

In the industry, many processes which are open-loop stable can be expressed as a first-order plus deadtime (FOPDT) model given by

$$G_p(s) = \frac{K_p \exp(-\theta s)}{\tau_p s + 1} \tag{2}$$

where K_p, τ_p and θ are the process gain, time constant and deadtime respectively.

Take note that, some of the processes which are open-loop unstable can be expressed by an unstable first-order plus deadtime (UFOPDT) model given as follows

$$G_p(s) = \frac{K_p \exp(-\theta s)}{\tau_p s - 1} \tag{3}$$

Figure 1 shows the standard single-loop feedback control. Here, F_r denotes the setpoint pre-filter, G_c the controller and P the process transfer functions; R, D_i, D_o and Y denote the setpoint, input disturbance, output disturbance and controlled variable signals respectively.

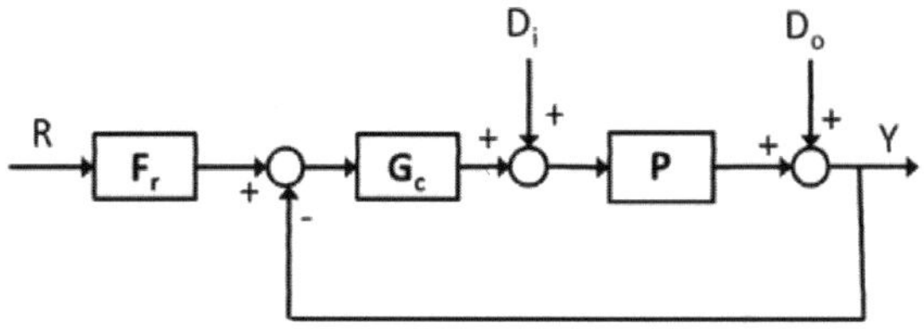

Figure 1. Standard single-loop feedback control structure.

For the standard single-loop control structure in figure 1, the closed-loop setpoint tracking and output disturbance transfer functions are given by (4) and (5), respectively.

$$H_r(s) = \frac{F_r(s)G_c(s)G_p(s)}{1+G_c(s)G_p(s)} \tag{4}$$

$$H_d(s) = \frac{1}{1+G_c(s)G_p(s)} \tag{5}$$

The closed-loop characteristic equation for both transfer functions are similar, i.e.:

$$1+G_c(s)G_p(s)=0 \tag{6}$$

Since the closed-loop transfer functions have a similar characteristic equation, this means that if the controller is stable for the setpoint tracking, then it should be stable also for the output disturbance rejection.

To perform the stability analysis on the closed-loop characteristic equation, it is convenient to first approximate the deadtime term as a rational transfer function, e.g., by using the 1/1 Padé approximation given as follows

$$\exp(-\theta s) \approx \frac{1-\alpha s}{1+\alpha s} \tag{7}$$

where $\alpha = 0.5\theta$.

3. SISO PID stability analysis

3.1. *Stability region determination*

Let us consider the stabilizing regions of PID controller for the two process types given as FOPDT and UFOPDT as Cases A and B respectively.

3.1.1. Case A: FOPDT. Based on the first-order Padé approximation of the deadtime, the closed-loop characteristic equation can be simplified to a polynomial as follows:

$$[\alpha\tau_I(\tau_p - K\tau_D)]s^3 + \tau_I[\alpha + \tau_p + K(\tau_D - \alpha)]s^2 + [\tau_I + K(\tau_I - \alpha)]s + K = 0 \tag{8}$$

Please note that the loop gain is $K = K_c K_p$. To establish a stabilizing region of PID controller, let us apply the two PID stability theorems established by Seer and Nandong [7]. By using the first part of PID stability theorem, we can construct a region of PID parameters which can meet the necessary criterion of Routh-Hurwitz stability.

It can be shown that via the first stability theorem, the ranges of derivative time, reset time and loop gain are given as follows

$$\tau_D > \alpha \tag{9}$$

$$\tau_I > \alpha \tag{10}$$

$$0 < K < \tau_p / \tau_D \tag{11}$$

To meet the sufficient criterion of Routh-Hurwitz stability, the second part of PID stability theorem [7] is applied. The sufficient condition for the closed-loop stability is given by

$$\tau_I > \max(\alpha, \tau_{I,\min_A}) \tag{12}$$

with the assumption that the derivative time and loop gain values are first set to be within the region established by the first part of stability theorem, i.e., as in (9) and (11). Note that $\tau_{I,min}$ is the lower limit on the reset time established via the sufficient criterion of Routh-Hurwitz stability – the second part of PID stability theorem.

$$\tau_{I,\min_A} = \frac{K\alpha}{1+K}\left[1 + \frac{\tau_p - K\tau_D}{\alpha + \tau_p + K(\tau_D - \alpha)}\right] \tag{13}$$

3.1.2. Case B: UFOPDT. By applying the first PID stability theorem; see [7], a region of PID parameter values fulfilling the necessary criterion of Routh-Hurwitz stability is established as follows:

$$\alpha < \tau_D < \tau_p \tag{14}$$

$$\tau_I > \tau_{I,\min 1_B} = \frac{\alpha \tau_p}{\tau_p - \tau_D} \tag{15}$$

$$\frac{\tau_I}{\tau_I - \alpha} < K < \frac{\tau_p}{\tau_D} \tag{16}$$

Provided that the derivative time and loop gain are within the region defined by the first stability theorem, to ensure closed-loop stability the second PID stability theorem is applied. This results in the following range for the reset time

$$\tau_I > \max\left(\tau_{I,\min 1_B}, \tau_{I,\min 2_B}\right) \tag{17}$$

Another lower limit of the reset time is given by the sufficient criterion of Routh-Hurwitz stability:

$$\tau_{I,\min 2_B} = \frac{K\alpha}{K-1}\left[1 + \frac{\tau_p - K\tau_D}{\tau_p - \alpha + K(\tau_D - \alpha)}\right] \tag{18}$$

4. Decentralized PID tuning procedure

4.1. Tuning relations

Let us introduce three dimensionless scaling parameters denoted as r_p, r_d and r_i. These parameters are used in the decentralized PID controller tuning, i.e., to obtain the values of K_c, τ_I and τ_D. All the tuning relations required for cases A and B are summarized in the following sections.

4.1.1. Tuning relations for case A. The following relations are used for tuning the decentralized PID control for Case A.

$$\tau_D = r_d \alpha, \quad r_d > 1 \tag{19}$$

$$K_c = \frac{r_p}{K_p}\left(\frac{\tau_p}{\tau_D}\right), \quad 0 < r_p < 1 \tag{20}$$

$$\tau_I = r_i\left(\frac{\tau_p}{\theta}\right)\left[\max\left(\alpha, \tau_{I,\min_C}\right)\right], \quad r_i > 1 \tag{21}$$

4.1.2. Tuning relations for case B. The proposed tuning relations for a class of open-loop unstable MIMO system are given as follows:

$$\tau_D = r_d\left(\tau_p - \alpha\right) + \alpha, \quad 0 < r_d < 1 \tag{22}$$

$$K_c = \frac{r_p}{K_p}\left[\frac{\tau_p}{\tau_D} - \frac{\tau_I}{\tau_I - \alpha}\left(1 - \frac{1}{r_p}\right)\right], \quad 0 < r_p < 1 \tag{23}$$

$$\tau_I = r_i\left(\frac{\tau_p}{\theta}\right)\left[\max\left(\tau_{I,\min 1_B}, \tau_{I,\min 2_B}\right)\right], \quad r_i > 1 \qquad (24$$

4.2. MIMO model assumptions
In order to use the case A tuning relations, we make the following assumptions.

- **Assumption A.1**: A class of $n \times n$ MIMO system P.
- **Assumption A.2**: All diagonal transfer functions $g_{ii}, i = 1, 2, \dots n$ can be approximated by the FOPDT model in (2) – consider direct controller pairings are used.
- **Assumption A.3**: All transfer function in the plant matrix (**P**) are open-loop stable, i.e., none of them with an unstable or integrating pole.
- **Assumption A.4**: The PID controller for any given *i-th* control loop is designed by using the corresponding diagonal transfer function g_{ii}, i.e., direct controller pairings are adopted.
- **Assumption A.5**: The time-constant to deadtime ratio of a diagonal transfer function $\left(\tau_{p_{ii}}/\theta_{ii}\right) > 1$.

Meanwhile, to use *Case B* tuning relations, the following assumptions are applied:

- **Assumption B.1**: A class of $n \times n$ MIMO system as represented by (25).
- **Assumption B.2**: All diagonal transfer functions $g_{ii}, i = 1, 2, \dots n$ can be approximated by the UFOPDT model in (3).
- **Assumption B.3**: All transfer functions in the plant matrix (P) are open-loop stable except for the diagonal transfer functions.
- **Assumption B.4**: The PID controller for any given *i-th* control loop is designed by using the diagonal transfer function g_{ii}, i.e., direct controller pairings are adopted.
- **Assumption B.5**: The ratio of time-constant to deadtime of a diagonal transfer function $\left(\tau_{p_{ii}}/\theta_{ii}\right) > 1$.

4.3. Tuning algorithm 1
This tuning algorithm is applied to Case A.
*Step 1: Arrange the plant matrix (**P**) so that the Relative Gain Array (RGA) gives direct controller pairings.*
Step 2: Use initial settings ($r_p = 0.2, r_i = 5, r_d = 2$).
Step 3:Performance Evaluation

- Method 1: Time-domain plots. From the step responses for setpoint tracking and output disturbance rejection, observe whether desired responses are achieved or not. If not, go back to step 2, and readjust the parameters.
- Method 2: Optimization-based technique to obtain optimal values of r_p, r_i and r_d that minimizes the total Integral Absolute Error (IAE) value.

The recommended ranges for scaling parameters are as follows:

$$0.1 \le r_p \le 0.6, 2 < r_i \le 8, 1.5 \le r_d \le 5 \tag{25}$$

4.4. Tuning algorithm 2

This tuning algorithm is applied to Case B.

*Step 1: Arrange the plant matrix (**P**) so that the Relative Gain Array (RGA) suggests direct controller pairings.*

Step 2: Use initial settings ($r_p = 0.2, r_i = 2, r_d = 0.1$).

Step 3:Performance Evaluation

Can either use Method 1 or Method 2 as mentioned in the algorithm 1.

Recommended ranges for the scaling parameters are given by

$$0.2 \le r_p \le 0.6, 1.5 < r_i \le 6, 0.01 \le r_d \le 0.25 \tag{26}$$

5. Numerical examples

5.1. Example 1

Consider an industrial distillation process (case A) reported in [8]. In a decentralized control design, it is crucial to first determine the right controller pairings. This can be done by applying the Relative Gain Array (RGA) analysis; see [9]. The diagonal RGA values are 1.0926, 0.1039 and 0.0983. These suggest that the direct controller pairings should be adopted, i.e., $U_1 - Y_1, U_2 - Y_2, U_3 - Y_3$ pairings where U_i and Y_i denote i-th input and output respectively.

Two running values obtained via the proposed Routh-Hurwitz (RH) method are compared with the SML-MSC tuning [2]. Here, we consider two sets of RH tuning: set 1 with $r_p = 0.15$, $r_i = 6$ and $r_d = 4$ while set 2 with $r_p = 0.3$, $r_i = 6$ and $r_d = 4$. The set 1 tuning leads to K_c = [-7.092, -5.428, -0.109], τ_I = [200, 24.3, 27.2], and τ_D = [1.42, 1.36, 3.18]. Meanwhile, the set 2 yields K_c = [-14.18, -10.86, -0.2185], τ_I = [200, 24.2, 29.4] and τ_D = [1.42, 1.36, 3.18]. To evaluate the closed-loop performances corresponding to the different tuning values, sequential step changes of 1 unit each in the setpoints of Y_1,Y_2 and Y_3 are applied. Figure 2 shows the comparative closed-loop responses under the different controller tunings. The proposed R-H tuning can provide improved performance in term of the total Integral Absolute Error (IAE) value over the SML-MSC tuning. Note that, the closer the loop gain to its upper limit (or as the tuning parameter r_p increases), the faster the closed-loop responses become. However, due to process interactions and deadtimes, there is an upper limit on the r_p above which the system becomes unstable. To ensure closed-loop stability, we recommend the value of r_p to be between 0.1 and 0.5. When the interactions among the control loops become more severe, a smaller value of r_p should be used to ensure closed-loop stability.

5.2. Example 2

In this example, we consider a 2x2 unstable system

$$\mathbf{P} = \begin{bmatrix} \dfrac{1.5e^{-3s}}{20s-1} & \dfrac{1.2e^{-2s}}{20s+1} \\ \dfrac{1.1e^{-5s}}{38s+1} & \dfrac{1.6e^{-5s}}{22s-1} \end{bmatrix} \tag{27}$$

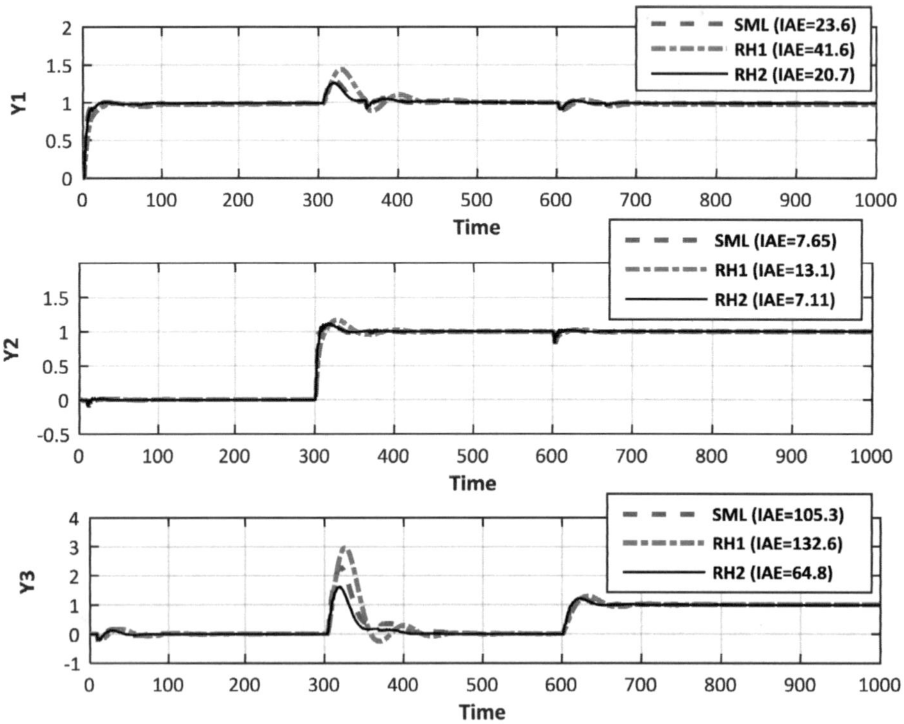

Figure 2. Setpoint responses for example 1.

For this unstable process (27), the R-H tuning (Case B) is adopted. Note that, many of the existing methods for the decentralized PID control tuning are not applicable to this type of unstable MIMO system. For an unstable system under PID controller, it is quite common to encounter a large overshoot in response to a setpoint change. To reduce this large overshoot, it is recommended to use a setpoint pre-filter which takes the form of

$$F_r = \frac{(\tau_I/\varepsilon)s+1}{\tau_I s+1}, \quad 1.5 \le \varepsilon \le 5 \tag{28}$$

where τ_I is the reset time of the given PID controller, and ε is a positive scaling parameter. The recommended range of the scaling parameter is between 1.5 and 5. The control performances are evaluated based on sequential step changes of 1 unit each in the setpoints of Y_1 and Y_2. Figure 3 shows the comparative closed-loop responses corresponding to two different tuning values: set 1 with $r_p = 0.6, r_i = 2, r_d = 0.05$ and set 2 with $r_p = 0.4, r_i = 2, r_d = 0.05$. Set 1 tuning leads to $K_c = [3.579, 2.642]$, $\tau_I = [31.9, 37.6]$ and $\tau_D = [2.425, 3.475]$ while set 2 tuning leads to $K_c = [2.615, 1.979]$, $\tau_I = [39.4, 47.4]$ and $\tau_D = [2.425, 3.475]$. For this system, it is not possible to increase the loop gain all the way close to its upper limit because of the process interactions. Also note that, reducing the loop gain so that it becomes too close to its maximum lower limit can cause the system to exhibit high oscillation. The closed-system can become unstable if the loop gain becomes too near to either its minimum upper limit or its maximum lower limit.

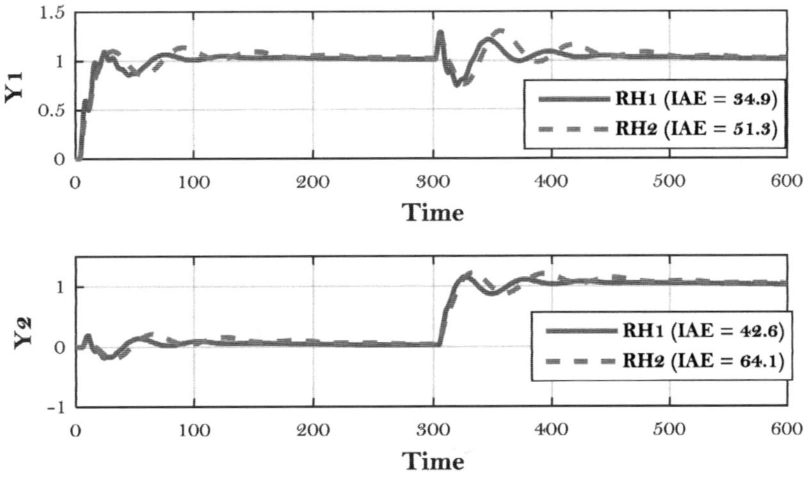

Figure 3. Responses for example 2.

6. Conclusion

A tuning method has been proposed via which we can simplify the tuning task from finding $3n$ PID parameters to finding only 3 common tuning parameters, namely r_p, r_i and r_d. This represents a great simplification for an otherwise very complicated tuning problem.

References

[1] Huang H, Jeng J, Chiang C and Pan W 2003 A direct method for multi-loop PI/PID controller design *Journal Process Contol* **13** pp 769- 86

[2] Nandong J 2015 Heuristic-based multi-scale control procedure of simultaneous multi-loop PID tuning for multivariable processes *Journal of Process Control* **35** pp 101-12

[3] Nandong J and Zang Z 2013 High-performance multi-scale control scheme for stable, integrating and unstable time-delay processes, *Journal of Process Control* **23** pp 1333-43

[4] Nandong J and Zang Z 2014 Inter-Communicative Decentralized Multi-Scale Control (ICD-MSC) Scheme: A New Approach to Overcome MIMO Process Interactions *Chemical Product and Process Modeling* **9** pp 165-78

[5] Silva G, Datta A and Bhattacharyya S 2002 New results on the synthesis of PID controllers," *IEEE Transactions of Automaic Control* **47** pp 241-52

[6] Ho W, Hang C and Cao L 1995 Tuning of PID controllers based on gain and phase margin specifications *Automatica* **31** pp 497-502

[7] Seer Q and Nandong J 2017 Stabilization and PID tuning algorithms for second-order unstable processes with time-delays *ISA Transactions* **67** pp 233-45

[8] Tyreus B Multivariable control system design for an industrial distillation column 1979 *Industrial & Engineering Chemistry Process Design and Development.* **18** pp 177-82

[9] Liceaga-Castro E, Liceaga-Castro J and Ugalde-Loo C 2005 Beyond the existence of diagonal controllers: from the relative gain array to the multivariable structure function *CDC-ECC'05, 44th IEEE Conf* (Seville Spain)

IOP Conf. Series: Materials Science and Engineering **458** (2018) 012052 doi:10.1088/1757-899X/458/1/012052

PID multivariable tuning system using BLT method for distillation column

Totok Ruki Biyanto[1], Nurul Atiqah Sordi[2], Nurulsapika Sehamat[2], Haslinda Zabiri[2,*]

[1] Department of Engineering Physics, Institut Teknologi Sepuluh Nopember (ITS) Surabaya 60111, East Java, Indonesia

[2] Chemical Engineering Department, Universiti Teknologi PETRONAS, 32610 Bandar Seri Iskandar, Perak Darul Ridzuan, Malaysia.

*haslindazabiri@utp.edu.my

Abstract. Most of the process industries are handling Multi-Input-Multi-Output (MIMO) system by using the classical Proportional-Integral-Derivative (PID) controller although there are wide researches in advanced control techniques. However, finding a good value for the controller parameters might be very challenging and best tuning approach need to be considered to obtain a better performance. In this paper, biggest-log modulus (BLT) tuning method is adopted to tune decentralized PID controller based on 2x2 MIMO system and the effectiveness of the tuning method is observed. A comparative study on the settling time, overshoot and integral absolute error (IAE) are evaluated when a step change in u1, u2 and disturbance is introduced to the system. The simulation result demonstrates that BLT method has a better performance for a disturbance change response compared to set point change response.

1. Introduction

Over last few decades, there are numerous research on advance control strategy involving model predictive control (MPC), linear quadratic control etc, yet Proportional Integral Derivative (PID) controller still extensively used in industry over the advance control strategy. PID is preferable over other controller scheme because of simplicity yet resulting most efficient control strategies [1]. This is proven as there are more than 90% of the industrial control loops being operated in the process industries either PI or PID mode [2-4].

However, finding the best tuning parameters for PID controller is still a major problem for the controller due to nonlinearity, time-variability, and time delay of processes [5]. [6] has stated that an improperly tuned PID controllers may resulted in more serious issue which may reduce the profitability and productivity of the plants. Hence, proper tuning is helpful to overcome this problem [7].

Typically, multiple inputs and outputs (MIMO) system is adopted in most of real industrial processes. MIMO system slightly differs compared with single-input single output (SISO) system as it is more complex and require a proper handling to the process interaction in the systems. Interaction process generally encountered when a change in a manipulated variable, affect the corresponding output variable and other output variables. Therefore, proper arrangement of input and output variable may aid in minimizing interaction between control loops [1]. Relative Gain Array (RGA) is one of the quantitative techniques in determining the correct pairing of manipulated and controlled variables.

IOP Conf. Series: Materials Science and Engineering **458** (2018) 012052 doi:10.1088/1757-899X/458/1/012052

In literature provided by Katebi [1], multivariable PID tuning method has been categorised into two main groups which is involve parametric and non-parametric model. BLT method is classified as one of parametric tuning method. Other than that, there are a few numbers of designing method develop for MIMO method that can be categorised into 5 groups which involve detuning, sequential loop closing, iterative or trial-and-error, simultaneous equation solving and independent methods [8]. Biggest-log-modulus (BLT) proposed by Luyben is one of the most famous method grouped in detuning category yet become preferable method compared to others due to its simplicity and practicality.

In this paper, the simplest form of MIMO system known as two-input two-output (TITO) system is adopted by implementing two PID controllers in PI mode. The previous paper discussed on tuning a multivariable process using BLT method focusing on servo task [9] and a systematic data-based method to tune PID controller based on the disturbance rejection [10]. The outcome of the paper exhibits that the controller tuned by using proposed method demonstrated a better performance based on disturbance rejection. Thus, in this study, a TITO system is emphasized on the composition control of distillate and bottom products in a distillation column. Proper tuning using BLT method on which serve as an extension of classical Ziegler-Nichols (Z-N) method is handled on the TITO system and the response for servo (set point) and regulatory (disturbance) performance are observed accordingly.

2. Multivariable System

Multivariable system is a system which involves multiple inputs and multiple outputs (i.e. shown in figure 1). Multivariable control involves the goal of maintaining several control variables at independent set points. A multivariable system will be more challenging to control if there is existence of cross couplings.

Transfer function matrix, $G_p(s)$ is given by equation 1 and each of the element of the transfer matrix $G_{ij}(s)$ is represented in a first order plus dead time (FOPDT) model, shown in equation 2.

$$G_p(s) = \begin{bmatrix} G_{11}(s) & G_{12}(s) \\ G_{21}(s) & G_{22}(s) \end{bmatrix} \tag{1}$$

$$G_{ij}(s) = \frac{K_{pij} e^{-\theta_{ij} s}}{(\tau_{ij} s + 1)} \; ; i = 1,2 \; j = 1,2 \tag{2}$$

In this study, decentralized controller is designed. The equation 3 indicates the design of PID controller in decentralized form.

$$G_c(s) = \begin{bmatrix} G_{c11}(s) & 0 \\ 0 & G_{c22}(s) \end{bmatrix} \tag{3}$$

PID controller mode is defined in $G_{c,ij}(s)$. Thus, in this study, the parallel form of PID controller is assume and the mode of PID controller is set as in PI mode. Therefore, the elements of the controller is defined as equation 4.

$$G_c(s) = K_c \left(1 + \frac{1}{\tau_I s}\right) \tag{4}$$

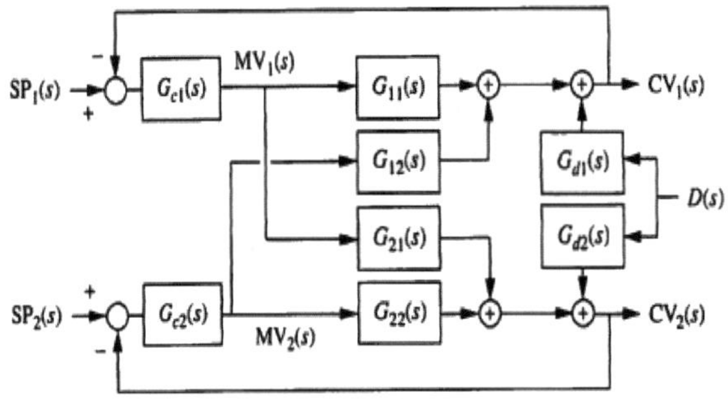

Figure 1. Block diagram of 2x2 with two single-loop controllers. [11]

3. PID tuning strategy

In this study, two PID controllers will be implemented to the distillation column model which is then properly tuned using BLT method and conventional Internal Model Control (IMC) method. In most multivariable control, the system is treat as a series of single-loop and further designs involving other methodologies is proceeded. One common approach is to detune controller parameters derived using SISO design approach.

3.1. Ziegler-Nichols Method

Ziegler and Nichols have described two techniques for tuning the parameter of P-, PI- and PID mode controllers. The open-loop method indicates in figure 2 has been chosen and calculated using the formula listed below (equation 5 to equation 9).

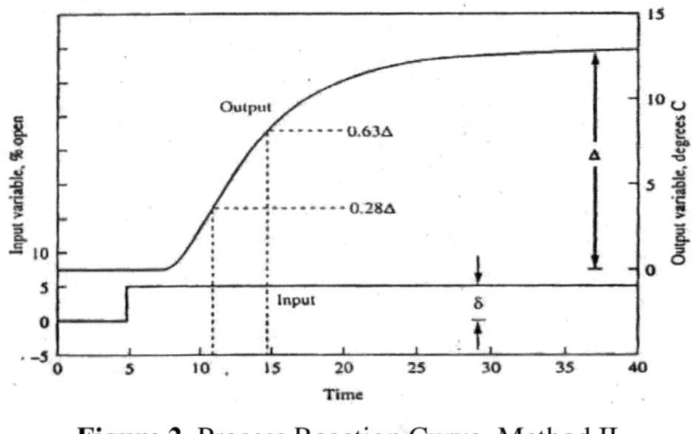

Figure 2. Process Reaction Curve, Method II

$$Y(\theta + \tau) = 0.632(\Delta) \tag{5}$$

$$Y(\theta + \tau/3) = 0.283(\Delta) \tag{6}$$

$$t_{28\%} = \theta + \frac{\tau}{3} \tag{7}$$

$$\tau = 1.5(t_{63\%} - t_{28\%}) \tag{8}$$

$$\theta = t_{63\%} - \tau \tag{9}$$

3.2. Biggest-Log Modulus (BLT) Method

BLT method is an extension of the classical Ziegler and Nichols method for multivariable case. Luyben (1986) has proposed this method and stated that there are 2N tuning parameters to be selected if PI controllers are considered which involves gains and reset times. The gains and reset times should be specified correctly or it will result in unstable system. Steps for BLT method are shows as follows;

3.2.1. Determine the controller settings. The controller settings are set by using Ziegler-Nichols open loop method for each individual loop using classical SISO as explained in subsection above. The corresponding tuning parameters is tabulated in table 1.

Table 1. PID tuning parameters for process reaction curve (method II)

Controller	*Gain* ,K_p	Integral *time*,τ_I	*Derivative Time*, τ_D
P	$\frac{\tau}{K\theta}$		
PI	$\frac{0.9\tau}{K\theta}$	$\frac{\theta}{0.3}$	
PID	$\frac{1.2\tau}{K\theta}$	$\frac{\theta}{0.5}$	0.5θ

3.2.2. A detuning factor F is assumed using equation 10 and equation 11. The F values usually greater than 1 and within the range of 1.5 to 4.

$$K_{CI} = \frac{K_{ZNI}}{F} \tag{10}$$

$$\tau_{ui} = \tau_{ZNi} \times F \tag{11}$$

The *F* factor is considered as a detuning factor which would be applied to all loops. The larger the value of *F* indicate the more stable of the system. (*For this simulation F=1.35 is chosen*)

3.2.3. Multivariable Nyquist plot of the scalar fraction is determined as follows in equation 12 to equation 14:

$$W_{(i\omega)} = 1 + Det\left[I + G_{m(i\omega)}B_{(i\omega)}\right] \tag{12}$$

$$G_m(s) = \begin{bmatrix} G_{11}(s) & G_{12}(s) \\ G_{21}(s) & G_{22}(s) \end{bmatrix} \tag{13}$$

$$B(s) = \begin{bmatrix} \frac{K_{c1}(\tau_{I,1}s + 1)}{\tau_{I,1}s} & 0 \\ 0 & \frac{K_{c2}(\tau_{I,2}s + 1)}{\tau_{I,2}s} \end{bmatrix} \tag{14}$$

3.2.4. Evaluate L_{cm}^{max} value. The value is estimated using equation 15

$$L_{cm}^{max} = max_{\omega}\left\{20log_{10}\left|\frac{W}{1+W}\right|\right\} \tag{15}$$

3.2.5. Repeat step 3.2.4 if the value of $L_{cm}^{max} \neq 2N$, whereby N is indicating the order of the system. In this study, TITO system is considered. Thus, iteration of F should be stopped as the value of $L_{cm}^{max} = 4db$ is obtained.

4. Simulation Results

The distillation column simulation was adapted from [12] to demonstrate the effectiveness and merits of the presented method on MIMO system. Equation 16 expressed the transfer function of the plant.

$$\begin{bmatrix} x_D(s) \\ x_B\ (s) \end{bmatrix} = \begin{bmatrix} \frac{0.0747e^{-3s}}{12s+1} & \frac{-0.0667e^{-2s}}{15s+1} \\ \frac{0.1173e^{-2.2s}}{11.7s+1} & \frac{-0.1253e^{-2s}}{10.2s+1} \end{bmatrix} \begin{bmatrix} F_R(s) \\ F_V\ (s) \end{bmatrix} + \begin{bmatrix} \frac{0.70e^{-5s}}{14.4s+1} \\ \frac{1.3e^{-3s}}{12s+1} \end{bmatrix} x_F(s) \tag{16}$$

A few steps changes are introduced to the simulation; (a) a set point change of x_D (U1), (b) a set point change of x_D (U2), and (c) a step change of disturbance. The performance of controller tuned using BLT method is compared with Internal Model Control (IMC) and will further discussed in the following section.

The corresponding PI controller parameters for both methods are listed in table 2. The simulation is completely free from steady state error for both tuning method.

Table 2. PI controller parameters.

Method	$K_{p,1}$	$K_{I,1}$	$K_{p,2}$	$K_{I,2}$
BLT	35.789	2.659	-27.207	-3.034
IMC	26.774	2.231	-20.351	-1.995

4.1. A step change of first input, U1 (set point of x_D) at time, t=50s

Both figures 3 (a) and (b) illustrate the servo responses for a step change given in U1 and the corresponding control actions (y1 and y2). Table 3 listed the performance analysis of both y1 and y2.

Aggressive oscillation and a higher peak response are observed when the system is tuned by BLT method for both top composition tracking(y_1) and bottom composition tracking (y_2). Nevertheless, the response is still under control as the percentage overshoot is still within 0% for y_1 response. Meanwhile, the conventional IMC method demonstrate a minimum oscillation but, it will produce sluggish response for both output accordingly. BLT detuned method is believable to improve the stability of the response but then it will yield a bit sluggish response as the value of K_c is reduced by the tuning factor, F. BLT method take the shortest time to return its desired steady state value and having the minimum IAE value for both y1 and y2 responses. The responses tabulated in Table 3 clearly indicates that BLT method have the best result for both y_1 and y_2 response in term of lowest IAE value and shorter settling time when a step change in u_1 is introduced.

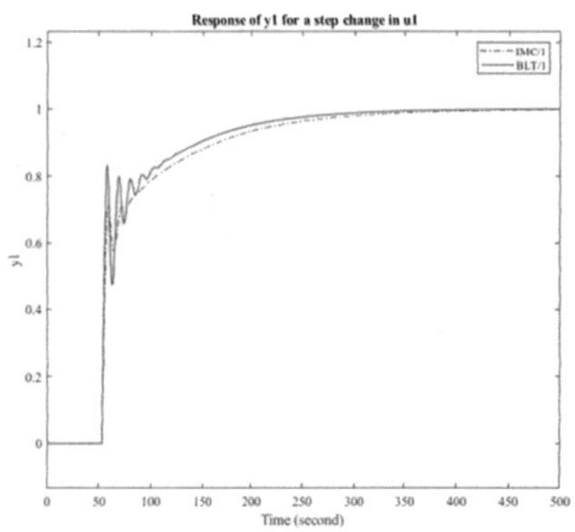

Figure 3 (a). Response of first output, y1 as a step change in first input, u1 is set.

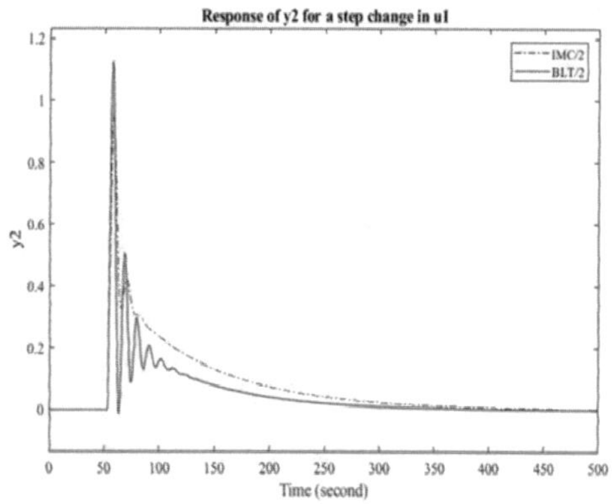

Figure 4 (b). Response of second output, y2 as a step change in first input, u1 is set.

Table 3. Performance data of y1 and y2 for two different tuning methods.

	y_1				y_2		
	Maximum peak	Overshoot (%)	Settling Time (s)	IAE	Maximum peak	Settling Time (s)	IAE
BLT	0.9980	0	441.14	30.53	1.1259	398.09	25.07
IMC	0.9958	0	498.42	36.19	0.9481	520.55	37.87

4.2. A step change of second input, U2 (set point of x_B) at time, t=50s

The response of a step change upon U2 is shown in figure 4 (a) and 4 (b) and table 4 tabulates the parameters on the performance of y1 and y2 on BLT and IMC based on decentralized PI controller. BLT method is observed to have a greater number of oscillations compared to the conventional IMC-PI method and resulted in a significant overshoot for the bottom composition tracking (y2). IMC method seems to be able to surpass the BLT method in reducing the instability of the responses and having less number of oscillation. Nevertheless, IMC method shows the slowest response to return to its desired steady state value for both y_1 and y_2 response compared to the BLT method. Among the listed methods, BLT method shows the best performance as it having a smaller IAE value and shortest settling time for both y_1 and y_2 response respectively.

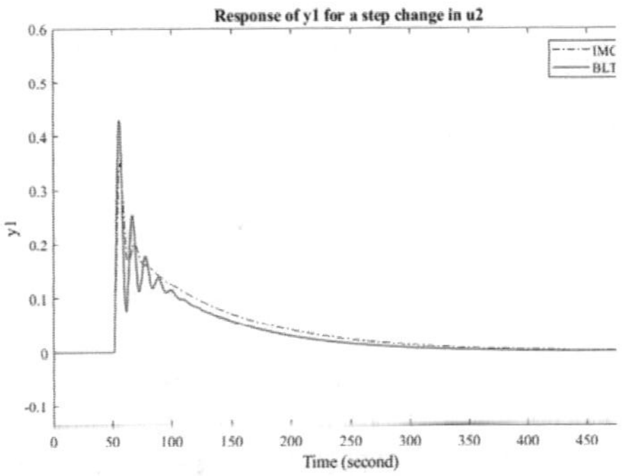

Figure 5 (a). Response of first output,y1 as a step change in second input, u2 is presented.

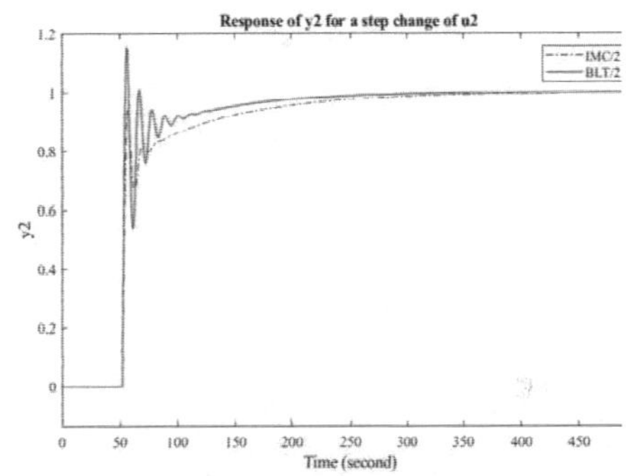

Figure 6 (b). Response of second output, y2 as a step change in second input, u2 is presented.

Table 4. Performance data of y1 and y2 for two different tuning approaches.

	y_1			y_2			
	Maximum peak	Settling Time (s)	IAE	Maximum peak	Overshoot (%)	Settling Time (s)	IAE
BLT	0.4315	484.27	16.24	1.1551	15.6253	341.53	16.60
IMC	0.3498	570.89	19.24	0.9972	0	436.45	24.13

4.3. A step change on disturbance input, at t=50s

Figure 5(a) and 5(b) demonstrate the response of y1 and y2 accordingly when a step change in disturbance is set to the system using two different methods. Performance analysis on both tuning method is tabulated in table 5.

The performance analysis for both tuning method as tabulated in table 5. BLT method shows a significant oscillation for both $y1$ and $y2$ response, but correspondingly the IMC method shows less oscillation. From the performance analysis table (Table 5), a faster response towards the new desired steady state value shows by BLT method after a step change at disturbance is introduced to the system.

An insignificant peak is shown by y1 and y2 response respectively. In regulator problem, BLT has a much the lowest settling time and IAE values compared to the servo problem (when a step input of U1 and U2 are introduced to the system) for both $y1$ and $y2$ response. These clearly verify that BLT method is one of the most suitable tuning approach for a PID controller mainly in handling with a disturbance change system.

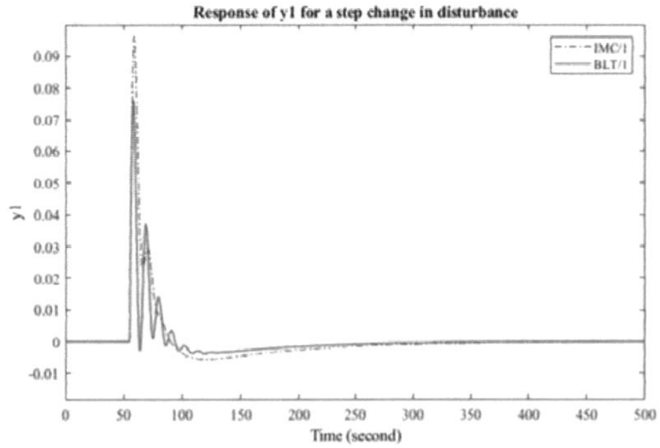

Figure 7 (a). Response of first output, y1 as a step change in disturbance is set.

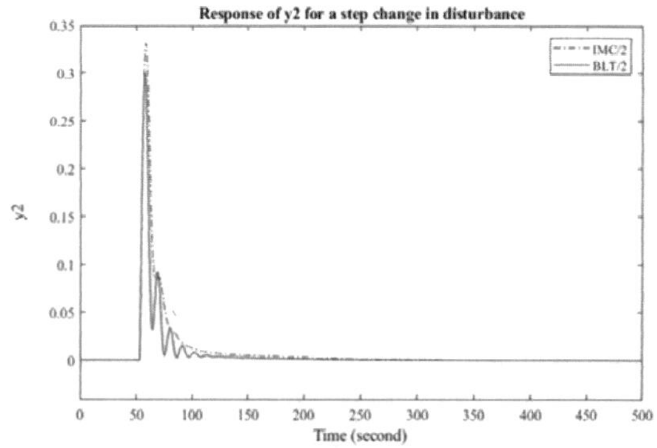

Figure 8 (b). Response of second output, y2 as a step change in disturbance is set.

Table 5. Performance data of y1 and y2 for two different tuning techniques.

	y_1			y_2		
	Maximum peak	Settling Time (s)	IAE	Maximum peak	Settling Time (s)	IAE
BLT	0.0758	320.84	1.055	0.3002	130.50	3.215
IMC	0.0962	362.36	1.657	0.3320	179.60	4.878

5. Conclusion

The application of multiloop control structure in various industrial and chemical contexts is widely implemented due to the simplicity and economic consideration for operation. In this paper, the response of First-order plus Time-delay (FOPTD) process is compared by using BLT method and conventional IMC tuning method that operates using PI controller mode. In this paper, BLT method is being adopted to the MIMO system to obtain the best performance of controller parameter of having a minimum IAE value. Although BLT demonstrate a greater number of oscillations in the simulation results, but BLT has a good performance on the regulator problem compared to servo problem. This is because BLT has the fastest time to remain to its desired steady state value for both top composition tracking (y1) and bottom composition tracking (y2) and the relatively minimum IAE value.

Acknowledgement

The Authors wish to thank Institut Teknologi Sepuluh Nopember and Universiti Teknologi PETRONAS for their financial support and knowledge to conduct the research. This support is gratefully acknowledged.

References

[1] Katebi R 2012 *PID Control in the Third Millennium: Lessons Learned and New Approaches (Advances in Industrial Control)* (Springer Verlag London Ltd) chapter 9 pp 255-80

[2] McMillan G 2012 *PID Control in the Third Millennium: Lessons Learned and New Approaches (Advances in Industrial Control)* (Springer Verlag London Ltd) chapter 14 pp 415-60

[3] Coelho L and Pessôa M 2011 A tuning strategy for multivariable PI and PID controllers using differential evolution combined with chaotic Zaslavskii map *Expert Systems with Applications* **38** (11) pp 13694–701

[4] Desborough L and Miller R 2002 Increasing Customer Value of Industrial Control Performance Monitoring—Honeywell's Experience *CHEMICAL PROCESS CONTROL-VI Assessment and New Directions for Research* **98** 326
[5] Zhang J, Zhuang J, Du H and Wang S 2009 Self-organizing genetic algorithm based tuning of PID controllers *Information Sciences* **179** (7) pp 1007-18
[6] O'Dwyer A 2012 *PID Control in the Third Millennium: Lessons Learned and New Approaches (Advances in Industrial Control)*. (Springer Verlag London Ltd) chapter1 pp. 3-60
[7] Wang Q, Zou B, Lee T and Bi Q 1997 Auto-tuning of multivariable PID controllers from decentralized relay feedback *Automatica* **33** (3) pp 319-30
[8] Nandong J 2015 Heuristic-based multi-scale control procedure of simultaneous multi-loop PID tuning for multivariable processes *Journal of Process Control* **35** pp 101-12
[9] Besta C and Chidambaram M 2015 Tuning of Multivariable PI Controllers by BLT Method for TITO Systems *Chemical Engineering Communications* **203** (4) pp 527-38
[10] Jeng J and Ge G 2016 Disturbance-rejection-based tuning of proportional–integral–derivative controllers by exploiting closed-loop plant data *ISA Transactions* **62** pp 312-24
[11] Luyben W L 2009 *Process modelling, simulation, and control for chemical engineers for dynamic performance (*Boston McGraw-Hill)
[12] Seborg D E, Edgar T F and Mellichamp D A 2004 *Process dynamics and control* (United States of America John Wiley & Sons, Inc)

Optimization of PID controller tuning parameters for multivariable system using Duelist algorithm

Totok Ruki Biyanto[1], **Nurulsapika Sehamat**[2], **Nurul Atiqah Sordi**[2], **Haslinda Zabiri**[2,*]

[1] Department of Engineering Physics, Institut Teknologi Sepuluh Nopember, 60111 Surabaya, East Java, Indonesia.
[2] Chemical Engineering Department, Universiti Teknologi PETRONAS, 32610 Bandar Seri Iskandar, Perak Darul Ridzuan, Malaysia.

* haslindazabiri@utp.edu.my

Abstract. In this paper, one of the most current new stochastic optimization method is proposed to tune decentralized proportional-integral-derivative (PID) controller for a multi-input-multi-output (MIMO) system. PID controller is still the most used controller in industrial application due to its easy application in control system. However, it still having problem to be tuned correctly especially for MIMO system. The decentralized controllers in this paper will be tuned simultaneously using duelist algorithm (DA) method and optimal value of controller parameters is obtained from a minimum of integral absolute error (IAE) value. The optimization of the objective function is based on the evaluation of IAE when a step change in u1 and u2 is introduced to the system. Form the simulation result, the proposed method is observed to have a better performance than other popular tuning method.

1. Introduction

Multi-input-multi-output (MIMO) system is the most encountered system in real industrial processes. A process in industry is becoming more complex to ensure the quality and quantity of the product is conserved as well as due to energy integration [1-3]. Hence, a more sophisticated way to control the system is needed to meet the requirements. A MIMO system is more complex than a single-input-single-output (SISO) system due to the increase in the number of control loops, thus increase the interactions between the loops [1, 4-5]. A manipulated variable could affect more than one controlled variable due to the loop interactions and increase the complexity of the system. An effective control system should be implemented to such system to ensure the efficiency of the operation and energy of the system [3].

Nowadays, there are many advanced controller has been developed in line with the increase in the complexity of a system. However, proportional-integral-derivative (PID) controller remained the most widely used controller in industrial application [6-7]. Over 90% of controller in industrial control application is using PID algorithm controller [1, 8]. The popularity gained by PID are due to its simple structure, good stability, reliable, robust performance, and wide operating condition and quite easy to be tuned on line [2, 9-14]. Those advantages make a PID controller easy to be operated by an operator or engineer. Despite the advantages, PID controller still having difficulties to be tuned and designed due to the interactions between the loops [7, 15].

There are many methods had been developed to tune a PID controller. One of the method is by optimization method. [8] had listed a few optimization methods had been used to tune a PID controller such as genetic algorithm, particle swarm optimization, tribes algorithm, harmony search, evolution strategy, ant colony and more. [16] had proposed to use multi-objective optimization design procedure in PID controller tuning application. [9-10, 17] had modified the genetic algorithm method to optimize PID parameters to find the suitable controller parameters with minimum integral absolute error (IAE). There is also a hybrid optimization where the author combined the differential evolution algorithm with chaotic Zaslavskii map to optimize PID controller parameters in a multivariable system [8].

In this paper, the most recent new optimization technique, which is duelist algorithm (DA) introduced by [18] is proposed to tune a decentralized PID controller for a multivariable system. Previously, DA has been implemented to simultaneously tune two PID controllers for a cascade control system [19]. It can be observed that DA optimization method did improved the performance of the PID controllers in the cascade control system. Thus, in this study, DA optimization algorithm will be implemented to two decentralized PID controller for a MIMO system of a distillation column to observed whether it can improve the performance of the controller or not.

2. Multivariable System

A multivariable system is a process that has either more than one input variable or output variable. A multivariable system could be difficult to control if there is an existence of cross couplings in the process. The cross coupling should be encountered well to avoid any problems to the control system. There are two main problems could be aroused due to the negligence of the cross couplings.

1. A change in an input could affect more than one output of the process.
2. The tuning of the PID controller will be difficult due to the interaction and the controller will be less stable.

In multivariable system, the system is control to ensure a change in a set point would not or mild affect other control variables.

A matrix of transfer function for a TITO system $G_p(s)$, can be writen as below in equation 1.

$$G_p(s) = \begin{bmatrix} G_{11}(s) & G_{12}(s) \\ G_{21}(s) & G_{22}(s) \end{bmatrix} \quad (1)$$

Let a First Order plus Dead Time (FOPDT) model represents the each element of the transfer function matrix $G_{ij}(s)$ in equation above. FOPDT can be written as shown in equation 2 below.

$$G_{ij}(s) = \frac{K_{pij}e^{-\theta_{ij}s}}{(\tau_{ij}s + 1)}; \; i = 1,2 \;\; j = 1,2 \quad (2)$$

PID controller design can be either a centralized or a decentralized controller. In this paper, a decentralized PID controller is designed. The structure for a decentralized controller can be defined as in equation 3 below:

$$G_c(s) = \begin{bmatrix} G_{c11}(s) & 0 \\ 0 & G_{c22}(s) \end{bmatrix} \quad (3)$$

$G_{c,ij}(s)$ is defined based on the mode of PID controller used. PID controller can be in either P mode, PI mode, PD mode or PID mode. PID controller for this paper is assumed to be in parallel form. PI mode of PID controller is used in this paper. The elements of the controller when PI mode is used is described as equation 4.

$$K_c\left(1 + \frac{1}{\tau_I s}\right) \quad (4)$$

Figure 1 illustrates a multivariable system with two decentralized PID controller.

IOP Conf. Series: Materials Science and Engineering **458** (2018) 012053 doi:10.1088/1757-899X/458/1/012053

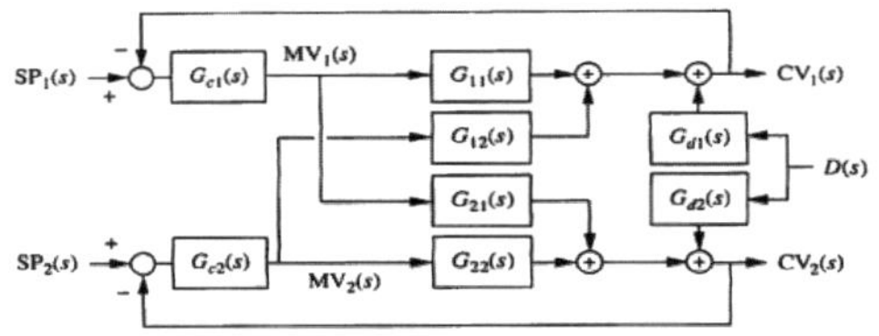

Figure 1. Block diagram of closed 2x2 system with two single control loop.

3. Optimization Using Duelist Algorithm

In this paper, the multivariable system of a distillation column is considered. In the system, the manipulated variables are reflux flow (MV1) and reboiler flow (MV2), meanwhile the controlled variable is the distillate (CV1) and bottom (CV2) product. There are two decentralized PID controller implemented in the system. The optimization of the four controller parameters ($K_{p,1}$, $K_{i,1}$, $K_{p,2}$ and $K_{i,2}$) will be done by simulation in MATLAB. A suitable value of each parameter will be obtained to have an optimal objective function for the system.

3.1. Objective Function Evaluation

In this paper, a simultaneous unit step change in set point value for u1 and u2 are set and allow an evaluation of Integral Absolute Error (IAE), which is defined in equation 5 up to the settling time. The IAE of the system is used as an objective function for the optimization and to be minimized by DA.

$$\mathrm{IAE} = \int_0^{\infty} |e(t)|dt = \int_0^{\infty} |r(t) - y(t)|\, dt \qquad (5)$$

The performance of the controller will be analysed based on the IAE, steady state error, settling time and maximum overshoot.

3.2. Optimization by Duelist Algorithm

Duelist Algorithm (DA) is a new stochastic optimization algorithm inspired by human fighting and learning capabilities. In DA, the individual of the population is known as duelist. The duelist will be fight to each other in order to determine the winner, loser and champion. The result from the fighting, either win or lose will determine what kind of action will be taken by the duelist. Figure 2 shows the sequence of DA method.

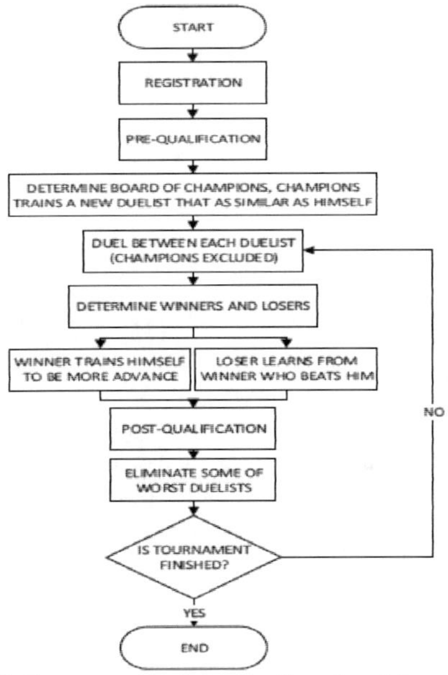

Figure 2. Sequence on how duelist algorithm (DA) is worked.

3.2.1. Registration. Each duelist is registered using a binary array which also known as skillset in DA while in genetic algorithm known as chromosome.

3.2.2. Pre-Qualification. A test taken by each duelist to measure their fighting abilities. Their abilities depend on their skillset.

3.2.3. Determination of Board of Champions. The best duelist for the fighting is kept in the Board of Champions. The champion will train other duelist to be as par as it in its fighting capabilities. The new duelist with the same abilities from previous champion will join the next game.

3.2.4. Duel between each Duelist. The fighting game is set randomly for each duelist. The winner and loser for very match is determined from the duelist's capabilities and luck. The duelist's luck is purely determined by a random function. This is to avoid local optimum. Algorithm shown below is the pseudocode used to determine the loser and the winner.

```
Required: Duelist A and B; Luck_Coefficient
A(Luck) = A(Fighting_Capabilities) * (Luck_Coefficient + (rand(0-1) * Luck_Coefficient));
B(Luck) = B(Fighting_Capabilities) * (Luck_Coefficient + (rand(0-1) * Luck_Coefficient));
If ((A(Fighting_Capabilities) + A(Luck)) <= (B(Fighting_Capabilities) + B(Luck)))
A(Winner) = 1;
B(Winner) = 0;
Else
A(Winner) = 0;
B(Winner) = 1;
End
```

3.2.5. Improvement by Duelist. The duelists will try to improve themselves after the match. There are three types of action or treatment taken by a duelist. For a loser, it will learn from the winner by copying the binary array of a winner. For a winner, it will try on something new by randomly manipulate its own array and for a champion, it will train other new duelist.

3.2.6. Elimination. Elimination is done based on the fighting capabilities. The worst duelist in the duel will be eliminated. Below are the details on DA optimization parameters:
Population = 20
MaxGeneration = 100
FightCapabilities = 20
Champion = 0.1
ProbLearning = 0.8
ProbInnovate = 0.1
Luckcoeff = 0.01
Dimension = 6

4. Results and Discussion

The plant used for the simulation is taken from Marlin (2000). Equation 6 and 7 show the transfer function of the plant for manipulated variable and disturbance respectively;

$$G_p = \begin{bmatrix} x_D(s) \\ x_B(s) \end{bmatrix} = \begin{bmatrix} \frac{0.0747e^{-3s}}{12s+1} & \frac{-0.0667e^{-2s}}{15s+1} \\ \frac{0.1173e^{-2.2s}}{11.7s+1} & \frac{-0.1253e^{-2s}}{10.2s+1} \end{bmatrix} \begin{bmatrix} F_R(s) \\ F_V(s) \end{bmatrix} \quad (6)$$

$$G_d = \begin{bmatrix} \frac{0.70e^{-5s}}{14.4s+1} \\ \frac{1.3e^{-3s}}{12s+1} \end{bmatrix} [x_F(s)] \quad (7)$$

The simulation is done for a few step change; set point change of x_D (u1), set point change of x_B (u2), simultaneous set point change of x_D (u1) and of x_B (u2) and step change of disturbance. The performance of the controller using DA method is being compared with other methods, which are conventional Internal Model Control (IMC) and biggest-log modulus tuning (BLT) method. Table 1 shows the controller parameters when being tuned using three different tuning methods. From the simulation, the steady state error for all response using those three different methods are observed to be zero.

Table 1. PID controller parameters of three different tuning methods.

Method	$K_{p,1}$	$K_{p,2}$	$K_{i,1}$	$K_{i,2}$
DA	10.210	-31.644	6.855	-4.394
IMC	26.774	-20.351	2.231	-1.995
BLT	35.789	-27.207	2.659	-3.034

4.1. A step change in first input, u1 (set point of x_D) at time, t=10s

Figures 3(a) and 3(b) show the responses given by y1 and y2, respectively, when a step change in u1 is set, while table 2 shows the performance analysis of the controller for different methods. From figures 3(a) and 3(b), the response shown by DA method is the best among the other two methods. When a step change of u1 is introduced to the system, PID controller optimized by DA shows the faster response to the desired value for y1 and y2, which is 1 and 0 respectively. Besides, both y1 and y2 responses when using DA optimization method illustrates a minimum oscillation, which leads to a more stable system towards the step change in u1. BLT method shows the highest oscillation among the three methods. IMC is seen to have a very minimum overshoot, however, the response is too sluggish for both output. The response obtained using DA is hence the best and DA method is observed to have the minimum value of IAE for both output, y1 and y2 compared to other methods.

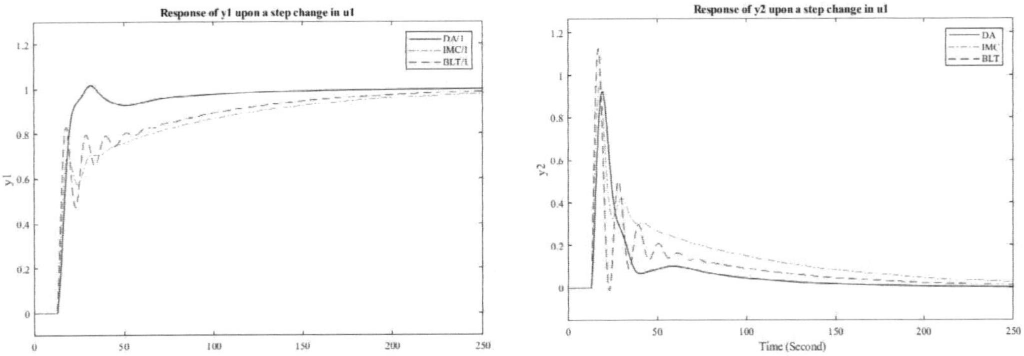

(a) (b)

Figure 3. Response of (a) first output, y1 and (b) second output, y2 when a step change of first input, u1 is introduced.

Table 2. Performance analysis of y1 and y2 for three different tuning methods.

	y1				y2		
	Maximum peak	Overshoot (%)	Settling Time (s)	IAE	Maximum peak	Settling Time (s)	IAE
DA	1.0177	1.7736	225.7905	12.03	0.9217	295.8903	17.38
IMC	0.9987	0	520.1091	36.53	0.9471	545.5867	38.23
BLT	0.9995	0	451.2056	30.63	1.1255	405.6164	25.16

4.2. A step change in second input, u2 (set point of x_B) at time, t=10s

Figures 4(a) and 4(b) show the responses of y1 and y2 accordingly when a step change in u2 is introduced to the system. Meanwhile, table 3 lists the performance analysis for all different methods. From the analysis, PID controller parameters optimized using DA method shows the fastest response to the set point value compared to BLT and IMC method for both output responses. DA method gives the shortest setting time. Minimum overshoot or peak is given by IMC method for both output responses but the system is very sluggish. IMC method has the longest settling time compared to other two methods. As can be observed from figures 4(a) and 4(b), DA method has a very minimum oscillation and the system can be seen to be very stable. However, in y2 response, DA method is observed to show a quite aggressive response towards the bottom composition tracking (y2) where there is a significant overshoot can be seen. DA method still shows the best performance among the three methods as the IAE value is very minimum and gives faster response.

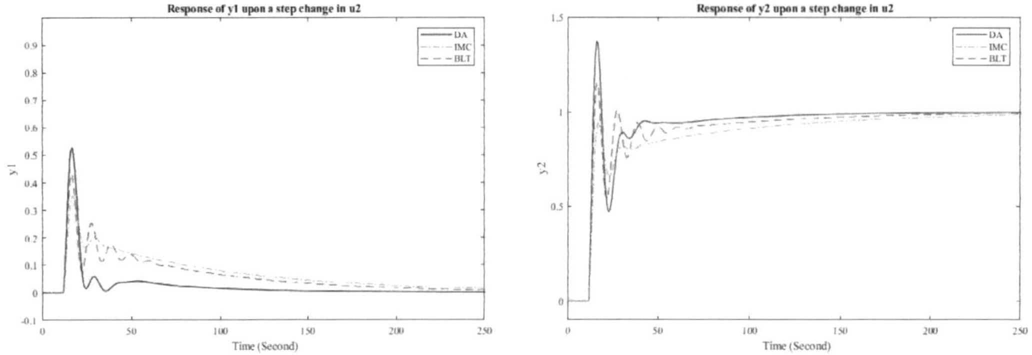

(a) (b)

Figure 4. Response of (a) y1 and (b) y2 when a step change of second input, u2 is introduced.

Table 3. Performance analysis of y1 and y2 for three different tuning methods.

	y1			y2			
	Maximum peak	Settling Time (s)	IAE	Maximum peak	Overshoot (%)	Settling Time (s)	IAE
DA	0.5272	242.6308	6.334	1.3736	37.3588	237.4746	13.3
IMC	0.3498	604.9694	19.44	0.9991	0	451.3486	24.35
BLT	0.4315	497.8624	16.31	1.1554	15.5375	346.6113	16.66

4.3. A step change in disturbance input at time, t=10s

Figures 5(a) and 5(b) show the responses of y1 and y2, respectively, using three different methods when a step change in disturbance is introduced to the system. Table 4 above shows the performance analyses for the method. DA is observed to give the fastest time to remain to its desired steady state value for both y1 and y2 when a disturbance change is set to the system. For y1 response, there is very insignificant oscillation observed when the disturbance is set to the system. The peak for all method is below 0.1 and 0.4 for y1 and y2 response accordingly. However, even though the overshoot is not very significant, the response obtained when DA is used is observed to remain as the best response. From table 4, it clearly states that the minimum IAE and peak belongs to DA method. Moreover, DA method has the fastest response to return to the desired steady state value.

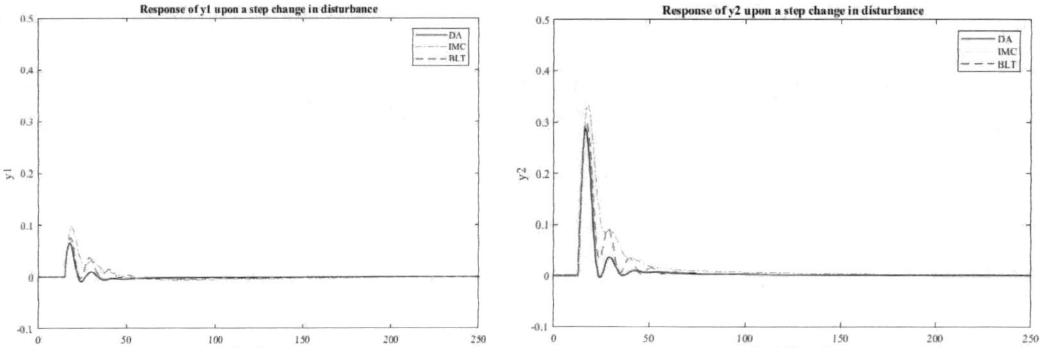

(a) (b)

Figure 5. Response of (a) first output, y1 and (b) second output, y2 when a step change of disturbance is introduced.

Table 4. Performance analysis of y1 and y2 for four different tuning methods.

	y1			y2		
	Maximum peak	Settling Time (s)	IAE	Maximum peak	Settling Time (s)	IAE
DA	0.0652	186.9565	0.5774	0.2868	118.1192	2.231
IMC	0.0962	372.2831	1.667	0.332	182.2883	4.889
BLT	0.0758	325.4664	1.059	0.3002	130.5139	3.218

5. Conclusion

In this paper, DA optimization method is being used to find the optimal PID controller parameters for a minimum IAE value. A MIMO system of a distillation column is considered for the simulation. From the result obtained, DA has proved that it could improve the performance of the PID controller. The system is less oscillated when a step change in either set point or disturbance is introduced to the system. It has a low overshoot and peak. Moreover, in all step changes, DA gives the shortest settling time for both y1 and y2 as well as the most minimum value of IAE. Thus, DA is seen to be one of a stochastic optimization method that could be used to tune and optimize a decentralized PID controller parameters for a MIMO system. DA method has a high potential to improve the performance of the controller towards the system when either there is a step changes in set point or disturbance is introduced to the system. Further improvement to the DA algorithm could be done to improve its ability in optimizing the controller parameters by reducing the overshoot.

Acknowledgments

The authors would like to show gratitude to the Institut Teknologi Sepuluh Nopember for providing the facilities and assistant to conduct the research. The Authors would also like to thank Universiti Teknologi PETRONAS for the support to complete the research.

References

[1] Xiong Q and Cai W J 2006 Effective transfer function method for decentralized control system design of multi-input multi-output processes *Journal of Process Control* **16** pp 773–84

[2] Katebi M R and Elkawafi M A 1995 Comparison of Robustness of Multivariable PID Tuning Methods *IFAC Proceedings Volumes* **28** pp 291–96

[3] Daoutidis P, Zachar M and Jogwar S S 2016 Sustainability and process control: A survey and perspective *Journal of Process Control* **44** pp 184–206

[4] Besta C S and Chidambaram M 2015 Tuning of Multivariable PI Controllers by BLT Method for TITO Systems *Chemical Engineering Communications* **203** pp 527–38

[5] Razali M, Wahab N A, Balaguer P, Rahmat M F and Samsudin S I 2013 Multivariable PID controllers for dynamic process *2013 9th Asian Control Conference (ASCC)*

[6] Chen D and Seborg D E 2002 PI/PID Controller Design Based on Direct Synthesis and Disturbance Rejection *Industrial & Engineering Chemistry Research* **41** pp 4807–22

[7] Nandong J 2015 Heuristic-based multi-scale control procedure of simultaneous multi-loop PID tuning for multivariable processes *Journal of Process Control* **35** pp 101–12

[8] Coelho L D S and Pessôa M W 2011 A tuning strategy for multivariable PI and PID controllers using differential evolution combined with chaotic Zaslavskii map *Expert Systems with Applications*

[9] Zhang J, Zhuang J, Du H and Wang S A 2009 Self-organizing genetic algorithm based tuning of PID controllers *Information Sciences* **179** pp1007–18

[10] Chang W 2007 A multi-crossover genetic approach to multivariable PID controllers tuning *Expert Systems with Applications* **33** pp 620–62

[11] Ghazali A B 2015 A PID de-tuned method for multivariable systems, applied for HVAC plant *IOP Conference Series: Materials Science and Engineering* **88** 012058

[12] Tan W, Chen T and Marquez H J 2002 Robust Controller Design and PID Tuning for Multivariable Processes *Asian Journal of Control* **4** pp 439–51

[13] Vu T N L and Lee M 2010 Multi-loop PI controller design based on the direct synthesis for interacting multi-time delay processes *ISA Transactions* **49** pp 79–86

[14] Jeng J C and Ge G P 2016 Disturbance-rejection-based tuning of proportional–integral–derivative controllers by exploiting closed-loop plant data *ISA Transactions* **62** pp 312–24

[15] Mokadam H R, Patre B M and Maghade D K 2013 Tuning of multivariable PI/PID controllers for TITO processes using dominant pole placement approach *International Journal of Automation and Control* **7** p 21

[16] Reynoso-Meza G, Sanchis J, Blasco X and Martínez M 2016 Preference driven multi-objective optimization design procedure for industrial controller tuning *Information Sciences* **339** pp 108–31

[17] Ganesh P and Chidambaram M 2005 Multivariable controller tuning for non-square systems by genetic algorithm *Proc. of 2005 Int. Conf. on Intelligent Sensing and Information Processing*

[18] Biyanto T R, Fibrianto H Y, Nugroho G, Hatta A M, Listijorini E, Budiati T and Huda H 2016 Duelist Algorithm: An Algorithm Inspired by How Duelist Improve Their Capabilities in a Duel *Lecture Notes in Computer Science Advances in Swarm Intelligence* pp 39–47

[19] Biyanto T R, Alfarisi M S, Afdanny N, Setiawan H and Hasan A 2016 Simultaneous optimization of tuning PID cascade control system using Duelist Algorithms *2016 International Seminar on Intelligent Technology and Its Applications (ISITIA)*

IAM: An Intuitive ANFIS-based method for stiction detection

Sean S Jeremiah[1], H Zabiri[1], M Ramasamy[1], B Kamaruddin[1], Teh W K[1] and A A A Mohd Amiruddin[1]

[1]Department of Chemical Engineering, Universiti Teknologi PETRONAS (UTP), 32610 Seri Iskandar, Perak, Malaysia.

Abstract. Stiction in control valves is an industry-wide problem which results in degradation of control performance. A new approach to detect the presence of stiction by utilising only the PV-OP data from control loops is proposed using an Adaptive Neuro-fuzzy Inferencing System (ANFIS). Intuitively, the error between the output of an FIS model developed with stiction and a process with stiction would be minimal. When benchmarked against seventeen well-known industrial control loop case studies, the Intuitive ANFIS-based Method (IAM) accurately predicts the presence or absence of stiction in 65% of loops tested.

1. Introduction

Modern chemical plants employ regulatory control strategies that require control loops to be far more integrated than those of yesteryear so that plant economy can be increased. However, control loops which function poorly can affect plant performance targets. Such loops may cause substandard product quality, increased utility wastage as well as increased downtime and frequency of maintenance. Poor control performance itself can be the result of improper controller tuning, plant-model mismatch, and actuator nonlinearity among others. Control valves are the most commonly used actuator and the well-known causes of control valve nonlinearity are backlash, saturation, and stiction [1, 2].

Control valve stiction is a particularly interesting fault. It progressively worsens as the mechanical components of the valve gradually deteriorate throughout their service life. Ideally, the controller output (OP) i.e. the desired valve position will be equal to the actual valve position (VP). However, the response of the control valve will begin to deviate from the required command of the controller due to mechanical wear or deposits on the valve stem. This, in turn, would cause oscillation which accelerates wear and further degrades control performance. The oscillation also tends to propagate downstream from the faulty loop to other control loops resulting in further deterioration of control performance.

Research in control valve stiction is mainly divided into modelling, detection, and quantification of stiction as well as compensation for valve stiction. The end goal is to mitigate the negative effects stiction has on the performance of a control loop. Since there can easily be hundreds or even thousands of control loops in process plants it would be difficult to diagnose every valve for the presence of stiction or detect when a loop abnormally enters a limit cycle. Only when stiction is identified can corrective measures be taken to restore control performance. This would be done either manually by manipulating the control loop parameters or by having the controller reconfigure itself automatically.

Detecting stiction is challenging because of two factors. Firstly, a sticky valve exhibits nonlinear behaviour such that the VP does not precisely map onto the OP. Thus, linear approximation methods of the process would fail. Secondly, aside from the OP, the process variable (PV), and the set-point

(SP), seldom is any other process data available to be utilised. The dynamics of the valve, i.e. the VP, would largely have to be inferred from the PV-OP data. Among the nonlinear system identification methods is a method that uses fuzzy logic and adaptive neural networks [3].

Fuzzy logic is an inferencing system that can be used to map an input space (the input variables) to an output space (the decision) using a set of IF-THEN rules. Each rule comprises of an antecedent which follows the "IF" and a consequent which follows the "THEN". Two main types of fuzzy rules exist: the Mamdani fuzzy rules and the Takagi-Sugeno (TS) fuzzy rules [4]. The key difference between them is that in Mamdani systems, the consequent part is based on fuzzy sets whereas in TS systems it is function-based. This makes TS systems especially useful for nonlinear system modelling.

Jang [3] showed how adaptive neural networks (ANN) can be used to overcome the limitation of ordinary TS systems where human knowledge and experience is needed to develop the fuzzy IF-THEN rule base and tune the membership functions to minimize output error during modelling. Using the backpropagation gradient descent algorithm, the ANN recursively recalculates the output error and sends it back to the input until the error falls below an acceptable threshold. This coupling of fuzzy logic and ANN is known as an Adaptive Neuro-Fuzzy Inferencing System (ANFIS).

This paper is organised in the following manner: Section 2 takes a glance at stiction detection methods and Section 3 breezes through ANFIS applied in stiction detection. Sections 4 and 5 show how the Intuitive ANFIS-based Method (IAM) is developed and used to detect stiction. In Section 6 are the results and the discussion. Sections 7 holds the conclusion.

2. Control Valve Stiction

A well-regarded definition of control valve stiction based on empirical observations is given by Choudhury et al [5]. Stiction is described by two variables; S which represents the sum of the deadband and the "stickband", and J which represents "slip-jump". When a force acts on a valve stem, it is initially held in place (stuck) by static friction. This causes an accumulation of potential energy. As the force increases, the friction is eventually overcome, and the valve stem slips and jumps due to an abrupt release of kinetic energy. The valve stem may then overshoot the position commanded by the controller thus requiring the controller to issue a signal for it to reverse direction. When the valve moves back and forth repeated, the process undergoes what is known as a limit cycle.

Stiction detection methods conventionally capitalise on the limit cycle pattern, the signal waveform shape, or employ some form of nonlinearity detection [6]. The method described in this paper is better classified as a model-based method. Daneshwar and Noh [7] have explored the use of fuzzy logic for stiction detection. They model the nonlinear dynamic part of a process with stiction using the OP and VP obtained from a smart valve. Arumugam and Panda [8] appear to have used ANFIS for detection but they do not clearly demonstrate how the ANFIS is trained nor how the amount of stiction present is determined.

3. ANFIS for System Identification

The essential mathematics behind fuzzy logic is given here with the goal of showing how fuzzy logic can be applied in system identification. This is a high-level overview of fuzzy logic taken from Mehran's [9] exposition of the subject. The basic structure of a Mamdani fuzzy rule can be represented by (1). When used for modelling, the TS fuzzy rule is written as (2).

$$\begin{aligned}&\text{IF} \quad u_1 \text{ is } M_1 \text{ AND } u_2 \text{ is } M_2 \text{ AND } u_3 \text{ is } M_3\\&\text{THEN} \quad y_1 \text{ is } M_a, y_2 \text{ is } M_b\end{aligned} \tag{1}$$

$$\begin{aligned}&\text{IF} \quad y(k) \text{ is } M_1 \text{ AND } y(k\text{-}1) \text{ is } M_2 \text{ AND } y(k\text{-}2) \text{ is } M_3 \text{ AND } u(k) \text{ is } M_4 \text{ AND } u(k\text{-}1) \text{ is } M_5\\&\text{THEN} \quad y(k+1) = F(y(k), y(k\text{-}1), y(k\text{-}2), u(k), u(k\text{-}1))\end{aligned} \tag{2}$$

where u_1 to u_3 are input variables; M_1 to M_5 are fuzzy input sets; AND is a logic operator; y_1 and y_2 are output variables; M_a and M_b are fuzzy output sets; $y(k)$, $y(k\text{-}1)$ and $y(k\text{-}2)$ are the output of the system at time k, k-1 and k-2 respectively; $u(k)$ and $u(k\text{-}1)$ are the input of the system at time k and k-1 respectively; $y(k+1)$ is the output of the system; and $F(\cdot)$ is an arbitrary function.

Fuzzy rules make up part of a fuzzy inferencing system called the rule base which can have many such rules. Together with the rule base, the database which defines the membership functions of the fuzzy sets make up the knowledge base. In ANFIS, ANN is used to develop the IF-THEN rules and membership functions. The decision-making unit performs inferencing operations whereas the fuzzification and defuzzification interfaces transform crisp input to degrees of match and then to crisp output as shown in figure 1.

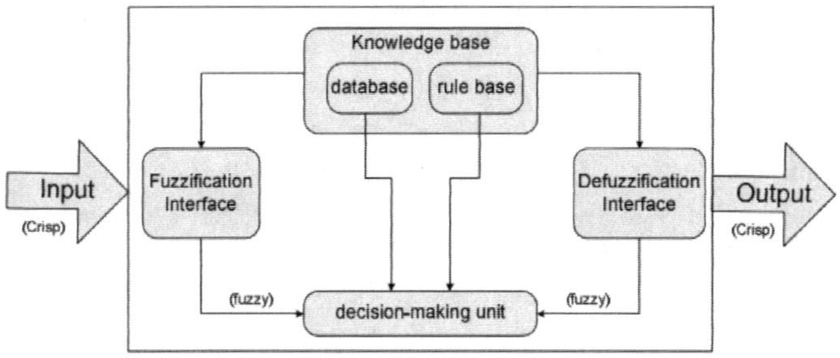

Figure 1. The structure of a fuzzy inferencing system (FIS) [3].

4. Development of the FIS model

To develop an FIS model for a process with stiction, simulated controller output (OP) data and process variable (PV) data are obtained from a feedback control process model with a sticky valve [10, 11]. The Choudhury valve stiction model is used in this scenario as it is sufficiently accurate. As per the Choudhury model, the stiction parameters are defined as "deadband plus stickband", S = 3 and "slip-jump", J = 1. Three thousand PV-OP data points are obtained over a period of 3,000 timesteps.

The simulated PV-OP data is scaled from a range of -1 to 1. Feature scaling overcomes the inability of the FIS to generate accurate output data if the range of input values differ significantly from the range of values used when modelling the FIS. The absolute value of the maximum and minimum values of the data sets may not be equal leading to an uneven spread of positive and negative values. This does not pose a problem for this simulation.

The following steps closely follow a tutorial example available on the MATLAB website for modelling a nonlinear dynamical system using ANFIS [12]. To determine which input variables should be used to develop the FIS model, a heuristic approach called sequential forward search (SFS) is used. Feature selection using this approach seeks to minimize the mean square error (MSE) by gradually adding features thus creating different permutations of the input variables.

Before running the SFS algorithm (seqsrch.m in MATLAB), the PV-OP data is delayed hence creating ten input candidates as shown in (3). The PV data is delayed by one to four timesteps and the OP data is delayed by one to six timesteps. The input candidates are then truncated leaving each with 2,994 datapoints.

$$\begin{bmatrix} Y_{k-1} \\ \vdots \\ Y_{k-4} \\ U_{k-1} \\ \vdots \\ U_{k-6} \end{bmatrix} = \begin{bmatrix} y_6 & y_7 & \cdots & y_{k-1} \\ y_5 & y_6 & \cdots & y_{k-2} \\ y_4 & y_5 & \cdots & y_{k-3} \\ y_3 & y_4 & \cdots & y_{k-4} \\ u_6 & u_7 & \cdots & u_{k-1} \\ u_5 & u_6 & \cdots & u_{k-2} \\ u_4 & u_5 & \cdots & u_{k-3} \\ u_3 & u_4 & \cdots & u_{k-4} \\ u_2 & u_3 & \cdots & u_{k-5} \\ u_1 & u_2 & \cdots & u_{k-6} \end{bmatrix} \quad (3)$$

The number of input candidates and the number of time steps to delay each dataset is chosen arbitrarily. To reduce the tendency for overfitting, conventional validation is implemented by using only the first half of the PV-OP data set for training. The second half is used for checking. Cross-validation is not used as sufficient data points are available for training and checking.

The SFS function is used to select the permutation of the best four input candidates. Four is the maximum number of input candidates which the function can determine. The number of membership functions (MFs) for SFS is set to three and the number of epochs is set to ten. Beyond three MFs, MATLAB warns that it may run out of memory as the rule base is large. Beyond ten epochs, the MSE does not decrease. An alternative feature selection method which could be used is exhaustive search (exhsrch.m). This method runs through all possible permutations of input candidates but is more time consuming and might not yield better results than SFS.

Once the SFS determines the optimal permutation of input candidates, the FIS is generated using the ANFIS GUI. The four input candidates selected by the SFS algorithm and the PV data are used. Again, the first half of the data is used for training and the other half for checking. A FIS model is generated using the grid partitioning method with three generalized bell-shaped MFs assigned to each of the four inputs. The output is assigned a constant membership function. The parameter estimation of the MFs is done using the hybrid method which combines backpropagation and least squares estimation. Error tolerance is set to 0.01 and training is run for thirty epochs.

5. Diagnosing Loops using the Intuitive ANFIS-based Method (IAM)

Once the FIS model is developed, other loops can be tested by comparing the PV data from loops being investigated and with the FIS output when it is fed with PV-OP data from the same loop.

Intuitively, since the FIS has been developed using stiction data, the FIS output will be similar to the PV data from a loop undergoing stiction.

To generate the output data using the FIS model, the PV-OP data from a loop being investigated is scaled from -1 to 1. Then, it is delayed by the same number of instances as the data used to generate the model. For example, if the input candidates used to generate the FIS were $u(t\text{-}1)$, $u(t\text{-}3)$, $y(t\text{-}5)$ and $y(t\text{-}6)$, the same input candidates are generated from the loop's PV-OP data. The input candidates are then fed to the FIS model to get the FIS output.

The FIS output is then compared with the loop PV data by calculating the mean absolute error (MAE) given by (4), where x loop PV data, y is the FIS output, and n is the number of data points.

$$\text{MAE} = \frac{\sum_{i=1}^{n}|y_i - x_i|}{n} \tag{4}$$

The MAE is the mean of the absolute distance between each point of the loop PV data with the corresponding point of the FIS output. Other error metrics such as the mean squared error (MSE) and the root mean square error (RMSE) were also tested during this study. However, they do not yield apparent performance benefits over the MAE. Furthermore, the MAE is conceptually simpler.

To select an appropriate error threshold which can be used to identify stiction, the ninety-four industrial control loops published by Jelali and Scali [13] were examined. By looking at the literature on stiction detection, the goal was to identify loops which have stiction present to a high degree of certainty. A complication that arises here is that the stiction detection methods have a high tendency of misdiagnosis. Therefore, between authors, little consensus exists on the presence of stiction in many of the control loops. This poses a problem for the threshold selection. If the threshold is set too high, it will result in too many false negatives. And if it is set too low, there will be too many false positives.

A timely publication of a review paper by Capaci and Scali [6] supported the selection of an appropriate threshold. They have identified five loops which are evidently present with stiction. (The nomenclature used when referring to data sets and detection methods henceforth is consistent with that used by Jelali and Huang [14] and later by Capaci and Scali [6].) The loops are CHEM 1, CHEM 10,

CHEM 12, CHEM 23, and CHEM 26 [14]. Using the method explained above, the MAE is obtained for each of these five loops.

Ultimately, the MAE of loop CHEM 10 which is 0.63 is selected as the threshold. The reason for this is three-fold. Firstly, Jelali and Scali have said that CHEM 23 and CHEM 26 are only likely to be having stiction. Secondly, all stiction detection methods reviewed by Capaci and Scali identify this loop to have stiction. Although some methods could not be used on CHEM 10, there are no false negatives. The YAMA and YAMA+ tests failed to identify stiction in CHEM 01 and CHEM 12 respectively. Thirdly, the MAE of CHEM 10 is the highest among the five loops indicating that this method needs to allow for that much of error to work.

One hundred and fourteen industrial control loops were available to the authors via the International Stiction Database (ISDB) [14] but only seventeen were analysed in this study because there is no consensus on whether the rest of the loops have been accurately labelled with stiction being present or otherwise.

Of the twenty-six loops listed by Capaci and Scali, the five loops used during the process of selecting the threshold were omitted from stiction analysis as including them would be self-serving. The four other loops omitted are CHEM 3, CHEM 33, CHEM 34 and PAP 7 because Capaci and Scali [6] could not reach a global consensus on those four loops. They recommend that additional tests be carried out to identify the type of nonlinearity present there.

6. Results and Discussion

The IAM can correctly indicate the presence (or absence) of stiction in eleven out of the seventeen loops tested. Of the six failures, one is a false negative (indicating stiction when it is not actually present) and the remaining five are false positives. The results are tabulated in table 1.

Table 1. Results of stiction analysis issued by the IAM

LOOP	STICTION[a]	MAE	RESULT	STATUS	OTHER[b]
CHEM 02	YES	0.22	YES	PASS	9
CHEM 05	*YES*	*1.03*	*NO*	***FAIL***	*9*
CHEM 06	YES	0.24	YES	PASS	9
CHEM 11	YES	0.23	YES	PASS	8
CHEM 24	YES	0.46	YES	PASS	10
CHEM 29	YES	0.50	YES	PASS	7
CHEM 32	YES	0.13	YES	PASS	8
MIN 01	YES	0.28	YES	PASS	9
PAP 02	YES	0.17	YES	PASS	9
PAP 05	YES	0.21	YES	PASS	7
CHEM 04	NO	0.96	NO	PASS	5
CHEM 13	*NO*	*0.25*	*YES*	***FAIL***	*9*
CHEM 14	NO	0.46	YES	**FAIL**	1
CHEM 16	NO	0.45	YES	**FAIL**	1
CHEM 58	NO	0.27	YES	**FAIL**	4
PAP 04	NO	0.74	NO	PASS	3
PAP 09	NO	0.26	YES	**FAIL**	8

[a] Presence of stiction according to Capaci and Scali **[6]**
[b] The number of other methods to accurately detect presence (or absence) of stiction. A total of fourteen other methods are compared.

The IAM was compared against fourteen other stiction detection methods. The comparison is tabulated in table 2. Of the fourteen methods, only the results for HAMM2 and HAMM3 tests were available for all seventeen loops. Some methods could not issue verdicts for certain cases due to various reasons and are considered to have failed to detect stiction in those cases. If no verdicts were made available for some loops, it is also considered that the methods failed there as the ISDB has been openly available since 2009.

Table 2. Synthesis of the verdicts issued by stiction detection methods

	IAM	RELAY [15]	HAMM2 [16]	BIC [10]	HIST [17]	HAMM3 [18]	CURVE [19]	SLOPE [20]	ZONES [20]	YAMA+ [21]	CORR [22]	YAMA [23]	AREA [24]	FUZ [7]	BRA [25]
NRI	**11**	12	12	11	10	10	9	9	9	9	8	7	5	4	1
False +ve	**1**	0	1	0	3	1	3	3	4	0	0	2	5	0	0
False -ve	**5**	4	4	4	2	6	2	0	1	1	2	1	1	0	0
UNC/NA /NIV/XX	**0**	1	0	2	2	0	3	5	3	7	7	7	6	13	16
NRI (%)	**65**	71	71	65	59	59	53	53	53	53	47	41	29	24	6

NRI: number of right indications; UNC: uncertain or unknown; NA: not applicable; NIV: verdict not issued; XX: not applicable to type of loop

Considering the number of right indications (NRI) alone, IAM having eleven NRI is on par with BIC. HIST and HAMM3 come close with ten NRI each. RELAY and HAMM2 have twelve NRI each but RELAY does not work on one loop. The IAM has five false negatives and one false positive. As seen in table 2, generally methods with a higher number of false positives tend to have a lower number of false negatives and vice versa.

Figures 2 and 3 are typical plots of the output generated during the tests. Looking at the failed loops, CHEM 5 and CHEM 13 are probably the most jarring as most other methods succeed here. Figure 2 shows the scaled PV vs. t plot for the test on CHEM 5 which is a flow control loop with stiction. The high MAE can be attributed to the generated PV being slightly out of phase from the actual and the wider range of values generated compared to the actual PV.

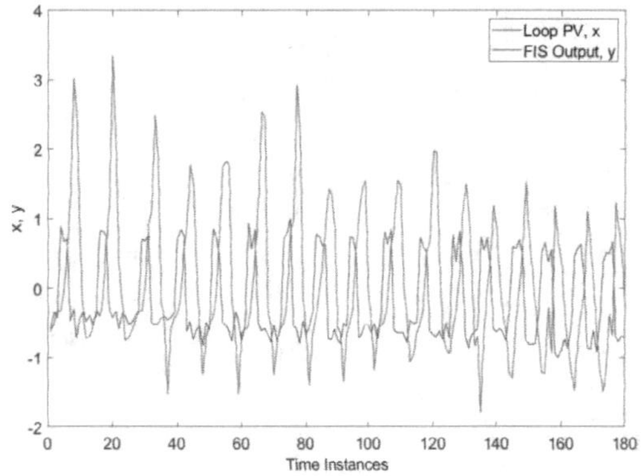

Figure 2. CHEM 5, a false negative.

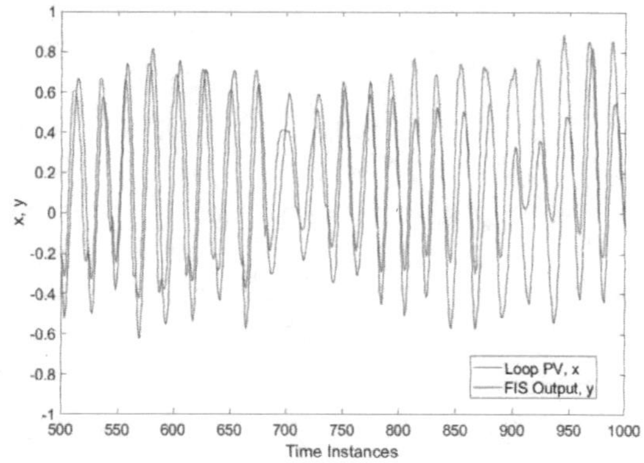

Figure 3. CHEM 13, a false positive.

7. Conclusion

This paper details a novel method to detect stiction using ANFIS. The IAM can correctly diagnose 65% of loops tested and does not require much pre-processing or filtering of the data. Further studies could investigate developing the FIS model using different sets of stiction data or by simultaneously using multiple FIS models do diagnose or even quantify stiction in a loop.

Acknowledgement

The Ministry of Science, Technology, and Innovation of Malaysia funded this work via the eScienceFund (Project No.: 03-02-02-SF0236).

References

[1] Choudhury M, Shah S and Thornhill N F 2008 *Diagnosis of Process Nonlinearities and Valves Stiction: Data Driven Approaches* (Berlin: Springer)

[2] Karra S, Jelali M, Karim M N and Horch A 2010 Detection of oscillating control loops *Detection and Diagnosis of Stiction in Control Loops: State of the Art and Advanced Methods* ed. Jelali M and Huang B (London:Springer) pp 61-100

[3] Jang J S R 1993 ANFIS: adaptive-network-based fuzzy inference system *IEEE Transactions on Systems, Man and Cybernetics* **23** pp 665-85

[4] Takagi T and Sugeno M 1985 Fuzzy identification of systems and its applications to modeling and control *IEEE Transactions on Systems, Man and Cybernetics* SMC-**15** pp 116-32

[5] Shoukat Choudhury M, Thornhill N F and Shah S 2005 Modelling valve stiction *Control Engineering Practice* **13** pp 641-58

[6] di Capaci R B and Scali C 2018 Review and comparison of techniques of analysis of valve stiction: from modeling to smart diagnosis *Chemical Engineering Research and Design* **130** pp 230-65

[7] Daneshwar M and Noh N M 2015 Detection of stiction in flow control loops based on fuzzy clustering *Control Engineering Practice* **39** pp 23-34

[8] Arumugam S and Panda R C 2014 Identification of stiction nonlinearity for pneumatic control valve using ANFIS method *International Journal of Engineering & Technology* **6** pp 570-78

[9] Mehran K 2008 *Takagi-Sugeno Fuzzy Modeling for Process Control*

[10] Choudhury M S, Shah S L and Thornhill N F 2004 Diagnosis of poor control-loop performance using higher-order statistics *Automatica* **40** pp 1719-28

[11] Shoukat Choudhury M A 2008 *Choudhury Stiction Model* http://teacher.buet.ac.bd/shoukat/download.html

[12] MATHWORKS (30-Nov) *Nonlinear System Identification:* https://www.mathworks.com/help/fuzzy/examples/nonlinear-system-identification.html

[13] Jelali M and Scali C 2010 Comparative study of valve-stiction-detection methods *Detection and Diagnosis of Stiction in Control Loops: State of the Art and Advanced Methods* ed. Jelali M and Huang B (Springer London) pp 295-358

[14] Jelali M and Huang B 2009 *Detection and Diagnosis of Sstiction in Control Loops: State of the Art and Advanced Methods* (Springer London)

[15] Scali C and Rossi M 2010 A relay-based technique for detection of stiction *Detection and Diagnosis of Stiction in Control Loops: State of the Art and Advanced Methods* ed. Jelali M and Huang B (Springer London) pp 165-181

[16] Lee K H, Ren Z and Huang B 2008 Novel closed-loop stiction detection and quantification method via system identification *ADCONIP Conf., Jasper* pp 4-7

[17] Horch A 2010 Stiction detection based on cross-correlation and signal shape *Detection and Diagnosis of Stiction in Control Loops: State of the Art and Advanced Methods* ed. Jelali M and Huang B (Springer London) pp 115-147

[18] Karra S and Karim M N 2009 Comprehensive methodology for detection and diagnosis of oscillatory control loops *Control Engineering Practice* **17** pp 939-56

[19] He Q P, Wang J, Pottmann M and Qin S J 2007 A curve fitting method for detecting valve stiction in oscillating control loops *Industrial & Engineering Chemistry Research* **46** pp 4549-60

[20] Dambros J n W, Farenzena M and Trierweiler J O 2016 Data-based method to diagnose valve stiction with variable reference signal *Industrial & Engineering Chemistry Research* **55** pp 10316-27

[21] Scali C and Ghelardoni C 2008 An improved qualitative shape analysis technique for automatic detection of valve stiction in flow control loops *Control Engineering Practice* **16** pp 1501-08

[22] Horch A 1999 A simple method for oscillation diagnosis in process control loops *Proc. of the 1999 IEEE Int. Conf. on Control Applications (Cat. No.99CH36328)* **2 pp** 1284-89

[23] Yamashita Y 2006 An automatic method for detection of valve stiction in process control loops *Control Engineering Practice* **14** pp 503-10

[24] Singhal A and Salsbury T I 2005 A simple method for detecting valve stiction in oscillating control loops *Journal of Process Control* **15** pp 371-82

[25] Brásio A S, Romanenko A and Fernandes N C 2015 Detection of stiction in level control loops *IFAC-PapersOnLine* **48** pp 421-26

IOP Conf. Series: Materials Science and Engineering **458** (2018) 012055 doi:10.1088/1757-899X/458/1/012055

System identification based proxy modeling of a reservoir under iWAG

B M Negash[1,*], **L D Tufa**[2] **and M Ramasamy**[2]

[1]Petroleum Engineering Department, Universiti Teknologi PETRONAS, 32610 Seri Iskandar, Perak Darul Ridzuan, Malaysia
[2]Chemical Engineering Department, Universiti Teknologi PETRONAS, 32610 Seri Iskandar, Perak Darul Ridzuan, Malaysia

*bmamo.negash@utp.edu.my

Abstract —The high computational cost associated with first principle based numerical reservoir models create a barrier for realizing their full potential. Currently, high fidelity proxy models, capable of mimicking the numerical reservoir models, are used to undertake some of the tasks. Nonetheless, large number of experimental simulations are required to generate modeling data. Moreover, experimental realizations are obtained by sampling static geological properties, hence, the resulting proxy model does not honor the dynamic nature of reservoirs, especially during immiscible water alternating gas injection (iWAG) process. In this study, a system identification based proxy modeling algorithm is developed for a reservoir under iWAG. A case study is used to evaluate the algorithm's performance in reproducing oil, gas and water production rates. By systematically designing the input pattern, only a single experimental run is used to generate modeling data. A blind validation and a cross plot analysis resulted in an average percentage fit of 89% and a tolerance level of up to 10%, respectively. The algorithm can have potential application for optimization of injection scenario and hence for automatic well control during iWAG.

Keywords—Numerical reservoir model; proxy model; water alternating gas injection; production prediction

1. Introduction

Different alternative methods have been studied in order to replicate the performance of full field numerical reservoir simulation models and at the same time decrease optimization time and operational cost [1]. Currently, proxy models are used to facilitate the full exploitation of the potential of reservoir simulators. These current approach of building a proxy model are based on either reduced order modeling [2], response surface based modeling [3] or surrogate reservoir modeling[4]. In addition to being expensive to build such types of proxy models, they have limitations. In reduced order modeling, the full numerical reservoir is reduced to a less complex and manageable model by upscaling the defining finite difference grid blocks and/or reducing the physics. Apart from being computationally expensive, the resulting simplified model may not truly represent the actual reservoir. In response surface based models, experimental simulation runs are undertaken at systematically selected design points to generate training and validation datasets. The design points are selected based on central composite design or Box-Behnken design of experiment techniques. Incases, where there are many sensitive independents reservoir parameters the number of simulation runs required will be very large and hence large number of simulation runs will be required. For instance, in order to design an experiment based on the Box-Behnken sampling technique for ten sensitive and independent reservoir parameters 1025 simulation runs are required. This results in expensive computations. In addition, response surface models that are in use today does not honor the dynamic nature of reservoirs, especially when injection of a secondary fluid such as water and gas is involved.

A more recent technique that is based on spatio-temporal data on a grid block basis and artificial neural network has claimed to reduce the number of runs to 15-20 [4, 5]. Even though the time and number of run has been significantly reduced, when compared with traditional proxy modelling, making 15-20 runs still requires a significant amount of time and effort. In addition, designing and organizing of a proper spatio-temporal data is a daunting and computationally expensive task. The degree of success, particularly in practical aspects, of these approaches remains to be argued. Moreover, several runs have to be performed before a reliable model is built [6]. Development of alternative, modest and yet powerful mathematical models allow quick reservoir performance evaluation with an acceptable accuracy [7]. In this research, we developed an algorithm for potential application during the third stage of numerical reservoir modeling and simulation for secondary recovery phase. The algorithm employs system identification principles to construct proxy models. The idea of using system identification for development of a proxy model is not a new concept. However, only few studies have demonstrated the suitability of system identification for development of a proxy model of a reservoir [6]. Moreover, the studies are concentrated on injection of only a single fluid. In this study, a case study approach is used to evaluate the performance of a system identification based proxy modeling algorithm in order to develop a proxy model for a reservoir undergoing iWAG in secondary recovery mode. The case study involves a benchmark reservoir model obtained from UNISIM research group in Brazil [8, 9].

2. Algorithm for proxy modelling of a reservoir under iWAG

The proposed system identification based workflow is discussed in this section. The workflow begins by designing a suitable injection pattern which is bounded by an upper and lower limit. Once a proxy model is developed it can be utilized for optimization of multiple injection scenarios [10].
The set of procedures included in the workflow are:

1. Design a suitable injection pattern and introduce it to numerical reservoir model input file
2. Run numerical reservoir simulation
3. Organize a dataset consisting of injection rate and production rates.
4. Split the dataset set into modeling and validation data set
5. Select an appropriate model structure (ARX, ARMAX, OE, BJ and their respective noise integrated versions.)
6. Select model order between 1 and 10.
7. Use the data reserved for modeling to estimate model parameters
8. Cross-validate the infinite step ahead simulated output and validation data.
9. If validation is not successful, then iterate by changing the selected model order.
10. If validation is not successful, even after exploring several model orders, then change the model structure and continue the iteration as in step 9.
11. If the validation is not successful, increase the number modeling data and repeat steps 6 to 10.
12. Repeat steps 6 to 11 until a valid proxy model is obtained.

3. System description: case study

The case study deals with developing an alternative system identification based proxy model using data collected from a reservoir model under iWAG. The reference model is a benchmark reservoir model developed by the "Research Group in Reservoir Simulation and Management (UNISIM)" [11, 12]. It was built in a high resolution geocellular model, using public data from Namorado Field, Campos Basin, Brazil. Geocellular models replicate the 3-D structure of a real reservoir, with the stratigraphic envelope, reservoir sublayers, and faults all represented in three dimensions. The original intention in developing the case was to create, test and compare new approaches related to reservoir development and management by research centers. The reservoir volume is meshed into smaller 3-D blocks, referred to as Grid blocks. The dimensions of each grid block are about 50 m (163 ft) thick and 100m X 100m (328 X 328 ft) areally. There are a total of 93960 such grid blocks with 81,58 and 20 blocks arranged in the x, y and z directions, respectively. The water-oil and gas-oil contacts are located at a depth of 3200m and 3000m, respectively. Initial oil originally in place is calculated to be 175000 m^3 and the initial reservoir pressure was 300bar. Figure 1 presents the top view of initial fluid distribution, well locations of the UNISIM reservoir model. Figure 2 displays the top view of fluid saturation at the end of simulation period. In both Figure 1 and Figure 2 100% water is represented by blue color, 100% gas is represented by red color and 100% oil is represented by green color. The reservoir model has 17 production wells which are optimally placed and perforated at locations that favor maximum oil production. This includes high permeability, high pore volume, and high oil saturations. In addition, 11 injector wells are introduced and optimized using streamline simulation options that indicates the streamline flows. Constrains on both injection and production wells were adopted from the original model.

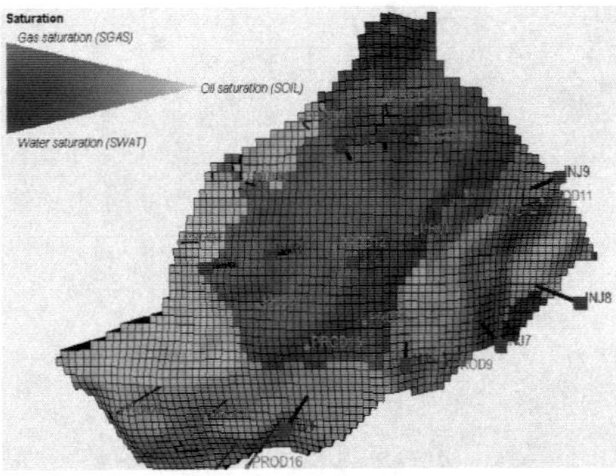

Figure 1. Initial fluid distribution of the UNISIM reservoir model

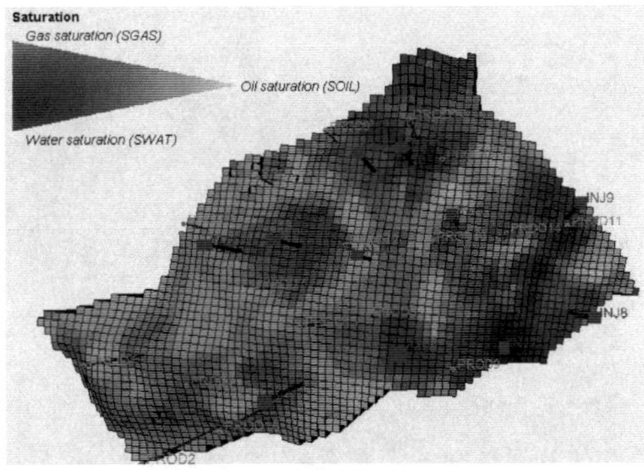

Figure 2. Fluid distribution of the UNISIM reservoir model at the end of simulation period

The reservoir model of Namorado field was modified so that all of the eleven injection wells inject water and gas alternatively while maintaining a PRBS form of injection pattern. Water and gas injection were allowed to alternate every 15 days. The resulting data were used to carryout system identification in a Two Input – Single Output (MISO) system mode where the total fluid production rate was considered as output and the total gas injection and total water injection rates were treated as two inputs to the system. Data preparation for system identification includes visual inspection and selection of an appropriate and representative dataset. Injection and production data were made available for system identification by splitting them into modeling and validation datasets. A simple model such as an ARX model is identified and its transient response is studied, then more complex model structures are investigated by using the order and time delay, obtained in the first stage, as

Table 1. Lower and upper bounds used in the design of a PRBS input during SWG.

Well Name	Gas injection Sm^3/day	Water injection Sm^3/day	Water injection Sm^3/day	Water injection Sm^3/day
	Lower bound		Upper bound	
Well 1	40000	1500	140000	4500
Well 2	45000	1750	145000	4750
Well 3	50000	2000	150000	5250
Well 4	55000	2250	155000	5550
Well 5	60000	2500	160000	5750
Well 6	65000	2750	165000	6000
Well 7	40000	3000	155000	6250
Well 8	45000	3250	160000	6500
Well 9	50000	3500	165000	6750
Well 10	55000	3750	165000	7000
Well 11	60000	4000	155000	7250

initial approximation. The use of simpler model allows to obtain insight into the order and delay of more complex model structures that are capable of identifying the system more accurately. In this study, the inputs, gas injection rates and water injection rates were designed as a PRBS switching between two bounds as shown in Table 1, and at the same time alternating every 15 days.

Injection of water and gas alternatively allows piston like displacement as displayed in Figure 3. This has the potential to effectively sweep the oil with only a small amount of residual oil.

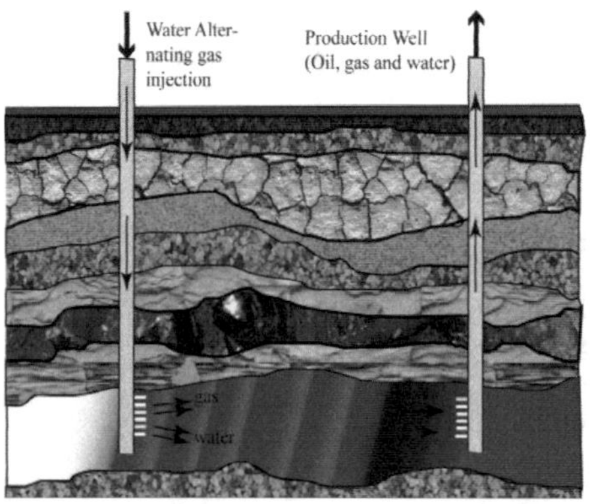

Figure 3. Physics of a reservoir flow during iWAG.

4. Results and Discussion

This section presents the results and discussions of modeling total oil, gas and water production rates, as a function of total gas injection and total water injection rates. The models obtained in this study can be useful for optimization of water and gas injection rates during iWAG.

4.1. Modeling oil production

This system has two inputs, namely total gas injection rate and water injection rate, and one output, viz. oil production rate. Figure 4 presents the modeling and validation data used in modeling oil production as a function of gas and water injection rates. A total of 2666 data at a frequency of 1/day were obtained from the reservoir simulator. The first 1500 data points were used for modeling and the next 10 data were removed to fulfill the requirements of predictive models. The remaining 1156 data points were used for cross validation.

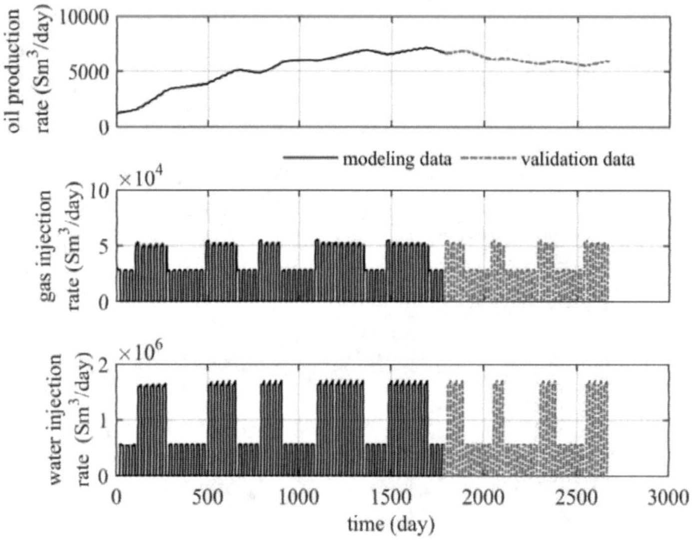

Figure 4. input and output data for modeling and validation of total oil production during iWAG injection.

The identification of a characteristic LTI model had started by identifying a simple ARX model. The model order and time delay of the identified ARX model was then used as initial approximation for further investigation into more complex model structures. More complex Polynomial LTI model structures (ARMAX, OE, and BJ) with and without noise integration had been thoroughly searched for a model that gives minimum error or best fit during validation. This was done by systematically changing the time delays and orders of the defining polynomials in both the deterministic and stochastic components of each model structures until a model is successfully identified. As a result, a discrete-time Box-Jenkins Polynomial with Noise Integration model that exhibits a one-time delay and fourth order in all the coefficient polynomials is found to be an optimal estimator in terms of the percentage fit obtained. The model was named BJI (4-4-4-4-1). The percentage fit to modeling and validation data were found to be 99.94% and 89.96%, respectively. The mean squared error (MSE) and Final prediction error (FPE) were 1.037and 1.085. The BJI (4-4-4-4-1) model is presented below:

$$y(t) = \left[\frac{B(q)}{F(q)}\right]u(t) + \left[\frac{C(q)}{D(q)(1-q^{-1})}\right]e(t) \tag{1}$$

where:

$B1(q) = [3.813 \quad 2.276 \quad -0.1771 \quad -0.508] * e^{-5}$

$B2(q) = [-1.734 \quad 6.035 \quad 5.963 \quad -0.641] * e^{-7}$

$C(q) = [-1.232 \quad -0.5181 \quad 0.9541 \quad -0.1607]$

$D(q) = [-2.37 \quad 1.403 \quad 0.3641 \quad -0.3927]$

$F1(q) = [-1.29 \quad 0.2356 \quad -0.4558 \quad 0.5118]$

$F2(q) = [-0.8214 \quad -0.819 \quad 0.1862 \quad 0.4544]$

A comparison plot of the identified BJI (4-4-4-4-1) model's infinite step simulated output and validation data is shown in Figure 5. It can be observed that the model has successfully represented oil production rate as a function of gas and water injection rates.

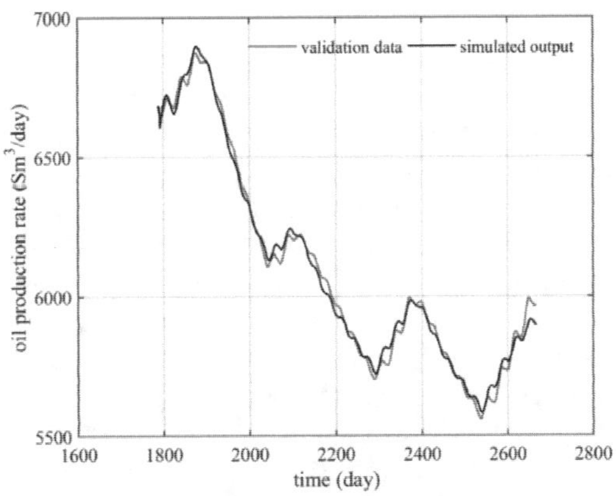

Figure 5. Validation of the BJI (4-4-4-4-1) mode.

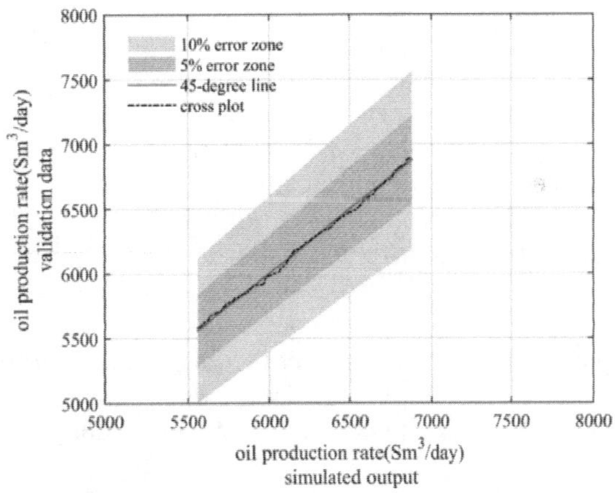

Figure 6. Cross plot for BJI (4-4-4-4-1).

Figure 6 presents a cross plot obtained by plotting simulated output vs. validation data on a linear and square plot. It can be seen that the 45-degree line closely matches the cross plot indicating that the model does, neither underestimated nor overestimated during validation.

4.2. Modeling gas production

The previous session outlined and discussed modeling and validation of total oil production rate as a function of total water and total gas injection. This section applies the same procedure to model total gas production as a function of total gas and total water injection rates. Figure 7 shows the input and output data collected at a sampling frequency of 1/day from year 2005 to end of simulation year. A total of 2666 data at a frequency of 1/day were obtained from the reservoir simulator. A simple visual inspection, shows that the first 500 data set is not representative of the dataset observed at a later time, hence, it was not considered during the modeling stage. The remaining data were used for identification. Data from 500 to 1500 were used for modeling and the next 10 data were removed to fulfill the requirements of predictive models. The remaining 1156 data points were used for cross validation.

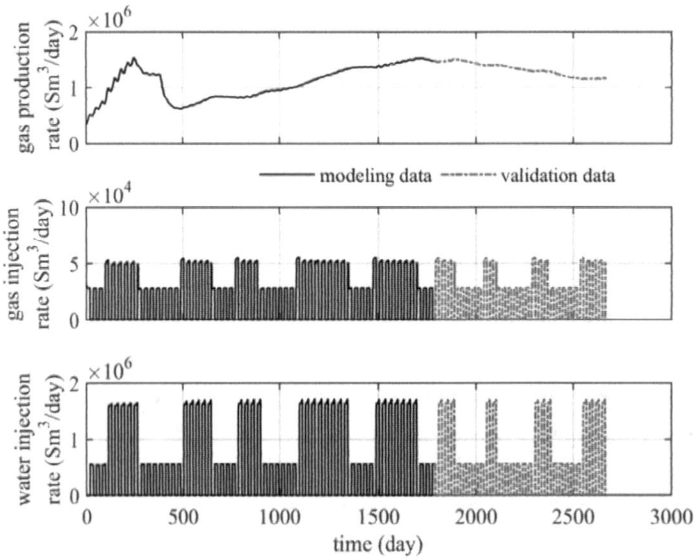

Figure 7. Input and output data for modeling gas production during iWAG injection.

At first an ARX and ARIX model structures, a simple model structures, were searched for by systematically changing their model order and time delay. The best estimating ARX or ARIX model's order and time delay were then utilized as initial approximation for further investigation into more complex model structures. of Consequently, a Box-Jenkins model of order seven in all the numerators and denominators and a one-time delay was found to produce best estimation in terms of calculated percentage fit. The model was named BJ (7-7-7-7-1) and it was cross-validated using data that has not been employed during the modeling stage. Figure 8 presents a comparison between validation data and simulated output of the identified model. The fit to estimation data was 99.86% and the fit to validation data was 91.06%. Final prediction error (FPE) and mean square error (MSE) were 6076 and 9892, respectively. By removing the first 500 points, the percentage fit has improved when compared to a similar identification process with all the data points included. The best match which captured the gas production rate dynamics during identification and while using the whole set of data was 60.71%. This case serves as an example of the importance of inspecting of input and output data prior to splitting the data into modeling and validation.

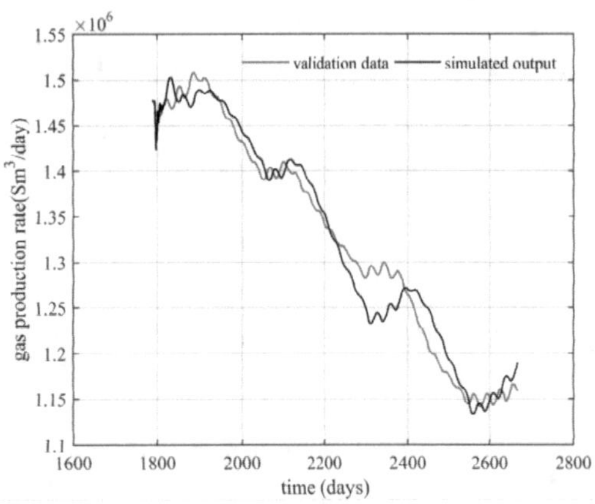

Figure 8. Validation of the BJ (7-7-7-7-1) model

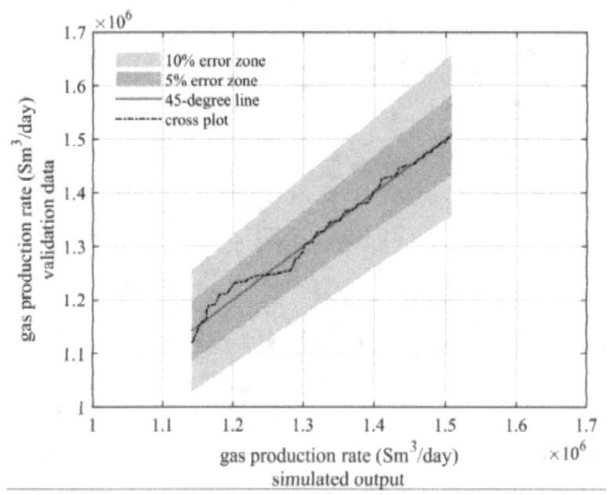

Figure 9. Cross plot analysis for model BJ (7-7-7-7-1).

The discrete time BJ (7-7-7-7-1) model is presented below:

$$y(t) = \left[\frac{B(q)}{F(q)}\right] u(t) + \left[\frac{C(q)}{D(q)}\right] e(t) \tag{2}$$

where

$B1(q) = [0.005366 \quad 0.01955 \quad -0.01959 \quad 0.0143 \quad -0.007743 \quad 0.00717 \quad -0.004015]$ $D(q) = [-2.176 \quad 1.547 \quad -0.8752 \quad 0.8192 \quad -0.4003 \quad 0.2673 \quad -0.1819]$

$B2(q) = [-40.53 \quad 70.24 \quad -29.34 \quad 32.58 \quad 66.9 \quad 32.84 \quad 1.513] * e^{-5}$ $F1(q) = [-2.421 \quad 2.024 \quad -0.6362 \quad -0.135 \quad 0.2282 \quad -0.0415 \quad -0.01869]$

$C(q) = [0.02011 \quad 0.03306 \quad 0.0005555 \quad -0.01909 \quad -0.01029 \quad 0.0004809 \quad 0.0006449]$ $F1(q) = [-2.01 \quad 1.199 \quad -0.9998 \quad 1.845 \quad -1.234 \quad 0.06368 \quad 0.1357]$

Figure 9 presents a cross plot analysis of an infinite step ahead prediction and validation data. The plot was made by first concatenating the two data and sorting them in ascending order. The plot is almost in line with the 45-degree line indicating that the identified mode has neither underestimated nor overestimated during validation.

4.3. Modeling water production

In addition to modeling oil and gas production, modeling water production is also an essential task. This section demonstrates the application of system identification in building a proxy model for capturing the relationship between total water production rate, total water injection rate and total gas injection rate during water alternating gas flooding. Data for identification were collected by proper excitation of the Namorado field's numerical reservoir model. The same simulation run that produced the gas and oil production rates in the previous two sessions also provided water production rate. Figure 10 presents the input and output data used for modeling and cross validation of a polynomial model. A total of 2666 data at a frequency of 1/day were obtained from the reservoir simulator. However, a simple visual inspection of the data revealed that the reservoir's response has only begun after 500 days of injection and production. This signifies that the first 500 data is not useful for identification. Hence, data points from 500 to 1500 were used for modeling and the next 10 data were removed to fulfill the requirements of predictive models. The remaining 1156 data points were used for cross validation.

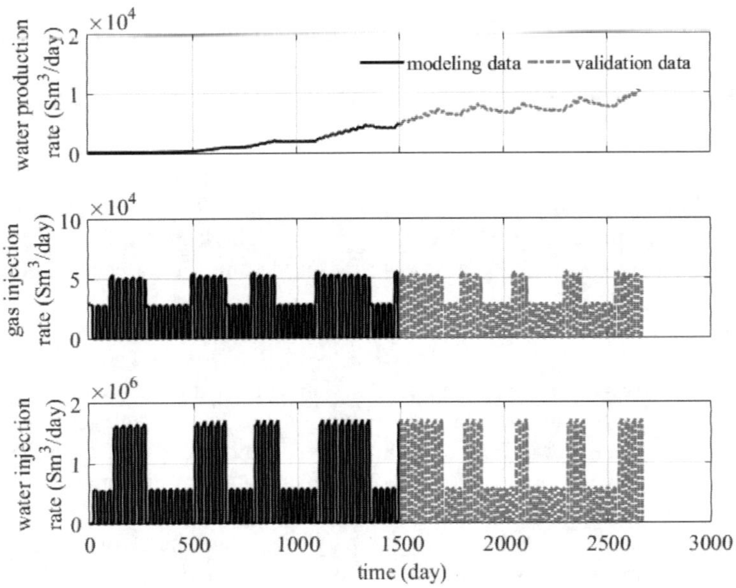

Figure 10. input and output data for modeling and validation of total water production during gas injection.

Identification of a suitable model has beggun by first exploring a simpler polynomial model structure, ARX, and its noise integrated version, ARIX. We found that an ARX model whose order is eight in both the numerators and denominators and a zero-time delay was the best estimating model in terms of the percentage fit exihibited. This model was named ARX (8-8-0). Using ARX (8-8-0) as initial estimate, more complex model structures were searched. We found a noise integrated Box-Jenkin model that has a model order of four in all the numerator and denominator polynomilas and a zero-time delay to best fit the validation data. This model was named BJI (4-4-4-4-0) and was validated using the validation data. The fit to estimation and independent validation data were found to be 99.86% and 60.52%. Moreover, the final prediction error and mean square error were calculated to give 2.727 and 425.2, respectively. Figure 11 presents a graphical comparison between validation data and simulated output. It is observed that the identified model has closely predicted total water production during validation. The structure and polynomial coefficients of BJI (4-4-4-4-0) model are shown below:

$$y(t) = \left[\frac{B(q)}{F(q)}\right]u(t) + \left[\frac{C(q)}{D(q)(1-q^{-1})}\right]e(t) \tag{3}$$

where:

$B1(q) = [0.0001924 \quad 0.0002775 \quad 0.0001607 \quad 3.495e^{-5}$

$D(q) = [-1.782 \quad -0.1408 \quad 1.731 \quad -0.7927]$

$B2(q) = [1.961 \quad -1.076 \quad 3.861 \quad -1.877] * e^{-6}$

$F1(q) = [-0.4819 \quad -0.9691 \quad 0.1753 \quad 0.3016]$

$C(q) = [-0.2918 \quad -1.438 \quad 0.3392 \quad 0.5495]$

$F2(q) = [-2.015 \quad 1.437 \quad -0.5185 \quad 0.09636]$

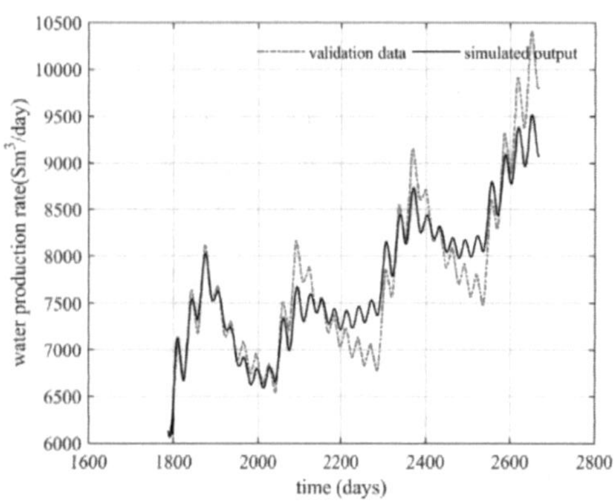

Figure 11. Validation of model BJ (7-7-7-7-0).

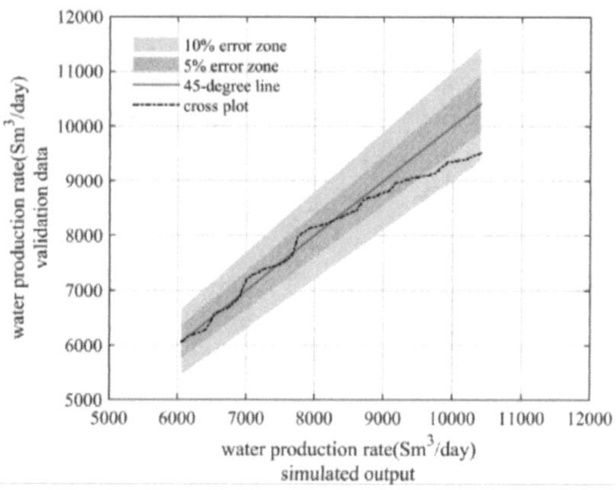

Figure 12. Cross plot analysis of model BJI (4-4-4-4-0).

Crossplot analysis was also conducted to find out whether the model has overestimated, underestimated or well estimated the output. This is realized by first sorting and plotting simulated output vs. validation data and comparing the plot with a straight 45-dgree line. Figure 12 shows that the model does not underestimate or overestimate the production of water at the early stage of the production.

5. Conclusion

The large computational time required to make simulation runs inhibit the full exploitation of the potential of numerical reservoir models. In this study, an algorithm that infuses system identification via LTI model structures and numerical reservoir models is proposed and evaluated. The resulting outcome is a proxy models to replace the role of numerical reservoir simulators. As opposed to current methods of proxy modeling, system identification based proxy models require only a single experimental run with and a pseudo random binary sequence form of injection pattern. The application is demonstrated by using a benchmark case study obtained from UNISIM, a research group established in 1996 and focused on studies related to numerical simulation and management of petroleum reservoirs. The case study involves a reservoir model with 18087 active grid blocks with 11 injectors and 19 producers which are optimally placed. Proxy models are developed to represent oil, gas and water production rates as a function of alternating water and gas injection rate. A percentage fitness test followed by whiteness test and cross plot analysis were used to measure the performance of the developed models. As a result, an average percentage fit of 89% is obtained indicating that the algorithm is reasonably efficient. The crossplot analysis also reveal that the developed proxy models are accurate and have a tolerance level of up to 10%. With this performance the algorithm can be useful for well setting and production optimization purposes.

References

[1] Zubarev D I 2009 Pros and cons of applying proxy-models as a substitute for full reservoir simulations *SPE Annual Technical Conference and Exhibition* (4-7 October 2009, New Orleans, Louisiana)

[2] Zangl G, Graf T and Al-Kinani A 2006 Proxy modeling in production optimization *SPE Europec/EAGE Annual Conference and Exhibition* (12-15 June 2006, Vienna, Austria)

[3] Negash B M, Vasant P M and Jufar S R 2018 Application of artificial neural networks for calibration of a reservoir model *Intelligent Decision Technologies* 12 (1) pp 67-79

[4] Alenezi F and Mohaghegh S 2017 Developing a Smart Proxy for the SACROC Water-Flooding Numerical Reservoir Simulation Model *SPE Western Regional Meeting* (23-27 April 2017 Bakersfield, California)

[5] Lutidze G and Younis R 2018 Proxy reservoir simulation model for IOR operations *Saint Petersburg 2018*

[6] Negash B M, Tufa L D, Ramasamy M and Awang M B 2017 System Identification Based Proxy Model of a Reservoir under Water Injection *Modelling and Simulation in Engineering* **2017** pp 1-10

[7] He Q 2016 *Investigating Continuously Updated History Matching Using Smart Proxy (Surrogate Reservoir Model)* (West Virginia University)

[8] Ljung L 1998 System identification *Signal analysis and prediction* (Springer) pp 163-173

[9] Nelles O 2013 *Nonlinear system identification: from classical approaches to neural networks and fuzzy models* (Springer Science & Business Media)

[10] Negash B M, Vel A and Elraies K A 2017 Artificial Neural Network and Inverse Solution Method for Assisted History Matching of a Reservoir Model *International Journal of Applied Engineering Research* **12** (11) pp 2952-62

[11] Gaspar A T, Santos A, Maschio C, Avansi G, Hohendorff Filho J and Schiozer D 2013 Study Case for Reservoir Exploitation Strategy Selection based on UNISIM-I Field ed: Campinas-SP, Brazil, http://www.unisim.cepetro.unicamp.br/unisim-i

[12] Gaspar A T, Avansi G D, Maschio C, Santos A A and Schiozer D J 2016 UNISIM-IM: Benchmark Case Proposal for Oil Reservoir Management Decision-Making *SPE Trinidad and Tobago Section Energy Resources Conference* (13-15 June, Port of Spain, Trinidad and Tobago)

Decontamination of benzene from aqueous solution by green functionalization of activated carbon

Hirra Anjum[1], Nirmala Gnanasundaram[2], Khairiraihanna Johari[1], Thanabalan Murugesan[1*]

[1]Department of Chemical Engineering, Universiti Teknologi PETRONAS (UTP), 32610 Seri Iskandar, Perak Darul Ridzuan, Malaysia.
[2]Department of Chemical Engineering, VIT University, Vellore, 632014 Tamil Nadu, India.

[*]murugesan@utp.edu.my

Abstract. Environmental pollution caused by industrial wastewater is at its alarming rate. Benzene falls under the category of those contaminants that most commonly found in produced water from chemical industries. To get rid of this carcinogenic compound, the present study proposed an environmentally benign method to modify the commercially activated carbon by means of 1-Hexyl-3-methylimidazolium bis (trifluoromethylsulfonyl)imide. This green functionalization attempts to simultaneously maximize the removal efficiency of adsorbent by elevating the surface area and pore volume of the modified adsorbent and reduce the overall cost of the process. The activated carbon in pristine and functionalized forms was characterized by using, N_2 adsorption/desorption, FTIR, Thermogravimetric analysis, Raman spectroscopy, XRD analysis and Scanning Electron Microscopy. Batch adsorption experiments were performed and modified activated carbon exhibited higher adsorption capacity compared to that of pristine AC. The Pseudo second order model and Langmuir isotherm model best fitted the data of the adsorption of benzene. Thus, modified activated carbon managed to enhance the benzene removal efficiency up to 87%.

1. Introduction

Aromatic compounds are most commonly used in chemical industries. They are often used as raw materials and solvents in manufacturing processes. Benzene is found to be the most dangerous among them. Because of its carcinogenic and mutagenic nature, it imparts severe health impacts on the environment. Thus, its removal is critical to ensure the safe water supply. A considerable effort has been made by the researchers to develop the most efficient method for the obliteration of benzene but the most widely used is the process of adsorption that can eradicate the unwanted volatile organic compounds even at very low concentrations [1].

There are several economically viable adsorbents namely, activated carbon, clay minerals, zeolites and polymeric resins that have been studied extensively. But activated carbon is found to be the most vital mesoporous adsorbent with high specific surface area and surface to volume ratio, complex surface characteristics. This surface chemistry is the one factor that ultimately defines the sorption capacity of the adsorbent. However, this surface chemistry is directly related to the surface groups. These surface groups can be easily modified by thermal, acidic and basic treatments [2, 3].

Furthermore, ionic liquids gained much attention as an alternative of acidic or alkaline modification due to their unique tuneable properties and diverse characteristics (extremely low vapor pressure, high thermal and chemical stabilities, non-flammability and high solvent capacity) and environmentally friendly nature. In order to ensure the covalent bonding, they act as a binder between the activated carbon and the other component [4]. We selected 1-Hexyl-3-methylimidazolium bis (trifluoromethyl-sulfonyl)imide. The use of bulky group in the conventional position enhances the dispersion ability of surfactant. However, methyl group avoids the deprotonation in the compound. Moreover, most commonly used ILs are imidazolium-based ionic liquids because of easy dispersion and longer chain length with high integrity and stability depending upon the anion used. On the other hand, hydrophobic groups increase the solubility of the surfactant in aqueous media, as perfluorinated sulfonyl imide forms hydrolytically stable hydrophobic ionic liquids with low viscosity and high electrical conductivity [5].

In this paper, the commercial activated carbon was chosen as an adsorbent for the sorption of benzene in aqueous solution. The modification of activated carbon was carried out via ionic liquid treatment in order to introduce oxygen bearing functional groups over the surface of graphitic structure and demonstrate the comparison with the help of detailed characterization, kinetics and isotherm studies revealed that activated carbon functionalized with ionic liquids exhibited a superior sorption capacity compared to the pristine AC.

2. Materials and Methods

Activated carbon (12.01g/mol, < 0.1 mmHg, -100 mesh particle size, Sigma Aldrich), 1-Hexyl-3-methylimadazolium bis(trifluoromethylsulfonyl)imide (447.42g/mol, Merck), ethanol (95.2%, 46.07g/mol) and acetonitrile (99.9%, 46.07g/mol).

Five grams of pretreated AC was dissolved in 500ml of ethanol and 5ml of 1-Hexyl-3-methylimidazolium bis (trifluoromethylsulfonyl) imide. Afterwards, 0.02g of Cetyltrimethylammonium (CTAB) was added in solution mixture as a surfactant with continuous stirring in an oil bath at 65°C for 3h. The oxidized AC (O-AC) was filtered and dried overnight at 60°C.

The surface area and pore volume was analyzed by BET surface area analyzer (Micromeritics ASAP 2020), the surface functional groups were evaluated by FTIR (SHIMADZU, 8400S), thermal stability was examined by Thermogravimetric analysis ((Perkin-Elmer TGA, Pyris I), the structural defects were probed by Raman spectroscopy (Horiba Jobin Yvon HR800), crystallographic structure was determined by XRD analysis ((Bruker, D8) and surface morphology was investigated by FESEM (Zeiss Supra 55 VP).

3. Results and Discussions

3.1. Pore Structure Characterization

N_2 adsorption/desorption isotherms of Pristine and modified activated carbon at 77K is illustrated in figure 1a. The initial curvature of the isotherm defines the monolayer coverage and rounded knee is the evidence of the presence of some micropores in the modified activated carbon [6]. The knee of O-AC is broader which indicates larger monolayer coverage as compared to that of raw sample. This large monolayer coverage implies that O-AC have much larger surface area than that of the P-AC.

Since surface area is the function of pore size and the number of pores. So, pore size distribution is presented as the evidence of the effect of chemical treatment on the performance of activated carbon after oxidation as depicted in figure 1b. It is evident that micropore size distribution is quite pronounced in the ionic liquid modified sample as compared to that of the un-functionalized sample which is due to the diffusion of the oxidizing agent through the graphitic structure as a matrix to create porosity in the framework for the removal of BTX [7].

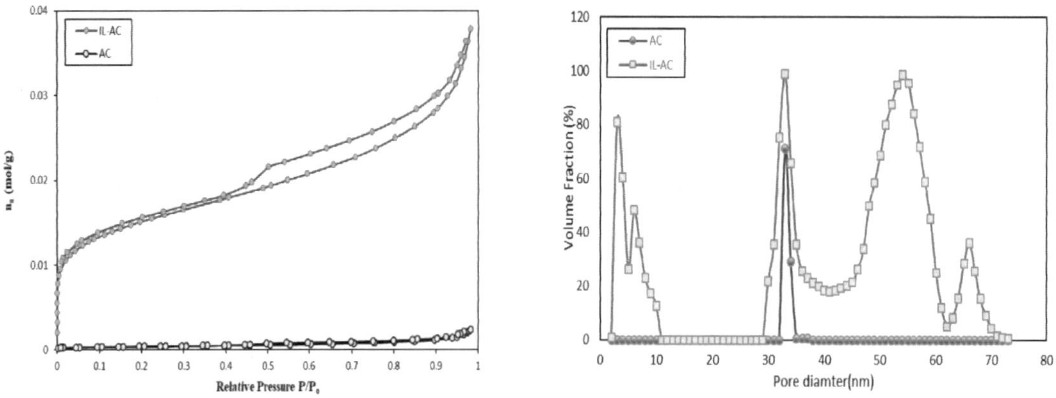

Figure 1. a) Adsorption/desorption isotherms. b) Particle size distribution of P-AC and O-AC.

Hence, the generation of these micropores is observed to be helpful in the enhanced sorption capacity of activated carbon.

3.2. FTIR

The IR spectra of P-AC and O-AC are exhibited in figure 2. It is clear that the peaks are more pronounced after modification with a combination of ionic liquid as compared to that of other modifications. The bands at 3400cm^{-1}, 3163.3cm^{-1}, 2961.7cm^{-1}, 2869.2cm^{-1} are related to –OH stretch from carboxylic acid groups, -CH stretch asymmetric and symmetric from CH_3 group respectively. The band 2369cm^{-1} is found in all peaks after modification, that is related to strongly hydrogen bonded COOH group [8].

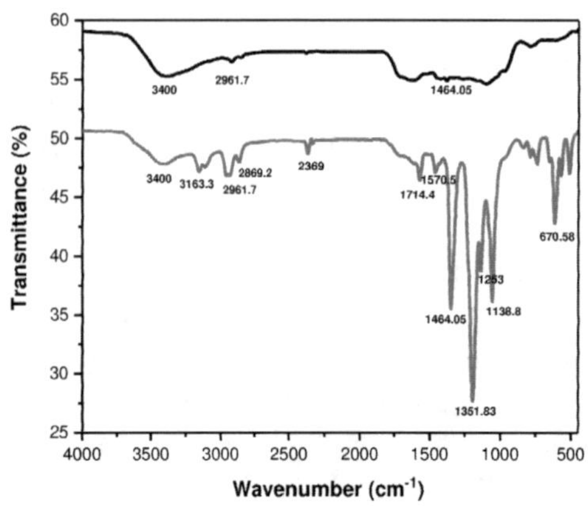

Figure 2. IR spectra of P-AC and O-AC.

The peaks at 1714.4cm^{-1} and 1570.5cm^{-1} signify the –C=O and OC=O stretch. However, the bands at 1351.83cm^{-1}, 1196.07cm^{-1},1138.8cm^{-1}, and 1058cm^{-1} are introduced in ionic liquid modified only and are an indication of the presence of (carboxylate anion) CH3COO-, (alcoholic) -OC, COH, and phenolic -OH groups respectively.

3.3. Thermogravimetric Analysis

Thermo-gravimetric analysis of the activated carbon before and after modification enabled ascertaining their thermal stability to state the optimum conditions for calculation. The plot (figure 3) exhibits that in the first region that is related to moisture evaporation, the thermal stability of pristine and ionic liquid treated activated carbon is 71.03% and 99.97% respectively. Even after degradation of hydroxyl groups the thermal stability of ionic liquid treated activated carbon is the higher than that of P-AC.

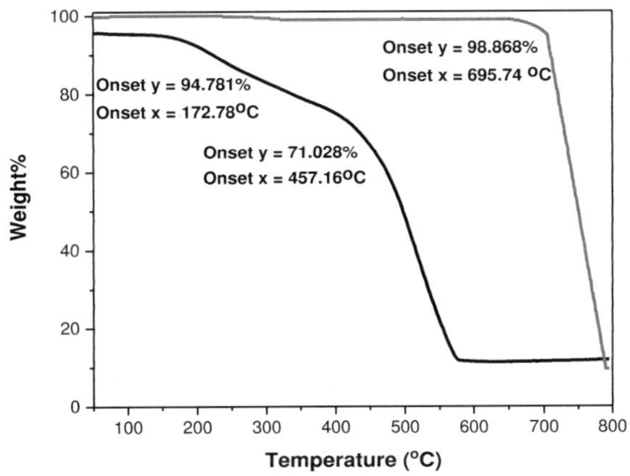

Figure 3. TGA curves of P-AC and O-AC.

The difference in the weight loss between pristine and oxidized AC can be assigned to the change in the hydrophilic nature due to the replacement of carboxylic groups on AC. In the second stage where rapid weight loss occurs due to decomposition of carbon, the thermal stability of pristine and ionic liquid treated activated carbon is 11.32% and 98.86%. The third weight loss is associated with decomposition of the chemical structure at around 795°C in which of pristine and ionic liquid treated activated carbon show 11.85%, and 9.44% stability. The remaining material in pristine activated carbon is most probably the catalyst that was removed in other oxidation treatments and improved the overall purity of the surface modified activated carbon. In conclusion, O-AC samples exhibit much higher thermal decomposition temperature which is the indication of a thermally stable combination of adsorbent. This increase in thermal stability might be due to increased intermolecular interactions and coordinating nature of 1-hexyl-3-methylimidazolium triflorosulfonylimide.

3.4. Raman Spectroscopy

The structural fingerprints are identified by the vibrational spectroscopic approach called as Raman spectroscopy (figure 4). The spectrum is enunciated in a form of the intensity of scattered light versus Raman shift (inverse of wavelength). The degree of functionalization (I_D/I_G ratio) can be determined by Lorentzian fitting areas of D and G band [9]. The degree of functionalization (I_D/I_G ratio) was found to be 1.00, 1.55 for pristine and ionic liquid treated activated carbon, respectively.

This finding reveals that attachment of the oxygen-bearing groups over the carbon lattice leads to an addition of more sp^3 hybridized carbon atoms, stipulating that higher quality functionalized activated carbon are formed. Consequently, the overall profile of modified sample suggests that even after chemical treatment the inherent hexagonal structure of activated carbon remains unmutilated.

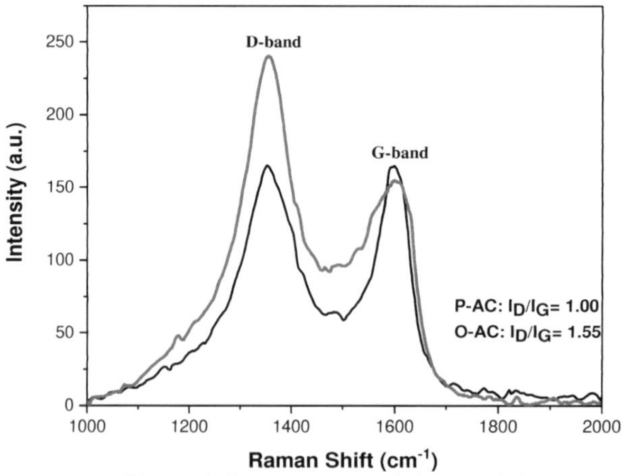

Figure 4. Raman spectra of P-AC and O-AC.

3.5. X-ray Diffraction

The phase analysis by x-ray diffraction of P-AC, and O-AC is exhibited in figure 5. In both samples, the broad peak at 26.8° and 25.4° is observed for P-AC and O-AC respectively. It is due to the reflections from planes can be noticed which is clear indication of the presence of graphitic ordering (amorphous carbon) in molecular planes[10]. The peak has become clearly broad in O-AC which is the validation of the presence of amorphous carbon in the structure. Other sharp peaks signify the presence of impurities in the samples. These peaks tend to attenuate in the O-AC plot and remain in micro range. From this (002) broad peak, interplaner distance (dhkl) was determined by utilizing Braggs law and average crystal size with the help of Deby- Scherrer equation [11, 12].

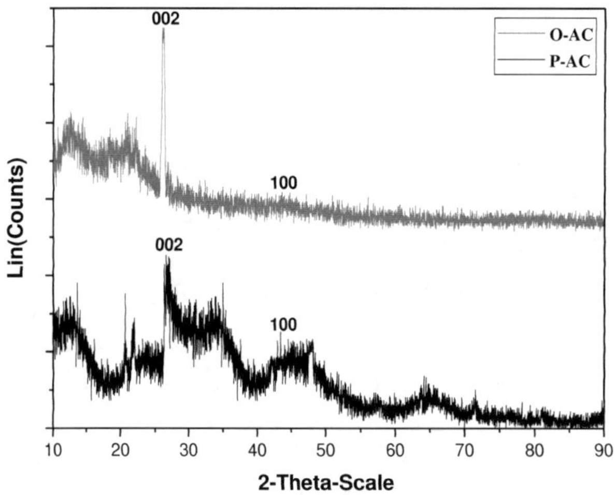

Figure 5. XRD peaks for P-AC and O-AC.

The crystalline dimensions of activated carbon before and after surface modification has been summarized in table 1. It reveals that with the increase in peak intensity at (002), O-AC exhibits the highest crystalline dimension as compared to that of other samples.

IOP Conf. Series: Materials Science and Engineering **458** (2018) 012056 doi:10.1088/1757-899X/458/1/012056

Table 1. Crystalline dimensions of pristine and oxidized activated carbon.

Crystalline dimensions	P-AC	O-AC
d_{hkl} (nm) $d_{hkl} = \frac{n\lambda}{2sin\theta_{B(hkl)}}$	0.332	0.3506
L_c (nm) $L_c = \frac{k\lambda}{FWHM_{hkl}\ cos\theta_{B(hkl)}}$	2.254	3.495
L_a (nm) $L_a = \frac{k\lambda}{FWHM_{hkl}\ cos\theta_{B(hkl)}}$	4.862	7.495
N_p $N_p = \frac{L_c}{d_{hkl}}$	6.789	9.968

Although, the calculated values are believed to be lower than the true crystallite dimensions but can be used as a relative estimate of these quantities [13]. Since, there is no significant difference in diffractogram of all plots, recommending that the surface modification process did not affect the microstructure of carbon material.

3.6. Scanning Electron microscope

The morphological behavior of pristine and modified activated carbon is summarized in figure 6. Two different magnifications (10μm and 1μm) were employed to detect the surface morphology before and after the functionalization. It is clear that the P-AC is comprised of irregular, heterogeneous surface before functionalization.

Figure 6. FESEM of P-AC and O-AC.

After the surface modification with the ionic liquid, no apparent damage to the structure is observed. The surface roughness has gradually been improved from ionic liquid treatment which is associated with the introduction of oxygen-containing functional groups. Furthermore, the amorphous nature of activated carbon has been approved by the introduction of small and larger pores in the structure after functionalization. These microspheres are believed to be responsible for the high specific surface area and total pore volume. O-AC showed a larger number of pores in their structures which implies the fact that other than preparation method, the morphology of activated carbon strongly depends upon the oxidizing agent used [14].

3.7. Adsorption Study

3.7.1. Adsorption kinetic study. Adsorption kinetic models are used to evaluate the rate of sorption mechanism and the efficacy of the sorbents for the removal of hydrophobic contaminants. In this study, the adsorption data of benzene by O-AC were fitted by three well known kinetic models i.e. pseudo-first-order, pseudo-second-order and intra-particle diffusion models.

Table 2. Parameters of kinetic models obtained of this study.

		Pseudo first order			Pseudo second order			Intra particle diffusion		
Sample	q_{exp} mg/g	$q_{e(cal)}$ mg/g	K_1 1/hr	R^2	$q_{e(cal)}$ mg/g	K_2 g/mg.hr	R^2	K_{ID} g/mg.hr$^{1/2}$	C mg/g	R^2
P-AC	87.5	155.85	0.0294	0.905	88.50	0.0140	0.998	18.76	82.69	0.9865
O-AC	184.00	150.66	0.048	0.9217	181.82	0.0644	0.9954	11.24	132.29	0.8289

It is clear from table 2 that pseudo second order model best suited for the sorption of benzene by O-AC. Moreover, the equilibrium adsorption capacity calculated by using this kinetic model is rationally comparable to sorption capacity attained from the experiments.

3.7.2. Adsorption Isotherm study. In order to design the adsorption system, adsorption isotherms are used. Adsorption equilibrium data of benzene using pure and O-AC was fitted with Langmuir, Freundlich, and D-R isotherm model.

Table 3. Parameters of isotherm models obtained from this study.

	Langmuir Isotherm			Freundlich Isotherm			Dubinin–Radushkevich isotherm			
Sample	Q_{max}	K_L	R^2	K_F	1/n	R^2	K_{DR}	q_s	E	R^2
P-AC	89.29	0.0102	0.9929	1.63	1.0731	0.9706	63.05	79.58	0.0890	0.9813
O-AC	188.68	0.0053	0.9941	11.82	4.1427	0.9155	145.83	171.16	0.0586	0.8961

It is evident from the table 3 that Langmuir model best describes the monolayer adsorption with maximum adsorption capacity of 188.68 mg/g while Freundlich model provides information about heterogeneous adsorption on adsorbent surface.

4. Conclusion

Commercial activated carbon is modified with 1-Hexyl-3-methyl imidazolium bis TFSI. As a result of surface modification high surface area and pore volume with a wide distribution of porosity are obtained within micro and mesopore ranges. O-AC exhibited the promising performance under the suggested modification route. Subsequently, this functionalization resulted in the introduction of oxygen-bearing complexes over the surface of graphitic carbon network structure which ultimately resulted in increased sorption capacity of modified AC. The obliteration of benzene was carried out in batch experiments.

Results demonstrated that the higher removal of benzene with O-AC occurs under similar the experimental conditions as with that of O-AC. Kinetic studies revealed that adsorption of benzene obeys a pseudo-second-order model. Adsorption isotherms study indicated that Langmuir isotherm models demonstrate very good fit with experimental data as 89.29mg/g and 188.68 mg/g for P-AC and O-AC respectively.

Acknowledgement
The authors gratefully acknowledge the financial support by Universiti Teknologi PETRONAS in the form of URIF project (0153AA-G19).

References

[1] Anjum H, Shamim M, Johari K and Murgesan T 2018 Adsorption of Benzene by "Green" functionalization of Montmorillonite *MATEC Web of Conferences* **150** p 2001
[2] Anjum H, Cheemat F, Gnanasundaram N, Arunagiri A and Murugesan T 2017 Impact of Surface Modification of Activated Carbon on BTEX Removal from Aqueous Solutions: A Review, *Air, Gas, and Water Pollution Control Using Industrial and Agricultural Solid Wastes Adsorbents,* Routledge pp 293–3123
[3] Anjum H and Murugesan T 2016 Effect of Functionalization Condition on Characterization of Carbonaceous Adsorbent *Procedia Engineering* **148** pp 1346–50
[4] Palomar J 2009 Adsorption of ionic liquids from aqueous effluents by activated carbon *Carbon* **47** (7) pp 1846–56
[5] Holbrey J D, Reichert W M and Rogers R D 2004 Crystal structures of imidazolium bis (trifluoromethanesulfonyl) imide 'ionic liquid'salts: the first organic salt with a cis-TFSI anion conformation *Dalton Transactions* **15** pp 2267–71
[6] Lu C, Su F and Hu S 2008 Surface modification of carbon nanotubes for enhancing BTEX adsorption from aqueous solutions *Applied Surface Science* **254** (21) pp 7035–41
[7] Phan N H 2006 Production of fibrous activated carbons from natural cellulose (jute, coconut) fibers for water treatment applications *Carbon* **44**(12) pp 2569–77
[8] Davis W 1999 Quantitative Fourier Transform Infrared spectroscopic investigation humic substance functional group composition *Chemosphere* **38**(12) pp 2913–28
[9] Tsai C and Chen C 2003 Characterization of bias-controlled carbon nanotubes *Diamond and Related Materials* **12**(9) pp 1615–20
[10] González P and Pliego-Cuervo Y 2013 Physicochemical and microtextural characterization of activated carbons produced from water steam activation of three bamboo species *Journal of analytical and applied pyrolysis* **99** pp 32–39
[11] Warren B 1941 X-ray diffraction in random layer lattices *Physical Review* **59**(9) p 693
[12] Carrott P 2001 Preparation of activated carbon fibres from acrylic textile fibres *Carbon* **39**(10) pp 1543–55
[13] Oberlin A, Bonnamy S and Lafdi K 1998 Structure and texture of carbon fibers *Carbon Fibers* 3rd Ed. Marcel Dekker, Inc, 270 Madison Avenue, New York, NY 10016, USA pp 85-159
[14] Tran H N, You S -J and Chao H -P 2017 Fast and efficient adsorption of methylene green 5 on activated carbon prepared from new chemical activation method *Journal of Environmental Management* **188** pp 322–336

Reduced tomography to investigate inhomogeneous solid in multiphase column

M F A Rahman, E M F E Chik, H Ithnin, L Dahing, M R Shari and H Hassan

Plant Assessment Technology Group, Industrial Technology Division
Malaysian Nuclear Agency, Bangi, 4300 Kajang, Selangor.

Abstract. Reduced tomography analyzed the data of cross section which no rotation and translation of the detectors and radioactive source. The main objective is investigating the industrial process in term of mechanical and process problems such as formation of catalyst/solid in multiphase packed bed column. A diameter of 0.2 m packed bed column was used in this study. The gas velocities were varied to get some movement of the particles. Multiple scanning was conducted using Industrial Process Computed Tomography instrumentation. However, only three detectors were used to collect and analyzed the data. The results were compared with other portable instruments with one source and three detectors manually. In addition, Monte Carlo modelling was conducted to get better results comparison. The results were successfully diagnostic any mechanical, and process problem such as maldistribution or channelling happen in the packed bed column.

1. Introduction

Industrial Process Computed Tomography usually using a lot of images to analyze the investigation process. The systems for computed tomography (CT) used many rotation and translation of the detector and radioactive source to get best image resolution. Many hydrodynamics studies have conducted to various types of column or pumps using non-invasive measurement techniques such as Gamma-ray CT to analyze gas hold distribution [1-3]. However, the CT systems take a lot of time, cost expensive and safety consideration. Industrial Computed tomography for process and mechanical diagnostic rarely used in refinery plants due to these shortcomings. A lot of factors such as column size, materials and height of the column obstruct the using of CT in the plants. Generally, the CT system is fixed and not yet has portable systems.

In industrial practice of gamma scanning for the packed bed column, multiple orientation of the scanning can determine any process and mechanical problems. It is also time consuming, many scanning orientation and radiation safety exposure. To overcome the shortcoming of CT and industrial practice of gamma scanning, the Reduced Gamma-ray Computed Tomography (RGCT) was introduced. RGCT can be defined as using the smallest number of scans and scanning orientation, less time of scanning and no movement of detector and source for translation and rotation. The ideas are to apply the concept of CT to any column in plants. This research aims to minimize the CT system operation and to overcome to industrial problems.

2. Experimental setup

Industrial Computed Process Tomography consists of a radioactive source of Cesium-137 (2.4 mCi) and 9 NaI scintillation detectors (figure 1). The inner diameter and height of the fluidized bed column are 0.2 m and 1.14 m, respectively. A gas inlet of 0.02 m diameter in the bottom part of the column used to supply the compressed air. The column was packed with glass bead of 0.003 m diameter and bio ball of 0.0385m as to mimic as catalyst in usual plant operation condition for multiphase system.

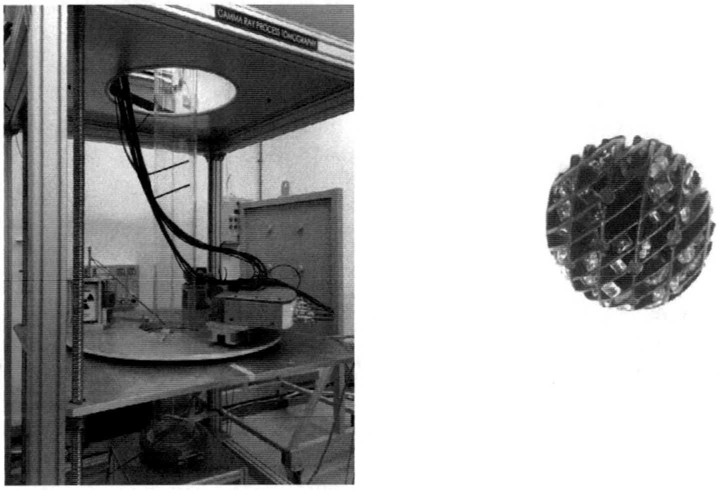

(a) (b)

Figure 1. (a) Industrial Computed Tomography System, (b) bio ball sample

The experiment was conducted to investigate distribution of the solid and gas holdup inside the multiphase column. Therefore, a range of gas velocities was applied to the fluidized bed. A multiple bio ball used to represent any corroded and degraded catalyst.

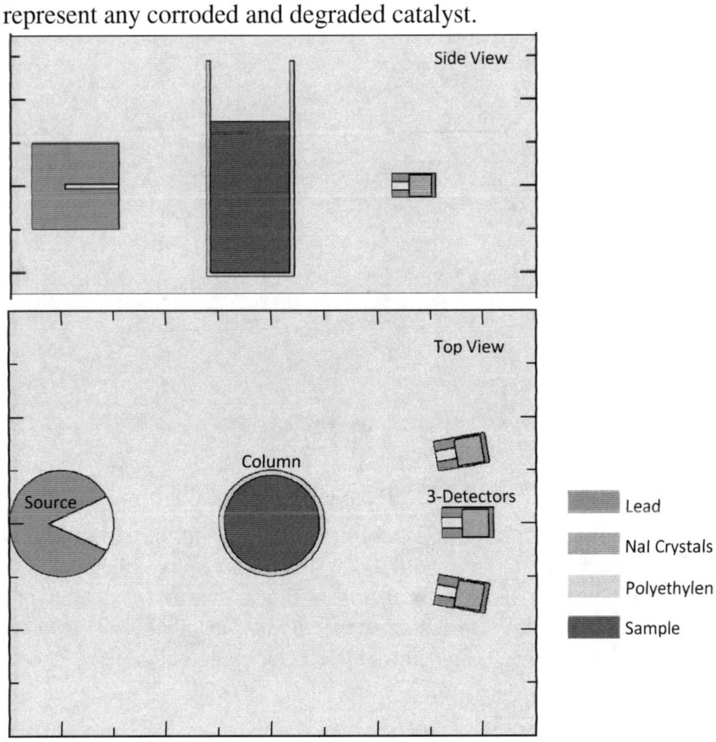

Figure 2. Model geometry of Industrial Computed Tomography for simulation study.

Multiple scans were conducted to represent of cross section data. However, only detector 3, 5 and 7 data are analyzed to represent the concept of RGCT. Graph was plotted to determine any process and mechanical problems. A series of comparison was conducted with portable systems, Gamma Spider systems and Monte Carlo simulation system. Gas holdup calculation is a subtract between static radiation counts and dynamics systems which involved the change of gas flow rate.

For comparison with simulation, Photon and Heavy Ion Transport System (PHITS) [4], code which implemented the Monte Carlo method is used. The geometry for the simulation is setup as in figure2. Simulation is calculating the gamma photon count at the three detectors from source pot through the column. In the simulation the material density inside the column is very from 0 to 8 g/cm^3 for simulating empty column at 0 and increasing density which represent different condition inside the column.

3. Results and discussions

Figure 3 demonstrates the results of the gas holdup versus gas flow rate. It was shown that at the beginning the gas holdup increases with the increase in gas flow rate. However, the gas holdup starts to decrease at the gas flow rate 180 SCFH for all detectors. This phenomenon can be described as the bio ball will always block the radiation signal as a result decrease the intensity counts.

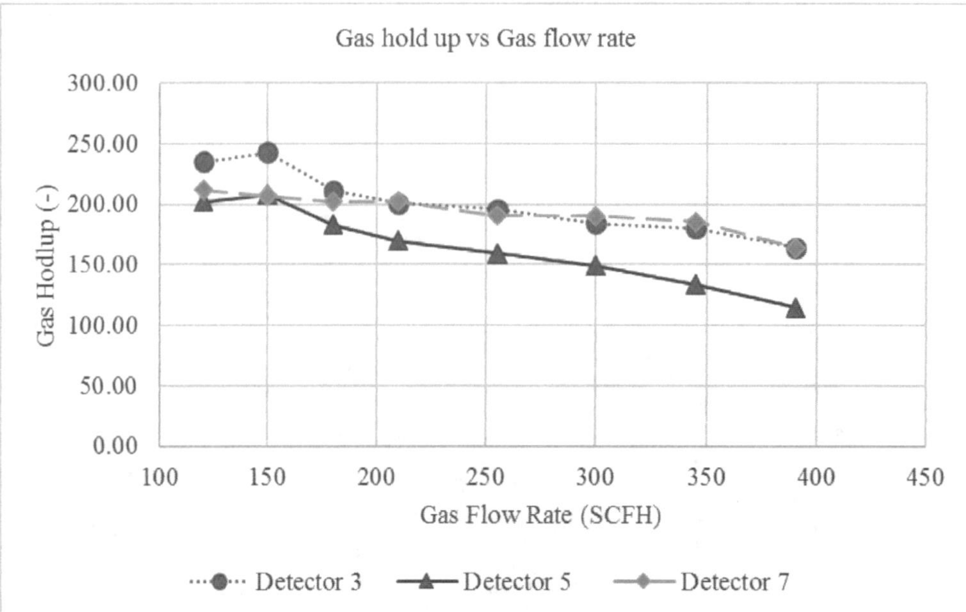

Figure 3. Gas Holdup vs Gas Flow Rate.

The gas holdup appeared some non-homogeneous solid distribution was observed during the operations. The findings can be compared with other instruments to get solid conclusions. Many factors involved of the findings such as the size of the solid, gas distributor design and columns sizes. Figure 4 shows the ideal solid distribution inside the column. The images were taken from Gamma Spider Instrumentation. Further analyses can be conducted from data collection.

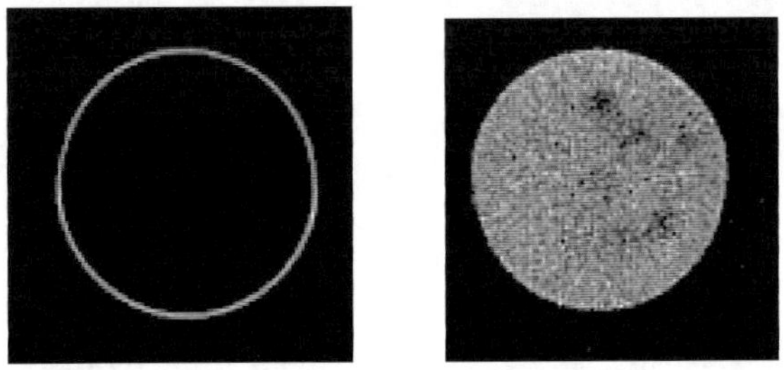

(a) (b)

Figure 4. (a) Empty column; (b) image of column with bio ball and glass bead.

Figure 5 shows the graph of photon counting at the three detectors namely, det-3, det-5, and det-7 versus density of material inside the column. The density is increased from 0 g/cm3 which represent empty column, and 1 g/cm^3 to 8 g/cm^3 to represent different condition inside the column. It can be seen the decreasing exponentially trend with increasing of density of the material inside the column. This is rightly followed the density dependent of photon attenuation. It can be seen both side detector has higher photon count compare to middle detector. This is due to path of the photon trajectory through the column in shorter compared to the middle detector which travel through the diameter of the column.

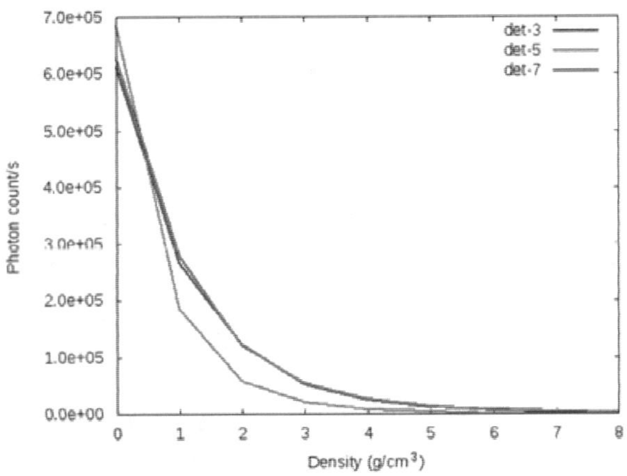

Figure 5. Graph of photon count versus density of the material inside column.

4. Conclusion

In conclusion, the experiments indicate possibility to reduce the number of detector, time and scanning orientation to investigate flow distribution of the gas inside the packed bed reactor. The output will be in graph mode of gas holdup profile in multiphase systems. The results also compared with portable instrumentation, Gamma Spider instrumentation and Monte Carlo simulation. However, further experiments should be conducted in real plant or bigger diameter process column to get best results.

Acknowledgments

The authors would like to thank the Dr. Jaafar Abdullah (retired) for developed the Industrial Process Gamma Tomography from IAEA budget. Very special thanks go to Malaysian Nuclear Agency and PAT research group for their generous support and cooperation.

References

[1] Abdelsalam E 2016 Scale-up Investigation and Hydrodynamics Study of Gas-Slod Fludized Bed Reactor using Advanced Non-invasive Measurement Techniques *Doctoral Dissertations*, Missouri Science and Technolgy Unverstiy, Rolla, Missouri, USA

[2] Schafer T, Bieberle A, Neumann M and Hampel U 2015 *Flow Measurement and Instrumentation* **46** pp. 262–267

[3] Patel A K and Thorat B N 2008 *Chemical Engineering Journal* **137** pp. 375–385

[4] Sato T, Niita K, Matsuda N, Hashimoto S, Iwamoto Y, Noda S, Ogawa T, Iwase H, Nakashima H, Fukahori T, Okumura K, Kai T, Chiba S, Furuta T and Sihver L 2013 *J. Nucl. Sci Technol.* **50** 9 pp 913-923

IOP Conf. Series: Materials Science and Engineering **458** (2018) 012058 doi:10.1088/1757-899X/458/1/012058

Investigating the effect of silver nanoparticles on carbon dioxide hydrates formation

Omar Nashed[1], Sadadine Mahamat Youssouf[1], Khalik M. Sabil[2], Azmi Mohd Shariff[1], Suriati Sufian[1], Bhajan Lal[1,*]

[1] Chemical Engineering Department Universiti Teknologi PETRONAS, 32610 Bandar Seri Iskandar Perak, Malaysia.
[2] Institute of Petroleum Engineering, School of Energy, Geoscience, Infrastructure and Society, 9 Heriot-Watt University Malaysia, No 1 Jalan Venna P5/2, Precinct 5, 62200 Putrajaya, 10 Malaysia

[*]bhajan.lal@utp.edu.my

Abstract. Gas hydrates have been considered as a promising approach for gas separation, storage, and transportation. Low hydrates formation rate has been identified as a major weakness of this technology. However, introducing hydrate promoter could accelerate the hydrates formation process. In this paper, silver nanofluids are investigated as kinetic CO_2 hydrates promoter by measuring the induction time, the initial rate of CO_2 hydrate formation, and the amount of CO_2 gas consumed at a pressure of 2.7MPa and temperature of 277.15 K. Silver nanofluids are prepared at a concentration range of 0.01 to 0.1 wt% of silver nanoparticles mixed with 0.08 SDS wt%. The experiments are conducted through high pressure reactor hydrate cell. The results show that nanoparticles enhance the initial hydrate formation rate and amount of gas consumption. However, no significant effect is observed on the induction time compared to SDS solution.

1. Introduction

The global emission of Carbon dioxide (CO_2) according to the Emission Database for Global Atmospheric Research has increased by 48% in the past two decades. CO_2 as an essential heat-trapping (greenhouse) gas, is released through human activities and natural processes such as burning fossil fuels, respiration, and deforestation [1]. There are few techniques used to capture CO_2 such as absorption, adsorption, membrane separation, and cryogenic distillation. However, these techniques have some drawbacks such as high cost, equipment faulting, solvent degradation, limited capacity, and high energy consumption [2]. Gas Hydrates or clathrates are referred to crystal structured solid that form from hydrogen-bonded water cages entraps small guest gas molecules such as methane, ethane or carbon dioxide formed at low temperature and high pressure. The guest gases are stabilized by Van der Walls force [3,4]. Gas hydrates cause a blockage in the pipeline so it is known as a serious problem for oil and gas industry [5-7]. Gas hydrates application has received huge attention recently by numerous researchers and scientists for the area of gas separation, storage and transportation. Hydrate based gas separation (HBGS) is a new technique in which CO_2 can be separated from other gases. CO_2 has different phase equilibrium than other gases allowing it to form hydrates crystal easier than other flue gases such as N_2. The advantage of hydrate based CO_2 separation is less energy consumption and the solvent used is water [8]. The solubility limitation of gases in water and the slow formation process of gas hydrates restrain the hydrate applications [8]. Therefore, chemical additives such as surfactants are used as a kinetic promoter to enhance the hydrate formation. However, no definitive solution has been reached for this problem, and more investigations are required. In 2006, Li et al. introduced copper nanoparticles as kinetic hydrates promoters [9]. Even though Nanoparticles do not affect the hydrate dissociation process notably, they reduce the induction time as well as increase the rate at which hydrates grow and storage capacity of the cages [9-12]. This ability is gained by increasing heat and mass transfer increases the solubility of gases in the water, expand contact of gas and liquid area, and supply nucleation sites for hydrate to form. Li et al. (2006), investigated the effect of copper nanoparticles on 1,1,1,2-Tetrafluoroethane used in air conditioning and refrigeration system. They found out that nanocopper have reduced the induction time of the hydrate [9]. Lee et al., studied the methane occupation of hydrates cages in the presence of TiO_2-Ag-SiO_2 and found out that the formation of methane hydrates rate is double in the presence of 1 wt% TiO_2-Ag-SiO_2 than that of methane in pure water [10]. Park et al., (2010), studied the effect of multi-walled carbon nanotube (MWCNTs) on the methane hydrate formation at concentration varies from (0.001-0.006 wt%). They reported as well that 0.004 wt.% MWCNT solution consumed the highest amount of methane as much as 300% compared to that in pure water [11]. Arjang et al., employed the silver nanoparticles as kinetic methane hydrate promoters [12]. They found that the induction time decreases by 85% and 73.9% respectively and gas consumption increased by 33.7% and 7.4% respectively. They also reported that SDS still batter than silver nanoparticle. Rahmati-Abkenar et al. also investigated triangular silver nanoparticles for methane hydrate promotion [13]. The results revealed triangular silver nanoparticles reduced the induction time up to 97% in comparison with pure water. Zhou et al. suggested that selected nanoparticles must be acid-resistant to be effective under acidic gas hydrate formation condition [14]. Hence, nanographite as kind of metalloid was studied as CO_2 hydrate promotor. The results showed that the formation of CO_2 hydrate was 80.8% faster in the presence of graphite nanoparticles than in pure water. Meanwhile the CO_2 consumption increased by 12.8%. Liu et al. reported that SDS coated Fe_3O_4 is not only able to shorten the induction periods but also increases the gas storage capacity of hydrates, particularly at low SDS concentrations [15]. They also mention that size of particle showed obvious influence on the promotion of SDS coated Fe_3O_4 to methane hydrate formation.

This paper performs stability study of silver nanoparticles in different SDS concentrations, and evaluate the performance of 0.01 to 1 wt. % silver nanofluid as a kinetic promoter for CO_2 gas hydrate formation by measuring the induction time, rate of CO_2 hydrate formation, and the amount of CO_2 gas consumed. High Pressure Reactor Hydrate Cell is used to run the experiment at temperature 277.15 K and pressure of 2.7MPa.

2. Experimental

2.1. *Materials*

Silver nanoparticles are supplied by US Research Nanomaterials Inc, with purity 99.9%, and particle size is 30-50nm. Sodium dodecyl sulfate (SDS) used to stabilize the nanofluid supplied by Merck. CO_2 gas with purity 99.9% is supplied by Air Product Sdn. Bhd.

2.2. *Apparatus and procedures*

To identify the optimum concentration of SDS used to stabilize the nanofluid, visual observation of the suspensions as well as nanoparticle size test using Brooklyn Haven 90 plus particle size analyser are used. High pressure reactor hydrate cell is used to study the kinetic of gas hydrates formation. A schematic representation of the reactor is shown in figure 1.

The High pressure reactor hydrate cell consists of a high-pressure stainless-steel vessel with an internal volume of 412 ml. The maximum working pressure for the vessel is 30 MPa. The vessel is immersed in a water bath to keep the temperature constant at the desired value. The temperature inside the vessel is monitored both in the gas phase and in the liquid phase by two thermocouples with an accuracy of ±0.01 °C. To achieve proper mixing in the liquid sample, a magnetic stirrer is placed in the vessel. The pressure inside the cell is measured with a pressure transducer.

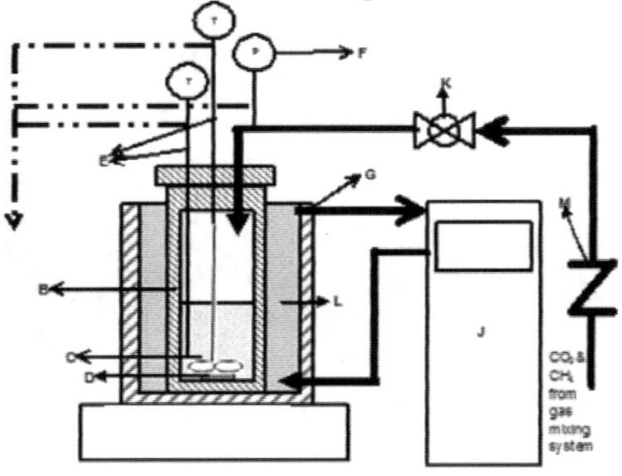

Figure 1. Pressure reactor hydrate cell schematics

Approximately 100 ml of sample is inserted into the vessel (high pressure reactor). The system then vacuumed, and the reactor is purged with gas twice to remove excess air inside of the reactor. Afterward, the reactor is submerged inside the cooling bath. The reactor is allowed to cool down to the temperature of 279.65 K that is 2 K higher than the hydrate equilibrium temperature. Finlay, sufficient CO_2 is introduced to the reactor up to 2.7 MPa. After that, the system is cooled down up to 277.15 K without stirring during the cooling period. The decrease in pressure is observed due to the dissolution of gas into the liquid mixture. When pressure regains stability and temperature of the system reached to 277.15 K, the stirrer is turned on. When hydrate formation occurs, the sudden pressure drop in the system is observed. The change in pressure and temperature is recorded every 10 seconds by data acquisition system. When the pressure of the system remains unchanged for 2 to 3 hours, this indicates that hydrate formation is completed, and the experiment is ended.

Three kinetic indexes are calculated to evaluate the silver nanoparticles as promoters. First, the induction time is taken from the starting moment of cooling up to hydrate formation indicated by the sudden drop in pressure. Secondly, the gas consumption rate is calculated using Eq (1).

$$r(t) = \frac{n_{g,\,i-1} - n_{g,\,i+1}}{(t_{i+1} - t_{i-1})} \quad (1)$$

where $n_{g,i-1}$ and $n_{g,i+1}$ are the mole number of gas in the gas phase at t_{i-1} and t_{i+1}, respectively. The amount of gas consumption is calculated using Eq (2).

$$\Delta n_{gas} = \frac{P_0 V_0}{Z_0 R T_0} - \frac{P_t V_t}{Z_t R T_t} \quad (2)$$

where P and T are the cell pressure and temperature, respectively. V is the volume of the gas inside the cell; R is the universal gas constant, and subscripts 0 is for the initial time and t is for a time equal to t. Z is the compressibility factor calculated by the Peng Robinson equation of state.

3. Results and Discussion

3.1. Stability study

The nanofluids are prepared by adding the silver and SDS to the water. Afterward, vigorous mixing applied for an hour using magnetic stirrer. Then, the nanofluid is placed in the ultrasonic bath for 60 minutes. For stability study, 0.1 wt% of silver nanoparticles are prepared with different SDS concentrations 0.03. 0.05, 0.08, and 1 wt%. At the initial state where time equal to zero all the samples look similar and no change was recorded until after two days (figure 2). The settlement increases and after seven days is shown clear that silver nanoparticle with 0.08 wt% SDS concentration is the most stable.

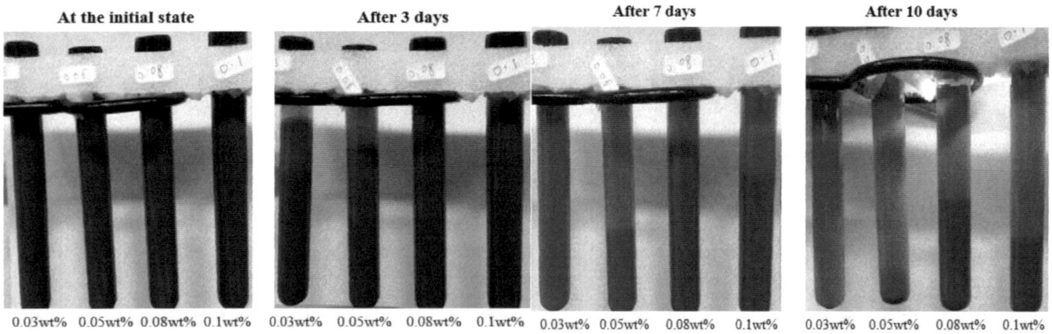

Figure 2. Stability of 0.1 wt. % Silver nanoparticles in different SDS concentrations.

Table 1. Particle Size of silver particles in different SDS concentration.

SAMPLE	Eff. Diam. (nm)
0.03SDS +0.1Ag	198.10
0.05SDS +0.1Ag	125.89
0.08SDS +0.1Ag	73.16
0.1SDS +0.1Ag	102.82

The particles size test also shows that the silver nanoparticles size is smallest at SDS concentration of 0.08 wt %. It indicates that, the nanoparticles tend to distribute better in the suspension and they do not aggregate, which results in better stability. Particle size for all different SDS concentration is shown in table 1 [14].

3.2. Kinetic promotion

After the most stable SDS concentration is determined, it used with three different silver nanoparticle concentration 0.01-0.1 wt%. The results obtained are illustrated in figures 3, 4 and 5.

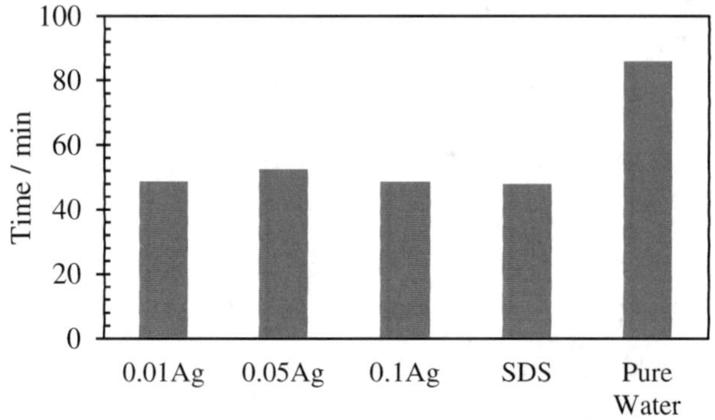

Figure 3. Induction time for 4°C Sub-cooling for pure water, 0.08 wt% SDS, and nanofluids.

Figure 3 illustrates the induction time, for pure water, 0.08 wt% SDS, and three different concentration of silver nanoparticle. It can be observed that all studied concentration have a promotional effect. In addition, there is no appreciable difference in the induction time between SDS solution and silver nanoparticles for average three experiment tests, with enhancement rate of 44% compared with water.

The initial rate of CO_2 gas consumption is plotted in figure 4. It shows that the pure water has the lowest consumption rate. 0.01 wt% Ag can enhance the initial rate by 417.4% and 15% compared with pure water and 0.08 wt% SDS, respectively. Then, it followed by SDS, 0.05 and 0.1 wt% as it shown in figure 4.

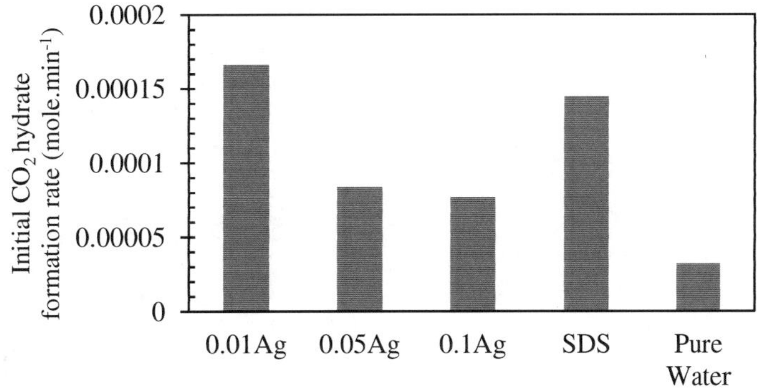

Figure 4. Initial CO_2 hydrate formation rate for pure water, 0.08 wt% SDS, and nanofluids.

The amount of CO_2 consumed is found to be the highest at a concentration of 0.1 wt% Ag. The experimental results show that the presence of Ag increased about 9.7, 1.8, and 11.4% in the amount of CO2 consumption for a period of 180 min for 0.01, 0.05, 0.1 wt%, respectively. The presence of a mixture of Ag + 0.08 wt% SDS enhance the CO2 consumption compared to 0.08 wt % SDS about 3.3 % and 5.1 % for 0.01 and 0.1 wt%, respectively.

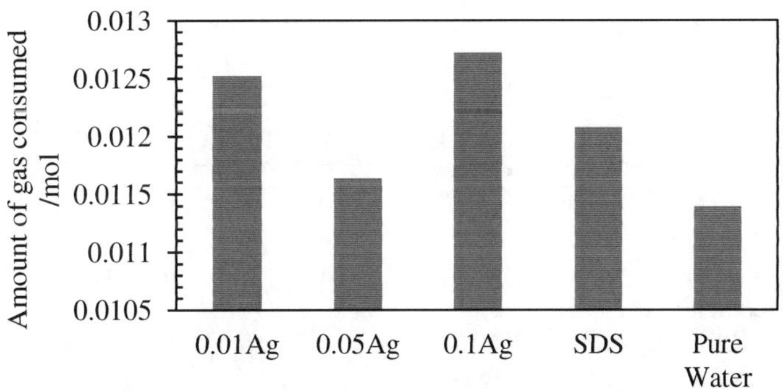

Figure 5. Amount of CO_2 Gas consumed at 180 minutes for pure water, 0.08 wt% SDS, and nanofluids.

The enchantment of gas hydrate formation by adding silver nanoparticles is attributed to its high thermal conductivity. Gas hydrates formation is an exothermic process that leads to increasing in the temperature, thus decreasing the driving force for hydrate formation. Therefore, the heat of formation can be removed from the system more effectively and much faster by using a higher thermal conductivity fluid. This will provide a more suitable temperature profile that is required for better nucleation. In addition, the presence of solid particles in the system, offer the nucleation sites for easier heterogeneous nucleation [13-15].

4. Conclusion

The silver nanofluid shows the best stability and smallest particle size at 0.08 wt % SDS. The kinetic study revealed that adding kinetic promoters shorten the induction time. However, no significant effect has been observed for Silver Nanoparticles. The initial rate of hydrates formation and amount of gas consumption are enhanced by adding silver nanoparticles. 0.01 wt% Ag enhance the initial rate by 417.4% and 15% compared with pure water and 0.08 wt% SDS. Meanwhile, 0.1 wt% show the highest gas consumption enhancement rate by 5.1 wt.

Acknowledgment
The authors thank Universiti Teknologi PETRONAS and CO_2 Research Centre for providing the facilities.

References

[1] Le Quéré C, Raupach M R, Canadell J G, Marland G, Bopp L, Ciais P, Conway TJ, Doney SC, Feely RA, Foster P and Friedlingstein P 2009 Trends in the sources and sinks of carbon dioxide *NAT GEOSCI* **2** pp 831–36
[2] Leung D Y, Caramanna G and Maroto-Valer MM 2014 An overview of current status of carbon dioxide capture and storage technologies *Renew Sust Energ Rev* **39** pp 426–43
[3] Koh C A, Sloan E D, Sum A K and Wu D T 2011 Fundamentals and applications of gas hydrates *Annual review of chemical and biomolecular engineering* **2** pp 237–57
[4] Sloan E D and Koh C 2008 Clathrate Hydrates of Natural Gases 3rd Ed. New York Marcel Dekker CRC press p 119
[5] Nashed O, Sabil K M, Ismail L, Japper-Jaafar A and Lal B, 2018 Mean induction time and isothermal kinetic analysis of methane hydrate formation in water and imidazolium based ionic liquid solutions *J Chem Thermodyn* **117** pp 147–54
[6] Nashed O, Dadebayev D, Khan M S, Bavoh C B, Lal B and Shariff A M 2018 Experimental and modelling studies on thermodynamic methane hydrate inhibition in the presence of ionic liquids. *J. Mol. Liq* **249** pp 886–91
[7] Bavoh C B, Nashed O, Khan M S, Partoon B, Lal B and Sharif A M 2018 The impact of amino acids on methane hydrate phase boundary and formation kinetics *J Chem Thermodyn* **117** pp 48–53
[8] Babu P, Linga P, Kumar R and Englezos P 2015 A review of the hydrate based gas separation (HBGS) process for carbon dioxide pre-combustion capture *Energy* **85** pp 261–79
[9] Li J, Liang D, Guo K, Wang R and Fan S 2006 Formation and dissociation of HFC134a gas hydrate in nano-copper suspension *Energ Convers Manage* **47** pp 201–10
[10] Lee J D, Kim H C, Kim Y S, Kim Y D and Lee M S 2007 Synthesis of nanosized TiO_2-Ag-SiO_2 sols by modified sol-gel method and their application for methane hydrate formation *In solid state phenomena* **124** pp 1059–62
[11] Park S S, Lee S B and Kim N J 2010 Effect of multi-walled carbon nanotubes on methane hydrate formation. *J. Ind. Eng. Chem.*, **16** (4) pp 551-55
[12] Arjang S, Manteghian M and Mohammadi A 2013 Effect of synthesized silver nanoparticles in promoting methane hydrate formation at 4.7 MPa and 5.7 MPa *Chem. Eng. Res. Des.* **91** pp 1050–54
[13] Rahmati-Abkenar M, Manteghian M and Pahlavanzadeh H 2017 Experimental and theoretical investigation of methane hydrate induction time in the presence of triangular silver nanoparticles *Chem. Eng. Res. Des.* **120** pp 325–32
[14] Zhou S d, Yu Y S, Zhao M M., Wang S L and Zhang G Z 2014 Effect of graphite nanoparticles on promoting CO_2 hydrate formation *Energy & Fuels* **28** pp 4694–98
[15] Liu, G Q, Wang F, Luo S-J, Xu D Y and Guo R-B 2017 Enhanced methane hydrate formation with SDS-coated Fe_3O_4 nanoparticles as promoters *J. Mol. Liq* **230** pp 315–321

IOP Conf. Series: Materials Science and Engineering **458** (2018) 012059 doi:10.1088/1757-899X/458/1/012059

Mixing analysis of nitrogen and hydrogen gases for ammonia synthesis via magnetic induction in monolithic channel

M Z Abdullah* and N A Mohamed Rashidi

Chemical Engineering Department, Universiti Teknologi PETRONAS, Bandar Seri Iskandar, Perak

*zamriab@utp.edu.my

Abstract. The capability of external magnetic force towards intensifying the dynamics of nitrogen (N_2) and hydrogen (H_2) in the monolithic channel is investigated through the computational fluid dynamics (CFD) approach. The ultimate aim of this study is to optimize the yield of ammonia (NH_3) in a microfluidic environment; where the reaction between N_2 and H_2 is attributed by the aid of the nanocatalyst array that are imbued onto support wires and filled inside the monolithic channel. The configuration of wire elements in the channel is significant to enhance prevalent mixing of the reactant gases, which thus increases the synthesized NH_3. Through ANSYS Fluent 15.0, the study shows that the monolithic channel that consists of 90° pitch-arranged wire elements and with less restrictive wire support at the entrance, middle and exit of the channel provides better mixing dynamics. The gases' velocity contour and its Lorentz force vector plots further indicates that the nanocatalyst should be densely grown at the locality after one-fifth the length of the channel from its entrance. The presence of magnetic field into the system provides not only to align the electron spin of the catalyst to enhance the reaction in the microfluidic environment, it also induces better prevalent mixing that implies to increasing the yield of NH_3.

1. Introduction

Ammonia (NH_3) is well recognized as a fundamental part in the industry for numerous applications such as plastic, pesticides, detergent and fertilizer. A high percentage of around 80% from the total production has been applied in the synthetic fertilizer industry, which leads to a remarkable market demand in the world of business [1]. Consumption of fertilizer has improved significantly in 2010 and is predicted to grow steadily at 2.0% annually from 2011 to 2015 [2].

The typical pathway of synthesizing NH_3 is through Haber-Bosch process. Even though the method through the route of Haber-Bosch creates huge impact on agricultural industry and the course of modern history [3], it implies to high operating conditions where such environment oblige to very cautious process system management that increases hazards and risks to the workers and plant facilities. The manufacturing of NH_3 requires high temperature (400°C - 500°C) and high pressure (150 bar – 300 bar); thus, is energy-intensive and costly in term of capital and operational expenditure, yet only able to synthesize 10-20% of NH_3 [4].

An alternative is discovered where with the aid of wires element in monolithic channel in microfluidic environment, high yield of NH_3 can be achieved at ambient condition [5]. This technique is perceived to enhance the production of NH_3 by exposing the reactant gases of N_2 and H_2 to high surface area-to-volume ratio, which have been demonstrated previously by having parallel and perpendicular obstruction to boost the dynamic mixing. The parallel and perpendicular placement leads to a favorable condition where the mixing index reaches approximately 90% with the desired ratio for the reactant gases to react [6]. As compared to simple square pitched arrangement, the possibility of dynamic mixing by having different configuration of wires support and pitch angle have been explored in this work to elevate the yield of NH_3.

Downsizing of this device has induced attractive features to the channel such as the installation time and time taken for the reaction to achieve steady state is decreased. In addition, the process safety risks and hazards may be reduced due to the process operation is at ambient conditions. By using smaller dimension channel, results shown that the yield attained and product selectivity is better [7]. The capability of this channel is noticed as it enables chemical reaction to be completed in reaction space several orders of magnitude smaller than conventional batch reactor [8].

The theory from First Law of Thermodynamics described that energy cannot be created or destroyed but can exist in different form [9]. Magnetohydrodynamic (MHD) has been recognized as a better alternative to synthesize NH_3 compared to Haber-Bosch process [10]. The important fragment of this research is to emphasize the application of conversion of magnetic energy to chemical energy instead of conversion of heat and mechanical energy to chemical energy in ammonia production. By coupling nanocatalyst with the new concept of magnetic induction, the catalytic activity and production output increased. Magnetic induction is interpreted as a response of material when a magnetic field is applied to a material. The relationship between magnetic induction and magnetic field is a property of material [11]. The mutual interaction between magnetic field, B, and a velocity field, u, arises partially as a result of the laws of Faraday and Ampere, and partially because of the Lorentz force experienced by a current-carrying body [12]. Magnetic characterization has been utilized in numerous application such as magnetic spectroscopy [13], superparamagnetic iron oxide nanoparticles loaded red blood cell [14], permanent magnet induction generator [15], magnetic ordered mesoporous carbon adsorbent [16].

Mixing plays an important role to elevate the reaction of N_2 and H_2 to produce NH_3. Primarily the significant is to reduce the inhomogeneity that leads to secondary effects such as undesired reaction and change in properties [17]. Great flow dynamic is preferred in a channel if intimate mixing is anticipated between two components [18]. Without alteration of geometry, mixing is hardly being accomplished in minute-scale system. Several patterns of channel were investigated formerly to address the significant of mixing such as serpentine-shaped channel [19], linked-twist map barriers [20], floor-grooved channel [21] and rhombic micromixer [22]. In this project, the simulation is done through configuring the wires and wires support assembly in parallel and perpendicular arrangement respectively inside the channel with the aid of external magnetic field that will induce dynamic mixing via chaotic advection to increase the efficiency of NH_3 production.

2. Methodology

The monolithic channels are designed using ANSYS Fluent 15.0 software package, where three configurations of geometries are developed via Design Modeler (figure 1). The simulation proceeds with mesh generation where discretization of elements for the flow domain is produced at dissimilar relevance centres. The flow properties are defined as in table 1, where the value of mole fraction, temperature, magnetic field and velocity were used to simulate the flow dynamics of the reactant gases. The analyses of mixing dynamics are done through Fluent-Post, where contour plots and flow profiles are investigated accordingly.

2.1. Geometry Development

Figure 1 (a) illustrates the monolith channel that consists of the outer channel (unshaded circle area) and the supporting wires array along its axial direction (shaded area). The outer channel has a dimension of 10 mm in diameter and 50 mm in length, while each of the wires in the assembly is of 1-mm diameter. The wires assembly begins at a distance on 3 mm from the inlet and ends at 3 mm prior to outlet. The distance between each wires to another in the assembly is 2 mm, with either 90° or 60° pitch, respectively. The wires are arranged in the longitudinal axis (YZ axis) in the channel with three supports that are located at the inlet, centre and outlet of channel to hold the wires in its position. Three configurations of geometries are developed with different parameters to observe the designs that could provide better flow dynamics of N_2 and H_2 gases throughout the monolithic channel. As the wire elements are in parallel position, there is no change in wires path nor intersection of the wires.

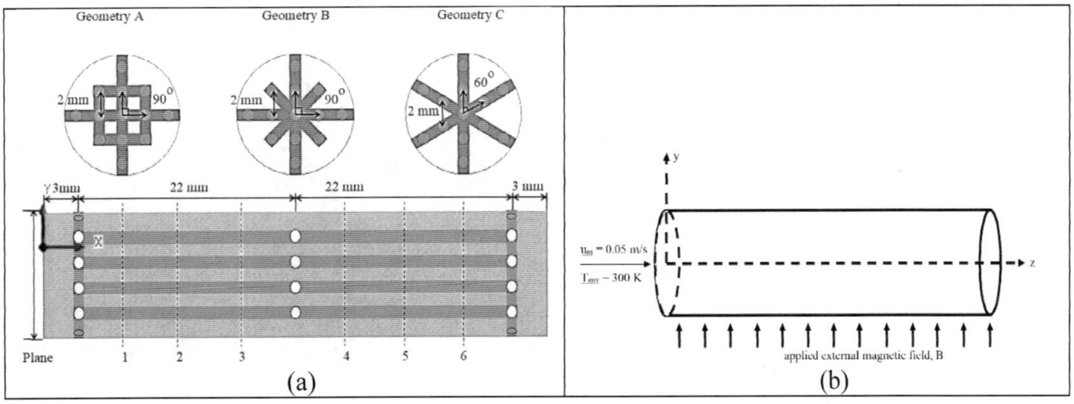

Figure 1. (a) General dimension of monolithic channel in radial and axial view. (b) Applied external field at x-direction.

2.1.1. Magnetic Induction Method. The process of analysing results is solved via ANSYS Fluent MHD Module where the magnetic induction method is applied in the simulation process. A laminar viscous model coupled with discrete phase model is applied. Conducting wall boundary conditions have been chosen for MHD boundary condition. No-slip boundary condition is also selected at monolithic channel wall, while the applied external magnetic field is chosen at constant value of 1.2 T at Y direction [24].

Table 1. Flow parameters setup.

Parameter	**Flow Properties**
Gas Inlet Velocity	0.05 ms^{-1}
Volume Fraction	N_2:H_2 = 0.25:0.75
T and P	25°C, 1 atm
MHD Model	Magnetic Induction, DC Field
Magnetic Force	1.2 T
Discrete Phase Model	Continuous Phase
Viscous Model	Laminar

2.1.2. Flow Properties Setup. Table 1 illustrates the flow properties that are selected for the CFD simulation. It is essential to highlight that the simulation is in a 'non-reactive mode', by which the actual reaction between N_2 and H_2 that is supposed to produce NH_3 is not considered. Hence, this study is purely fluid dynamics with no reaction involve.

3. Results and Discussion

The velocity of both nitrogen and hydrogen gases are analysed in the form of contour plots in radial (XY) and axial (YZ) directions, shown in figure 2 and figure 3, respectively. Each radial arrangement consists of six equidistance planes to observe the mixing dynamics of both gasses for Geometry A,

Geometry B and Geometry C. The planes are located at the position of 5, 11, 17, 27, 33 and 39 mm from the inlet of the monolithic channel, respectively. The magnitude of Lorentz force, on the other hand, is illustrated in isometric view as vector plot shown in figure 4.

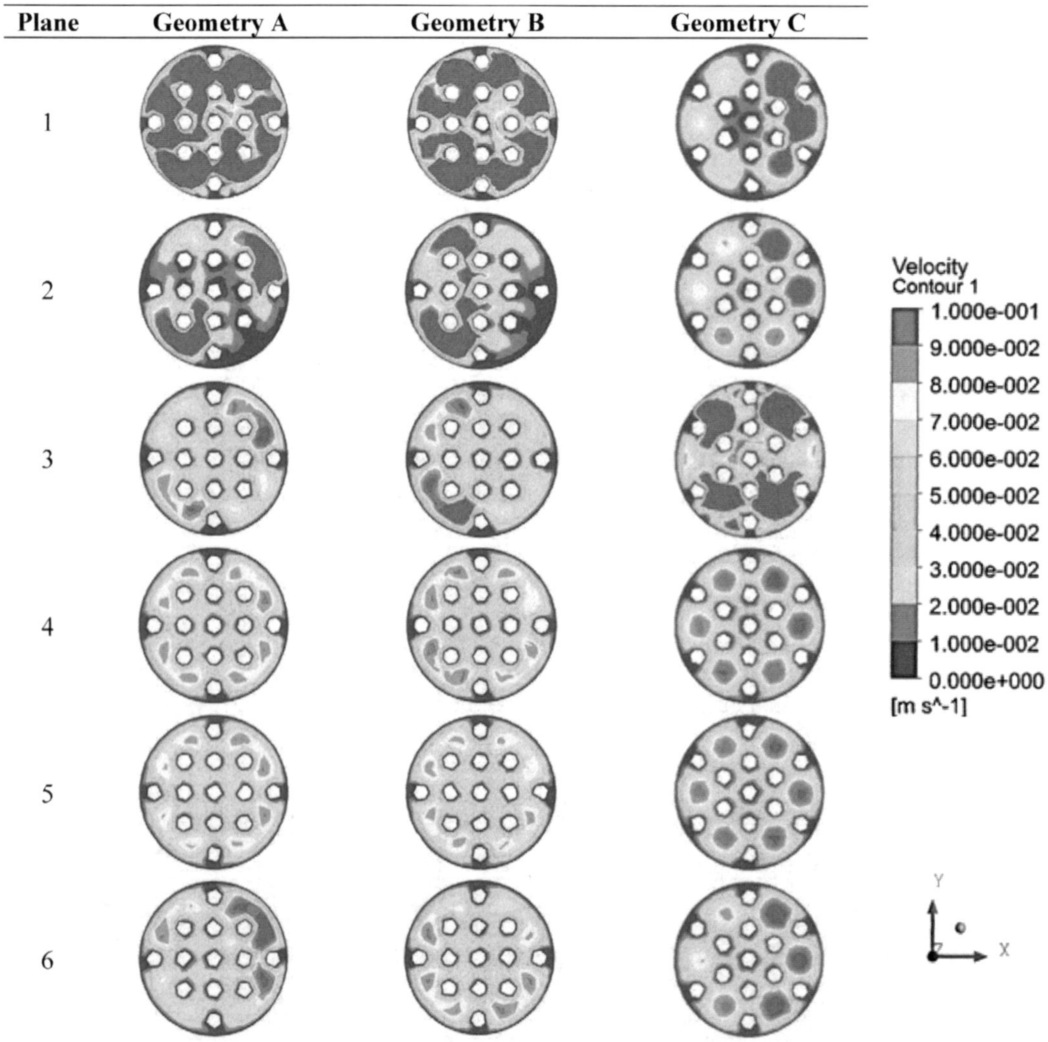

Figure 2. Velocity contours for geometries A, B and C at XY-axis.

Figure 2 displays the profiles of the gases velocity in radial direction. An active dynamic mixing is anticipated when velocity contours agree closely to the set value of 0.05 m/s. Analysing the contours, Geometry A and B illustrated nearly similar dynamic behaviour throughout the channel, albeit slight higher velocities are seen at some portion within the vicinity of the wall. This is mainly ascribed to the similarities of wire elements configuration that both geometries possess, where they are arranged in the same 90°-pitch to one another.

The gases circulating near the wall also speed at higher magnitude compared to the central radial core of the channel, which is due to the external magnetic field applied at Y-direction [25]. At Plane 1, Geometry A presented faster moving gas due to the greater restrictive flow area subjected by the preamble square-shaped support at the entrance section. At Geometry C, reactant gases flow at higher speed at the outer core of the channel compared to the inner radial side and continues throughout the longitudinal direction of the flow.

Through these observation, it is implied that the placement of the nanocatalyst that would be supported on the wire elements should be optimized starting from the distance of 11-mm onwards from the entrance (begins from Plane 2) in order to enhance higher reaction for the yield of NH_3. In addition, nanocatalyst should either be densely packed at the central core of the channel compared to its near-wall due to its lower velocities, or be grown at taller height than the rest of the channel to induce more collision between reactant gases and the nanocatalyst that will increase the synthesis of NH_3.

In figure 3, velocity of N_2 and H_2 can be visualized at an altered angle where the contour is obtained at the point Z = 0 on YZ-axis, showing the axial centreline of the movement for both gases. High velocity of gases are seen at the inlet and outlet of the channel as there are no wire elements at the said area.

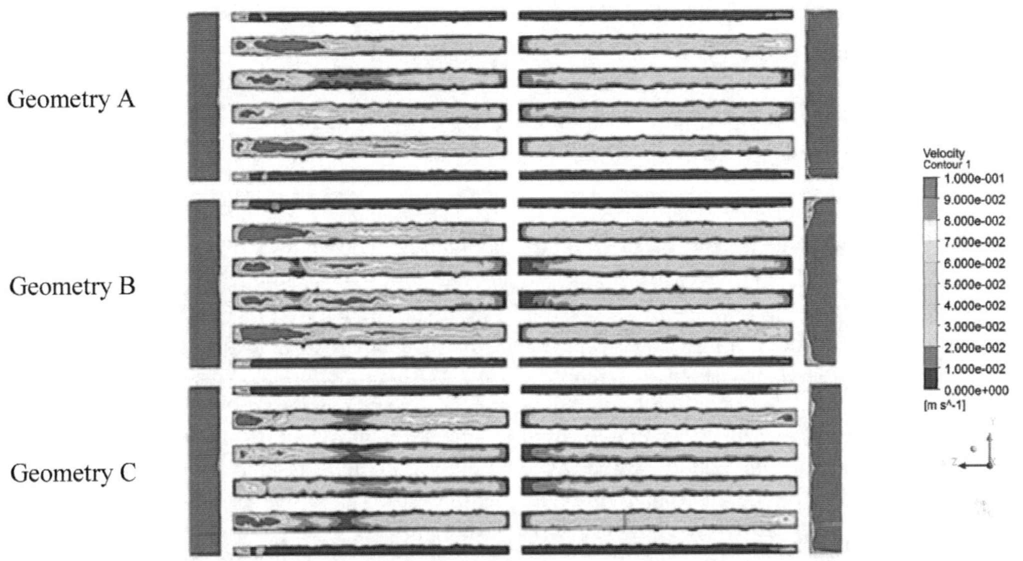

Figure 3. Velocity Contours for geometries A, B and C at YZ-axis.

As the gases pass through the first wire supports, larger boundary length is required in order the velocity to settle towards the set value of 0.05 m/s, hence they are showing irregular contour up to 11 mm from the inlet (upon reaching the analysed plane 2). Once surpassing this mixing length, the gases show good mixing dynamics where the mixture velocities lie in the range of 0.04 m/s to 0.05 m/s, within the region of the set value.

This implies that greater dynamics would be achieved once the gases flow after more than one-fifth the length of the monolithic channel, which further purported to the possibility of optimizing the dense-packed nanocatalyst to be imbued onto the wire element from this locality onward in order to enhance the yield of NH_3.

The purpose of magnetic induction is to align the electron spins of the catalysts, where this method has produced ammonia gas when reaction is done in the electromagnetic field (produced by Helmholtz coil) [5]. The possibilities of having better mixing process attributed by external magnetic induction is analyzed through the vector plot shown in Figure 4. The resemblance of the Lorentz force vectors of the three geometries throughout the monolithic channel length implies that the magnetic force impacted more towards the free spaces within the channel that allows gases to flow. This is given by the green-colored vector that shows the flow domain that is not restricted by the wire support, whereas the blue-colored indicates the wire support at the entrance, middle and exit section of the channel.

Due to the configuration of Geometry B that has less restrictive wire support at its three axial locations (YZ axis) compared to the other two geometries, a slight higher force is exerted on this geometry instead.

This is proven by the denser and thicker vectors shown in figure 4 for Geometry B along the channel when it is induced by 1.2 T magnetic field. This further suggests that dynamic mixing would occur much greater in this geometry and should be the one to be utilized in order to enhance the yield of NH_3.

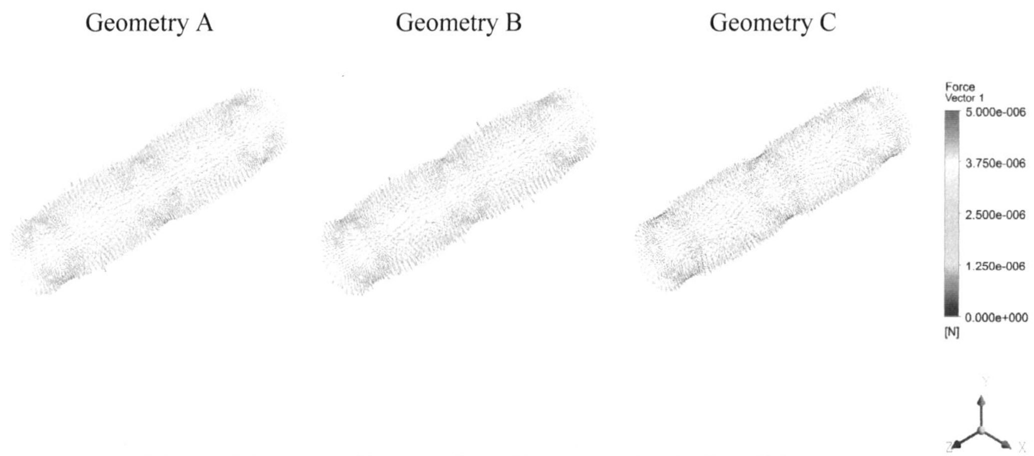

Figure 4. Vectors of Lorentz force for geometries A, B and C.

4. Conclusions

The aid of magnetic element in a minute-scale monolithic channel with modified wires arrangement were configured using laminar viscous, discrete phase and MHD model via ANSYS Fluent simulation. Geometry B, which consist of 90°-pitch arranged wire elements and has less restrictive wire support at the entrance, middle and exit of the channel provides better mixing dynamics as shown in the gases' velocity contour plots and its Lorentz force vector plot. From this arrangement, the placement of nanocatalyst that will be imbued onto the wire would be optimized to be densely packed and grown at the distance from 1/5 the length of the channel and onwards in order to enhance the synthesis of ammonia in a microfluidic environment.

Acknowledgement

The authors would like to acknowledge Universiti Teknologi PETRONAS for the provision of computational facilities, and the Ministry of Higher Education for funding the project.

References

[1] Egenhofer C and Schrefler L 2014 For a Study on Composition and Drivers of Energy Prices and Costs in Energy Intensive Industries *The Case of The Chemical Industry-Ammonia* pp 19–20
[2] FAO, Food and Agriculture Organization of the United Nations Rome 2011 *Current World Fertilizer Trends and Outlook to 2015*
[3] Erisman J W, Sutton M A, Gallowy J, Klimont Z and Winiwarter W 2008 *Nature Geoscience* **1** pp 636–39
[4] Puspitasari P, Razak J A and Yahya N 2012 *American Institute of Physics* **1482** pp 605–10
[5] Yahya N, Puspitasari P, Koziol K, Mohd Zabidi N A and Othman M F 2010 *Journal of Basic and Applied Science* **10** pp 60–64
[6] Abdullah M Z, Kashif S and Tan L S, 2014 *Characterizing the Mixing Dynamics of N_2 and H_2 Gases in a Monolithic Microchannel Impacted by the Wires Assembly*, unpublished
[7] Kashif S, Abdullah M Z and Ku Shaari K Z 2016 *Procedia Engineering* **148** pp 1266–73
[8] Asano Y, Togashi S, Tsudome H, Murakami S 2010 *International Society for Pharmaceutical Engineering* **30** pp 1–9
[9] Nguyen N and Wu Z 2005 *Journal of Micromechanics and Microengineering* **15** pp 1–16

[10] Struchtrup H 2014 *Thermodynamics and Energy Conversion* (London: Springer-Verlag Berlin Heidelberg)
[11] Yahya N, Puspitasari P and Noordin N H 2013 *Defect and Diffusion Forum* **334-335** pp 329–36
[12] Spaldin N A 2011 Magnetic Materials: Fundamentals and Applications, 2nd. ed., (United Kingdom: University Press, Cambridge)
[13] Davidson P A 2001 An introduction to Magnetohydrodynamics (Cambridge University Press)
[14] Weaver J B and Rauwerdink A M 2010 *The Effect of Molecular Binding on the Phase of MSB Measurements* pp 19–25
[15] Markov D, Boeve H, Gleich B, Antonelli A, Sfara C and Magnani M 2010 SPIO *Nanoparticles Encapsulation into Human Erythrocytes for MPI Application* pp 26–31
[16] Sharma P, Bhatti T S and Ramakrishnan K S S 2011 *Journal of Engineering Science and Technology* **6** (3) pp 332–338
[17] Nejada N F, Shamsb E and Amini M K 2015 *Journal of Magnetism and Magnetic Materials* **390** pp 1–7
[18] Mills P L, Quiram D J and Ryley J F 2007 *Chemical Engineering Science* **62** pp 6692–7010
[19] Liu R B, Stremler M A, Sharp K V, Olsen M G, Santiago J G, Adrian R J, Aref H and Beebe D J 2000 *Journal of Microelectromechanical Systems* **9** pp 190–197
[20] Wang R, Lin J and Li H 2007 *Chaos, Solitons and Fractals* **33** pp 1362–66
[21] Zhang Z, Yim C H, Lin M and Cao X 2008 *Biomicrofluidics* **2** pp 1–9
[22] Chung C K and Shih T R 2007 *Microfluid Nanofluid* **4** 4 pp 19–425
[23] Nurrahman R, Shuib A S, Abdullah M Z and Ku Shaari K Z 2013 *Science and Engineering Research* pp 369–373
[24] Alqasem B, Yahya N, Qureshi S, Irfan M, Rehman Z U and Soleimani H 2017 *Material Science and Engineering: B* **217** pp 49–62
[25] Altintas A and Ozkol I 2014 *Journal of Applied Fluid Mechanics* **8** pp 507–14
[26] Ranganath G 2002 *Mysterious Motions and Other Intriguing Phenomena in Physics* (India: Sangam Book Ltd)

Semi-clathratic impact of tetrabutylammonium hydroxide on the carbon dioxide hydrates

Khor Siak Foo[*1,2], **Muhammad Saad Khan**[*2,3], **Bhajan Lal**[*2,3], **Suriati Sufian**[2]

[1]Murphy Sarawak Oil Co., Ltd., Tower 2, Petronas Twin Towers, Kuala Lumpur City Centre, 50088, Kuala Lumpur.
[2]Chemical Engineering Department, Universiti Teknologi of PETRONAS, Bandar Seri Iskandar, 32610, Perak, Malaysia.
[3]CO_2 Research Centre (CO_2RES), Bandar Seri Iskandar, 32610, Perak, Malaysia.

*khor_siak_foo@murphyoilcorp.com; *khansaaad@gmail.com; *bhajan.lal@utp.edu.my

Abstract. In the present experimental study, the phase behaviour of Tetrabutylammonium hydroxide (TBAOH) on Carbon dioxide (CO_2) hydrates is evaluated via the T-cycle method. All the experiments performed in high-pressure hydrate reactor for various concentrations of TBAOH (1, 5 and 10 wt%) at moderate pressure and temperature ranges of 2.0-4.50 MPa and 276.0-285.0 K, respectively. The hydrate dissociation enthalpy (ΔH_{diss}) is also calculated for all the studied systems via Clausius Clapeyron equation. Results revealed that the TBAOH able to shift the hydrate liquid vapour equilibrium (HL_wVE) curve towards higher temperature and lower pressure regions hence provide the promotional impact on all the studied concentrations. 10 wt% TBAOH able to heighten hydrate promotional impact ($\Delta \bar{T}$) up to 1.15 K. Additionally, enthalpy (ΔH_{diss}) data revealed that the promotional impact of TBAOH attributed due to the semi-clathratic nature of TBAOH. Thus, it can be concluded that the studied TBAOH in the present work has the immense potential for the industrial applications like natural gas separation and gas storage.

1. Introduction

Gas hydrates are non-stoichiometric crystal-like inclusion compounds formed under thermodynamically favourable conditions and stabilized via hydrogen bonding network of water. In gas hydrates, the water molecules are acted as host molecule while smaller molecules gases like methane (CH_4), ethane (C_2H_6), Carbon dioxide (CO_2) and others are entrapped (as a guest molecule) within the hydrate structures. Usually, there are three types of hydrate structures are often found which known as structure I (sI), structure II (sII) and structure H (sH) hydrates [1]. The type of hydrate structure mainly depends upon the presence of guest molecules together with thermodynamic condition accounted for hydrate formation.

Earlier, gas hydrate often saw as a perennial problem in flow assurance. The mitigation of gas hydrates accounted for almost 70 percent of all the flow assurance problems which leads towards the spending of 100s of Million dollars ($) without any permanent solutions [2]. Therefore, much research on gas hydrates focused on mitigation of gas hydrate via conventional [3] and unconventional inhibitors [2,4–6]. The conventional thermodynamic hydrate inhibitors include alcohols like methanol, ethylene glycol together with conventional salts for instance sodium chloride (NaCl) and potassium chloride (KCl) [7]. The unconventional inhibitors include various novel chemicals that possess relatively superior environmentally friendly properties especially lower vapour pressure. The prime examples are amino acids [4] and ionic liquids [4,8–10].

Recently, researchers receiving more attraction on developing novel hydrate based technologies, applicable to the energy and environmental fields. Since hydrate formation is the tedious process, which involve high pressure, low temperature alongside lower rate of hydrate formations that causes hindrances for their potential applications. Therefore, for the numerous applications of gas hydrates, thermodynamic gas hydrate promoters (THP) are inserted to decrease the equilibrium hydrate formation pressure and to increase the hydrate formation temperatures [11]. Shahnazar and Hasan [12] intensely reviewed the formation studies of gas hydrate promoters to evaluate the emergence of the novel applications of gas hydrate. Energy recovery, gas separation gas storage and transportation alongside seawater desalination are the potentials applications which triggered gas hydrate to be in the continual research during recent years.

Ammonium based hydrate promoters take up an essential part of the formation of clathrate and semi-clathrate hydrate, for instance, tetra-n-alkyl ammonium salts formed some unusual hydrate structures even at atmospheric pressure. Contrasting from clathrate hydrates where water atoms form hydrate cages, and small guest molecules like CH_4 or CO_2 entrapped inside the hydrate lattice structure. In semi-clathrate hydrates, the anion of halide uproots the water particle in the system, shaping Hydrogen bond with water molecules, then the tetra-n-alkylammonium cation also participated in hydrate crystalline structure [13].

Initially, Karimi et al. [14] thermodynamically evaluated the tetrabutylammonium hydroxide (TBAOH) as hydrogen (H_2) hydrate promotor for the storage of hydrogen gas. Results revealed that TBAOH could shift HL_wVE boundary towards higher temperature and pressure regions around 0-40 MPa pressures. They further reported that the promotional impact of TBAOH found to be pressure dependent as the pressure increased the promotional impact also found higher due to more hydrate enclathration. Similar hydrate promotional behaviour was also observed by the Su et al. [15] with CH_4 gas. TBAOH was able to promote CH_4 hydrate at higher-pressure conditions (4.46–16.05 MPa). Previously, Nashed et al. [16] reported the physical-chemical properties of TBAOH; they further reported that in the presence of the TBAOH, the gas uptake of CO_2 gas significantly enhanced compared to deionized water. Recently, Khan et al. [17,18], evaluated the impact of alkyl chain on the AILs and reported that TBAOH able to promote hydrate formation in CO_2 and CH_4 enrich hydrte systems.

Therefore, here the focus of this study is to report the impact of TBAOH on the HL_wVE boundaries of CO_2 hydrates at different concentrations (1, 5 and 10 wt%) for the first time. The experiments are performed at the moderate pressure and temperature range of 277 K-286 K and 2.0 MPa- 4.5 K respectively. The average hydrate promotional temperature ($\Delta Ŧ$) values of all the studied systems are also reported. The Clausius Clapeyron equation is employed to determine the enthalpy of hydrate dissociation (ΔH_{diss}) for all the studied systems. Also, an attempt is also made to define the hydrate promotional mechanism of TBAOH in the presence of CO_2 hydrates.

2. Methodology

2.1. Materials

The table 1 provide the details of the materials used in this work. The all chemicals are employed without any further purification. The gravimetrical analytical balance is used for all the sample preparations.

2.2. Details of the Experimental Apparatus and Method

The high-pressure hydrate reactor is used in this study. The details of the experimental setup and methods can be found elsewhere [19,20]. The HL_wVE data of pure $CO_2 + H_2O$ and CO_2 + TBAOH + H_2O hydrate systems are measured via the isochoric T-cycle with step heating method. The complete details of the experimental method can also be found in others places [1,4,21,22].

Table 1. Materials used for this gas hydrates study.

No	Chemical Name	Symbol	Purity	Chemical Structure
1	Carbon dioxide	CO_2	99.95 mole %	O=C=O
2	Water	H_2O	Deionized	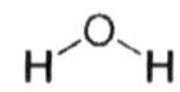
3	Tetrabutylammonium hydroxide	TBAOH	40 wt% aqueous solution	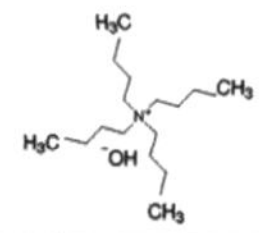

2.3. HLwVE data and Enthalpy (ΔH_{diss}) Analysis

In this study, the average promotional temperature (ΔT) is calculated via Equation 1 determined by Xia and co-workers [23].

$$\overline{T} = \frac{\sum \Delta T}{n} = \frac{\sum_{i=1}^{n}(T_{0,pi} - T_{1,pi})}{n} \quad (1)$$

where, $T_{0,\ pi}$ and $T_{1,\ pi}$ represents the hydrate equilibrium temperatures of CO_2 gas for deionized water and an aqueous solution of TBAOH respectively. The values of both dissociation temperatures attained at the uniform pressure *pi,* and *n* denotes the number of considered pressure points.

The dissociation enthalpies (ΔH_{diss}), of gas hydrates, are obtained through the Clausius Clapeyron equation (see Equation 2) which also used in preceding studies [4,24–26].

$$\frac{\partial \ln P}{\partial \frac{1}{T}} = -\frac{\Delta H_{diss}}{zR} \quad (2)$$

where T and P are the equilibrium temperature and pressure, R denotes the universal gas constant, z represents the compressibility factor of CO_2 gas (calculated via Peng Robinson Equation of state), and ΔH_{diss} represents the enthalpy of CO_2 hydrates for all the studied systems.

3. Results and Discussions

The experimental characterization and analysis of CO_2 hydrate in the presence of TBAOH is presented in this section.

3.1. HLwVE Analysis of CO_2+TBAOH+H_2O hydrates

As mentioned afore very inadequate data offered on the thermodynamic promotional impact of ILs on CO_2 hydrate systems[5]; most of the system dealt with hydrate inhibition impacts [5,27]. The phase behaviour of the various concentrations (1, 5 and 10 wt%) of CO_2 + TBAOH + H_2O system is presented in figure 1.

The acquired results expose that all the studied concentrations of TBAOH can enhance the CO_2 hydrate equilibrium curves to a higher temperature and lower pressure regions, which reveals the thermodynamic promotional impact. It can be evident from the results that the hydrate promotional influence is composition-dependent. As the quantity of TBAOH increasing from 1 – 10 wt%, the promotional impact is also increased (see figure 1). The prior AILs studies revealed that shorter alkyl chain AILs (like TMAOH, TMACl, TEAOH, TEAI and TPrAOH) were performed as hydrate inhibitors [20,21,24]. The hydrate promotional behaviour of TBAOH is observed perhaps attributed due to the longer chain of AIL. It should be noted that this promotional impact might cause due to the semiclathratic nature of TBAOH, which is evaluated in next section (see section 3.2). For further quantifying the hydrate promotional impact, the ΔŦ values of different concentrations of TBAOH are depicted in figure 2. For different concentrations of CO_2 + TBAOH + H_2O systems; 1 wt% showed the very less ΔŦ value of 0.04 K, while higher concentrations of 5 and 10 wt% are found to be enhanced the ΔŦ values up to 0.37 K and 1.15 K, respectively.

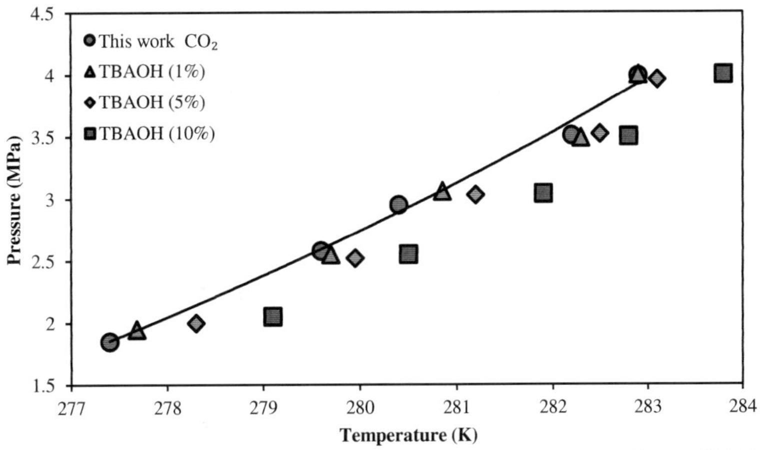

Figure 1. HL_wVE data for 1 wt%, 5 wt% and 10 wt% concentrations of TBAOH.

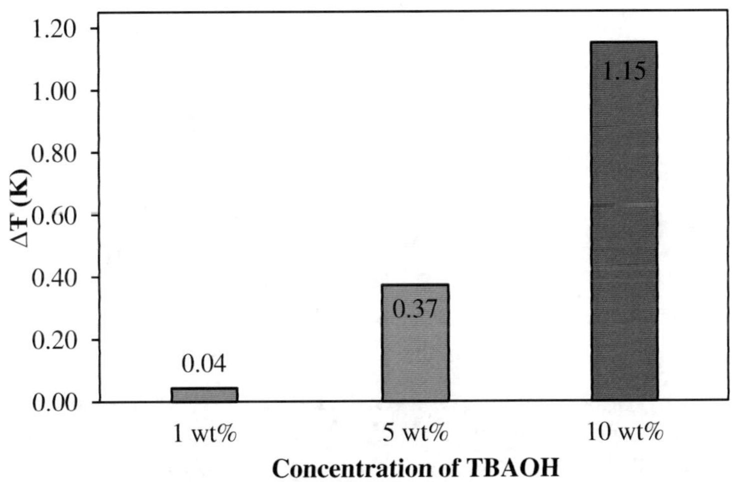

Figure 2. Average promotional temperature (ΔŦ) for different concentrations (1, 5 and 10 wt%) of CO_2 + TBAOH + H_2O systems.

3.2. Enthalpy of dissociation (ΔH_{diss}) of CO_2+TBAOH+H_2O hydrates

To investigate the semi-clathratic influence of TBAOH on the cage occupancy, the Clausius Clapeyron equation is used to calculate the ΔH_{diss} of CO_2 hydrates in the presence and absence of TBAOH. The calculated ΔH_{diss} of CO_2 +TBAOH + H_2O systems are presented in table 2.

The average ΔH_{diss} of H_2O + CO_2 is 64.73 kJ/mol lies within the range of CO_2 hydrate enthalpy data which means structure I hydrate formed [3]. Referring to the table 2, it can be observed that the enthalpies of all the studied TBAOH + CO_2 + H_2O systems appear very different to the pure CO_2 hydrate. This behaviour proposes that the presence of TBAOH have a substantial impact on the ΔH_{diss} of the system and therefore participate in the CO_2 hydrate structure and cages occupancy. Hence, semi-clathrate impact of TBAOH found on CO_2 hydrate structure which is in line with the preliminary data

[14,15]. The enthalpy results further rationalize the promotional influence of TBAOH, which predominantly comes owing to higher alkyl chain together with its semi- clathrate effect. The probable mechanism of semi-clathrate nature of TBAOH further underpinning via the structural representation of TBAOH as observed in figure 3.

Table 2. Calculated molar enthalpies of hydrate dissociation, ΔH_{diss} (kJ/mol) of CO_2 hydrate in the presence of CO_2 + TBAOH +H_2O solutions

Pressure (MPa)	TBAOH concentration in the solution			
	0 wt%	1 wt%	5 wt%	10 wt%
4.0	57.481	59.664	62.538	63.648
3.5	61.071	63.647	66.521	68.078
3.0	65.052	66.807	70.390	71.842
2.5	67.655	70.527	74.226	75.516
2.0	72.435	74.575	77.902	79.131
Overall	**64.739**	**67.044**	**70.315**	**71.643**

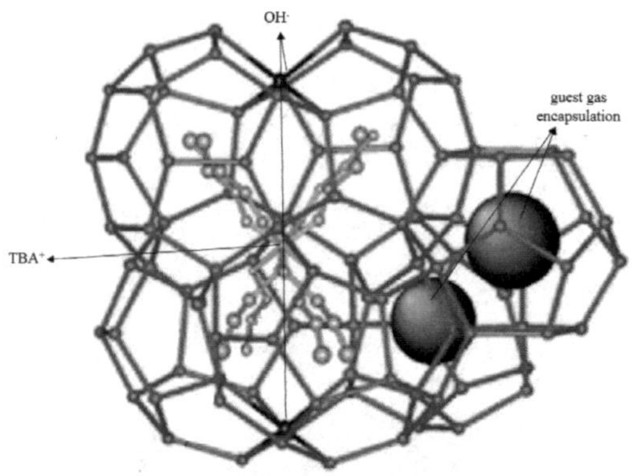

Figure 3. The graphical representation of TBAOH molecule in the semi-clathrate structure.

4. Conclusions

In this work, the HL_wVE measurement for CO_2 + TBAOH + H_2O is reported for different concentrations (1, 5 and 10 wt%). The attained results revealed that the occurrence of TBAOH increases the HLwVE phase boundary towards higher temperature and lower pressure regions. The promotional impact of TBAOH seems to be concentration dependent since 10 wt% can shift the ΔŦ value up to 1.15K. The hydrate promotional impact of TBAOH found perhaps attributed due to the presence of large cation (TBA^+) which induce semi-cathartic behaviour. Furthermore, the molar enthalpies of dissociation for the CO_2 + TBAOH +H_2O hydrate systems suggested that TBAOH actively participated in the hydrate crystalline structure and its impacts on the hydrate promotion caused due to its semi-clathratic nature. Therefore, the outcome of this study would highlight the importance of TBAOH for designing the hydrate based technological applications such as hydrate-based desalination of saline water, storage of natural gas and designing of separation processes for gas phase separations.

Acknowledgments

Authors wishing to acknowledge the chemical engineering department, Universiti Teknologi PETRONAS for providing laboratory facilities throughout the study. Authors also would like to thanks, Universiti Teknologi PETRONAS for providing financial support during this project.

References

[1] Khan M S, Lal B, Partoon B, Keong L K, Bustam M A and Mellon N B 2016 Experimental Evaluation of a Novel Thermodynamic Inhibitor for CH_4 and CO_2 Hydrates *Procedia Eng.* **148** pp 932–40

[2] Xiao C and Adidharma H 2009 Dual function inhibitors for methane hydrate *Chem. Eng. Sci.* **64** pp 1522–27

[3] Sloan E D and Koh C A 2008 *Clathrate Hydrates of Natural Gases* vol 87, ed Third (Boca Raton; London; New york: CRC Press Taylor & Francis)

[4] Bavoh C B, Nashed O, Saad Khan M, Partoon B, Lal B and Sharif A M 2018 The Impact of Amino Acids on Methane Hydrate Phase Boundary and Formation kinetics *J. Chem. Thermodyn.* **117** pp 48–53

[5] Tariq M, Rooney D, Othman E, Aparicio S, Atilhan M and Khraisheh M 2014 Gas hydrate inhibition: A review of the role of ionic liquids *Ind. Eng. Chem. Res.* **53** pp 17855–68

[6] Bavoh C B, Lal B, Nashed O, Khan M S, Lau K K and Bustam M A 2016 COSMO-RS: An ionic liquid prescreening tool for gas hydrate mitigation *Chinese J. Chem. Eng.* **24** pp 1619–24

[7] Sakthivel S, Chhotaray P K, Velusamy S, Gardas R L and Sangwai J S 2015 Synergistic effect of lactam, ammonium and hydroxyl ammonium based ionic liquids with and without NaCl on the surface phenomena of crude oil/water system *Fluid Phase Equilib.* **398** pp 80–97

[8] Tariq M, Atilhan M, Khraisheh M, Othman E, Castier M, Garcia G, Aparicio S and Tohidi B 2016 Experimental and DFT Approach on the Determination of Natural Gas Hydrate Equilibrium with the use of Excess N_2 and Choline-Chloride Ionic Liquid as an Inhibitor *Energy and Fuels* **30** pp 2821–32

[9] Tariq M, Connor E, Thompson J, Khraisheh M, Atilhan M and Rooney D 2016 Doubly dual nature of ammonium-based ionic liquids for methane hydrates probed by rocking-rig assembly *RSC Adv.* **6** pp 23827–36

[10] Khan M S, Liew C S, Kurnia K A, Cornelius B and Lal B 2016 Application of COSMO-RS in Investigating Ionic Liquid as Thermodynamic Hydrate Inhibitor for Methane Hydrate *Procedia Eng.* **148** pp 862–69

[11] Kumar A, Bhattacharjee G, Kulkarni B D and Kumar R 2015 Role of Surfactants in Promoting Gas Hydrate Formation *Ind. Eng. Chem. Res.* **54** pp 12217–32

[12] Shahnazar S and Hasan N 2014 Gas hydrate formation condition: Review on experimental and modeling approaches *Fluid Phase Equilib.* **379** pp 72–85

[13] Hashimoto S, Tsuda T, Ogata K, Sugahara T, Inoue Y and Ohgaki K 2010 Thermodynamic Properties of Hydrogen + Tetra-n-Butyl Ammonium Bromide Semi-Clathrate Hydrate *J. Thermodyn.* **2010** pp 1–5

[14] Karimi A A, Dolotko O and Dalmazzone D 2014 Hydrate phase equilibria data and hydrogen storage capacity measurement of the system H_2+tetrabutylammonium hydroxide+H_2O *Fluid Phase Equilib.* **361** pp 175–80

[15] Su Y, Searles D J and Wang L 2017 Semiclathrate hydrates of methane + tetraalkylammonium hydroxides *Fuel* **203** pp 618–26

[16] Nashed O, Koh J C H and Lal B 2016 Physical-chemical Properties of Aqueous TBAOH Solution for Gas Hydrates Promotion *Procedia Eng.* **148** pp 1351–56

[17] Khan M S, Lal B, Keong L K and Sabil K M 2018 Experimental evaluation and thermodynamic modelling of AILs alkyl chain elongation on methane riched gas hydrate system *Fluid Phase Equilib.* **473** pp 300–309

[18] Khan M S, Bavoh C B, Partoon B, Nashed O, Lal B and Mellon N B 2018 Impacts of ammonium based ionic liquids alkyl chain on thermodynamic hydrate inhibition for carbon dioxide rich binary gas *J. Mol. Liq.* **261** pp 283–90

[19] Idris A, Man Z, Maulud A S and Khan M S 2017 Effects of Phase Separation Behavior on Morphology and performance of Polycarbonate Membranes *Membranes (Basel)* **7** pp 1–18

[20] Khan M S, Partoon B, Bavoh C B, Lal B and Mellon N B 2017 Influence of

tetramethylammonium hydroxide on methane and carbon dioxide gas hydrate phase equilibrium conditions *Fluid Phase Equilib.* **440** pp 1–8

[21] Khan M S, Bavoh C B, Partoon B, Lal B, Bustam M A and Shariff A M 2017 Thermodynamic effect of ammonium based ionic liquids on CO_2 hydrates phase boundary *J. Mol. Liq.* **238** pp 533–39

[22] Muhammad Saad Khan, Yaqub S, Manner N, Karthwathi N A, Qasim A, Mellon N B and Bhajan Lal 2018 Experimental Equipment Validation for Methane (CH_4) and Carbon Dioxide (CO_2) Hydrates *IOP Conf. Ser. Mater. Sci. Eng.* **344** pp 1–10

[23] Xiao C, Wibisono N and Adidharma H 2010 Dialkylimidazolium halide ionic liquids as dual function inhibitors for methane hydrate *Chem. Eng. Sci.* **65** pp 3080–87

[24] Nashed O, Dadebayev D, Khan M S, Bavoh C B, Lal B and Shariff A M 2018 Experimental and modelling studies on thermodynamic methane hydrate inhibition in the presence of ionic liquids *J. Mol. Liq.* **249** pp 886–91

[25] Yusof Z A M, Ahmed I, Khan M S, Hussain S A, Hussain A, Mutalib I bin A, Balkhair K S and Albeirutty M H 2015 Thermal Evaluation of Diesel/Hydrogen Peroxide Fuel Blend *Chem. Eng. Technol.* **38** pp 2170–80

[26] Khan M S, Ahmed I, Lal B, Idris A-A, Albeirutty M H, Ayoub M and Sufian S binti 2018 Physicochemical and FTIR Study of Diesel-Hydrogen Peroxide Fuel Blend *IOP Conf. Ser. Mater. Sci. Eng.* **344** 012026 pp 1-9

[27] Khan M S, Lal B, Bavoh C B, Keong L K and Bustam A 2017 Influence of Ammonium based Compounds for Gas Hydrate Mitigation : A Short Review *Indian J. Sci. Technol.* **10** pp 1–6

IOP Conf. Series: Materials Science and Engineering **458** (2018) 012061 doi:10.1088/1757-899X/458/1/012061

Improvement of energy density and energy yield of oil palm biomass by torrefaction in combustion gas

Y Uemura[1*], V Sellappah[1], T H Trinh[1], M Komiyama[1], S Hassan[2], K Tanoue[3]

[1] Center for Biofuel and Biochemical Research, Institute for Sustainable Living, Department of Chemical Engineering, Universiti Teknologi PETRONAS, 32610 Seri Iskandar, Perak, Malaysia

[2] Department of Mechanical Engineering, Universiti Teknologi PETRONAS, 32610 Seri Iskandar, Perak, Malaysia

[3] Department of Mechanical Engineering, Yamaguchi University, Tokiwa-dai, Ube, Yamaguchi 755-8611, Japan

[*] yoshimitsu_uemura@utp.edu.my

Abstract. This study investigates torrefaction of oil palm empty fruit bunch (EFB) and palm kernel shell (PKS) in a fixed-bed reactor in biomass combustion gas, which consisted of 8-9 vol% O_2 and 13-14 vol% CO_2. Torrefaction experiments were carried out at 473-573K for 30 min. The mass and energy yields, and properties of torrefied biomass were investigated. Torrefaction in the presence of combustion gas resulted in a decrease in mass yield and an increase in the energy density and carbon content. For both the biomass, the highest energy yield and density was obtained by torrefaction at 473 K and 573 K in combustion gas, respectively. This study has proved that use of combustion gas has advantages in terms of product yield as well as energy saving. Production capacity of torrefied biomass at oil mill was also estimated.

1. Introduction

Torrefaction is a promising technique to improve some inferior properties of biomass residue, such as lower energy density, shorter shelf life and poor grindability. Many fundamental studies, in which torrefaction was carried out in inert gas such as nitrogen, have been done [1-3], and have proved that torrefaction can improve the inferior properties listed above.

Even though torrefaction can improve the inferior properties, biomass solid fuel, the products from the process, is rather cheap in selling price. Therefore, intensive efforts are required to minimize the investment and utility costs. The authors have proposed a process shown at the bottom of figure 1 for this purpose [4]. The background of this issue was detailed elsewhere [4]. The top illustration represents the representative current process, which was drawn based on the process proposed elsewhere [5]. In the proposed process at the bottom of figure 1, boiler flue gas is utilized directly as torrefaction gas and energy source. By comparing with the representative existing process [5], one can see the advantages of the proposed process which include less complicated pipings and free from additional equipment and auxiliary energy source for providing thermal energy for torrefaction.

In this study, a comprehensive investigation for torrefaction of empty fruit bunch (EFB) and palm kernel shell (PKS) has been conducted to validate torrefaction in combustion gas. The mass yield, energy density, and energy yield were determined for torrefaction in combustion gas at 473, 523 and 573 K. Torrefaction in nitrogen gas was also done for comparison at all the temperatures. In our previous report

[4], only one type of biomass was used, and torrefaction in nitrogen was carried out at only one temperature.

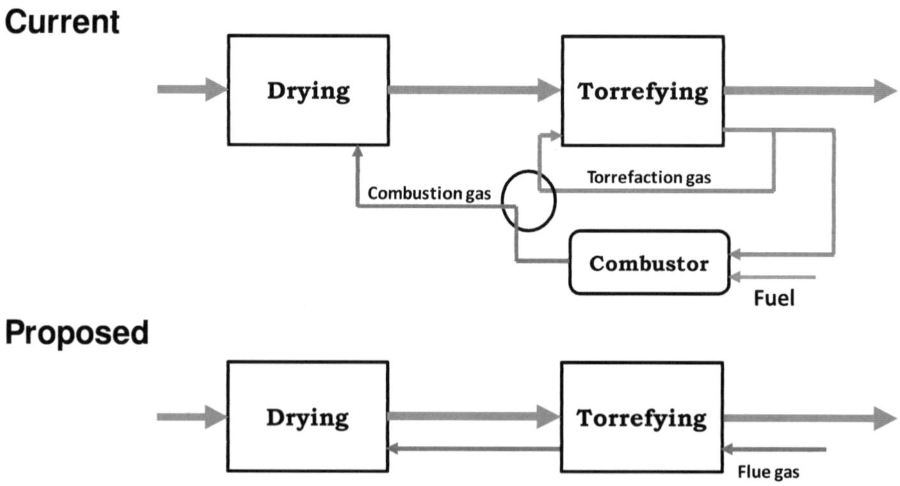

Figure 1. Comparison of the torrefaction process proposed in this study with existing one.

2. Experimental

2.1. Material

EFB and PKS collected from a nearby palm oil mill were dried in an oven at 378 K for 24 h. EFB were chopped into smaller chips with dimensions of 2.0 cm x 0.5 cm x 1.0 cm. PKS was used without further size adjustment. The elementary and chemical analyses result of EFB and PKS used in this study are listed in table 1.

Table 1. Properties of biomass used in this study.

Biomass	EFB	PKS
HHV [MJ/kg]	15.2	19.6
Ultimate analysis [wt%]		
Carbon	40.6	50.4
Hydrogen	5.1	6.5
Nitrogen	1.2	0.7
Oxygen	38.1	36
Moisture content	7.4	3.8
Ash content	7.6	2.7
Chemical composition [wt%]		
Hemicellulose	22.1	18.2
Cellulose	59.7	33.2
Lignin	18.1	48.6

2.2. Experimental setup and procedure

The experimental setup consisted of a combustion chamber for producing combustion gas and a torrefaction reactor (figure 2) The chamber and reactor were stainless steel-made vertical tubular columns with internal diameters of 0.10 m and 0.13 m, respectively. A prescribed amount of EFB (100 g) or PKS (434 g) was torrefied in combustion gas or N_2 at 473, 523 or 573 K for 30 min. The combustion gas generated contained 13-14 vol% of carbon dioxide and 8-9 vol% of oxygen (nitrogen balance). The details were reported elsewhere [4].

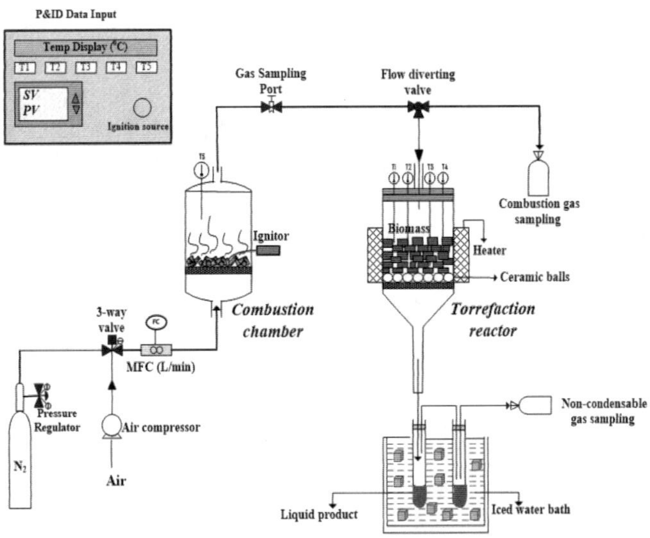

Figure 2. Experimental setup for torrefaction under combustion gas atmosphere.

2.3. Data processing

The equations employed in quantifying the mass yield (Y_m), calorific value (CV) ratio (CV_{ratio}) and energy yield (Y_e) from the torrefaction result are shown below:

$$Y_m = \frac{m_f}{m_i} \quad (1)$$

$$CV_{ratio} = \frac{CV_f}{CV_i} \quad (2)$$

$$Y_e = \frac{CV_f \, m_f}{CV_i \, m_i} \quad (3)$$

where m denotes the mass of biomass, CV_{ratio} CV ratio, subscript i at initial stage, subscript f after torrefaction.

3. Results and Discussion

Figure 3 shows the mass yield, CV (calorific value) ratio and energy yield of the torrefied biomass under nitrogen and combustion gas atmosphere at 473, 523 and 573 K for EFB (left) and PKS (right). The energy density (or CV ratio) of biomass torrefied in combustion gas is higher than that in nitrogen gas at all the temperatures regardless of biomass type. The energy yield of biomass torrefied in combustion gas is higher than that in nitrogen gas at the lowest temperature of 473 K. This trend is reversed at higher temperatures of 523 and 573 K. As the authors already reported, the severity of torrefaction in the presence of oxygen or carbon dioxide is more significant than that in inert gas [6], this trend may be attributed to partial oxidation of the biomass by the oxidative atmosphere.

After all, if higher energy density is preferable, torrefaction at 573 K in combustion gas is recommended. On the other hand, if higher energy yield is preferable torrefaction at 473 K in combustion gas is recommended.

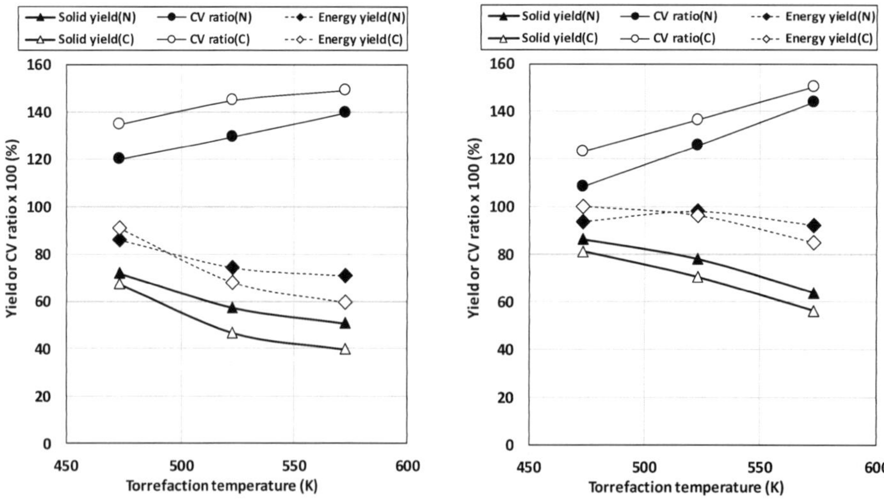

Figure 3. Torrefaction result of EFB (left) and PKS (right) in nitrogen (N) and combustion gas (C).

The calorific value of biomass is the most important parameter to determine the energy density and energy yield if the biomass is used as a solid fuel. The authors once reported that the carbon content of lignocellulosic biomass is the major factor to determine the calorific value for biomass torrefied in inert gas [7]. But no report has been found about this issue for torrefaction under combustion gas. To investigate this point, the calorific vale is plotted against the carbon content in figure 4. The data consist of the four groups of torrefaction results: PKS in nitrogen, PKS in combustion gas, EFB in nitrogen and EFB in combustion gas. All the data are approximated by only one straight line (y = 1.31x + 24.3) regardless of type of torrefaction gas, as shown in figure 4. From the present study, it has been proved that the type of torrefaction gas has little effect on the relationship between the calorific value and carbon content.

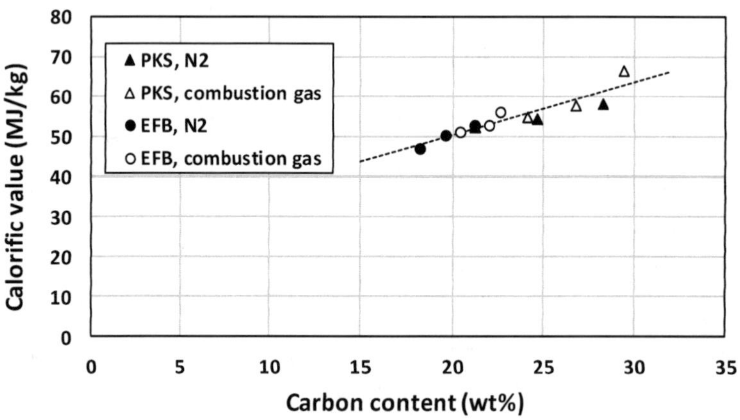

Figure 4. A plot of higher calorific value against carbon content.

Finally, estimation of torrefied biomass production capacity at an average-sized oil mill in Malaysia was carried out based on the mass yield data obtained from this study. 'An average-sized oil mill' is defined as a mill of which oil production is the average in Malaysia. Torrefied biomass production at an average-sized palm oil mill was calculated for the following four cases:

(1) Torrefaction of EFB in combustion gas at 573 K for higher energy density,
(2) Torrefaction of EFB in combustion gas at 473 K for higher energy yield,
(3) Torrefaction of PKS in combustion gas at 573 K for higher energy density, and
(4) Torrefaction of PKS in combustion gas at 473 K for higher energy yield.

The result is summarized in figure 5. The limiting factor was set as the thermal energy supplied by 10% of the total monocarp fiber generated at the mill.

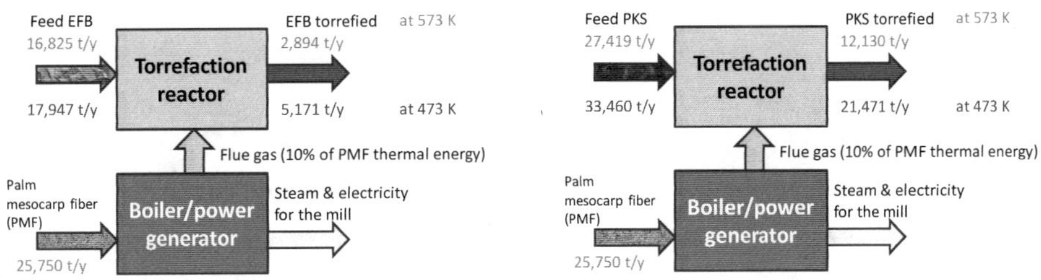

Figure 5. Estimation of torrefied EFB (left) and PKS (right) production at an average-sized oil mill.

Torrefied PKS production ranges from 12 to 21 thousand t/y, while 3 to 5 thousand t/y for EFB as shown in figure 5. The poor productivity of torrefied EFB is due to the higher moisture content of raw feed EFB (57 wt%). This may be improved if sunlight drying is applied to pre-drying of raw EFB [8].

4. Conclusions

In the present study, the effect of combustion gas on torrefaction behavior of empty fruit bunch (EFB) and palm kernel shell (PKS) at a temperature range of 473-573 K was investigated. Combustion gas atmosphere resulted in lower mass yield and higher calorific value ratio in comparison with nitrogen gas atmosphere. The energy yield behaved in a little more complex way. For both the biomass, the highest energy yield was obtained by torrefaction at 473 K in combustion gas. The highest energy density was obtained at 573 K in combustion gas. The relationship between the calorific value and carbon content was well correlated by only one line regardless of torrefaction atmosphere. Finally, production capacity for torrefied EFB and PKS was estimated at an average-sized palm oil mill based on the mass and energy balance.

Acknowledgement

This work was supported by the Mitsubishi Corporation Education Trust Fund, Japan, and PRGS Grant from the Ministry of Higher Education, Malaysia.

References

[1] Arias B, Pevida C, Fermoso J, Plaza M G, Rubiera F and Pis J J 2008 Influence of torrefaction on the grindability and reactivity of woody biomass *Fuel Processing Technology* **89** pp 169–75
[2] Prins M J, Ptasinski K J and Janssen F J J G 2006 Torrefaction of wood Part 2. Analysis of products *J. Anal. Appl. Pyrolysis* **77** pp 34–40
[3] Bridgeman T G, Jones J M, Shield I and Williams P T 2008 Torrefaction of reed canary grass, wheat straw and willow to enhance solid fuel qualities and combustion properties. *Fuel* **87** pp 844–856

[4] Uemura Y, Sellappaha V, Thanh T H, Hasan S and Tanoue K 2017 Torrefaction of empty fruit bunches under biomass combustion gas atmosphere *Bioresource Technology* **243** pp 107–117
[5] http://articles.extension.org/pages/72903/terra-green-energy:-pioneers-in-torrefaction-for-commercial-co-combustion
[6] Uemura Y, Saadon S, Osman N, Mansor N and Tanoue K 2015 Torrefaction of oil palm kernel shell in the presence of oxygen and carbon dioxide *Fuel* **144** pp 171–179
[7] Uemura Y, Omar W N, Tsutsui T, Subbarao D and Yusup S 2010 Relationship between calorific value and elementary composition of torrefied lignocellulosic biomass *J. of Appl. Sci.* **10** (24) pp 3250–56
[8] Imanishi H, Uemura Y and Niiyama H 2006 Utilization of Water Hyacinth as a renewable energy source in tropical areas having canal system *Proceedings for the Renewable Energy* 2006, P-B-3

IOP Conf. Series: Materials Science and Engineering **458** (2018) 012062 doi:10.1088/1757-899X/458/1/012062

Process simulation of integrated palm oil mill, refinery and oleochemical processes

H Kamarden[1], K Kidam[1,2], H Hashim[1], M H Hassim[1], S S Shahlan[1], N Ngadi[1], A Johari[1], M J Kamaruddin[1] and J Y Ten[a]

[1]Department of Chemical Engineering, Faculty of Chemical and Energy Engineering, Universiti Teknologi Malaysia, Malaysia.
[2]Institute for Oil & Gas, Universiti Teknologi Malaysia, Malaysia.

kamarizan@cheme.utm.my

Abstract. Processes involved in the palm oil industry can be categorized into three main operations; upstream operation (planting and milling), midstream operation (physical/chemical refining), and downstream operation (palm-based oleochemical processing). For the last few years, some industrial key players have emphasized a change from stand-alone operations towards an integrated manufacturing facility. This has become the basis idea for integration of palm-based production processes. It includes material integration, shared energy and utility resources and waste-to-feed utilization. Integrated process however, involves a complex network; thus, the process has to be simulated to analyze its feasibility in terms of process throughput and material, energy and utility consumption. Hence, this paper addresses the basis to simulate the integrated palm oil process using process simulator Aspen Plus V8.6 for its dynamic capabilities in adjusting process variables to meet the required capacity for a target product. This study is critical in providing a basis and overview of the typical processes in palm oil industry and the possible interrelations associated with the proposed integrated facility for palm oil-based production.

1. Introduction

For the past few years, there has been emphasis on a transformation from stand-alone operations towards integrated manufacturing facility. The integrated manufacturing facility is intended to produce a more complete range of value-added products from the raw fruits/seeds to consumers' usage [1].

In current practice, upstream operation involves planting, harvesting and processing of fresh fruit bunch (FFB) to obtain the crude palm oil (CPO). As for the midstream operation, it involves further refining of CPO to obtain refined, bleached and deodorized palm oil (RBDPO), and to separate the RBDPO into palm stearin and palm olein. Lastly, the downstream operation deals with oleochemical processing such as fractionation (separation) of palm fatty acids distillate (PFAD) from the refining section. Many plantation companies in Malaysia manage both upstream and midstream operations, but they are mostly operated separately.

Recently, the East Coast Economic Region of Malaysia (ECER) in 2010 has promoted the investment for Palm Oil Industrial Cluster (POIC) to be constructed at Gebeng Industrial area in Pahang [2]. The POIC is planned to facilitate the upstream (plantation, milling and milled products) and downstream (food-based, oleo and biodiesel) of the palm oil industries.

This recent revelation shows the possibility of integrating the palm oil-based processes for sustainable production of palm oil upstream and downstream products. However, the literature review conducted has found no evidence of integrated palm oil process that addresses the possibility of integrating the processes in production of palm oil and palm-derived products. Therefore, via process simulation, the process throughput, operating variables, and material, energy and utility consumption can be analyzed for both stand-alone process unit and integrated process unit.

2. Research approach

Processes involved in palm oil industry can be categorized into three main section; milling, refining (physical/chemical) and oleochemical processing. To carry out this analysis, the involved processes are briefly described in the following section.

2.1. Process description

2.1.1. Oil Extraction (Milling Process). Milling process involves oil extraction from the palm FFB. The raw fruits are fed into a sterilizer and then sent to a threshing drum where the fruits are separated from the bunches. The fruits then sent to digesters followed by pressing in a screw presser. The pressed oil is pumped to the clarification tank for oil separation. The separated oil containing fibrous solids is further separated in high speed centrifuge and then sent to vacuum dryer [3]. The desired products from the milling process are crude palm oil (CPO) and kernel nuts; while the waste consists of palm oil mill effluent (POME), empty fruit bunch (EFB) and fruit fiber.

2.1.2. Oil Refining (Degumming, Bleaching and Deodorizing Process). There are three stages in CPO refinery: degumming, bleaching and deodorizing of oil. In degumming, the CPO feed is treated with phosphoric acid (H_3PO_4) with a concentration of 85% and demineralized hot water [4]. In bleaching, the degummed oil is treated with bleaching earth in a vacuum bleaching mixer before entering the vacuum bleacher. In deodorization, the degummed and bleached oil is treated at high temperature of between 240°C to 260°C, under vacuum condition (2-4 mmHg) with direct steam injection of 2.5-4.0 % of oil weight. The products from oil refining process are refined, bleached and deodorized palm oil (RBDPO) and palm fatty acid distillate (PFAD), while the waste consists of spent bleaching earth (SBE).

2.1.3. Oleochemical Processing (*Palm Oil Fatty Acid Fractionation*). Oleochemical process involves various production of specialty products. Fractionation of palm fatty acid distillate (PFAD) from the oil refining process is considered in this analysis. The operation consists of four main process units: drying column, pre-cut column, C16/18 column and fatty acid evaporator. Drying column is used to distill out the water content from PFAD feed. The first fractionation column (pre-cut column) is used to drawn off a C12-C14 fatty acids as by-product. The second fractionation column (C16/18 column) is producing a C16-C18+ fatty acids as final product.

2.2. Process Integration Approach

The aim of process integration is to integrate the use of materials and energy and to minimize the generation of emissions and wastes. Hence, the proposed integrated process is depicted in figure 1, where the whole manufacturing process of upstream, midstream and downstream processes in palm oil industry, are considered as a big picture of an integrated system with interconnecting processing units inclusive of process, utility and waste streams. There are two levels of integration in the proposed scheme. Firstly, the interconnection of material stream from one process unit to the other. Another integration level is the interconnection of centralized utility system (CUS) with each process plant unit, where the supply of utilities (water, electricity, steam and vacuum) to the process plant unit is centralized.

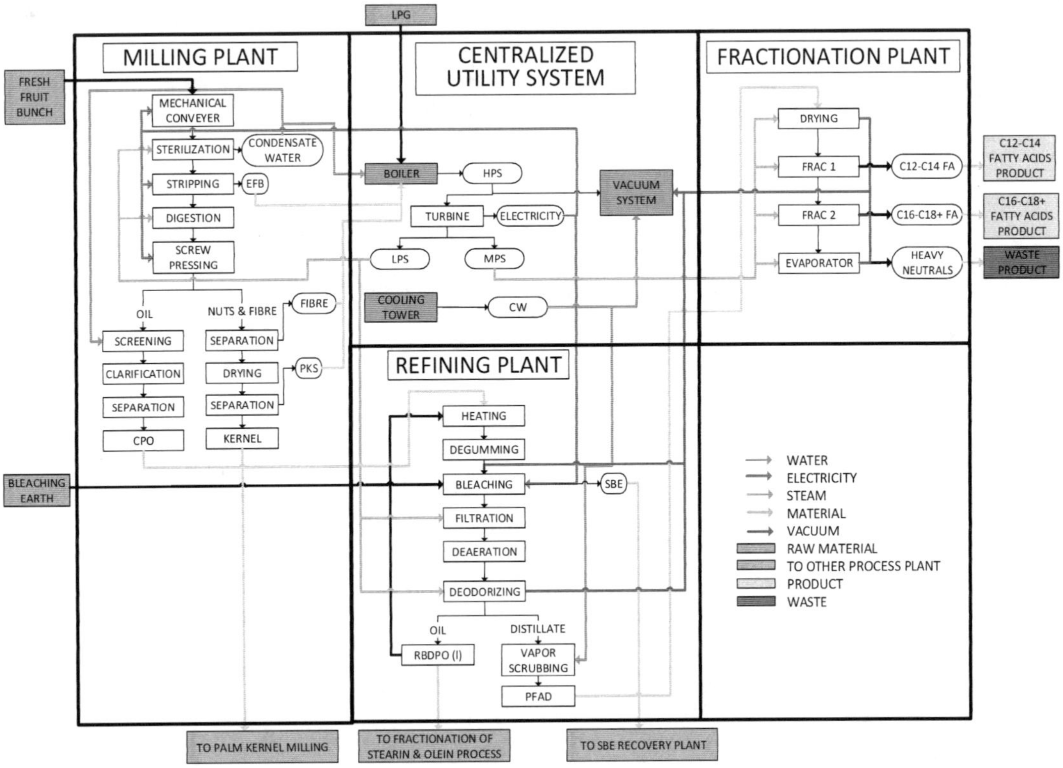

Figure 1. Proposed integrated process of milling, refining and fractionation process.

2.3. *Simulation Procedure*

The simulation of production process for upstream, midstream and downstream operations in palm oil industry is performed using the process simulator Aspen Plus V8.6. The main process data of the simulation are presented in table 1. For upstream process, the process units involved in milling are mostly the mechanical equipment that handle the solids (FFB). Solid model is employed for milling process simulation where the physical properties and chemical composition of the solid feed are obtained from [5] and [6] respectively. Data on chemical composition of the liquid feed (mesocarp oil of the FFB) is obtained from [7] and [8]. Feed to midstream and downstream operation comes from the subsequent product, starting from milling section, where its product (CPO) is the feed to the refinery section and product from refinery (PFAD) is then fed to the fractionation section.

3. Results and discussion

In this study, the integration is based on consideration of the whole manufacturing process of the analysis; milling, refining and fractionation of palm fatty acid distillate (PFAD). From a big picture, these three operations have the possibility for interconnection of the process units, utility and waste streams. Through preliminary analysis of mass balance data for each operation, they share similar requirement on utilities (electricity, steam, cooling water and hot water). Abundance waste stream from milling operation such as EFB, mesocarp fibers, palm kernel shell (PKS) and condensate water from sterilizer which is part of the mixture effluent from milling process (POME)) is a potential feed for generating utilities. These wastes can be used to fuel up the boiler to generate steam. The steam generated can be used for both steam supply to the process and for turbine operation to generate electricity, which are among the utilities required to operate all the three processes.

Table 1. Main process data for process simulation of palm oil processes integration.

Features	Value	Features	Value
Milling		***Fractionation***	
Feed composition		Dryer	
Solid fraction	35.11% cellulose, 25.26% hemicellulose, 28.01% lignin, 5.68% extractive, 5.93% ash	Temperature/ Pressure No. stages	~ 30 mbar / ~ 128°C 3 stages
Liquid fraction	0.2% C12, 1.1% C14, 44.1% C16, 4.5% C18, 39.2% C18:1, 10.5% C18:2, 0.4% C20:0	Column 1 Temperature/ Pressure	 ~ 140°C / ~ 7 mbar (top), ~ 230°C (bottom)
		No. of stages	20 stages
Feed flowrate	45,000 kg/hr	Product	C12 - C14
Steam	LPS; 10080 kg/hr		
Product	CPO	Column 2	
Refining		Temperature/ Pressure	~ 140°C/ ~ 6 mbar (top column), ~ 230°C (bottom)
Feed	CPO from milling		
Temperature & Pressure	90-110°C / Vacuum (8 mbar)		
BE flowrate	2% of feed	No. of stages	20 stages
H_2PO_4	85% conc., 0.2% of feed	Product	C16 – C18 +

Table 2. Composition of the main process stream calculated by simulation.

		FFB	CPO	RBDPO	PFAD	C12-C14	C16-C18+
T (°C)		***30***	***201.5825***	***353.3637***	***130***	***166.2392***	***153.0815***
P (bar)		***1***	***0.1973847***	***0.9973684***	***0.9973684***	***0.024673***	***0.011843***
Mass Flow (kg/hr)		***45000***	***10274.58***	***6972.9***	***3309.084***	***50.00038***	***2000***
CELLULOS	Cellulose	27.00%	-	-	-	-	-
XYLAN	Hemicellulose	19.42%	-	-	-	-	-
LIGNIN	Lignin	21.55%	-	-	-	-	-
EXTRACT	Extractives	4.37%	-	-	-	-	-
ASH	Ash	4.56%	-	-	-	-	-
H2O	Water	0.00%	0.08%	0.00%	0.34%	0.02%	0.01%
Lauric	C12:0	0.05%	0.19%	0.10%	0.38%	0.00%	0.00%
Myristic	C14:0	0.25%	1.07%	0.80%	1.65%	0.11%	0.00%
Palmitic	C16:0	10.19%	43.94%	41.82%	48.14%	73.79%	33.00%
Stearic	C18:0	1.04%	4.52%	5.20%	3.07%	1.34%	2.88%
Oleic	C18:1	9.06%	39.30%	41.00%	35.48%	17.96%	49.15%
Linoleic	C18:2	2.43%	10.51%	10.49%	10.48%	6.66%	14.96%
Heavy neutrals	C20:0	0.09%	0.39%	0.36%	0.47%	0.12%	0.00%
H3PO4	Phosphoric Acid	-	-	0.24%	-	-	-

Table 3. Energy and utility consumption calculated by simulation.

	Milling	Refining	Fractionation
Energy (kJ/hr)	8.00E+06	9.10E+06	1.86E+06
Vacuum duty (kJ/hr)	-5.28E+05	-2.42E+06	-5.13E+05
Steam (kg/hr)	10080	308.46	-
Hot water (kg/hr)	4937.63	-	-
Cooling Water (kg/hr)	-	2154.24	-

Based on table 2, it is found that the weight percent (wt%) of desired products from each process (milling, refining and fractionation) after the integration, is as according to current production of their stand-alone operation. However, the feed basis of the process integration performed is based on a stand-alone milling operation, where the capacity of a normal operational milling plant is 45 t/hr of FFB (based on observation and data collection from site visits). Therefore, the end-operation (fractionation process) is only able to produce a very small amount as compared to its current plant capacity of ~500kg/hr of C12-C14 fatty acids (current simulated production is only 50 kg/hr) and ~14,5000kg/hr of C16-C18+ fatty acids (current simulated production is only 2000 kg/hr). For instance, if the throughput is set from backwards, to accommodate a larger fractionation process capacity, the feed of PFAD required from refinery and CPO from milling sections will be larger. A larger capacity of milling for example, will generate a larger amount of waste that can be used to produce more amount of steam and more electricity generation for the operational consumptions. These changes can be simulated in Aspen Plus to obtain the value of energy and utility consumption (table 3) as compared to the waste generated and its conversion to energy in terms of steam and electricity generation capacity.

4. Conclusion

Process integration is an important branch of process engineering that refers to the application of system-oriented, thermodynamics-based and integrated approaches for the analysis, synthesis and retrofit of a process plant. Since the goals of process integration are to integrate the use of materials and energy and to minimize the generation of emissions and wastes, hence integration of upstream, midstream and downstream operations in palm oil-based industry is relevant to minimize the resources by direct transfer of material from one process unit to another and sharing of energy supplies from a centralized utility system as previously depicted in figure 1. Hence, by using the process simulation tools, a basis and overview of the typical processes in palm oil industry and the possible interrelations within the proposed integrated facility can be observed and analysed virtually for further details and retrofit of process design to realize a performance target. For future work, a comprehensive integration of palm oil processes based on current capacity of downstream operation can be done to analyse its effect to the throughput of raw material required at the upstream operation and the impact to overall integrated process, in terms of waste generation, raw material and energy and utility consumption.

Acknowledgment

We would like to thank UTM for the financial support to conduct this research under vot number 15H62.

References

[1] Anderson D 2005 A Primer on Oils Processing Technology *Edible Oil and Fat Products: Processing Technologies* ed F Shahidi 6th Ed (New Jersey, United States of America: John Wiley & Sons, Inc.) Part **5** pp 1-56

[2] East Coast Economic Region Development Council 2010 *Palm Oil Industrial Cluster- Sustaining The Innovative Growth of The Palm Oil-Based Industry* Retrieved June 27, 2014, from http://www.ecerdc.com.my/wp-content/uploads/2014/06/4-ECER-Palm-Oil-Industrial-Cluster.compressed.pdf

[3] Department of Environment 1999 *Industrial Processes & the Environment of Crude Palm Oil Industry* Department of Environment, Ministry of Science, Technology and Environment of Malaysia.

[4] Mohd Zin R 2006 *Process Design in Degumming and Bleaching of Palm Oil* Centre of Lipid Engineering and Applied Research (CLEAR) Universiti Teknologi Malaysia.

[5] Poku K 2002 Smal-Scale Palm Oil Processing in Africa *FAO Agricultural Services Bulletin* **148** pp 1–56

[6] Saka S, Munusamy M V, Shibata M, Tono Y and Miyafuji H 2008 Chemical Costituents of the Different Anatomical Parts of the Oil Palm (Elaeis guineensis) for Their Sustainable

Utilization *Natural Resources and Energy Environment, JSPS-VCC Program on Environmental Science, Engineering and Ethics (Group IX) Kyoto, Japan: Kyoto University & Universiti Sains Malaysia* pp 19–34

[7] Lim T K 2012 Elaeis guineensis *Edible Medicinal and Non-Medicinal Plants* (Springer Science and Business Media) **1** pp 157–168

[8] Obahiagbon F I 2012 A Review: Aspects of the African Oil Palm (Elaeis guineesis jacq.) and the Implications of its Bioactives in Human Health *American Journal of Biochemistry and Molecular Biology* **2** (3) pp 106–119

Appendices

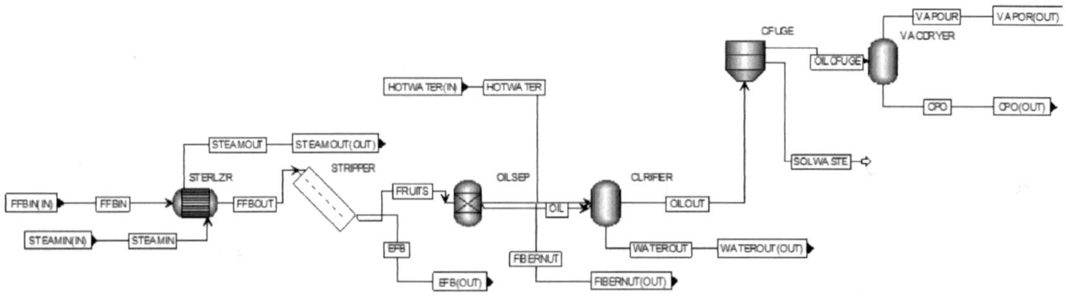

Appendix A. Upstream process simulation (milling process flow diagram).

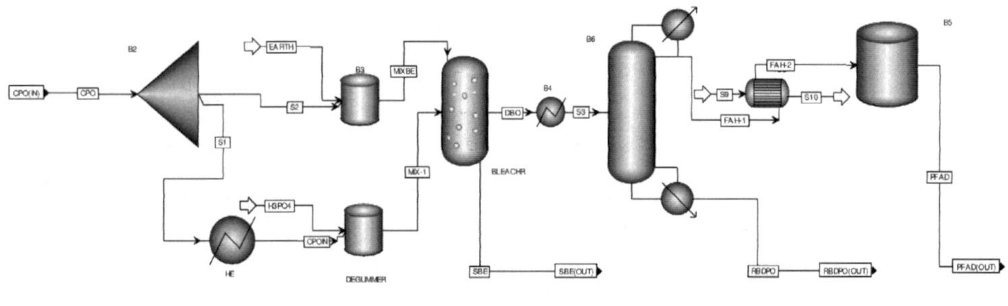

Appendix B. Midstream process simulation (refinery process flow diagram).

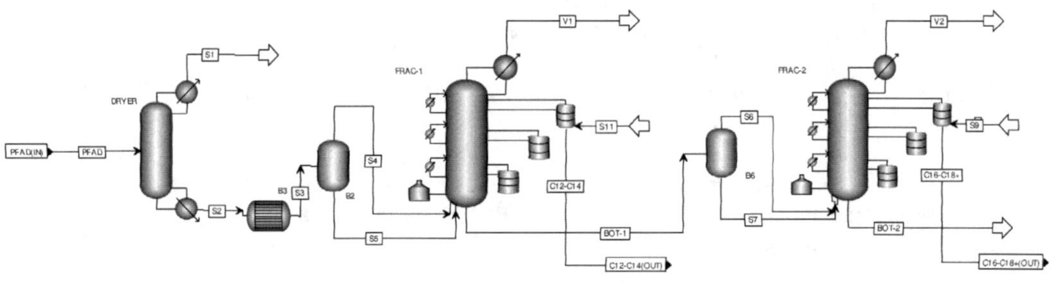

Appendix C. Downstream process simulation (fractionation process flow diagram).

Kinetics of CO_2 reaction with N-methyldiethanolamine and aminobutanol using stopped flow technique

Abdelbaki Benamor[1]*, Nafis Mahmud[1], Mustafa Nasser[1] and Hazim Qiblawey[2]

[1] Gas Processing Centre, College of Engineering, Qatar University, 2713, Doha, Qatar
[2] Department of Chemical Engineering, College of Engineering, Qatar University, 2713, Doha, Qatar

*E-mail: benamor.abdelbaki@qu.edu.qa

Abstract. Removal of carbon dioxide (CO_2) from natural gas streams is mandatory to avoid pipeline corrosion, increase the heating value of the gas and reduce the gas volume in case of liquefied natural gas (LNG). The use of N-Methyldiethanolamine (MDEA) solution combined with other rate promoters such as Piperazine (PZ) is a common practice in Gas treatment technology. In this work, the use of Aminobutanol (AB) mixed with MDEA for the removal of (CO_2) from natural gas streams is presented. The reaction kinetics of (CO_2) with mixtures of MDEA and AB was investigated using stopped flow technique. The experiments were performed over a temperature range of 293 to 313 K and solution concentration of 0.5 and 1 moles/l in different MDEA/AB proportions. Obtained kinetics data were interpreted using zwitterion mechanisms for the primary amine AB. The individual rate constants of the participating reactions were regressed and their corresponding activation energies were estimated.

1. Introduction

Natural gas is a vital source of energy. It consists of methane with other hydrocarbons and non-hydrocarbon components. Compositions of natural gas varies in accordance to its formation process and it may contain diluents like CO_2 and N_2 [1]. The presence of carbon dioxide within natural gas stream causes an increase in the gas volume, reduces its heat value and may enhance corrosion within the pipelines and treatment plants. Therefore, CO_2 need to be removed to meet the required pipeline or LNG specifications [2]. Amine based absorption technology is a recognized and industrially implemented technology to remove CO_2 [3]. Many variants of chemical solvents based on primary, secondary and tertiary amines are available in the market with distinguished performances. However, several drawbacks such as low absorption rate, solvent degradation, high regeneration energy and severe equipment corrosion are encountered with their use.

Blended amines have been suggested for the absorption of acid gases [4]. As a common practice in the industry, tertiary amines are mixed with faster reacting primary, secondary amines or Piperazine that act as rate promoters. The addition of small amounts of promoters to slow reacting tertiary amines can enhance the rate of absorption of CO_2 to a large extent without appreciably changing the stripping characteristics [5].

Amino butanol, a primary amine, have recently been investigated by Idris et al [6] as a possible solvent for CO_2 capture. The equilibrium solubility of carbon dioxide in aqueous solutions of 4-amino-1-butanol at low partial pressures and two different concentrations (5 molar and 30 wt. %) and a temperature of 40 °C was presented [6]. It was found that CO_2 solubility in amino butanol is higher than its solubility in MEA. Benamor et al [7] recently presented kinetics data involving the reaction of amino butanol with CO_2 over a temperature range varying from 293 to 313 K. The kinetics data were interpreted using two different reaction mechanisms and compared to several primary and secondary amines. It was found that aminobutanol, has a rate constant lower than that of MEA but still higher than those of secondary amine, DEA, and sterically hindered amine, AMP.

In this work, the reaction kinetics of carbon dioxide with aqueous mixtures of MDEA and AB using stopped flow technique is presented for the first time. The temperature was varied from 298 to 313 K while the amine total concentration was varied from 0.5 to 1 mole/l in different proportions of AB/MDEA.

2. Reaction Mechanism and Rate Models

2.1. Reaction of CO_2 with MDEA

The overall reaction of CO_2 and MDEA in aqueous solution is well documented and can be written as [8]:

$$CO_2 + MDEA + H_2O \xrightarrow{k_{MDEA}} MDEAH^+ + HCO_3^- \quad (1)$$

Its pseudo-first-order reaction rate is given as:

$$R_{CO_2\text{-}MDEA} = k_{MDEA}[CO_2][MDEA] \quad (2)$$

Beside reaction 1, the reaction of bicarbonate formation in water is given by

$$CO_2 + OH^- \rightleftharpoons HCO_3^- \quad (3)$$

However, the contribution of this reaction is minimal and is usually neglected and is not considered here.

2.2. Reaction of CO_2 with Aminobutanol

The reaction of CO2 with primary amine aminobutanol is well explained [7]. According to the zwitterion mechanism, this proceeds first by the formation of an intermediate zwitterion, which then undergoes a deprotonation by all bases present in the solution resulting in carbamate formation according to the following Eqs (4, 5):

$$CO_2 + RR'NH \underset{k_{-1}}{\overset{k_1}{\rightleftharpoons}} RR'NH^+COO^- \quad (4)$$

$$RR'NH^+COO^- + B \underset{k_{-b}}{\overset{k_b}{\rightleftharpoons}} RR'NCOO^- + BH^+ \quad (5)$$

The rate of reaction of CO_2 in aqueous solutions of 2-amino-1-butanol can be described as:

$$R_{CO_2} = -k_{ov}[CO_2] = -\frac{k_1[CO_2][AB]}{1+\left(k_{-1}/\left(\sum k_{b_i}[B_i]\right)\right)} \quad (6)$$

k_{bi} denote the deprotonation rate constants of the zwitterion by any base, such as $[H_2O]$, $[OH^-]$ or [AB]. Eq. (6) can be re-arranged in the following form:

$$k_{ov} = -\frac{[AB]}{1/k_2 + \left(1/\left(k_a[AB] + k_{hyd}[OH^-] + k_w[H_2O]\right)\right)} \quad (7)$$

where, the new constants are defined as $k_a = k_2 k_{AM}/k_{-1}$, $k_{hyd} = k_2 k_{OH^-}/k_{-1}$, $k_w = k_2 k_{H_2O}/k_{-1}$

In a single-step termolecular mechanism [9] as represented by Eq. (8) and illustrated in figure 1, only one step is involved in the reaction.

$$CO_2 + RR'NH...B \rightleftharpoons RR'NHCOO^-...BH^+ \tag{8}$$

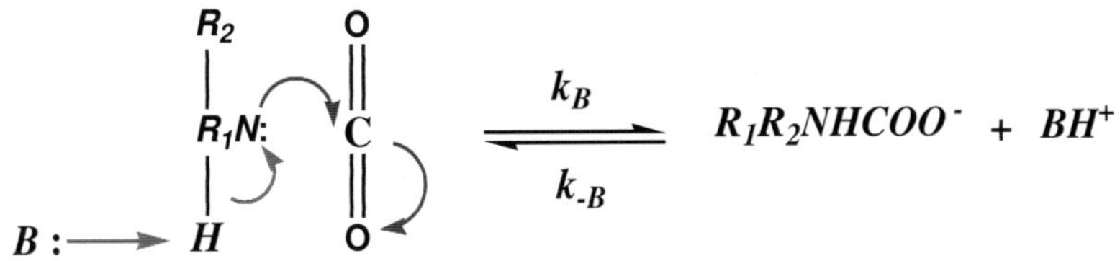

Figure 1. Schematic drawing of single-step termolecular reaction mechanism [8]

According to this mechanism, the CO_2 reaction rate expression is a second order with respect to amine, it is very similar to the one derived by Crooks and Donnellan [8]:

$$Am\cdots B + CO_2 \leftrightarrow AmCOO^- \cdots BH^+ \tag{9}$$

Analysis of the rate expression above shows that termolecular mechanism is equivalent to the asymptotic limit of the zwitterion mechanism when deprotonation is the rate-determining step: $1 \ll (k_{-1}/\sum k_{bi}[B_i])$

In this case Eq (6) becomes:

$$\begin{aligned} R_{CO2} &= k_{ov}[CO_2] = [CO_2][AB]\{\sum k_{bi}[B_i]\} \\ &= [CO_2][AB]\{k_a[AB] + k_w[H_2O] + k_{hyd}[OH^-]\} \end{aligned} \tag{10}$$

In general and regardless of the employed mechanism, a carbamate and a protonated base are the accepted products of the CO_2-amine reaction. In this present work, only zwitterion mechanisms is applied to interpret the obtained kinetics data.

2.3. Reaction of CO_2 with Mixtures of MDEA and Aminobutanol

For blends of MDEA and AB, the overall reaction rate with CO_2 is considered as the sum of reaction rates of CO_2-MDEA and CO_2-AB; hence:

$$R_{CO_2} = R_{CO_2\text{-}AB} + R_{CO_2\text{-}MDEA} \tag{11}$$

Using zwitterion mechanism, Equation 11 can be written as:

$$R_{CO_2} = -\left(k_2[AB] \Big/ \left(1 + \left(k_{-1}/\sum_i k_{b_i}[B_i]\right)\right) + k_{MDEA}[MDEA]\right)[CO_2] \tag{12}$$

$$\text{Or } R_{CO_2} = (k_{AB} + k_{MDEA}[MDEA])[CO_2] = k_{ov}[CO_2] \tag{13}$$

$$\text{where } k_{AB} = -\frac{k_2[AB]}{1 + \left(k_{-1}/\left(k'_{AB}[AB] + k'_{OH^-}[OH^-] + k'_{MDEA}[MDEA] + k'_{H_2O}[H_2O]\right)\right)} \tag{14}$$

which can be written as:

$$k_{AB}=-\frac{[AB]}{\frac{1}{k_2}+\left(1/\left(\frac{k_2k'_{AB}}{k_{-1}}[AB]+\frac{k_2k'_{OH^-}}{k_{-1}}\left[OH^-\right]+\frac{k_2k'_{MDEA}}{k_{-1}}[MDEA]+\frac{k_2k'_{H_2O}}{k_{-1}}[H_2O]\right)\right)} \tag{15}$$

For simplicity, new constants, k_a, k_{hyd}, k_b and k_w are defined as $k_a=k_2k'_{AB}/k_{-1}$, $k_{hyd}=k_2k'_{OH^-}/k_{-1}$, $k_b=k_2k'_{MDEA}/k_{-1}$, $k_w=k_2k'_{H_2O}/k_{-1}$. Then, Eq 15 becomes:

$$k_{AB}=\frac{[AB]}{(1/k_2)+\left(1/\left(k_a[AB]+k_{hyd}\left[OH^-\right]+k_b[MDEA]+k_w[H_2O]\right)\right)} \tag{16}$$

3. Materials and Methods

Reagent grade N-methyldiethanolamine (MDEA) with a mass purity of 99 % and 2-amino-1-butanol (AB) with a mass purity of 99% purchased from Sigma-Aldrich were used in this work. CO_2 solutions were prepared by bubbling CO_2 (> 99.99 % pure) in deionized water (DI) that was previously degassed using pure N_2. The concentration of CO_2 in the water was measured with a gas chromatograph (GC-6890, Agilent). This solution was then diluted by adding fresh water until the amine concentration used was at least 20 times higher than that of CO_2 to ensure that the pseudo first order condition with respect to CO_2 is always met [10]. The concentrations of amine mixture were checked by titration using standard 0.1M hydrochloric acid solution. The homogeneous reaction rate between CO_2 dissolved in water and amine solution was measured using stopped-flow apparatus (Hi-Tech Scientific Ltd., UK, Model SF-61DX2). Temperature was controlled by Lauda water bath model Alpha RA8 within ± 0.1 K. In all cases, the change of the amine concentration within the run time was very small and can be considered negligible for very low CO_2 loading (such as our case) as was demonstrated by Kierzkowska-Pawlak et al [11]. The reaction was monitored by measuring the conductivity, 'Y', change as function of time as described by Knipe et al. [12]. The change in the conductivity with respect to time was fitted to a first-order exponential equation in the form:

$$Y=-A.\exp(-k_{ov}.t)+Y_\infty \tag{17}$$

where k_{ov} is the pseudo first-order rate constant. Throughout the experiments, the average of three runs was taken to obtain consistent k_{ov} values. In all case, reproducibility error for k_{ov} was less than 5%.

4. Results and discussion

The experimental data of the observed rate (k_{ov}) of the CO_2 reaction with methyldiethanolamine (MDEA) and aminobutanol (AB) were obtained by fitting the conductivity-time curves to Eq (13). The rate constant k_{AB} was calculated by taking into account Eqs (13) and (14), Values of k_{MDEA} were taken from a previous work [13]. The obtained overall reaction rate constants, k_{ov}, were plotted against temperatures for different total concentrations (Figures 2 and 3). As observed in all runs, the overall rate constants (k_{ov}) increased with increased temperature and increased proportion of aminobutanol in the blend.

Upon applying the power law kinetics by plotting the CO_2-AB rate constant against the concentrations of aminobutanol (figure 4), an average exponent of 1.08 was obtained, which affirms that the pseudo first order regime prevails. Therefore, within the experimental concentration range, the reaction can be analyzed via the zwitterion mechanism [14].

IOP Conf. Series: Materials Science and Engineering **458** (2018) 012063 doi:10.1088/1757-899X/458/1/012063

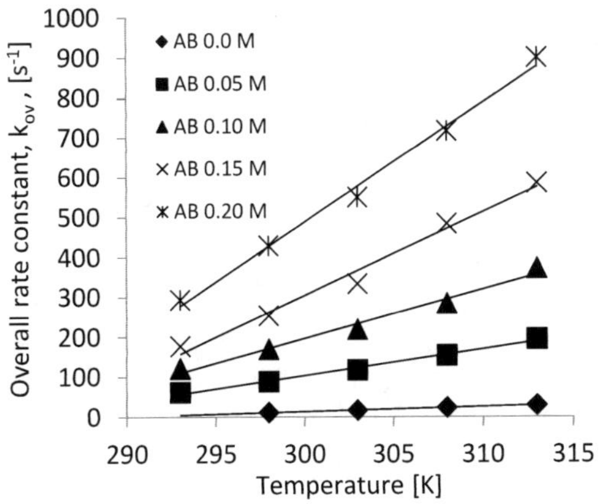

Figure 2. Rate constant, k_{ov}, versus temperature for total 1 molar solution

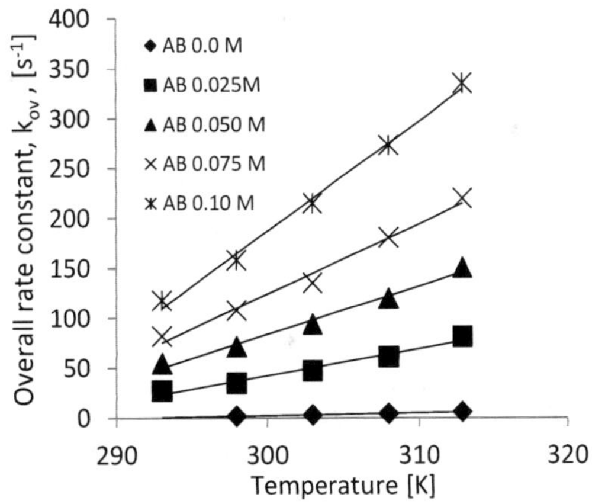

Figure 3. Rate constant, k_{ov}, versus temperature for total 0.5 molar solution.

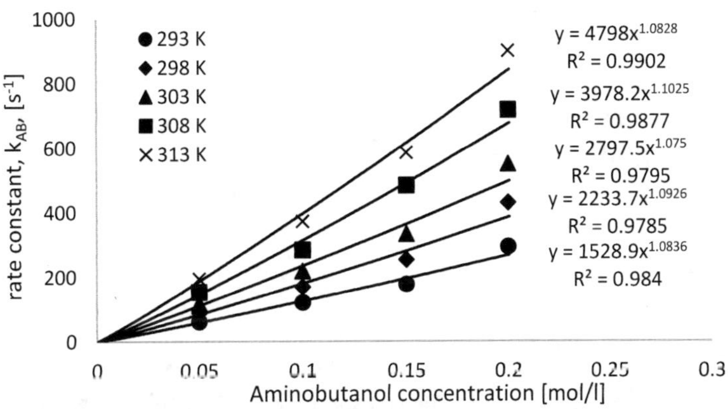

Figure 4. Rate Constant (k_{AB}) as a function of Arginine concentration at different temperatures

Experimental rate constants, k_{AB}, data were fitted to Eq (16) to extract the individual blocks of rate constants described in this equation. The concentrations of water molecules (H_2O) were calculated by mass balance while hydroxyl ions (OH^-) were estimated from the relation given by Astarita et al [10].

$$[OH^-] = \sqrt{\frac{K_W}{K_{Pi}}[AM]} \tag{18}$$

In this equation, amine concentration [AM] was considered to be the concentrations of MDEA and Arg independently. In this case, the total [OH-] was taken to be the sum of [OH-] ions produced by both MDEA and AB. The water dissociation constant, Kw, and protonation constant, K_{pi}, for MDEA and AB were taken from the literature [15-17]. Using Excel solver® nonlinear regression technique, experimental k_{AB} values were fitted to Eq (17) taking into account the species concentrations, H_2O, AB, OH- and MDEA previously calculated. The generated rates constants are summarized in table 1. values of k_2 were taken from a previous work [7].

Using these generated rate constants, the k_{AB} reaction rate constant values were predicted using Eq (16) and were plotted against the experimental ones as shown in figure 5. It is very clear that the adopted rate model along with extracted individual rate constants represent very well the experimental results

Table 1. Reaction rate constants at different temperatures

Rate constant	293 K	298 K	303 K	308 K	313 K
k_a (m^6 mol^{-2} s^{-1})	2045.0	2580.9	3270.3	5544.0	6192.9
k_{hyd} (m^6 mol^{-2} s^{-1})	891.6	4144.7	5537.5	9134.8	12552.0
k_b (m^6 mol^{-2} s^{-1})	8.0	7.6	9.9	17.5	22.0
k_w (m^6 mol^{-2} s^{-1})	23.4	24.4	33.8	37.7	45.7

with an average absolute deviation, AAD, of 6.8 %, which validated the choice of the kinetics model used to interpret the data represented by Eq (12). Furthermore, these results confirm the contribution of AB, OH- and MDEA, AB species in the base-catalytic formation of carbamate.

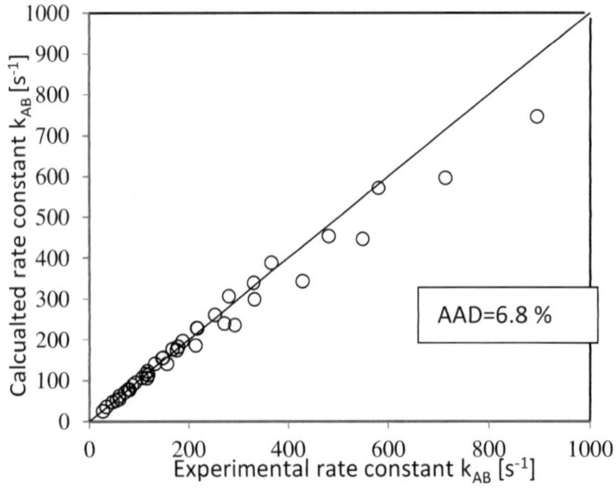

Figure 5. Calculated versus experimental rate constants

The individual rate constants in table 1 were plotted as function of temperature as shown in figure 6 and the results are summarized in table 2. The activation energy (E_a) of each reaction derived from the Arrhenius plots along with the pre-exponential coefficient of each rate constant are given in table 2.

Table 2. Summarized kinetics rate constants for CO_2-MDEA-AB reaction

Rate	$k_{AB}=[AB]/\left(\left(1/k_2\right)+\left(1/\left(k_a[AB]+k_{hyd}[OH^-]+k_b[MDEA]+k_w[H_2O]\right)\right)\right)$		
	Lnk_0	Ea (kJ/mol)	Equation
k_a (m^6 mol^{-2} s^{-1})	26.23	4423.0	k_a = 2.46 x 10^{11}*exp(-5464/T)
k_{hyd} (m^6 mol^{-2} s^{-1})	19.68	5211.1	k_{hyd} = 3.52 x 10^8*exp(-5211.1/T)
k_b (m^6 mol^{-2} s^{-1})	45.49	11217.0	k_b = 5.75 x 10^{19}*exp(-11217/T)
k_w (m^6 mol^{-2} s^{-1})	14.22	3256.9	k_w = 1.50 x 10^6*exp(-3256.9/T)

The E_a for AB, OH-, MDEA and H_2O catalytic carbamate formation showed that the contribution of water molecules to the overall formation of carbamate (14.22 kJ/mole) is the most important followed by that of Hydroxyl ions (19.68 kJ/mole) and aminobutanol molecules (26.23 kJ/mole). While the contribution of MDEA molecules to this reaction (45.49 kJ/mole) were found to be the least influential.

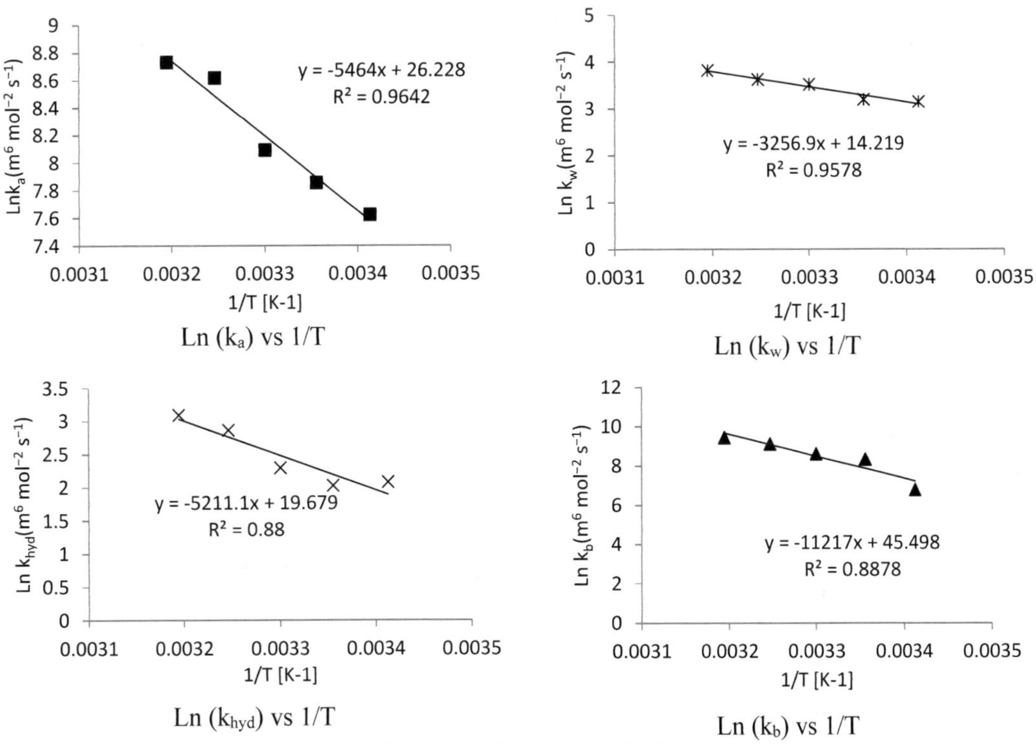

Figure 6. Arrhenius plots of CO_2-AB-MDEA rate constants.

5. Conclusion

The reaction kinetics of CO_2 with AB in aqueous solutions was studied for the first time using stopped-flow apparatus. The kinetics measurements were performed for concentrations ranging from 0.1 to 0.3 M of AB and a temperature range of 293 to 313 K. Using nonlinear regression technique, the overall rate constants of CO_2-AB were fitted to two different kinetic models derived from zwitterion termolecular mechanisms respectively. Among the two rate models adopted to interpret the kinetics data, the zwitterion mechanism was found to be very successful in accurately representing the experimental data with an AAD of 6.1 %. Subsequent rate constants and their temperature dependency along with their respective activation energies were derived. Analysis on the contribution of OH^-, AB and H_2O to the catalytic carbamate formation pathway showed that the contribution of H2O to the overall formation of carbamate (E_a = 91.96 kJ/mole) is more important, followed by the contribution of OH- (E_a = 104.41 kJ/mole). The contribution of AB was found to be the smallest (E_a = 201.57 kJ/mole). Comparison between the Kinetics of CO_2–AB reaction and those of CO_2 with MEA, MDEA, DEA and AMP showed that the CO_2-AB reaction is slower that of CO_2 with MEA but faster than those of CO_2 with MDEA, DEA and AMP, respectively.

Acknowledgement

This paper was made possible by an NPRP Grant #7-1154-2–433 from the Qatar National Research Fund (a member of Qatar Foundation). The statements made herein are solely the responsibility of the authors.

References

[1] Faramawy S, Zaki T and Sakr A A E 2016 Natural gas origin, composition, and processing: A review *Journal of Natural Gas Science and Engineering* **34** pp 34–54

[2] Peters L, Hussain A, Follmann M, Melin T and Hägg M B 2011 CO2 removal from natural gas by employing amine absorption and membrane technology—A technical and economical analysis *Chemical Engineering Journal* **172** pp 952–60
[3] Qi R and Henson M A 1998 Optimization-based design of spiral-wound membrane systems for CO2/CH4 separations *Separation and purification technology* **13** pp 209–25
[4] Chakravarty T, Phukan U K and Weiland R H 1985 REACTION OF ACID GASES WITH MIXTURES OF AMINES *Chemical Engineering Progress* **81** pp 32–6
[5] Liang Z, Rongwong W, Liu H, Fu K, Gao H, Cao F, Zhang R, Sema T, Henni A, Sumon K, Nath D, Gelowitz D, Srisang W, Saiwan C, Benamor A, Al-Marri M, Shi H, Supap T, Chan C, Zhou Q, Abu-Zahra M, Wilson M, Olson W, Idem R and Tontiwachwuthikul P 2015 Recent progress and new developments in post-combustion carbon-capture technology with amine based solvents *International Journal of Greenhouse Gas Control*
[6] Idris Z, Peresunko N, Jens K J and Eimer D A 2015 Equilibrium solubility of carbon dioxide in aqueous solutions of 3-amino-1-propanol, 4-amino-1-butanol and 5-amino-1-pentanol at low partial pressures *Fluid Phase Equilibria* **387** pp 81–7
[7] Benamor A, Mahmud N, Nasser M S and Tontiwachwuthikul P 2018 Reaction Kinetics of Carbon Dioxide with 2-Amino-1-butanol in Aqueous Solutions Using a Stopped-Flow Technique *Industrial & Engineering Chemistry Research* **57** pp 2797–804
[8] Crooks J E and Donnellan J P 1988 Kinetics of the formation of N, N-dialkylcarbamate from diethanolamine and carbon dioxide in anhydrous ethanol *Journal of the Chemical Society, Perkin Transactions 2* pp 191–4
[9] da Silva E F and Svendsen H F 2004 Ab initio study of the reaction of carbamate formation from CO2 and alkanolamines *Industrial & engineering chemistry research* **43** pp 3413–8
[10] Astaria G, Savage D W and Bisio A 1983 Gas treating with chemical solvents (John Wiley)
[11] Kierzkowska-Pawlak H, Siemieniec M and Chacuk A 2013 Investigation of CO2 and ethylethanolamine reaction kinetics in aqueous solutions using the stopped-flow technique *Chemical Papers* **67** pp 1123–9
[12] Knipe A C, McLean D and Tranter R L 1974 A fast response conductivity amplifier for chemical kinetics *Journal of Physics E: Scientific Instruments* **7** p 586
[13] Benamor A, Al-Marri M J, Khraisheh M, Nasser M S and Tontiwachwuthikul P 2016 Reaction kinetics of carbon dioxide in aqueous blends of N-methyldiethanolamine and glycine using the stopped flow technique *Journal of Natural Gas Science and Engineering* **33** pp 186–95
[14] Alper E and Bouhamra W 1994 Kinetics and mechanisms of reaction between carbon disulphide and morpholine in aqueous solutions *Chemical Engineering & Technology* **17** pp 138-40
[15] Littel R J, Bos M and Knoop G J 1990 Dissociation constants of some alkanolamines at 293, 303, 318, and 333 K *Journal of Chemical and Engineering Data* **35** pp 276–7
[16] Mahmud N, Benamor A, Nasser M S and Tontiwachwuthikul P 2018 Carbamate Formation and Amine Protonation Constants in 2-Amino-1-Butanol–CO2–H2O System and Their Temperature Dependences *Journal of Solution Chemistry* **47** pp 262–77
[17] Edwards T, Maurer G, Newman J and Prausnitz J 1978 Vapor-liquid equilibria in multicomponent aqueous solutions of volatile weak electrolytes *AIChE Journal* **24** pp 966–76

IOP Conf. Series: Materials Science and Engineering **458** (2018) 012064 doi:10.1088/1757-899X/458/1/012064

Optimization of multistage cross current extraction by iterative dynamic programming

Tazien Rashid[1], Thanabalan Murugesan[2], Sadiq Hussain[3]

[1]Chemical Engineering Department, NFC Institute of Engineering and Fertilizers Research, Faisalabad, Pakistan.

[2]Department of Chemical Engineering, Universiti Teknologi PETRONAS, Bandar Seri Iskandar, Tronoh, 32610, Perak Malaysia.

[3]NFC Institute of Engineering and Technology, Multan, Pakistan.

Abstract: In this study the optimal extraction of lignin and solvent distribution in a multistage cross current extraction cascade system is treated for the solution of constrained non-linear maximization problem. The objective was to simultaneously minimize the solvent loss and maximize solvent convergence in a finite cascade of extractors. Different possible loops are calculated through iterative dynamic programming, and the loops are restricted to certain objective functions. An analytical solution is developed which enables the amount of the make-up solvent to be chosen such that the total profit is maximized. The tabular results are represented on a general graph, giving the optimal combination and number of stages for different solvent distribution conditions.

E-mail:trtazien8@hotmail.com

1. Introduction

Multiple optimum solutions of a multistage problem are already employed by the chemical and biochemical industries [1-3]. In the chemical manufacturing processes many criteria exist for the selection of the best optimum conditions under which to design and operate a project. It may be desired to increase the yield of final product, or to maintain its quality above a certain minimum, or more generally it is desired to increase the sales value and to reduce the operating costs for any operation [1,4]. An established practice to optimize the multistage problems is to apply the dynamic programming technique [5]. In dynamic programming various possible decisions are made at each individual stage so as to reduce the expected cost. The serial computation is employed in this optimization approach in logic that the decisions of the next stage are dependent upon the results of the previous stages involved [4, 5]. A mathematical model needs to be developed for such a process involving an objective function, which deals with the variables and the constraints for each individual stage. The constraints are subjected to certain inter stage dependencies.

Lagrange multipliers are traditionally used for constrained external problems. However, the Lagrangian multipliers method involves the differential functions and the multiplier approach becomes computationally unmanageable. These difficulties can be overcome by the use of dynamic programming. In dynamic programming multivariable optimization problems are solved individually into a series of one variable optimization problems each of these involving a small number of total variables. The individual one variable problem may be solved using standard methods of differential calculus or using simple search procedures.

In literature quite, a few applications of dynamic programming to chemical engineering problems are being reported. Aris et al., (1961) has solved problems concerned with reactor design and a problem of cross-current extraction cascade [6]. Similarly, Nemhauser et al., (1961) utilized dynamic programming to study the optimization of separation processes in detail [7]. In a study, Bellman et al., (1961) has proposed the basic theory of dynamic programming [8] by studying multistage decision processes in detail. Later, Rudd & Blum, (1962) extended the dynamic programming approach of Aris et al., (1961) and determined the optimum operating conditions for cross-current extraction with product recycle. Their work involved the iterative table entries and back substitutions [9]. These concepts were implemented on a multistage recycle system with recycle back to the first stage in detail by using the same dynamic approach by Van Cauwenberghe, et al., [10]. Optimization of multistage separation processes was explored by Nemhauser (1963) by using dynamic programming approach [7, 11, 12], they reported the mass transfer effect in solid separation problems in particular. Hence the use of dynamic programming is appropriate for the design in the chemical industry where the objective function for a complicated system can often be obtained by dividing the overall system into a series of stages. Optimizing the resulting simple stages can lead to the optimal solution for the original complex problem.

The present work uses a similar approach for analyzing the solvent distribution problem in the case of a cascade of four equal-sized extraction tanks. Based on our previous research Rashid et al., [13], it is evident that due to high efficiency of pyridinium formate ([Py][For]) a protic ionic liquid i.e. (≥ 90 % lignin extraction), it is economical to use it for further extraction cycles pertaining to the make-up solvent, in order to achieve more economical utilization of the solvent. All possible considerations have been taken into account and the final design choice of optimal solvent distribution can be made from a purely economic point of view. For this case the solvent distribution in a multistage cross current extraction cascade system is treated for the solution of constrained non-linear maximization problem. Different possible routes are calculated through iterative dynamic programming, and the routes are restricted to certain objective functions. In this particular arrangement, an analytical solution is developed which enables the optimum route and the amount of the make-up solvent to be chosen such that the total profit is maximized.

2. Materials

The starting material used for the current study include, pyridinium formate (synthesized as previously established procedure Rashid et al., [14]), Fresh biomass feed empty fruit bunch (EFB) samples were supplied by FELCRA, Nasaruddin Oil Palm Mill, Bota, Perak, Malaysia and stored at ≈ 5 °C. The raw biomass was washed with 2 % detergent solution in order to remove oil and greases. The washed biomass samples were dried under natural sun light for 24 hrs before being grinded and crushed using power cutting mill (Pulverizette 25). Subsequently, the grinded biomass samples were sieved to attain the smallest particle sizes of (0.1-0.3mm). All the chemicals were used as received. Triple distilled water was used for all washing purposes.

2.1 Methodology

2.1.1 General Schematics and Concept of Cross Current Extraction Cascade

The solvent [Py][For] and biomass feed continuously pass through different extraction stages in order to achieve the desired specification which is the solvent should achieve the saturation level and the biomass is nearly free of lignin. In chemical industry a common interest is the optimization of an objective function by the most suitable choice of the independent design variables. An objective function can be defined in which the individual costs or profits from each sub-unit in the complex process may be summed and the total optimized with respect to the design or operating variables using the common optimizing techniques [3, 4]. These concepts can be employed for the efficient concept of bio refinery which comprises of the efficient separation of oil palm biomass constituents as well as the efficient utilization of the solvent. To evaluate this particular need some sets of experiments were performed to collect data. Experiments up to two iterations containing (stage 1, 2 and 3) were conducted to collect data for efficiency graph (figure 2) and solvent loss in each stage. All the experiments were carried out on optimized conditions obtained from the second order polynomial equation for lignin extraction from our previously published work Rashid et al., [13]. To begin with, it was first considered as the operation of a single stage (figure 1(a)). The solute "lignin" is to be extracted using fresh solvent "S_o", in a series of "n" extraction tanks with "R_n" and "E_n" be the concentration of solute in the Raffinate and Extract streams leaving the nth stage. Based on this a general scheme of cross current extraction cascade is shown in figure 1(b). The biomass feed "F_o" has been introduced in first extraction stage. The concentration of the biomass feed entering in first stage has comparatively high initial lignin contents "C_o" as compared to second and third extraction stages. The fresh solvent "S_o" due to high extraction efficiency does not achieve the allowable lignin capacity (determined experimentally (0.710 g/ml) Rashid et al., [14]) through first extraction stage so the exit solvent "S_1" of this stage enters in the second stage which has comparatively low lignin contents "C_1".

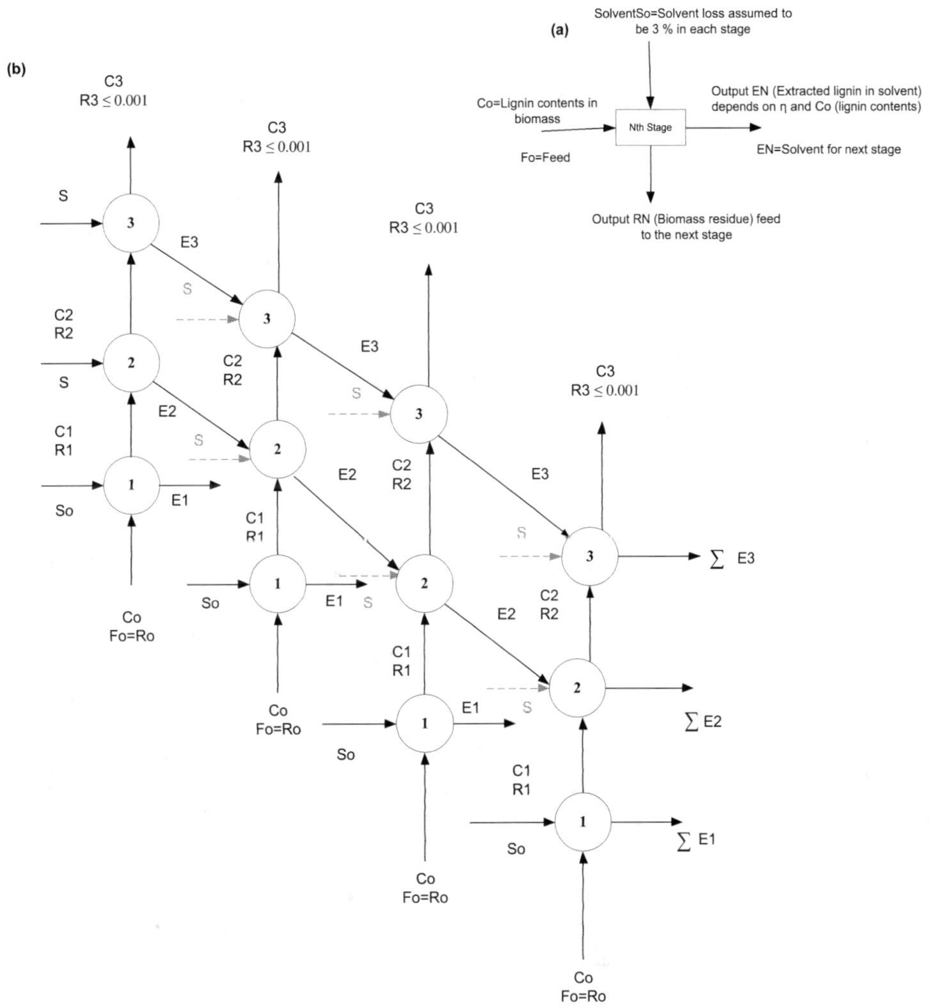

Figure 1. General schematic of cross current extraction cascade.

A certain amount of fresh solvent "S_o" was added to make the established wt % constant. The lower lignin contents "C_1" allows the solvent to extract the maximum remaining lignin contents to "C_2". This continues until the solvent achieves the saturation at 0.710g/ml. The scheme in Figure 1(b) is used for the selection of optimal solvent distribution for lignin extraction considering various routes. Various possible routes are systematically determined for optimal solvent utilization and is presented in results and discussion section. The mathematical model is formulated as a set of different equations and the details are discussed in the following sections;

2.1.2 Objective Function for Optimal Solvent Distribution

An objective function that serves as a comparison between the various routes suggested above is required. The formulation of the objective function and the notations used are detailed in this section. The extract "E_n" and raffinate "R_n" streams are in equilibrium, then; $E_n = R_n$. The problem is the maximization of a specified function f_n to the final state by an optimal choice of the decision θ_n.

$$E_n = f_n (E_{n-1}, \theta_n) \qquad n = 1, 2, N \tag{1}$$

where "θ_n" is the decisions that may be made for the solvent distribution for the operation of nth stage. The profit "P" obtained in each stage will be taken as being given by the sales income from the lignin extracted less the cost of the extracting solvent [Py][For] used [4, 15]. Thus if "λ_S" is the unit cost of the added solvent ($/ml) and "$\lambda_L$" is the unit income from the sale of the extracted lignin ($/gm), the profit from the nth stage may be written as;

$$\psi = \lambda_L \sum_{n=1}^{N} E_{n+1} - \lambda_S (V_o) \tag{2}$$

In actuality, the feed as well as the solvent used is all directly fed into the first extractor. There is a drop in the amount of solvent used between the first and second stages and then it rises slowly from the second to the nth stage. This extra amount of solvent used, "V_L" which is independent of the number of extractors in the system, also gives rise to an additional cost, which must be subtracted from the profit, this solvent loss was determined experimentally which is almost 3% for each stage, hence the objective function becomes,

$$\psi = \lambda_L \sum_{n=1}^{N} E_{n+1} - \lambda_S (V_o - V_L) \tag{3}$$

Extraction from Biomass, the equation governing this process is

$$R_n = F_{n^*} C_n \tag{4}$$

the initial concentration of lignin in biomass is C_o, then

$$C_{n+1} = [C_n - (\eta_{n+1} * C_n)] \tag{5}$$

The efficiency is a function of E_n (gm/ml), experimentally obtained Figure 2;

$$\eta_n = - 0.0021 * E_n^2 - 228 * E_n + 92.455 \tag{6}$$

As the solvent has a saturation limit of 0.710 g/ml; before reaching that limit it has a high efficiency in extracting lignin. In a given stage there is a combination of streams coming from different previous stages. The volume of solvent at stage "N" can be calculated as:

$$V_{n+1} = V_n - (V_n * W) \tag{7}$$

Where "W" represents the % loss of solvent (i.e amount left (ml) of the initial solvent used in each stage) achieved between stages which was determined experimentally. With a material balance for lignin around stage n (where n = 1, 2, 3) we get;

$$E_{n+1} = R_n * \eta_{n+1} / V_n + E_n \tag{8}$$

The solution has been implemented by using solver add-in tool, Microsoft Excel (nonlinear solving method). for the calculation at every stage and also for determining the optimal route to get the maximum profit using dynamic programming analysis of Rudd et al., [9]. The routes are based on the best selection of make-up solvent. The optimum route was considered which gives the minimum solvent loss and maximum lignin removal.

3. Results and discussion

The biomass feed introduced in the first stage enters at comparatively high lignin contents as the second and third extraction stages. The lignin contents of the biomass feed EFB were determined experimentally by TAPPI T222 om-02 method which were found to be 17.8 %. The first stage operates at 17.8 % lignin contents, the second stage operates at 1.42 % lignin contents and stage three operates at 0.11 % lignin contents as ≈ 90% of lignin is being extracted in each consecutive stage. The solvent in the first stage due to its high efficiency does not achieve the allowable lignin saturation ($\sum E_n+1 \approx 0.710$ g/ml) so the exit solvent is reused in the second and third stages. The lower initial lignin contents in stage two and three allows to remove the maximum remaining quantity of lignin contents but the solvent still has not achieved its saturation concentration. Hence it is economical to use it for further extraction cycles pertaining to the make-up solvent, in order to achieve more economical utilization of the solvent. The systematic determination of routes for optimal solvent utilization is shown in figure 1(b). Stage 1, (figure 1(b)) serves as a root stage of the cross current extraction cascade. The Feed "F_o"has to pass through stage 1 before passing through the stage 2 and 3. The possible routes for solvent saturation are shown in figure 1(b).

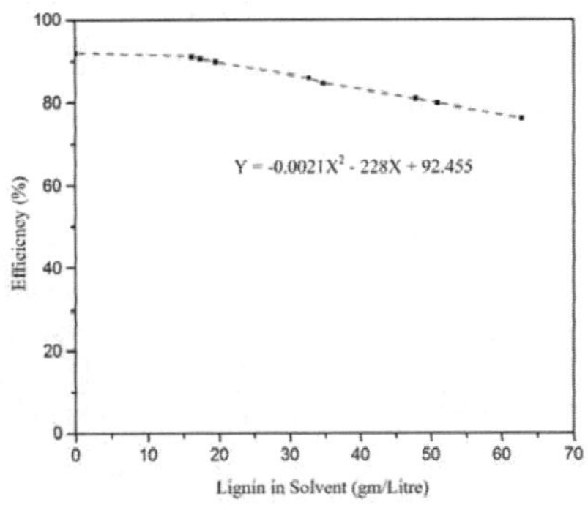

Figure 2. Efficiency vs amount of lignin in solvent.

The routes are based on the best selection of make-up solvent, for example route #1 follows fresh solvent at each stage 1 and no make-up solvent is involved in this scheme, and iterations are run until the constraints ($\sum E_1 \approx 0.710$ gm/ml) are achieved. In route #2 fresh solvent enters at stage 2 and passes downstream through stage 2 with makeup solvent, route # 3 involves the entry of fresh solvent at stage 3 and passes downstream with make-up solvent. The route # 4 involves the fresh solvent from stage 2 and passes through next loop's stage 2 and 3 respectively. The last route #5 involves the fresh solvent enters at stage 1 and 2 and the stage 3 involves make up solvent. All the routes iterations were run until the constraints ($\sum E_n+1 \approx 0.710$ g/ml) are achieved.

3.1. Selection of Optimal Route

Arranging stages into cascades allows more separation or less energy input than is possible in a single stage. The solution of the mathematical model formulated has been implemented by using solver add-in tool, Microsoft Excel Microsoft Corp. for the calculation at every stage and also for determining the optimal route to get the maximum profit using dynamic programming analysis of Rudd et al., [9]. For the route # 1, fresh feed "Fo" with "Co" lignin content enters at stage1, as the fresh feed contacts with fresh solvent in all stages so there is no drop-in extraction efficiency "ηn". The computations for the route # 1 are given in table 1.

Table 1. Computation for "E_n" for route # 1 for single iteration

Stage	V_o(ml)	F_o (gm)	C_n (%)	E_n (gm/ml)	η_n	R_n (gm)
			Route # 1			
1	80.00	7.894	0.1780	0.01612	0.9201	0.0938

For the route # 2 feed "F_1" after first extraction and reduced lignin contents "C_1" enters in stage 2, fresh solvent enters into stage 2 and follows downstream through stage 2 with make-up solvent. As the solvent is reused for multiple extractions of stage 2 so there is a drop in % extraction efficiency "η_n". This drop in" η_n" can be calculated from equation (6). The computations for the route # 2 are given in table 2.

Table 2. Computation for "E_n" for route # 2 for single iteration

Stage	V_o(ml)	F_o (gm)	C_n (%)	E_n(gm/ml)	η_n	R_n (gm)
			Route # 2			
1	80.00	7.894	0.1780	0.0162	0.9201	0.0938
2	66.85	6.597	0.0142	0.0013	0.9201	0.0074

For the route # 3 the same process is considered but the stage 2 is replaced by stage 3. The calculations for the route # 3 are given in table 3.

Table 3. Computation for "E_n" for route # 3 for single iteration

Stage	V_o(ml)	F_o (gm)	C_n (%)	E_n (gm/ml)	η_n	R_n (gm)
			Route # 3			
1	80.00	7.894	0.1780	0.0162	0.9201	0.09389
3	66.30	6.543	0.0011	0.0001	0.9201	0.00059

For the route # 4 it is considered that the feed "F_1" after first extraction and reduced lignin contents "C_1" enters at stage 2, fresh solvent enters into stage 2 and this solvent enters further into next stage 2 and 3 of the next loop (Figure 1(b)) with make-up solvent. The computations for the route # 4 are listed in table 4.

Table 4. Computation for "E_n" for route # 4 for single iteration

Stage	V_o(ml)	F_o (gm)	C_n (%)	E_n (gm/ml)	η_n	R_n (gm)
			Route # 4			
1	80.00	7.894	0.178	0.0162	0.9201	0.09389
2	66.85	6.597	0.0142	0.0013	0.9201	0.00740
2	66.85	6.597	0.0142	0.0013	0.9215	0.00740
3	66.30	6.543	0.0011	0.0001	0.9186	0.00059

The route # 5 is a conventional one in which the feed passes through stage 1,2 and 3 respectively. The fresh solvent enters at stage 1 and 2 and the stage 3 is working with make-up solvent (table 5).

Table 5. Computation for "E_n" for route # 5 for single iterition

Stage	V_o(ml)	F_o (gm)	C_n (%)	E_n (gm/ml)	η_n	R_n (gm)
			Route # 5			
1	80.007	7.894	0.1780	0.0162	0.9201	0.09389
2	66.857	6.597	0.0142	0.0013	0.9201	0.00740
3	66.309	6.543	0.0011	0.0001	0.8847	0.00059

The solvent loss, E_n and R_n concentrations for all of the routes and stages are determined. In table 6 it can be observed that route # 5 which is the most conventional route gives the maximum convergence but on the other hand it can be seen that the solvent loss and number of stages are maximum per iteration. The optimum cost for the individual stage and process will be higher, hence route # 5 is rejected. From the table 6, the feasible solvent distribution and combination of stages seems to be route # 2, which produces an equal profit as route # 5, showing that route # 2 cascade is a fairly good approximation. This table also shows that the extraction is more complete, if more stages are used. Although this requires more solvent and more loss of solvent consequently.

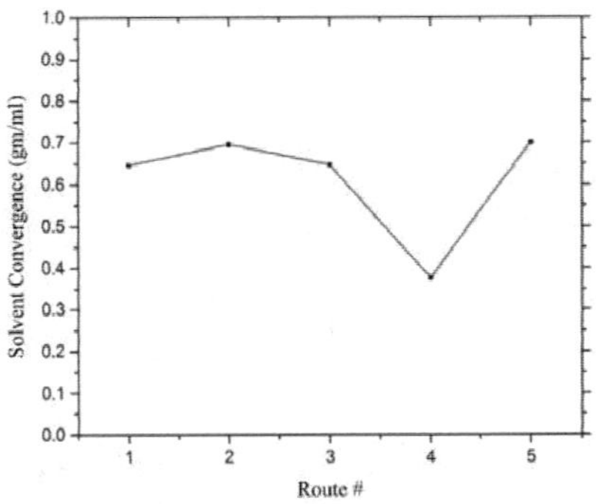

Figure 3. Solvent convergence plot for various routes.

It can be suggested that, in order to attain optimal operation fresh solvent should be utilized in the first stages and should be reused during further downstream. From table 6 it can also be concluded that route #1 requires less solvent and hence the solvent loss is minimum, but on the other hand the raffinate concentration in the exit stream is comparatively higher than other routes which is also another parameter to consider for bio-refinery concept. Hence the selection of the route is dependent on the concentration of R_n for lignin extraction as it should be minimum in the remaining biomass residue.

Table 6. Overview of various studied routes

Cascade Route	Convergence E_n (gm/ml)	Solvent Loss (ml)	Profit ($)	No.of Iterations	Stages Iteration
Route # 1	0.646	96	108	10	4
Route # 2	0.696	176	113	10	8
Route # 3	0.647	176	107	10	8
Route # 4	0.377	168	88	10	8
Route # 5	0.700	255	113	10	12

From figure 4 it is clear that the profit is a function of solvent loss as greater the solvent loss lesser is the profit achieved [10]. Conversely it is evident from figure 4 that solvent convergence is higher for the longer route as more lignin is available to be extracted and hence the profit is increased. Pertaining to this fact it can be observed from figure 4 that the route # 5 is giving maximum solvent loss but instead the profit is found to be increased.

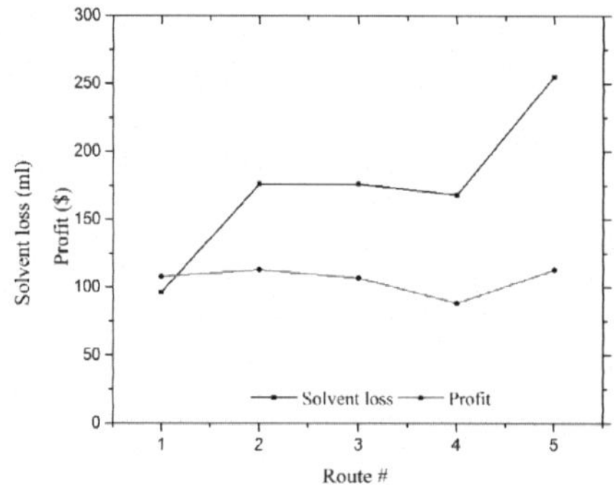

Figure 4. Profit and solvent loss relation for lignin crosscurrent extraction cascade.

This may be due to the reason that more lignin is available to be extracted in route # 5 and hence the profit is greater, but this increase in profit is very small that it cannot be considered on the cost of the solvent loss. Hence, from figure 3 and figure 4 it is evident that route # 2 and 3 are comparable but due to higher convergence achieved by route # 2, it is selected as optimum route.

4. Conclusions

A multistage cross current extraction cascade system is treated for the solution of constrained non-linear maximization problem for optimum solvent distribution. Different possible routes are calculated through iterative dynamic programming and the route which gave maximum solvent saturation with minimum solvent losses was suggested for further large-scale applications. This is of a great significance from an industrial viewpoint, as a significant amount of solvent and material and energy could be reduced if such a process is implemented.

References

[1] Levenspiel O 1999 *Chemical Reaction Engineering* 3rd ed (John Wiley & Sons)
[2] de Andrade Lima L and Hodouin D 2006 Simulation study of the optimal distribution of cyanide in a gold leaching circuit *Minerals Engineering* **19** pp 1319-27
[3] Moreira N and Malcata F 1996 On the optimum distribution of enzyme feed in a cascade of CSTR's performing an enzyme-catalyzed reaction with deactivation *Bioprocess and Biosystems Engineering* **15** pp 301-6
[4] Oi K, Sayama H and Takamatsu T 1978 Multimodality analysis of a class of multistage allocation problems *Journal of Optimization Theory and Applications* **25** pp 33-48
[5] Antonio J K, Tsai W K and Huang G M 1991 A highly parallel algorithm for multistage optimization problems and shortest path problems *Journal of Parallel and Distributed Computing* **12** pp 213-22
[6] Aris R 2000 The Optimal Design of Chemical Reactors A Study in Dynamic Programming **3** (Elsevier)
[7] Nemhauser G L 1961 Ph.D. thesis (Northwestern University)
[8] Bellman R 1961 Dynamic Programming (New Jersey)
[9] Rudd D and Blum E 1962 Optimum cross-current extraction with product recycle *Chemical Engineering Science* **17** pp 277-81
[10] Van Cauwenberghe A 1967 Optimization of multistage recycle system *Chemical Engineering Science* **22** pp 193-200
[11] Mitten L G and Nemhauser G L 1963 Multistage optimization *Chem. Eng. Progress* **59** (1) pp 52-60
[12] Mitten L G and Nemhauser G L 1963 Optimization of multistage separation processes by dynamic programming *The Canadian Journal of Chemical Engineering* **41** pp 187-94

[13] Rashid T, Gnanasundaram N, Appusamy A, Kait C F and Thanabalan M 2018 Enhanced lignin extraction from different species of oil palm biomass: Kinetics and optimization of extraction conditions *Industrial Crops and Products* **116** pp 122-36
[14] Rashid T, Kait C F, Regupathi I and Murugesan T 2016 Dissolution of kraft lignin using Protic Ionic Liquids and characterization *Industrial Crops and Products* **84** pp 284-93
[15] McEwan M W and Beveridge G S G 1965 Process optimization at the design stage with an objective function linear in profit and investment *Chemical Engineering Science* **20** pp 987-995

CFD simulations of convective heat transfer in actively heated tube bundle

Ola Karar, Sampath Emani, Ramasamy Marappa Gounder * and Hilmi Mukhtar

Chemical Engineering Department, Universiti Teknologi PETRONAS, 32610, Bandar Seri Iskandar, Perak D.R., Malaysia

marappagounder@utp.edu.my

Abstract. Active heating is used as a flow assurance technique to prevent the huge temperature drop of crude oil while transporting from wells to facilities, which creates a friendly environment for hydrate formation and solid depositions. In the present study, convective heat transfer in tube bundle with three internal tubes was investigated through computational fluid dynamics simulation. A three dimensional model for existing bundle was developed. The model described the heat transfer occurring inside the bundle, aiming to heat up or at least maintain the production fluid temperature. The heat transfer from heating pipe to the whole bundle can be visualized through the temperature contours and the temperature gradient at the outlets. The effect of heating medium temperature and flow velocity are studied in the developed model.
Keywords - Active heating; tube bundle; computational fluid dynamics

Nomenclature

A	Area [m^2]
C_p	Specific heat [J/kg.K]
H	Source term
g	Gravitational force [m/s^2]
K	Thermal conductivity [W/m.K]
M	Mass flowrate [Kg/s]
n	Number of tubes
Nu	Nusselt number
P	Pressure [Pa]
Pr	Prandtl number
Q	Rate of heat transfer [w]
Ra	Rayleigh number
T	Temperature [K]
v	Velocity [m/s]
ρ	Density [kg/m^3]
τ	Shear Stress [Pa]
μ	Dynamic viscosity [$N.s/m^2$]

Subscripts

f	Film

1. Introduction

The presence of hydrocarbons, water and other impurities, in crude oil eases the formation of hydrates as crude temperature falls below the hydrate formation temperature when transported from wellhead to facilities through flowlines and risers. Hydrates are solid-crystalline structured compounds formed when methane and other light hydrocarbons are trapped within water crystals under high pressure and low temperature. Once formed, hydrates continue to grow and aggregate and act as flow obstruction that causes production losses and environmental issues with high costs of maintenance. A study reported three incidents associated with hydrate plugs where the elbow was ruptured and resulted in loss of lives and over USD 7 million in capital cost [1].

Another flow related issue is wax deposition which restricts the crude oil flow and forms a three dimensional structure known as crude oil gel [2]. Wax precipitates from the oil when the temperature falls below wax appearance temperature (WAT) and transport with the production fluid or build up on the pipe surface.

Offshore harsh environment has increased the need to find ways to mitigate the issues of hydrate formation, wax and solid depositions which was later introduced as flow assurance by the oil conglomerate Petrobras in the 1990's [3]. Flow assurance techniques guarantee safe, economic and reliable flow throughout the whole field life. The main concern is to maintain the temperature of the oil above hydrate formation temperature and WAT.

1.1. Flowline Thermal Management

1.1.1 Chemical Injection. Pour point depressants (PPDs) are added to waxy crude oil at an optimum concentration to lower its pour point, viscosity, yield strength and the wax deposition rates [4]. PPDs once injected in the pipeline, co-precipitate with the wax crystals and disturb its structure or hinder the wax particles to grow further. Low dosage hydrate inhibitors (LDHIs) are more cost-effective compared to conventional thermodynamic inhibitors [5]. Lower concentration of LDHI is needed compared to the conventional thermal inhibitors. Thus, it reduces the injection rates and the volume required. However, there are some challenges associated with using LDHI as it is non-recoverable [5].

1.1.2 Insulation. Insulation of flow lines and risers also reduces the heat loss to seawater; therefore, it keeps the crude oil warm and prevents hydrate formation. Using insulated tubing system will significantly decrease the volume of inhibitors injected into the system [6]. Materials such as polypropylene, polyethylene and polyurethane are used for flowline insulation depending on the required thermal conductivity [7]. Insulation techniques have improved starting from the conventional pipe in a pipe system filled with gas or vacuum in the annulus. However, the double-walled pipe is not economic for longer pipes

1.1.3 Direct and indirect heating. In addition to the aforementioned methods, thermal hydrate management includes electrical heating and hot fluid circulation; in the first method, heat is directly generated through electrical cables where anodes are used to provide electrical communication with sea water. The other method is indirect heating through hot fluid circulation, where the heat is transferred through convection and radiation. Hot fluid circulation can be used with the conventional pipe in a pipe configuration, the heating medium flows in the annulus of the carrier pipe to maintain or heat up the production fluid temperature and prevent hydrate formation. Another configuration is using heat circulation in tube bundle which consists of multiple internal pipes enclosed together in one carrier pipe. The space between the internal pipes can be filled with air, nitrogen or water [8]. Most common heating fluids are water and glycol solution. Glycol-water solutions proved to have a better efficiency in heat transfer, lower corrosion and good microbiological control properties [9].

Throughout the past decays, tube bundles proved to be the most economical, efficient and practical flow assurance strategy for hydrate and wax mitigation through combining hot fluid circulation with cost effective insulation. Moreover, the carrier pipe or outer jacket provides a good resistance to the seawater hydrostatic pressure [10].

Several studies have contributed in solving the problems of flow assurance in industry using indirect active heating method. However, heat transfer processes employed in some situations and the heat transfer coefficient predictions require improvement. Thus, the present study aims to investigate the heat transfer in 3 internal pipes tube bundle through computational fluid dynamics modelling. The effect of the heating medium temperature and flowrate on total heat gained by the production fluid was studied in the present work.

2. Convective heat transfer in tube bundles

Throughout the years, both internal fluid flow within an enclosure and external convection in a body surrounded by an infinite fluid medium grasped the interest of researchers in heat transfer area. Among the early theoretical models for natural convection was the correlation developed by Churchill and Churchill [11] for vertical plates, horizontal cylinders and spheres, which proposed an expression for Nusselt number (for both laminar and turbulent flow regimes) as in:

$$Nu^{0.5} = Nu_0^{0.5} + \left[\frac{Ra/300}{\left[1+\left(\frac{0.5}{Pr}\right)^{9/16}\right]^{16/9}} \right]^{1/6} \quad (1)$$

which gives satisfactory results for all Ra > 10^{-4}

Another simplified correlation describing convective heat transfer in tube bundles was developed by Tillman [12] for in-line arrays of tubes

$$Nu_{\mathrm{f}} = 0.057\, {Ra_{\mathrm{f}}}^{0.5} \quad (2)$$

and equation (3) for staggered arrays

$$Nu_{\mathrm{f}} = 0.067\, {Ra_{\mathrm{f}}}^{0.5} \quad (3)$$

Churchill and Churchill correlation was then taken as a basis for an experimental and numerical study on tube bundle active heating by Liu [13]. Convective heat transfer between the three internal tubes enclosed in a carrier pipe filled with water was investigated and Nusselt number of a given tube surface was expressed in terms of Rayleigh number for all the tubes and Prandtl number from the expression.

$$Nu_{\mathrm{i}} = c\, {Ra_{\mathrm{i}}}^{1/3}\, [f(\mathrm{Pr})]^{1/3}\, (\,\Pi_{j=1}^{n}\, Ra_{\mathrm{j}}\,)^{d}] \quad j \neq i \quad (4)$$

where

$$[f(\mathrm{Pr})]^{1/3} = 0.8829 - 0.0013628\,(T_{\mathrm{f}} - 21.85) \quad (5)$$

A study was conducted for Gulf of Mexico field to investigate the thermal interaction between the pipes in a bundle system and the effect on overall thermal performance and fluid arrival temperature [14]. The bundle had 3 insulated flow lines encased in steel, it was reported that using active heating, the entire bundle is performing as counter flow tube heat exchanger and computer model was developed to simulate multiphase flow in the bundle. The results have shown that doubling the flow rate of the heating medium increased the crude oil temperature by 6.8℃ [14]. Thermal performance of tube bundle depends on bundle configuration, production fluid and heating medium temperature and characteristics, flow properties and the fluid flowing in the annulus of the tube.

Theoretical and experimental investigations on heat transfer analysis through tube bundles are available in literature for steady and transient states. Zabaras and Zhang [15] studied the steady state and transient behavior for tube bundle system using six different configurations. Overall heat transfer coefficients were determined and the cool down performances of the six configurations were compared to the single pipe in pipe. However, the natural convection in nitrogen filling the bundle annulus was excluded from calculations.

Sallehud-Din [16] developed CFD model to predict heat transfer in 3 meter long tube bundle with 4 internal pipes enclosed in a sleeve pipe with spacers at both ends to hold the internal pipes in place. The model provided steady state calculations for the tube bundle in horizontal, inclined and vertical positions and it was reported that increasing the inclination angle will increase heat loss to surroundings and thus, reduce the efficiency of the tube bundle [16].

Danielson and Brown [17] developed steady state analytical model to predict the thermal performance of a 3-lines tube bundle and temperature threshold for hydrate mitigation. An OLGA simulation was performed and the data from the analytical model was used to evaluate the number of grids and overall heat transfer coefficient.

3. Flow circuit

The one-meter long bundle consists of two production pipes (204 mm OD) and one heating pipe (101.52 mm OD) enclosed in a carrier pipe (453.43 mm ID). A stainless steel tube (507.38 mm inside diameter) then surrounds the carrier pipe. Tubes are kept in position between two 25mm thick acrylic resin (Perspex) sheets. The sheets are tied together by tie rods external to the bundle.

Figure 1 presents the flow circuit, where water is continuously added to the internal tubes and the outer cavity. The storage tanks are filled with water which flows into respective tubes through pumps, temperature controllers are fitted to the tanks associated with the internal tubes to maintain the desired temperature and when exiting the tubes, water is fed back to the tanks.

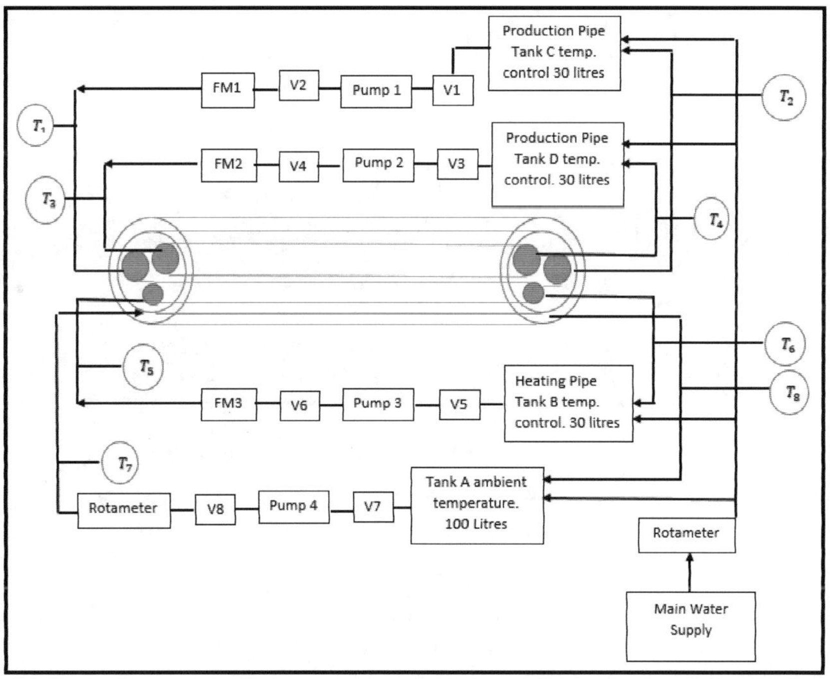

Figure 1. Tube bundle flow circuit.

4. Computational fluid dynamics study

CFD is used in simulating various flow and complex heat transfer problems. The governing equations for the model are solved together for each node in the developed mesh to investigate the system thermodynamic behaviour. ANSYS fluent was used to simulate the flow inside the tube bundle and predict the heat transfer.

4.1. Geometry and Meshing

A three-dimensional model of the prescribed tube bundle is developed. Steady state pressure based solver was used with K-epsilon turbulence model. Material flowing inside the internal tubes was set to be water. Figure 2 shows the developed geometry using design modeller.

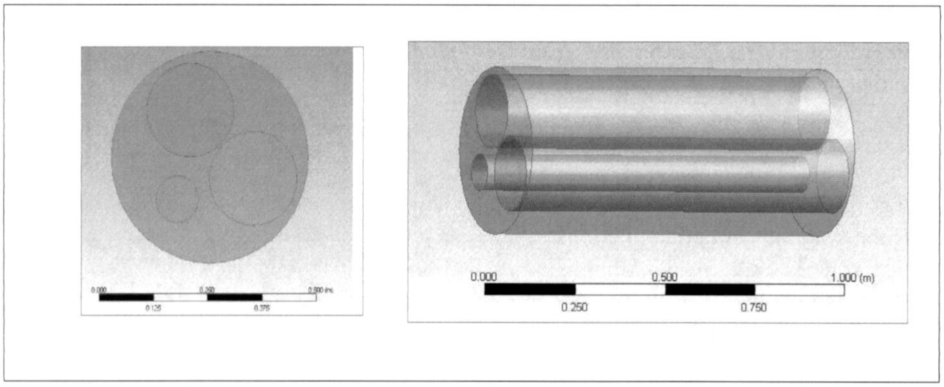

Figure 2. Three dimensional model of the tube bundle

Tetrahedral mesh was generated for the tube bundle with a total elements number of 0.46 million where governing equations are solved at each node. Boundary conditions were set to be velocity inlet at the tubes inlet and outlet at tubes end was set as pressure outlet. Figure 3 shows the developed grid.

Figure 3. Developed mesh

4.2. Governing Equations

The flow inside the tube bundle is governed by Navier Stokes equation which represents the heart of fluid modeling, derived from conservation of mass and momentum. The non-linear differential equations are solved together for each node in the developed mesh. Conservation of mass is represented by continuity equation.

$$\frac{\mathrm{d}\rho}{\mathrm{d}t} + \nabla \cdot (\rho v) = 0 \tag{6}$$

Conservation of momentum:

$$\frac{\mathrm{d}(\rho v)}{\mathrm{d}t} + \nabla . (\rho v v) = -\nabla p + \rho g + \nabla \cdot (\tau) \tag{7}$$

IOP Conf. Series: Materials Science and Engineering **458** (2018) 012065 doi:10.1088/1757-899X/458/1/012065

Conservation of energy:

$$\frac{d(\rho\, C_p T)}{dt} + \nabla.\left(\rho\, C_p u\, T\right) = \nabla.\left(k\Delta T\right) + H \quad (8)$$

Total heat transferred within a tube is calculated from:

$$Q = \dot{m}\, C_p\, (T_{inlet} - T_{outlet}) \quad (9)$$

Steady state pressure based solver was used in the developed model, energy equation was enabled to predict the heat transfer rates and Standard k-epsilon viscous model was used for better convergence. Simulations were run for 1000 iterations. Fluid flowing inside the 4 tubes was set to water to later be validated with the experimental results.

5. Results

CFD simulations predict the heat transfer inside the tube bundle and the outlet temperature of all the streams were obtained. Convective heat transfer takes place where the hot water in heating pipe loses heat to the annulus fluid and further heat or at least maintain the production fluid temperature. Figure 4 shows the temperature contours for the annulus in between the tubes in 4 planes starting from inlets till the end of the bundle. Results were used to calculate the overall heat transfer rate inside the tubes to further investigate the effect of the heating medium temperature and flow velocity on the total heat gained by the production pipe.

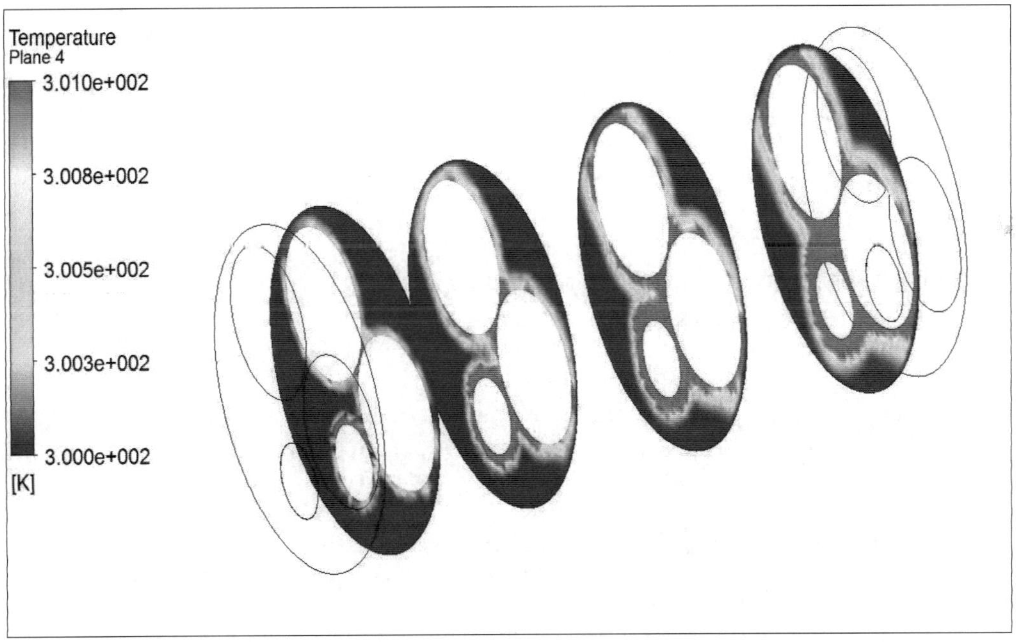

Figure 4. Temperature contours in planes along the bundle.

5.1. Effect of heating medium temperature

Simulations were run with different inlet temperatures for the heating line (323 – 363 K) and the rest of the tubes were kept at constant temperature of 313 K. It can be observed that increasing the inlet temperature gives higher heat transfer rates and consequently, higher outlet temperature of the production fluid. However, the length of the tube bundle limits the time for heat transfer to occur fluid inside the tube, thus, the temperature gradients are low. Total heat gained by production pipe at different heating medium inlet temperature is shown in figure 5.

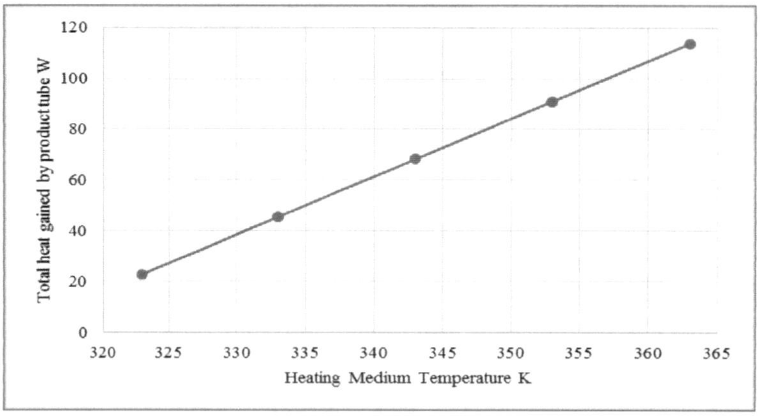

Figure 5. Total heat gained by production tube at different heating inlet temperatures.

5.2. *Effect of heating medium flow velocity*

The study covered 5 cases of heating medium flow velocity (0.03, 0.04, 0.05, 0.06 and 0.07 m/s). Low flow velocities were simulated to provide enough time for heat transfer to occur as the bundle is one meter in length. Higher flow velocities gave higher temperature gradient and thus higher overall heat transfer rates. Figure 6 shows the graph for the heating medium flow velocities against the heat gained by production tube.

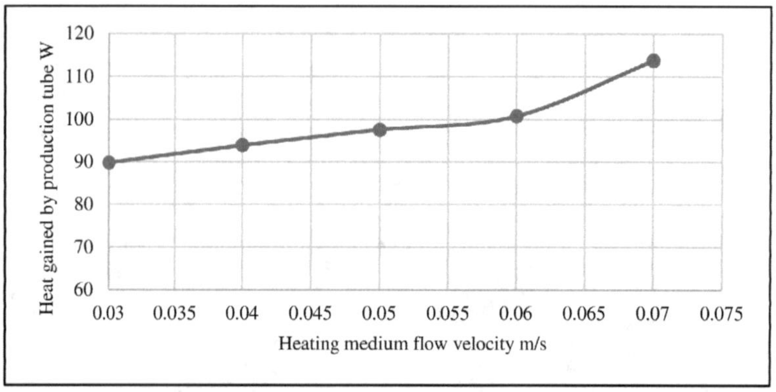

Figure 6. Total heat gained by production tube at different flow velocities.

6. Conclusion

A three dimensional CFD model was developed to predict the heat transfer rates in a tube bundle with three internal pipes. The outlet temperatures of all the streams to evaluate the heat transfer rates inside the tubes were obtained. Cases with different inlet temperatures and flow velocities of the heating pipe were run and it was observed that the higher the heating medium inlet temperature and flow velocity the more heat gained by production pipe and thus, higher arrival temperature of production fluid.

Acknowledgments

Authors of the study are grateful to Universiti Teknologi PETRONAS for providing support and facilities through the research grant YUTP 0153AA-E07.

References

[1] Davalath J and Barker J 1995 Hydrate inhibition design for deepwater completions *SPE Drilling & Completion* **10** pp 115–121

[2] Ellison B, Gallagher C, Frostman L and Lorimer S 2000 The physical chemistry of wax, hydrates, and asphaltene *Offshore Technology Conf.* 11963 Texas, USA

[3] Camargo R, Gonçalves M, Montesanti J, Cardoso C and Minami K 2004 A perspective view of flow assurance in deepwater fields in Brazil *Offshore Technology Conf.* 16687 Texas, USA

[4] Thant M 2012 Experimental investigation of natural convection heat transfer in bundle pipeline PhD Thesis Imperial College

[5] Huo H, Wang R, Ni H, and Liu Y 2014 An Experimental Study on the Synergetic Effects of Kinetic and Thermodynamic Gas Hydrate Inhibitors *Petroleum Science and Technology* **32** pp 1940-47

[6] Rubel M and Broussard D 1994 Flowline insulation thermal requirements for deepwater subsea pipelines *SPE Annual Technical Conf. and Exhibition* Texas, USA pp 193-201

[7] Bai Y and Bai Q 2012 *Subsea Engineering Handbook* (Gulf Professional Publishing)

[8] Thant M, Sallehud-Din M, Hewitt G, Hale C and Quarini J 2011 Mitigating Flow Assurance Challenges in Deepwater Fields using Active Heating Methods *SPE Middle East Oil and Gas Show and Conf.* 140997 Bahrain

[9] Brown L 2002 Flow Assurance: A π Discipline *Offshore Technology Conf.* 14010 Texas, USA

[10] Goodlad M 2013 Bundle technology for the future *SPE Offshore Europe Oil and Gas Conf. and Exhibition* Aberdeen, UK

[11] Churchill S and Churchill R 1975 A comprehensive correlating equation for heat and component transfer by free convection *AIChE Journal* **21** pp 604-606

[12] D'Orazio A and Fontana L 2010 Experimental study of free convection from a pair of vertical arrays of horizontal cylinders at very low Rayleigh numbers *International Journal of Heat and Mass Transfer* **53** pp 3131-42

[13] Liu L, Choudhury T, Hu B, Duggan N, Richardson S and Hewitt G 2006 Experimental studies of heat transfer in a tube bundle model *Proc. of the Institution of Mechanical Engineers Part E Journal of Process Mechanical Engineering* **220** pp 151-159

[14] Brown T, Clapham J, Danielson T, Harris R and Erickson D 1996 Application of a transient heat transfer model for bundled multiphase pipelines *SPE Annual Technical Conf. and Exhibition* Denver, Colorado pp 377-393

[15] Zabaras G and Zhang J 1998 Bundle-Flowline Thermal Analysis *SPE Journal* **3** pp 363-372

[16] Sallehud-Din M 2012 Prediction of Natural Convection Heat Transfer in a Pipeline Bundle PhD Thesis Imperial College London

[17] Brown L, Clapham J, Belmear C, Harris R, Loudon A and Maxwell S 1999 Design of Britannia's Subsea Heated Bundle for a 25 Year Service Life *Offshore Technology Conf.* Houston, Texas

IOP Conf. Series: Materials Science and Engineering **458** (2018) 012066 doi:10.1088/1757-899X/458/1/012066

A kinetic study of CO_2 sorption improvement in the CA-CNTs mixed matrix membrane

E T S Wong[1], Z A Jawad[1] *, B L F Chin[1], S K Wee[2]

[1] Department of Chemical Engineering, Faculty of Engineering and Science, Curtin University Malaysia, CDT 250, 98009 Miri, Sarawak, Malaysia
[2] Department of Petroleum Engineering, Faculty of Engineering and Science, Curtin University Malaysia, CDT 250, 98009 Miri, Sarawak, Malaysia

Abstract. Greenhouse emission notably carbon dioxide (CO_2) in the atmosphere have been increasing significantly over the last century. Hence, there is a need to address this issue to minimize the adverse environmental effects. Membrane separation is one solution that could potentially help in this endeavour. Among the many types of membranes available, mixed matrix membrane (MMM) is the most promising with the potential to surpass the many deficiencies present in both inorganic and polymeric membranes. In this study, results confirmed that incorporation of functionalized CNTs increases CO_2 permeance. Hence, the different loadings of MWCNTs-F were investigated to find the optimum gas separation performance. From this optimization study, excellent MMM performances in terms of CO_2 permeance were shown in 0.1wt% loading of MWCNTs-F. This superior performance was attributed to a good homogeneous dispersion of MWCNTs-F in the CA matrix which consequently enlarged the free volumes between the polymer chains and improved the polymer nano particulate interface. Apart from this, a kinetic sorption study showed improvements in the CO_2 solubility coefficient over previous related work at 14.9601 x 10^{13} cm^3(STP)/cm^4.cmHg. Hence, the usage of lower acetyl content is proven to be capable of increasing the CO_2 solubility coefficient.

1. Introduction

Over the years, greenhouse gasses (GHGs) such as carbon dioxide (CO_2) have been increasing significantly in the earth's atmosphere [1]. This has serious implications to the Earth from global warming to weather changes [2]. CO_2 gas is the main subject matter among these GHGs as it is released in large quantities compared to rest [2]. Hence, a few incentives have been taken to reduce the release of CO_2 into the environment [3].

The usage of membranes in separation is one that could possibly help in this endeavour as it addresses the many issues plaguing commercial CO_2 capture [4]. Some of the polymeric materials available are cellulose acetate (CA), polyimide (PI), polysulfone (PSF), polyethersulfone (PES), and polycarbonates (PC). Commercially, only PI and CA have been mainly used for CO_2/CH_4 and CO_2/N_2 separations [5]. This is due to the excellent gas separation performance of these 2 polymers [6].
Mixed matrix membrane (MMM) which is made up of a polymer matrix and an inorganic filler is currently the most promising way forward. [7]. Carbon nanotubes (CNTs) were proved to be able to transport light gases many times faster than any other known microporous material [8]. However, the poor distribution and weak interfacial interaction of the inorganic fillers within the polymer matrix leads to the formation of unselective channels within the membranes [9]. Thus, there is need to enhance the compatibility between the inorganic fillers and the polymer matrix within the MMMs.

The main objective of this research study is to synthesise MMM from cellulose acetate (CA) polymer matrix of 39% acetyl group and CNTs with high CO_2 permeance and solubility coefficients for the first time. From previous works, CA polymer containing a high acetyl content was documented to have high CO_2 pemeance and sorption coefficients. However, it is known to easily plasticize upon exposure to high CO_2 pressures. Hence, in this study CA membranes with a lower acetyl group (39%) would be fabricated and compared in terms of CO_2 permeance and sorption coefficients. Results reveals that a lower acetyl group (39%) improves the membrane's CO_2 permeance and sorption coefficients.

2. Experimental

2.1 Fabrication of Cellulose Acetate Membrane (CA-M)

The cellulose acetate membrane (CA-M, Table 1) was prepared based on the method published in previous work [10].

2.2 Fabrication of Mixed Matrix Membrane (CA–CNTs)

Pristine MWCNTs (MWCNTs-P) was functionalized using β-CD employing Chen's soft cutting method [11]. A set quantity of MWCNTs-P or Functionalized MWCNTs (MWCNTs-F) was added to the CH_3 COOH and H_2O before it is sonicated, as shown in Table 1. Stirring of the suspension was then instigated for 4 hr for better particle distribution of the MWCNTs. After that, CA was added gradually into the suspension of 55°C and stirred for 3 hr till it is dissolved completely followed by the synthesis method described in previous published work [10].

Table 1. The composition of the membranes fabricated.

Sample Description	Polymer	Solvent mixture (ratio 70:30)		MWCNTs		
	CA (wt%)	CH_3COOH (wt%)	H_2O (wt%)	Total Filler[a] (wt%)	Solid base MWCNTs[b] (wt%)	Solid base β-CD[c] (wt%)
CA-M	15	63	27	0.000	0.000	0.000
MMM-0.05P	15	62.993	26.997	0.050	0.005	0.000
MMM-0.1F	15	63.780	26.910	0.100	0.010	0.300
MMM-0.2F	15	62.544	26.805	0.200	0.021	0.630
MMM-0.05F	15	62.890	26.953	0.050	0.005	0.152

[a]Total filler = embedded MWCNTs into CA polymer. [b]Solid base MWCNTs = total filler x CA)/ (1- total base filler (1+ 30)). [c]Solid base β-CD = 30 x solid base MWCNTs

2.3 Kinetic Sorption Study

Membrane sorption experiments were carried out using a previous published procedure [12].

3. Results and Discussions

3.1 Integration of MWCNTs into CA Polymer.

3.1.1 Membrane Characterization. Scanning Electron Microscope (SEM) used to observe the physical structure and morphology of the cellulose acetate (CA) membrane, MMM-0.05P and the MMM-0.05F. Based on Figure 1a-b, it can be seen that the CA membrane has a smooth surface with a compact polymer layer near to the membrane/air interface. However, when pristine Multi-Walled Carbon Nanotubes (MWCNTs-P) were embedded into the CA in Figure 1c-d, it was found to have almost a nonexistent compact layer as compared to the neat CA membrane.

This was due to the MWCNTs-P being agglomerated within the polymer matrix from the strong van der Waals forces and their hydrophobic nature [13]. This inferred that the MWCNTs-P had a stronger interaction between each other in contrast to the polymer matrix [14]. This phenomenon was schematically described by Ahmad et al. [10].

Conversely, Figure 1e-f demonstrates a thinner compact layer at the membrane/air interface as compared to the pure CA membrane. No agglomeration was present in the polymer matrix indicating a better interaction between the CNTs with the CA. The beta cyclodextrin (β-CD) used to functionalize the MWCNTs (MWCNTs-F) helped eliminate agglomeration by shortening the MWCNTS and removing the van der Waals forces between them [15].

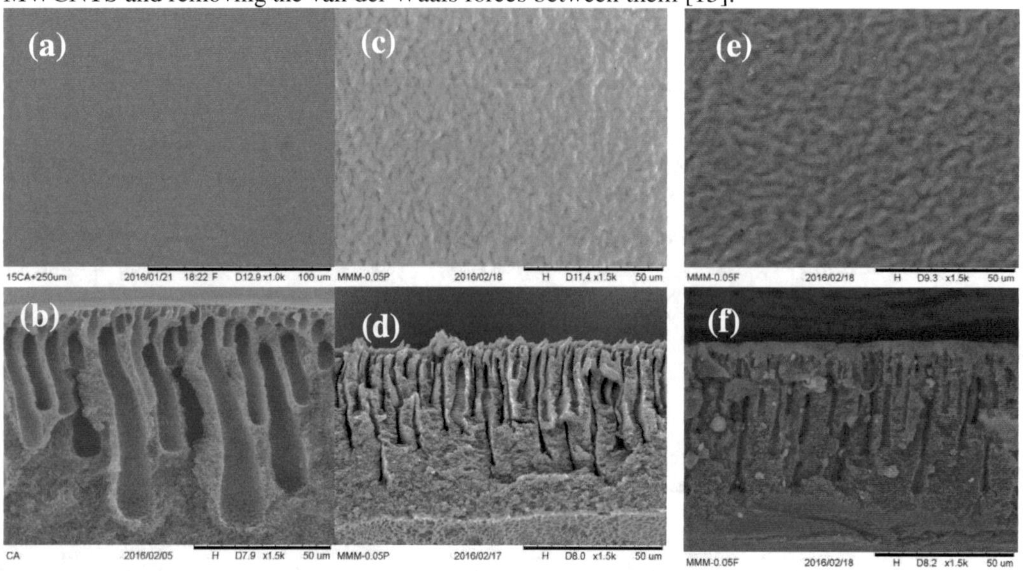

Figure 1. SEM morphologies of surface and cross section of (a-b) CA membrane and mixed matrix membrane with (c-d) MWCNTs-P (MMM-0.05P) and (e-f) MWCNTs-F (MMM-0.05F).

*3.1.2 Membrane Performance towards CO_2.*The gas separation performance for membranes was first evaluated by conducting a single gas permeation test followed by a kinetic sorption study. From this single gas permeation test, the permeance of the membranes towards CO_2 could be ascertain, enabling optimized MWCNTs loading for greater membrane separation. Gas permeation properties of MMMs incorporated with pristine and functionalized CNTs were initially investigated and compared to the neat CA membrane. It was observed that the CO_2 permeance of the MMM-0.05P (270.242 + 0.069 GPU) and MMM-0.05F (374.513 + 0.102 GPU) were higher than the pure CA membrane (131.580 + 0 GPU), as presented in Figure 2.

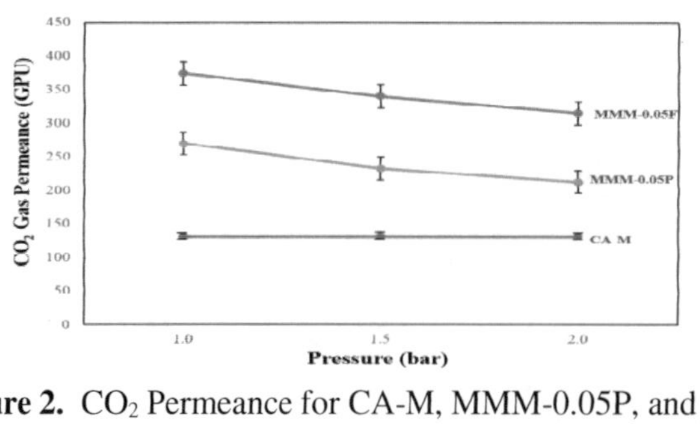

Figure 2. CO_2 Permeance for CA-M, MMM-0.05P, and MMM-0.05F

Based on Figure 2, a higher CO_2 permeance was obtained when MWCNTs-P were integrated into the CA membrane results from more free volume space available in the polymer matrix [15]. The disruption in the polymer chain packing due to the interaction between polymer segmental chains and CNTs ultimately leads to more void spaces in the membrane formed [16]. Thus, an increase in the overall permeance is a direct consequence of these reasons.

Meanwhile, when MWCNTs-F are embedded into CA polymer matrix, the CO_2 permeance continues to increase due to the functionalizing of CNTs [10, 12]. Functionalization shortens and opens up the ends of the MWCNTs, creating an open ended structure covered with a uniform coating of β-CD. Fast transport of gases are possible by the forming of these smooth nanochannels CNTs walls [17]. This inadvertently causes favourable CO_2 permeance improvements [15]. Apart from this, the functionalized CNTs have better CO_2 permeance due to the availability of the hydroxyl functional group on the β-CD. This helps improve the solubility of CO_2 in the MMM from the strong interaction present between the CO_2 molecules and the functional groups on the MWCNTs-F [18].

3.1.3 Transient Sorption and Diffusivity of the Membrane. Subsequently, the membrane performance was also determined through the Kinetic Sorption Study. This was because the gas transport though dense membrane are normally governed through the solution-diffusion mechanism [19]. Effectively, this emphasizes the solubility of specific gasses and its diffusion through the membrane. Hence, gas separations are dependent not only on the diffusion capabilities of the gasses but also on the physical and chemical interactions between the gas species and the polymers affecting the quantity of gas that can be accumulated within the polymer matrix. For a deeper and clearer understanding on the transport phenomena of CO2 through the MMMs, the kinetic coefficients were calculated and ascertained [12].

Based on Figure 3, the time spent to reach its equilibirum adsorption was in the order of CA-M > MMM-0.05F > MMM-0.05P. The reason behind this difference was the increment in the membrane's capability to adsorb more CO_2, as calculated from the TGA results (0.46365 mg for CA-M, 0.05883 mg for MMM-0.05P, and 0.074 mg for MMM-0.05F). Hence, a greater adsorption capacity would definitely incur a longer equilibrium time as compared to a smaller one.

Another aspect compared through the sorption study was the solubility coefficient of the CO_2 gas through the neat CA membrane, MMM-0.05P and MMM-0.05F. Referring to Figure 4, MMM-0.05P has the lowest CO_2 solubility coefficient among all the membranes. This was caused mainly by the agglomeration of CNTs in the polymer matrix due to the strong van der Waals forces among the pristine CNTs [13]. The embedment of MWCNTs-P into the CA polymer affects long-term stability of the dope along with the membrane's phase separation kinetics becoming one of the causes of agglomeration to be found in the membrane. Dope stability is concerned about the MWCNTs particles capability of homogeneously dispersing in a solution. Factors that influence the stability include the affinity of the MWCNTs with other constituents of dope and rate of particle settling [20]. At the same time, the solubility coefficient of CO_2 has been improved by more than a factor of 2.6, from 0.9399 x 10^{13} cm^3 (STP)/cm^4.cm Hg (CA-M) to 2.4968 x 10^{13} cm^3 (STP)/cm^4.cm Hg (MMM-0.05F) when functionalized MWCNTs were embedded within the CA polymer matrix. This was due to the dispersed CNTs within the polymer matrix when β-CD was used to functionalize the MWCNTs that might help to improve the CO_2 solubility of the gas molecules through strong interactions and prevents agglomeration [18]. On the other hand, the implementation of the kinetic sorption study enabled a comparative analysis of the incorperation of MWCNTs on the membrane's CO_2 diffusion cofficient (1.4 X 10^{-11} cm^2/s for CA-M, 3.0 X 10^{-11} cm^2/s for MMM 0.05P, and 1.5 X 10^{-11} cm^2/s for MMM 0.05F). Referring to these results, numerically there are significant changes in the membranes ability to diffuse CO_2 . This would infer that the embedding of CNTs does affect the CO_2 diffusion cofficient especially for MMM-0.05P.

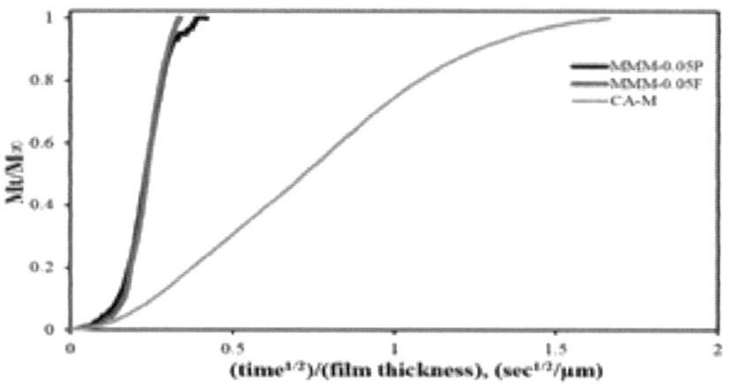

Figure 3. CO_2 Uptake Curves for CA-M, MMM-0.05P, and MMM-0.05F

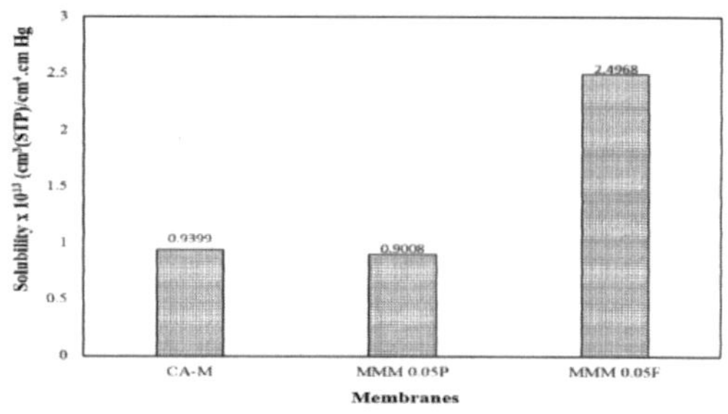

Figure 4. CO_2 solubility coefficient of CA-M, MMM-0.05P, and MMM-0.05F

3.2 Effect of Concentration Loadings of Functionalized MWCNTs

3.2.1 Membrane Characterization. As mentioned earlier, SEM characterization would enable correlative MWCNTs loading effects towards membrane structure and evaluating the successfulness of the polymer-CNTs integration. Figures 5 displays the surface and cross section micrographs of MMM-0.05F, MMM-0.1F and MMM-0.2F. Structurally, the MMM-0.1F displays a smooth and defect free membrane structure with a dense skin layer responsible for enhanced gas selectivity capabilities as compared to MMM-0.05F and MMM-0.2F. Meanwhile, MMM-0.05F and MMM-0.2F were shown to possess a compact and a porous structure surface in Figures 5a,b,e and f.

Furthermore, MMM-0.1F also displays a very compact and thin matrix near the membrane/air interface supported by a finger structure. MWCNTs-F macromolecular densification which overcomes the development of surface defects and agglomerated structures is the reason behind all the aforementioned better membrane structural buildup [21]. Vertical directed membrane/air interface densification led to the formation of a thin dense-skin layer suppressing the pin hole surface defects. Lastly, a better polymer–particle interaction and compatibility was induced from the strong adhesion between the CA polymer and MWCNTs-F. All of this contributes in forming a defect free skin layer [21]. However at high loadings of 0.2 wt%, there is an increased chance of agglomeration. This was tentatively observed through Figure 5b-c and schematically described by Ahmad et al. [10].

The casting dope solution viscosities for different functionalized MWCNTs loadings were calculated (282.183 ± 0.65cP for CA-M, 512.750 ± 7.00cP for MMM-0.05F, 554.475 ± 5.02cP for MMM-0.1F, and 992.333 ± 1.44cP for MMM-0.2F). Based on these results, the viscosity of the casting solution increased with increased the MWCNTs loading which is in line with the SEM images (Figure 5). At such a high viscosity of 992.333 ± 1.44cP was at 0.2 wt% CNTs. Where, the membrane morphology would have a highly dense and compact packing layer (Figure 5f). This was due to the diffusion of the phase inversion was suppressed as the rheological hindrance from the agglomeration of CNTs causing delayed exchange between the solvent and non-solvent [22].

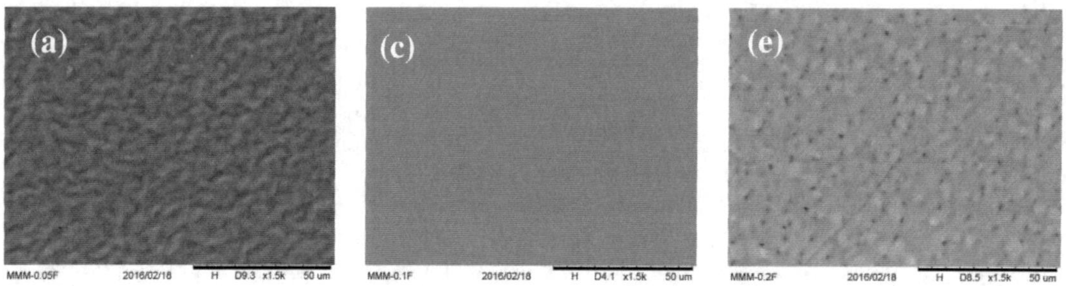

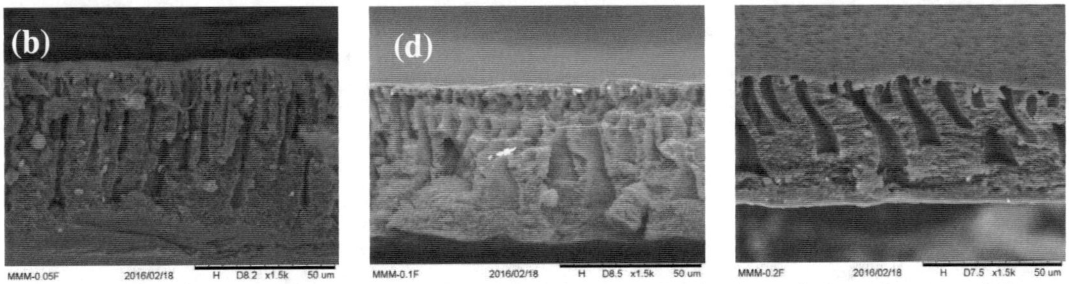

Figure 5. SEM morphologies of surface and cross section of mixed matrix membrane at different MWCNTs-F loadings of (a-b) 0.05 wt%, (c-d) 0.1 wt%, and (e-f) 0.2 wt%.

*3.2.2 Membrane Performance towards CO_2.*The MMMs with 0.05 wt%, 0.1 wt% and 0.2 wt% MWCNTs-F underwent the single gas permeation test to find the relationship between CNTs loadings and the CO2 permeance. Figure 6 showed that MMM-0.1F possesses the greatest CO2 permeance of 1017.288 + 0.613 GPU as compared to MMM-0.05F (374.513 + 0.102 GPU) and MMM-0.2F (117.184 + 0.068 GPU). This remarkable difference could be attributed to the good interfacial interaction between MWCNTs and polymer matrix [23]. In addition, it might be due to increase the availability of fast smooth nanochannels for gas transport along with better solubility from more β-CD in the membrane [18].

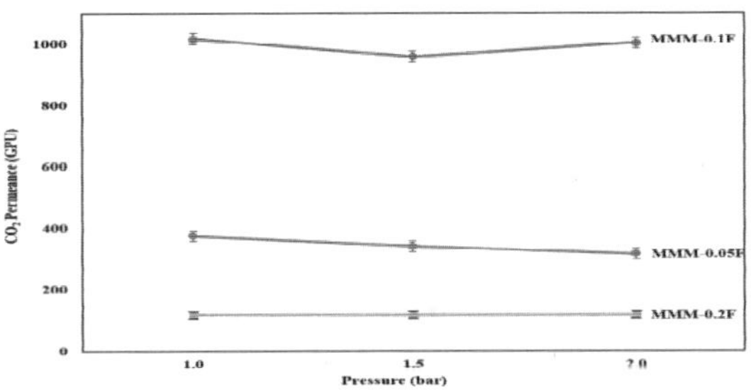

Figure 6. CO_2 Permeance for MMM-0.05F, MMM-0.1F, and MMM-0.2F.

3.2.3 Transient Sorption and Diffusivity of the Membrane. Likewise, the effect of the CNTs loading on the membrane performance was also evaluated through the Kinetic Sorption Study. Based on Figure 7, it can be seen that MMM-0.1F and MMM-0.2F reaches its CO_2 adsorption equilibrium slower than MMM-0.05F. This was due to the increment in the membrane's capability to adsorb CO_2 quantified from the difference in weight measured by the TGA (0.074 mg for MMM-0.05F, 0.78521 mg for MMM-0.1F, and 0.71894 mg for MMM-0.2F). A higher adsorption potential would naturally prolong the membranes from achieving its maximum adsorption state.

Based on Figure 8, the MMM-0.1F had the highest CO_2 solubility coefficient of 14.9601 x 10^{13} cm^3 (STP)/cm^4.cm Hg as compared to MMM-0.05F (2.4968 x 10^{13}) cm^3 (STP)/cm^4.cm Hg and MMM-0.2F (1.4648 x 10^{13}) cm^3 (STP)/cm^4.cm Hg. This was due to the well dispersed inorganic filler and the controlled amount of functionalized MWCNTs within the CA polymer matrix which further improve the MMM separation performance [21, 24, 25].

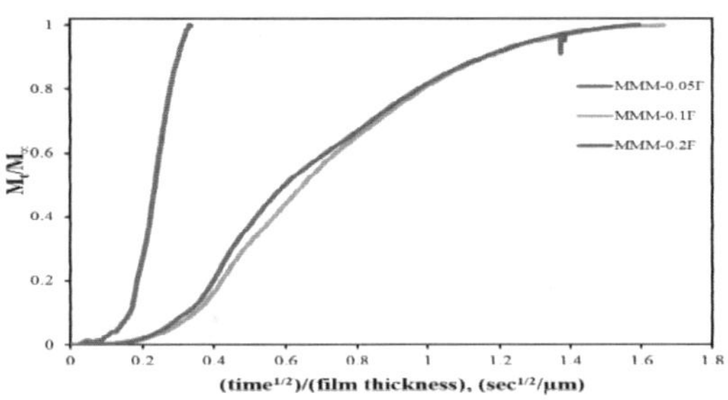

Figure 7. CO_2 Uptake Curves of CO_2 Permeance for MMM-0.05F, MMM-0.1F, and MMM-0.2F

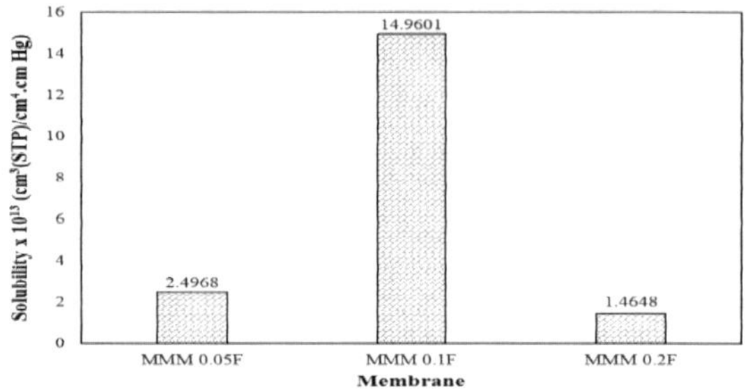

Figure 8. CO_2 solubility coefficient of CO_2 Permeance for MMM-0.05F, MMM-0.1F, and MMM-0.2F.

In this study, the CO_2 diffusion cofficients membrane diffusivity were also determined (1.5 x 10^{-11} cm^2/s for MMM 0.05F, 6.8 x10^{-12} cm^2/s for MMM 0.1F, and 8.0 x 10^{-12} cm^2/s for MMM 0.2F). Based on these results, it can be observed that CO_2 diffusion cofficient was the highest when 0.05% of MWCNTs was embedded into the CA polymer matrix. Meanwhile, it reduced when the MWCNTs concentration was increased to 0.1 wt%. These results have proved that the CO_2 diffusion cofficient was mainly affected by the membrane's solubility.

In fact, diffusivity does not contribute much for improving CO_2 permeance. Conversely, in this study the solubility can be effectively increased the CO_2 permeance. In short, these results proved that the driving force in the transport of CO_2 is to be controlled by the solubility and in direct agreement with a closely related kinetic adsorption study [12].

Lastly, this present work was compared to another research work Jawad et al. [12] as depicted in Table 2. Based on Table 2, the low acetyl content of 39% improved the CO_2 permeance to 1017.29 GPU and solubility coefficient to 14.960 x 10^{13} cm^3 (STP)/cm^4.cmHg. This is in agreement with Puleo et al (1989) which conclude that a lower acetyl group content would reduce the plasticization along with permeance and solubility coefficients [26].

Table 2. Summary of the permeation properties achieved by the present work compared to other research works.

References	Polymer	Filler	P_{CO2} (GPU)	CO_2 solubility coefficient (cm^3(STP)/cm^4.cmHg)	CO_2 diffusion coefficient (cm^2/s)	Acetyl Content
Present work	CA	MWCNTs-F 0.10 wt%	1017.29	14.960 x 10^{13}	6.8 x10^{-12}	39%
Jawad et al. [12]	CA	MWCNTs-F 0.10 wt%	741.67	198.352x10^{11}	3.70x10^{-11}	54.6-56%

4. Conclusion

In summary, the optimal MWCNTs-F loading was ascertained to be 0.1 wt% as it has the highest CO_2 permeance of 1017.288 $\pm$ 0.613 GPU and CO_2 solubility coefficient through the membrane along with a smooth and defect morphology. Solubility was improved as the loading increased reaching its highest values of 14.9601 x 10^{13} cm^3 (STP)/cm^4.cm Hg due to the presence of more β-CD in the MMMs. However, at higher loadings the solubility reduced due to the agglomeration of CNTs in the polymer matrix. Kinetic adsorption studies also uncovered that the main transport of CO_2 gasses in MMMs are determined by its solubility. Hence, greater emphasis in enhancing membrane solubility coefficient towards CO_2 can significantly create superior gas separation perfomance and apply the solution-diffusion mechansim.

Acknowledgements
The authors would like to offer their appreciation to the Ministry of Higher Education Malaysia (MOHE) for approving Fundamental Research Grant Scheme (FRGS) (MOHE Ref No: FRGS/1/2015/TK02/CURTIN/03/01) and Cost Centre: 001048. We would also like to thank LRGS USM (Account No: 304/PJKIMIA/6050296/U124) and Curtin Cost Centre: 001047.

References

[1] Haszeldine R S 2009 *Science* **325** 1647
[2] Speight J G 2007 *Natural Gas: A Basic Handbook*. Elsevier Science
[3] Leung D Y C, Caramanna G and Maroto-Valer M M 2014 *Renew. Sust. Energ. Rev.* **39** 426
[4] Baker R W 2002 *Ind. Eng. Chem. Res.* **41** 1393
[5] Han S H and Lee Y M 2011 Recent High Performance Polymer Membranes for CO_2 Separation. In *Membrane Engineering for the Treatment of Gases:Gas-Separation Problems with Membranes* (vol 1) The Royal Society of Chemistry Chapter 4 pp 84-124
[6] Drioli E, Barbieri G and Pullumbi P 2011 *Membrane Engineering for the Treatment of Gases*. Royal Society of Chemistry
[7] Cong H, Radosz M, Towler B F and Shen Y. 2007 *Sep. Sci. Technol.* **55** 281
[8] Skoulidas A I, Ackerman D M, Johnson J K and Sholl D S 2002 *Phys. Rev. Lett.* **89** 185901
[9] Mahajan R and Koros W J 2000 *Ind. Eng. Chem. Res.* **39** 2692
[10] Ahmad A L, Jawad Z A, Low S C and Zein S H S 2014 *J. Membr. Sci.* **451** 55
[11] Chen J, Dyer M J and Yu M F 2001 *J. Am. Chem. Soc.* **123** 6201
[12] Jawad Z A, Ahmad A L, Low S C, Chew T L and Zein S H S *J. Membr. Sci.* **476** 590
[13] Sahoo N G, Rana S, Cho J W, Li L and Chan S H 2010 *Prog. Polym. Sci.* **35** 837
[14] Ma P C, Kim J K and Tang B Z 2007 *Composites Science and Technology* **67** 2965
[15] Sanip S M, Ismail A F, Goh P S, Soga T, Tanemura M and Yasuhiko H 2011 *Sep. Purif. Technol.* **78** 208
[16] Cong H, Zhang J, Radosz M and Shen Y 2007 *J. Membr. Sci.* **294** 178
[17] Sanip S M, Ismail A F, Goh P S, Ng B C, Abdullah M S, Soga T, Tanemura M and Yasuhiko H. 2010 *Nano.* **05** 195
[18] Kim S, Jinschek J R, Chen H, Sholl D S and Marand E 2007 *Nano Letters* **7** 2806
[19] Paul D R and Yampol'skii Y P 1993 *Polymeric Gas Separation Membranes* Taylor & Francis
[20] Ismail A F, Rahim N H, Mustafa A, Matsuura T, Ng B C, Abdullah S and Hashemifard S A 2011 *Sep. Purif. Technol.* **80** 20
[21] Aroon M A, Ismail A F, Montazer-Rahmati M M and Matsuura T 2010 *J. Membr. Sci.* **364** 309
[22] Choi J H, Jegal J, Kim W N 2006 *J. Membr. Sci.* **284** 406
[23] Khan M M, Filiz V, Bengtson G, Shishatskiy S, Rahman M M, Lillepaerg J and Abetz V 2013 *J. Membr. Sci.* **436** 109
[24] Khan M M, Filiz V, Bengtson G, Shishatskiy S, Rahman M and Abetz V 2012 *Nanoscale Res. Lett.* **7** 504
[25] Kim S, Chen L, Johnson J K and Marand E 2007 *J. Membr. Sci.* **294** 147
[26] Puleo A, Paul D and Kelley S 1989 *J. Membr. Sci.* **47** 301

IOP Conf. Series: Materials Science and Engineering **458** (2018) 012067 doi:10.1088/1757-899X/458/1/012067

Extractive separation of benzene and cyclohexane using 1-butyl-3-methylimidazolium acetate

M. Zulhaziman M. Salleh*[1], Mohamed K. Hadj-Kali[2], M. A. Hashim[1], Emad Ali[2]
[1] University of Malaya Centre for Ionic Liquids (UMCiL), Department of Chemical Engineering, Faculty of Engineering, University of Malaya, Kuala Lumpur, Malaysia.
[2] Chemical Engineering Department, King Saud University, P.O Box 800, Riyadh 11421, Saudi Arabia.
*Corresponding author: zulhaziman@siswa.um.edu.my

Abstract. Separation of benzene and cyclohexane is a challenging process in petrochemical industry due to the difficulty to separate them by a simple distillation. Recognizing ionic liquid as a potential alternative, this work investigated the performance of 1-butyl-3-methylimidazolium acetate ([C_4Mim][Ac]) as an extracting solvent in the separation of benzene and cyclohexane via liquid–liquid extraction. The ternary liquid–liquid equilibria for this IL with benzene and cyclohexane was investigated at 25°C and 1 atm, with feed concentration of benzene ranging from 10 to 60 wt %. Good agreement was achieved for the tie-lines obtained from the COSMO-RS model and experimental data. The NRTL model was successfully employed to correlate the experimental tie lines. The performance of [C_4Mim][Ac] was compared with organic solvent and other ILs reported in literature. The selectivity and distribution ratio of benzene were in the range of 4.17–5.47 and 0.51–0.72, respectively. At equilibrium, no IL was present in the cyclohexane layer and the concentration of cyclohexane in the IL layer was very low. This observation indicated that there was minimum cross-contamination between the phases and less energy was required for solvent recovery.

1. Introduction

The separation of benzene and cyclohexane by conventional distillation is difficult and challenging because they have nearly equal boiling points, with the difference of only 0.64°C. This condition requires an advanced distillation process such as the azeotropic or extractive distillations [1]. Despite their application in industry, both processes suffer from some disadvantages such as process complexity, high capital and operating costs, and high energy consumption [2, 3]. Moreover, these processes are carried out by adding a third compound as the entrainer, in which the removal of the third compound from the distillate causes even higher process complexity and cost. In addition, azeotropic distillation and extractive distillation are only suitable for high (> 90%) and medium (65–90 %) concentration of benzene in the feed, respectively.

Liquid–liquid extraction (LLE) is widely used in the chemical industry because it is a simple process that can be operated at mild conditions. In the case of benzene-cyclohexane mixture, LLE is a suitable method to separate low concentrations of benzene (20–65 %). However, there is still no industrial process available for the concentrations less than 20 wt %. While the pervaporation process was reviewed as a potential technique, it was still limited by the challenges encountered during economic evaluation [2]. Several reports have investigated the performance of industrial organic solvents in LLE process such as ethylene glycol, tetra-ethylene glycol, sulfolane, and *N*-

methylpyrrolidone as extractants [4-6]. The results revealed that their application is associated with several challenges because these solvents are usually volatile, toxic and flammable.

Interestingly, ionic liquids (ILs) have been extensively studied for wide applications of chemical processes. In the last decades, the potential of ILs in many industrial applications has become more diverse due to their appreciable advantages such as being non-flammable, non-volatile, environmentally benign, and thermally stable over a wide range of temperature. ILs have also been recognized for their ability to break the azeotropic properties in many mixtures such as ethanol-water, alcohol-alkane, aromatic-aliphatic, ketone-alkane, and others [7]. For the IL–benzene–cyclohexane system, the reports of new data for ternary phase equilibria proved the feasibility of using ILs as the extracting solvents. This is reflected from earlier studies involving some ILs such as $[C_4Mim][BF_4]$[8], $[C_4Mim][SCN]$[8], $[C_4Mim][MeSO_4]$[9], $[C_1im][DBP]$[10] $[C_2Mpyr][EtSO_4]$[11] and many more. Despite the huge number of ILs available, $[C_4Mim][Ac]$ is particularly an interesting IL because it is an affordable IL that consists of a common cation-anion pair. $[C_4Mim][Ac]$ has been extensively studied in the process of CO_2 capture [12, 13]. Reports showed that the carboxylation reaction can occur between $[C_4Mim][Ac]$ and CO_2, leading to the high solubility of CO_2 in the IL [14, 15]. It is known that CO_2 and benzene are both non-polar compounds. In addition, as observed in COSMO-RS programme, both compounds also showed similar σ-profiles and σ-potential. Thus, it can be hypothesized that benzene is also highly soluble in $[C_4Mim][Ac]$, which will produce high selectivity of IL in LLE. This is the first study that investigates the performance of $[C_4Mim][Ac]$ as an extracting solvent to separate benzene-cyclohexane mixture.

2. Methodology

2.1 Chemicals and materials

The following chemicals were obtained as follow and used without further purification: benzene (99.7 %) from Merck, cyclohexane (99.5%) from Merck and $[C_4Mim][Ac]$ (95.0 %) from Sigma Aldrich. For the determination of ternary composition in extract and raffinate phases at equilibrium, the NMR analysis was performed using deuterated chloroform with purity ≥99.8% (stabilized with silver).

2.2 Liquid–liquid extraction experiment

The feed mixture containing 10 wt % benzene in cyclohexane was firstly prepared inside a 20-mL screw-capped scintillation vial using an analytical balance with accuracy of ±0.0001 g. This procedure was repeated for other concentrations of benzene (i.e., 20, 30, 40, 50, and 60 *wt* %) in the feed. Based on the binary composition, the total weight of the feed mixture was fixed at 2 g. The $[C_4Mim][Ac]$ IL was then mixed into every feed mixture in a weight ratio of 1:1. Parafilm tape was used to seal each vial and avoid the component loss due to evaporation. The vials were then placed in an incubation shaker at 25°C and 1 atm, where spring clamps were used to tighten them upon shaking. The shaking was subsequently conducted at 200 rpm for 6 h. After stopping the mixing procedure, the mixture was left for approximately 12 h to reach equilibrium. This period is sufficient as we have conducted the settling time study to ensure the equilibrium state was fully reached.

2.3 Determination of ternary molar compositions

A drop (±0.035 mL) of sample was taken out from the extract and raffinate layers using a micropipette. For the extract layer, a few bubbles were purged from the micropipette tip to avoid cyclohexane contamination from the raffinate. This drop of sample was then dissolved in ±0.7 mL of deuterated chloroform placed inside an NMR tube. The sample and solvent in the tube were shaken carefully to form homogenous mixture. Each tube was tightly sealed with parafilm to avoid chemical loss. The ^1H NMR spectrometer of Bruker 400 MHz was used to measure the peak of hydrogen molecules in each component and calculate the molar fraction of each component in both layers.

IOP Conf. Series: Materials Science and Engineering **458** (2018) 012067 doi:10.1088/1757-899X/458/1/012067

2.4 Experimental selectivity and distribution ratio

The distribution ratio of benzene (D_B) and the solvent selectivity (S) of [C_4Mim][Ac] were used to evaluate the extraction performance by using Equations 1 and 2.

$$D_B = \frac{x_B^1}{x_B^2} \tag{1}$$

$$S = \frac{x_B^1}{x_B^2} \Big/ \frac{x_C^1}{x_C^2} \tag{2}$$

where x_B and x_C are the concentrations of benzene and cyclohexane, respectively. The superscripts 1 and 2 represent the extract and raffinate phase, respectively.

3. Results and discussion

3.1 Ternary LLE

The LLE data for the ternary system of [C_4Mim][Ac] + benzene + cyclohexane is tabulated in Table 1, and the corresponding ternary plot is depicted in Figure 1. As seen in these results, the tie lines obtained from experimental and COSMO-RS approaches showed a good agreement, with the RMSD value of 1.59 %. The tie lines also showed negative slope and they become more negative at higher concentrations of benzene in the feed. Interestingly, there was no trace of IL found in cyclohexane-rich layer. This indicates a favourable extraction process because the solvent cross contamination can be avoided. The concentration of cyclohexane in bottom layer was also very low in which indicates an easier regeneration of IL. Based on the measured tie lines, it was observed that the immiscibility region in [C_4Mim][Ac] is slightly higher than that in [C_2Mim][Ac] from our previous work [16]. This shows that the higher alkyl length of IL resulted into higher mutual solubility between IL and benzene/cyclohexane. This is mostly contributed by the stronger dispersive interaction between the three components at higher alkyl chain of imidazolium cation. This trend is consistent with another study using N-methylimidazole [C_1im] and N-ethylimidazole [C_2im] as the solvents [10].

Table 1. Molar composition of the tie-lines with the distribution ratio and selectivity data for benzene (1) + cyclohexane (2) + [C_4Mim][Ac] (3) at 25 °C and 1 atm.

Raffinate phase			**Extract phase**			***D***	***S***
x_1	x_2	x_3	x_1	x_2	x_3		
0.078	0.922	0.000	0.056	0.122	0.822	0.72	5.47
0.163	0.837	0.000	0.102	0.087	0.810	0.63	6.01
0.250	0.750	0.000	0.158	0.068	0.773	0.63	6.94
0.345	0.655	0.000	0.190	0.083	0.727	0.55	4.35
0.447	0.553	0.000	0.241	0.085	0.674	0.54	3.50
0.545	0.455	0.000	0.280	0.056	0.664	0.51	4.17

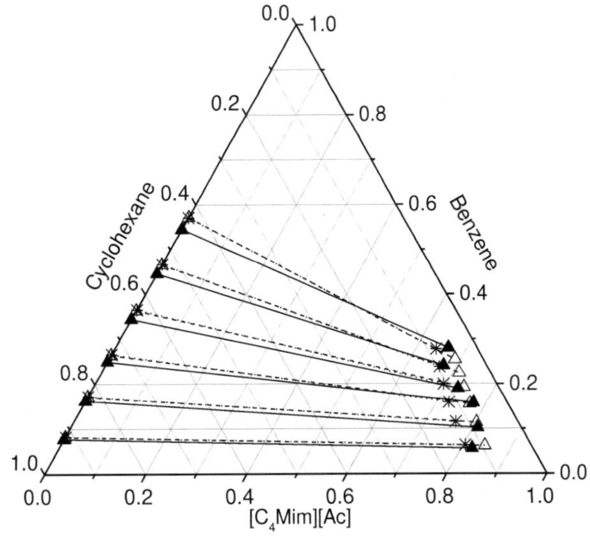

Figure 1. Ternary LLE diagram for $[C_4mim][Ac]$ + benzene + cyclohexane at 25 °C and 1 atm: –▲–, experimental; --Δ--, COSMO-RS; ··×··, NRTL.

3.2 Selectivity and distribution ratio

Figure 2 shows the extractive performance of $[C_4mim][Ac]$ in view of distribution ratio and selectivity. The values of selectivity were higher than unity which confirms the possibility of extraction. In general, the distribution ratio and selectivity slightly decreased at higher mole fraction of benzene in raffinate layer. Referring to our previous work [16], the benzene distribution ratio for the system using $[C_4Mim][Ac]$ was higher than the one with $[C_2Mim][Ac]$. This is because at higher cation alkyl length, the ability of IL to dissolve benzene and cyclohexane increased. The selectivity, in contrast, decreased at higher cation alkyl length. Selectivity defines the ability of an IL to selectively extract benzene from cyclohexane, rather than extracting cyclohexane altogether. The longer alkyl chain of cation creates additional interaction between the IL and cyclohexane in the non-polar region, and therefore reduces the selectivity.

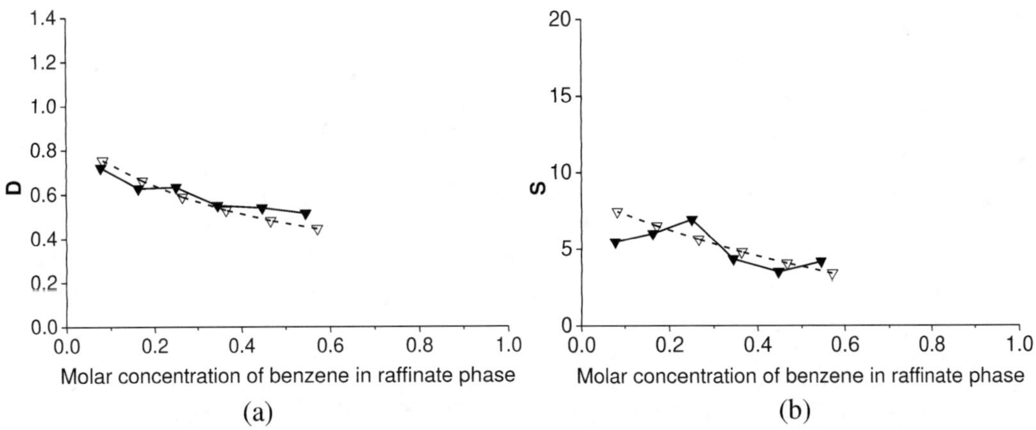

Figure 2. The distribution ratio (a) and selectivity (b) with respect to concentrations of benzene in the raffinate layer. The full symbol with solid line represent the experimental data, while the empty symbol with dashed line indicates the COSMO-RS predictions.

3.3 Consistency test

For the experimental results, the consistency tests were conducted using the Hand's correlation. This correlation was used to obtain the following Hand's equation [17].

$$ln\left(\frac{x''_{Bz}}{x''_{Cyc}}\right) = a + b\,ln\left(\frac{x'_{Bz}}{x'_{IL}}\right) \quad (3)$$

where x_{Cyc}, x_{Bz}, and x_{IL} represent the concentrations of cyclohexane, benzene, and IL, respectively. The superscripts $'$ and $''$ refer to the extract and raffinate phases, respectively. The value of parameters a, b, and R^2 for each ternary system are 1.442, 2.558 and 0.983, respectively. The linearity of each plot (i.e. the value of R^2 close to unity) indicates the excellent degree of consistency for the ternary LLE tie lines reported in this work.

3.4 NRTL correlation

In LLE calculations, an isothermal liquid-liquid flash is solved at a given temperature and pressure to obtain phase compositions. The flash calculation consists of the Equations 4, 5 and 6.

Material Balance : $x_i - (1-\omega)x_i^{L1} - \omega x_i^{L2} = 0, \quad i = 1, N_c$ (4)

Equilibrium Equation : $x_i^{L1}\gamma_i^{L1} - x_i^{L2}\gamma_i^{L2} = 0, \quad i = 1, N_c$ (5)

Equation of Summation : $\sum_i x_i^{L1} - \sum_i x_i^{L2} = 0$ (6)

In these equations, ω is the liquid-liquid splitting ratio; x_i, the amount of component i in the mixture; x_i^{L1}, the amount of component i in liquid phase $L1$; x_i^{L2}, the amount of component i in liquid phase $L2$; and N_C, the number of constituents of the liquid phases. The parameters γ_i^{L1} and γ_i^{L2} are the activity coefficients of component i in $L1$ and $L2$, respectively.

In this work, the activity coefficients were evaluated using the non-random two-liquid (NRTL) model [18]. This model was found to be useful for correlating the experimental LLE data of systems containing ILs without requiring any special modification. For a multicomponent system, the activity coefficient of component i is given by the general expression, shown in Equation 7:

$$ln\,\gamma_i = \frac{\sum_j \tau_{ji} G_{ji} x_j}{\sum_j G_{ji} x_j} + \sum_j \frac{G_{ji} x_j}{\sum_k G_{kj} x_k}\left(\tau_{ij} - \frac{\sum_k \tau_{kj} G_{kj} x_k}{\sum_k G_{kj} x_k}\right) \quad (7)$$

with: $ln\,G_{ij} = -\alpha_{ij}\tau_{ij}$, $\alpha_{ij} = \alpha_{ji}$ and $\tau_{ii} = 0$

where τ_{ij} and τ_{ji} are binary interaction parameters and α_{ij} is the non-randomness parameter.

The binary interaction parameters τ_{ij} and τ_{ji} were estimated by minimizing the root mean square deviation (RMSD) between calculated and experimental solubilities of each constituent in each phase using the Simulis® software package [19]. The parameter α_{ij} in the NRTL model measures the non-randomness in the mixture; i.e. the mixture is said to be completely random when α_{ij} is zero. In this work, α_{ij} was fixed equal to 0.20 for all binary combinations. The choice of $\alpha_{ij} = 0.2$ is made following successful fitting of NRTL model for ternary LLE data for systems containing ILs and deep eutectic solvents as reported in our previous works [16, 20, 21]. The RMSD value between experimental and NRTL calculation was 2.1 %, which indicates that NRTL correlation represents the experimental data very well. This excellent fitting is also observable in Figure 1. The values of NRTL binary interaction parameters regressed for this ternary system are shown in Table 2. In order to conserve coherence between this work and the previous one applied to deep eutectic solvents for the same binary mixture, the binary interaction parameters between benzene and cyclohexane, despite the IL used, were taken from our previous work [21] without any adjustment.

Table 2. NRTL parameters for the ternary system $[C_4Mim][Ac]$ + benzene + cyclohexane.

i	j	τ_{ij}	τ_{ji}
Benzene	Cyclohexane	-189.57	453.65
Benzene	$[C_4Mim][Ac]$	4401.56	62.90
Cyclohexane	$[C_4Mim][Ac]$	1913.19	144.24

3.5 Comparison with other solvents

The efficiency of extraction process can be evaluated from its distribution ratio of solute (benzene for this case) and the selectivity of solvent. Figure 3 shows the comparison of these values from our work with some other reported works that used different types of ILs. Sulfolane [22] was included as a benchmark to represent the performance of common organic solvents used in industry.

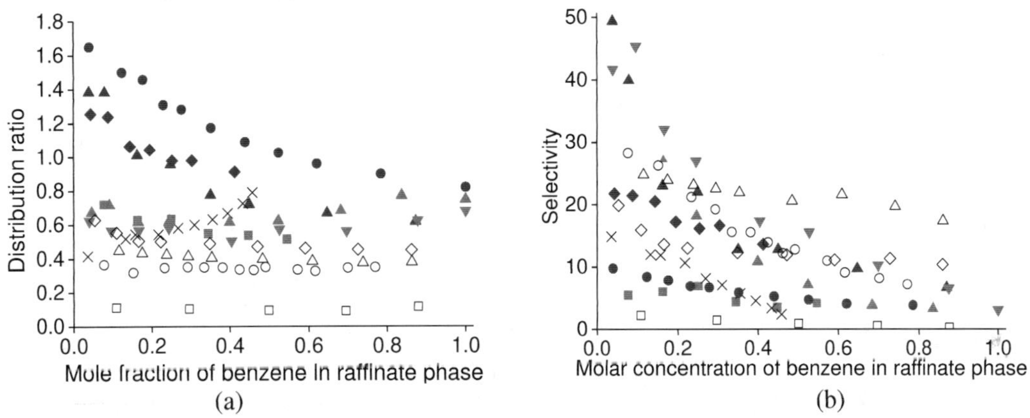

Figure 3. Distribution ratio (a) and selectivity (b) of the reported ILs with respect to mole fraction of benzene in raffinate phase. The symbols are: ■–$[C_4Mim][Ac]$(this work); ♦–$[C_4Mim][AlCl_4]$ [23]; ▲–$[C_4Mim][PF_6]$ [24]; ●–$C_4Mim][TF_2N]$ [25]; ▼–$[C_4Pyr][BF_4]$ [8]; ▲–$[C_4Mim][BF_4]$ [8]; □–$[Mim][BF_4]$ [26]; Δ–$[C_4Pyr][EtSO_4]$ [27]; ◊–$[C_2Mim][EtSO_4]$ [28]; ○–$[C_4Mpyr][BF_4]$ [29]; and x–sulfolane [22].

It was observed that at low concentration of benzene in raffinate phase ($x_B < 0.3$), the distribution ratio for $[C_4Mim][Ac]$ was slightly higher than that with sulfolane. Even so, the selectivity of sulfolane is very low. Therefore, any IL that has better or comparable value of distribution ratio with respect to that in sulfolane will have more effective extraction. As seen in Figure 3, the distribution ratio of $[C_4Mim][Ac]$ were generally in the intermediary region, as similar as most of the other reported ILs. The distribution ratio for $[C_4Mim][Ac]$ is higher than $[Mim][BF_4]$ [26], $[C_4Mpyr][BF_4]$ [29], $[C_2pyr][EtSO_4]$ [27], $[C_2Mim][EtSO_4]$ [28], and $[C_4pyr][BF_4]$ [8]. In addition, $[C_4Mim][Ac]$ also have comparable distribution ratio with $[C_4Mim][BF_4]$ [8] and $[C_4pyr][BF_4]$ [8].

It was noteworthy that three ILs i.e. $[C_4Mim][TF_2N]$ [25], $[C_4Mim][PF_6]$ [24] and $[C_4Mim][AlCl_4]$ [23] showed exceptionally higher distribution ratio than this work. However, their halide constitution must be considered carefully to ensure less effect towards the environment and process safety.

Besides, each of them could also bring significant drawbacks upon the application as the extracting solvent. Firstly, although the extraction using [C_4Mim][Tf_2N] showed the highest distribution ratio, its selectivity was relatively much lower than the other ILs. Next, for the extraction using [C_nMim][PF_6] (n = 4, 5, 6) [24], the high distribution ratio would be interrogated by the cost of operation because it is a highly expensive type of IL. Finally for [C_4Mim][$AlCl_4$], it was reported that this IL is instable in humid air, thus creates difficulty in process handling. The selectivity of [C_4Mim][Ac] in this work is generally less than the other ILs. This could be possibly tolerated because the solvent capacity, which is directly related to distribution ratio, was reported to have much greater influence on the cost of production than does the selectivity [30].

4. Conclusion

In this work, the extractive performance of [C_4Mim][Ac] was investigated for the separation of benzene from its mixture with cyclohexane. Increasing cation alkyl chain resulted in increased distribution ratio and reduced selectivity. There was no IL presence in cyclohexane layer and the concentration of cyclohexane in IL layer was very low. Good agreement was also observed for the tie lines between COSMO-RS, experimental and NRTL. This study indicates that apart from being available commercially at affordable price, [C_4Mim][Ac] also showed good efficiency in the extraction of benzene from benzene–cyclohexane mixture.

Acknowledgement

The authors extend their appreciation to the International Scientific Partnership Program ISPP at King Saud University for funding this research work. The authors also express their gratitude to UMCiL at the University of Malaya for facilities and other supports given along this work.

References

[1] Dechant J 1990 *Acta Polym.* **41** 361
[2] Garcia Villaluenga J P and Tabe-Mohammadi A 2000 *J. Membr. Sci.* **169** 159
[3] Johann G Stichlmair and Fair J R 1998 *Distillation: Principles and Practices* (Wiley-VCH) pp 246
[4] Ali S H, Lababidi H M S, Merchant S Q and Fahim M A 2003 *Fluid Phase Equilib.* **214** 25
[5] Al-Sahhaf T A and Kapetanovic E 1996 *Fluid Phase Equilib.* **118** 271
[6] Wang W, Gou Z and Zhu S 1998 *J. Chem. Eng. Data* **43** 81
[7] Pereiro A B, Araújo J M M, Esperança J M S S, Marrucho I M and Rebelo L P N 2012 *J. Chem. Thermodyn.* **46** 2
[8] Zhou T, Wang Z, Ye Y, Chen L, Xu J and Qi Z 2012 *Ind. Eng. Chem. Res.* **51** 5559
[9] Domínguez I, Calvar N, Gómez E and Domínguez Á 2013 *J. Chem. Eng. Data* **58** 189
[10] Wang R, Li C, Meng H, Wang J and Wang Z 2008 *J. Chem. Eng. Data* **53** 2170
[11] González E J, Domínguez I, González B and Canosa J 2010 *Fluid Phase Equilib.* **296** 213
[12] Shiflett M B, Drew D W, Cantini R A and Yokozeki A 2010 *Energy Fuels* **24** 5781
[13] Barrosse-Antle L E and Compton R G 2009 *Chem. Commun.* 3744
[14] Besnard M, Cabaço M I, Chávez F V, Pinaud N, Sebastião P J, Coutinho J A and Danten Y 2012 *Chem. Commun.* **48** 1245
[15] Besnard M, Cabaco M I, Vaca Chávez F N, Pinaud N l, Sebastião P J, Coutinho J o A, Mascetti J l and Danten Y 2012 *J. Phys. Chem. A* **116** 4890
[16] Zulhaziman M S, Hadj-Kali M K, Hashim M A and Mulyono S 2018 *J. Mol. Liq.* **266** 51
[17] Hand D B 1929 *J.Phys. Chem.* **34** 1961
[18] Renon H and J M Prausnitz 1968 *AIChE J.* **14** 135
[19] Simulis® thermodynamics. [cited 2017; Available from: http://www.prosim.net/]
[20] Hizaddin H F, Hadj-Kali M K, Ramalingam A and Ali Hashim M 2016 *J. Chem. Thermodyn.* **95** 164

[21] Zulhaziman S, Wazeer I, Mulyono S, El-blidi L, Hashim M A and Hadj-Kali M K 2017 *J Chem. Thermodyn.* **104** 33
[22] Chen J, Li Z and Duan L 2000 *J. Chem. Eng. Data* **45** 689
[23] Lyu Z, Zhou T, Chen L, Ye Y, Sundmacher K and Qi Z 2014 *Chem. Eng. Sci.* **115** 186
[24] Zhou T, Wang Z, Chen L, Ye Y, Qi Z, Freund H and Sundmacher K 2012 *J. Chem. Thermodyn.* **48** 145
[25] Calvar N, Domínguez I, Gómez E and Domínguez Á 2011 *Chem. Eng. J.* **175** 213
[26] Sakal S A, Shen C and Li C-x 2012 *J.Chem. Thermodyn.* **49** 81
[27] Gómez E, Domínguez I, Calvar N, Palomar J and Domínguez Á 2014 *Fluid Phase Equilib.* **361** 83
[28] González E J, Calvar N, González B and Domínguez Á 2010 *J. Chem. Eng. Data* **55** 4931
[29] Abu-Eishah S I and Dowaidar A M 2008 *J. Chem. Eng. Data* **53** 1708
[30] Zahed A H, Gashghari M A 1992 *Eng. Sci.* **4** 101

IOP Conf. Series: Materials Science and Engineering **458** (2018) 012068 doi:10.1088/1757-899X/458/1/012068

Phase identification of natural gas system with high CO_2 content through simulation approach using Peng-Robinson model

Ali Qasim, Azmi Mohd Shariff, Suriati Sufian, Muhammad Ayoub, Bhajan Lal*
Department of Chemical Engineering, Universiti Teknologi PETRONAS, 32610 Seri Iskandar, Perak, Malaysia.

*bhajan.lal@utp.edu.my

Abstract: The phase behavior for natural gas mixtures with high CO_2 content and heavier hydrocarbons up to pentane for cryogenic separation has been obtained through simulation using Peng-Robinson (PR) model. The bubble point and dew point curves which define the liquid-vapor and solid-vapor domains for separation are obtained by generating the pressure-temperature phase diagram. The simulations are performed using Aspen HYSYS. The validation for the use of PR model is done by comparing the results of CO_2 freeze-out temperature predictions for PR and Soave-Redlich-Kwong (SRK) model with the experimental data available in the literature.

1. Introduction

The energy consumption and demand throughout the world is increasing. Also the increase in population calls for an increase in energy usage. Natural gas serves as a clean and sustainable source of energy, so for this reason it is widely employed in industrial and household applications [1, 2]. It was reported in 2010 that natural gas fulfills almost a quarter of the world's total energy requirements [3]. Although natural gas is considered a clean fuel in comparison with the fossil fuels but raw natural gas contains impurities such as CO_2, water and other hydrocarbons. Regarding natural gas reservoirs, it is estimated that almost 33% of these reservoirs are acrid. In Malaysia, some of the reserves contain more than 70% of CO_2 in natural gas [4, 5]. In these impurities, removal of CO_2 is of utmost importance as it significantly decreases the calorific content of natural gas and causes erosion in delivery pipelines [6, 7].

In order to predict the phase behavior of natural gas mixture, a number of equation of states are used for modeling and simulation purposes such as Peng-Robinson (PR), Soave-Redlich-Kwong (SRK) and Benedict-Webb-Rubin-Starling (BWRS). Among these equations, Peng-Robinson equation of state model is suitable to be used in oil, gas and petrochemical applications. It is extensively used to model VLE behavior of natural gas in cryogenic gas processing systems. Peng-Robinson equation of state also holds high degree of efficiency and reliability for its use in low pressure and higher temperature conditions [8, 9]. The cryogenic separation method, among the other CO_2 removal processes, is a resourceful technique for CO_2 capture. Cryogenic separation method involves the removal of CO_2 impurity at lower temperatures of up to -120°C or below [10]. In hybrid cryogenic separation technique, conventional distillation columns and packed beds are incorporated and separation of the components can be performed in both liquid-vapor and solid-vapor regions of the phase [11].

In this work, the phase behavior of two natural gas mixtures with high CO_2 content is determined through simulation using the suitable equation of state that predicts it more accurately, which in this case is PR equation of state. The investigation of phase behavior is done to determine the vapor, liquid and sold boundaries for cryogenic based separation. Simulations are performed in Aspen HYSYS. The phase behavior obtained through the use of PR and SRK equations of states for binary gas mixture containing CO_2 and CH_4 are compared with the experimental data for validation.

2. Methodology

The composition of the natural gas feed gas mixtures is shown in Table 1. It is based on the maximum available amount of higher hydrocarbons that exists in Malaysian wells [12]. Also, in both mixtures, more than half of the total amount of gas constitute of CO_2.

Table 1. Natural gas feed streams composition [12].

Component	Composition (Mol %)	
	Gas Mixture 1	Gas mixture 2
CH_4	30	20
CO_2	60	70
C_2H_6	5	5
C_3H_8	2.8	2.8
C_4H_{10}	1.2	1.1
C_5H_{12}	1	1

The VLE behavior of gas mixtures has been predicted through the generation of pressure-temperature phase diagram using Aspen HYSYS 8.0. Bubble point and dew point curves along with the CO_2 freeze out line construe the boundaries for liquid-vapor and solid-vapor regions. Figure 1 shows the simulation environment where the feed enters the separator and is separated in vapor and liquid states.

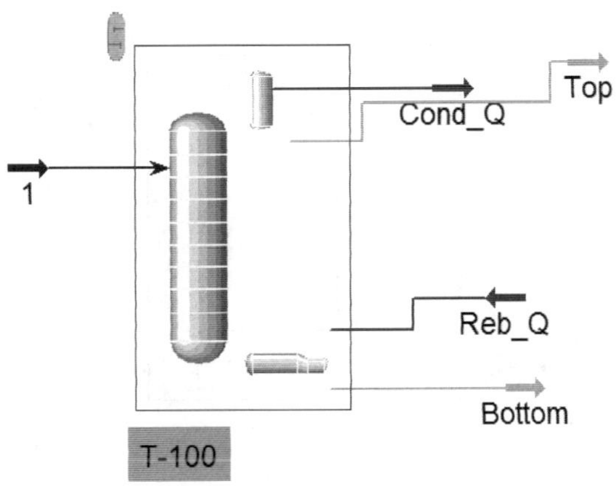

Figure 1. HYSYS simulation for phase identification.

In order to determine the accuracy of the software and suitability of the equation of state, the Solid-Liquid-Vapor (SLV) locus is obtained through simulation for both the PR and SRK models. It is obtained at the same pressure and temperature conditions that were used experimentally in previous study. Donnelly et al. had reported the experimental data for SLV locus of CO_2-CH_4 binary gas mixture which is available in literature [13]. The comparison between the predicted and experimental SLV locus is reported in Table 2 of the next section along with the discussion.

3. Results and discussion

The simulation results for phase separation of the selected compositions of natural gas mixtures are obtained by using PR and SRK equation of states. In case of gaseous mixture composed of hydrocarbons, previous studies suggest the use of PR equation of state for the accurate prediction and description of vapor-liquid, vapor-solid and liquid-solid phase behavior [14, 15]. Some studies also involved SRK model to investigate the thermodynamic behavior of natural gas for the optimization of a cryogenic system in a large scale natural gas processing plant [16]. In this work, for validation purposes, the SLV locus predicted by these equilibrium relationships through simulation is compared with the available experimental data. The experimental range of pressure at which the CO_2 freeze out temperatures and SLV locus are determined vary from 9 bar to 38 bar. Donnelly et al. determined experimentally the three phase locus describing the equilibria wherein solid CO_2 came in coexistence with the liquid and vapor phases. Their experimental setup consisted of a glass-windowed pressure cell in an alcohol bath used to confine the system [13]. For validation, Table 2 shows the comparison between experimental set of values at which CO_2 freeze out occurs and the phase exists in liquid, vapor and solid states with the predicted values for SLV locus considering both the PR and SRK equation of states.

Table 2. Comparison of experimental and predicted CO_2 freeze out temperatures [13].

P (bar)	Experimental Temp (°C)	PR Model Predicted Temp (°C)	SRK Model Predicted Temp (°C)	AAE for PR Model (%)	AAE for SRK Model (%)
9	-57.7	-60	-57.4	3.846	0.602
10.7	-58.3	-57.8	-57.8	0.914	0.928
12.8	-57.5	-58.2	-58.2	1.217	1.163
18	-59.1	-59.3	-59.3	0.225	0.205
24.3	-59.7	-60.6	-60.6	1.47	1.48
29.7	-60.5	-61.7	-61	1.89	0.706
31.2	-61.1	-62.1	-61	1.618	0.208
32	-61.3	-62.3	-61	1.484	0.659
38	-61.3	-63.7	-61	3.765	0.654
39.5	-61.6	-64	-61	3.784	1.1
40.2	-62.2	-64.2	-61	3.179	1.983
41.8	-63	-64.6	-61	2.449	3.277
44.2	-62.7	-65.4	-61	4.177	2.847
44.6	-63	-65.5	-61	3.877	3.275
44.9	-64.1	-65.6	-61	2.234	4.949
46.9	-64.7	-66.3	-61	2.438	5.764
47.8	-66.9	-66.7	-61	0.365	8.891
46.4	-69.7	-67.8	-61	2.678	12.522
44.7	-71.1	-69.7	-61	1.909	14.232
43	-72.7	-71.6	-61	1.507	16.197
37.5	-78.6	-78.1	-61	0.578	22.419
				MAE=2.172%	MAE=4.955%

The average absolute error in terms of percentage for the predicted temperature values for both the models is calculated and mean absolute error (MAE) is also determined. The mean absolute error for SRK model is about 5 % while for PR model, it is only 2.172 %. The comparison between the results predicted by PR model and SRK model clearly shows that the ability of PR model to predict the actual phase behavior is much better as compared to that of SRK model. The results also show the imprecision of SRK model to predict the phase behavior and the actual CO_2 freeze out temperatures at higher pressures and the model is not suitable to simulate the phase behavior for natural gas in high pressure applications. So the phase identification for the feed gas mixtures mentioned in Table 1 is done by using PR equation of state and the Figure 2 shows the boundaries of liquid, vapor and solid phases for both the gas mixtures.

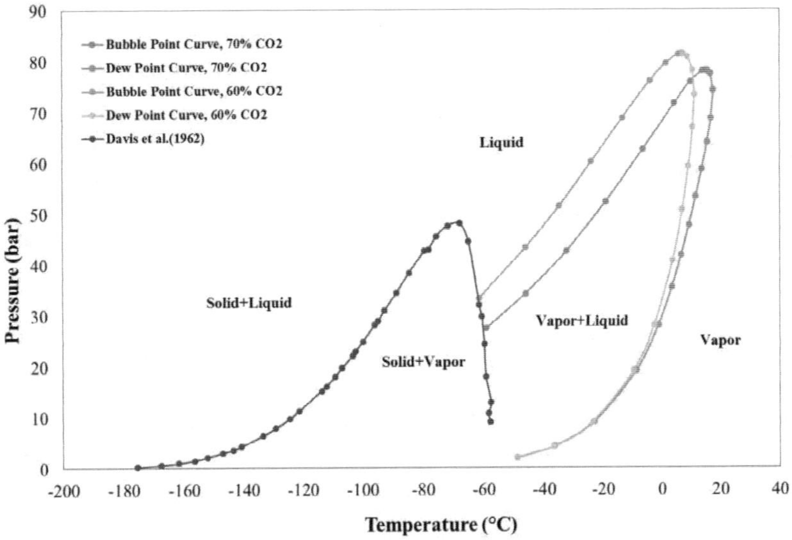

Figure 2. Phase regions for natural gas mixtures.

From the above figure it can be noticed that the liquid vapor (L-V) region starts from about 0°C to around -60°C and solid-vapor (S-V) domain lies in the temperature range starting from about -60°C to -120°C. Bubble point and dew point curves are obtained in order to determine the phase behavior.

Figure 2 also indicates that from the temperature range of 40°C to 0°C, the gas mixture is in vapor state. It is also be noticed that the vapor-liquid region lies between the temperature range of 0 to -60°C and beyond -60°C, the phase enters into solid-vapor region. The solid-vapor region has been worked upon by Davis et al. and it is reported that in this region only CO_2 exists in solid form while CH_4 is all vapor as the freezing point of methane is much lower than that of CO_2 [17]. At very low temperatures as -120°C and at both lower and higher pressures, phase enters into solid-liquid region. It is feasible to do separation in liquid-vapor and solid-vapor regions. The energy requirements for separation in liquid-solid region is high as more energy is needed to lower the temperatures to enter this region. Also the solidification of methane is avoided in this region as it is possible at very lower temperatures therefore it is not recommended to separate CO_2 in solid-liquid region.

4. Conclusion

The liquid-vapor and solid-vapor phase regions are identified for natural gas systems with high CO_2 content through simulation using PR model. These domains are determined in order to separate natural gas components at lower temperatures for cryogenic applications. PR equation of state was selected due to more accurate prediction of the phase behavior of natural gas with the data available in literature in comparison to that of SRK model. PR model predicts the VLE phase behavior and CO_2 freeze out temperatures more accurately.

Acknowledgement

The authors would like to acknowledge Universiti Teknologi PETRONAS grant YUTP-0153AAE16 for providing the necessary facilities and financial support for this research project work.

References

[1] Fantazzini D, Höök M and Angelantoni A 2011 Global oil risks in the early 21st century *Energy Policy* **39** 7865
[2] Speight J G *Natural gas: a basic handbook* 2007 Elsevier
[3] Rufford T, Smart S, Watson G, Graham B, Boxall J and Da Costa J D 2012 The removal of CO_2 and N_2 from natural gas: a review of conventional and emerging process technologies *J. Petrol. Sci Eng.* **94** 123
[4] Partoon B, Sabil K M, Roslan H, Lal B and Keong L K 2016 Impact of acetone on phase boundary of methane and carbon dioxide mixed hydrates *Fluid Phase Equilib.* **412** 51
[5] Khan M S, Bavoh C B, Partoon, Lal B, Bustam M A and Shariff A M 2017 Thermodynamic effect of ammonium based ionic liquids on CO_2 hydrates phase boundary *J. Mol. Liq.* **238** 533
[6] Parrish W R and Kidnay A J 2006 *Fundamentals of natural gas processing* CRC Press Taylor & Francis Group
[7] Abdullah M Z and Qasim A 2016 Parametric analysis of carbon dioxide adsorption on nanoporous activated carbon using computational approach Procedia *Eng.* **148** 1416
[8] Hart A and Gnanendran N 2009 Cryogenic CO_2 capture in natural gas *Energy Procedia* **1** 697
[9] Xu G, Li L, Yang Y, Tian L, Liu T and Zhang K 2012 A novel CO_2 cryogenic liquefaction and separation system *Energy* **42** 522
[10] Song C, Kitamura Y and Li S 2014 Optimization of a novel cryogenic CO_2 capture process by response surface methodology (RSM) *J. Taiwan Inst. Chem. Eng.* **45** 1666
[11] Maqsood K, Mullick A, Ali A, Kargupta K and Ganguly S 2014 Cryogenic carbon dioxide separation from natural gas: a review based on conventional and novel emerging technologies *Rev. Chem. Eng.* **30** 453
[12] AbulHassan A 2015 Study on desublimation based cryogenic packed beds for CO_2 removal from natural gas PhD, Universiti Teknologi Petronas
[13] Donnelly H G and Katz D L 1954 Phase equilibria in the carbon dioxide–methane system *Ind. Eng. Chem.* **6** 511
[14] Williams C A, Zana E N and Humphrys G E 1980 Use of the Peng-Robinson equation of state to predict hydrocarbon phase behavior and miscibility for fluid displacement in *SPE/DOE Enhanced Oil Recovery Symposium*
[15] Wu J and Prausnitz J M 1998 Phase equilibria for systems containing hydrocarbons, water, and salt: An extended Peng– Robinson equation of state *Ind. Eng. Chem. Res.* **37** 1634
[16] Cavenati S, Grande C A and Rodrigues A E 2006 Separation of $CH_4/CO_2/N_2$ mixtures by layered pressure swing adsorption for upgrade of natural gas *Chem. Eng. Sci.* **61** 3893
[17] Davis J A, Rodewald N and Kurata N F 1962 Solid-liquid-vapor phase behavior of the methane-carbon dioxide system American *Inst. Chem. Eng.* **8** 537

Fundamental studies on extraction of vanadyl oxide tetraphenyl porphyrin (VO TPP) presence in heavy oil model using toluene assisted ionic liquids

N F Salleh[1], N Ishak[1], M S H Ruslan[1,3], P C Mandal[2], and S Yusup[1].

[1] Department of Chemical Engineering, Universiti Teknologi PETRONAS, 32610 Seri Iskandar, Perak, Malaysia.
[2]Titas Gas Transmission and Distribution Co. Ltd, 105 Kazi Nazrul Islam Avenue, Kawran Bazar, Dhaka 1215, Bangladesh.
[3] Email: hazwan.ruslan@utp.edu.my

Abstract. The demand on petroleum resources are increasing in recent years but the production of conventional oil is depleted. Hence, the unconventional oil seem to be an alternative. Unfortunately, the value of heavy oils are lower than conventional oils because of the content of contaminants such as heavy metals. These heavy metals cause problems during the upgrading process. It tends to accumulate during distillation, poisoning the catalyst and will cause injurious to the wall of furnace. The aim for this work is to remove Vanadyl Oxide Tetraphenyl Porphyrin (VO TPP) from heavy oil model using toluene-assisted ionic liquids (ILs). Temperatures used are 30°C, 60°C, and 90°C at mixing time of 30 minutes, 60 minutes, and 90 minutes. Result shows 1-ethyl-3-methylimidazolium octyl sulfate (EMIM OS) have higher efficiency of VO TPP removal compared to 1-butyl-3-methylimidazolium octyl sulfate (BMIM OS). Although EMIM OS removed VO TPP higher than BMIM OS but the removal was decreased at 90°C. The optimum extraction achieved when using EMIM OS is about 45 % removal at 30°C and 30 minutes. Thus, EMIM OS is suggested to be used as it capable to remove VO TPP at minimal temperature and lowest mixing time which low operating cost.

1. Introduction

The amount of conventional oil is depleting day by day with increasing market demand. Study shows that world conventional production after 2025 is likely to be quite flat and the gap between the production and demand is large and need to be replace by unconventional oil as alternative [1]. Heavy oil resources from around the world are mostly located Venezuela, Canada, United State, Middle East and Russia [2]. Unfortunately, heavy crude oil has low value in market price due to its contaminants, high viscosity and density which make the production process more difficult [3]. Heavy oil is unconventional oil with American Petroleum Institute (API) gravity 10° and 20° [4].

In global, the heavy oil and bitumen resources is estimated around 6.2 trillion barrels, according to The World Energy Outlook from International Energy Agency. However, this heavy oil is not well commercialized as it having high amount of impurities such as heavy metals. Heavy metals found in heavy oil are chromium, copper, nickel and vanadium. The most abundant and undesired heavy metals are nickel and vanadium [5]. Mandal et al, stated that the most abundant trace elements in heavy oil are vanadium and nickel which identified at level ranging up to 200 ppm to 2000 ppm [6]. Approximately, 2% to 54% of vanadium and 1% to 47% of nickel in crude oil occur as extractable metalporphyrins (MPs) [6]. Previously, several technologies have been used to remove the heavy metals in heavy crude oil. Hydro treatment process which is the alternative for delayed coking gave disadvantages in removing heavy metals [7]. This costly method only able to produce 50-60% of useful residual oil, and the remainder contain high sulphur fuel oil. Same goes to vis-breaking and solvent deasphalting where both of these methods are inefficient and costly [7]. Due to uneconomical and unpractical existence methods, researchers still exploring new method to develop green technology to remove heavy metals from crude oil.

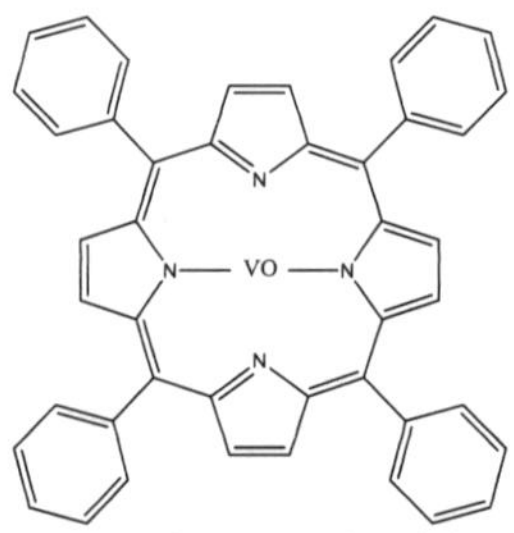

Figure 1. Molecular structure of vanadyl oxide tetraphenyl porphyrin

Ikyerene et al. reports on removing metal ions such as nickel, vanadium, and sulphur from heavy oil using natural zeolite. The result revealed that zeolite is good to remove a specific metal ion from the crude oil. However, this method is ineffective as zeolites deactivated due to deposition of heavy compounds in crude oil besides having low efficiency of metal ion removal [8]. In 2012, Ameur et al. proposed salting out separation method to remove metal ions from heavy crude oil. The result shows that 10 % (vol) of NaCl provided the best window for separation at 100 ppm heavy oil, while 20 vol % was needed for 1000 and 2000 ppm heavy oil. The advantage of this technique was able to extract highly charged molecules. The downside of this technique is it consumes high cost and the efficiency of removal is low [9]. Wang et al. was using a series of cationic starches with different degrees of substitution synthesized on removal of nickel and vanadium from Iranian and Shengli crude oils by a microwave-dry process. As the result, they found that the optimum conditions were 200 mg/L of CS4, 300 W microwave power and 5 minutes microwave time. Under these conditions, the removal efficiencies of nickel and vanadium were approximately 60 % and 80 % respectively. The disadvantage of this technique are it costly as it consumed high energy, and the temperature control is very difficult [10].

The objectives of this study is VO TPP extraction from model heavy using toluene-assisted ionic liquids (ILs). The molecular structure of VO TPP can be found in Figure 1. Subcritical toluene is a condition where the temperature of toluene is between boiling point temperature (110 °C) and critical point temperature (318.64 °C). The use of toluene is because it is highly diffusive and a good solvent to asphaltene where most organometallic species in heavy oil are agglomerated. ILs get attention among researchers and widely used in industry nowadays as it is known as green solvent. The uniqueness of ILs are it is environmentally friendly, good solubility, thermally stable, negligible vapor pressure and noncorrosive [11]. Earlier studies show that ILs has been applied in many areas. There are electrochemistry [12], chemical engineering [13], and extraction purpose such as glycerin extraction from palm oil based biodiesel [3].

IOP Conf. Series: Materials Science and Engineering **458** (2018) 012069 doi:10.1088/1757-899X/458/1/012069

2. Methodology

2.1. *Material and Apparatus*

Vanadyl Oxide Tetraphenyl Porphyrin (VO TPP), 1-ethyl-3-methylimidazolium Octyl Sulfate (EMIM OS), and 1-butyl-3-methylimidazolium Octyl Sulfate (BMIM OS) were purchased from Avantis Laboratory Supply, Malaysia. The other chemical being used was toluene. The chemicals were used without further treatment. VO TPP is purple crystal with empirical formula $C_{44}H_{24}N_4VO$ and molecular weight of 679.66 g/mol, EMIM OS having molecular weight of 320.45 g/mol and BMIM OS having molecular weight of 348.502 g/mol. Toluene was used as a solvent.

2.2. *Equipment used*

The experiments were performed in 25 mL autoclave reactor invented by Shanghai Yanzheng Experiment Instrument Co., Ltd., China, and designed for maximum temperature of 230°C and pressure of 3 MPa (gauge). The reactor is made of stainless steel, which is durable, and gas-tight. Meanwhile, inner chamber of the reactor is polytetrafluoroethylene (PTFE) structure to minimize corrosion.

The analysis was carried out on UV-visible spectrophotometer Perkin Elmer Lambda 25 model. 12.5 mm × 12.5 mm × 45 mm cuvette was used and the reference used was toluene. Lambert-Beer Law was applied in order to compute the concentration of porphyrin and its hydrogenated intermediate compounds.

2.3. *Generating Calibration Curve*

Metal compound solution was prepared by dissolving 0.015 g of VO TPP crystal in 50ml of toluene at the temperature of 50°C with 200 rpm. 150 ml of toluene was further added in three occasions, 50 ml each occasion over certain period. Once VO TPP was completely dissolved in toluene, the solution was transferred into 250ml volumetric flask.

For construction of the calibration curve, five standard solutions were prepared. 0.5 ml, 1.0 ml, 1.5 ml, 2.0 ml and 2.5 ml was measured from the metal compound solution and transferred into different 25 ml volumetric flask. Toluene was added into the volumetric flask up to the 25ml mark. All the standard solutions are analyze using UV-Visible Spectrophotometer (UV-Vis) to determine the concentration of porphyritic reactant as well as the reactant intermediate in the samples. The analysis was carried out on UV-Visible spectrophotometer Perkin Elmer Lambda 25 model with 200 – 700 range of wavelength. 12.5 × 12.5 × 45 mm cuvette was used, and the reference used was toluene. Lambert-Beer Law was applied to compute the concentration of porphyrin and its hydrogenated intermediate compounds. The graph of absorbance against wavelength was shown in Figure 2.

2.4. *Experimental Procedure*

0.5 ml of ionic liquid and 5 ml of heavy oil model was loaded into the reactor. The reactor was placed on top of hot plate and the experiment condition is set. Once the parameter was set, the reactor is left on the hot plate for few minutes to equalize the temperature. After that, IL and heavy oil model were mixed for a certain period of time at 150 rpm. After reaction, the sample was cool at room temperature. Then, sample was sent for analysis. Percentage of VO TPP removal were determined using the following equation.

$$Removal\ (\%) = \frac{C_o - C_t}{C_o} \times 100 \qquad (1)$$

where C_o is initial VO TPP concentration (0.06 g/L). Meanwhile, C_t is concentration of VO TPP after extraction.

3. Results and Discussion

3.1. Calibration Curve

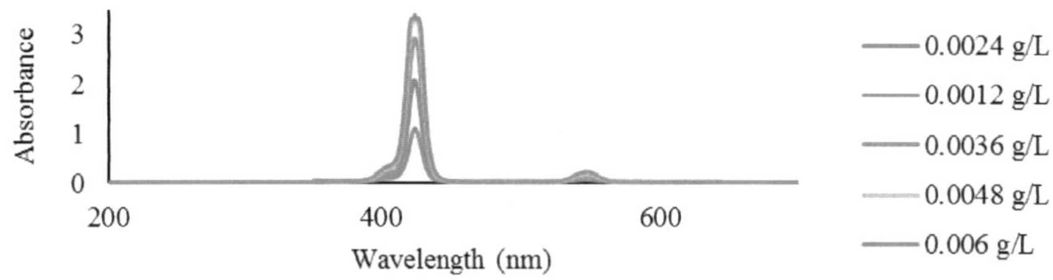

Figure 2. Absorption peak of five different samples with different concentrations for calibration curve.

Figure 2 shows the absorbance value of five different samples with different concentration. From this absorbance, a straight line graph of calibration curve was plotted. The graph obtained is linear with the value of R^2 = 0.9995. The equation obtained from the graph is y=40.417 x - 0.0011. From this equation, the final concentration for each sample can be calculated through its absorbance value.

The porphyrin macrocycle is greatly conjugated, and a number of resonance forms can be recognized using UV-visible spectrum. Soret band (*S* band) and several *Q* bands are the most significant characteristics of electrons in UV-Vis absorption spectrum as stated by Mandal et al. [6]. Soret band at approximately 400 nm refer to all type of free-metal porphyrins while *Q* bands refer to particular MPs complex when the metal replaced two hydrogen atoms in the centre of the porphyrin ring. Different MPs has different value of absorbance for Q band as stated in previous studies. Mandal et al. [14] stated the wavelength of Ni-TPP found using UV-visible spectrum is at 527.5 nm. In another report, Mandal et al. [15] stated the characteristic absorption peak for nickel etioporphyrion (Ni-EP) is at peak of 550 nm. Meanwhile, Mandal et al. [16] carried out another study on vanadyl etioporphyrin (VO EP).For this experiment, the Soret band obtained from Figure 2 is at 420 nm. Meanwhile *Q* band formed for VO TPP is at wavelength of 548 nm when nickel was replaced the two hydrogen atoms at the center of the tetraporphyrin ring. In a study reported by Mandal et al. [6], the absorption wavelength of Soret band is at 422 nm meanwhile VO TPP is at 548 nm which close to the result obtained.

3.2. VO TPP Removal

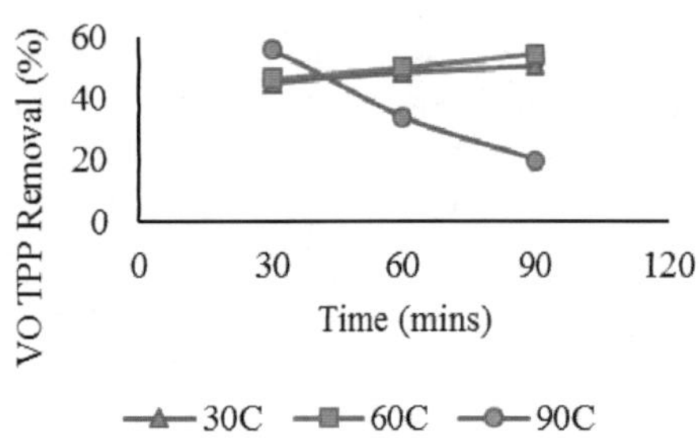

Figure 3. Effect of temperature and mixing time on VO TPP removal using EMIM OS

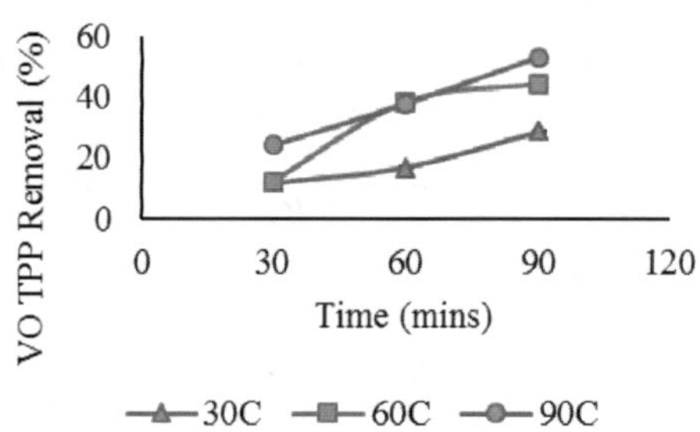

Figure 4. Effect of temperature and mixing time on VO TPP removal using BMIM OS

The experiment was conducted at different temperature which are 30°C, 60°C and 90°C to study the effect of temperature on efficiency of VO TPP removal using two ILs, BMIM OS and EMIM OS. Based on the results obtained in Figure 3 and Figure 4, EMIM OS and BMIM OS show opposite trends to each other. In Figure 3, at mixing time of 30 minutes (mins), VO TPP removal is increased up to 57 % as temperature increased from 30°C to 90°C by using EMIM OS. While, at 60 mins and 90 mins, the removal of VO TPP increased from 50 % to 55 % when temperature at 30 °C and 60 °C but it drastically declines below 35 % to 20 % when the temperature reached 90 °C. The possible reason for the result is that when increasing the temperature, the EMIM OS becomes more miscible with toluene. Then, decreasing its selectivity, thus made EMIM OS extract more toluene instead of vanadium.

In the other hand for BMIM OS, as the temperature increase from 30°C to 90°C, the removal of VO TPP using BMIM OS gradually increases. Based on Figure 4, it shows that when the temperature increased, the removal of VO TPP can be reach up to 25 %, 40 % and 55 % at mixing time 30 mins, 60 mins and 90 mins respectively. This clearly shows that increase in temperature can enhance the removal of VO TPP when BMIM OS was used. This is because at high temperature, the metal ion transfer becomes faster as stated by Germani et al. At low temperature, the high viscosity of BMIM OS reduces the rate of mass transfer between the two phases during liquid-liquid extraction. By increasing the temperature, the viscosity of BMIM OS decreases and the frequency of collision between ionic liquid and heavy oil model increases, thus increasing the extraction rate.

A study on the removal of vanadium from crude oil at temperature of 50°C and extraction time of two hours was reported in 2013 [17]. The ILs used were 1-ethyl-3-methylimidazolium (EMIM) and 1-butyl-3-methylimidazolium (BMIM) as cation with several anions. The results revealed that only EMIM bis(trifluoromethylsulfonyl)imide was capable to remove vanadium at 7.3 %. Besides, they also carried out another experiment to remove vanadium from residue [18]. They used BMIM hexafluorophosphate, BMIM bromide, and EMIM ethyl sulphate at 150°C and was mixed overnight. As the result, 40 % of vanadium was removed by BMIM hexafluorophosphate, 11 % of vanadium was removed by BMIM bromide and 23 % of vanadium was removed by EMIM ethyl sulfate. They did another set of experiment were done using EMIM trifluoroacetate and EMIM chloride at 120°C for 2 hours of mixing time [19]. It was revealed that both ILs capable to remove vanadium for approximately 22% From these previous studies, BMIM cation-based and EMIM cation-based ILs have the capability to remove vanadium but their performance depends on source of petroleum fraction likes vacuum gas oil, crude oil and residue.

3.3. Effect of Mixing Time on Percentage of VO TPP Removal

Effect of mixing time was examined at 30 minutes (mins), 60 mins and 90 mins at respective temperature and stirred at 150 rpm. Figure 3 and Figure 4 illustrates the efficiency of VOTPP removal as the mixing time increased. Result shows that when mixing time increases, the percentage of VO TPP removal increases by using BMIM-OS. At temperature of 30°C, the removal of VO TPP increased to 30 %. Meanwhile at 60°C and 90°C, the removal increased to 45 % and 54 % respectively. The performance of EMIM OS to remove VO TPP are not as good as BMIM OS. When mixing time increased, the removal increased only for temperature of 30°C and 60°C. When the temperature was extended to 90°C, the removal dropped drastically from 50% to 20% at mixing time of 30 mins to 90 mins. Even though the extraction efficiency seems to increase with time, the increase in efficiency was not very significant. It can be concluded that mixing time influence the extraction efficiency on selectivity of the ions. Further study on mechanism should be accomplishing to have deep understanding on the metal ions transfer process as ILs consists of both anion and cation.

Messadi et al. [19] studied the extraction mechanism of cation exchange and ion-pair extraction for task-specific ionic liquids with coordinating anion for heavy metal ions extraction. In their studies, they generated a cationic ester derivative of glycine betaine with bis(trifluoroethylsulfonyl)imide and dicyanamide anions in order to extract copper ions from water and the experiment was done at room temperature. They reported that the extraction process involved both cation exchange and ion-pairing. It was found that the transfer extent was determined by hydrophobicity of organic cation of IL phase through the cationic exchange process together with ion-pairing extraction mode. The elongation of the alkyl chain on the imidazolium moiety from 5 to 10 carbon atoms encourages a gradual change from cationic exchange to ion-pairing transfer mode and caused progressive decrease of the extraction yields.

3.4. Capability of Ionic Liquids

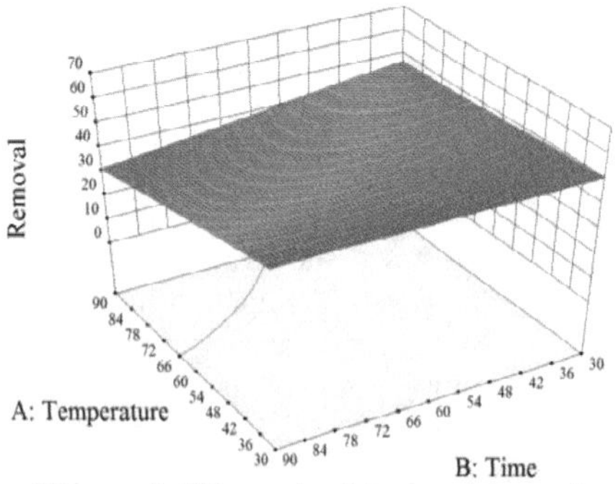

Figure 5. 3D graph of factor time and temperature toward VO TPP removal using EMIM OS

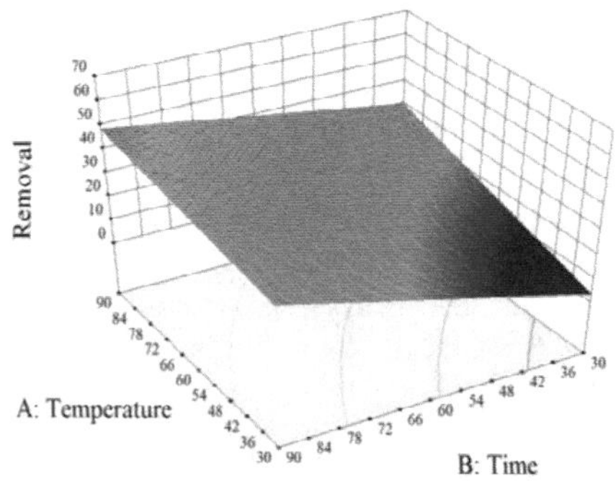

Figure 6. 3D graph of factor time and temperature toward VO TPP removal using BMIM OS

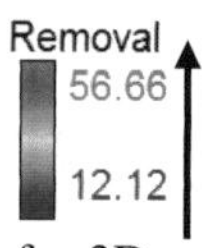

Figure 7. Color indicator for 3D graph of VO TPP removal

Table 1. ANOVA Table.

Source	Sum of Squares	DF	Mean Square	F value	Prob > F
Model	3017.882	6	502.980	6.906	0.0031
A	51.834	1	51.834	0.712	0.4168
B	245.346	1	245.346	3.369	0.0936
C	1036.339	1	1036.339	14.230	0.0031
AB	118.349	1	118.349	1.625	0.2287
AC	704.107	1	704.107	9.668	0.0099
BC	861.908	1	861.908	11.835	0.0055
Residual	801.1034	11	72.828		
Cor Total	3818.985	17			

Based on Figure 5 and Figure 6 represent 3D graph of VO TPP removal for BMIM OS and EMIM OS with the factors of temperature and mixing time. Meanwhile Figure 7 shows the colour indicator for the 3D graph found in Figure 5 and Figure 6. In Figure 7, the colour range from blue to green then to red indicated the increasing of VO TPP removal. Based on Figure 5 and Figure 6, it shows that there were interactions between the ionic liquid and heavy oil thus allow extraction process to occur Figure 5 shows maximum removal of VO TPP by EMIM OS is at 30°C with almost 50% removal meanwhile Figure 6 shows maximum removal of VO TPP by BMIM OS for almost 50% at temperature of 90°C. In the other hand, for mixing time factors, EMIM OS extracted almost 32% of VO TPP at 90 minutes while BMIM OS extracted almost 32% of VO TPP at 30 minutes. Due to that, BMIM OS is suggested to be used for extraction process as it gave optimum extraction of VO TPP with only minimal energy required and lowest mixing time, which require low operating cost. Other than that, BMIM OS is has good potential for VO TPP extraction because of the alkyl group length. According to Meindersma et al. [20], the shorter alkyl group length leads to higher toluene selectivity. Since the extraction is non-selective, EMIM OS ionic liquid tends to extract more toluene compare to VO TPP.

Based on Table 1, the Prob > F value of the model possess less than 0.05 which mean the model is significant. In this experiment, the parameters effecting the removal of VO TPP in decreasing order are type of ILs (C), mixing time (B) and temperature (A). It can be seen in Table 1 that the sum of squares for type of ILs is the highest value which is 1036.339 followed by mixing time with 245.346 and lastly temperature with 51.834 of sum of squares. From Table 1, it shows that when type of ILs interacted with time (BC) and temperature (AC) possess less than 0.05 of P value. These interaction can be seen in 3D surface graph in Figure 5 and Figure 6. Figure 5 shows the 3D surface graph of VO TPP removal using EMIM OS. From the graph, it concluded that when treatment time increased from 30 minutes to 90 minutes, the removal of VO TPP decreased. Meanwhile, when temperature increased from 30 °C to 90 °C, the removal of VO TPP was increased. 3D surface graph of BMIM OS can be found in Figure 6. It shows that when temperature increased from 30°C to 90°C and treatment time increased from 30 minutes to 90 minutes, the VO TPP removal decreased. But when time interacted with temperature without type of liquids (AB), the P value is more than significant value which means that type of ionic liquid plays an important in removing VO TPP in this experiment. The optimized condition obtained through this model is using EMIM OS at temperature of 30 °C and mixing time of 30 mins. Equations generated to predict the respond at any given mixing time and temperature, as shown in equation below:

EMIM OS: $Removal\ (\%) = 48.92 + (0.12 \times t) + (0.07 \times T) - (4.27 \times 10^{-3} \times t \times T)$
BMIM OS: $Removal\ (\%) = -30.79 + (0.69 \times t) + (0.58 \times T) - (4.27 \times 10^{-3} \times t \times T)$

where t is mixing time (minutes) and T is temperature (°C).

4. Conclusion

In conclusion, the project was able to study the capability of EMIM OS and BMIM OS to remove vanadium from heavy oil model. Somehow, the extraction was non-selective as both toluene and VO TPP are being extracted by IL. Further analysis should be done to confirm the behaviour of the extraction. Besides that, it shows that extraction temperature and mixing time can affect the extraction efficiency of vanadium. For optimization, it is preferable to use EMIM OS to extract VO TPP with only minimal temperature and lowest mixing time. This can provide good extraction efficiency at low operating cost.

Acknowledgement
The authors would like to thank Yayasan Universiti Teknologi PETRONAS (UTP) (cost centre: 0513AA-E33) for the financially support, Department of Chemical Engineering, UTP and Centre of Research in Ionic Liquids (CORIL).

References

[1] Greene D J, Hopson J L and Li J Running 2004 Out of and Into Oil: Analyzing Global Oil Depletion and Transition Through 2050 *J. Transport. Res. Rec.* **1880** 1

[2] Gupta A "Countries With The Biggest Oil Reserves", *Hydrocarbon Technology* Retrieved on 24 January 2018
[3] Hayyan M, Mjalli F S, Hashim M A and AlNashef I M 2010 A Novel Technique for Separating Glycerine From Palm Oil-Based Biodiesel using Ionic Liquids. 2010 *Fuel. Process. Technol.* **91** (1) 116-120.
[4] Speight J G 2007 *The Chemistry and Technology of Petroleum*. (New York: CRC Press) p 14.
[5] Ali M F and Abbas S. A Review of Methods for Demetallization of Residual Fuel Oils 2006 *Fuel. Process. Technol.* **87** 573
[6] Mandal P C, Coto M and Sasaki M 2014 Removal of Nickel and Vanadium from Heavy Oils using Supercritical Water *J. Jpn. Pet. Inst.* **57** (1) 12
[7] Rajendran A, Ragupathy D, Priyadarshini M, Magesh A, Jaishankar P and Madhavan M 2011 Effective Extraction of Heavy Metals from Their Effluents Using Some Potential Ionic Liquids as Green Chemicals. *E-J. Chem.* **8** (2) 697
[8] Ikyereve R E, Nwankwo C and Mohammed 2014 A Selective Removal of Metal Ions from Crude Oil using Synthetic Zeolites *Int. J. Sci. Res.* **4**(5) 1
[9] Ameur Z O and Husein M M 2012 Salting-out Induced Aggregation for Selective Separation of Vanadyl-oxideTetraphenylporphyrin from Heavy Oil *Energy Fuels* **26** 4420
[10] Wang S, Xu X, Yang J and Gao J 2011 Effect of the Carboxymethyl Chitosan on Removal of Nickel and Vanadium from Crude Oil in the Presence of Microwave Irradiation *Fuel. Process. Technol.* **92** 486
[11] Sowmiah S, Srinivasasedikan V, Tseng M C and Chu Y H 2009 On the Chemical Stabilities of Ionic Liquids *Mol.* **74** 3780
[12] MacFarlane D R, Forsyth M, Howlett P C, Pringle J M, Sun J, Annat G, Neil W and Izgorodina E I 2007 Ionic Liquid in Electrochemical Devices and Processes: Managing Interfacial Electrochemistry *Acc. Chem. Res.* **40** (11) 1165
[13] Jenck J F, Agterberg F and Droescher M J 2004 Products and Processes For Sustainable Chemical Industry: A Review of Achievements and Prospects *Green Chem.* **6** 544
[14] Mandal P C, Wahyudiono, Sasaki M and Goto M 2011 Nickel Removal from Nickel-5, 10, 15, 20-tetraphenylporphyrine using Supercritical Water in Absence of Catalyst: A Basic Study *J. Hazard. Mater.* **187** 600
[15] Mandal P C, Wahyudiono, Sasaki M and Goto M 2012 Nickel Removal from Nickel Etioporphyrin (Ni-EP) using Supercritical Water in the Absence of Catalyst *Fuel. Process. Technol.* **104** 67
[16] Mandal P C, Wahyudiono, Sasaki M and Goto M 2012 Non-catalytic Vanadium Removal From Vanadyl Etioporphyrin (VO-EP) using A Mixed Solvent of Supercritical Water and Toluene: A kinetic Study *Fuel* **92** 288
[17] Serban M, Bhattacharyya A, Bussche K M V, Nicholas C P and Bennion W K Process for Removing Metals from Resid. US8608950B2. Issued Dec.17, 2013
[18] Serban M, Bhattacharyya A, Bussche K M V, Nicholas C P and Bennion W K Process for Removing Metals from Resid. US8608950B2. Issued Dec.17, 2013
[19] Messadi A, Mohamadou A, Boudesocque S, Dupont L and Guillon E 2013 Task-Specific Ionic Liquid with Coordinating Anion for Heavy Metal Ion Extraction: Cation Exchange versus Ion-pair Extraction *Sep. Purif. Technol.* **107** 172
[20] Meindersma G W, Podt A (J G) and Haan A B 2005 Selection of Ionic Liquids For The Extraction of Aromatic Hydrocarbons From Aromatic / Aliphatic Mictures *Fuel. Process. Technol.* **87** 59

Structural Analysis on the Effect of Base-Catalysed Delignification Process Parameters on Palm Oil Empty Fruit Bunches Fibres using Glycome Profiling

Noor Haida Sebran[1], Law Poh Gaik[1], Azlan Shah Hussain[1]*

[1]PETRONAS Research Sdn. Bhd., Lot 3288 & 3289, Off Jalan Ayer Itam, Kawasan Institusi Bangi, 43000 Kajang, Selangor Darul Ehsan, MALAYSIA

azlanshah_hh@petronas.com.my

Abstract. Malaysian Oil Palm Empty Fruit Bunches (EFB) fibre was used as a model substrate and was subjected to base-catalysed delignification technique as a function of different temperature, time and base concentration. The resultant substrates were analysed using glycome profiling technique to ascertain the degree of delignification effectiveness and position of the residual lignin in the structure of the pre-treated biomass. The study indicated that the base-catalysed delignification technique removes all the outer lignin structure of the fibre due to the accessibility of the base to the surface of the biomass. Base wash concentration using sodium hydroxide (NaOH) and temperature are the two main factors that determine the effectiveness of lignin removal and increment of surface area accessibility (SAA). The pre-treated biomass that contains the lowest amount of residual lignin content of 6.19 wt.% which corresponds to 74.47% lignin removal shows complete removal of the outer layer of lignin structure but leaves the inner lignin structure intact. This study suggests that the complex interaction between different cell wall constituents plays a much bigger role in order to achieve efficient delignification.

1. Introduction

Biomass resources, in particular, agricultural wastes or forestry residues posed significant challenges due to their recalcitrant nature, varying degree of constituents' content and complex intermolecular hydrogen-bond networks. In the review made by Zhao *et al.*, pretreatment techniques shall exert influence on accessible surface area, biomass relevant-factors such as pore size, chemical composition (such as lignin and carbohydrate content) and cellulose crystallinity [1]. Reviews made by Zhao *et al.* and Kumar *et al.* provide an extensive body of knowledge on the various physico-chemical techniques of treating and conditioning such feedstock prior to its valorisation to fuels or chemicals [1, 2]. Martin *et al.* provides a review which summarises the limitation of such techniques as it contributes to the formation of non-beneficial side products that could render subsequent process to be less effective such as short-chained mono and di acids, ferulic acid and even monosaccharides [3].

Biomass recalcitrance is a term that is being used to describe the interaction between cell-wall microstructure which essentially provide the strength and structure to any plant [4-6]. Cell-wall is essentially a microfibril structure that consists of vascular bundles built from polymeric cyclodextrin (or cellulose) and surrounded by lignin-hemicellulose matrix. The same hemicellulose contains ferulated functionality which provides site for cross-bonding between the lignin and carbohydrates, also known as lignin-carbohydrate complex [7-9]. Due to this structural heterogeneity and intertwining of different polymeric chemical compositions, it is fairly difficult to deconstruct any biomass.

Glycome profiling is a high-throughput technique that employs glycan-directed monoclonal antibodies (mAbs) to monitor structural changes of plant cell wall [10-11]. The technique allows in situ visualisation of specific substrates, or epitopes as it is called, using specific molecular probes post selective chemical extractions with increasing severity. This essentially allows large and wide range monoclonal antibodies to be utilised to detect and identify diverse features of cell wall polysaccharides [12].

Interactions between lignin, hemicellulose and pectin with cellulose and their contribution to overall biomass recalcitrance has been widely illustrated using Poplar or *populus sp.* as model biomass. One of the striking features of poplar, as observed by DeMartini *et al.*, is the existence of cross-linked pectin, arabinogalactan and various loosely-bounded xylans [13-14]. Removal of gene controlling pectin synthesis in poplar has been proved to contribute to reduced recalcitrance of the plant cell wall whilst maintaining the amount of polysaccharide as per the native species [15]. Variation of lignin content, as measured in part with syringyl and guaicol ratio (S/G ratio) was exemplified in poplar, of which lower lignin content would have favourable effect on the sugar hydrolysis [16].

Using Poplar as model biomass, the effect of pretreatment techniques such as hydrothermal, dilute acid and ammonia explosion (AFEX™) on plant cell wall deconstruction can be semi-quantified using glycome profiling. Hydrothermal technique has shown that outer-layer xylans is more soluble in this condition with additional benefit of incremental average pore diameter [17-20]. More closely-bounded xylans in the macrofibrils is much harder to remove, as exemplified by removal of xylan epitopes or subsets with harsher extraction techniques. Similar observations were made by Singh *et al.* using corn stover as substrate [6].

This work is a continuation of the work first published herein [21]. The work shall describe the use of glycome profiling to investigate the changes on the plant cell wall and structural modification induced by base catalysed delignification process using palm oil biomass as substrate. As far as we are concern, no work of similar nature has been done on palm oil biomass. Three (3) independent parameters, namely, temperature, base concentration and time were varied with final lignin content of the cellulose-rich material as the dependent parameter.

2. Materials and Methodologies

2.1 Materials and Chemicals

Empty Fruit Bunches (EFB) fibres were sourced from a local mill located in Selangor, Malaysia. The fibres were already processed and pre-cut to 2 mm average fibre length. Further grinding was done on lab scale grinder (IKA Labortechnik) with sieve of 1 mm to get the desired fibre length. All chemicals used, sodium hydroxide (NaOH, Fisher Chemical), ammonium oxalate ($C_2H_8N_2O_4$, Acros Organics), sodium carbonate (Na_2CO_3, Acros Organics), potassium hydroxide (KOH, J.T. Baker) and sodium chlorite ($NaClO_2$, Acros Organics) were used as received unless stated otherwise.

2.2 Brunauer-Emmett-Teller (BET) Gas Adsorption Method

Method described by Sing *et al.* (1985) was used to determine surface area of the biomass [22]. Adsorbed materials from the surface of our samples were eliminated by heating the samples under vacuum at a temperature of 353.15K for 10 hours before analysis. The adsorption/desorption experiments were carried out at liquid nitrogen temperature of 77.15 K and were measured on a Micromeritics ASAP 2020 automatic surface area and pore radius distribution analyzer. Quantity of adsorbed and desorbed gas was calculated from changes in pressure using pressure transducers. BET analysis was carried out to determine total surface area of our samples.

2.3 Sodium Hydroxide (NaOH) Pretreatment of Empty Fruit Bunches (EFB) Fibres
The NaOH pretreatment or 'based-catalyst delignification' of EFB was followed the procedure published by Zawawi *et al.* [21]. NaOH is one of the strongest base catalyst and it is capable to break the linkage between lignin and hemicellulose in lignin-carbohydrate complexes (LCC). The NaOH pretreatment of untreated EFB fibres was carried out in a 1.5 L PARR reactor (PARR USA). The concentrations of NaOH used was 0.5 wt.% to 5.5 wt.%, residence time was 60 to 120 min and stirring rate was 500 rpm. These experiments were conducted at a temperature range of 373 K to 473 K.The composition of the pre-treated and untreated EFB samples were then analysed using Van Soest method and ELISA (enzyme-linked immunosorbent assay). The methodology for Van Soest method was adopted from work by Goering and Van Soest (1970), Van Soest *et al.*(1991) and Chaves *et al.*(2002) [23-25].

2.4 Total Sugar Estimation and Glycome Profiling using ELISA
Sequential extraction of the cell walls (AIR) samples and glycome profiling was carried out as described by Pattathil *et al.* [12] to isolate fractions enriched in various cell wall components. Sequential extraction involved different reagents (50 mM ammonium oxalate, 50 mM sodium carbonate, 1 M KOH, 4 M KOH, three additions of 0.125 g of sodium chlorite and 4 M KOH post chlorite). The supernatants were labelled as 'Ammonium Oxalate Extract', 'Carbonate Extract', '1 M KOH Extract', '4 M KOH Extract' and post chlorite – '4 M KOH PC fraction'. All extracts were dialyzed (using 3,500 Da molecular weight cut-off tubing, spectrum Laboratories Inc. CA, USA) against four changes of deionized water (with an approximate sample to water ratio of 1:60) for 48 h at room temperature and freeze-dried individually. All extracts were dissolved in deionized water at a concentration of 0.2 mg/mL in a 15 mL centrifuge tube. The total sugar estimation was done using a phenol-sulphuric acid colorimetric assay [26, 27]. The extract samples were diluted to a sugar concentration of 0.2 mg/mL. ELISA plates (Costar 3598) were coated with 50 μL per well of extract and allowed to evaporate to dryness overnight in a ventilated incubator at a temperature of 310 K until dry. The glycome profiling analyses involved ELISA-based screening of EFB biomass materials against seven plant cell wall glycan-directed monoclonal antibodies (mAbs). Negative controls consisting of water blanks without samples and antigen were included in all assays and their absorbance subtracted from all samples. The ELISA responses of the monoclonal antibodies binding intensities to each extract were presented as heat maps. Monoclonal antibodies were obtained as hybridoma cell culture supernatants from the main suppliers of plant cell wall glycan-directed mAbs such as CarboSource, PlantProbes and BioSupplies. There were seven mAbs used in the ELISA including xylan (LM10, CCRC-M113), xyloglucan (LM15, CCRC-M96, CCRC- M102), (1→3)-β-glucan (400-2) and (1→3, 1→4)-β-glucan (400-3).

3. Results and Discussion

3.1 Surface Area Measurement (BET) and Compositional Analysis
Table 1 and Table 2 summarize the BET results and compositional analysis of six (6) different samples subjected to increasing severity of delignification process. BET surface area measurements revealed that EFB pretreated with high concentrated NaOH solution resulted in a more porous material with higher surface area compared to those untreated materials or those treated in less concentrated NaOH solutions. The average pore diameter of sample E is 1.6 times larger compared to sample UN (untreated). Surface area of sample E further confirmed this observation as it is 2.37 times larger than sample UN. Rezende *et al.* confirmed that when sugarcanes bagasse samples were pretreated with NaOH concentrations between 1-4%, no significant morphological differences were observed between the samples [28]. Therefore, in order to increase assessable surface area as well as larger average pore

diameter of EFB samples with low lignin content, it is crucial to subject the EFB fibers to harsh pretreatment conditions of 5.5 wt.% NaOH solutions at 473K for 1 hour.

The lower BET surface area of sample A to D as compared to untreated sample is due to the dehydration step of which it removes water and collapse the open pores. The same observation was made by Kaewprasit *et al.* [29].

Table 1. BET results for pretreated and untreated EFB samples

Sample code	Pretreatment	BET surface area (m^2/g)	Adsorption average pore diameter (Å)
UN	Untreated EFB	1.2795	67.0955
A	Base 0.5 wt.%, 423 K, 120 min	1.2152	61.6034
B	Base 1.0 wt.%, 423 K, 120 min	0.9755	36.0061
C	Base 1.75 wt.%, 423 K, 90 min	0.8016	45.4848
D	Base 3.0 wt.%, 423 K, 120 min	1.7422	46.8362
E	Base 5.5 wt.%, 473 K, 60 min	3.0393	108.0183

Table 2 shows the composition of hemicellulose, cellulose and lignin by using the Van Soest method. The hemicellulose content for all pre-treated samples varied between 8.83 wt.% to 18.41 wt.% while untreated samples gave hemicellulose content at 26.76 wt.%. Sample E that was pretreated with the highest base concentration gave the highest cellulose content at 84.53 wt.% and followed by sample D and C at 71.41 wt.% and 70.86 wt.%, respectively. The cellulose content for other pretreatments including untreated EFB samples were varied in the range of 41.26 wt.% to 66.40 wt.%. Sample E showed the lowest lignin content of 6.19 wt%. This is followed by samples D, C, B and A at 8.69 wt.%, 9.55 wt.%, 14.06 wt.% and 18.20 wt.%, respectively. Untreated EFB samples gave the highest lignin content at 24.25 wt.%. Based on the overall results, sample E with a pretreatment parameters of NaOH concentration 5.5 wt.% at 473 K for 60 min showed the best performance with highest cellulose content and lowest lignin content as compared with other pretreatment parameters. Study by Lima *et al.* reported that higher NaOH concentration at 4.0 wt.% is adequate to keep lignin dispersed in the pretreatment liquor but with lower NaOH concentration of less than 1.0 wt.%, it is not sufficient to open the cellulose microfibrils and to expose the crystalline surfaces [30]. The study conducted by Rezende *et al.* using sugarcane bagasse samples further confirmed the effectiveness of higher NaOH concentration in lowering the lignin content of biomass [28].

Table 2. Analysis of EFB samples using Van Soest method

Sample code	Pretreatment	Van Soest method (wt.%)			
		Hemicellulose	Cellulose	Lignin (AIL)	Grand total
UN	Untreated EFB	26.76	41.26	24.25	92.27
A	Base 0.5 wt.%, 423 K, 120 min	5.60	66.40	18.20	90.20
B	Base 1.0 wt.%, 423 K, 120 min	18.41	62.04	14.06	94.51
C	Base 1.75 wt.%, 423 K, 90 min	10.15	70.86	9.55	90.56
D	Base 3.0 wt.%, 423 K, 120 min	16.55	71.41	8.69	96.65
E	Base 5.5 wt.%, 473 K, 60 min	8.83	84.53	6.19	99.55

*AIL = Acid Insoluble Lignin

3.2 Glycome Profiling of EFB Fibres

Recent studies on the usage of glycome profiling technique as a primary semi-quantitative tool to analyse the plant biomass and its components have been published [10, 12-13]. Glycome profiling technique uses a set of extractions to solubilize the different carbohydrate materials from the plant cell wall. It depends on how these components are bound into the walls. All untreated and pre-treated EFB samples were further investigated for the changes on the plant cell wall and structural modification especially on lignin distribution using this technique. The extracted samples released from each untreated and pretreated EFB samples by reagents were loaded onto the ELISA plates and were screened against the monoclonal antibodies. Several plant cell wall glycan-directed monoclonal antibodies (mAbs) were selected to recognize epitopes (specific glycan subcultures) present in the extracts. Figure 1 shows the overall heat maps for pretreated and untreated EFB samples.

Sequential extraction using ammonium oxalate, sodium carbonate, 1 M KOH, 4 M KOH, chlorite and 4 M KOH post chlorite are selected since they have the capability to separate plant cell wall components [31]. These extracts are selectively enriched with various classes of cell wall glycans based on the tightness with which they are integrated into the cell walls [10]. Based on Figure 1, xylan epitopes represent the most abundant recognizable hemicellulosic epitopes in all pretreated and untreated EFB samples. Xylans were extracted out under all extraction conditions as indicated by the binding patterns with strong significant binding of mAbs such as LM 10 with the sample extracts. This observation was supported by the work done by Simmons *et al.* which reported that xylan was the most dominant non-cellulosic polysaccharides that binds to cellulose in the cell wall. The interaction of xylan with cellulose is expected to have the most profound effect on the cell wall characteristics with higher xylan folded onto cellulose surfaces [32] which later increased the cellulose content. Based on our study, this ELISA results rationalize the Van Soest analysis results which showed higher cellulose content in samples A to E and untreated EFB sample as compared to their respective hemicellulose and lignin content. The study conducted by Pattathil *et al.* further confirmed that the plant cell wall is comprised largely of cellulose microfibers that are embedded in a macromolecular matrix of glycans [17].

Figure 1 shows other hemicellulosic epitopes, such as fucosylated xyloglucan (CCRC-M102) and non-fucosylated (CCRC-M96) were less abundantly present based on weak/moderate binding with samples extracts of pretreated and untreated EFB, indicating that xyloglucans are not completely solubilized during the pretreatment even at the high base concentration of NaOH (sample E). Both fucosylated and non-fucosylated xyloglucans were found to be released more in the 4 M KOH treatment for pretreated and untreated EFB samples (Figure 1). Among these xyloglucans, the presence of fucosylated xyloglucan epitopes were in higher proportions compared to non-fucosylated xyloglucans epitopes. The relative distribution of solubilized xylans and xyloglucans were found to be comparable among the pretreatments (from lower to higher base concentration of NaOH), indicating that all pretreatments solubilize similar pools of hemicellulosic glycans. β-glucan were also present in less abundance based on weak/moderate binding of this mAbs with the sample extracts of pretreated and untreated EFB (Figure 1).

This study explains the composition, extractability and relative proportions of cell wall components such as hemicellulose, cellulose, pectin and lignin in EFB fibres. This is to ensure that our hypothesis on alkaline pretreatment is accurate especially on lignin removal. More tightly bound lignin components of the cell wall materials are removed by using a chlorite treatment, leaving polysaccharides into the generated extract. Based on heat maps for chlorite extraction treatment in Figure 1, the removal of lignin in the EFB pretreatment will give the strong binding between the mAbs and sample extracts. However, glycome profiling results shows differently with weak and moderate binding occurred between the mAbs and samples A to E including untreated EFB extracts, indicating that these lignin-associated cell wall glycans were still trapped in the extract of samples after the chlorite treatment. Chlorite extract of pretreated and untreated EFB contained xylans and β-glucan as denoted by the weak binding of mAbs with sample extracts. No xyloglucans were present in the chlorite treatment (Figure 1). The ELISA results of lignin content were comparable with the Van Soest analysis of lignin content at 6.19 wt.% to 18.20 wt.% that shows considerable amount of lignin still remained tightly bound in the samples even after the harsh alkaline pretreatment (e.g 5.5 wt.% of base concentration). The post chlorite alkaline extract (4 M KOH PC) was used to further solubilize the remaining polysaccharides after lignin removal to make it more extractable. Based on Figure 1, the extracts produced in 4 M KOH PC was found comparable in composition to other alkaline extracts (4 M KOH) with proportions of hemicellulosic polysaccharides, xyloglucans and xylans. The study done by DeMartini *et al.* and Venketachalam *et al.* confirmed that the results were similar for the post chlorite extracts with the significant amounts of various epitopes in the cell wall [10, 33].

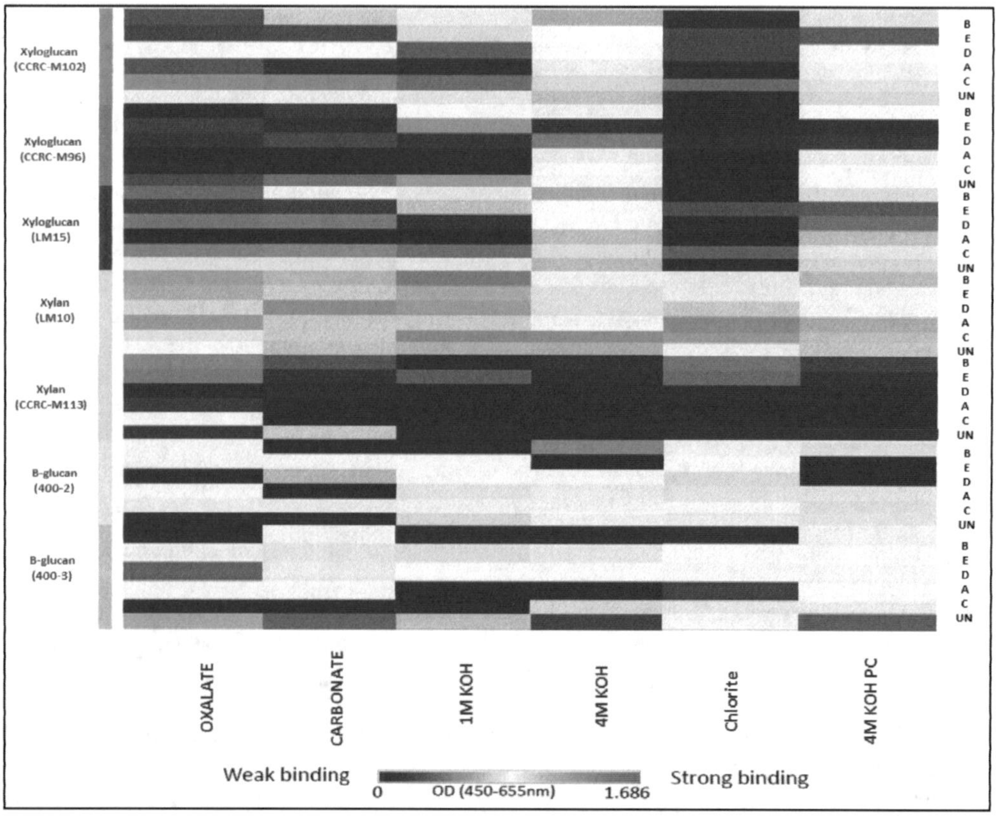

Figure 1. Glycome profiling of sequential cell wall extracts of pretreated EFB and untreated EFB. Binding intensities of antibodies to the extracted cell wall glycans, as measured in ELISAs, are depicted as heat maps with blue to red colours representing no binding to strongest binding, respectively. The reagents used for extraction are labelled at the bottom of the heat map.

4. Conclusion

BET surface area measurement further confirmed that EFB samples that undergone harsh alkaline pretreatment has the highest surface area among all 6 samples tested. The changes in the cell wall structure of pretreated EFB especially for higher alkaline pretreatment was due to the solubilisation and removal of lignin and lignin-associated carbohydrates during the pretreatment. Glycome profiling technique further confirmed our observations made throughout the chemical analysis using Van Soest method especially on delignification efficiency. It showed that lignin is still present in the structure of EFB fibres whilst most of the outer lignin layer of the EFB sample was removed after harsh alkaline pretreatment, indicating a tightly bound secondary lignin surrounding inner fibrils of cellulose. Sodium hydroxide pretreatment with base concentration more than 1.0 wt.% was able to reduce lignin content to less than 10.0 wt.%. This pretreatment can be improved by using combined pretreatments such as physical (mechanical)-chemical, physico-chemical or biological-chemical processes.

Acknowledgement

The authors would like to acknowledge PETRONAS and PETRONAS Chemical Group Bhd. (PCGB) for funding this work.

References

[1] Zhao X, Zhang L and Liu D 2012 *Biofuel Bioprod Bior* **6(4)** 465-482
[2] Kumar P, Barrett D M, Delwiche M J and Stroeve P 2009 *Ind Eng Chem Res* **48** 3713–29
[3] Martin C and Jonsson L J 2016 *Bioresour Technol* **199** 103-112
[4] Foston M and Ragauskas A J 2012 *Ind Biotechnol* **8** 191-208
[5] Himmel M E, Ding S Y, Johnson D K, Adey W S, Nimlos M R, Brady J W and Foust T D 2007 *Science* **315** 804-807
[6] Singh S, Cheng G, Sathitsuksanoh N, Wu D, Varanasi P, George A, Balan V, Gao X, Kumar R, Dale B E, Wyman C E and Simmons B A 2015 *Front Energy Res* **2(62)** 1-12
[7] Rubin E M 2008 *Nature*, **454** 841-845
[8] Khandelwal M, and Windle A 2014 *Carbohydr Polym* **106** 128-131
[9] Conley K, Godbout L, Whitehead M A and van de Ven, T G M 2015 *Carbohydr Polym* **135** 285-299
[10] DeMartini J D, Pattathil S, Avci U, Szekalski K, Mazumber K, Hahn M G and Wyman C E 2011 *Energ Environ Sci* **4** 4332-39
[11] Pattathil S, Avci U, Baldwin D, Swennes A G, McGill J A, Popper Z, Botten T, Albert A, Davis R H, Chennareddy C *et al.* 2010 *Plant Physiol* **153** 514-525
[12] Pattathil S, Avci U, Miller J S and Hahn M G 2012 *Biomass conversion: Methods and Protocol* vol **908**, ed M E Himmel (New York: Humana Press) p 61-72
[13] DeMartini J D, Pattathil S, Miller J S, Li H, Hahn M G and Wyman C E 2013 *Energ Environ Sci* **6** 898-909
[14] Vogel J 2008 *Curr Opin Plant Biol* **11** 301-307
[15] Chung D, Pattathil S, Biswal A K, Hahn M G, Mohnen D and Westpheling J 2014 *Biotechnol Biofuels* **7** 147-158
[16] Studer, M H, DeMartini J D, Davis M F, Sykes R W, Davidson B, Keller M, Tuskan G A and Wyman C E 2011 *Proc Natl Acad Sci* **108** 6300-05
[17] Pattathil S, Hahn M G, Dale B E and Chundawat S P S 2015 *J Exp Bot* **66** 4279-94
[18] Meng X, Wells Jr T, Sun Q, Huang F and Ragauskas A 2015 *Green Chem* **17** 4239-46
[19] Singh S, Varanasi P, Singh P, Adams P D, Auer M and Simmons B A 2013 *Biomass Bioenerg* **54** 276-283
[20] Jing M, Zhe J, Jia C C, Xia Z, Kim Y S and Feng X 2015 *Planta* **242** 327-337
[21] Zawawi A Z, Law P G, Sebran N H, Othman J and Hussain A H 2017 *Biomass Convers Biorefin* https://doi.org/10.1007/s13399-017-0246-x
[22] Sing K S W, Everett D H, Haul R A W, Moscou L, Pierotti R A, Rouquerol J and Siemieniewska T 1985 *Pure Appl Chem* **57(4)** 603-619
[23] Goering H K and Van Soest P J 1970 *Agricultural Research Service (ARS) Handbook. No. 379.* Washington, DC.
[24] Van Soest P J, Robertson J B and Lewis B A 1991 *J Dairy Sci* **74** 3583–97
[25] Chaves A V, Waghorn G C and Tavendale M H 2002 *Proc N Z Grassland Assoc* **64** 129-133
[26] Dubois M, Gilles D A, Hamilton J K, Rebers P A and Smith F 1956 *Anal Chem* **28** 350–356
[27] Masuko T, Minami A, Iwasaki N, Majima T, Nishimura S and Lee Y C 2005 *Anal Biochem* **339(1)** 69-72

[28] Rezende C A, deLima M A, Maziero P, deAzevedo E R, Garcia W and Polikarpov I 2011 *Biotechnol Biofuels* **4(54)** 1-18
[29] Kaewprasit C, Hequet E, Abidi N and Gourlo J P 1998 *J Cotton Sci* **2** 164-173
[30] Lima M A, Lavorente G B, Hana K P da Silva, Bragatto J, Rezende C A, Bernardinelli O D, deAzevedo E R, Gomez L D, McQueen-Mason S J, Labate C A and Polikarpov I 2013 *Biotechnol Biofuels* **6(75)** 1-17
[31] Zhu X, Pattathil S, Mazumder K, Brehm A, Hahn M G, Dinesh-Kumar S P and Joshi C P 2010, *Mol Plant* **3(5)** 818-833
[32] Simmons T J, Mortimer J C, Bernardinelli O D, Poppler A, Browns S P, deAzevedo E R, Dupree R and Dupree P 2016 *Nat Commun* **7** 1-9
[33] Venketachalam S, Pattathil S, Kandemkavil S and Sharma S 2013 *J Text Sci Eng* S1:001 pp 1-4

Evaluation of Tetraethylammonium Chloride on Methane Gas Hydrate Phase Conditions

Z Kassim[1,2], M S Khan[1], B Lal[1], B Partoon[1], A M Shariff[1]

[1]*Research Centre for CO_2 Capture (RCCO_2C), Chemical Engineering Department, Universiti Teknologi PETRONAS, Bandar Seri Iskandar, 32610, Perak, Malaysia*

[2]*PETRONAS Research Sdn Bhd, Kawasan Institusi Bangi, Kajang, 43000, Selangor, Malaysia*

E-mail: zamzila_kassim@petronas.com.my

Abstract: In this work, hydrate liquid vapor equilibrium conditions for methane (CH_4) gas is measured in the absence and presence of different aqueous TEACl concentrations. The temperature cycle (T-cycle) method has been used for the hydrate equilibrium temperature of CH_4 + H_2O + TEACl system measurement within the pressure ranges of 3.90 – 8.30 MPa. Results reveal that, for 10 wt% aqueous TEACl average of suppression temperature (ΔT) (inhibition effect) appeared of 0.9 K for CH_4 hydrate. Additionally, the enthalpy of dissociation for methane gas hydrate with and without the present of inhibitor is calculated using Clausius-Clapeyron equation.

1. Introduction

In oil and gas industry, gas hydrate is one of the major challenges as exploration and production facilities are moving towards application of game-changer technologies with operating condition at extremely low temperature and high pressure in such conducive condition for hydrate formation [1,2]. Gas hydrates are crystalline compound consist of water cage structure and stabilized with gases as guest molecules under thermodynamically driven conditions [3]. Normally, small guest molecules occupying the water cages are light hydrocarbons such as methane (CH_4) and propane (C_3H_8), besides other gas carbon dioxide (CO_2) and hydrogen sulfide (H_2S). There are three gas hydrate structures namely structure I (sI), structure II (sII) and hexagonal structure H (sH) primarily subject to the size of the guest molecules. Gas hydrate issues in offshore production facilities has caused plugging and interrupted production line thus causing disastrous economic, inefficient operation and safety issue [2,3]. Occurrence of hydrate at low temperature and high pressure environment has made offshore production in deeper oceans such as North Sea, Gulf of Mexico and Alaska exposed to hydrate plugging issues. Remediation due to hydrate flow assurance issues, industry required exorbitance cost annually depending on the severity and location. The conventional hydrate remediation strategy includes dehydration of free water, depressurization, insulation or heat tracing and chemical inhibition [3]. However, chemical inhibition is considered as the most efficient hydrate prevention strategy existed in the industry [3]. Chemical inhibitor that works by shifting hydrate curve to the left were identified as thermodynamic hydrate inhibitor (THI). Among known THI are methanol and mono ethylene glycol (MEG) and they were used at very high concentration 20-50% by volume depending on aqueous phase in the system [4]. As the gas production and exploration moves to deeper seas large amount of THI need to be injected as the pressure and temperature conditions become more conducive for hydrate formation. Although THI is effective, factors such as economic, operational health & safety and product quality has forced oil and gas industry to develop low dosage hydrate inhibitor (LDHI) that commonly injected at very low dose 0.1- 2.0wt % [5]. There are two types of LDHI namely kinetic hydrate inhibitor (KHI) and anti-agglomerant (AA). KHI typically a water soluble polymer delayed the hydrate formation by slowing down the growth of hydrate crystallization in the pipelines and processes unit while AA inhibitors allowed hydrate formation, this surface active molecules attach to and disperse hydrate particles that can still be transported in slurry form [6]. When tiny hydrate particle began to form, AA will bind to particles and makes it slightly hydrophobic, thus prevented from hydrate cluster to grow [7]. However, AA performance are only effective at low to moderate water cut resulting industry to further research more effective hydrate inhibitor. Despite being in the industry for years, both THI and KHI (as well as AA) are still far from satisfactory to give an economic solution [8]. In previous research activity, it is admitted that the combined effect of thermodynamic and kinetic inhibitors would be able to promise an improved solution to the current issue [5]. Recently, more efforts has been conducted to define effective operability of new-kind inhibitors than the current commercial chemical inhibitors [9].

Among the earliest discovery of a new category of inhibitors, known as ionic liquids (ILs) is by Xiao and Adidharma [10] and Xiao et al [4]. It has been found that ILs can powerfully function as both THI and KHI. As of now, commonly studied are imidazolium bases ionic ILs [4], [11], [12]; therefore insufficient work is done to study the effects of other group of ILs as hydrate inhibitor, especially ammonium-based ionic liquid (AILs) [13]. Tariq et al. [13] have evaluated that AILs worked as dual function hydrate inhibitor. Govinda et al. [14] has studied the applications of AILs in several research area such as carbon capture and storage, fuel cells, corrosion inhibition included flow assurance applications. AILs have been claimed to possess a better eco-friendly properties as compared to imidazolium based ILs group as AILs are commonly found to be used as the food stabilizers [15]. Li et al. [16] have studied that tetramethylammonium chloride (TMACl) as gas hydrate inhibitor and it is found TMACl produce better THI in comparison to imidazolium group ILs. Consequently, Keshavarz et al [17] has assessed the influence of tetraethylammonium chloride (TEACl) with other imidazolium-based ILs and discovered that all investigate ILs shows inhibition performance as high as 0.9 K [17–20].

As limited understanding in literature on the inhibition performance of AILs,` this study are obliged to further investigate the thermodynamic inhibition performance of one of the AILs group named tetraethylammonium chloride (TEACl) on the hydrate equilibrium conditions of methane (CH_4) hydrates at 1, 5 and 10 wt% concentrations of TEACl.

2. Methodology

2.1 Material

Materials used in this study are TEACl (95%) and acquired from Merck milli-pore company Germany. CH_4 gas is supplied by Air Products PTE LTD. Ultra-purified deionized water is used to dilute the TEACl to intended concentration.

2.2 Experimental apparatus

Experimental set-up consists of fixed volume Saphhire glass cell is used in this work for determining the hydrate phase equilibrium conditions of CH_4 gas in the absence and presence of TEACl. The set-up was furnished with an internal cell with 60 cm^3 in volume with design pressure up to 20 MPa and design temperature in the range of 253 – 333 K. The temperature, pressure and volume are measured in evert two seconds with an accuracy of ±0.1 K , ±0.01 MPa and ±0.001 cm^3, respectively. The glass cell is submerged in a ethanol coolant bath where temperature is controlled by means circulating the coolant in Lauda + DLK45 thermostats. The scientific equipment setup used in this work, depicted in Figure 1. The temperature in the equilibrium cell both in liquid and gas phase are measured by means of platinum-resistance (Pt100) thermometer probes and it is calibrated by Laboratoire National d'Essaies Paris. The pressure in the equilibrium cell is measured by stable pressure transducer pressure calibrated by Armines-Mines Paristech Research Centre. The glass cell is equipped with magnetic stirrer in order to promote homogenous state of sample, which can be operated at maximum operating speed of 2000 rpm. Tank 1 or Tank 2 is used as a collecting tank for the gas sample prior entering Tank 3 which act as mixing tank. The gas sample is pressurized to the cell via Haskel AG-303 air driven gas booster compressor. Liquid water is transferred into the equilibrium cell using the manual syringe pump provided in this set up. Alternatively, images and video of sample in the equilibrium cell can be observed via a camera attached at the front of the thermostat bath.

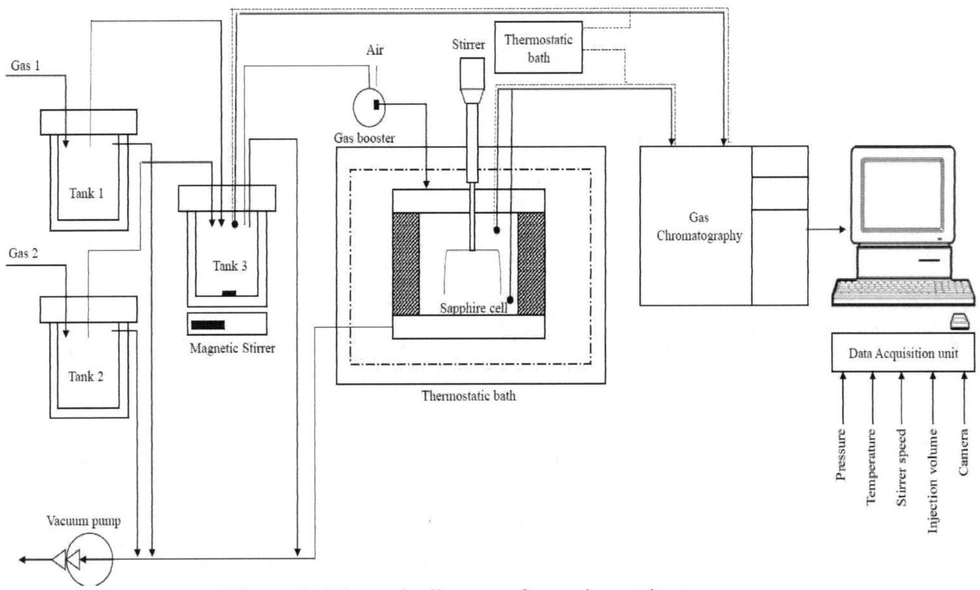

Figure 1. Schematic diagram of experimental set-up.

2.3 Experimental Procedure

Thermodynamic equilibrium measurements for CH_4 gas in this work used a traditional T-cycle method. Prior to the measurements, Sapphire cell is systematically washed with ultra-purified deionized water and subsequently vacuumed to ensure the cell is air free. The equilibrium cell is first charged with 10 ml of aqueos sample of TEACl followed by gas sample. The cell then is charged with CH_4 gas and pressurized to the desired conditions (within the range of 3.9 – 8.3 MPa). Once the system stabilized, the mixture is stirred at 300 rpm. The temperature is reduced by applying rapid cooling method to enhance hydrate formation. The lenght of each cooling step varies from 2 to 4 hours maximum. Hydrate formation is oberved both visually along with the sudden pressure drop. Once the gas mixture has fully encapsulated and hydrate has fully formed with no further pressure drop being observed, the equlibrium cell then is heated slowly for complete dissociation. Pressure-temperature trace is plotted in a graph, the first point where heating curve meets with the cooling curve is taken as hydrate equilibrium conditions.

3. Results and discussion

3.1 Measurement of Hydrate Equilibrium Pressure and Temperature of Methane Hydrate

Table 1 shows the experimental results from the measurement of CH_4 hydrates equilibrium pressure and temperature in the absence and presence of TEACl. The experimentals were measured from 3.9 to 8.3 MPa pressure range.

Table 1. Hydrate equilibrium pressure and temperature of CH_4 hydrates with and without the TEACl

System	**TEACl Concentration in the solution**							
	0		**1wt%**		**5 wt%**		**10 wt%**	
	T (K)	**P (MPa)**	**T (K)**	**P (MPa)**	**T (K)**	**P (MPa)**	**T (K)**	**P (MPa)**
CH_4+TEACl +H_2O	277.25	3.97	277.05	3.95	277	3.97	276.8	3.95
	280.15	5.16	279.9	5.1	279.5	5.1	279	5.1
	282.15	6.62	281.9	6.5	281.6	6.5	281.15	6.7
	284.15	8.3	284.1	8.2	283.8	8.1	283.2	8.2

The hydrate equilibrium pressure and temperature of CH_4+H_2O and CH_4+TEACl+ H_2O system at different concentrations of aqueous solution are presented in Figure 2. From Figure 2, it is evident that the presence of TEACl at different concentrations shift the CH_4 hydrate formation curves to higher pressure and lower temperature condition, which indicates the influence of TEACl is resulting a hydrate inhibition effect. Interestingly, the inhibition performance are found to be concentration driven, an increase in TEACl concentrations enhance the THI performance of TEACl. Figure 3 describes the influence of ΔT over experimental pressure of studied TEACl system. It is also observed that, the inhibitions effect is more potent at pressure value of 5 MPa for 5 and 10 wt% of TEACl concentrations.The findings are in line with the earlier AILs studies [21,22]. The average suppresion temperatures (ΔT) for 1, 5 and 10 wt% TEACl is 0.2 K, 0.45 K and 0.9 K, correspondingly.

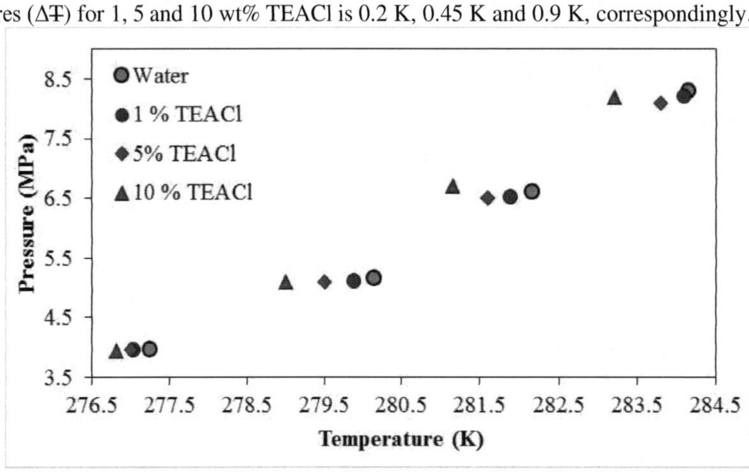

Figure 2. Hydrate phase equilibrium conditions for CH_4+TEACl+H2O at 1, 5 and 10 wt%.

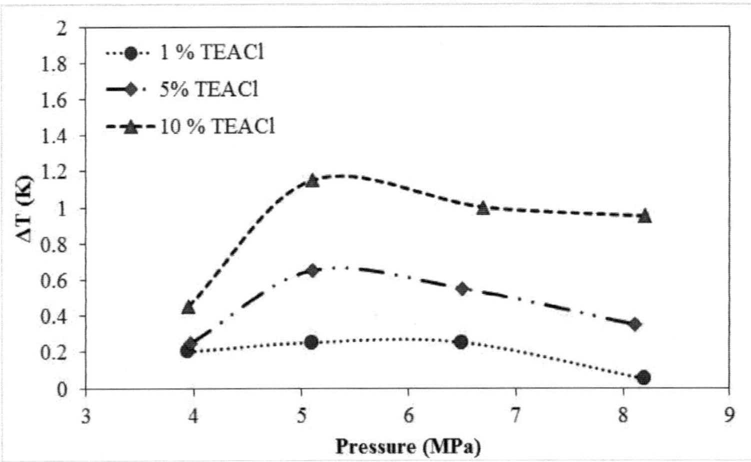

Figure 3. Average suppression temperature (ΔT) of CH_4+TEACl+H2O at 1, 5 and 10 wt%.

The inhibition effects of TEACl on CH_4 hydrate in this work is compared to TEACl and other works related to ammonium based ionic liquid which; tetramethylhydroxide (TMAOH). The comparison of previous work is as per presented in Figure 4. In Figure 4, at 10 wt%, TEACl study in this work is found to be low in inhibition performance in comparison to studied AILs reported by Khan et al [22] at pressure of 5 MPa. However, the inhibition effect of 10 wt% TEACl is almost equivalent to the value testified by others [17] at pressure values higher than 5 MPa.

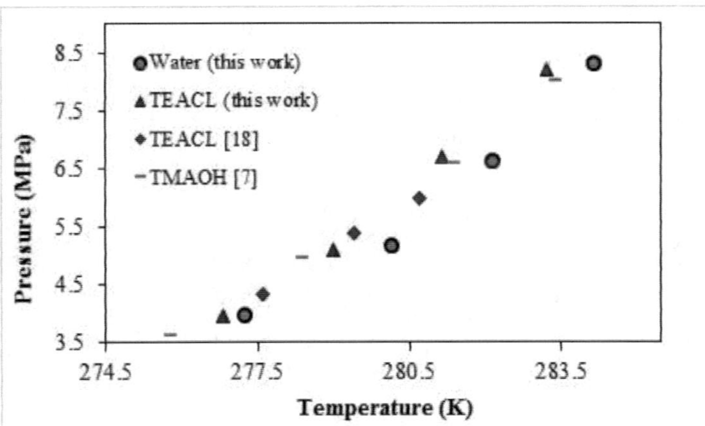

Figure 4. Hydrate equilibrium condition of CH_4 + TEACl + H_2O system with different type of ammonium based ionic liquids at 10 wt% concentration.

3.2 Enthalpy of dissociation

As hydrate forming and dissociation reaction The heat required to decompose hydrate cyrstals into liquid and hydrate-forming gas phase is known as dissociation enthalpy. The enthalpy of dissociation (ΔH_{diss}) would rely on the type of gas-guest molecule, size of the moleculs and type of hydrate structure [23]. For example, typical sII hydrate formers such as propane occupy large cage and was reported with enthalpy of dissociation of 129 kJ/mol [19]. As for CH_4 gas which formed sI hydrate, it is occupying small cavity with ΔH_{diss} = 58.88 kJ/mol as reported in various literature [17,26].

In this work the enthalpy of dissociation is calculated by using the Clausius-Clapeyron equation [19–21, 23–25]. Enthalphy of dissociation is predicted by utilising the CH_4 hydrate equilibrium pressure and temperature measured in this study, while z is gas compressibility factor and R is universal gas constant.

$$\frac{d\ln P}{d\left(\frac{1}{T}\right)} = \frac{\Delta H_{diss}}{zR} \qquad \text{Eq. 1}$$

The calculated values for ΔH_{diss} for CH4 + TEACl + H2O system is as per listed in Table 2. Assumption is made that the amount of hydrate phase is insignificant when the system is at equilibrium [17]. ΔH_{diss} value presented in this table of systems without the presence of TEACl is comparable to CH_4 hydrate entalphy data, 58.88 kJ/mol [17,26]. It has been observed, for system with 1 and 5 wt% of TEACl presence in CH_4 hydrate, the ΔH_{diss} is similar as sI hydrates enthalpy value. This evidently shown as at low concentrations TEACl is not inhabiting the hydrate cavity. Increased in TEACl concentrations to 10 wt% shows the increasing of ΔH_{diss} as a result of TEACl molecules entering the large water cavity. Nevertheless, the ΔH_{diss} values calculated in this work are much less as compared to typical sII hydrate formers such as propane, therefore it indicates sI methane hydrate is competing the hydrate cage occupancy with TEACl.

Table 2. Average dissociation enthalpy for CH_4 hydrates with and without the presence of TEACl aqueous solution

TEACl concentrations (wt %)	ΔH_d (kJ/mol)
0	60.700
1	58.367
5	58.557
10	64.877

4. Conclusions

In this work, an experimental investigation for hydrate equilibrium conditions of CH_4 hydrate in the absence and presence of various concentrations of aqueous TEACl has been carried out. . In the presence of TEACl in the CH_4 gas, it has been observed that the hydrate formation pressure is increase. . The conditions are causing by the disruptions of water activity during hydrate crystalline structure formation by TEACl, thus decreasing hydrate phase equilibrium conditions. The average suppression temperature ($\Delta \bar{T}$) reduction in the CH_4 systems are up to 0.9 K in the presence of 10 wt% aqueous TEACl, which can be considered useful in ILs perspectives. In addition, Clausius-Clapeyron equation was used to quantify the enthalpy of dissociation for TEACl + CH_4 + H_2O hydrate system which are found that there is chance of TEACl is occupying large cavities of TypeII hydrate at 10 wt% (higher) TEACl concentration.

Acknowledgement
This research is part of a project funded under grant number 01253C1-034 and 0153AB-F07 funded by PETROLIAM RESEARCH FUND (PRF). The authors would like to express their sincere gratitude to University Teknologi PETRONAS and PETRONAS Research Sdn Bhd for providing laboratory facilities, knowledge and support throughout the project.

References
[1] Sloan Jr, Dendy; Koh C S A 2010 *Natural gas hydrates in flow assurance* (Gulf Professional Publishing)
[2] Sloan Jr D 1991 Natural Gas Hydrates *J. Pet. Technol.* **43**
[3] Sloan Jr D K C 2007 *Clathrate Hydrates of Natural Gases* (CRC Press Taylor & Francis)
[4] Xiao C, Wibisono N and Ã H A 2010 Dialkylimidazolium halide ionic liquids as dual function inhibitors for methane hydrate *Chem. Eng. Sci.* **65** 3080–7
[5] Kelland M A 2006 History of the development of low dosage hydrate inhibitors *Energy Fuels* **20** 825–47
[6] Marie L. Zanota; Christophe, Dicharry; Alain G 2005 Hydrate Plug Prevention by Quaternary Ammonium Salts *Energy Fuels* **19** 584–90
[7] Aman Z M, Haber A, Ling N N A, Thornton A, Johns M L and May E F 2015 Effect of Brine Salinity on the Stability of Hydrate-in-Oil Dispersions and Water-in-Oil Emulsions *Energy Fuels* **29** 7948–55
[8] Demirbas A 2010 *Methane Gas Hydrate* (Springer London)
[9] Peng X, Hu Y, Liu Y, Jin C and Lin H 2010 Separation of ionic liquids from dilute aqueous solutions using the method based on CO 2 hydrates *J. Nat. Gas Chem.* vol 19 pp 81–5
[10] Xiao C and Adidharma H 2009 Dual function inhibitors for methane hydrate *Chem. Eng. Sci.* **64** 1522–7
[11] Ebrahim M, Nasrollah H, Abareshi B, Ghotbi C, Taghikhani V, Jalili A H, Nasrollah M E H, Abareshi B, Ghotbi C and Taghikhani V 2013 Investigation of Six Imidazolium-Based Ionic Liquids as Thermo-Kinetic Inhibitors for Methane Hydrate by Molecular Dynamics Simulation *Proceedings of 2nd National Iranian Conference on Gas Hydrate* p 12
[12] Avula V R, Gardas R L and Sangwai J S 2014 An improved model for the phase equilibrium of methane hydrate inhibition in the presence of ionic liquids *Fluid Phase Equilib.* **382** 187–96
[13] Tariq M, Rooney D, Othman E, Aparicio S, Atilhan M and Khraisheh M 2014 Gas hydrate inhibition: A review of the role of ionic liquids *Ind. Eng. Chem. Res.* **53** 17855–68
[14] Govinda V, Venkatesu P and Bahadur I 2016 Molecular interactions between ammonium-based ionic liquids and molecular solvents: current progress and challenges *Phys. Chem. Chem. Phys.* **18** 8278–326
[15] Hough W L 2010 *Functional Ionic Liquids for use in Pharmaceutical Applications* (Tuscaloosa, Alabama: University of Alabama)
[16] Li X-S, Liu Y-J, Zeng Z-Y, Chen Z-Y, Li G and Wu H-J 2011 Equilibrium Hydrate Formation Conditions for the Mixtures of Methane + Ionic Liquids + Water *J. Chem. Eng. Data* **56** 119–23
[17] Keshavarz L, Javanmardi J, Eslamimanesh A and Mohammadi A H 2013 Experimental measurement and thermodynamic modeling of methane hydrate dissociation conditions in the presence of aqueous solution of ionic liquid *Fluid Phase Equilib.* **354** 312–8
[18] Khan M S, Bavoh C B, Partoon B, Lal B, Bustam M A and Shariff A M 2017 Thermodynamic effect of ammonium based ionic liquids on CO2 hydrates phase boundary *J. Mol. Liq.* **238** 533–9
[19] Nashed O, Dadebayev D, Khan M S, Bavoh C B, Lal B and Shariff A M 2018 Experimental and modelling studies on thermodynamic methane hydrate inhibition in the presence of ionic liquids *J. Mol. Liq.* **249** 886–91
[20] Mellon B, Nashed O, Dadebayev D, Khan M S, Bavoh C B, Lal B and Shariff A M 2018 Experimental and modelling studies on thermodynamic methane hydrate inhibition in the presence of ionic liquids *J. Mol. Liq.* **249** 886–91

[21] Tariq M, Connor E, Thompson J, Khraisheh M, Atilhan M and Rooney D 2016 Doubly dual nature of ammonium-based ionic liquids for methane hydrates probed by rocking-rig assembly *RSC Adv.* **6** 23827–36
[22] Khan M S, Partoon B, Bavoh C B, Lal B and Mellon N B 2017 Influence of tetramethylammonium hydroxide on methane and carbon dioxide gas hydrate phase equilibrium conditions *Fluid Phase Equilib.* **440** 1–8
[23] Kwon T, Kneafsey T J and Rees E V L 2011 Thermal Dissociation Behavior and Dissociation Enthalpies of Methane À Carbon Dioxide Mixed Hydrates 8169–75
[24] Partoon B, Sabil K M, Roslan H, Lal B and Kok L 2016 Fluid Phase Equilibria Impact of acetone on phase boundary of methane and carbon dioxide mixed hydrates *Fluid Phase Equilib.* **412** 51–6
[25] Thermodynamics J C, Sabil K M, Nashed O, Lal B, Ismail L and Japper-jaafar A 2015 Experimental investigation on the dissociation conditions of methane hydrate in the presence of imidazolium-based ionic liquids *J. Chem. Thermodyn.* **84** 7–13
[26] Nasir Q, Lau K K, Lal B and Sabil K M 2014 Hydrate Dissociation Condition Measurement of CO2-Rich Mixed Gas in the Presence of Methanol/Ethylene Glycol and Mixed Methanol/Ethylene Glycol + Electrolyte Aqueous Solution *J. Chem. Eng. Data* **59** 3920–6

Thermokinetics of SO_3H-functionalized dicationic ionic liquids: Effect of anions

A N Masri[1,2,4], M I Abdul Mutalib[1,2], W Z N Yahya[1,2], N F Aminuddin[1,2], J-M Leveque[3],

[1] Department of Chemical Engineering, Universiti Teknologi PETRONAS, 32610 Seri Iskandar, Perak, Malaysia.
[2] Centre of Research in Ionic Liquids (CORIL), Universiti Teknologi PETRONAS, 32610 Seri Iskandar, Perak, Malaysia.
[3]LCME/CISM, Universite Savoie Mont-Blanc, 73376 Le Bourget du Lac Cedex, France.
[4]Corresponding author: asiahnusaibah88@gmail.com

Abstract. In this study, the kinetics of thermal degradation of three ionic liquids (ILs) were investigated using non-isothermal thermogravimetric analysis to investigate the thermal stability of ILs with various anion. Di(4-sulfonic acid) dibutyltriethylene diammonium di(hydrogen sulfate, ([DABCODBS][HSO_4]$_2$), di(4-sulfonic acid) dibutyltriethylene diammonium di(trifluoromethanesulfonate) ([DABCODBS][CF_3SO_3]$_2$) and di(4-sulfonic acid) dibutyltriethylene diammonium di(chloride) ([DABCODBS][Cl]$_2$) were subjected to thermogravimetric analysis (TGA) in the temperature range of 50-800°C at four heating rates of 10, 15, 20 and 25°C/min using nitrogen at a flow rate of 20 L/min as purge gas. The average activation energies (E_a) calculated using Kissinger-Akahira-Sunose (KAS), Flynn-Wall-Ozawa (FWO) and Starink methods are 163.49, 173.33 and 162.58 kJ/mol for [DABCODBS][HSO_4]$_2$, 119.59, 130.41 and 119.07 kJ/mol for [DABCODBS][CF_3SO_3]$_2$, and 199.72, 210.50 and 198.63 kJ/mol for [DABCODBS][Cl]$_2$, respectively. The results presented through all kinetic methods are in close agreement with each other and showed a decreasing thermal stability trend of [DABCODBS][Cl]$_2$ > [DABCODBS][HSO_4]$_2$ > [DABCODBS][CF_3SO_3]$_2$. The results also showed that intermediates were formed for [DABCODBS][CF_3SO_3]$_2$ and [DABCODBS][HSO_4]$_2$ until achieved total decomposition, while there was no intermediate formed for the pyrolysis kinetic of [DABCODBS][Cl]$_2$. This study shows the effect of nature of anion towards the thermal stability and decomposition mechanism of ILs.

1. Introduction

Ionic liquids (ILs) have become an attractive candidate as effective and environmental-friendly materials investigated in the last decade. ILs are molten salts in liquid form at temperature below 100°C [1] and are composed of a cation and an anion [2]. Their properties can be tuned by combination of organic cations with various inorganic or organic anions. Through these combination, a large amount of salts can be synthesized for numerous possible applications [3,4]. Their remarkable physicochemical properties such as high thermal stability and negligible vapor pressure solved the issue related to volatility and flammability [5,6]. They are reviewed as prospective materials due to their inherent characteristics such as being less corrosive, ease of separation, recyclable, applicable for continuous process and less production of waste water. Therefore, the number of reactions and purification steps could be reduced [7,8]. Their distinctive properties encouraged their application as heat transfer fluids, lubricants, propellants, solvents, catalyst and additives [9,10]. However, these applications are often implemented in high temperature condition. This may result to decomposition of materials which lead to significant hazard cause by gas generation, inflammation and rapid detonation [11]. Therefore, it is essential to study the thermal behavior and degradation temperature of ionic liquids for commercial scales application at wide temperature range [12]. ILs thermal stability can be affected by several factors such as structure of cation and anion, length of alkyl chain, functional group and impurities [13,14]. Hence, several techniques have been applied to assess the thermal stability of materials. Among them, thermogravimetric analysis technique is often used for its simplicity. The thermogravimetric analysis technique can be classified into several groups such as ramping technique, isothermal technique and isoconversional technique (integral, differential and incremental) [9,15]. Through these techniques, the activation energy can simply be calculated through Kissinger-Akahira-Sunose (KAS) [16], Flynn-Wall-Ozawa (FWO) [17,18] and latest Starink [19] methods. These methods also allow determination of the chemical steps of the degradation process and kinetic parameters for each step. The kinetic parameters can be calculated using the thermogravimetric (TG) curve of various heating rate under nitrogen or air flow [20]. In ILs field, this thermokinetic study has just recently being explored for a better understanding on the thermal stability of ILs. Such study includes exploration on thermal stability of ILs of various cations such as imidazolium [21,22], benzimidazolium [23], pyrrolidinium [13,24,25], piperidinium [12] and phosphonium [13]. However, the exploration on thermal properties and kinetic parameters covered only limited ILs. The effect of structure of cation and anion towards the thermokinetics behavior has yet to be compared in a single study. In our previous study, we have reported DABCO based ILs as potential catalysts for production of biodiesel [10]. In this present work, isoconversional thermal analysis were conducted to evaluate the thermal stabilities of the three DABCO-based ionic liquids with variety of anions, namely di(4-sulfonic acid) dibutyltriethylene diammonium di(hydrogen sulfate, ([DABCODBS][HSO_4]$_2$), di(4-sulfonic acid) dibutyltriethylene diammonium di(trifluoromethanesulfonate) ([DABCODBS][CF_3SO_3]$_2$) and di(4-sulfonic acid) dibutyltriethylene diammonium di(chloride) ([DABCODBS][Cl]$_2$). The kinetic parameters were determined by three isoconversional methods including KAS, FWO and Starink methods.

2. Experimental

2.1. Reagents and materials

All chemicals were commercially available. They were used without further purification unless otherwise stated. ^{1}H NMR spectra were recorded on a Bruker 400MHz spectrometer using D_2O as solvent.

2.2. Synthesis of ionic liquids

1,4-diazabicyclo[2.2.2]octane (DABCO) was reacted with 1,4-butane sultone in ethanol as solvent at 55°C for 3 days, forming white precipitate. The precipitate was then filtered, washed with solvents and dried. The white powder product was the zwitterion precursor. Double equivalent of sulphuric, trifluoromethanesulfonic and hydrochloric acid were also reacted on the same zwitterion precursor. After that, the product was dried to remove water, giving solid ILs at ambient temperature. The structure of the ILs were shown in **Figure 1**. a) **Di(4-sulfonic acid) dibutyltriethylene diammonium di(hydrogen sulfate), ([DABCODBS][HSO_4]$_2$)**: ^{1}H NMR (δ ,ppm): 1.74 (m, 2×2H), 1.92 (m, 2×2H), 2.89 (t, 2×2H), 3.53 (t, 2×2H), and 3.92 (s, 6×2H). b) **Di(4-sulfonic acid) dibutyltriethylene diammonium di(trifluoromethanesulfonate), ([DABCODBS][CF_3SO_3]$_2$)**: ^{1}H NMR (δ ,ppm): 1.74 (m, 2×2H), 1.92 (m, 2×2H), 2.89 (t, 2×2H), 3.53 (t, 2×2H), and 3.92 (s, 6×2H). c) **Di(4-sulfonic acid) dibutyltriethylene diammonium di(chloride), ([DABCODBS][Cl]$_2$)**: ^{1}H NMR (δ ,ppm): 1.75 (m, 2×2H), 1.93 (m, 2×2H), 2.91 (t, 2×2H), 3.55 (t, 2×2H), and 3.93 (s, 6×2H).

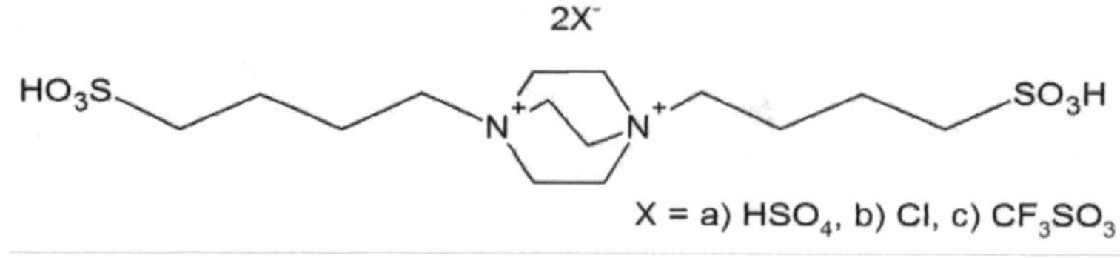

Figure 1. Structures of the DABCO based ionic liquids

2.3. Thermokinetic study

The thermal behaviour of ILs was measured using Perkin-Elmer, Pyris V-3.81 thermal gravimetric analyser (TGA). The samples were placed in sealed aluminium pan under nitrogen flow of 20 L/min in the temperature range of 50-800°C at four heating rate of 10, 15, 20 and 25°C/min. The kinetic parameters was determined based on the data of thermogravimetric graphs following the approach of Kissinger-Akahira-Sunose (KAS) method, Flynn-Wall-Ozawa (FWO) method and Starink method. These methods were mathematically expressed by equation (1), (2) and (3) for KAS, FWO and Starink methods, respectively.

Kissinger-Akahira-Sunose (KAS) method:
$$\ln\left(\frac{\beta}{T^2}\right) = \ln\left(\frac{AR}{E_a g(\alpha)}\right) - \frac{E_a}{RT} \quad (1)$$

Flynn-Wall-Ozawa (FWO) method:
$$ln\beta = \ln\left(\frac{AE_a}{Rg(\alpha)}\right) - 5.331 - \frac{1.052E_a}{RT} \quad (2)$$

Starink method:
$$\ln\left(\frac{\beta}{T^{1.92}}\right) = \ln\left(\frac{AR^{0.92}}{g(\alpha)E_a^{0.92}}\right) - 0.312 - \frac{1.008E_a}{RT} \quad (3)$$

In these equations, T was temperature (K), β was heating rate (K/min), A was pre-exponential factor (min^{-1}), R was ideal gas constant (8.314 J/mol K), E_a is activation energy (kJ/mol), α is conversion factor and g(α) is function of conversion factor. Based on the equation, A and E_a were calculated using the intercept and slope values, respectively.

3. Results and discussion

Dynamic thermogravimetric analysis was executed to evaluate the decomposition temperatures of DABCO-based ILs with various anions, namely [DABCODBS][HSO_4]$_2$, [DABCODBS][CF_3SO_3]$_2$ and [DABCODBS][Cl]$_2$ ILs. **Figure 2** illustrates the thermogravimetric (TG) and derivative thermogravimetric (DTG) curves of all three ILs with different anions for various heating rates which are 10, 15, 20 and 25 °C/min, respectively. Both TG and DTG curves for each IL show constant pattern behavior. This is accredited to similar reaction mechanism for each individual IL regardless of heating rate [26]. The shifting of curve towards right side has shown the increase of thermal decomposition. The increment trending of thermal decomposition with increasing heating rate as thermodynamics and decomposition kinetics influenced the thermal decomposition rate [27].

The thermal decomposition of ILs is usually represented by the onset temperature of the TG curve. The onset temperature can be extracted from the TG curve by the intersection of the mass at 100% with the tangent of the curve. Based on **Figure 2**, the comparison of the first onset temperature of TG curve for all three ILs shows a decreasing temperature (onset) trend in sequence of [DABCODBS] [Cl]$_2$ > [DABCODBS][HSO_4]$_2$ > [DABCODBS][CF_3SO_3]$_2$ with the onset temperature range of 351.4-366.2°C, 155.1-166.1°C and 140.1-151.8°C, respectively. Even though ILs with halide anions are known to have low thermal stability due to their higher basicity [28], in this case, IL with Cl^- anion has highest onset temperature. The high thermal stability of ILs with Cl^- anion can be due to less steric hindrance of ILs as the anion is less bulky compare to the other two ILs [29]. Besides that, additional sulfonyl group in HSO_4^- and $CF_3SO_3^-$ anions also lower the thermal stability of the ILs [30].

In general, the ILs may decomposed by defragmentation to a range of volatile degradation product of simply revert back to starting materials [31,32]. The decomposition mechanisms are highly dependent on the structure of ILs. In TG and DTG curves of all three ILs, [DABCODBS][Cl]$_2$ curves depict that one-step decomposition mechanism is taking place while [DABCODBS][HSO_4]$_2$ and [DABCODBS][CF_3SO_3]$_2$ curves show more than one step decomposition mechanism is involved. The small mass loss at low temperature of the first onset temperature for [DABCODBS][HSO_4]$_2$ and [DABCODBS][CF_3SO_3]$_2$ TG curves can be accredited to the evaporation and decomposition of the ILs. The multi-step decomposition mechanism also may be due to separate decomposition of cation and anion. Liang et al (2013) has also determined multi-step decomposition reaction for their cyano-functionalized imidazolium-based ILs which has HSO_4^- anion [33].

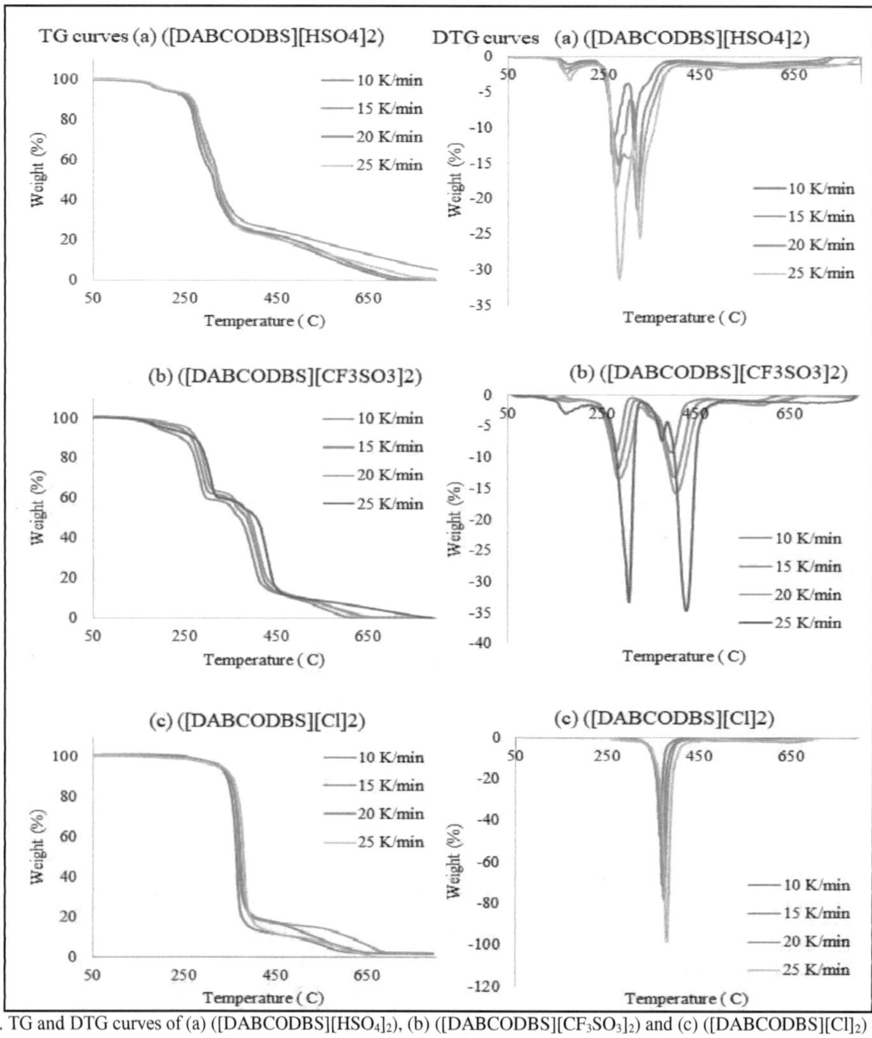

Figure 2. TG and DTG curves of (a) ($[DABCODBS][HSO_4]_2$), (b) ($[DABCODBS][CF_3SO_3]_2$) and (c) ($[DABCODBS][Cl]_2$) at various heating rate.

The thermal decomposition kinetic parameters for ILs, namely, activation energy, E_a and pre-exponential factor, A can be calculated through DTG and TG approaches. Through DTG approach, the value of T_{max} which is the peak temperature in DTG curve is extracted and used in the mathematical model for thermokinetic study. However, this approach is only suitable for one-step thermal decomposition mechanism [34]. Therefore, in this study, TG approach is used instead. In TG approach, the kinetic parameters are calculated distinctively for each fraction conversion, α. The trend for kinetic parameters is then compared to define number of step for thermal decomposition mechanism. In order to compare the kinetic parameters for thermal decomposition of various ILs, the average value is calculated. For this research, KAS, FWO and Starink methods are used to calculate the kinetic parameters using equation (1), (2) and (3), respectively as mentioned in experimental section. From the equations, kinetic parameters are derived as equation (4) and (5) for KAS method, equation (6) and (7) for FWO method and equation (8) and (9) for Starink method:

$$E_a = slope \times R \tag{4}$$

$$A = \frac{\alpha \times E_a \times e^{intercept}}{R} \tag{5}$$

$$E_a = \frac{slope \times R}{1.052} \tag{6}$$

$$A = \frac{\alpha \times R \times e^{intercept+5.331}}{E_a} \tag{7}$$

$$E_a = \frac{slope \times R}{1.008} \tag{8}$$

$$A = \left(\frac{E_a}{R}\right)^{0.92} \times e^{slope+0.312} \tag{9}$$

Figure 3, 4 and **5** depict the thermokinetics plot of KAS, FWO and Starink methods and activation energy, E_a at various conversion, α for [DABCODBS][HSO_4]$_2$, [DABCODBS][CF_3SO_3]$_2$ and [DABCODBS][Cl]$_2$ ILs, respectively. The results presented through all kinetic methods for all ILs are in close agreement with each other. For [DABCODBS][HSO_4]$_2$ and [DABCODBS][CF_3SO_3]$_2$ ILs activation energy, E_a plot in **Figure 3d** and **4d**, respectively, there are slight increase of E_a at lower conversion degree, α. This results may be due to evaporation or partial thermal decomposition of ILs at lower temperature. After that, the E_a trending fluctuate showing the multi-step thermal decomposition mechanism of ILs where intermediates are formed during decomposition process due to complexity of IL structure [35]. For [DABCODBS][Cl]$_2$ IL activation energy, E_a plot in **Figure 5d**, the E_a trending is almost constant at $\alpha > 0.1$. This result is attributed to a one-step thermal decomposition mechanism of IL where no intermediate is formed during the decomposition process.

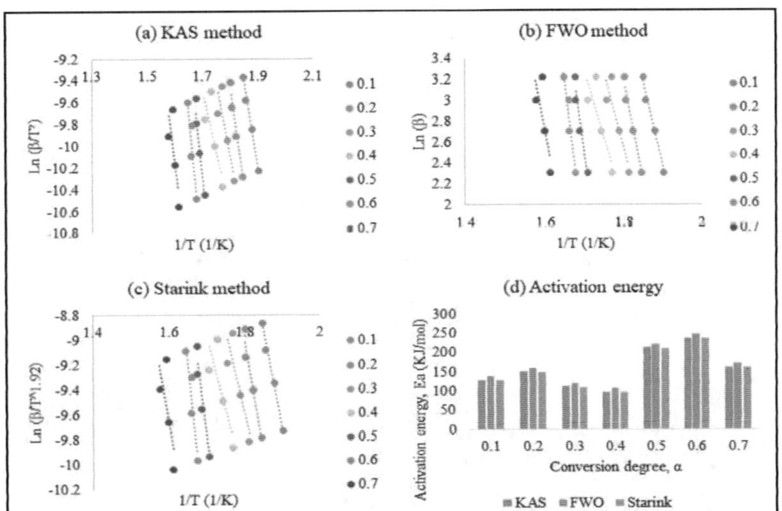

Figure 3. Thermokinetics of ([DABCODBS][HSO_4]$_2$) using KAS (a), FWO (b) and Starink (c) method, and also the activation energy profiles (d).

The average kinetic parameters are then calculated at conversion degree, α = 0.3 – 0.7 as suggested by literatures [36,37] and the values are listed in **Table 1**. The average activation energy (E_a) calculated by KAS, FWO and Starink methods for [DABCODBS][HSO_4]$_2$ are 163.49 kJ/mol, 173.33 kJ/mol and 162.58 kJ/mol. For [DABCODBS][CF_3SO_3]$_2$, the Ea values are 119.59 kJ/mol, 130.41 kJ/mol and 119.07 kJ/mol respectively. For [DABCODBS][Cl]$_2$ the Ea values are 199.72 kJ/mol, 210.50 kJ/mol and 198.63 kJ/mol. While pre-exponential factor, presented in term of Log A calculated by KAS, FWO and Starink methods are 13.96 min^{-1}, 14.13 min^{-1} and 14.33 min^{-1} for [DABCODBS][HSO_4]$_2$, 8.96 min^{-1}, 9.42 min^{-1} and 9.34 min^{-1} for [DABCODBS][CF_3SO_3]$_2$, and 15.85 min^{-1}, 15.87 min^{-1} and 16.21 min^{-1} for [DABCODBS][Cl]$_2$, respectively. All the values calculated using the three methods show good correlation. The highest activation energy for thermal decomposition was found for [DABCODBS][Cl]$_2$ and the lowest was [DABCODBS][CF_3SO_3]$_2$. It can be deduced from the results, the decreasing thermal stability trend of [DABCODBS][Cl]$_2$> [DABCODBS][HSO_4]$_2$ > [DABCODBS][CF_3SO_3]$_2$.

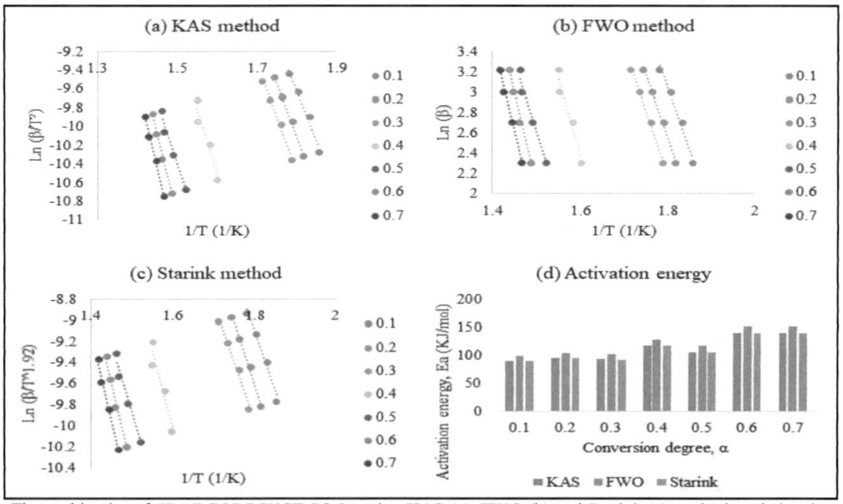

Figure 4. Thermokinetics of ([DABCODBS][CF_3SO_3]$_2$) using KAS (a), FWO (b) and Starink (c) method, and also the activation energy profiles (d).

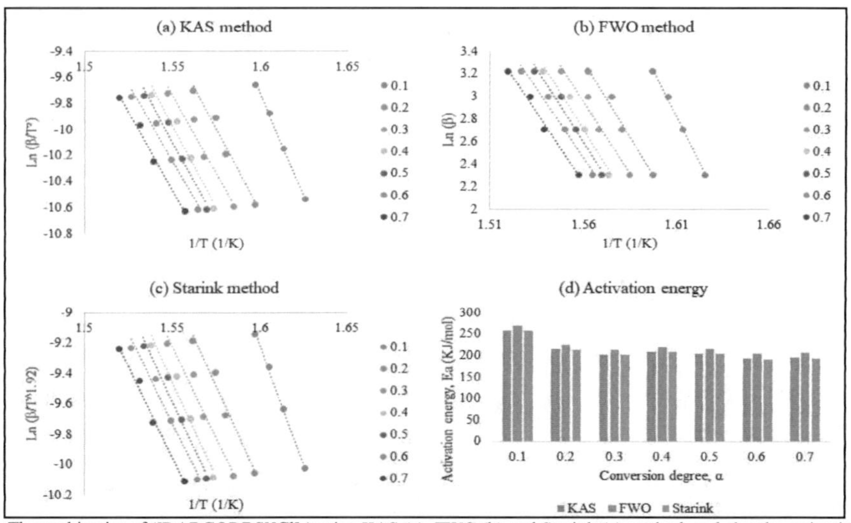

Figure 5. Thermokinetics of ([DABCODBS][Cl]$_2$) using KAS (a), FWO (b) and Starink (c) method, and also the activation energy profiles (d).

Table 1. Average activation energy, E_a and pre-exponential factor in term of Log A for ILs using KAS, FWO and Starink methods

Method	KAS		FWO		Starink	
Kinetic parameters	E_a (kJ/mol)	Log A (min^{-1})	E_a (kJ/mol)	Log A (min^{-1})	E_a (kJ/mol)	Log A (min^{-1})
[DABCODBS][HSO_4]$_2$	163.49	13.96	173.33	14.13	162.58	14.33
[DABCODBS][CF_3SO_3]$_2$	119.59	8.96	130.41	9.42	119.07	9.34
[DABCODBS][Cl]$_2$	199.72	15.85	210.50	15.87	198.63	16.21

4. Conclusion

Thermokinetic study of DABCO-based IL with three types of anions, namely [DABCODBS][HSO_4]$_2$, [DABCODBS][CF_3SO_3]$_2$ and [DABCODBS][Cl]$_2$ ILs has been executed using TG approach following KAS, FWO and Starink method. The results determined from all the three methods are in close agreement with each other. The ILs are thermally stable before reaching an onset temperature range of 155.1-166.1 °C for [DABCODBS][HSO_4]$_2$ IL, 140.1-151.8 °C for [DABCODBS][CF_3SO_3]$_2$ IL and 351.4-366.2 °C for [DABCODBS][Cl]$_2$ IL. The average kinetic parameters (E_a and Log A) are calculated for KAS, FWO and Starink methods. The results also has shown a decreasing thermal stability trend of [DABCODBS][Cl]$_2$ > [DABCODBS][HSO_4]$_2$ > [DABCODBS][CF_3SO_3]$_2$. Intermediates are formed for [DABCODBS][CF_3SO_3]$_2$ and [DABCODBS][HSO_4]$_2$ until achieved total decomposition, while there is no intermediate formed for the pyrolysis kinetic of [DABCODBS][Cl]$_2$.

Acknowledgement
The authors acknowledged the Yayasan UTP grant, Centre of Research in Ionic Liquids (CORIL) and Chemical Engineering Department of Universiti Teknologi PETRONAS for their supports in conducting this study.

References

[1] Masri AN, Abdul Mutalib MI, Leveque JM A Review on Dicationic Ionic Liquids: Classification and Application 2016 *Ind. Eng. Manage.* **5** 197
[2] Hossain MI, Samir BB, El-Harbawi M, Masri, AN, Abdul Mutalib MI, Hefter G and Yin CY Development of a novel mathematical model using a group contribution method for prediction of ionic liquid toxicities 2011 *Chemosphere* **85** 990-994
[3] Godajdar BM, Kiasat AR and Hashemi MM Synthesis, characterization and application of magnetic room temperature dicationic ionic liquid as an efficient catalyst for the preparation of 1,2-azidoalcohols 2013 *J Mol Liq* **183** 14-19.
[4] Bui TLT, Korth W and Jess A Influence of acidity of modified chloroaluminate based ionic liquid catalysts on alkylation of iso-butene with butane-2 2012 *Cat Comm* **25** 118-124.
[5] Man Z, Elsheikh YA, Bustam MA, Yusup S, Abdul Mutalib MI, Muhammad N (2013). A Brønsted Ammonium Ionic Liquid-KOH Two-stage Catalyst for Biodiesel Synthesis from Crude Palm oil 2013 *Industrial Crops and Products* **41** 144-149.
[6] Mohammad Fauzi AH and Saidina Amin NA Optimization of Oleic Acid Esterification Catalyzed by Ionic Liquid for Green Biodiesel Synthesis 2013 *Energy Conv. Management* **76** 818-827.
[7] Sheldon RA, Lau RM, Sorgedrager MJ and Van Rantwijk F Biocatalysis in ionic liquids 2002 *Green Chem* **4** 147-151.
[8] Lin YC, Chen SC, Wu TY, Yang PM, Jhang SR and Lin JF Energy saving and rapid transesterification of jatropha oil using a microwave heating system with ionic liquid catalyst 2015 *J Taiwan Inst Chem Eng* **49** 72-78.
[9] Maton C, De Vos N, Steven CV Ionic Liquid Thermal Stabilities; Decomposition Mechanism and Analysis Tools 2013 *Chem. Soc. Rev.* **42** 5963-5977.
[10] Masri AN, Abdul Mutalib MI, Aminuddin NF, Leveque JM Novel SO3H-functionalized dicationic ionic liquids – A comparative study for esterification reaction by ultrasound cavitation and mechanical stirring for biodiesel production 2018 *Sep. Pur. Tech.* **196** 106-114.
[11] Rowe SM Thermal stability: A review of methods and interpretation of data 2002 *Org Proc Res Dev* **6** 877-833.
[12] Wang J, Wang M, Yang X, Zou W and Chen X Crystal structure and thermal decomposition kinetics of 1-(pyridinium-1-yl)propane-(1-methylpiperidinium) bi[bis(trifluoromethanesulfonyl)imide], [PyC3Pi][NTf2]2 2015 *Chi J Chem Eng*
[13] Parajo JJ, Villanueva M, Otero I, Fernandez J and Saldago Thermal stability of aprotic ionic liquids as potential lubricants. Comparison with synthetic oil bases 2017 *J Chem Thermodynamics*
[14] Siedlecka EM, Czerwicka M, Stolte S and Stepnowsi P Stability of ionic liquids in application conditions 2011 *Curr Org Chem* **15** 1974-1991.
[15] Simon P Isoconversional methods – Fundamentals, meaning and application 2004 *J Therm Anal Cal* **76** 123-132.
[16] Sbirazzuoli N, Girault Y and Elegant L Simulations for evaluation of kinetic methods in

differential scanning calorimetry. Part 3. Peak maximum evolution methods and isoconversional methods 1997 *Thermochim Acta* **293** 25-37.
[17] Flynn JH and Wall LA A quick, direct method for the determination of activation energy from thermogravimetric data 1966 *J Pol Sci B: Pol Let* **4** 5 323-328.
[18] Ozawa T A new method of analyzing thermogravimetric data 1965 *Chem Soc Jpn* **38** 11 1881-1886.
[19] Starink MJ The determination of activation energy from linear heating rate experiments: a comparison of the accuracy of isoconversion methods 2003 *Thermochim Acta* **404** 163-176.
[20] Al-Maydama H, El-Shekeil A, Khalid MA and Al-Karbouly A Thermal degradation behavior of some polydithiooxamide metal complexes 2006 *Ecl Quim* Sao Paulo **31** 1 45-52.
[21] Khan AS, Man Z, Bustam MA, Chong FK, Ullah Z, Sarwono A and Wilfred CD Pyrolysis kinetics of 1-propyronitrile imidazolium trifluoroacetate ionic liquid using thermogravimetric analysis 2016 *Pro Eng* **148** 1332-1339.
[22] Blake DM, Moens L, Rudnicki D and Pilath H Lifetime of imidazolium salts at elevated temperatures 2005 *J Sol Energy Eng* **128** 54-57.
[23] Ullah Z, Bustam MA, Man Z and Khan AS Thermal stability and kinetic study of benzimidazolium based ionic liquid 2016 *Pro Eng* **148** 215-222.
[24] Quraishi KS, Bustam MA, Krishnan S, Khan MI, Wilfred CD and Leveque JM Thermokinetics of alkyl methylpyrrolidinium [NTf2] ionic liquids – Effect of alkyl chain on thermal stability 2017 *J Therm Anal Calorim* **129** 261-270.
[25] Salgado J, Parajo JJ, Fernandez J and Villanueva M Long-term thermal stability of some 1-butyl-1-methylpyrrolidinium ionic liquids 2014 *J Chem Thermodynamics* **74** 51-57.
[26] Liang R, Yang M and Xuan X Thermal stability and thermal decomposition kinetics of 1-butyl-3-methylimidazolium dicyanamide 2010 *Chin J Chem Eng* **18** 5 736-741.
[27] Efimova A, Pfutzner L and Schmidt P Thermal stability and decomposition mechanism of 1-ethyl-3-methylimidazolium halides 2015 *Thermochim Acta* **604** 129-136.
[28] Ngo HL, LeCompte K, Hargens L and McEwen AB Thermal properties of imidazolium ionic liquids 2000 *Thermochim Acta* **357** 97-102.
[29] Kosmulski M, Gustafsson J and Rosenholm JB Thermal stability of low temperature ionic liquids revisited 2004 *Thermochim Acta* **412** 47-53.
[30] Amarasekara AS and Owereh OS Thermal properties of sulfonic group functionalized Bronsted acidic ionic liquids 2011 *J Therm Anal Calorim* **103** 3 1027-1030.
[31] Kroon MC, Buijs W and Peters CJ Quantum chemical aided prediction of the thermal decomposition mechanisms and temperatures of ionic liquids 2007 *Thermochim Acta* **465** 40-47.
[32] Rooney D, Jacquemin J and Gardas R Thermophysical properties of ionic liquids 2009 *Top Curr Chem* **290** 185-212.
[33] Liang WD, Li HF, Gou GJ and Wang AQ Study of preparation and thermal stability of cyano-functionalized imidazolium type ionic liquids 2013 *Asian J Chem* **25** 9 4779-4782.
[34] Kissinger HE Reaction kinetics in differential thermal analysis 1957 *Anal Chem* **29** 1702-1706.
[35] Abd-Elghany M, Klapotke TM and Elbeih A Investigation of 2,2,2-trinitroethyl-nitrocarbamate as a high energy dense oxidizer and its mixture with nitrocellulose (thermal behavior and decomposition kinetics) 2017 *J Anal App Pyrolysis* **128** 397-404.
[36] Abd-Elghany M, Klapotke TM, Elbeih A and Zeman S 2017 Investigation of different thermal analysis techniques to determine the decomposition kinetics of ε-2,4,6,8,10,12-hexanitro-2,4,6,8,10,12-hexaazaisowurtzitane with reduced sensitivity and its cured PBX *J Anal App Pyrolysis* **126** 267-274.
[37] Yan QL, Zeman S, Zang TL and Elbeih A 2013 Non-isothermal decomposition behavior of fluorel bonded explosives containing attractive cyclic nitramines *Thermochim Acta* **574** 10-18.

Application of Electrolyte Based Model on Ionic Liquids-Methane Hydrates Phase Boundary

Muhammad Saad Khan[1,2], Cornelius B. Bavoh[1,2], Bhajan Lal[*1,2], Lau Kok Keong[1,2], Nurhayati Bt Mellon[1], Mohamad Azmi Bustam[1,2] Azmi Mohamad Shariff[1,2]

[1]Chemical Engineering Department, Universiti Teknologi of PETRONAS, Bandar Seri Iskandar, 32610, Perak, Malaysia.

[2]CO_2 Research Centre (CO2RES), Bandar Seri Iskandar, 32610, Perak, Malaysia.

*bhajan.lal@utp.edu.my

Abstract. In the current study, the phase behaviour of selected methane (CH_4) hydrate-Ionic Liquids (ILs) systems were predicted via Dickens and Quinby-Hunt model. The model chosen is an electrolyte-based model; therefore easily accommodate ILs that are molten salts. The experimental hydrate vapour liquid Equilibrium (HL_wVE) data of ILs was extracted from various literature sources for validation of the applied model. The overall predicted results suggested that the studied predictive model found in line with the experimental literature data for almost all the studied systems. The maximum deviation observed from the pyrrolidinium family of IL which also found to less than 0.67 K. Apparently it can be concluded that the selective model could applicable for accurate prediction of thermodynamic hydrate phase boundaries of methane hydrates in the presence of ILs. Since hydrate experimentations are very time-consuming, accurate thermodynamic predictions of hydrate phase are very crucial for exploring various hydrate-based technologies like flow assurance, natural gas recovery and gas storage and transportations.

1. Introduction

Gas hydrate formation is one of the foremost impedance faced by the offshore oil and gas exploration and production facilities. The problem is even getting harder since petroleum exploration is a move to more in-depth off-shores, which provide further thermodynamically favourable conditions leads to more close hydrate encounters. According to Xia et al. [1], nearly 70 per cent of all the flow assurance problems accounted for due to the hydrate formations. To avoid hydrate formations, oil and gas industry spends 100s of Million dollars ($) without any permanent solutions [1,2].

Gas or clathrates hydrates are non-stoichiometric ice-type enclosure solids, produced when small gas molecule (guest) trapped inside the hydrogen bonded water structure (host) under thermodynamically favourable conditions. Gas hydrates stabilized, owing to the existence of an active hydrogen-bonded network of water in hydrate structure. Methane (CH_4), ethane (C_2H_6), carbon dioxide (CO_2) are among the typical hydrate formers [3]. Often, three types of conventional gas hydrate structures are found which recognised as structure I (sI), structure II (sII) and structure H (sH) hydrates which depend upon the thermodynamic conditions and the presence of guest molecules [4].

There are four conventional methods available for mitigation of gas hydrates, which includes thermal heating, depressurization, dehydration and chemical inhibition. In most of the scenarios, chemical inhibition is the only viable method available for hydrate mitigations [5]. There are three types of chemicals inserted as hydrate inhibitors; namely: Thermodynamic hydrate inhibitors (THIs), Kinetic hydrate inhibitors (KHIs) and Anti Agglomerates (AAs) [6,7]. THIs shifts the hydrate formation conditions towards lower-temperature and higher-pressure regions. The conventional thermodynamic hydrate inhibitors include alcohols like ethylene glycol (EG), methanol besides traditional salts for example potassium chloride (KCl) and sodium chloride (NaCl) [8]. The unconventional THI include different chemicals, which are relatively more environmentally friendly in properties particularly with lower vapour pressure like ionic liquids [9–12].

Since hydrate experimentation is very complicated and time-consuming process; it is practically not possible to experimentally investigate the all the potential chemicals (ILs). Besides, the commercial hydrate prediction soft wares (such as CSMGem, PVTSim, etc.) are very limited and could only incorporate the commercial inhibitors (like methanol, EG or NaCl) which restrain their applications to the newer inhibitors (ILs). ILs are a molten salts having melting points less than 100 °C. The ILs are the combination of organic part (electropositive) cation and inorganic part anion (electronegative). The ILs could influence THI behaviour uniquely; they possess strong electrostatic force (coulombic charge) together with enhanced hydrogen-bonding ability with water molecules. For that reasons, ILs are efficiently able to act as dual-functional hydrate inhibitors. However, the most of the ILs studies mainly focused on THI impact only.

There are many families of ILs; like imidazolium, pyrrolidinium, morpholinium, ammonium etc. The gas hydrate-ILs research mainly focused on imidazolium-based ILs [7]. Relatively less literature has published over the different families of ILs.

Therefore, at this point, the emphasis of this study is to apply the electrolyte based (Dickens and Quinby-Hunt) model on the selected reported literature data of CH_4 hydrates. The model predictions (HL_wVE boundaries) are stretched on the different ILs-families to evaluate the applicability of the studied model. Therefore, this work is expected to efficiently predict the HL_wVE data of ILs-CH_4 system which ultimately save time, cost, and to support the appropriate selection of more efficient ILs for THI mitigation.

2. Methodology

2.1. Model Theory

Mainly most of the hydrate thermodynamic models determine the activity of additive in the aqueous phase. The THI behaviour induced due to the disturbance occurred in the activity of water. Therefore, to calculate the HL_wVE points of CH_4 hydrates in the presence of ILs, the electrolytes based model offered by Dickens and Quinby-Hunt [13] was used in this study. Since ILs are molten salts, hence the considered model excellently worked for electrolytes (salts) [14]. The applied model was the adaptation of the Pieroen [15] model; and earlier used for the electrolytes modelling like ionic liquid and amino acids solutions on the CH_4 hydrate phase boundary by Partoon et al. [16], Javanmardi et al. [17,18] and other researchers [6,19,20]. Recently, the existing model was also successfully applied on the CO_2 and CH_4 riched binary mixtures (30-70 and 70-30) of CH_4-CO_2 gases in the presence of ammonium based ionic liquids (AILs) [20,21]. The model has fundamentally established from the classical thermodynamic theory, which assumed that the amount of gas in the water phase is negligible, and vice versa. Also, the effect of ILs on CH_4 HL_wVE boundary was sole to decrease water activity (a_w), and at small temperature ranges, the hydrate enthalpy of dissociation (ΔH_d) remained constant. The derivation details had provided by Dickens and Quinby-Hunt [13] and Pieroen [15]. Based on this model, the effect of ILs on the methane hydrate dissociation temperature can be denoted as;

$$In\ a_w = \frac{\Delta H_d}{nR}\left[\frac{1}{T_w} - \frac{1}{T_{IL}}\right] \tag{1}$$

where,

$$In\ a_w = \frac{\Delta H_{FUS(i)}}{R}\left[\frac{1}{T_{f(i)}} - \frac{1}{T_f}\right] \tag{2}$$

where a_w denotes water activity, n was the methane hydrate hydration number (5.75 [22]), ΔH_d represented the dissociation enthalpy of methane (59.966 kJ/mol), R denoted as the universal gas constant, T_{IL} and T_w represented the hydrate formation temperatures in water + ILs solution and pure water, respectively. CSMGem software was used to calculate the water- methane hydrate dissociation temperatures. $\Delta H_{FUS(i)}$ was the heat of fusion of ice (6.008 kJ/mol), $T_{f(i)}$ and T_f were the freezing point temperatures of water (273.15 K) and water + aqueous ILs solutions, respectively. T_f was calculated as suggested by Dickens and Quinby-Hunt [13] by using a cryoscopic constant for water as 1.853 K·kg/mol. Hence, merging Eq. (1) and (2), narrated the temperature offset of methane hydrate phase condition and the temperature of the ice-water equilibrium condition in any IL solution at constant pressure as follows;

$$\left[\frac{1}{T_w}-\frac{1}{T_{IL}}\right]=\frac{n\Delta H_{FUS(i)}}{\Delta H_d}\left[\frac{1}{T_{f(i)}}-\frac{1}{T_f}\right] \quad (3)$$

Therefore, Eq. (3) can be used to calculate T_{IL}, the hydrate formation temperature in the presence of ILs. The model predictions average absolute error (AAD) is calculated using Eq. (4).

$$AAE=\frac{1}{m}\sum_{i=1}^{m}\left|T_{Exp.}-T_{Cal.}\right|_i \quad (4)$$

2.2. Material

Table 1 provided the details of the selected (literature) ILs used in this work. These literature HL_wVE data have chosen for the model application since they covered the different types (families) of ILs. The typical concentration of ILs was selected as 10 wt%. Overall, modelling of 19 ILs-CH_4 systems (see Table 1) were performed in this work.

3. Results and Discussions

The proposed electrolyte model had used to predict the influence of ILs on the HL_wVE conditions for CH_4 hydrates. The authors would like to mention that the considered model had been successfully used for different electrolyte system in the literature to predict the dissociation temperature within the acceptable deviation range (Line). However, for validation purpose the model had also applied for common salts system, i.e., NaCl+CH_4 and KCl+CH_4 by the literature reported data of Mohammadi *et al.* [27]. For the modelling, the constants and model parameters have kept uniform with recently published article of Nashed *et al.* [19]. The model predicted results were found in acceptable agreement with the salts literature data as evident in the following Figure 1 and Table 2.

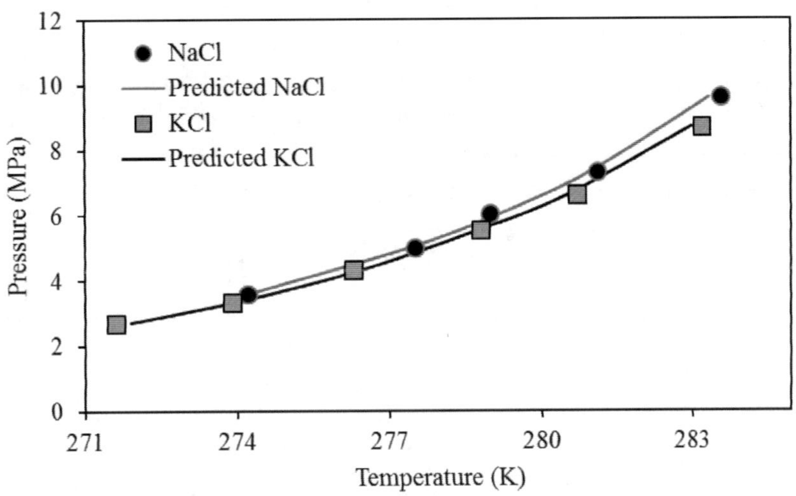

Figure 1: Experimental and Model predicted HLwVE data for 5 wt% salts (NaCl and KCl) solutions for validation.

Table 1. Details of the reported aqueous ILs solutions used (10 wt%) in this work.

S. No.	Name of ILs	Symbol	Author
1	1-[2-Hydroxyethyl]-3-methylimidazolium chloride	[OH-EMIM][Cl]	Sabil et al. [23]
2	1-[2-Hydroxyethyl]-3-methylimidazolium bromide	[OH-EMIM][Br]	Sabil et al. [23]
3	1-butyl-3-methylimidazolium bromide	[BMIM][Br]	Sabil et al. [23]
4	1-butyl-3-methylimidazolium chloride	[BMIM][Cl]	Sabil *et al.* [23]
5	1-butyl-3-methylimidazolium perchlorate	[BMIM][ClO_4]	Sabil *et al.* [23]
6	1-butyl-3-methylimidazolium dicyanamide	[BMIM][$N(CN)_2$]	Sabil *et al.* [23]
7	1-butyl-3-methylimidazolium hydrogen sulfate	[BMIM][HSO_4]	Sabil *et al.* [23]
	1-butyl-3-methylimidazolium trifluoromethane sul ona	[BMIM][CF SO]	Sabil *et al.* [23]
9	1-butyl-3-methylimidazolium methylsulfate	[BMIM][CH_3SO_4]	Sabil *et al.* [23]
10	1-ethyl-3-methylimidazolium chloride	[EMIM][Cl]	Xiao *et al.* [24]
11	1-ethyl-3-methylimidazolium bromide	[EMIM][Br]	Xiao *et al.* [24]
12	1-propyl-3-methylimidazolium iodide	[PMIM][I]	Xiao *et al.* [24]
13	1-butyl-3-methylimidazolium chloride	[BMIM][Cl]	Xiao *et al.* [24]
14	1-butyl-3-methylimidazolium bromide	[BMIM][Br]	Xiao *et al.* [24]
15	1-butyl-3-methylimidazolium iodide	[BMIM][I]	Xiao *et al.* [24]
16	1-[2-hydroxyethyl]-1- methylpyrrolidinium tetrafluoroborate	[HEMP][BF_4]	Kim *et al.* [25]
17	1-butyl-1-methylpyrrolidinium tetrafluoroborate	[BMP][BF_4]	Kim *et al.* [25]
18	1-ethyl-1-methylimidazolium tetrafluoroborate	[EMIM][BF_4]	Kim *et al.* [25]
19	Tetramethylammonium hydroxide	TMAOH	Khan *et al.* [26]

Table 2: Experimental and Model predicted HLwVE data analysis for 5 wt% salts (NaCl and KCl) solutions.

System	Temperature range (K)	No. data point	AAE (K)	R^2
		5 wt% aqueous salts solutions		
NaCl	274.2-283.3	5	0.13	0.998
KCl	271.9-283.0	6	0.15	0.999

It can be evident from the Figure 1 and Table 2 that the considered model was perfectly applicable for salt solution, therefore; it effectively be applied on other electrolytes solutions such as ILs solutions. Moreover, Figure 2 presented the experimental and predicted HLwVE data points of 10 wt% IL-CH_4 systems from literature. Additionally, Table 3 tabulated the model predicted range, AAD and correlation (R^2) values of each considered ILs.

IOP Conf. Series: Materials Science and Engineering **458** (2018) 012073 doi:10.1088/1757-899X/458/1/012073

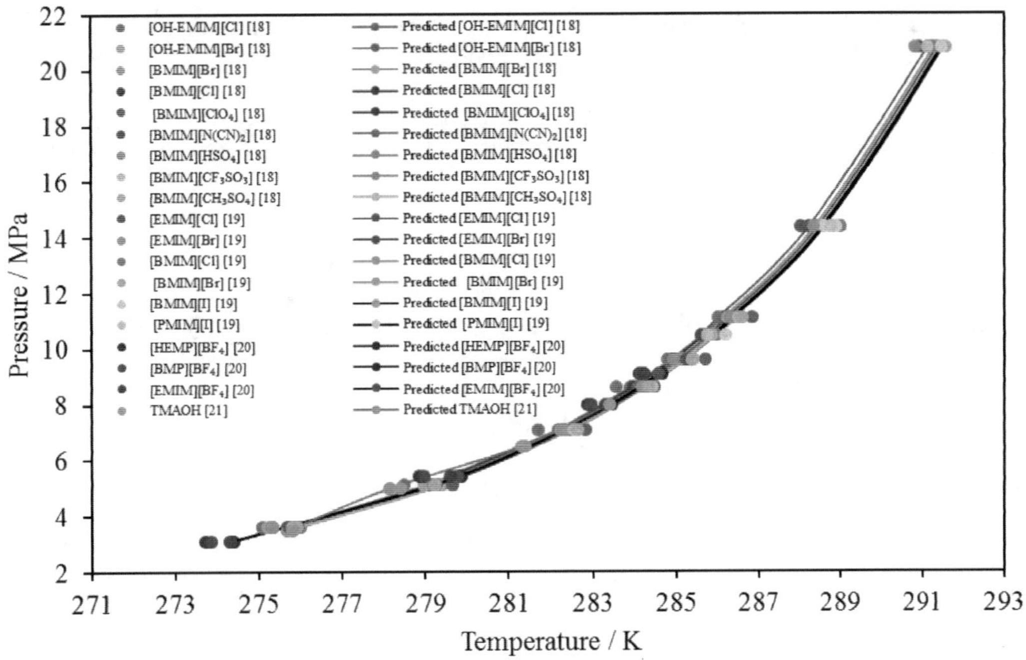

Figure 2. Experimental and Model predicted HL_wVE data for 10 wt% aqueous ILs solution.

Table 3. Predicted methane HL_wVE points in the presence aqueous ILs solutions

ILs	No. data point	Temperature range (K)	AAE (K)	R^2
[OH-EMIM][Cl]	06	279.77-287.45	0.24	0.996
[OH-EMIM][Br]	06	278.91-286.65	0.23	0.997
[BMIM][Br]	06	278.72-285.59	0.55	0.997
[BMIM][Cl]	06	279.91-286.77	0.3	0.995
[BMIM][ClO_4]	06	279.37-286.62	0.18	0.998
[BMIM][$N(CN)_2$]	06	275.82-286.32	0.15	0.999
[BMIM][HSO_4]	06	275.90-286.41	0.19	0.998
[BMIM][CF_3SO_3]	06	276.00-286.51	0.11	0.999
[BMIM][CH_3SO_4]	06	275.93-286.44	0.13	0.999
[EMIM][Cl]	03	285.59-291.14	0.18	0.998
[EMIM][Br]	03	285.80-291.35	0.26	0.999
[PMIM][I]	03	285.96-291.53	0.25	0.989
[BMIM][Cl]	03	285.73-291.29	0.30	0.987
[BMIM][Br]	03	285.89-291.45	0.14	0.998
[BMIM][I]	03	285.90-291.47	0.21	0.998
[HEMP][BF_4]	04	274.34-284.63	0.67	0.998
[BMP][BF_4]	04	274.37-284.67	0.60	0.998
[EMIM-BF_4]	04	274.29-284.58	0.14	0.999
TMAOH	04	275.65-283.38	0.13	0.999

Consulting to the Figure 2, it is evident that the model predicted data found in good agreement with the experimental literature data. It was clearly observed from the Table 3 that only pyrrolidinium based

ILs, i.e., [HEMP][BF_4] and [BMP][BF_4] slightly deviated with AAE values of 0.667 and 0.660, respectively. Furthermore, good correlations obtained with most of the R^2 values < 0.99 (with 95 % confidence level) were observed among the model predicted and experimental HL_wVE points. This validates the applicability of the studied model on the CH_4 hydrate and thus endorses the application of the electrolyte model for the prediction of the behaviour of the different ILs for other gas hydrate formers too.

4. Conclusions

In this work, the electrolyte-based model is applied to CH_4-ILs hydrate systems. Dickens and Quinby-Hunt Model predicted HL_wVE data revealed that the used model found in acceptable agreement with the experimental literature data of CH_4 hydrates in the presence of aqueous ILs solutions. Generally, for most of the ILs, the AAE values are found to be less than 0.25 K; while maximum AAE value (0.67) obtained for pyrrolidinium based IL ([HEMP][BF_4]) perhaps due to their higher alkyl chain length and molecular weights. Therefore, the suggesting model can be efficiently applied as a tool to predict new ILs with the aim to accomplish the desired THI inhibition for the petroleum industry.

Acknowledgements

Authors wishing to acknowledge PRSB for providing financial assistance via grant numbers 0153AB-F07 and 0153C1-024. Moreover, authors also would like to acknowledge the chemical engineering department, Universiti Teknologi PETRONAS for providing laboratory facilities throughout the study.

References

[1] Xiao C and Adidharma H 2009 Dual function inhibitors for methane hydrate *Chem. Eng. Sci.* **64** 1522–7
[2] Bavoh C B, Lal B, Nashed O, Khan M S, Lau K K and Bustam M A 2016 COSMO-RS: An ionic liquid prescreening tool for gas hydrate mitigation *Chinese J. Chem. Eng.* **24** 1619–24
[3] Muhammad Saad Khan, Yaqub S, Manner N, Karthwathi N A, Qasim A, Mellon N B and Bhajan Lal 2018 Experimental Equipment Validation for Methane (CH_4) and Carbon Dioxide (CO_2) Hydrates *IOP Conf. Ser. Mater. Sci. Eng.* **344** 1–10
[4] Khan M S, Lal B, Partoon B, Keong L K, Bustam M A and Mellon N B 2016 Experimental Evaluation of a Novel Thermodynamic Inhibitor for CH_4 and CO_2 Hydrates *Procedia Eng.* **148** 932–40
[5] Khan M S, Bavoh C B, Partoon B, Lal B, Bustam M A and Shariff A M 2017 Thermodynamic effect of ammonium based ionic liquids on CO_2 hydrates phase boundary *J. Mol. Liq.* **238** 533–9
[6] Bavoh C B, Partoon B, Lal B, Gonfa G, Foo Khor S and Sharif A M 2017 Inhibition Effect of Amino Acids on Carbon Dioxide Hydrate *Chem. Eng. Sci.* **171** 331–339
[7] Tariq M, Rooney D, Othman E, Aparicio S, Atilhan M and Khraisheh M 2014 Gas hydrate inhibition: A review of the role of ionic liquids *Ind. Eng. Chem. Res.* **53** 17855–68
[8] Sakthivel S, Chhotaray P K, Velusamy S, Gardas R L and Sangwai J S 2015 Synergistic effect of lactam, ammonium and hydroxyl ammonium based ionic liquids with and without NaCl on the surface phenomena of crude oil/water system *Fluid Phase Equilib.* **398** 80–97
[9] Bavoh C B, Nashed O, Saad Khan M, Partoon B, Lal B and Sharif A M 2018 The Impact of Amino Acids on Methane Hydrate Phase Boundary and Formation kinetics *J. Chem. Thermodyn.* **117** 48–53
[10] Tariq M, Atilhan M, Khraisheh M, Othman E, Castier M, Garcia G, Aparicio S and Tohidi B 2016 Experimental and DFT Approach on the Determination of Natural Gas Hydrate Equilibrium with the use of Excess N2 and Choline-Chloride Ionic Liquid as an Inhibitor *Energy and Fuels* **30** 2821–2832
[11] Khan M S, Lal B, Bavoh C B, Keong L K and Bustam A 2017 Influence of Ammonium based Compounds for Gas Hydrate Mitigation : A Short Review *Indian J. Sci. Technol.* **10** 1–6
[12] Khan M S, Liew C S, Kurnia K A, Cornelius B and Lal B 2016 Application of COSMO-RS in Investigating Ionic Liquid as Thermodynamic Hydrate Inhibitor for Methane Hydrate *Procedia Eng.* **148** 862–9
[13] Dickens G R and Quinby-Hunt M S 1997 Methane hydrate stability in pore water: A simple theoretical approach for geophysical applications *J. Geophys. Res.* **102** 773–83
[14] Kirkwood J G 1934 Theory of solutions of molecules containing widely separated charges with special application to zwitterions *J. Chem. Phys.* **2** 351–61
[15] Pieroen A P 1955 Gas hydrates- approximate relations between heat formation, composition and equilibrium temperature lowering by inhibitors *Rec. Trav. Chim* **74** 955–1002

[16] Partoon B, Wong N M S, Sabil K M, Nasrifar K and Ahmad M R 2013 A study on thermodynamics effect of [EMIM]-Cl and [OH-C_2MIM]-Cl on methane hydrate equilibrium line *Fluid Phase Equilib.* **337** 26–31

[17] Javanmardi J, Moshfeghian M and Maddox R N 2001 An accurate model for prediction of gas hydrate formation conditions in mixtures of aqueous electrolyte solutions and alcohol *Can. J. Chem. Eng.* **79** 367–73

[18] Javanmardi J, Moshfeghian M and Maddox R N 1998 Simple Method for Predicting Gas-Hydrate-Forming *Energy & Fuels* **28** 219–22

[19] Nashed O, Dadebayev D, Khan M S, Bavoh C B, Lal B and Shariff A M 2018 Experimental and modelling studies on thermodynamic methane hydrate inhibition in the presence of ionic liquids *J. Mol. Liq.* **249** 886–91

[20] Khan M S, Bavoh C B, Partoon B, Nashed O, Lal B and Mellon N B 2018 Impacts of ammonium based ionic liquids alkyl chain on thermodynamic hydrate inhibition for carbon dioxide rich binary gas *J. Mol. Liq.* **261** 283–90

[21] Khan M S, Lal B, Keong L K and Sabil K M 2018 Experimental evaluation and thermodynamic modelling of AILs alkyl chain elongation on methane riched gas hydrate system *Fluid Phase Equilib.* **473** 300–9

[22] Anderson G K 2004 Enthalpy of dissociation and hydration number of methane hydrate from the Clapeyron equation *J. Chem. Thermodyn.* **36** 1119–27

[23] Sabil K M, Nashed O, Lal B, Ismail L and Japper-Jaafar A 2015 Experimental investigation on the dissociation conditions of methane hydrate in the presence of imidazolium-based ionic liquids *J. Chem. Thermodyn.* **84** 7–13

[24] Xiao C, Wibisono N and Adidharma H 2010 Dialkylimidazolium halide ionic liquids as dual function inhibitors for methane hydrate *Chem. Eng. Sci.* **65** 3080–7

[25] Kim K-S, Kang J W and Kang S-P 2011 Tuning ionic liquids for hydrate inhibition *Chem. Commun.* **47** 6341–3

[26] Khan M S, Partoon B, Bavoh C B, Lal B and Mellon N B 2017 Influence of tetramethylammonium hydroxide on methane and carbon dioxide gas hydrate phase equilibrium conditions *Fluid Phase Equilib.* **440** 1–8

[27] Mohammadi A H, Afzal W and Richon D 2008 Gas hydrates of methane, ethane, propane, and carbon dioxide in the presence of single NaCl, KCl, and $CaCl_2$ aqueous solutions: Experimental measurements and predictions of dissociation conditions *J. Chem. Thermodyn.* **40** 1693–7

IOP Conf. Series: Materials Science and Engineering **458** (2018) 012074 doi:10.1088/1757-899X/458/1/012074

Effect of The natural Green Materials on Methane Hydrate Formation Kinetics

Sana Yaqub, Bhajan Lal *, Nurhayati Bt. Mellon and Suriati binti Sufian

Chemical Engineering Department, Universiti Teknologi PETRONAS,
32610 Bander Seri Iskandar, Perak Darul Ridzuan, Malaysia
*bhajan.lal@utp.edu.my

Abstract. The kinetic effect of resin (terpene) and ajinomoto (monosodium glutamate) for methane hydrate formation was reported. The induction time of 0.2wt% natural green materials was measured in a sapphire hydrate reactor at 12°C at 9.5MPa. The induction time of all materials were found to be less than pure water. In the presence of resin induction time decreased 20% more than pure water and the initial rate of hydrate formation was enhanced. Resin enhanced the rate of methane hydrate formation 17% more than pure water. However, compared to SDS the rate of hydrate formation by both natural materials was less. Moreover, natural materials consumed more methane during hydrate formation than pure water. In addition, resin promote methane hydrate formation and could act as suitable kinetic promoter for the natural gas separation, transportation and storage.

1. INTRODUCTION

Gas hydrate are crystalline, ice like anomalies, formed in oil and gas transmission lines by the concomitance of hydrogen bonded water molecules and former gases (methane, ethane etc.) at high pressure and low temperature [1, 2]. The formation of gas hydrate disturbs the gas transmission and slows down the production. However, recently natural gas hydrates (NGHs) have been gained much attention as a potential next generation energy source [3-6]. Generally gas hydrate formation take placed by entrapping the former gas molecules within the hydrate cages. The gas molecules form hydrogen bond with water molecules until nuclei of detectable size is formed. This agglomeration of nuclei at water-gas interface results in the formation of thin hydrate film on the surface which severely slow the rate of hydrate formation [3]. In order to improve interfacial interactions between gas and water various chemical additives are used which improves the gas hydrate storage and generation.

Synthetic chemical additives like SDS, sodium tetradecyl sulfate (STS) and sodium hexadecyl sulfate (SHS) are used as kinetic promoters. These additives enhance hydrate formation rate by increasing gas solubility and supporting micelle formation [7-9]. Tetrahydrofuran (THF) and acetone are used as thermodynamic promoter and enhance the hydrate formation by shifting the phase boundary to the right side [10-13]. Undesirably these synthetic surfactants and promoters are toxic in nature and can affect the aquatic life. Studies have been performed to evaluate the performance of natural green materials such as: chitosan [14], tapioca starch [15], amino acids[16] and antifreeze proteins [17]as kinetic hydrate inhibitor (KHI). However, the behaviour of natural materials as kinetic promoters are still unsearched.

Therefore, in order to prevent environmental hazards due to the excessive usage of toxic materials in pipeline, biodegradable and environment friendly chemicals are required to investigate. Consequently the effect of natural green materials such as: ajinomoto, and resin on methane hydrate kinetic is investigated in this work. Ajinomoto is the non-essential amino acid and similar to amino acids have zwitterions behaviour in water. While resin are the highly viscous materials from the pine tree and principally composed of terpenes. The chemical structures of additives are shown in Fig. 1.

Materials	Chemical Structure
Resin (Pine tree, terpenes)	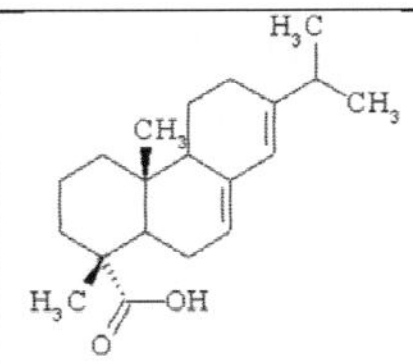
Ajinomoto (monosodium glutamate)	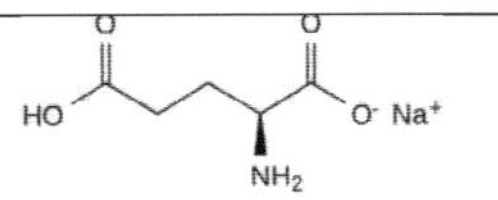
Sodium dodecyl sulfate (SDS)	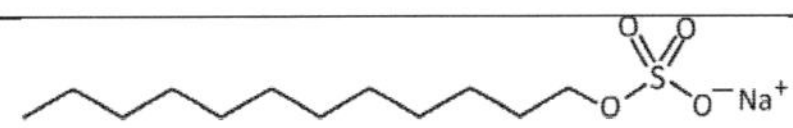

Fig.1. Chemical structures of studied natural green materials

2. EXPERIMENTAL

2.1. Materials

The chemical structure of the natural green materials, utilized in this work is shown in Fig. 1. These natural materials were used without any refining. Methane gas with purity of 99.995% was supplied by Gas Walker Sdn Bhd, Malaysia. Aqueous samples of additives were prepared using deionised water.

2.2. Apparatus

A hydrate sapphire cell or a rector as illustrated in Fig. 2 was used for kinetic hydrate measurements. The volume of transparent sapphire cell is 29.46ml. The sapphire hydrate reactor setup can operate up to 20 MPa in the temperatures range of 253.15-338.15 K. For temperature and pressure measurement during experiments two Pt100 thermocouples and a GP-M250 Keyence Japan Pressure transducer is used. The sapphire cell is place in a thermo-stated liquid bath that uses ethylene glycol as coolant. The temperature of liquid bath is controlled by Lauda Alpha RA 24 thermostats. The adequate stirring in sapphire is provided by the magnetic torque stirrer. A gas booster compressor is used to introduce methane into the sapphire. The aqueous natural green material solution is pumped into the cell via a manual hand pump. The temperature and pressure in the cell is continuously recorded in every 10sec by a data acquisition system with accuracy of ±0.15 K and ±0.01 MPa respectively.

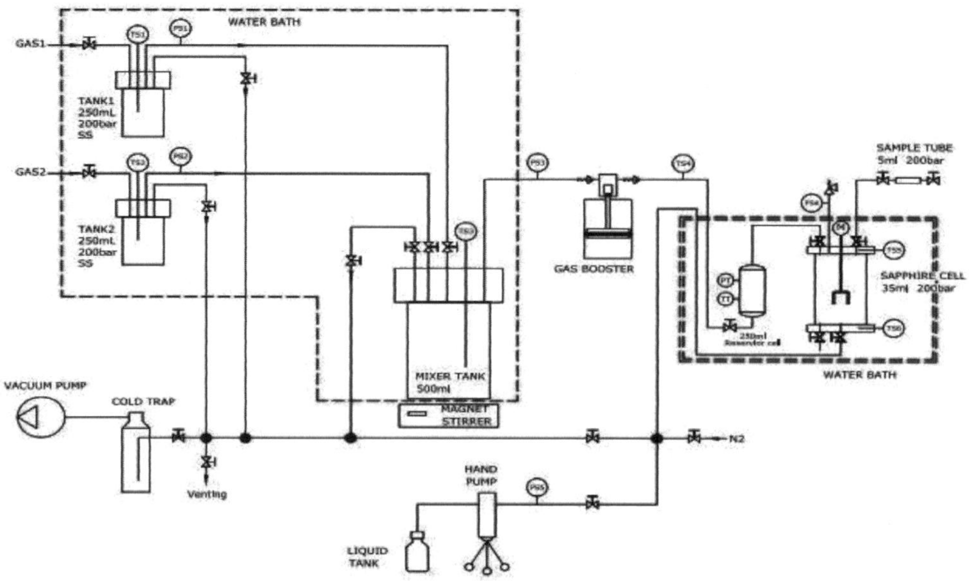

Fig.2 Experimental setup for gas hydrate kinetic measurements.

2.3. Kinetic measurements

There are several steps in order to perform kinetic measurements on sapphire hydrate reactor. Prior to experiment the reactor was washed thoroughly and vacuumed for 1 hr. Then reactor was filled with 18 ml of aqueous solution of natural green materials through injection pump. The temperature of reactor was set almost 2-3K, above the equilibrium temperature of methane. Once the system reaches the desired temperature, the reactor was pressurized with 9.5 MPa of pure methane. In order to get stable physical conditions, the system was kept at rest up to 20 min. Then the system temperature was reduced to the experimental temperature with the cooling rate of 4 K/hr for hydrate to form and stirrer was turned on. In sapphire hydrate reactor, the data is acquired by an interface that records the variations in pressure and temperature at every 10 s. In all kinetic experiments the experimental temperature was kept constant at 274.15 K. Kinetic measurement experiments generally take 2-3 hrs to complete. In terms of physical conditions, constant pressure and temperature indicates completion.

Kinetic measurement experiments gave three types of data set which includes induction time, moles of methane consumed and initial rate of hydrate formation. The induction time was delineate as mentioned in previous studies [18]. The real gas equation was used to determine the moles of methane consumed. The compressibility factor (z) in real gas equation was calculated from Peng–Robinson equation of state. The gradient of graph, plotted between moles of methane consumed and time was used to estimate the initial rate of hydrate formation.

3. RESULT AND DISCUSSION

The evaluated kinetic parameters such as induction time, initial rate of hydrate formation and the total moles of methane consumed are determined and used as kinetic inhibition indicators.

To guarantee the reproducibility, the experiments were repeated three times. To evaluate the effect of additives on the methane hydrate nucleation the induction time was determined by the pressure drop. The induction time is the time at which nuclei of detectable size is formed. Fig. 3 shows that in the presence of ajinomoto and resin the induction time was reduced to 14.8 and 13.8 min respectively at the sub-cooling temperature of 12 °C.

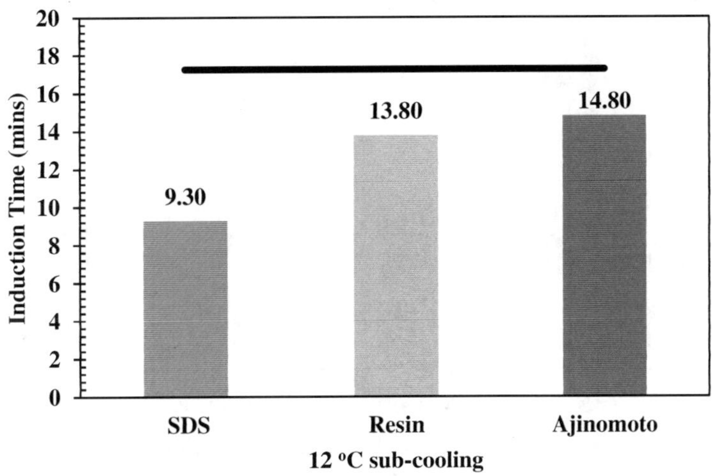

Fig.3 The induction time of methane hydrate in the presence of 0.2wt% concentration of natural green materials and commercial gas hydrate promoter, SDS.

The initial rate of hydrate formation in the presence of ajinomoto and resin is illustrated in Fig.4. Results showed that compared to pure water both additives attractively enhanced the initial rate of hydrate formation. Resin enhances the hydrate formation rate 17% more than pure water while ajinomoto slightly increased the rate of hydrate formation. However, in the presence of commercial kinetic hydrate promoter i.e. SDS the rate of hydrate formation was not much enhanced. As kinetic hydrate inhibition is probabilistic in nature, thus use of induction time and rate of hydrate formation individually be disingenuous. Therefore, the total moles of methane consumed were utilized to examine the effect of ajinomoto and resin on methane hydrate formation. The total methane consumed is taken as a point where the pressure became constant after hydrate formation. It is observed in Fig.5 that compared to pure water presence of ajinomoto significantly increased the total moles of methane consumed. However, compared to SDS both additives consumed less methane. The average methane consumed in the presence of ajinomoto was 4% less than average methane consumed in SDS hydrae formation. Compared to water in the presence of ajinomoto the methane consumed was increased 24%.

The promotional effect of ajinomoto and resin is similar to the adsorption mechanism of surfactants. Ajinomoto and resin hindered the accumulation of hydrate at the liquid interface thus support the displacement of more methane in to the liquid phase. Therefore in the presence of ajinomoto and resin the mole consumption of methane increased. The similar mechanism of methane hydrate promotion in the presence of arginine was observed by Bavoh et al.[16]

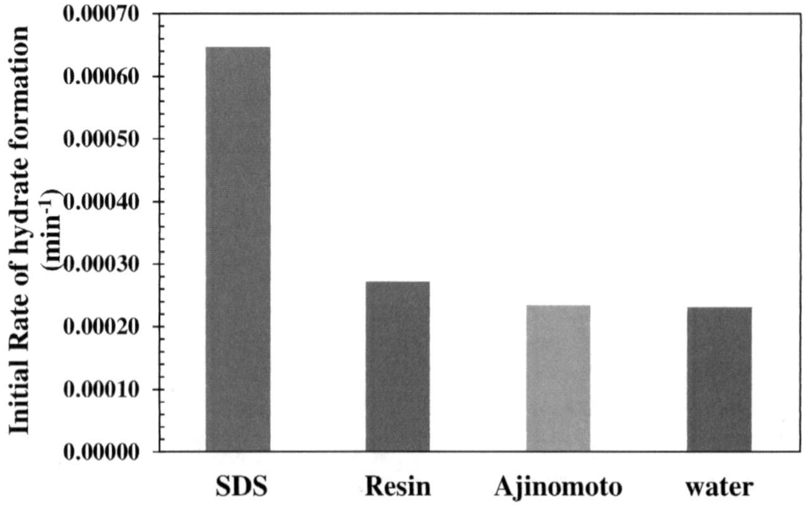

Fig.4 Initial rate of methane hydrate formation

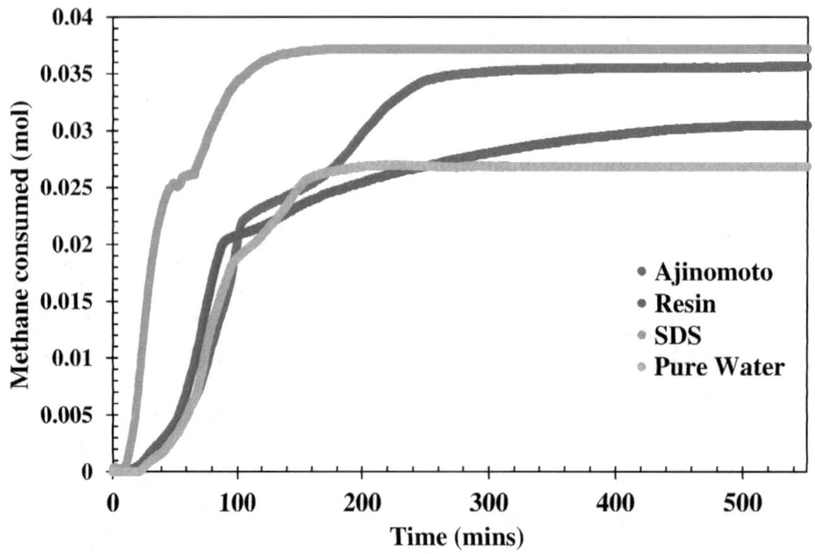

Fig.5 Moles of methane consumed during hydrate formation

4. CONCLUSIONS

The effect of ajinomoto and resin on the kinetics of methane hydrate formation is studied in sapphire hydrate reactor. It was found that compared to pure water both ajinomoto and resin decreased the induction time of hydrate formation. Furthermore, both additives enhanced the rate of hydrate formation. It was noticed that in the presence of ajinomto and resin the total moles of methane consumed are more than pure water. However, in the systems containing ajinomoto and resin, the rate of hydrate formation and moles of methane consumed was dwindled compared to comercial kinetic promotor i.e. SDS. Ajinomoto showed the hightest average methane hydrate promotion impact of about 24%. In addition, resin promote the methane hydrate formation and can act as suitable kinetic promoter for the natural gas separation, transportation and storage.

Acknowledgment
The Universiti Teknologi PETRONAS through YUTP Research grant from the Ministry of Higher Education, Malaysia, supported this work.

References

[1] K. M. Sabil, O. Nashed, B. Lal, L. Ismail, and A. Japper-Jaafar, "Experimental investigation on the dissociation conditions of methane hydrate in the presence of imidazolium-based ionic liquids," *J Chem Thermodyn.,* vol. 84, pp. 7-13, 2015.

[2] E. D. Sloan, "Fundamental principles and applications of natural gas hydrates," *Nature,* vol. 426, pp. 353-363, 2003.

[3] A. Arora, S. S. Cameotra, R. Kumar, C. Balomajumder, A. K. Singh, B. Santhakumari, *et al.*, "Biosurfactant as a promoter of methane hydrate formation: thermodynamic and kinetic studies," *Scientific reports,* vol. 6, p. 20893, 2016.

[4] Y. Liu, J. Hou, H. Zhao, X. Liu, and Z. Xia, "A method to recover natural gas hydrates with geothermal energy conveyed by CO_2," *Env.,* vol. 144, pp. 265-278, 2018.

[5] Y. Makogon, S. Holditch, and T. Makogon, "Natural gas-hydrates—A potential energy source for the 21st Century," *J. Pet. Sci. Technol.,* vol. 56, pp. 14-31, 2007.

[6] S.-Y. Lee and G. D. Holder, "Methane hydrates potential as a future energy source," *Fuel Process. Technol.* vol. 71, pp. 181-186, 2001.

[7] P. Gayet, C. Dicharry, G. Marion, A. Graciaa, J. Lachaise, and A. Nesterov, "Experimental determination of methane hydrate dissociation curve up to 55 MPa by using a small amount of surfactant as hydrate promoter," *Chem. Eng. Sci.,* vol. 60, pp. 5751-5758, 2005.

[8] J. Zhang, S. Lee, and J. W. Lee, "Kinetics of methane hydrate formation from SDS solution," *Ind. Eng. Chem. Res,* vol. 46, pp. 6353-6359, 2007.

[9] Y. Tian, Y. Li, H. An, J. Ren, and J. Su, "Kinetics of Methane Hydrate Formation in an Aqueous Solution with and without Kinetic Promoter (SDS) by Spray Reactor," *J. Chem.,* vol. 2017, 2017.

[10] B. Partoon and J. Javanmardi, "Effect of mixed thermodynamic and kinetic hydrate promoters on methane hydrate phase boundary and formation kinetics," *J. Chem. Eng. Data,* vol. 58, pp. 501-509, 2013.

[11] B. Partoon, S. N. A. Malik, M. H. Azemi, and K. M. Sabil, "Experimental investigations on the potential of SDS as low-dosage promoter for carbon dioxide hydrate formation," *Asia-Pac. J. Chem. Eng.,* vol. 8, pp. 916-921, 2013.

[12] B. Partoon, K. M. Sabil, H. Roslan, B. Lal, and L. K. Keong, "Impact of acetone on phase boundary of methane and carbon dioxide mixed hydrates," *Fluid Phase Equilib.,* vol. 412, pp. 51-56, 2016.

[13] B. Partoon, O. Nashed, Z. Kassim, K. M. Sabil, J. Sangwai, and B. Lal, "Gas Hydrate Equilibrium measurement of methane+ carbon dioxide+ tetrahydrofuran+ water system at high CO_2 concentrations," *Procedia Eng.,* vol. 148, pp. 1220-1224, 2016.

[14] Y. Xu, M. Yang, and X. Yang, "Chitosan as green kinetic inhibitors for gas hydrate formation," *J. Nat. Gas Chem.,* vol. 19, pp. 431-435, 2010.

[15] J. D. Lee, H. Wu, and P. Englezos, "Cationic starches as gas hydrate kinetic inhibitors," *Chem. Eng. Sci.,* vol. 62, pp. 6548-6555, 2007.

[16] C. B. Bavoh, O. Nashed, M. S. Khan, B. Partoon, B. Lal, and A. M. Sharif, "The impact of amino acids on methane hydrate phase boundary and formation kinetics," *J Chem Thermodyn.,* vol. 117, pp. 48-53, 2018.

[17] V. K. Walker, H. Zeng, H. Ohno, N. Daraboina, H. Sharifi, S. A. Bagherzadeh, *et al.*, "Antifreeze proteins as gas hydrate inhibitors," *Can. J. Chem.*, vol. 93, pp. 839-849, 2015.
[18] C. B. Bavoh, B. Lal, L. K. Keong, M. binti Jasamai, and M. binti Idress, "Synergic Kinetic Inhibition Effect of EMIM-Cl+ PVP on CO_2 Hydrate Formation," *Procedia Eng.*, vol. 148, pp. 1232-1238, 2016.

Improvement on plant thermal efficiency in MLNG Satu via MLNG Molecule Allocation Program (M-MAP)

Siti Farhana Binti Mohd Shaari[1,3], Putri Sophiazura Binti Megat Nor Hashim[1,3], Izyan Farhana Binti A Kaher[2,3], Nazatul Sheema Binti Che Din[2,3]

[1]Technical Services Department (Technology Section)
[2]Technical Services Department (Process Control Section)
[3]Malaysia LNG Sdn Bhd, Bintulu, Sarawak, Malaysia

Abstract. It has always been an aspiration for a Liquefied Natural Gas (LNG) plant to be able to monetize natural gas to its design plant capacity and best plant efficiency. Being able to achieve design capacity means project investment is paid off, whereas operating at the best efficiency means operating cost is optimum. Malaysia LNG Sdn Bhd (MLNG SATU) is consistently operated at design nameplate capacity post MLNG Rejuvenation and Revamp (MRR) project in 2005. Existing process control and optimization program has been helping the plant to consistently achieve maximum plant capacity within constraints on operational target and equipment performance. MLNG Molecule Allocation Program (M-MAP) was developed in 2016 to improve current program by looking at the plant thermal efficiency aspect, defines as Process Efficiency (PE). Leveraging on good existing plant data acquisition system and effective process control and optimization program in MLNG SATU, M-MAP has successfully expanded from a simple engineering calculation to a program that enables MLNG SATU production to achieve at its best efficiency. Developed in bi-phases, the Phase 1 of M-MAP has improved the overall efficiency of MLNG SATU by 1.3%. The Phase 2 of the M-MAP is currently under implementation; expecting another step in the overall efficiency of MLNG SATU within 0.5%.

Keywords: steam turbine driven LNG plant, process efficiency, molecule allocation program.

1. Introduction

Malaysia LNG Sdn Bhd (MLNG SATU) is a sea cooling water, steam turbine driven LNG plant with annual LNG production of 8.4 metric tonne per annum (mtpa). MLNG SATU consists of three identical modules namely Module 1, 2 and 3. Each of modules in general can be divided into three main areas which are gas treatment, liquefaction and utilities.

MLNG SATU had its first online plant data acquisition system in 2002 which enables the users to monitor plant condition in real time basis. In 2008, the first process control and optimization program was in place in MLNG SATU covering optimization of liquefaction process, named "Main Cryogenic Heat Exchanger Advanced Process Control". In 2013, the program was extended and includes a function that enables module selection and ability to maximize production of selected modules at minimum allowable feed gas pressure to the modules, named "Landing Pressure Automation". The program also includes process optimization in gas treatment area.

In 2016, MLNG SATU explored on possibility to further maximize LNG for a given feed by improving plant efficiency. There is a great potential to further enhance the value from the existing LNG assets using new industrial internet of things (IIoT) and data analytics technology [2]. A program named MLNG SATU Molecule Allocation Program (M-MAP) was initiated with two objectives (a) to determine the best combination of production target between LNG modules that able to achieve high plant efficiency of MLNG SATU (b) to identify efficiency gap for each of the modules and factors contributed to the gap in order to allow for rectification towards improving plant efficiency. The program combines data analytics of plant efficiency using real-timed plant data with existing MLNG Satu Advanced Process Control.

To achieve these objectives, a performance indicator named; Process Efficiency (PE) is introduced in M-MAP. PE is defined as the ratio of energy of total useful products (in gigajoules) with energy of total feed (in gigajoules) (See Equation 1). Examples of useful products in MLNG SATU are LNG, fuel gas and refrigerant whereas feed in MLNG SATU is in the form of natural gas (See Table 1 for the typical feeds and products in MLNG Satu). Higher PE means more useful products with higher energy were being generated for a given feed. PE is defined or calculated as follows:

$$Process\ Efficiency\ (PE) = \frac{Energy\ of\ Total\ Useful\ Product}{Energy\ of\ Total\ Feed}\ X\ 100\% \tag{1}$$

Table 1. Typical list of feeds and useful products in MLNG Satu.

Feeds	Useful Products
Natural Gas	Liquefied Natural Gas
Natural Gas Liquid	Refrigerants
Steam (As Energy Input to Compressor Driver or as an heating medium)	Liquefied Petroleum Gas
Electricity (As Energy Input to Pump Driver)	Fuel Gas

This paper describes the framework of M-MAP and its associated methodologies for in-situ LNG process efficiency performance analysis. Combining with existing process control and optimization program, the framework offers elements for monitoring, diagnosis and simple prognosis that helps MLNG SATU in achieving higher LNG production and overall plant efficiency in day to day operation.

2. Methodology

MLNG SATU Molecule Allocation Program (M-MAP) consists of two phases with the objectives to determine the best combination of production target between LNG modules that able to achieve high plant efficiency of MLNG SATU and to determine efficiency gap in each module and process parameters that contribute to the gap.

The heart of M-MAP lies in its MS Excel embedded formula file. M-MAP MS Excel incorporates a detailed calculation of overall and individual's mass and energy balance which will be used in calculating PE value in real-timed basis within the desired time interval. Defining MLNG SATU's process boundary is crucial when developing the mass and energy balances of the plant to prevent "over-counting" or "under-counting" of products and feeds that can lead to misleading of PE calculation as compared to actual plant efficiency of each module in MLNG SATU.

Once mass and energy balances are established, PE value can be calculated using the Equation 1. For data analysis purpose, MS Excel generates a correlation between PE and LNG production to find the equation's slope. The slope is named as PE Slope.

PE Slope is crucial as an input to determine the best combination of production targets between modules towards higher overall plant efficiency in MLNG SATU. The slope of graph between these two parameters is the indicator to show the magnitude of the impact to PE as LNG production increases or decreases. Those modules with higher slope (i.e. PE/LNG production) should be given priority to increase production as the impact to PE improvement is larger than those with lower slope. When production has to be reduced, those modules with lower slope will have the priority in order to minimize the degree of reduction in overall PE. M-MAP MS Excel, based on the example given in Figure 1 will propose the sequence of priority as X >>Y>>Z when it comes to production increment and Z >>Y>>X when the production reduces.

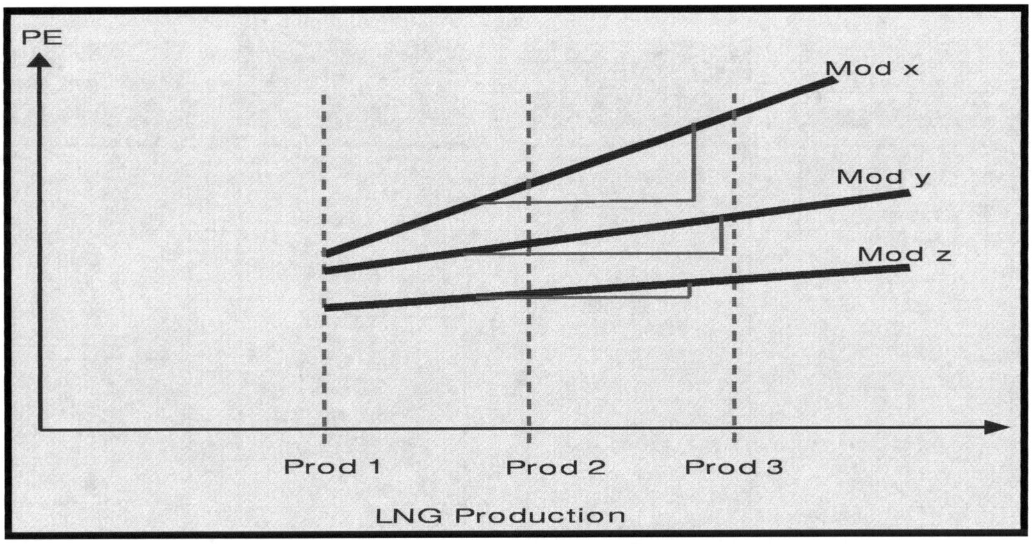

Figure 1. Typical correlation between PE and LNG production.

M-MAP MS Excel file is then made available in a relay server that is located in between plant network and business network. By enabling information from the business network to be pulled by the plant network, it sanctions the communication between the M-MAP MS Excel file to Advanced Process Control (APC) Application. The option of having optimum operating parameters supported by an Advanced Process Control system is more and more applied and reported cost benefits are straightforward in terms of a better utilization of train capacity and smoother operation leading to a better efficiency [1]. This whole network combining M-MAP MS Excel and APC is named as M-MAP Real Time Optimizer (RTO). Once M-MAP RTO starts running, MS Excel file will automatically retrieve the plant data from existing plant data acquisition system using Visual Basic for Applications (VBA) scripting and calculate the real-timed PE value of each module at respective LNG production band.

In any application that has the functionality of giving inputs to APC for plant operation, the function for bad data filtration is vital to prevent APC from causing operation upsets. In M-MAP, in the case of bad plant data inputs due to instrumentation under maintenance, the MS Excel File is designed with an "error handler" function. This function will alert the unavailability of good plant data input to the user for rectification and will temporarily replace the real-timed plant data with the design values.

In addition, LNG production band is created. It will be used to group the PE values within the same production band for every time interval and calculate the average of PE values before developing the real-timed correlation with LNG production. This is one of the way to ensure that the PE values used in developing the correlation are truly representing the actual plant PE during the desired time interval.

Once the real-timed PE slope values is generated upon the establishment of the correlation, it will be compared with the previously generated slope. To prevent too frequent changes in module priority during the implementation, the real-timed PE slope values will only be transferred to Distributed Control System (DCS) when the difference in slope with the previous one is significant.

The PE slope values will be transferred to DCS via a script file located in the DCS server. M-MAP RTO will automate the module's priority selection within the desired time interval by comparing each modules' real timed PE slope values. The outcome from M-MAP RTO will be the input to "Landing Pressure Automation" and "Main Cryogenic Heat Exchanger Advanced Process Control" which determines which module production to be adjusted. Figure 2 shows an illustration of M-MAP RTO data transfer choronology.

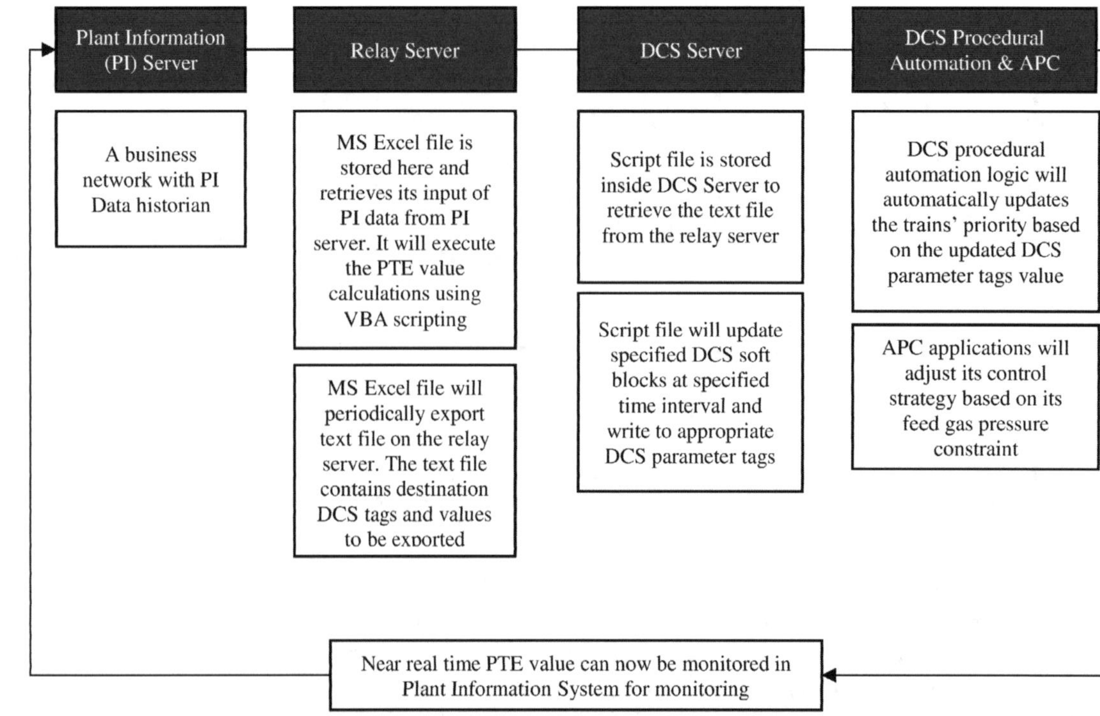

Figure 2. M-MAP RTO data transfer chronology.

While Phase 1 leverages on different PE between modules to determine the best combination of production to obtain higher overal efficiency of MLNG SATU, Phase 2 is established with the intent to reduce the gap of PE between modules. PE Gap Analysis Tool is introduced to identify reasons for the differences in PE and subsequently area for improvements to minimize the gap.

The tool is embeded inside M-MAP MS Excel File. The tool compares the daily energy performance of specific module with the best process efficiency performance that was achieved historically by any modules in MLNG SATU for the same LNG production range for the past one year, as illustrated in Figure 3 below. The benchmark values will be changed once the major gaps for a particular module has been rectified, normally during major turnaround or when the improvement has successfully brought the actual performance to or beyond the benchmark values.

	Module X Efficiency Gap	
LNG Production range	Same Production Band	Same Production Band
Process Efficiency (%)	Current PE	Best PE
	INPUT	
Descriptions	**Actual Energy**	**Energy Benchmark**
Supply1	Higher than benchmark	Benchmark value
Supply2	Healthy	
....	Healthy	
Steam	Higher than benchmark	
Fuel Gas	Higher than benchmark	
	OUTPUT	
Descriptions	**Actual Energy**	**Energy Benchmark**
Product1	Lower than benchmark	Benchmark value
Product2	Higher than benchmark	
Product3	Higher than benchmark	
Product4	Higher than benchmark	
....	Healthy	
LNG Refrigerant	Healthy	

Figure 3. Illustration of PE Gap Analysis Tool

In PE Gap Analysis Tool, the energy inputs (feed) and energy outputs (products) for particular module are listed separately between the input column and the output column. For energy inputs, the gap is defined when the actual energy is higher than the benchmark. Whereas, for energy outputs, PE Gap Analysis Tool introduces a new performance indicator which is "Low Heating Value (LHV) ratio" to identify the energy loss. The LHV ratio is defined as:

$$LHV\ Ratio = \frac{Higher\ Energy\ Products}{Leaner\ Energy\ Product} \quad (2)$$

The ratio of high energy products over lean energy products is compared to identify the energy loss. From PE point of view, higher energy products for the same energy inputs is more desirable, hence the tool looks for gap when low LHV ratio is identified.

Once the gaps of particular energy inputs and outputs identified, the tool allows for the users to evaluate the process parameters that contribute to the gaps, such as pressures, temperatures, flows and composition. The benchmarked values for identified process parameters were also made available for the users to quickly identify the most plausible process parameters that contribute to the gap.

3. Results and Discussions

The implementation of M-MAP Phase 1 and Phase 2 has helped to increase overall process efficiency in MLNG SATU and at the same time minimise the PE gap among all three modules. Figure 4 shows improvement in overall process efficiency in MLNG SATU by 1.3% that is equivalent to 189 tonne/day feedgas saving, based on 6-month evaluation between pre and post implementation of M-MAP Phase 1.

IOP Conf. Series: Materials Science and Engineering **458** (2018) 012075 doi:10.1088/1757-899X/458/1/012075

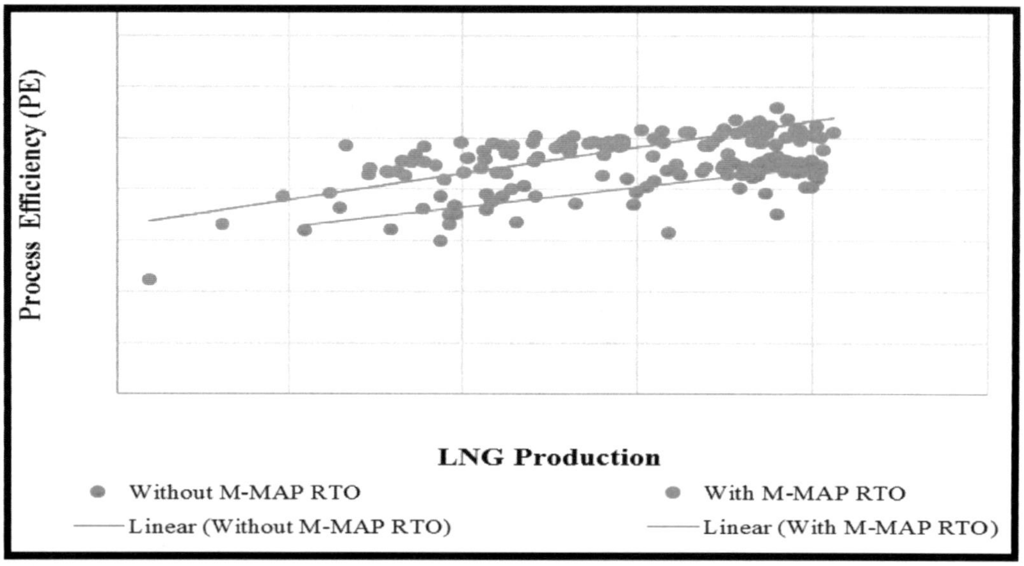

Figure 4. Improvement in overall process efficiency in MLNG SATU post M-MAP Phase 1.

As for Phase 2, several area of improvements to close the gaps in PE have been identified and actions have been planned for gap closure. Below are some of the summary of evaluation and findings obtained from the tool.

Higher steam consumption (energy input) for steam turbine in all Module X, Y, Z as compared to the benchmark by 20%, 10% and 5% respectively (see Figure 5). The tool identifies higher pressure at the surface condenser of steam turbine has led to higher steam flow to the turbine in order to compensate the loss in steam enthalpy for a given compressor power. Higher pressure at the surface condenser can be due to many reasons; among others are insufficient cooling medium or underperformance of ejector to remove noncondensibles medium from the system.

High steam consumption can be resolved by improving the cooling duty of the surface condenser and make good of ejector performance. It is projected that if the steam consumption can be reduced as per benchmark, an improvement of PE by 0.24%, energy saving equivalent to 33 tonne/day of natural gas upon resolution of this issue.

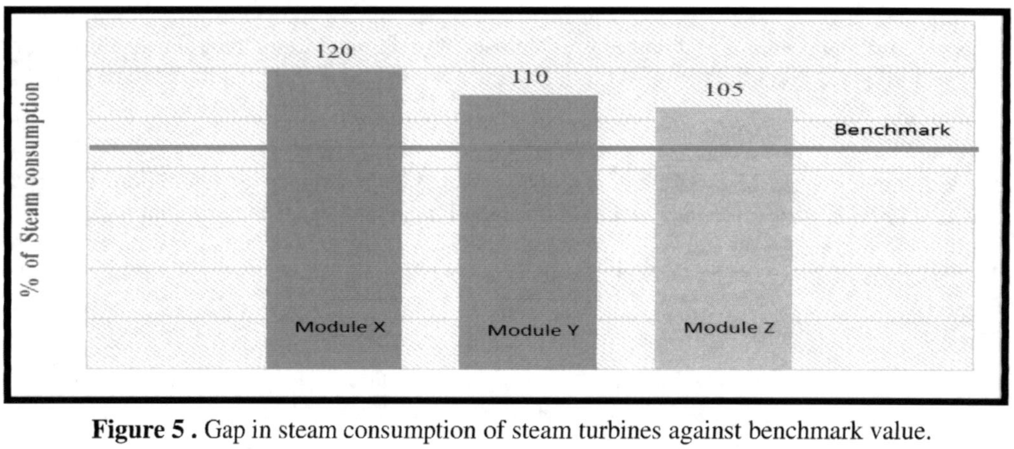

Figure 5 . Gap in steam consumption of steam turbines against benchmark value.

Module Y has higher steam consumption in heat exchanger by 15% as compared to benchmark (see Figure 6). This can be due to higher heating demand from process stream or inefficency in heat transfer due to fouling. The evaluation found that this can be resolved by further optimizing steam flow consumption in Module Y with a more robust control strategy while waiting for a suitable window for maintenace. If the steam consumption in the heat exchanger of Module Y can be reduced at least as the benchmark values, it is projected an improvement of PE by 0.05% in average upon resolution of this issue, equivalent to 6.5 tonne/day of feedgas energy saving.

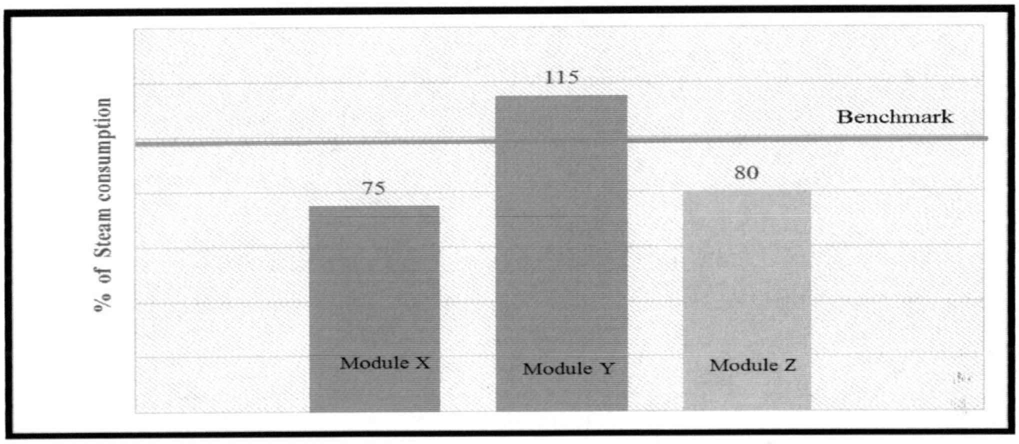

Figure 6. Gap in steam consumption of heat exchanger against benchmark value.

Module Y and Z have lower LHV Ratio by 75%-80% as compared to benchmark which indicates that the modules produce more lean energy products as compared to higher energy products. For example, producing fuel gas (methane dominant) as a product will lead to lower LHV ratio as compared to producing more LPG (propane dominant) for the same feed. In this case, Module Y and Z provide higher fuel gas export due to demand from other plants. From MLNG SATU PE point of view, exporting natural gas to other plant as fuel gas as opposed to producing LNG will reduce the process efficiency of these modules as fuel gas has lower energy content as compared to energy of LNG. Since April 2018, efforts has been done to control the treated natural gas supply from MLNG Satu modules by optimizing the fuel gas usage by other plants, hence, more natural gas from Module Y and Module X can be processed as LNG. The reduction of hot gas has improved the PE by 0.67 %, equivalent to 90 tonne/day of feedgas energy saving.

4. Conclusions

Molecule Allocation Program (M-MAP) was developed with the intention to have the best production and plant efficiency for MLNG SATU. The program has proven able to determine the best combination of production target among LNG modules that leads to high plant efficiency. Not only that, M-MAP is also capable to identify efficiency gap in each modules and the process parameters that contribute to the gap. This has led to several resolution initiatives created for MLNG SATU in 2017; approached either operationally or as part of maintenance activity. It is envisaged for similar LNG process efficiency performance monitoring and analysis to be embedded in other LNG Plants as well in order to maximize the realization of benefits to Petronas LNG Complex (PLC) as a whole.

Acknowledgments
The authors are also immensely grateful for the opportunity given by Malaysia LNG Sdn. Bhd to develop and implement M-MAP in MLNG Satu. The authors would also like to acknowledge assistance and support given by Operation, Engineering and Instrumentation Team of MLNG Satu for sharing their pearls of wisdom during the course of research and implementation.

References

[1] Chrystophe T and Denis C 2009 *Improving Energy Efficiency of LNG Plants*, **5**
[2] Jaleel V, David M and Kelly K 2017 *Realizing the Full Value of Industrial Internet and Data Analytics in LNG Facilities*, **1**

Reliability of Principal Component Analysis and Pearson Correlation Coefficient, for Application in Artificial Neural Network Model Development, for Water Treatment Plants

Jayaweera CD, Aziz N. *
School of Chemical Engineering, Engineering Campus, Universiti Sains Malaysia, Seri Ampangan,14300 Nibong Tebal, Seberang Perai Selatan, Penang, Malaysia
chamanthidj@gmail.com, chnaziz@usm.my*

Abstract. The Pearson Correlation Coefficient (PCC) and Principal Component Analysis (PCA) are methodologies commonly used for linear variable selection. PCC has been extensively used for variable selection, due to its simplicity and as it assists in recognizing the degree of correlation between input and output variables. Meanwhile, PCA has been used for recognizing variables that have high variances influencing the output variable. However, the use of linear forms of variables selection methodologies in non-linear modelling such as artificial neural networks (ANN) is questionable. In this work, the acceptability of PCC and PCA in variable selection for ANN modelling of the coagulation process in water treatment, is analysed. ANN models, aiming to predict coagulant dosage, treated water (TW) turbidity, TW pH and residual Aluminium, were developed. In order to compare the validity of inputs selected via PCC and PCA, an exhaustive search strategy of variable selection was carried out. The results showed that using the variables selected using PCA did not contribute in improving ANN model development. Meanwhile, variables selected by PCC were successfully used for all ANNs developed, except for TW pH prediction. The results also demonstrated that PCC and PCA are incapable of capturing collective effects of variables, on the output parameter.

1. Introduction

Coagulation process is primarily undertaken in water treatment plants for removal of turbidity. Alum is one of the most commonly used coagulants. The required dosage of coagulant is conventionally determined by jar tests, which consume time, chemicals and thus, is costly and not feasible to be used effectively, to respond to rapid changes of turbidity. Therefore, research has been carried out, to develop Artificial Neural Network (ANN) models, using past data of coagulant dosages and other factors such as pH, turbidity and alkalinity to approximate required dosages of coagulant with adequate accuracy.

The majority of studies that had developed ANN models for coagulation have adopted Multi-Layer Perceptron (MLP) architecture, as it had yielded in satisfying results. Theoretically, the Levenberg-Marquardt algorithm is considered a more superior learning algorithm for MLP models, than the back propagation algorithm, due to its ability to escape local minima.

Data preprocessing is one of fundamental aspects employed to ensure the validity of data used. [1] studied the effect of preprocessing and concluded that models trained with preprocessed data had a rapid learning rate and a smaller terminal error.

The Pearson Correlation coefficient (PCC) is a frequently used tool in the selection of input parameters in modeling coagulation to predict the optimal coagulant dosage [2,3]. PCC is used to determine the most related variables to the output parameter and eliminate inter-related input parameters. [2,3] had used a combined approach of using PCC and an exhaustive search strategy to arrive at the optimal model predicting the coagulant dosage. [2] have recognized that historical alum dosage (Al(t-1)) as an inherent factor in developing their model, thus this factor is adopted in this work. PCC has been used for the determination of input parameters for the prediction of the optimal coagulant dosage, but it is yet to be used for models predicting treated water quality parameters. The use of predictive models for monitoring treated water qualities is advantageous over the use of online measuring instruments, as it spares the cost spent on infrastructure and maintenance.

Principal Component Analysis (PCA) is another method used for selection of inputs as frequently as PCC [4,5]). The variables that account for the highest variance in the data could be recognized using PCA, which often mandatorily contribute to increase accuracy of a prediction model. Thus, PCA is used in this research to identify the variables with the highest variance.

Although PCC and PCA are often used in input parameter determination successfully, they are linear methods, and therefore the use of them in non-linear models is questionable as pointed out by [6]. Consequently, the objective of this work is to investigate the acceptability of using PCC and PCA for input parameter selection in the development of ANN models for coagulant dosage, Treated Water (TW) turbidity, TW pH and residual Aluminium prediction.

2. Procedure

One year's data corresponding to year 2005 from the Segama water treatment plant, Sabah, was used for the development of ANN models to predict alum dosage, TW turbidity, TW pH and residual Aluminium concentration. The input variables were selected from raw water quality parameters that include RW Color, RW Turbidity, RW Alkalinity, RW Total Dissolved Solids (TDS) and RW pH. Input variables were selected based on raw water quality parameters for simplicity and robustness. Raw water and treated water quality ranges are given in table 1.

Table 1 - Water quality ranges

	Raw water					Treated water						
	pH	Turb (NTU)	Col (HU)	TDS	Alkalinity (mg/l)	pH	Turb (NTU)	Color (HU)	TDS	Alkalinity (mg/l)	Al residual (mg/l)	Coagulant dosage
Min	6.6	11	36	60	50	6.5	0.19	0	90	20	0.01	20
Max	7.9	3405	656	160	170	7.5	5.24	6	170	100	0.2	170

Matlab 2017 was used to develop the feed-forward ANN models. The number of neurons in the hidden layer was decided based on the approach used by [7], where a maximum of 2I+1 hidden neurons is used, I being the number of input variables. The transfer function used for the input and output layers were tansig and purelin, respectively. Training data constituted of 60% of the data, 20% of data was used for validation, and the remaining 20% of the data was used for testing. Data division was carried out randomly using the Matlab's default option. The networks were trained using the Levenberg-Marquardt algorithm. The best model was based on the lowest test mean squared error and highest regression coefficient (R).

As indicated in the flowchart shown in Fig. 1, input parameter selection for developing ANN models was carried out using an exhaustive search strategy of the best combinations of RW variables. The selected input parameters were compared to input parameters selected using PCC and PCA. All values were normalized to carry out the principal component analysis, by dividing each variable by its maximum value. The exhaustive search strategy is a technique used for input selection where the optimum model is selected from all possible combinations of input parameters. It should be noted that it was intended to develop a model with minimal number of variables. In this work, four variables were found sufficient to achieve the acceptable model performance.

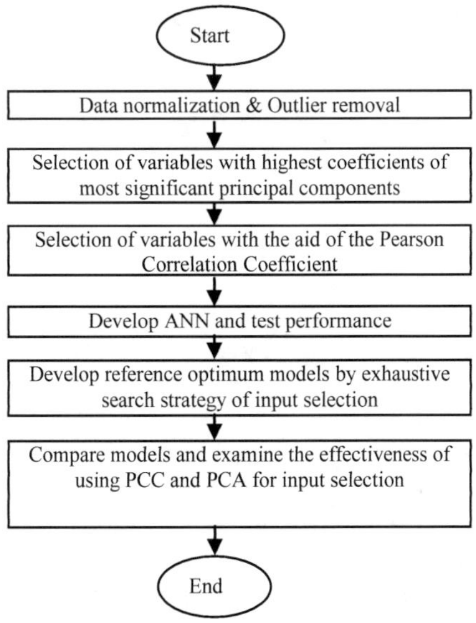

Fig. 1 – Research Methodology

3. Results and discussion

As shown in Table 2, the cumulative variance of the first two components is 89% and therefore they sufficiently represent the variability of the entire variable space. Thus, the first two principal components are chosen to represent the data set.

Table 2 - Principal components

Principal Component	Eigen vectors	Percentage total Variance	Cumulative variance
1	0.038	67	0.67
2	0.013	22	0.89
3	0.004	7	0.96
4	0.001	2	0.98
5	0.001	2	1

Table 3 shows the coefficients of each variable obtained for each principal component, from the analysis carried out. According to table 3, RW TDS and RW Alkalinity are the most significant variables, since they have high values for coefficients in both principal components. Consequently, both of them are deemed to be chosen input parameters as per the PCA.

Table 3-Coefficients of variables of significant principal components

Variables	PC1	PC2
RW Color	-0.071	-0.080
RW Turbidity	-0.105	-0.183
RW Alkalinity	0.728	-0.682
RW TDS	0.668	0.703
RW pH	0.080	0.025

An exhaustive search of the best combination of input variables was carried out by testing the performance of models with different combinations of the available input raw water quality parameters. The results of this search are given in table 5. It could be noted that the variables identified by the PCA are not among the best combinations in any of the models shown in table 5. Therefore, it is evident that PCA is not an efficient method for input parameter selection for the neural network models developed.

Pearson correlation coefficients of each raw water quality with respect to the output variable of each model are shown in table 4.

Table 4 - Pearson correlation coefficients

Output	Alum dosage	RW Alkalinity	RW Color	RW pH	RW TDS	RW turbidity	Residual Al
TW Turbidity	-0.86	-0.13	0.48	0.18	-0.33	0.99	Insignificant
TW pH	0.165	0.3587	-0.1056	0.917	0.4154	-0.1773	Insignificant
Residual Al	0.1054	-0.02	0.1097	0.062	0.012	0.1307	1
Alum dosage	1	-0.180	0.481	-0.292	-0.407	0.8536	0.107224

, RW parameters that have high correlation with respect to the output parameter are taken as input parameters, when using PCC. For example, the Pearson correlation coefficients relating Alum dosage with RW turbidity and RW color are relatively higher. Therefore RW turbidity and RW color are considered as input parameters to the model predicting Alum dosage. This result obtained by PCC agrees with the combination shown in table 5. According to table 4, input parameters for the model predicting TW turbidity are, RW turbidity, Alum dosage and RW color, which also agree with the combination shown in table 5. Similarly, input parameters for models predicting residual Al based on table 4 agree with the combination shown in table 5.

Table 5 – Architecture and input parameters used in optimum models

Model	Input parameters	Architecture
TW turbidity	Alum dosage, RW turbidity, RW color, TW turbidity(t-1)	4-9-1
TW pH	RW pH, Alum dosage, TW pH(t-1)	3-7-1
Residual Al	RW turbidity, RW color, Alum dosage, Residual Al (t-1)	4-9-1
Alum dosage	RW color, RW turbidity, Alum dosage (t-1)	3-7-1

The choice of input parameters shown in table 5 could be backed by the analytical studies done on coagulation in water treatment [8,9]). Although NOM is not measured using color, NOM is usually responsible for color. NOM affects the turbidity removal as it interferes with the coagulation process by forming complexes with Alum, thus explaining the influence of RW color in models predicting TW turbidity and Alum dosage. However this cannot be confirmed as corresponding data on Total Organic Carbon or UV-254 are unavailable.

Although pH is considered an important parameter affecting turbidity removal, residual aluminium concentration and alum dosage [8-11], the models developed do not demonstrate such a correlation. The lack of influence of pH on the models developed in this work, could be explained by how pH affects the coagulation mechanism of a coagulant. Charge neutralization is the dominant mechanism for alum at pH less than 6.5, at which the form of the prevalent monomer species highly depend on the pH, and therefore majorly affect turbidity removal and alum demand. The dominant mechanism switches to sweep flocculation at pH greater than 6.5 due to hydrolysation of monomer species [12], during which the effect of pH on alum demand and turbidity removal is lesser. According to table 1, the pH range of the water for the duration considered was 6.5 to 7.5. Therefore, sweep flocculation can persist throughout the range of data considered. The decisive value of pH affecting the residual Aluminium content is 6 [15], as Al exists in soluble forms at pH less than 6. At higher pH, the solubility of Al declines and it is easier to reduce the residual Al concentration.

Table 6 demonstrates the performance of the TW pH model with different combinations of input variables. It could be noted in table 6 that the mean squared error of models developed using RW TDS and RW Alkalinity is higher than the MSE of models developed with RW turbidity and alum dosage. However the results of the PCC suggest that RW Alkalinity and RW TDS demonstrate a significantly higher correlation to TW pH than the others, thus revealing a limitation of PCC.

Table 6 - Performance of models predicting TW pH

		Training	Validation	Testing
RW pH, Turbidity	**R**	0.8946	0.894	0.8982
	MSE	0.00808	0.0079	0.008093
RW pH, Color	**R**	0.88846	0.88739	0.88641
	MSE	0.008567	0.008249	0.008859
RW pH, Alkalinity	**R**	0.86353	0.85766	0.85617
	MSE	0.01018	0.01084	0.01088
RW pH, TDS	**R**	0.883	0.8884	0.8885
	MSE	0.008783	0.008675	0.008665
RW pH, Alum Dose	**R**	0.89146	0.88855	0.8993
	MSE	0.008096	0.008972	0.007915

The performance of the optimum models in table 5 is shown in table 7. The correlation coefficient of the turbidity model is poorer compared to other models, as the fluctuation of RW turbidity and TW turbidity is high and therefore it is difficult to establish a solid correlation.

Table 7 - Performance of optimum models

Optimum prediction model		Training	Validation	Testing
TW turbidity	**R**	0.77352	0.75268	0.93648
	MSE (mg/l)	0.01558	0.045833	0.0069
TW pH	**R**	0.9814	0.9788	0.98443
	MSE	0.001476	0.00171	0.001221
Residual Al	**R**	0.847	0.8998	0.958
	MSE	0.00014	0.00011	9.30E-05
Alum Dosage	**R**	0.9644	0.9636	0.9667
	MSE (mg/l)	17.22	17.45	12.76

In order to investigate the suitability of PCC for input parameter selection, this study was extended using an exhaustive search of higher number of input variables, including treated water parameters in the choice for input parameters for the prediction of alum dosage. The results obtained are demonstrated in table 8.

Table 8 - Optimum combinations of input variables

No. of variables	Input variables
3	RW turbidity, RW color, Alum (t-1)
4	RW turbidity, RW Alkalinity, TW turbidity, Alum (t-1)
5	RW turbidity, RW color, TW pH, TW color, Alum (t-1)
6	RW turbidity, RW color, TW pH, TW color, residual Aluminium, Alum (t-1)
7	RW turbidity, RW color, RW alkalinity, TW pH, TW turbidity, TW color, Alum (t-1)
8	RW turbidity, RW color, TW pH, TW turbidity, TW color, TW alkalinity, TW TDS, Alum (t-1)
9	RW turbidity, RW color, RW TDS, TW pH, TW turbidity, TW color, TW alkalinity, TW TDS, Alum (t-1)

It could be noted that under the combination of four variables, RW color is not considered as an input parameter, and it has been replaced by RW Alkalinity and TW turbidity. Based on Table 4, TW turbidity has a high correlation with alum dosage, but RW alkalinity shows low correlation with alum dosage. Therefore PCC was incapable of capturing the collective effect of the two variables on the alum dosage. However, it has been reported that alkalinity has a direct influence on the alum dosage [18]. The requirement for alum increases with decreasing alkalinity, due the corresponding decrease in the concentration of nucleating agents that stimulates coagulation. Therefore, a major drawback of the PCC is its inability to capture collective effects of input parameters on the output parameter.

However, several studies had been successfully carried out using PCC for input selection [2, 3]. Table 9 demonstrates that the performance of the model developed using PCC (model 1) and the model from table 8 with four variables (model 2) only differs slightly. Therefore PCC is a reliable tool for developing models with reasonable performance.

Table 9 - Test performance of models predicting alum dosage

	Correlation coefficient	Mean squared error (mg/l)
Model 1	0.9686	12.78
Model 2	0.9706	11.69

The regression plots of the two models shown in figure 2, confirms the results in table 8, as the plot of model 1 has a few more scattered points than model 2.

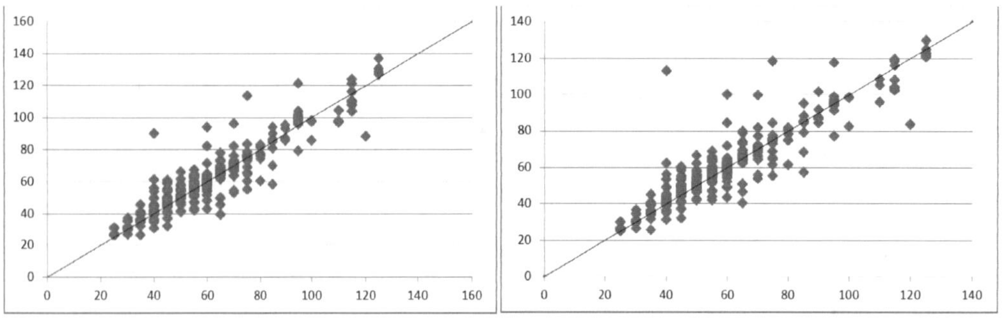

Fig 2 – Regression plots of model 1 (right) and model 2 (left)

4. Conclusion

Input parameters for models predicting coagulant dosage, TW turbidity, TW pH and residual Aluminium were determined using an exhaustive search of input variables, where the best combination of inputs were selected by testing the performance of ANN models. Input variables selected by principal component analysis and Pearson correlation coefficient, were compared with the best combination of variables selected from the exhaustive search. Variables selected by the PCA did not agree with the best combination of variables in any model. However the inputs selected by PCC agreed with models predicting coagulant dosage, TW turbidity and residual Aluminium. It was found that PCC was not able in selecting optimum parameters for the model predicting pH. It was also shown that PCC was not able to capture the collective influences of variables on the output parameter. However, all models developed in this study are capable of predicting the optimum coagulant dosage and treated water qualities with reliable accuracy. It should be noted that these models are valid for plants maintaining pH at neutral ranges, with moderately alkaline water, the source of which is not exposed to many industrial processes.

Acknowledgement

The financial support from Universiti Sains Malaysia through postgraduate conference fund to the first author and the cooperation of Sabah Water Supply Department and LDWS for supplying the Segama Water Treatment Plant data are greatly acknowledged.

References

[1]. Joo, D. (2000). The effects of data preprocessing in the determination of coagulant dosing rate. *Water Research*, 34(13), pp.3295-3302.

[2]. Wu, G. and Lo, S. (2010). Effects of data normalization and inherent-factor on decision of optimal coagulant dosage in water treatment by artificial neural network. *Expert Systems with Applications*, 37(7), pp.4974-4983.

[3]. Wu, G. and Lo, S. (2008). Predicting real-time coagulant dosage in water treatment by artificial neural networks and adaptive network-based fuzzy inference system. *Engineering Applications of Artificial Intelligence*, 21(8), pp.1189-1195.

[4]. Lamrini, B., Benhammou, A., Le Lann, M. and Karama, A. (2005). A neural software sensor for online prediction of coagulant dosage in a drinking water treatment plant. *Transactions of the Institute of Measurement and Control*, 27(3), pp.195-213.

[5]. Hernandez H. and Le Lann M. (2006), Development of a Neural Sensor for On-Line Prediction of Coagulant Dosage in a Potable Water Treatment Plant in the Way of Its Diagnosis, *Advances in Artificial Intelligence - IBERAMIA-SBIA* , pp. 249-257.

[6]. R. May, G. Dandy and H. Maier, "Review of Input Variable Selection Methods for Artificial Neural Networks", 2011.

[7]. H. Maier (2004), Use of artificial neural networks for predicting optimal alum doses and treated water quality parameters, *Environmental Modelling & Software*, 19(5), pp. 485-49.

[8]. A. Matilainen, M. Vepsäläinen and M. Sillanpää (2010), Natural organic matter removal by coagulation during drinking water treatment: A review, *Advances in Colloid and Interface Science*, 159(2), pp. 189-197.

[9]. R. Jiao, R. Fabris, C. Chow, M. Drikas, J. van Leeuwen and D. Wang (2016), Roles of coagulant species and mechanisms on floc characteristics and filterability, *Chemosphere*, 150, pp. 211-218.

[10]. Z. Yang, B. Gao and Q. Yue (2010), Coagulation performance and residual aluminum speciation of Al2(SO4)3 and polyaluminum chloride (PAC) in Yellow River water treatment, *Chemical Engineering Journal*, 165(1), pp. 122-132.

[11]. A.Robenson, S. Shukor and N. Aziz (2009), Development of Process Inverse Neural Network Model to Determine the Required Alum Dosage at Segama Water Treatment Plant Sabah, Malaysia.

[12]. C. Kim and M. Parnichkun, MLP, ANFIS, and GRNN based real-time coagulant dosage determination and accuracy comparison using full-scale data of a water treatment plant, *Journal of Water Supply: Research and Technology - Aqua*, vol. 66, no. 1, pp. 49-61, 2016.

[13]. E. Towler, B. Rajagopalan and R. Summers (2009), Using Parametric and Nonparametric Methods to Model Total Organic Carbon, Alkalinity, and pH after Conventional Surface Water Treatment, *Environmental Engineering Science*, 26(8), pp. 1299-1308.

[14]. F. N. Valentin (2009), An Hybrid Neural Network Based System for Optimization of Coagulant Dosing in a Water Treatment Plant, *Citeseerx.ist.psu.edu.*

[15]. K. Griffiths and R. Andrews (2011), The application of artificial neural networks for the optimization of coagulant dosage, *Water Science & Technology: Water Supply*, 11(5), p. 605.

[16]. M. Kennedy, A. Gandomi and C. Miller (2015), Coagulation modeling using artificial neural networks to predict both turbidity and DOM-PARAFAC component removal, *Journal of Environmental Chemical Engineering*, 3(4), pp. 2829-2838.

[17]. S. Dharman, V. Chandramouli and S. Lingireddy (2012), Predicting Total Organic Carbon Removal Efficiency and Coagulation Dosage Using Artificial Neural Networks, *Environmental Engineering Science*, 29(8), pp. 743-750.

[18]. J. Gregor, C. Nokes and E. Fenton (1997), Optimising natural organic matter removal from low turbidity waters by controlled pH adjustment of aluminium coagulation, *Water Research*, 31(12), pp. 2949-2958.

A study on blowdown of pressurized vessel containing $CO_2/N_2/H_2S$ at cryogenic conditions

Umar Shafiq[1], Azmi M Shariff[1*], Muhammad Babar[1] and Abulhassan Ali[2]

[1]CO_2 Research Centre (CO2RES), Universiti Teknologi PETRONAS, Bandar Seri Iskandar, 32610 Perak, Malaysia
[2]Department of Chemical Engineering, University of Jeddah, Jeddah, Saudi Arabia

*azmish@utp.edu.my

Abstract. At times, CO_2 capture and sequestration from the gaseous mixture have to be done at high pressure and high concentration of CO_2 in vessels and pipelines. During depressurization of CO_2 mixture from high pressure vessel, the chances of solidification increase and the mixture undergo sudden complex phase behavior changes that can lead to accidental events. By avoiding solidification and controlling the complex phase transitions, the chances of accident and unscheduled breakdowns can be minimized. In this paper, a simulation study is presented for the depressurization of highly pressurized vessel containing the ternary mixture of $CO_2/N_2/H_2S$. The simulation is performed over different blowdown valves with orifice sizes of 4.325 mm, 6.325 mm, and 8.325 mm using BLOWDOWN software package in Aspen HYSYS® V9 to predict the phase behavior of mixture and changes in the temperature and pressure. The results indicate that the chances of an accident are minimized by avoiding the solidification during depressurization process using blowdown valve with an orifice diameter of 4.325 mm.

Keywords: Depressurization, Accident, CO_2 Capture, Solidification, Pressure Relief Valve, Phase Transitions

1. Introduction

In recent years, the development of technologies for the economical CO_2 capturing and sequestration (CCS) from various industrial sources such as fossil fuel-fired thermal power plants and other chemical process plants has attracted remarkable attention [1; 2]. Considering the fact that highly concentrated CO_2 is asphyxiating and highly toxic, an adult can be exposed to 5000 ppm CO_2 only up to 1 min [3]. The CCS is considered as one of the most effective method to mitigate the continuously rising CO_2 concentration in the atmosphere [4; 5]. Another application of CCS involves depressurization of CO_2 in the reservoir to enhance the oil recovery [6]. The transportation of a huge quantity of CO_2 at high pressure via several kilometers long pipelines passing through populated areas increases the chances of accidents due to very low-temperature that can lead to the solid formation. The chances of accidents can be reduced by safety assessment of pressurized pipelines or vessels [7-11].

Safety assessment of the pressurized pipelines (or vessel) transporting a massive quantity of CO_2 involves the accurate prediction of the fundamentally important data, such as outflow parameters of pipeline in the event of a rupture. The mass discharge rate, fluid phase, the temperature of the mixture, and initial pressure within the pipeline are some of the important outflow parameters [7; 12; 13]. In case of any accident or failure event, further damage can be avoided by blowdown of the highly pressurized vessel or pipeline. The blowdown of the highly pressurized vessel containing CO_2 is very complex due to some specific challenges related to CO_2 transport. First, CO_2 has a different critical point (7.38 MPa at 31.1°C) and triple point (0.518 MPa at −56°C) as illustrated in Figure 1 which highlights that CO_2 transport is possible in the dense liquid state. Second, the transported CO_2 is generally impure [14]. Depending upon the fuel source and capturing process, CO_2 might contain oxygen, water, nitrogen, SO_x, methane, and other impurities. These two factors introduce considerable challenges in modelling and simulation studies since the presence of even small traces of impurities may significantly affect the thermodynamic and transport properties of the gaseous mixture [15; 16]. The composition of the gaseous mixture used for simulation by Aihara and Misawa [17] is presented in Table 1.

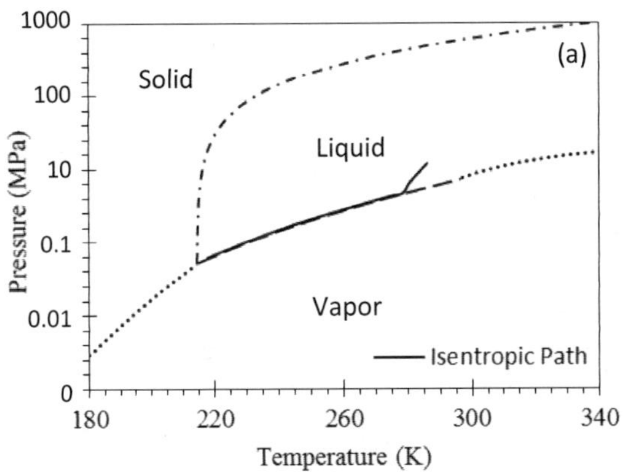

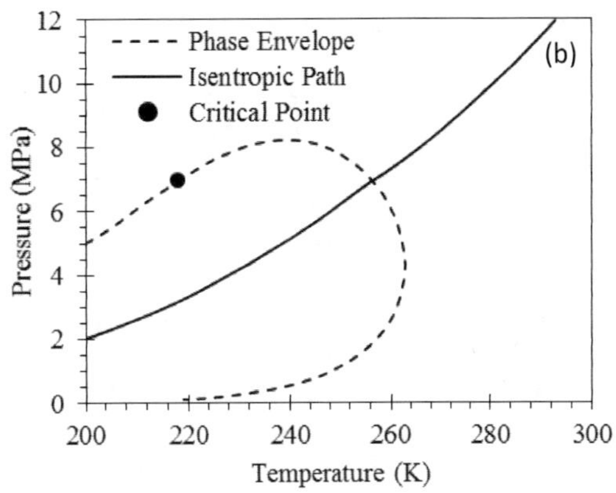

Figure 1. Isentropic depressurization of pure and impure CO_2 from 12 MPa at 293K. **(a)** CO_2 is in dense liquid state until it reaches the saturation line and then triple point, **(b)** Natural gas is in dense gaseous phase until it reaches the two-phase area

Table 1. Gaseous mixture composition [17]

Component	**(Mole %)**
CH_4	88.9
C_2H_6	6.2
C_3H_8	2.5
i-C_4H_{10}	0.4
n-C_4H_{10}	0.6
i-C_5H_{12}	0.1
n-C_5H_{12}	0.1
n-C_6H_{14}	0.1
N_2	0.3
CO_2	1.0

For dynamic simulation study, different models are developed for the study of depressurization of mixture over time for the oil and gas industry. Some of the significant models are LEKCON [18], BLOWDOWN [19], SAFIRE [13], OLGA [20], RELEASE [21], PHAST [22], and BLOWSIM [23-25]. In this study, BLOWDOWN is used because it doesn't assume the thermodynamic equilibrium and can simulate three phases. Density dependency can be modelled using the equation of state (EoS) based upon extended corresponding state principle which is more accurate than cubic EoS, however, it requires more run-time.

Data for depressurization of CO_2 or CO_2-N_2 mixture is available in the literature. Simulation studies or experiments were conducted by researchers previously. The results unveil that even small amounts of impurities will affect the depressurization path. Therefore, the aim of this research is the simulation study of depressurization of CO_2 with added impurities of N_2-H_2S to observe the effect of orifice size on depressurization path. To conduct this study, BLOWDOWN package in Aspen HYSYS® V9 is used as a simulation tool.

2. Research Methodology

Optimum blowdown time requires a delicate balance between the maximum admissible depressurization time, minimum fluid, and wall temperatures that may be safely considered [26]. To overcome issues regarding blowdown, the accurate predictions of these parameters are crucial. The traditional approach has been to use material more than required also known as 'overdesign'. However, this approach has become less attractive because of the rapidly decreasing profit margins of the oil and gas industry [24]. Therefore, to calculate optimum blowdown parameters, different simulations were performed to get most suitable blowdown parameters with minimum cost and hazards. In this work, three simulation-based case studies are performed to investigate the effect of orifice size and time on the temperature, pressure and phase behavior of a ternary mixture of CO_2, N_2, and H_2S.

To study the solid formation, initially the mixture is considered at the cryogenic condition. Since the mixture is already at very low temperature (cryogenic conditions), the Joule-Thomson effect with sudden depressurization causes the spontaneous drop in temperature that leads to the formation of solids. These case studies helped to observe phase behavior changes with time and orifice size. The mixture inside the vessel was at −40°C and 40 bar pressure. The cryogenic conditions helped to observe the worst-case scenario as well as planned blowdown of the CO_2 gaseous mixture. The simulation parameters and conditions used for this study are given in Table 2.

Table 2. Simulation parameters and conditions

Composition CO_2	82.4 Mole %
N_2	15.3 Mole %
H_2S	2.3 Mole %
Initial temperature	-40 °C
Initial pressure	40 bar
Vessel orientation	Vertical
Orifice diameter	4.325 mm (1st Case)
	6.325 mm (2nd Case)
	8.325 mm (3rd Case)
Leakage direction	Top
Ambient temperature	25 °C
Back pressure	1.01325 bar

For this study, a 1.524 m long and 0.273 m wide vessel with 25 mm thickness was used as given in Table 3. The tan-to-tan length of the vessel is 1.524 m, while the thickness of walls and head of the vessel is constant. To simulate the process, the BLOWDOWN package in Aspen HYSYS® V9 is used as a simulation tool. Peng-Robinson [27] EoS is selected as a property package. After selecting Blowdown utility, blowdown valve is linked to the top of the vessel. A feed stream at −40 °C and 40 bar is connected to the vessel. Vessel conditions and dimensions are defined along with ambient temperature and pressure. First, the blowdown process is converged for 4.325 mm orifice size followed by 6.325 mm and 8.325

mm. The vessel at high pressure (40 bar) is depressurized to atmospheric pressure with different orifice sizes. Larger the orifice size, smaller the depressurization time required to reach the atmospheric pressure.

Table 3. Dimensions of the vessel

Inner diameter	0.273 m
Length	1.524 m
Wall thickness	25 mm

Pressure-time, temperature-time profiles and phase behavior of mixture during blowdown process is generated. Pressure-time profiles display the depressurization path followed by the mixture during blowdown process. Temperature-time profiles are the dynamic temperature change of vapors inside the vessel. Pressure-Temperature graphs are the phase behavior of ternary mixture during the blowdown process. These profiles are created to see the effect of orifice size on the temperature and pressure inside the vessel. To observe the chances of solid formation, frost line for the ternary mixture is generated using Aspen HYSYS®. This will indicate that the mixture inside the vessel is entering the solid region from vapor-liquid region. To avoid solidification, the mixture should not pass the frost line, otherwise the solid formation starts that may cause the blockage.

3. Results and Discussions

Figure 2 represents the pressure-time profile for depressurization path followed by the mixture for orifice sizes of 4.325 mm, 6.325 mm and 8.325 mm. Orifice with the highest size is the quickest to reach the lowest pressure while depressurization through orifice 4.325 mm took more than 250 seconds to reach minimum pressure.

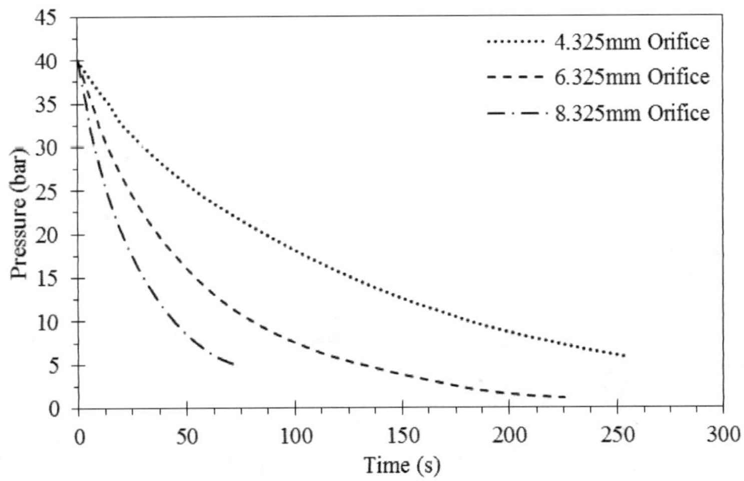

Figure 2. Main vessel pressure variations with time during depressurization

The mixture inside the vessel with an orifice size of 8.325 mm starts solidifying after 20 seconds when it reaches a pressure below 20 bar. Similarly, mixture inside the vessel with an orifice size of 6.325 mm takes 40 seconds to form solids. However, for 4.325 mm orifice size, the mixture remains in the vapor-liquid region throughout the process. During the depressurization process, the pressure suddenly decreases according to the Joule-Thomson effect that leads to changes in the phase behavior of mixture in the trivial time interval. This abrupt change in phase behavior enhances the solidification which is interconnected with the pressure stabilization [28].

Dynamic change in temperature of vapor inside the vessel is shown in Figure 3. Temperature variations manifest a sudden fall in temperature in earlier time of depressurization process and then temperature rises with time. This is because, at the initial conditions, the vessel was full of mixture and highly pressurized but with the passage of time, the mass of the mixture inside the vessel decrease and so does the pressure. Pressure drop can be observed from 40 bar to 5 bar within 60 seconds with orifice size 8.325 mm and temperature also falls below −70°C. After 20 seconds the mixture inside the vessel with 8.325 mm orifice size enters the solidification region while at the same time, the mixture is in vapor-liquid phase for other orifice sizes. Similarly, the mixture inside vessel with orifice size 6.325 mm starts forming CO_2 solids just after 40 seconds. However, mixture inside the vessel with orifice size 4.325 mm successfully reaches the minimum pressure without entering into the solidification region. In this research, 4.325 mm orifice size is found to be compatible for a successful blowdown of the vessel. But a further brief study is required to understand the complex phase behavior changes during depressurization process in order to avoid any accident.

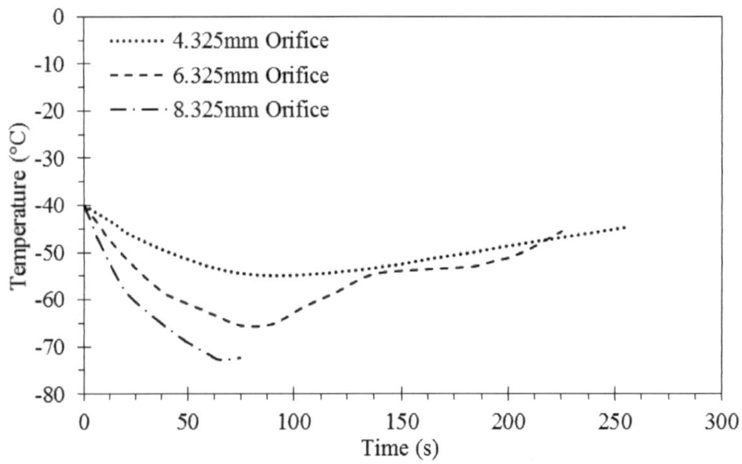

Figure 3. Temperature variations of vapor with time during depressurization

The prediction of solidification and phase behavior of ternary mixture during depressurization is illustrated in Figure 4. The vapor temperature inside the vessel is also reported along with the vapor-liquid and solidification region. Both phases are differentiated with frost line at which the formation of solid starts. At the beginning of the blowdown, the mixture is in vapor-liquid region but with the passage of time, it can enter the solidification region. Therefore, it can be observed in Figure 4 that the vapor temperature for 6.325 mm and 8.325 mm orifice diameter below 20 bar pressure falls below the freezing temperature. This can lead to the formation of solids of CO_2 because of swift decrease in temperature with an abrupt decrease in pressure. However, the orifice with diameter 4.325 mm indicates no chances of solidification because the mixture remains in the vapor-liquid region throughout the process. Although it takes more time to evacuate the mixture from the vessel with smaller orifice size but chances of accident are minimized.

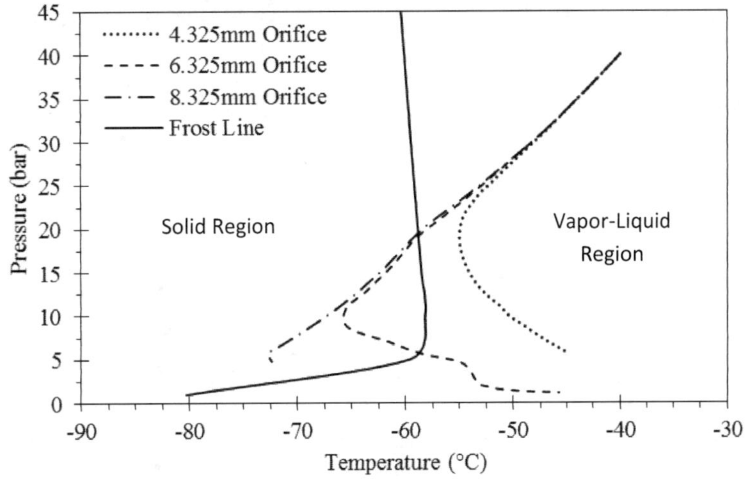

Figure 4. Phase behavior of ternary mixture during depressurization

4. Conclusions

A simulation study of depressurization of highly pressurized vessel containing ternary mixture is carried out aimed to investigate the temperature and pressure changes and phase behavior of gas mixture over three different orifice diameters i.e. for 4.325 mm, 6.325 mm, and 8.325 mm. The results show that the orifice with diameter 4.325 mm can prevent the solidification during depressurization and reduce the chances of accidents while other two orifice sizes can provoke solidification leading to the blockage or any accidental event in the pipelines. In future, blowdown process can be expanded to natural gas industry to avoid hazards in case of emergency or planned shutdown.

Acknowledgement

The authors would like to extend their most profound gratitude to the CO_2 Research Centre (CO2RES), Universiti Teknologi PETRONAS (UTP), Malaysia, and Hyundai Heavy Industries (HHI), South Korea, for the provision of research grant and on-campus state of the art facilities to accomplish this research.

References

[1] Kruse H and Tekiela M 1996 Calculating the consequences of a CO_2-pipeline rupture *Energy Convers. Manag.* **37** 1013
[2] Dall'Acqua D, Terenzi A, Leporini M, D'Alessandro V, Giacchetta G and Marchetti, B A 2017 New tool for modelling the decompression behaviour of CO_2 with impurities using the Peng-Robinson equation of state *Appl. Energy* **206** 1432
[3] Wilday A, McGillivray A, Harper P and Wardman M 2009 A comparison of hazard and risks for carbon dioxide and natural gas pipelines *IChemE Symposium Series*. pp. 392-398
[4] Sahu G C, Bandyopadhyay S, Foo D C, Ng D K and Tan, R R 2014 Targeting for optimal grid-wide deployment of carbon capture and storage (CCS) technology *Process Saf. Environ.* **92** 835
[5] Yu Y, Li Y, Lu H, Yan L and Zhang Z 2011 Performance improvement for chemical absorption of CO_2 by global field synergy optimization *Int. J. Greenh. Gas Con.* **5** 649
[6] Houston, J 2009 Petrobras tests CO_2 reinjection for Santos basin pre-salt Offshore (Tulsa) **69** 24
[7] Mahgerefteh H, Saha P and Economou I G 1999 Fast numerical simulation for full bore rupture of pressurized pipelines *AIChE Journal* **45** 1191
[8] Koornneef J, Spruijt M, Molag M, Ramirez A, Faaij A and Turkenburg W 2009 Uncertainties in risk assessment of CO_2 pipelines *Energy Procedia.* **1** 1587

[9] Mahgerefteh H, Fairweather M, Falle S, Melheim J, Ichard M, Storvik I, Taraldset O J, Skjold T, Economu I and Tsangaris D 2011 CO_2pipehaz: quantitative hazard assessment for next generation CO_2 pipelines *Inst. Chem. Eng.* 606

[10] Mahgerefteh H, Jalali N and Fernandez M I 2011 When does a vessel become a pipe? *AIChE Journal.* **57** 3305

[11] Park A, Ko Y and Lim, Y A 2007 Numerical Simulation of Rapid Depressurization in Pressure Vessels Incorporating Nucleate Boiling of Hydrocarbon Mixture *ASME 36th International Conference on Ocean, Offshore and Arctic Engineering*, American Society of Mechanical Engineers pp. V001T001A019-V001T001A019

[12] Oke A, Mahgerefteh H, Economou I and Rykov Y 2003 A transient outflow model for pipeline puncture *Chem. Eng. Sci.* **58** 4591

[13] Cumber P 2001 Predicting outflow from high pressure vessels *Process Saf. Environ.* **79** 13

[14] De Visser E, Hendriks C, Barrio M, Mølnvik M J, de Koeijer G, Liljemark S and Le Gallo Y 2008 Dynamis CO_2 quality recommendations *Int. J. Greenh Gas Con.* **2** 478

[15] Li H, Jakobsen J P, Wilhelmsen Ø, and Yan J 2011 PVTxy properties of CO_2 mixtures relevant for CO_2 capture, transport and storage: review of available experimental data and theoretical models *Appl. Ener.* **88** 3567

[16] Li H, Wilhelmsen Ø, Lv Y, Wang W and Yan J 2011 Viscosities, thermal conductivities and diffusion coefficients of CO_2 mixtures: Review of experimental data and theoretical models *Int. J. Greenh. Gas Con.* **5** 1119

[17] Aihara S and Misawa K 2009 Numerical simulation of unstable crack propagation and arrest in CO_2 pipelines The First international Forum on the Transportation of CO_2 in Pipelines.

[18] Woodward J L 1990 An integrated model for discharge rate, pool spread, and dispersion from punctured process vessels *J. Loss Prevent Proc.* **3** 33

[19] Richardson S and Saville G Blowdown of LPG pipelines 1996 *Process Saf. Environ.* **74** 235

[20] Bendiksen K H, Maines D, Moe R and Nuland, S 1991 The dynamic two-fluid model OLGA: Theory and application SPE production engineering **6** 171

[21] Johnson D W and Woodward J L, RELEASE: A Model with Data to Predict Aerosol Rainout in 2009 Accidental Releases (with CD-ROM)

[22] Witlox H W and Bowen P J 2002 Flashing liquid jets and two-phase dispersion: a review prepared by Det Norske Veritas Ltd for the Health and Safety Executive HMSO

[23] Mahgerefteh H, Saha P and Economou I G 2000 Modeling fluid phase transition effects on dynamic behavior of ESDV AIChE journal **46** 997

[24] Mahgerefteh H and Wong, S M A 1999 numerical blowdown simulation incorporating cubic equations of state Computers & Chemical Engineering **23** 1309

[25] Mahgerefteh H, Falope G B and Oke A O 2002 Modeling blowdown of cylindrical vessels under fire attack AIChE journal **48** 401

[26] Api R 1990 Recommended practice 521

[27] Peng D-Y and Robinson D B A 1976 New two-constant equation of state *Ind. Eng. Chem. Fundam.* **15** 59

[28] Eggers R and Green V 1990 Pressure discharge from a pressure vessel filled with CO_2 *J. Loss Prevent Proc.* **3** 59

IOP Conf. Series: Materials Science and Engineering **458** (2018) 012078 doi:10.1088/1757-899X/458/1/012078

CFD simulations of gas-liquid-liquid three-phase co-current flow in horizontal pipe by tracking volume fractions using VOF model

Muhammad Waqas Yaqub and Rajashekhar Pendyala*

Chemical Engineering Department, Universiti Teknologi PETRONAS, 32610 Seri Iskandar, Perak Darul Ridzuan, Malaysia.

*Corresponding author: rajashekhar_p@utp.edu.my

Abstract. The gas-liquid-liquid three-phase co-current flow is frequently observed in upstream petroleum pipelines. The identification of three-phase flow patterns is important to characterize three-phase flow dynamics. The three-phase flow patterns change with the pipe diameter, superficial velocities and physical properties of the fluids. Computational fluid dynamics (CFD) analysis of three-phase flow through 0.1524 m diameter and 3.5 m long horizontal pipe has been conducted. The superficial velocity ranges of gas-oil-water were 0.5-4, 0.08-0.32 and 0.08-0.32 m/s respectively. The volume of fluid (VOF) model was used to simulate the three-phase interfacial structures. According to the results, the three-phase flow patterns have been successfully simulated. Three-phase flow regime map has been constructed to analyze the flow pattern transition with the superficial velocities. The three-phase flow pressure drop increased with the superficial velocities of oil and water. However, the drop in pressure first increased and then decreased with the increase in superficial velocity of the gas.

1. Introduction

The fluid mixture produced from petroleum reservoirs mainly consists of gas, crude oil, and water. This fluid mixture is transported to the processing facilities through pipelines. Thus, gas-oil-water flow co-currently in upstream pipelines. Three-phase flow dynamics such as pressure drop, liquid holdup, and volumetric phase fractions are associated with the flow pattern. The liquid holdup and pressure drop are the common flow assurance problems. Various studies have been conducted to identify three-phase flow patterns. Three-phase flow patterns are differentiated based on gas-liquid and liquid-liquid interfacial relationship. Commonly occurring gas-liquid flow patterns are annular, plug, churn, slug, and stratified flow [1-4]. The oil-water could flow in the form of separate layers or as a homogeneous liquid mixture or one liquid as a continuous phase and the other one as dispersed phase [5, 6].

The three-phase flow pattern is the function of the diameter of the pipe, superficial velocities and physical properties of the fluids. The pipe diameter significantly influences the three-phase flow pattern and resultant pressure drop. Increase in the pipe diameter decreases the flow velocity which leads to the reduced pressure drop [7]. Some studies on three-phase flow to identify the flow patterns and the pressure drop are available in the literature. However, most of the previous studies were conducted with the small diameter pipes. The size of the commercial scale pipes is much bigger than laboratory scale pipes used in experiments for the identification of flow pattern. Therefore, the identification of realistic flow patterns and pressure drop is needed. The advancement in the processing ability of computers has made the CFD, a robust tool to study the flow dynamics in the desired range of parameters and geometries. The CFD analysis on two-phase flow has been carried out by Ekambara et al. [8], Bramara et al. [9] and Mazumder et al. [10]. However, the CFD analysis of three-phase flow in pipes is not available in the literature. The analysis was intended to simulate the three-phase flow patterns in the 0.1524 m internal diameter horizontal pipe at various gas, oil, and water superficial velocities.

2. Methodology

2.1. Geometry and Meshing
The geometry of the pipe was created in the Ansys design modeler tool. The length and internal diameter (ID) of the pipe were 3.5 m and 0.1524 m respectively. The minimum pipe length to diameter ratio (L/D) for turbulent flow to get the fully developed velocity profile was 10 [11] and the L/D ratio of the created pipe geometry has the (L/D) ratio of 23 to ensure the fully developed flow and to study the variations of flow parameters after the flow developing length as shown in Figure 1 (a). The results of CFD analysis strongly depends on the type of mesh employed [12]. The structured hexahedral mesh was preferred due to improved orthogonal quality and lower skewness as shown in Figure 1 (b). The mesh independence analysis was conducted at four different number of mesh cells by plotting the maximum absolute pressure for each mesh as shown in Figure 2.

(a)

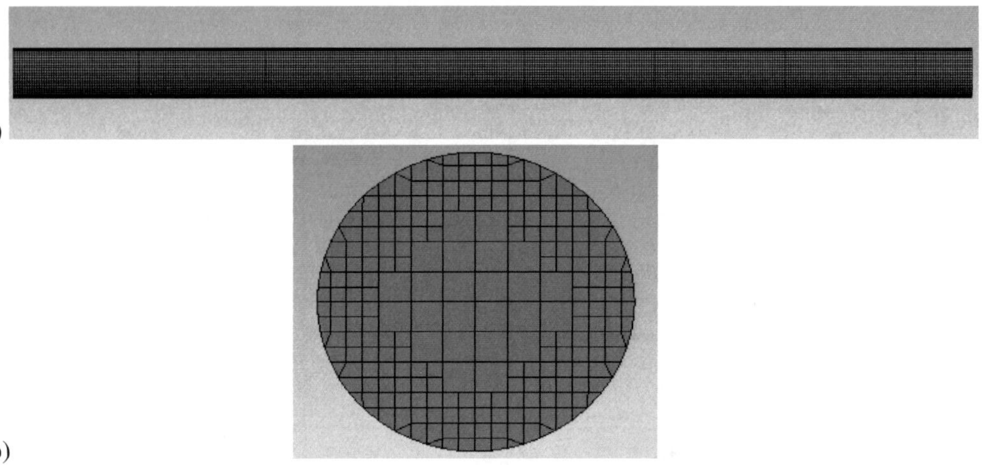

(b)

Figure 1. (a) Meshed horizontal pipe geometry, **(b)** Cross section view of the pipe mesh.

2.2. Boundary conditions and physical properties of fluids
The appropriate boundary conditions were chosen for a practical solution. The physical properties of the phases are given in table 1. The operating gauge pressure was set at zero as the flow system was considered to run at atmospheric pressure (101.32 *kPa*). The walls of the flow domain were considered to have no-slip condition and wall roughness was set as a smooth wall. The mixture velocity was given at the inlet according to the different set of superficial velocities in table 2. The volumetric phase fractions of oil and water based on the superficial velocities of each phase is calculated and specified at the inlet given in table 2. Air was declared as primary phase due to its higher phase fraction compared to other phases. Total 20 different combinations of gas-oil-water have been simulated to analyze the change in flow patterns from low to high superficial velocities. The surface tension effect was neglected due to the higher inertial forces than the surface tension force described by Weber number more than 1 [13].

2.3. Multiphase model
The volume of fluid VOF has been used to reveal the interfacial relationship between the phases. The VOF model solves a set of momentum equation to model the flow of immiscible fluids. The VOF model can model two or more immiscible fluids by the solution of a single set of momentum

equation. The implicit body force was used to include the gravitational effect in the flow along Y-axis. The momentum equation for VOF model is given in equation (1) [13].

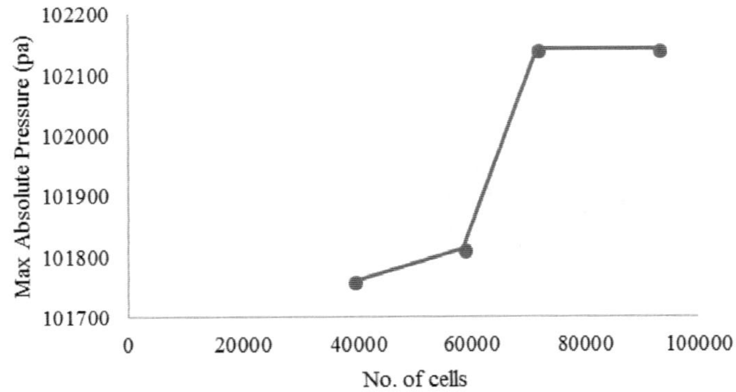

Figure 2. Mesh sensitivity analysis

Table 1. Physical properties of fluids

Sr No.	Name	Phase	Density (kg/m^3)	Viscosity (kg/m·s)
1	Air	Gas	1.225	0.00001789
2	Oil	Liquid	840	0.0084
3	Water	Liquid	1000	0.001

$$\frac{\partial}{\partial t}(\rho\bar{v})+\nabla.(\rho\bar{v}\bar{v}) = -\nabla p+\nabla.[\mu.(\nabla\bar{v}+\nabla\bar{v}^T)]+\rho\bar{g}+\bar{F} \quad (1)$$

Where g and t are the acceleration due to gravity and time respectively and ρ, p, μ and v are the density, pressure, dynamic viscosity and velocity of the phase k respectively. The term T is the stress tensor and F is the external force. The continuity equation for the volumetric fraction of one or more phases is given as.

$$\frac{1}{\rho_q}\left[\frac{\partial}{\partial t}(\alpha_q\rho_q)+\nabla.(\alpha_q\rho_q\vec{v}_q)=S_{\alpha q}+\sum_{p=1}^{n}(\dot{m}_{pq}-\dot{m}_{qp})\right] \quad (2)$$

Where, a and ρ are the volume fraction and density of the qth phase respectively and t is the time. The source term is zero by default is S_{aq}. The transfer of mass p to q and from q to p are denoted by term $\dot{m}_{pq}$ and $\dot{m}_{qp}$ respectively.

2.4. Turbulence model

The flow conditions adopted in the flow domain clearly indicates the occurrence of turbulence in the flow. Therefore, the simulation of turbulence was necessary to get the accurate results. The *k-epsilon* turbulence model was used, which is described by the following equations [14].

$$\rho v_j\frac{\partial k}{\partial x_j}=\frac{\partial}{\partial x_j}\left(\frac{\mu_t}{\sigma_k}\frac{\partial k}{\partial x_j}\right)+\mu_t\frac{\partial v_j}{\partial x_i}\left(\frac{\partial v_i}{\partial x_j}+\frac{\partial v_j}{\partial x_i}\right)-\rho\varepsilon \quad (3)$$

$$\rho v_j\frac{\partial \varepsilon}{\partial x_j}=\frac{\partial}{\partial x_j}\left(\frac{\mu_t}{\sigma_\varepsilon}\frac{\partial \varepsilon}{\partial x_j}\right)+C_1\mu_t\frac{\partial v_j}{\partial x_i}\left(\frac{\partial v_i}{\partial x_j}+\frac{\partial v_j}{\partial x_i}\right)-C_2\frac{\varepsilon}{k}\rho\varepsilon \quad (4)$$

Where, k represents the turbulent kinetic energy and the dissipation rate of k is denoted by ε. $C1$, $C2$, σ_ε, and σ_k are constants.

2.5. Solver and time step

Pressure based solver has been used. The algorithm used for the solution of Navier-Stokes equations was SIMPLE. The second-order upwind scheme was used for solving the momentum equation and

Table 2. Superficial velocities and volume fractions of fluids

	Gas superficial velocity (m/s)	Oil Superficial velocity (m/s)	Water superficial velocity (m/s)	Mixture velocity (m/s)	Oil volume fraction	Water volume fraction
1	0.5	0.08	0.08	0.66	0.121	0.121
2	1	0.08	0.08	1.16	0.068	0.068
3	2	0.08	0.08	2.16	0.037	0.037
4	3	0.08	0.08	3.16	0.025	0.025
5	4	0.08	0.08	4.16	0.019	0.019
6	0.5	0.16	0.16	0.82	0.195	0.195
7	1	0.16	0.16	1.32	0.121	0.121
8	2	0.16	0.16	2.32	0.074	0.074
9	3	0.16	0.16	3.32	0.048	0.048
10	4	0.16	0.16	4.32	0.037	0.037
11	0.5	0.24	0.24	0.98	0.244	0.244
12	1	0.24	0.24	1.48	0.162	0.162
13	2	0.24	0.24	2.48	0.096	0.096
14	3	0.24	0.24	3.48	0.068	0.068
15	4	0.24	0.24	4.48	0.053	0.053
16	0.5	0.32	0.32	1.14	0.280	0.280
17	1	0.32	0.32	1.64	0.195	0.19
18	2	0.32	0.32	2.64	0.121	0.121
19	3	0.32	0.32	3.64	0.087	0.087
20	4	0.32	0.32	4.64	0.068	0.068

Geometric reconstruction scheme was used to solve volume fraction equations to minimize numerical diffusion. Transient calculations were conducted to analyze the change in interface structure between the phases with the time. VOF model gives an excellent feature of the variable time step, which increases or decreases the time step size with the change in the size of the cell to meet the global courant number criteria. Thereby, decreases the computational time. The simulations were conducted for 8 seconds to allow the flow to become completely stable.

3. Results and Discussion

3.1. Three-phase flow patterns

The contours of three-phase horizontal flow in the pipe along the X-Y plane is shown in Figure 3. Three-phase plug, stratified, stratified wavy and stratified rolling wave flow patterns have been observed. Intermittent plugs of gas and liquid were moving in the upper section of the pipe as shown in the contours of plug flow in Figure 3. Most of the pipe cross-section was filled with the liquid because of the high volumetric phase fractions of oil and water. In such conditions where the volumetric fraction of gas is lower than the total liquid volume fraction, the gas phase is driven by the liquid phase [3]. At low gas superficial velocity, the oil and water were not completely mixed during plug flow as shown by the cross-section Figure 4 (c) and (d). However, increase in superficial velocity of gas increased the oil-water mixing shown in Figure 4 (g) and (h). According to the simulation results, the stratified flow was the highly occurring flow pattern in three-phase co-current horizontal flow in the pipe. The gas-liquid interface was straight and smooth. During stratified flow, liquid phases flow in separate layers or in the form of a mixture as shown in

	Gas-oil-water superficial velocities (m/s)	Three-phase flow in horizontal pipe	Gas-Liquid Flow Pattern
3(a)	0.5-0.08-0.08		Stratified wavy
3(b)	0.5-0.16-0.16		Stratified wavy
3(c)	0.5-0.24-0.24		Plug
3(d)	0.5-0.32-0.32		Plug
3(e)	1-0.08-0.08		Stratified wavy
3(f)	1-0.16-0.16		Stratified wavy
3(g)	1-0.24-0.24		Plug
3(h)	1-0.32-0.32		Plug
3(i)	2-0.08-0.08		Stratified wavy
3(j)	2-0.16-0.16		Stratified wavy
3(k)	2-0.24-0.24		Stratified wavy
3(l)	2-0.32-0.32		Stratified
3(m)	3-0.08-0.08		Rolling wave
3(n)	3-0.16-0.16		Stratified
3(o)	3-0.24-0.24		Stratified
3(p)	3-0.32-0.32		Stratified
3(q)	4-0.08-0.08		Rolling wave
3(r)	4-0.16-0.16		Stratified
3(s)	4-0.24-0.24		Stratified
3(t)	4-0.32-0.32		Stratified
		Gas Oil Water 	

Figure 3. Contours of the X-Y plane of the horizontal pipe

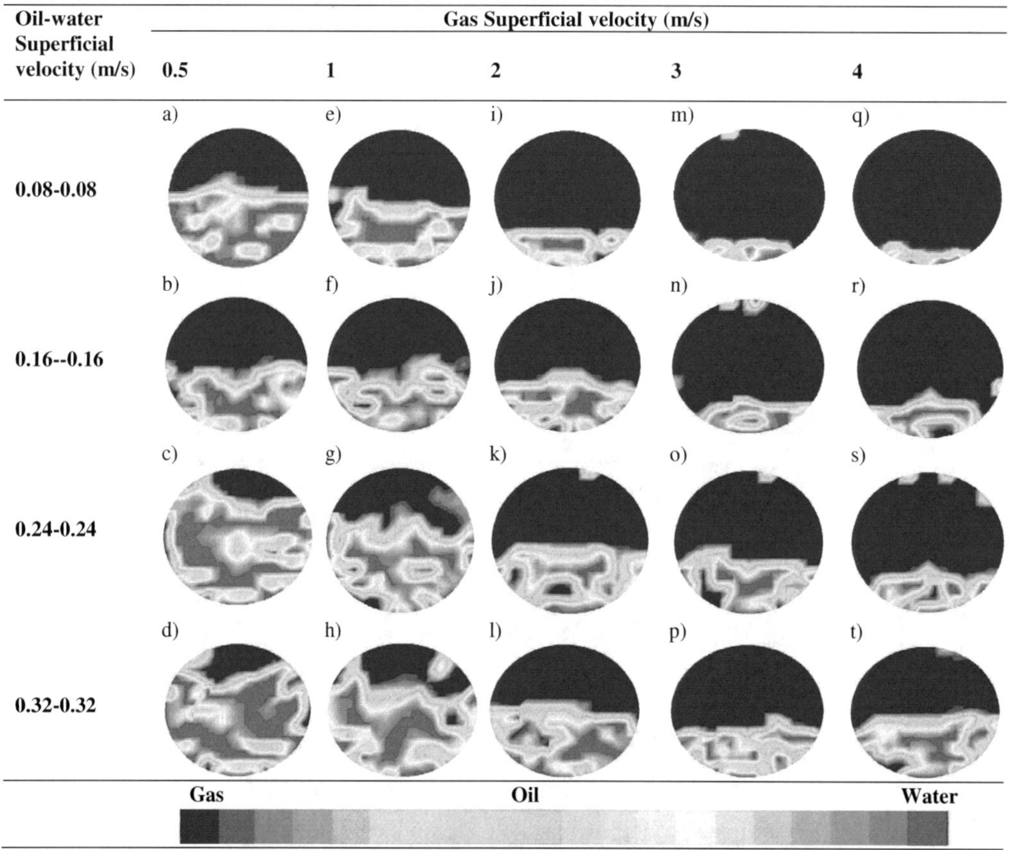

Figure 4. Contours of cross-section of three-phase flow in horizontal pipe

the contours of the cross-section in Figure 4. The three-phase stratified flow can be further subdivided based on the changes occur at the gas-liquid interface at various superficial velocities of fluids. In stratified wavy flow, the gas-liquid interface has the wave-like appearance as shown in Figure 3.

The stratified wavy flow pattern observed at low and moderate superficial velocities of gas and liquids. The pressure exerted by the gas on the liquid surface tends to push the liquids in the forward direction, which resulted in the creation of waves at the gas-liquid interface. The oil-water were flowing in the separate layers during the stratified wavy flow pattern at low superficial velocities of gas and liquids. However, the liquid layer changed to mixture flow with the increase in the superficial velocities of gas and liquids as shown by the cross-section contours in Figure 4. At a high superficial velocity of the gas and low superficial velocities of liquids, the stratified wavy flow pattern changed into the stratified rolling waves. This is due to the increase in pressure of gas at the interface, at which the interfacial waves changed to the rolling waves. Thereby created splashes at the gas-liquid interface. At high gas superficial velocity, more turbulence was induced in the liquid phases, which made a homogeneous mixture as shown in Figure 4 (k), (l), (o), (p) and (s).

3.2. Flow regime map

The flow regime map indicating three-phase flow patterns at various superficial velocities of gas and liquids is shown in Figure 5. At low gas and liquid superficial velocity, the gas-liquid adopted stratified wavy interface. Plug flow has been observed at low gas and high liquid superficial velocity. At a moderate superficial velocity of the gas, increase in the superficial velocity of the total liquid, shifted the stratified wavy flow to smooth stratified flow. Increase in the superficial velocity of the gas at a low superficial velocity of liquid changed the flow from stratified wavy to rolling wave. At high superficial velocities of gas and liquids, the flow pattern was smooth stratified. Separate layers of oil and water were observed at low superficial velocities of gas and liquids. Increase in the superficial velocities of gas and liquid increased the mixing in the liquid phase, due to which a complete homogeneous oil-water mixture was obtained. The three-phase flow regime map of Lee et al. [3] with 0.1 m internal diameter pipe showed that the stratified flow only occurred at low gas and liquid velocities. However, the comparison shows that the slug flow in Lee et al. [3] map is replaced by the stratified flow. The replacement of slug flow by smooth stratified flow is due to the increase in the diameter of the pipe as compared to the pipe used in the work of Lee et al. [3].

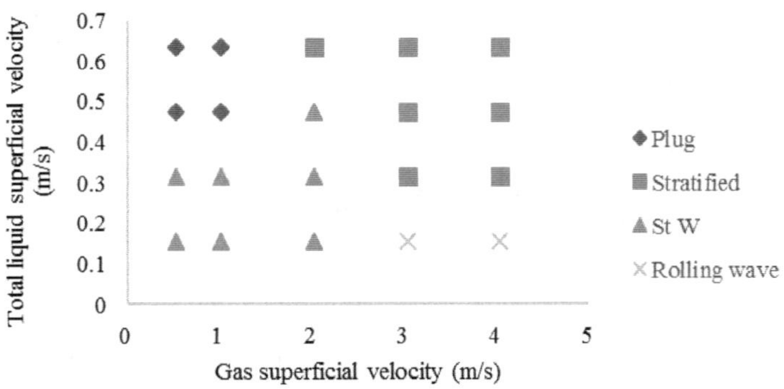

Figure 5. Three-phase horizontal flow regime map

3.3. Pressure drop

According to the literature, the pressure drop increased with the increase in superficial velocities of gas and liquid [15, 16]. The effect of gas on pressure drop was more profound than the liquid phases [17]. However, the trend of the pressure drop with the increase in the gas flow rate was slightly different as shown in Figure 6. The pressure drop increased with the superficial velocities of liquid. However, the pressure drop first increased with superficial velocity of the gas and then started decreasing with further increase in superficial velocity of the gas. This behavior is described by the governing flow pattern at respective superficial velocity of the gas. The pressure drop increased up to the gas superficial velocity of 2 m/s. The flow pattern at this superficial velocity was stratified wavy.

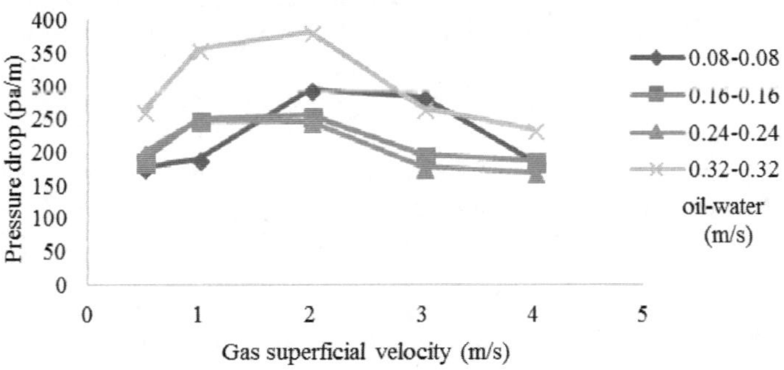

Figure 6. Variation of pressure drop with the gas velocity at different oil-water velocities

The reduction in pressure drop occurred when the type of flow pattern shifted from stratified wavy to smooth stratified flow at high superficial velocities of gas and total liquid. The pressure drop was reduced to almost initial level when the flow pattern is smooth stratified. At high superficial velocities of gas and total liquid, the momentum of the fluids was high enough to minimize the slip between the phases. The reduced slip, and smooth gas-liquid interface in stratified flow resulted in the reduced pressure drop.

4. Conclusions

CFD simulations of three-phase flow through a 0.1524 m diameter pipe were conducted. The superficial velocity ranges of gas-oil-water were 0.5-4, 0.08-0.32 and 0.08-0.32 m/s respectively. The flow patterns at various superficial velocities have been successfully simulated with VOF model. The observed flow patterns were stratified, stratified wavy, stratified rolling wave and plug flow. The flow regime map has been constructed to analyze the flow regime transition with the superficial velocities of fluids. Pressure drop increased with the increase in superficial velocity of liquids. However, pressure drop first increased and then decreased with increase in the gas flow rate. This behavior is because of following flow pattern at the given superficial velocities of fluids. The pressure drop increased due to the plug and stratified wavy flow and then decreased when the flow pattern changed to smooth stratified flow.

Acknowledgment

This work is supported by Chemical Engineering Department of Universiti Teknologi PETRONAS. The financial assistance is provided by YUTP 0153AA-E11.

References

[1] Açikgöz M, Franca F and Lahey R 1992 An experimental study of three-phase flow regimes *Int. J. Multiphase Flow* **18**(3) 327
[2] Woods G et al. 1998 Three-phase oil/water/air vertical flow *Chem. Eng. Res. Des.* **76**(5) 571
[3] Lee A, Sun J and Jepson W 1993 Study of flow regime transitions of oil water-gas mixtures in horizontal pipelines. in *The Third International Offshore and Polar Engineering Conference*. International Society of Offshore and Polar Engineers
[4] Qiang W et al. 2016 Experimental tomographic methods for analysing flow dynamics of gas-oil-water flows in horizontal pipeline *J. Hydrodyn.* Ser. B **28**(6) 1018
[5] Keskin C, Zhang H Q and Sarica C 2007 Identification and Classification of New Three-Phase Gas/Oil/Water Flow Patterns. in *SPE Annual Technical Conference and Exhibition*. Society of Petroleum Engineers.
[6] Spedding P, Donnelly G and Cole J 2005 Three phase oil-water-gas horizontal co-current flow: I. Experimental and regime map *Chem. Eng. Res. Des.* **83**(4) 401
[7] Cimbala J M and Çengel Y A 2014 *Fluid mechanics: fundamentals and applications*.
[8] Ekambara K et al. 2008 CFD simulation of bubbly two-phase flow in horizontal pipes *Chem. Eng. J.* **144**(2) 277
[9] Bhramara P et al. 2009 CFD analysis of two phase flow in a horizontal pipe–prediction of pressure drop *Momentum* **10** 476
[10] Mazumder Q H and Siddique S A 2011 CFD analysis of Two-phase Flow Characteristics in a 90 degree Elbow. *The J. of Comp. Multiphase Flows* **3**(3) 165
[11] Bergman T L and Incropera F P 2011 *Fundamentals of heat and mass transfer* John Wiley & Sons.
[12] Hernandez-Perez V, Abdulkadir M and Azzopardi B 2011 Grid generation issues in the CFD modelling of two-phase flow in a pipe T*he J. of Comp. Multiphase Flows* **3**(1) 13
[13] Fluent A 2013 *ANSYS fluent theory guide* 15.0. Inc, Canonsburg, PA
[14] Launder B E and Spalding D B 1972 *Mathematical models of turbulence* Academic press.

[15] Wang H Q, Wang Y and Zhang L 2011 Characteristics of pressure gradient fluctuation for oil-gas-water three-phase flow based on flow pattern. in *App. Mechs. and Mat.* Trans Tech Publ.
[16] Shmueli A, Unander T and Nydal O 2015 Characteristics of Gas/Water/Viscous Oil in Stratified-Annular Horizontal Pipe Flows. in OTC Brasil *Offshore Technology Conference*
[17] Bannwart A et al. 2009 Experimental investigation on liquid–liquid–gas flow: flow patterns and pressure-gradient *J. of Pet. Sci and Engg* **65**(1) 1

Improved vanadium barrier properties of perfluorinated sulfonic acid membranes for vanadium redox flow battery

S S Sha'rani[1,2], E Abouzari-Lotf[2,3*], M M Nasef[2,4], A Ahmad[2,3], T M Ting[5] and R R Ali[1]

[1]Malaysia-Japan International Institute of Technology, International Campus, Universiti Teknologi Malaysia, Jalan Semarak 54100, Kuala Lumpur, Malaysia.
[2]Advanced Materials Research Group, Center of Hydrogen Energy, Universiti Teknologi Malaysia, Jalan Semarak 54100, Kuala Lumpur, Malaysia.
[3]Department of Chemical Engineering, Univeristi Teknologi Malaysia, 81310 Johor Bahru, Malaysia.
[4]Chemical Engineering Department, Universiti Teknologi Petronas, 32610 Seri Iskandar, Perak, Malaysia.
[5]Nuclear Agency (Nuclear Malaysia) Bangi, 43000 Kajang, Selangor, Malaysia.

*ebrahim@utm.my

Abstract. The search for highly vanadium selective membrane for vanadium redox flow battery (VRFB) is fast growing to accelerate the commercialization of VRFB. Currently, VRFB is challenged by high vanadium permeability through the membranes lowering the battery performance. In this paper, a simple and effective strategy for improving the vanadium barrier permeability of the membranes, without altering the desired proton conductivity, was reported by applying layer(s) of polyelectrolytes. A commercial perfluorinated sulfonic acid membrane, GN115 was modified by layer of positively charged poly(diallyldimethylammonium chloride) (PDDA) or consecutive layers of PDDA and negatively charged poly(sodium styrene sulfonate) (PSS). Upon surface modification, the membranes exhibited a reduced vanadium permeability with a slight decreased in proton conductivity values compared with the pristine GN115 and N117. As a result, higher coulombic efficiency was obtained in VRFB single cell with the modified membranes. This confirms that the addition of the polyelectrolyte layers can suppress the vanadium permeability of the membranes.

1. Introduction

The vanadium redox flow battery (VRFB) is one of the best energy storage techniques that is appealing for promoting of renewable energy. Particularly, VRFB offers various attractive benefits, which meet the grid storage market demands such as long life cycle, deep discharge capacity, low environmental effect and low cost [1]. In VRFB, ion exchange membrane (IEM) plays an imperative role in deciding the battery performance. IEM is a polymer separator in which the positive and negative electrolytes of the cell are separated. IEM also serves as a transport medium for protons to complete the circuit during the passage of the current [2, 3]. To date, most of the IEMs employed in VRFB did not meet most of the requirements to prompt the commercialization of this battery. An ideal membrane for VRFB must have not only low vanadium permeability, high conductivity, small area resistance, excellent chemical and mechanical stability in the vanadium electrolyte medium but also low cost [4, 5].

There are various types of commercial polymeric membranes that were applied in VRFB applications. Such membranes involve a wide range of polymers ranging from perfluorinated sulfonic acids (PFSA)s [6-8] to hydrocarbon-based ones [9]. PFSA membranes such as Nafion from DuPont, have been widely used in VRFB and demonstrated a good performance. However, this membrane was found to be prone to high vanadium ion crossover leading to continuous capacity decay and lower efficiency of VRFB [10]. Thus, various research efforts have been exerted to solve the permeability issue by employing innovative IEMs or modifying the available PFSA membranes [11-13]. Applying polyelectrolyte layer(s) of opposite charge on the surfaces of PFSAs was reported to directly reduce the vanadium permeability without altering desired ion conductivity [14]. This could be due to the Donnan exclusion phenomenon as the positive force in the surface repulse the vanadium ions. Such modifications were applied on various types of membranes including commercial PFSAs. However, the modification was not considered widely since adding more cost to expensive PFSAs. Therefore, applying of polyelectrolyte layers on low-cost PFSAs could be considered as a commercially viable alternative for VRFB application.

In this work, a strategy for improving the performance of VRFB by applying layer(s) of polyelectrolytes on new alternative cheaper commercial PFSA membrane named as GN115 is proposed. The GN115 membrane was modified by applying layer of polycation or two layers of positively charged poly(diallyldimethylammonium chloride) (PDDA) and negatively charged poly(sodium styrene sulfonate) (PSS) polyelectrolytes consecutively. The vanadium permeability and proton conductivity of the modified membranes were evaluated in comparison with Nafion 117 membrane. The performance of the modified membranes was preliminarily tested in a VRFB single cell.

2. Methodology

2.1. Preparation of surface modified membranes

Two different surface modified composite membranes were prepared in this study. The PFSA modified with polycation was prepared by immersing the GN115 in the poly(diallyldimethylammonium chloride) (PDDA, 35 wt. % in H_2O, Mw <100,000) for 10 minutes at room temperature. To remove unbounded PDDAs, surface modified membranes with polycation (GN115-(PDDA)) was rinsed thoroughly in a deionized water and dried in a vacuum oven. Second type of surface modified membrane was prepared by applying polyanion layer onto surface of GN115-(PDDA). Poly(sodium styrene sulfonate) (PSS, 25 wt. % in H_2O, Mw ~ 1,00,000) was added by immersing of the GN115-(PDDA) for 10 minutes at room temperature followed by washing with deionized water and drying. The developed membrane was denoted as GN115-(PDDA/PSS).

2.2 Characterization and analysis

To investigate the existence of the PDDA and PSS on the composite membranes, the infrared (IR) absorption spectra were recorded on a FTIR spectrophotometer (Perkin Elmer, USA) in the region of 600 - 4000 cm^{-1}.

To evaluate the effectiveness of the modified membranes, proton conductivities and the vanadium permeability of the membranes were measured. The through-plane ($\sigma_{|}$) and in-plane ($\sigma_{||}$) conductivity of the membranes was measured by using a homemade cell with two disc stainless steel electrodes that were polished frequently and four-point probe of Bekk Tech conductivity cells (BT-112), respectively. A DC conductivity tester (Kickstart SMU 2400) was used to measure the conductivity of the membranes. The conductivity (σ) of the membrane was calculated using equation (1) [9]:

$$\sigma(S/cm) = \frac{L\ (cm)}{R(\Omega)xS(cm^2)} \quad (1)$$

where, R is the resistance obtained from the slope of voltage versus current plot, L is the distance between the two inner probes and S is the area of the membranes. The denominator of the equation represents the area of resistance, *AR*. Anisotropic conductivity ratio was calculated from in-plane and through-plane conductivity values following equation (2) [9]:

$$\sigma_{\parallel}/_{\mid} = \frac{\sigma_{\parallel}}{\sigma_{\mid}} \quad (2)$$

The vanadium permeability of the membranes was measured using the same method described in ref [8]. The membrane with an effective area of 1.72 cm^2 was clamped between the two half-diffusion cell (Perme Gear, Inc.) having two reservoirs. The left reservoir of was filled with 1.0 M $VOSO_4$ + 2.5 M H_2SO_4 and the right one was filled with 1.0 M Na_2SO_4 + 2.5 M H_2SO_4. The samples were collected from the right reservoir at pre-determined time intervals. The concentration of V(IV), $C_{V(IV)}$ was measured using UV-Vis spectrophotometer (UV-1800 Shimadzu) and was calculated using equation (3) [9]:

$$C_{V(IV)} = \frac{SP}{V_R T} C_L(t - t_0) \quad (3)$$

where, V_R is the volume of the reservoir, C_L is the concentration of vanadium ion in the left reservoir of the cell, *P* is the permeability, *T* is the thickness of the membrane, *S* is the effective area of the membrane and *t* is the diffusion time.

A single cell test was carried out to compare the performance of the surface modified membranes with N117. Carbon felt was activated prior to setup the single cell system as described elsewhere [15]. The VRFB single cell was assembled by clamping the membrane with two carbon felts (C200 AvCarb) (3 cm x 3 cm), two PVC plates and two rubber sheets to provide the necessary pressure. A solution of 1.5 M V^{3+}/V^{4+} in 3.0 M H_2SO_4 was used as electrolytes feeding the positive and negative tanks of the cell at 10 mL min^{-1} (Thermo Scientific FH100). The experiment was carried with a battery analyser (BST8-CDS, MTI Corp.) at room temperature. A schematic view of the VRFB single cell is shown in Figure 1.

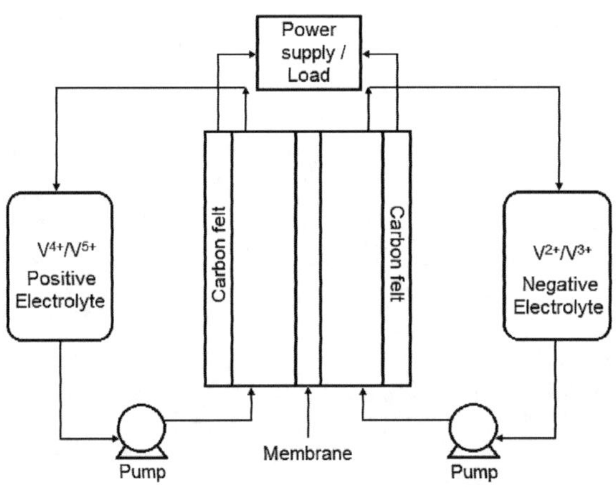

Figure 1. The schematic representation for the VRFB single cell

The coulombic efficiency (CE), voltage efficiency (VE), energy efficiency (EE) for each complete cycles were determined from equations 4, 5 and 6, respectively [5].

$$CE = \frac{I_{constant} t_{discharge}}{I_{constant} t_{charge}} \; x\, 100\% \quad (4)$$

$$VE = \frac{V_{discharge}}{V_{charge}} \; x\, 100\% \quad (5)$$

$$EE = CE \; x \; VE \quad (6)$$

3. Results and discussion

Two types of PFSA membranes were obtained by surface modification with monolayer of PDDA and two consecutive layers of polycation and polyanion following the scheme presented in Figure 2. The modified membranes are denoted as GN115-PDDA and GN115-PDDA/PSS, respectively.

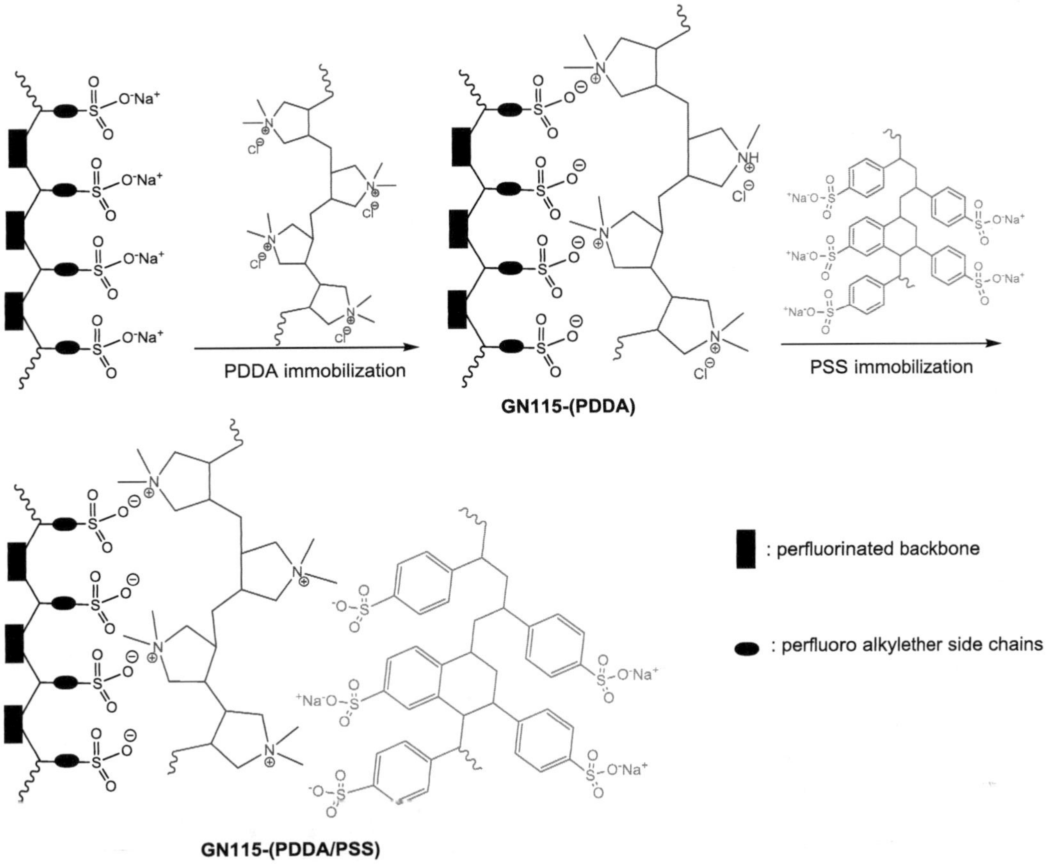

Figure 2. Pathway for the preparation of surface modified membranes

FTIR spectra of the developed membranes together with pristine GN115, PDDA and PSS are shown in Figure 3. A typical PFSA membrane, the C-F bonds stretching can usually be seen at around 1000 to 1360 cm^{-1}. Meanwhile, the peaks that correspond to C-O-C and S-O groups can be observed at 976 cm^{-1} and 1060 cm^{-1}, respectively [16-18]. The peak of C-H bond in the case of PFSA type membrane could hardly be seen since the structure is usually fluorinated [19]. Upon modification, a visible difference can be obviously seen in the modified membranes. The PDDA show some peaks at 2938 and 2865 cm^{-1} correspond to the stretching vibrations of methyl groups [20]. Meanwhile, PSS spectra shows some stretching vibrations of –CH and –CH_2 groups at 2925 and 2853 cm^{-1} respectively [21]. The corresponding peaks of $^{+}NR_4$ (1644 cm^{-1}) and C=C (1637 cm^{-1}) is clearly visible in the zoomed region [21, 22]. Therefore, it was confirmed that the PDDA and PSS bilayers were added on the surface of the GN115 membrane upon modification.

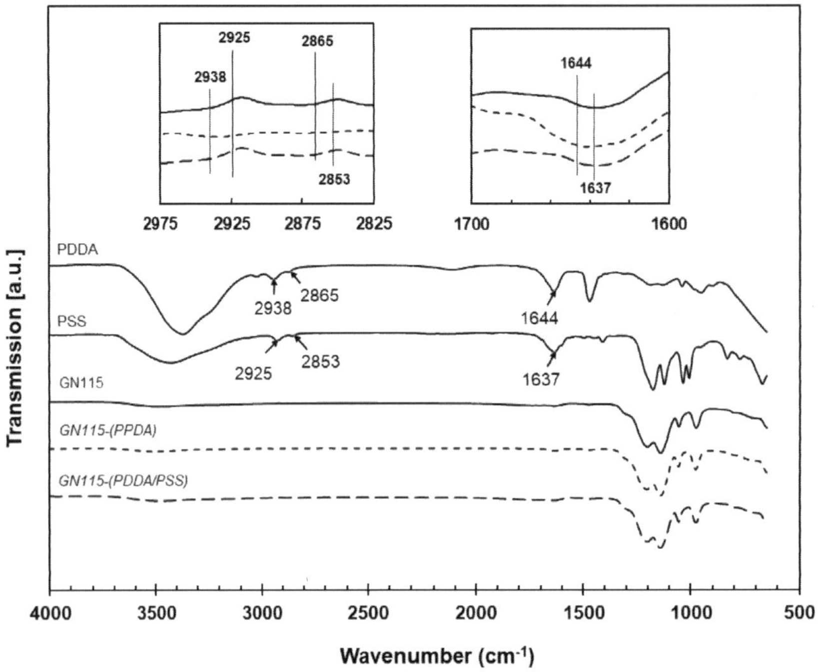

Figure 3. FTIR spectra of GN115, PDDA, PSS and resulting modified membranes.

The permeability of the GN115 membranes were measured before and after modification with polyions. It was found that the membranes modified with PDDA and PDDA/PSS of about 23.5x10^{-7} cm^{2} min^{-1} layers exhibited a lower permeability compared to that of the pristine membrane (26.05x10^{-7} cm^{2} min^{-1}). This is because, the additional PDDA, polycation create a repulsive charge that hindered the vanadium crossover. Such crossover reduction can be also attributed to the more complicated channels that were created by the surface modification with polycation and polyelectrolyte [23].

The proton conductivity and area resistance of the membranes were measured before and after modification and the obtained data is presented in Table 1. As can be seen the proton conductivity of the membranes decreased slightly for about 8 to 8 % with a little increase in the area resistance upon modification with both PDDA and PDDA/PSS. Nonetheless, the values of the area resistance of these modified membranes are still smaller than the reference Nafion117 membrane. Therefore, it is expected that the modified membranes remain suitable for making better performance in VRFB giving a higher voltage efficiency. All membranes showed anisotropic behaviour since different values for through-plane and in-plane conductivity were obtained.

Table 1. Proton conductivies and area resistance of the membranes before and after modification

Membrane	Thickness, µm	Proton conductivity, σ (mS/cm)		Area Resistance[1](Ω cm^{2})
		Through-plane ($\sigma_{\perp}$)	In-plane ($\sigma_{\parallel}$)	
GN115	115	96.1	96.0	0.12
GN115-(PDDA)	116	90.4	92.2	0.13
GN115-(PDDA/PSS)	119	89.4	90.6	0.13
Nafion 117	177	88.0	82.6	0.20

[1] Based on through-plane conductivity

The electrochemical properties of the newly developed membranes were further evaluated by charge-discharge test in the VRFB cell as shown in Figure 4. The experiment was carried out at 40 mA cm^{-2} current density for 10 cycles under same operating conditions. It was observed that the modified membranes, GN115-PDDA and GN115-PDDA/PSS have higher CE compared to N117 and pristine GN115 membranes. This value reflects the vanadium permeability pattern emphasizing that the permeability decreased upon modification. The VE of GN115 membrane was observed to be higher than Nafion117 membrane since the area resistance of the membrane is 40% higher than that of GN115 membrane. For the modified GN115 membranes, the value of VE is quite close to pristine GN115 and this because there was no apparent difference in the area resistance. The EE of the battery was determined by multiplying the CE and VE and the obtained data showed that the EE of the pristine and modified GN115 membranes at 40 mA cm^{-2} is higher than the reference Nafion 117 membrane. This remarkable improvement shown by the modified membranes confirm the suppression of vanadium crossover in the cell.

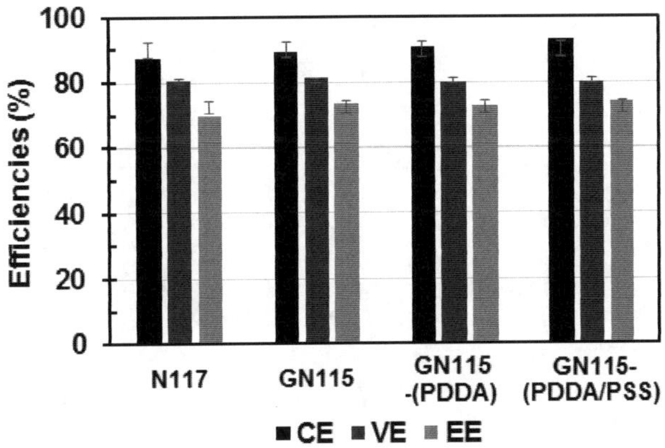

Figure 4. Efficiencies of the N117, GN115, GN115-PDDA, and GN-PDDA/PSS membranes at 40 mA cm^{-2}.

Figure 5 displays the cycling curves and the change of CE, VE and EE of GN115-PDDA and GN115-PDDA/PSS membranes. The preliminary performance of the modified membranes showed a stable performance at 40 mA cm^{-2} as the cycling curves were almost consistent for 10 cycles. This is due to the extra repulsive forces created between the positively charged PDDA and vanadium ion [23]. Hence, it can be concluded that the incorporated polyelectrolyte layers of the modified membranes are stable in VRFB. Comparing the performance of the modified membranes, the GN115-PDDA/PSS revealed a higher CE with lower VE than the pristine GN115 and N117. This indicates that the vanadium crossover can be reduced by the additional polyanion on the membrane surface.

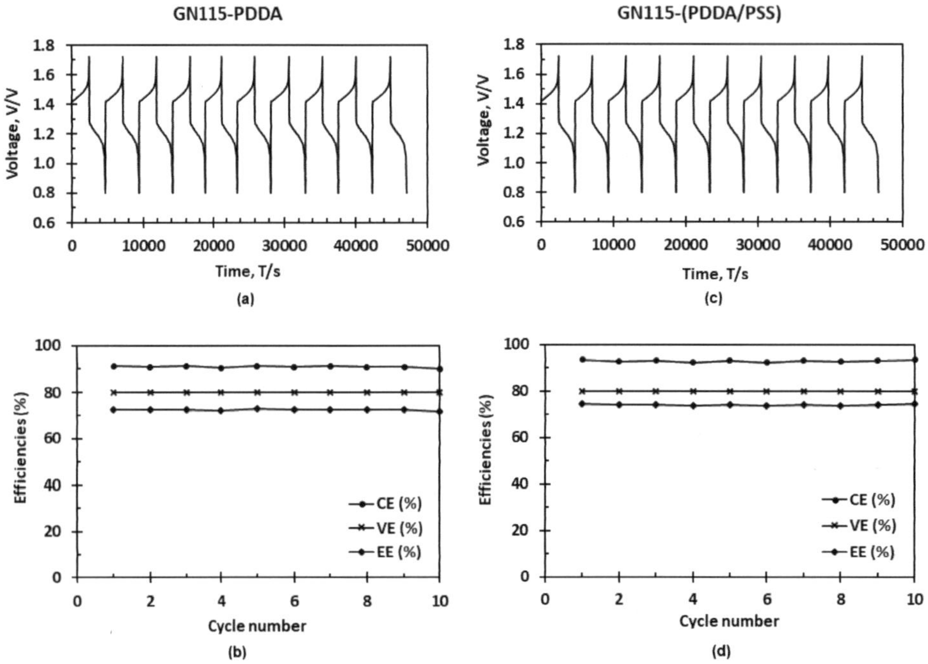

Figure 5. (a) Cycling curves and (b) efficiencies of GN115-(PDDA), (c) cycling curves and (d) efficiencies of GN115-(PDDA/PSS)

4. Conclusion

Modification of GN115 membrane was successfully carried out using PDDA polycation and PSS polyanion to enhance the barrier properties against vanadium crossover in VRFB. The monolayer and bilayer modified membranes showed lower vanadium permeability reasonable area resistance compared to Nafion 117 membranes. The cycling performance test showed that better efficiencies with PDDA and PDDA/PSS modified membranes compared to N117 and the pristine GN115. The negatively charged PSS further reduce the vanadium crossover through the membrane by Donnan effect. The modified membranes also exhibit a good stability in VRFB environment. It can be concluded that the adopted strategy was effective in improving the resistance to vanadium crossover of the commercial PFS GN115 membrane.

Acknowledgment

The research was primarily supported by Universiti Teknologi Malaysia and ministry of higher education under Fundamental Grants (vote no. #4F878) and joint partnership research program of Innovative Research Universities of Australia and Malaysia Research University Network (IRU-MRUN) (vote no. #00M84). The authors wish to thank General Energy Co., Ltd (China) for providing the samples.

References

[1] Minke C, and Turek T 2018 Materials, system designs and modelling approaches in techno-economic assessment of all-vanadium redox flow batteries – A review *J. Power Sources* **376** 66

[2] Luo X, Lu Z, Xi J, Wu Z, Zhu W, and Chen L 2005 Influences of permeation of vanadium ions through PVDF-g-PSSA membranes on performances of vanadium redox flow batteries *J. Phys. Chem. B* **109** 20310

[3] Sun C X, Chen J, Zhang H M, Han X and Luo Q T 2010 Investigations on transfer of water and vanadium ions across Nafion membrane in an operating vanadium redox flow battery *J. Power Sources* **195** 890

[4] Li X, Zhang H, Mai Z, Zhang H and Vankelecom I 2011 Ion exchange membranes for vanadium redox flow battery (VRB) applications *Energy & Env. Science.* **4** 1147
[5] Prifti H, Parasuraman A, Winardi S, Lim TM and Skyllas-Kazacos M 2012 Membranes for redox flow battery applications *Membranes.* **2** 275
[6] Zhang L, Ling L, Xiao M, Han D, Wang S and Meng Y 2017 Effectively suppressing vanadium permeation in vanadium redox flow battery application with modified Nafion membrane with nacre-like nanoarchitectures *J. Power Sources* **352** 111
[7] Cao L, Kronander A, Tang A, Wang D and Skyllas-Kazacos M 2016 Membrane permeability rates of vanadium ions and their effects on temperature variation in vanadium redox batteries *Energies* **9**
[8] Pezeshki A, Tang Z, Fujimoto C, Sun C, Mench M and Zawodzinski T 2015 Full cell study of diels alder poly(phenylene) anion and cation exchange membranes in vanadium redox flow batteries *J. Electrochem Society* **163** A5154
[9] Sha'rani S, Abouzari Lotf E, Ahmad A, Wan Ibrahim W, El-sayed Nasef M and Rasit Ali R 2016 Stability and performance evaluation of ion-exchange membranes for vanadium redox flow battery *Jurnal Teknologi* **78** 7
[10] Hoi S, Kim T and Cha S 2017 Patents on Membranes Based on Non-Fluorinated Polymers for Vanadium Redox Flow Batteries *Recent Patents on Nanotechnology* **11** 123-129
[11] Huang S, Yu H and Lin Y 2017 Modification of Nafion® Membrane via a Sol-Gel Route for Vanadium Redox Flow Energy Storage Battery Applications *J. Chem.* **2017** 1
[12] Lin C, Yang M and Wei H 2015 Amino-silica modified Nafion membrane for vanadium redox flow battery *J. Power Sources* **282** 562
[13] Luo Q, Zhang H, Chen J, Qian P and Zhai Y 2008 Modification of Nafion membrane using interfacial polymerization for vanadium redox flow battery applications *J. Memb. Sci.* **311** 98
[14] Teng X, Yu C, Wu X, Dong Y, Gao P and Hu H, 2017 PTFE/SPEEK/PDDA/PSS composite membrane for vanadium redox flow battery application *J. Mater. Science* **53** 5204
[15] Etesami M, Abouzari-Lotf E, Ripin A, Mahmoud Nasef M., Ting T, Saharkhiz A and Ahmad A 2018 Phosphonated graphene oxide with high electrocatalytic performance for vanadium redox flow battery *Int. J. Hydrog. Energy* **43** 189
[16] Feng K, Hou L, Tang B and Wu P 2015 Does thermal treatment merely make a H2O-saturated Nafion membrane lose its absorbed water at high temperature? *Phys. Chem. Chem. Phys* **17** 9106
[17] Jaumann E, Steinwand S, Klenik S, Plackmeyer J, Bats J, Wachtveitl J and Prisner T 2017 A combined optical and EPR spectroscopy study: azobenzene-based biradicals as reversible molecular photoswitches. *Phys. Chem. Chem. Phys.* **19** 17263
[18] Liang Z, Chen W, Liu J, Wang S, Zhou Z, Li W, Sun G and Xin Q 2004 FT-IR study of the microstructure of Nafion® membrane *J. Membr. Sci.* **233** 39
[19] Grosse A J, Nunes K C, Komsiyska L and Wittstock G 2016 Layer-by-layer modification of Nafion membranes for increased life-time and efficiency of vanadium/air redox flow batteries *J. Membr. Sci.* **510** 259
[20] Wang Y, Wang S, Xiao M, Han D, Hickner M and Meng Y 2013 Layer-by-layer self-assembly of PDDA/PSS-SPFEK composite membrane with low vanadium permeability for vanadium redox flow battery *RSC Advances* **3** 15467
[21] Mulijani S, Dahlan K and Wulanawati A 2014 Sulfonated Polystyrene Copolymer: Synthesis, Characterization and Its Application of Membrane for Direct Methanol Fuel Cell (DMFC) *IJMME* 36
[22] Cho E and Won J 2016 Novel composite membrane coated with a poly(diallyldimethylammonium chloride)/urushi semi-interpenetrating polymer network for non-aqueous redox flow battery application *J. Power Sources* **335** 12
[23] Xi J, Wu Z, Teng X, Zhao Y, Chen L and Qiu X 2008 Self-assembled polyelectrolyte multilayer modified Nafion membrane with suppressed vanadium ion crossover for vanadium redox flow batteries. *J. Mater. Chem.* **18** 1232

Synthesis and characterization of imidazolium ILs based chitosan–tripolyphosphate microparticles

Rizwan Safdar[1], Abdul Aziz Omar[1], A. Arunagiri[2], Murugesan Thanabalan[1]*,

[1]Chemical Engineering Department, Universiti Teknologi PETRONAS, Bandar Seri Iskandar 32610, Perak Darul Ridzuan Malaysia.
[2]Chemical Engineering Department, National Institute of Technology, Tiruchirappalli –620 015 Tamil Nadu, India.

*murugesan@utp.edu.my

Abstract. In this work, two imidazolium based ILs, 1–butyl–3– methylimidazolium hydrogen sulfate [bmim][HSO_4] and 1–butyl–3– methylimidazolium acetate [bmim][Ac] were blended with Chitosan (CS) to synthesize the CS–IL–TPP microparticles. The prepared microparticles were characterized by Zetasizer, FTIR, TGA, SEM and elemental analysis (EDX). The addition of these ILs resulted in controlled sized microparticles which possessed good thermal stability. The elemental analysis indicated that the better crosslinking of tripolyphosphate (TPP) with CS was achieved in the presence of IL, [bmim][Ac] as compared with [bmim][HSO_4] and CS–TPP formulation alone. The better crosslinking might result in the formation of stable and compact microparticles which could have a better control over the drug release applications. Therefore, this kind of IL might be suitable for improving the characteristics of chitosan as a drug carrier which could ultimately provide controlled release of unstable drugs.

Keywords—Chitosan, Ionic Liquid, microparticles, Drug delivery

1. Introduction

Chitosan (CS) is water–immiscible polysaccharide which can be obtained by the deacetylation of Chitin at 100–160 °C by using 40 to 50 % aqueous alkaline solution. This polysaccharide undergoes protonation when dissolved in weak acids. Acetic acid is one of the weak acids which is mostly used for its solubilization and it possess positive charge which is considered as one of its important properties. The presence of free amino groups in CS plays an important role in cross–linking with other anions enabling it suitable for various applications such as food and textile industry, wastewater treatment and pharmaceutical applications. Chitosan (CS) has been used with a number of polymers/materials to improve the characteristics of drugs [1, 2]. There are several factors which can affect the release profiles of drugs such as nature of polymer/drug, the method of preparation, particle size and stability of particles etc are the important factors which affect the release profiles of drugs [3–5]. Despite the great potential of CS, it is less stable in acidic pH as well as high temperature medium. This deficiency of CS creates a barrier for its commercialization in drug formulation [6]. Therefore, investigations are required which may improve the characteristics of CS. In this regard, one of the ways is to optimize and improve the synthesis process of CS microparticles (MPs). Finding the best conditions in terms of Molecular weight (MW) Concentration of CS, initial pH of CS solution, speed and time of mixing, binder type/concentration, the mass ratio of CS–binder may result in enhanced stability of particles. The second way is to find and investigate the new materials which may enhance the stability of these particles.

Ionic liquids have emerged as alternative solvents for various applications. Their unique properties can tailor the shortcomings of various drugs. Though the purity, toxicity and unknown biodegradability of several ILs is a big issue for their use in pharmaceutical applications, on the other side, the toxicity of ILs does not obstruct their use for pharmaceutical applications. Fortunately, some reports have indicated the possibility of synthesizing non–toxic ILs. There are several FDA approved counterions such as anionic salts (bromide, chloride, phosphate, mesylate, citrate, titrate and acetate and cationic salts (sodium, ammonium, N–methylglucamine, diethylamine, ethylenediamine) [7]. The toxicity study of 1–butyl–3–methylimidazolium chloride [bmim][Cl] showed that this IL did not possess the acute toxic effects (EC50≈13μM) on marine algae and enzymes such as Acetylcholinesterase. Some other studies also supported some of the imidazolium ILs in terms of low/negligible toxicity towards Caco–2 cells. The specified test on rats (LD50) also indicated the low toxicity (LD50=1300 mg/kg) of imidazolium ILs [8].

Dorota et al (2013) prepared the oil–water emulsion by using hydrophilic IL, 1–hexyl–3–methylimidazolium chloride [hmim][Cl] and water–oil emulsion by using hydrophobic IL, 1–butyl–3–methylimidazolium hexafluorophosphate [bmim][PF_6]. The use of these ILs resulted in a stable emulsion with low cytotoxicity and good microbial activity showing the potential of these ILs to be used as preservatives [9]. The use of these ILs resulted in a stable emulsion with low cytotoxicity and good microbial activity showing the potential of these ILs to be used as preservatives. Moniruzzaman et al (2010) investigated the imidazolium based ILs for micro–emulsion formation in drug delivery and reported that IL–oil micro–emulsion increased the topical and transdermal drug delivery six times higher than normal formulation. Extension to this work resulted in the formation of stable IL–micelle formation with stronger shelf life as compared with typical drug formulation [10–12]. Sometimes these solvents can be a good alternative to enhance the solubility, absorption and bioavailability of several drugs through oral route as compared with volatile, flammable and toxic polar solvents such as pyridine, N, N–Dimethylformamide (DMF) and Dimethyl sulfoxide (DMSO) [8]. Therefore, ILs in combination with CS could be a good approach to alter the release characteristics of various drugs. In the present work imidazolium based CS–IL–TPP MPs have been synthesized and their characterization is conducted.

2. Methodology

2.1. Materials

Chitosan (CS) with high purity and HMW (kDa) (CS) was purchased from Sigma Aldrich, Sodium tripolyphosphate (TPP) with purity ≥ 94.00 % from SRL Chemicals, Hydrochloric acid (HCL) with purity ≥ 37 % and Sodium hydroxide (NaOH) with purity ≥ 98.00 % were purchased from the Merk. Acidic acid was purchased from R and M Chemicals. The Ionic Liquids were purchased from the Merk. The stock solutions of TPP, HCL and NaOH were prepared by using distilled water, while the CS stock solution was prepared in 1.0 % acetic acid.

2.2. Synthesis of CS– IL–TPP Microparticles

The CS–IL microparticles (MPs) were prepared by using ionic gelation method by using TPP as a crosslinker [13]. In detail, the CS stock solution was prepared by dissolving CS in acetic acid (1 % w/v). Further, this solution was diluted with distilled water to make a concentration of (0.05 % w/v). The ILs, [bmim][HSO_4] (0.05 % w/v) and [bmim][Ac] (0.05 % w/v) were added to this solution and stirred for 4 hours to obtain the homogeneous mixtures. The pH of both CS–[bmim][HSO_4] and CS–[bmim][Ac] solutions were maintained at pH=4.80 by using 1.0 M NaOH solution. Afterward, the TPP (0.05 % w/v) solution was prepared in distilled water. The concentration of CS (0.05 % w/v), initial pH of CS (pH=4.80), concentration of TPP (0.05 % w/v), initial pH of TPP (pH=9.30), CS–TPP mass ratio (5:1) and operating parameters such as rpm (700) and stirring time (60 min) were kept constants.

The CS–IL–TPP MPs were prepared by adding the CS–IL solutions into the TPP solution at room temperature. After preparation, these MPs were collected by centrifugation at rpm 4000 and freeze dried.

2.3. Characterization of CS–IL–TPP Microparticles
The properties such as particle size, zeta potential and surface morphology have significance importance in controlling the release profile of various drugs therefore, it is necessary to characterize the modified MPs. This could provide a better understanding of modified MPs before they could use for drug release application. The freeze–dried MPs were dispersed in the deionized water, sonicated for 30 minutes at room temperature and particle size and zeta potential was estimated by using the Malvern Zetasizer Nano ZS.

FTIR analysis was conducted to study the structure of synthesized CS–IL–TPP MPs. This analysis was carried out by using SHIMADZU FTIR–8400S in the range of 4000–400 cm^{-1} for synthesized CS–IL–TPP MPs as well as for pure CS and TPP. Thermal degradation of the prepared MPs was studied by using a software based and autosampler Thermogravimetric analyzer (Perkin Elmer TGA 4000). For this purpose, a temperature range of 20–40010 °C was selected and % mass loss was recorded versus heat rate (10 °C/min). Scanning Electron Microscopy (SEM ZEISS) was used to study the morphology of the prepared MPs. Further, elemental analysis (EDX) was carried out to find the presence of elements after modification.

3. Results and Discussion
The particle size and zeta potential of CS–[bmim][HSO_4]–TPP MPs were found in the range of 800 ± 90 nm and 15.0 ± 0.50 mV respectively. On the other hand, the particle size and zeta potential of CS–[bmim][Ac]–TPP MPs were found in the range of 900 ±105 nm and 4.98 ± 0.70 mV. In both cases, the NPs were also produced but the average sizes of particles are reported. The particle size of later MPs is bigger than the former, while the zeta potential value is less. In comparison to zeta potential of CS (0.05 % w/v), the zeta potential of prepared MPs is shifted to low positive values giving a confirmation of the modification of these particles.

FTIR analysis for the pure CS, TPP, [bmim][HSO_4] and [bmim][Ac] based MPs was carried out in the range of 4000–400 cm^{-1}. In case of pure CS, the significant peaks were found in the regions 3600–3000 cm^{-1}, 1700–1500 cm^{-1} and 1190–850 cm^{-1}. The broader peak at wavelength 3370 cm^{-1} was due to the O–H stretching caused by the hydroxyl group. The small peak at wavelength 2871 cm^{-1} is due to the result of C–H vibration, whereas the peaks at the 1019 cm^{-1} indicate the C–O stretching. Similarly, another small peak was found at 1665 cm^{-1}. The IR spectra of the TPP showed two significant peaks, one at 890 cm^{-1} and second at 1140 cm^{-1} representing the P–O–P and P–O stretching respectively. The IR spectra of CS solution (0.05 % w/v) was taken as a reference to see the effects of ILs on the structural changes of MPs. The main peaks in this spectrum were found at 3440cm^{-1}, 1640 cm^{-1} and 1084 cm^{-1}. In case of CS–[bmim][HSO_4]–TPP and CS–[bmim][Ac]–TPP MPs, the major peaks were almost at the same wavelength as that of CS solution. However, some shifting was found in the fingerprint regions which can be seen in Figure 1. In case of CS–[bmim][HSO_4]–TPP MPs, the peaks in fingerprint region shifted from reference CS solution (1084 cm^{-1}) to 1087 cm^{-1} as well as from 893 cm^{-1} to 888 cm^{-1} respectively. In case of [bmim][Ac]–TPP MPs, these peaks from were shifted 1084 cm^{-1} to 1077 cm^{-1} and from 893 cm^{-1} to 897 cm^{-1} respectively.

The morphology of the CS–TPP MPs, CS–[bmim][HSO_4]–TPP MPs and CS–[bmim][Ac]–TPP MPs were studied by the SEM analysis. The images of the prepared MPs are given in Figure 2. Freeze drying is usually used for the long–term stability of particles, but it affects the shape of particles. Similar effects were seen in present work as prepared MPs somehow possessed uneven shapes after freeze drying.

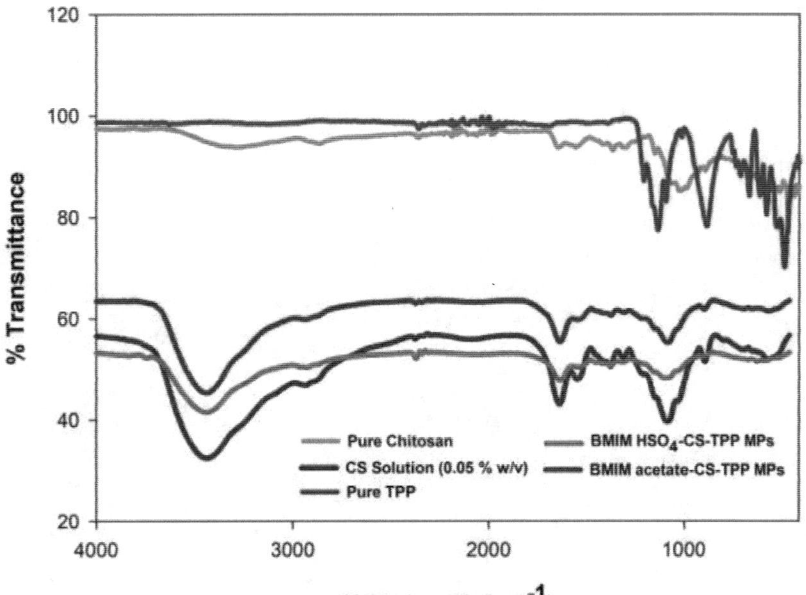

Figure 1. FTIR spectra of pure TPP, CS, aqueous CS and IL–CSS–TPP MPs

These results can be supported by the findings of Leoner et al (2008) [14], who investigated the effects of freeze drying on particle shape. Overall, the CS–TPP MPs, CS–[bmim][HSO_4]–TPP MPs and CS–[bmim][Ac]–TPP MPs had a similar structure with a fact that CS–[bmim][HSO_4]–TPP MPs possessed more agglomeration than other combinations. The reason could be low positive zeta potential which resulted in less stable particles. Overall, SEM analysis somehow showed the bigger sizes as compared with Zetasizer. As mentioned earlier, the reason could be the agglomeration of particles which formed clusters beside each other and at such magnification (100 X) appeared as bigger particles. While the analysis by using the Zetasizer showed reasonable particle size, the possible reason in this case could be the sonication technique which broke down the agglomerated particles into their actual sizes.

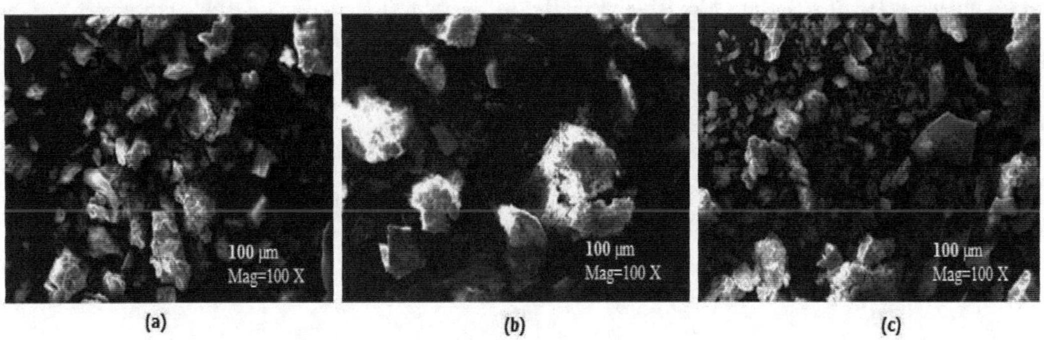

Figure 2. SEM Images for: (a), CS–TPP MPs; (b), CS–[bmim][HSO_4]–TPP MPs and (c), CS–[bmim][Ac]–TPP MPs

The cloud images and elemental analysis (EDX) are presented in Figure 3. The percentage of elements Carbon (C), Oxygen (O), Nitrogen (N) and Phosphorous (P) was found in the order of CS–

IOP Conf. Series: Materials Science and Engineering **458** (2018) 012080 doi:10.1088/1757-899X/458/1/012080

[bmim][Ac]–TPP MPs > CS–[bmim][HSO_4]–TPP MPs > CS–TPP MPs, CS–[bmim][HSO_4]–TPP MPs > CS–TPP MPs > CS–[bmim][Ac]–TPP MPs, CS–TPP MPs=CS–[bmim][HSO_4]–TPP MPs > CS–[bmim][Ac]–TPP MPs and CS–[bmim][Ac]–TPP MPs > CS–TPP MPs > CS–[bmim][HSO_4]–TPP MPs respectively.

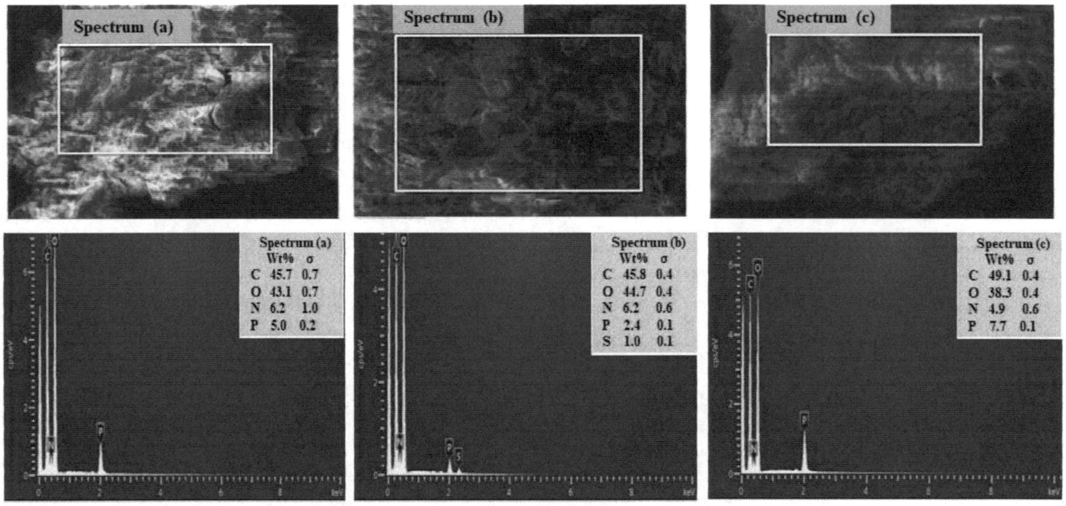

Figure 3. Electron Images and elemental analysis for: (a), CS–TPP MPs; (b), CS–[bmim][HSO_4]–TPP MPs and (c), CS–[bmim][Ac]–TPP MPs

The considerable sulfur (S) contents are only found in CS–[bmim][HSO_4]–TPP MPs. From the results it was found that the phosphorous contents are higher in the microparticles prepared by the [bmim][Ac] as compared with other studied combinations. This means that better crosslinking of Chitosan and TPP was achieved by using this IL, ultimately resulting in compact particles. The compact and mechanically stable particles are essential to control the release mechanism of drugs.

The TGA results (Figure 4) indicated that the CS–TPP MPs are more stable than pure CS, CS–[bmim][HSO_4]–TPP MPs and CS–[bmim][Ac]–TPP MPs. The % mass loss of these MPs within temperature range 40–215 °C was found in the order of CS–[bmim][HSO_4]–TPP MPs > CS–[bmim][Ac]–TPP MPs > pure CS > CS–TPP MPs. For the temperature range 216–260 °C, % mass loss was found in order of CS–[bmim][Ac]–TPP MPs > CS–[bmim][HSO_4]–TPP MPs > pure CS > CS–TPP MPs. Finally, for the temperature range 261–400 °C, the % mass loss was found in the order of CS–[bmim][HSO_4]–TPP MPs > CS–[bmim][Ac]–TPP MPs > pure CS > CS–TPP MPs. Overall, all the microparticles exhibited good thermal stability which supports their use as carrier materials for in vitro drug release applications.

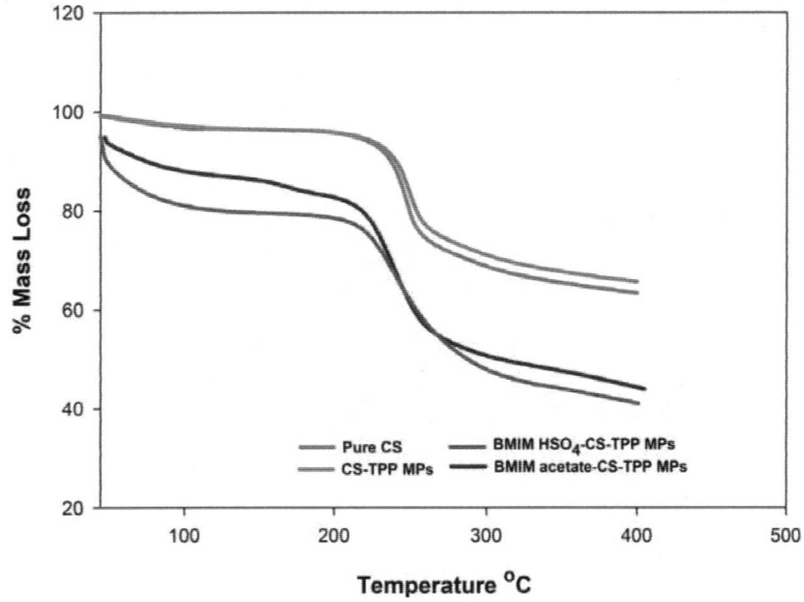

Figure 4. TGA of pure CS, CS–TPP MPs, CS–[bmim][HSO_4]–TPP MPs and CS–[bmim][Ac]–TPP MPs

4. Conclusion

In this study two imidazolium based ILs were used for the preparation of IL–CS–TPP microparticles. The prepared MPs were characterized in terms of size, zeta potential, morphology and thermal stability. The results concluded that the prepared MPs possess good thermal stability. Use of IL, [bmim][Ac] resulted in more stable and compact CS–TPP MPs which could be useful in altering the shortcomings of drugs and better release profiles can be obtained. Further research in this area can bring improvement in drug formulations.

Acknowledgment

Authors would like to thank YUTP project (0153AA-H 44) for providing the financial support and also like to thank Universiti Teknologi PETRONAS for providing Graduate Assistance for Mr Rizwan Safdar.

References

[1] Zheng C H, Gao J Q, Zhang Y P and Liang W Q 2004 A protein delivery system: Biodegradable alginate-chitosan-poly(lactic-co- glycolic acid) composite microspheres *Biochem. Biophys. Res. Commun.* **323** 1321

[2] Chakravarthi S S and Robinson D H 2011 Enhanced cellular association of paclitaxel delivered in chitosan-PLGA particles *Int. J. Pharm.* **409** 111

[3] Agnihotri S A, Mallikarjuna N N and Aminabhavi T M 2004 Recent advances on chitosan-based micro- and nanoparticles in drug delivery *J. Control. Release* **100** 5

[4] Thanoo B C, Sunny M C and Jayakrishnan A 1992 Cross- linked Chitosan Microspheres: Preparation and Evaluation as a Matrix for the Controlled Release of Pharmaceuticals *J. Pharm. Pharmacol.* **44** 283

[5] Xu Y and Du Y 2003 Effect of molecular structure of chitosan on protein delivery properties of chitosan nanoparticles *Int. J. Pharm.* **250** 215

[6] Zargar V, Asghari M and Dashti A 2015 A Review on Chitin and Chitosan Polymers:

Structure, Chemistry, Solubility, Derivatives, and Applications *ChemBioEng Rev.* **2** 204
[7] Shamshina J L, Barber P S and Rogers R D 2013 Ionic liquids in drug delivery *Expert Opin. Drug Deliv.* **10** 1367
[8] Moniruzzaman M and Goto M 2011 Ionic liquids: Future solvents and reagents for pharmaceuticals *J. Chem. Eng. Japan* **44** 370
[9] Dobler D, Schmidts T, Klingenhöfer I and Runkel F 2013 Ionic liquids as ingredients in topical drug delivery systems *Int. J. Pharm.* **441** 620
[10] Moniruzzaman M, Tamura M, Tahara Y, Kamiya N and Goto M 2010 Ionic liquid-in-oil microemulsion as a potential carrier of sparingly soluble drug: Characterization and cytotoxicity evaluation *Int. J. Pharm.* **400** 243
[11] Moniruzzaman M, Kamiya N and Goto M 2010 Activation and stabilization of enzymes in ionic liquids *Org. Biomol. Chem.* **8** 2887
[12] Moniruzzaman M, Kamiya N and Goto M 2010 Ionic liquid based microemulsion with pharmaceutically accepted components: Formulation and potential applications *J. Colloid Interface Sci.* **352** 136
[13] Calvo P, Remunan-Lopez C, Vila-Jato J L and Alonso M J 1997 Novel hydrophilic chitosan-polyethylene oxide nanoparticles as protein carriers *J. Appl. Polym. Sci.* **63** 125
[14] Leonor I B, Baran E T, Kawashita M, Reis R L, Kokubo T and Nakamura T 2008 Growth of a bonelike apatite on chitosan microparticles after a calcium silicate treatment *Acta Biomater.* **4** 1349

Empirical Relationship between Hardness and Tensile Properties of High Pressure Torsion-processed Al 6061

Fauziana Lamin[1,2], Ahmad Kamal Ariffin[1], Intan Fadhlina Mohamed[1]

[1] Mechanical and Materials Engineering Department, Faculty of Engineering and Built Environment, Universiti Kebangsaan Malaysia, 43600 Bangi, Selangor, Malaysia.

[2] Vehicle Safety and Biomechanics Research Centre, Malaysian Institute of Road Safety Research, Lot 125-135, Jalan TKS 1, Taman Kajang Sentral, 43000 Kajang, Selangor, Malaysia.

Email: fauziana1215@gmail.com

Abstract. Improved strength is the ultimate goal of advanced material processing technique, including high-pressure torsion (HPT). To evaluate this goal, it requires determination of strength properties of the processed sample. While tensile test involves destructive test procedures, hardness is a closely related mechanical property in evaluating strength. This illustrates the need for a reliable hardness-tensile properties relationship. This paper determined an empirical relationship between hardness and several tensile properties i.e. ultimate tensile strength, Young modulus and tangent modulus, of HPT-processed Al 6061 samples. As results, an acceptable linear correlation for each property was obtained with a unique relationship. The degree of agreement between the studied properties and hardness was discussed. With recent advanced in computational method, the derived empirical relationships could optimise the HPT processing and aid the sample behaviour control and prediction.

1. Introduction

High pressure torsion (HPT) is a material strengthening technique through severe plastic deformation (SPD). The key SPD features includes introduction of a significant hydrostatic pressure and excessive strain with insignificant change in the overall sample dimensions [1–3]. As shown in Figure 1, HPT technique obtains these features by applying a simultaneous high compressive force and torsional straining. By restriction of material free flow, these forces subsequently induce micro-mechanical strengthening mechanisms to the processed sample, including grain size, precipitation, dislocation and solid-solution hardening [4].

Factors such as material and processing parameters affect the strengthening output. Therefore, evaluation of the strengthening level achieved by the specific HPT-processed materials is necessary. Uni-axial tensile test is a basic procedure to examine mechanical properties. However, engineers and researchers often conduct hardness test as a quick check on the strength properties [5]. It is comparatively non-destructive, fast, inexpensive and more effective for small-scale samples. This definitely requires an acceptable hardness-strength relationship. Previous researcher, Gasko and Roserberg [6] verified a greater possibility of tensile strength prediction by hardness measurement, as compared to yield strength, on dual phase (ferrite-martensite) steel.

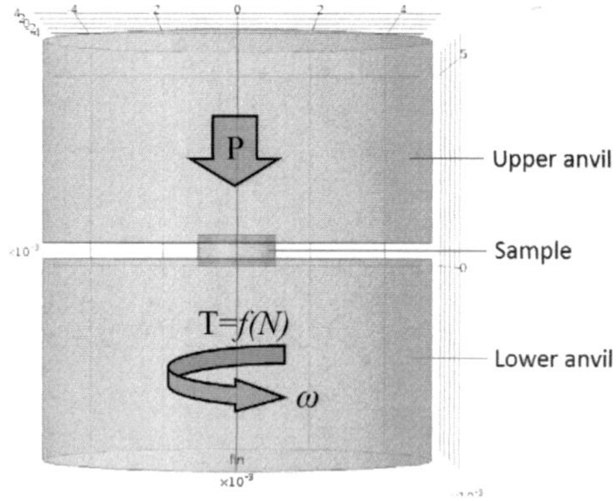

Figure 1. Schematic diagram of a HPT process where a thin disc sample subjected to a simultaneous pressure (P) and torsion (T), until n^{th} number of revolutions (N) at a determined rotational speed (ω).

Some hardness-strength linear correlations have been previously reported [5–7]. The simplest developed correlation is the three-time general hardness-strength (H-σ) relationship, as shown in Equation (1).

$$H = f(3\sigma) \tag{1}$$

Zhang et al. [5] presented a hardness-strength relationship unique classification by material type and processing technique. In view of processing technique, it was shown that the relationship of Cu–Zn alloys deformed by Equal Channel Angular Pressing (ECAP) is $H \approx 3\sigma$ and $H < 3\sigma$ when subjected to HPT. In both cases, σ represents the yield or tensile strength due to the limited work-hardening ability of the SPD-processed material. However, these generic relationships predict a very wide range of strength properties, particularly for the HPT materials.

Later, based on findings presented by number of researchers, Tiryakioğlu et al. [7] concluded that the hardness-strength relationship is better expressed in the form of non-zero y-intercept, as shown in Equation (2).

$$H = \beta_1\sigma - \beta_2 \tag{2}$$

in which β_1 is the slope and β_2 is the y-intercept of the fitted linear correlation.

The discussed inconsistency led to the examination of a hardness-strength correlation for a specific processed material. While a number of literature reported on steel [6,8-10], to date, inadequate studies describing the specific correlation on aluminium alloy exists. This paper aims to obtain an empirical relationship between hardness and tensile properties of high pressure torsion-processed Al 6061.

2. Methodology

This paper studies an Al 6061 with material composition listed in Table 1. Disc with diameter of 10 mm and thickness of 1 mm was prepared. The pre-HPT samples were solution treated (ST) at 530°C for 4 hours in an air atmosphere and then immediately quenched into ice water. Each disc was processed by HPT using procedure as reported in previous literature [4], [11], at room temperature under quasi-constrained conditions with an applied pressure of P = 6 GPa for revolution up to 5, and a rotation speed of ω = 1 rpm. The full revolution samples were then subjected to subsequent ageing at100°C in the air atmosphere for certain periods of time up to 15360 minutes, which is equivalent to 10.5 days.

Mechanical properties of aluminium alloy 6061 subjected to HPT was evaluated. Two mechanical tests were conducted which are uni-axial tensile test and hardness test. Figure 2 illustrates the locations for hardness measurements and tensile specimen on the 10 mm diameter HPT disc sample.

Table 1. Material composition of the studied Al 6061

Measurement	Material							
	Mg	Si	Cu	Fe	Cr	Zn	Ti	Al
Composition (wt%)	0.96	0.59	0.29	0.29	0.02	0.01	0.01	Balance

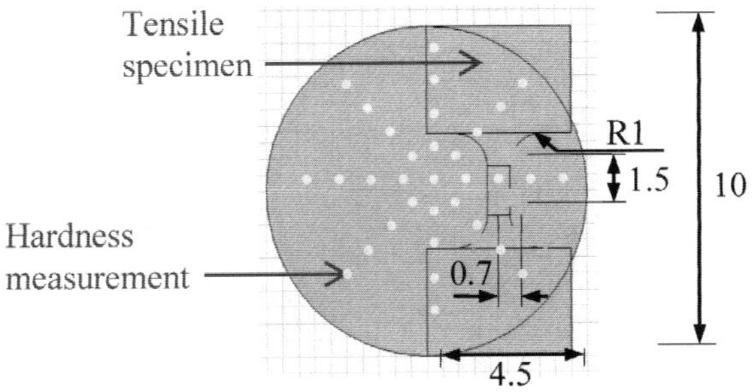

Figure 2. Locations for hardness measurements and tensile specimen on a HPT disc sample.
Note: Units: mm

The tensile test specimen having 1.5 mm gauge length, 0.7 mm width and 0.5 mm thickness were cut from the HPT sample using electrical discharge machining (EDM). Tensile tests were carried out at room temperature with an initial strain rate of 2 x $10^{-3}s^{-1}$. Tensile test generates properties such as yield stress, Young modulus, tangent modulus and ultimate tensile strength.

Whereas hardness tests, which measures the surface resistance against plastic deformation was conducted using Vickers test method. Vickers micro hardness was evaluated by application of a 100 g load for a dwell time of 15 s at equal distances of 1 mm from the disc centre along several radial directions. An average hardness value was calculated for each sample. Regression analysis was then conducted using the property values obtained by the tensile and hardness test.

3. Results and Discussions

Mechanical properties of the processed samples were presented in the first subsection. This will be followed by findings on the empirical relationship between hardness and tensile properties.

3.1. Mechanical properties of Al 6061

Figure 3 shows the engineering stress against engineering strain plot for the samples processed by HPT for 5 revolutions and subsequently aged at 100°C for several ageing durations. The stress-strain curve for pre-HPT sample was also included for comparison. Flow stress of the HPT-processed material exhibits work hardening while reducing some ductility. It can be clearly observed that the yield strength and tensile strength were significantly increased as compared to the ST sample. A similar behaviour of the HPT stress flow was reported in various materials including pure metals [12-14], alloys [15-16] and composite [17]. The properties were also affected by ageing duration.

Table 2 summarises the hardness and tensile properties of the studied samples. It can be seen that the highest strength of 622.2 MPa and hardness of 166 Hv were achieved in the 5 revolutions sample with subsequent ageing at 100°C for 15360 minutes. The minimum values of both strength and hardness are 534.0 MPa and 161.6 Hv respectively for 5 revolutions and 15 minutes aged sample. When there was an increase/decrease in tensile strength, the values of hardness increased/decrease accordingly. These patterns indicate that hardness proportionally affects the tensile properties. The following subsections discuss this properties relationship.

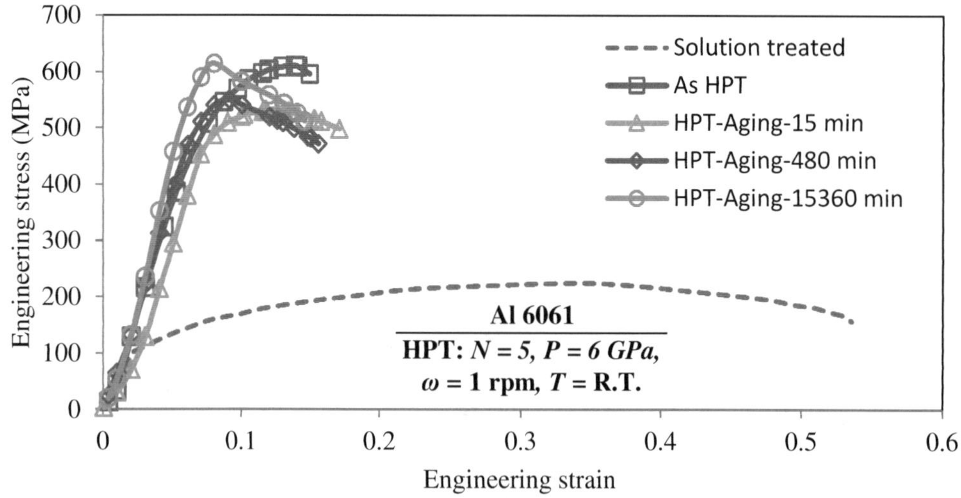

Figure 3. Stress-strain curves for the samples processed by HPT for 5 revolutions and subsequently aged at 100°C for several aging durations.

Table 2. Hardness and tensile properties of the studied samples

Number of HPT revolution, N	Process	Hardness, H (Hv)	Tensile strength, σ_{UTS} (MPa)	Young modulus, E (MPa)	Tangent modulus, E_t (MPa)
$N = 0$	Solution treated	61.35	225.6	5000.0	250.0
$N = 5$	As-HPT	163.9	613.0	8625.0	3333.3
	HPT-Aging 15 min	161.6	534.0	8333.3	1428.6
	HPT-Aging 480 min	162.9	552.9	8947.4	2555.6
	HPT-Aging 15360 min	166.0	622.2	10526.3	4117.6

3.2. Hardness and tensile strength relationship

Ultimate tensile strength (σ_{UTS}) is the maximum stress withstand by a material while being loaded. Comparison between pre and post-HPT showed that H and σ_{UTS} values were increased from the pre-HPT to the post-HPT. The pre-HPT (as annealed) data is referring to other published data, which are Al 3% Mg alloy [18] and Al 33% Cu alloy [19]. H values for the pre-HPT samples are ranging between the 50 Hv to 100 Hv. This value was significantly increased to the range between 150 Hv and 250 Hv after sample was subjected to HPT. The process also raises σ_{UTS} from less than 100 MPa to the upper range, between 450 MPa to 850 MPa.

Figure 4 shows the plot of hardness (H) against σ_{UTS}. Observation on the HPT-processed Al 6061 indicates that H correlates linearly with σ_{UTS}. Comparison was also conducted on other HPT-processed aluminium alloy published data, which are Al 4 wt% Cu [4] and 5483 Al–5%Mg alloy [20]. As results, a similar linear correlation pattern was observed on these additional data.

In terms of H- σ_{UTS} relationship, a distinctive pattern can be observed between the pre and post-HPT materials. It can be seen that the as annealed data are scattered along the three-time general hardness-strength relationship, as written in Equation (1). However, the work-hardened HPT materials are significantly deviate from the three-time general relationship. The deviation, by means of hardness to strength ratios, is between the range of 0.25 and 0.36.

The HPT-processed materials still obey the linear relationship, with a lower gradient that yield a specific value of y-intercept. Table 3 presents the H- σ_{UTS} relationship and the respective correlation coefficient for the observed materials. A strong linear correlation was obtained for each material with the minimum and maximum coefficient of 0.84 and 0.96, respectively. This result confirms that the HPT-processed aluminium alloys follows the linear non-zero y-intercept H- σ_{UTS} relationship model, as expressed in Equation (2). This model is consistent with finding reported by Salazar-Guapiriche et al. [21], particularly on Al 7010 alloy.

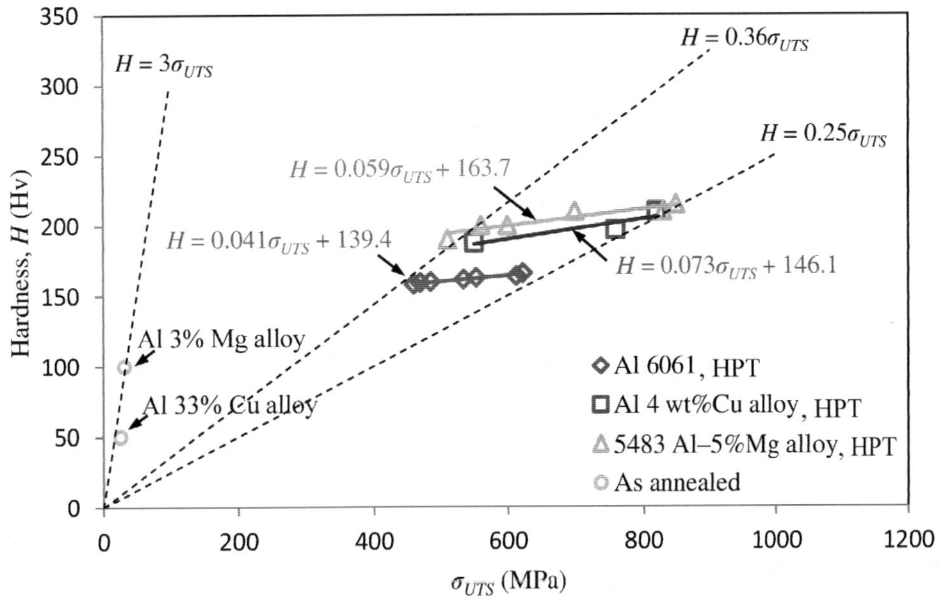

Figure 4. Hardness (H) against σ_{UTS} for several aluminium alloys.

Table 3. Hardness-tensile strength linear correlation of aluminium alloys

Aluminium Alloy	Relationship, $H = \beta_1\sigma_{UTS} + \beta_2$	Correlation coefficient, R^2
AlMg1SiCu (This study)	$H = 0.042\sigma_{UTS} + 139.3$	0.957
Al 4 wt% Cu alloy [4]	$H = 0.073\sigma_{UTS} + 146.1$	0.844
5483 Al–5%Mg alloy [18]	$H = 0.059\sigma_{UTS} + 163.7$	0.865

3.3. Hardness and modulus relationship

Young modulus, E is the ability of a material to withstand elastic deformation with increasing forces, which determined by the slope of stress-strain curve in the elastic region. On the other hand, rate of stress changes over plastic strain defines tangent modulus, E_t. It represents the material resistance to deform plastically. Both moduli describe material rigidity or stiffness. High E or E_t value describe low material potential to be elastically or plastically deformed, respectively.

Figure 5 shows the graph of modulus versus hardness for samples processed by HPT through 5 revolutions and aged at several ageing durations, as summarised in Table 3. It can be observed that with increasing hardness, the Young modulus and tangent modulus increase. This pattern demonstrates that the hardness is related to the modulus parameters. It is a linear function of both the elastic and plastic parts. The linear regression for each modulus was indicated in the corresponding graph.

In comparison, difference in tangent modulus for samples that have undergone dissimilar hardening treatments is more apparent than Young modulus. This behaviour was described by the higher H- E_t linear correlation slope, 606, compared to the H-E, 475.7. For a clearer comparison, Figure 6 shows the plot of normalised modulus ($\acute{E} = E/E_{max}$ and $\acute{E}_t = E_t/_{t,max}$) against hardness. This graph explains the response of E and E_t against changes in H, in a similar data range. Modulus data that correspond to the same hardness value are referring to the similar sample that has undergone specific process. It can be clearly observed that the gradient of normalised tangent modulus against hardness is steeper than the Young modulus. This result implies that hardness has a greater detection of changes in the E_t than E.

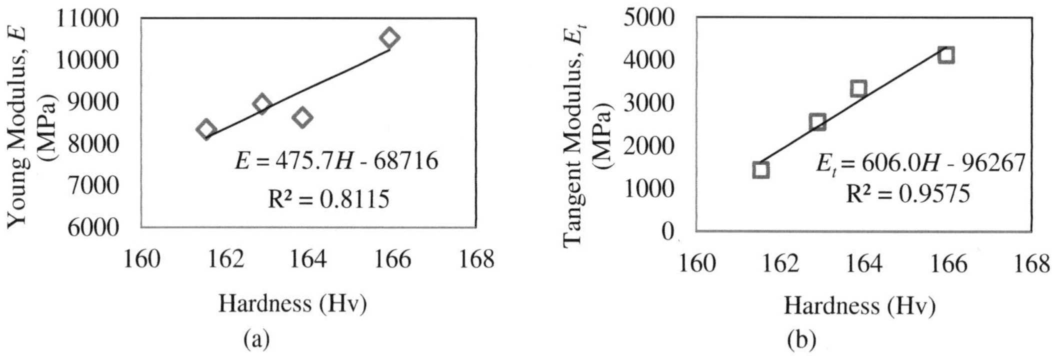

Figure 5. Correlation of different modulus against hardness: (a) Young modulus, (b) tangent modulus.

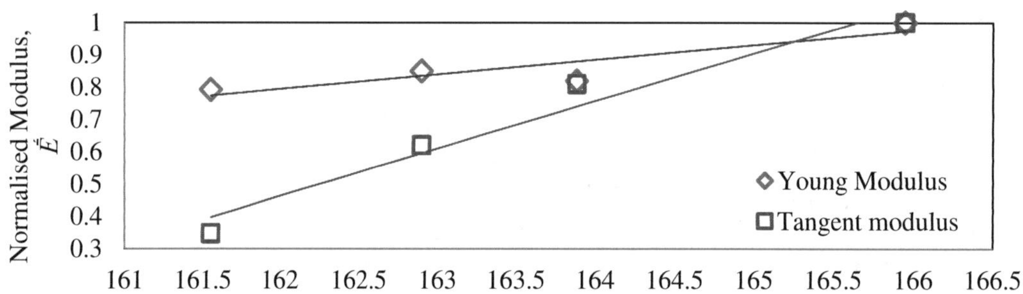

Figure 6. The response of E and E_t against changes in H.

The observed trend may due to the hardness and tangent modulus measurement itself, which correspond to the plastic deformation. Hardness, which is a measure of a material's flow resistance to dislocations movement, describes the degree of plastic deformation in materials. Meanwhile, tangent modulus describes material potential to be plastically deformed. This could also be the factor of better tangent modulus approximation by hardness. The high value of correlation coefficient, R^2 for H- E_t relationship, which is 0.96, reflects that H and E_t has an excellent agreement.

These correlations will benefit to determine an appropriate material model in computational mechanics. In order to simulate an HPT process for a particular sample, the material plasticity model requires a value of E and E_t as the parameter input. Through a simpler test, by measuring hardness, the value of E and E_t for the Al 6061 can be appropriately determined by the developed correlation relationship. An accurate model allows a better prediction of the material behaviour with HPT loading.

4. Conclusion

The empirical relationship between tensile properties and hardness of HPT-processed Al 6061 has been developed. This study showed an increase in material strength and rigidity with the increasing hardness. The HPT-processed aluminium alloys results including Al 6061, Al 4 wt% Cu and 5483 Al–5%Mg alloy, indicate that the strength correlates linearly over a hardness range with a non-zero y-intercept pattern. The calculated correlation are $H = 0.042\sigma_{UTS} + 139.3$ for Al 6061 alloy, $H = 0.073\sigma_{UTS} + 146.1$ for Al 4 wt% Cu alloy and $H = 0.059\sigma_{UTS} + 163.7$ for 5483 Al–5%Mg alloy. Hardness of the studied Al 6061 alloy also correlates well with Young modulus and tangent modulus. The relationship is linear, which are $E = 475.7H - 68716$ and $E_t = 606.0H - 96267$ for the respective modulus. Higher gradient of the E_t-H relationship, 606.0, as compared to 475.7 for the E-H relationship showed that hardness has a better detection of changes in tangent modulus, with a strong correlation of 0.96. The similar nature of plastic deformation measurement in hardness and tangent modulus might be the reason. These findings confirm that hardness is an appropriate alternative indication of the underlying mechanical properties of the HPT-processed Al 6061 alloys.

Acknowledgments

The author(s) gratefully acknowledges Malaysia Public Service Department for the Federal Training Award (HLP) scholarship.

References

[1] A. Azushima *et al.*, "Severe plastic deformation (SPD) processes for metals," *CIRP Ann. - Manuf. Technol.*, vol. 57, no. 2, pp. 716–735, 2008.

[2] M. Borodachenkova, W. Wen, and A. M. de B. Pereira, "High- Pressure Torsion- Experiments and Modeling," in *Severe Plastic Deformation Techniques*, Intech, 2017, pp. 93–112.

[3] S. Panda, L. S. Toth, J. Fundenberger, O. Perroud, and J. Guyon, "Analysis of heterogeneities in strain and microstructure in aluminum alloy and magnesium processed by high-pressure torsion," *Mater. Charact.*, vol. 123, pp. 159–165, 2017.

[4] I. F. Mohamed, Y. Yonenaga, S. Lee, K. Edalati, and Z. Horita, "Age hardening and thermal stability of Al-Cu alloy processed by high-pressure torsion," *Mater. Sci. Eng. A*, vol. 627, pp. 111–118, 2015.

[5] P. Zhang, S. X. Li, and Z. F. Zhang, "General relationship between strength and hardness," *Mater. Sci. Eng. A*, vol. 529, no. 1, pp. 62–73, 2011.

[6] M. Gaško and G. Rosenberg, "correlation between hardness and tensile properties in ultra-high strength dual phase steels – short communication," vol. 18, pp. 155–159, 2011.

[7] M. Tiryakio, J. S. Robinson, M. A. Salazar-guapuriche, Y. Y. Zhao, and P. D. Eason, "Hardness – strength relationships in the aluminum alloy 7010," *Mater. Sci. Eng. A*, vol. 631, pp. 196–200, 2015.

[8] O. Takakuwa, Y. Kawaragi, and H. Soyama, "Estimation of the Yield Stress of Stainless Steel from the Vickers Hardness Taking Account of the Residual Stress," vol. 2013, no. October, pp. 262–268, 2013.

[9] E. J. Pavlina and C. J. Van tyne, "Correlation of Yield Strength and Tensile Strength with Hardness for Steels Correlation of Yield Strength and Tensile Strength with Hardness for Steels," *J. Mater. Eng. Perform.*, vol. 17, pp. 888–893, 2008.

[10] J. T. Busby, M. C. Hash, and G. S. Was, "The relationship between hardness and yield stress in irradiated austenitic and ferritic steels," vol. 336, pp. 267–278, 2005.

[11] I. F. Mohamed *et al.*, "Grain refinement and microstructure evolution in aluminum A2618 alloy by high-pressure torsion," *J. Teknol.*, vol. 78, no. 6–9, pp. 149–152, 2016.

[12] A. Hohenwarter, "Incremental high pressure torsion as a novel severe plastic deformation process: Processing features and application to copper," *Mater. Sci. Eng. A*, vol. 626, pp. 80–85, 2015.

[13] S. Naghdy, L. Kestens, S. Hertele, and P. Verleysen, "Evolution of microstructure and texture

in commercial pure aluminum subjected to high pressure torsion processing," *Mater. Charact.*, vol. 120, pp. 285–294, 2016.

[14] H. Shahmir and T. G. Langdon, "Characteristics of the allotropic phase transformation in titanium processed by high-pressure torsion using different rotation speeds," *Mater. Sci. Eng. A*, vol. 667, pp. 293–299, 2016.

[15] E. A. Lukyanova *et al.*, "Strengthening of age-hardenable WE43 magnesium alloy processed by high pressure torsion," *Mater. Lett.*, vol. 170, pp. 5–9, 2016.

[16] A. Korneva *et al.*, "Grain refinement of intermetallic compounds in the Cu-Sn system under high pressure torsion," *Mater. Lett.*, vol. 179, pp. 12–15, 2016.

[17] S. N. Alhajeri, K. J. Al-Fadhalah, A. I. Almazrouee, and T. G. Langdon, "Microstructure and microhardness of an Al-6061 metal matrix composite processed by high-pressure torsion," *Mater. Charact.*, vol. 118, pp. 270–278, 2016.

[18] H.-J. Lee, J.-K. Han, S. Janakiraman, B. Ahn, M. Kawasaki, and T. G. Langdon, "Significance of grain refinement on microstructure and mechanical properties of an Al-3% Mg alloy processed by high-pressure torsion," *J. Alloys Compd.*, vol. 686, pp. 998–1007, 2016.

[19] M. Kawasaki, J. Foissey, and T. G. Langdon, "Development of hardness homogeneity and superplastic behavior in an aluminum-copper eutectic alloy processed by high-pressure torsion," *Mater. Sci. Eng. A*, vol. 561, pp. 118–125, 2013.

[20] P. Bazarnik, B. Romelczyk, Y. Huang, M. Lewandowska, and T. G. Langdon, "Effect of applied pressure on microstructure development and homogeneity in an aluminium alloy processed by high-pressure torsion," *J. Alloys Compd.*, vol. 688, no. October, pp. 736–745, 2016.

[21] M. Salazar-Guapuriche, Y. Y. Zhao, A. Pitman, and A. Greene, "Correlation of Strength with Hardness and Electrical Conductivity for Aluminium Alloy," *Mater. Sci. Forum*, vol. 519–521, pp. 853–858, 2006.

IOP Conf. Series: Materials Science and Engineering **458** (2018) 012082 doi:10.1088/1757-899X/458/1/012082

Optimisation of corncob based heterogeneous acid catalysed biodiesel synthesis using response surface methodology

Z E Tang[1], S Lim[1,*], Y L Pang[1] and H C Ong[2]

[1]Department of Chemical Engineering, Lee Kong Chian Faculty of Engineering and Science, Universiti Tunku Abdul Rahman, 43000 Selangor, Malaysia.
[2]Department of Mechanical Engineering, Faculty of Engineering, University of Malaya, 50603 Kuala Lumpur, Malaysia.

*stevenlim@utar.edu.my

Abstract. Biodiesel, which is also known as fatty acid methyl ester (FAME) can be produced through esterification reaction of vegetable oil or animal fats catalysed by heterogeneous acid catalyst. The objective of this study was to synthesise a corncob derived carbon based heterogeneous acid catalyst functionalised by the arylation of 4-benzenediazonium sulfonate (4-BDS) for biodiesel production by using palm fatty acid distillate (PFAD) as feedstock for the esterification reaction. Subsequently, the biodiesel production reaction was optimised by using response surface methodology (RSM). RSM was employed to study the interaction between the primary factors: reaction time (2.5-6.5 h), temperature (60 to 100 ºC), oil to methanol molar ratio (1: 13 to 1: 29) and catalyst loading (5 to 11 wt.%) in esterification reaction. The five-level, four factors central composite design (CCD) consisted of 30 experiments was chosen in this investigation. The predicted optimum reaction conditions was 6.48 h reaction time, 89.21 ºC reaction temperature, 1 to 21.94 molar ratio of PFAD to methanol and 11 wt.% catalyst loading with 85.94% of predicted biodiesel yield. The actual optimum biodiesel yield of 83.48% was successfully achieved at the corresponding optimum operating conditions which proved the validity of the statistical optimisation model.

1. Introduction

Since 2014, world energy consumption has been increasing by more than 1% yearly due to the rapid development of world economics [1]. Following behind the industrial sector, transportation sector contributes the second highest energy consumption and it was predicted that energy demand in transportation sector will continue to grow rapidly in the next 35 years [2]. For the transportation sector, non-renewable fossil fuels is the largest energy source based on the current trend. U.S. Energy Information Administration reported that about 92.63% of the energy consumed in transportation sector originates from the petroleum derived fuels in year 2017. However, these non-renewable fossil fuels have drawbacks where combustion of fossil fuels will release large amount of greenhouse gases such as carbon dioxide into the atmosphere which will then lead to global warming.

Biodiesel, which is also known as fatty acid methyl ester (FAME), is a renewable energy that can be produced from vegetable oils or animal fats through transesterification reaction. Transesterification is also known as methanolysis where the vegetable oils or animal fats will react with alcohol in the presence of base catalyst to form FAME. However, development of biodiesel production is facing a large obstacle due to its high production cost. About 70% of the production cost is associated with the usage of high quality refined oil with low free fatty acid (FFA) content as feedstock [3]. These low

FFA content oils are required to prevent saponification reaction between the oil and base catalyst [4]. Nonetheless, usage of high FFA unrefined oil is able to significantly decrease the production cost incurred by the crude feedstock used. Biocatalyst (enzyme) such as lipase was capable to be used with oil feedstock with high FFA content without promoting saponification reaction [5]. However, enzymatic catalysts are usually expensive while enzyme catalysed reaction requires long reaction time.

In this context, low cost, non-toxic and reusable heterogeneous acid based catalyst will be more favourable without the drawbacks. Many researchers had utilised different types of biomass to synthesis carbon-based catalysts such as bamboo [4], coffee residue [6], de-oiled waste cake [7], glucose [8], microalgae residue [9], oil palm trunk and sugar bagasse [10] and had been proven to be very promising for biodiesel production. For this study, corncob was selected as the raw material as they are produced in abundance and easily available in Malaysia at all time [11]. Although oil palm waste is a more popular commercial plantation in Malaysia, researches based on this waste has been relatively saturated. To the best knowledge of the authors, very few researches on biodiesel synthesis had conducted by using corncob waste as the carbon catalyst precursor. Therefore, the feasibility of corncob to synthesis the heterogeneous catalyst for biodiesel production was being investigated in this study.

RSM which comprises of mathematical and statistical methods can be used to optimise the response of the study. Through RSM, the optimum operating conditions can be predicted based on the mathematical models at reduced number of experimental runs than the full factorial design required to carried out and in return, reducing the cost and time consumed [12]. Alejandro et al. had investigated the effect of catalyst mass, methanol to oil molar ratio and temperature on the FAME yield by using three factors central composite design (CCD) and obtained an optimised yield of 99.15% [13]. The effect of catalyst loading (sodium hydroxide and sulfuric acid) and methanol to oil molar ratio on FAME yield was performed by Anand et al. using CCD and found that methanol to oil molar ratio has a significant effect on the FAME yield [14].

A four factors five levels CCD was used in the RSM carried out in the present work to study the combination effects of operating conditions (reaction temperature, time, catalyst loading and methanol to oil molar ratio) to the FAME yield obtained through esterification reaction. CCD is a surface response design that consist of full factorial design with lesser number of runs. It requires five levels design with additional star points as compared to three levels design to provide rotatability that ensure the variance would stay constant at all points equidistance from the centre [13]. In this study, esterification reaction catalysed by the corncob derived carbon based heterogeneous acid catalyst sulfonated through arylation of 4-benzenediazonium sulfonate (4-BDS) was investigated to obtain the maximum FAME yield and the optimum operating conditions based on the mathematical model from RSM.

2. Experimental

2.1. Materials

Corncob was obtained from the local market in Selangor, Malaysia. Palm fatty acid distillate (PFAD) was obtained from local edible oil production company. Sulfanilic acid, hydrochloric acid 37%, phosphoric acid 85%, ethanol 95%, methanol (≥99.9% purity) and n-heaxane (≥96% purity) were obtained from Merck. Sodium nitrite (98.5% purity) was obtained from Acros Organics. Internal standard (methyl heptadecanoate, 99%) and external standards (methyl palmitate, 99% methyl stearate, 99% methyl oleate and 99% methyl linoleate) were all obtained from Fluka Chemie, Germany for gas chromatography (GC) quantification.

2.2. Preparation of activated carbon

Collected corncob was washed, cut into smaller pieces and dry overnight in the oven at 80 °C. Corncob powder after ground was then impregnated with 30 v/v% phosphoric acid as the activating agent at the ratio of 1: 6 corncob to 30 v/v% phosphoric acid weight ratio for 24 h. Impregnated corncob was later subjected to thermal treatment by carbonisation in Carbolite furnace at 900 °C for 2 h at 5 °C /min. The activated carbon synthesised was denoted as CCAC900.

2.3. Synthesis of catalyst

Corncob derived carbon acid catalyst was synthesised through sulfonation method by arylation reaction of 4-BDS with activated carbon. In this process, 4-BDS will be first synthesised and directly followed by sulfonation of activated carbon with 4-BDS. Firstly, sulfanilic acid was dispersed into 300 mL of 1M HCl in a 500 mL round bottom flask in an ice bath and stirred continuously, followed by the dropwise addition of 1M sodium nitrite. The mixture was then stirred for 1 h at temperature 3 – 5 °C. Next, the solution was filtered by using vacuum pump to collect the precipitate formed, which was the 4-BDS. 4-BDS was then added into a mixture of 60 mL ethanol and 200 mL of deionised water at 3 – 5 °C. It was then followed by addition of CCAC900 and 100 mL 30 v/v% phosphoric acid and stirred for 30 min. CCAC900 will then be functionalised to form catalyst. The catalyst synthesised was denoted as Cat900.

2.4. Esterification Reaction

Esterification reaction was carried out by using PFAD and methanol as the reactant and Cat900 as the catalyst for the biodiesel production. In this reaction, 10.0 g of PFAD was kept constant and used as the basis to calculate the methanol to oil (PFAD) ratio. When the reaction is completed, the catalyst was filtered out by using filter paper and the methanol was evaporated and the remaining product (biodiesel and unreacted PFAD) was kept for analysis.

2.5. Catalyst characterisation and product analysis

Surface morphology CCAC900 and Cat900 was studied by using field emission scanning electron microscope (FESEM) JEOL, JSM-6710F at different magnifications. The functional group attached on the sample was studied by using FT-IR (Perkin Elmer Spectrum RX)-1with KBr pellet technique at wavelength range of 400 to 4000 cm-1. Biodiesel yield was investigated by using gas chromatography (GC) Perkin Elmer Claurus 500 equipped with FID detector with Nukol^{TM} capillary column at helium gas flow rate 3 ml/min, injector temperature 250 °C of and detector temperature 220 °C. The initial oven temperature was set at 110 °C then increased to 220 °C at 10 °C/min. Biodiesel yield was calculated following equation (1) below:

$$Y = \frac{\sum C_{ME} \times m_{product}}{m_{PFAD}} \times 100\% \quad (1)$$

In equation (1), Y is the FAME yield (%), C_{ME} is the total concentration of methyl esters $\left(\frac{\text{g}}{\text{g product}}\right)$, $m_{product}$ is the product mass (g product) and m_{PFAD} is the PFAD feed mass (g).

2.6. Design of Experiment

Esterification reaction in the presence of the corncob derived acid catalyst to produce biodiesel was conducted according to the factorial design by using the response surface methodology (RSM) with Design Expert 7.0.0. A four factors and five levels central composite design (CCD) was employed in the design of the experiment. In this study, temperature (40 – 120 °C), time (0.5 – 8.5 h), catalyst loading (2 – 14 wt.%) and methanol to oil molar ratio (5:1 – 37:1) were the targeted independent variables to be investigated and FAME yield (%) from the esterification reaction was the dependent variable, which is the response of the experiment. Table 1 shows the variables and levels of the experimental design in this study. The data obtained in this study was fitter into the second order polynomial expression as show in equation (2).

$$R = \beta_0 + \sum_{i=1}^{4} \beta_i X_i + \sum_{i=1}^{4} \beta_{ii} X_i^2 + \sum_{i=1}^{4} \sum_{j=i+1}^{4} \beta_{ij} X_i X_j \quad (2)$$

In equation (2), R is the response factor, X_i is the independent factor, β_0 is the intercept, β_i is the linear coefficient, β_{ii} is the quadratic coefficient and β_{ij} is the cross products coefficient.

Table 1. Variables range and levels for response surface methodology.

Independent Variables	Symbols	Levels				
		$-\alpha$	-1	0	+1	$+\alpha$
Temperature (°C)	A	40	60	80	100	120
Time (h)	B	0.5	2.5	4.5	6.5	8.5
Catalyst Loading (wt.%)	C	2	5	8	11	14
Methanol to oil molar ratio	D	5	13	21	29	37

3. Results and discussion

3.1. Characterisation of catalyst

Figure 1 shows the SEM images of the surface morphology of CCAC900 at different magnification of 2000x and 10000x. As shown in the SEM image, CCAC900 had uneven and rough surface with a porous structure. The black holes observed in the SEM image represented the pores formed during activation and carbonisation [15]. The pores formed on the activated carbon is able to give good access to the reactant such as free fatty acid [6].

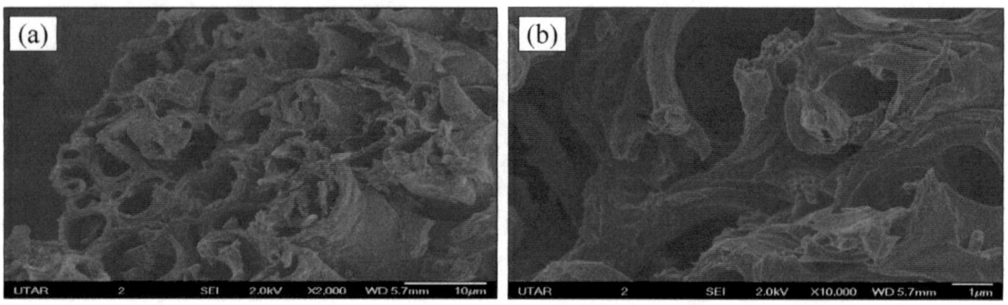

Figure 1. SEM images of CCAC900 at magnification of (a) 2000x and (b) 10000x.

The presence of active site $-SO_3H$ was confirmed by performing the FT-IR analysis. Figure 2 shows the FT-IR spectra for CCAC900 (black curve) and Cat900 (red curve) from wavelength of 400 to 4000 cm^{-1}. The FT-IR spectra for CCAC900 had a broad peak at wavelength of 1126 cm^{-1} is associated with C=O stretching in acids, alcohols and phenols [7, 16]. Peak appeared at wavelength 2377 cm^{-1} may be due to the adsorption of carbon dioxide from the atmosphere on the surface of activated carbon [6]. On the other hand, the red curve in figure 2 shows the FT-IR spectra of the sulfonated activated carbon, Cat900. Overall, Cat900 exhibited similar spectra with the CCAC900. The main difference of the spectra between Cat900 and CCAC900 is that peak was observed at wavelength 1167 and 1031 cm^{-1} in Cat900 indicating asymmetric –SO3 stretching and O=S=O symmetric stretching modes, respectively and this is in accordance to the results reported by several literatures [4, 17, 18]. The presence of the $-SO_3$ and O=S=O stretching modes confirmed the successful attachment of sulfonic group, $-SO_3H$ onto the carbon framework as the active site for esterification reaction.

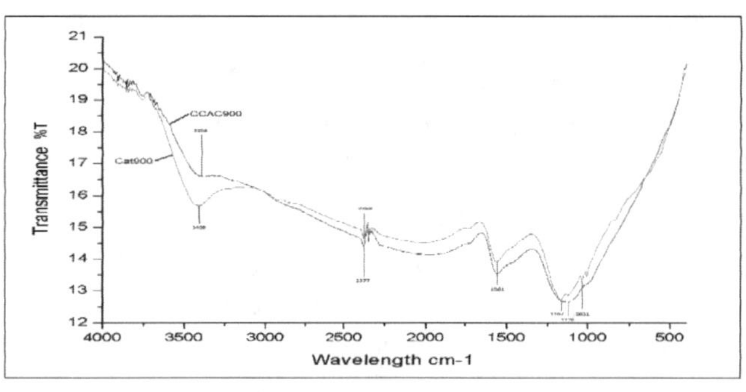

Figure 2. FT-IR spectra of activated carbon CCAC900 and catalyst Cat900.

3.2. Quadratic Regression Model

The response (biodiesel yield) was calculated from the quadratic regression equation as shown in equation (3).

$$Y = 76.41 + 4.03A + 5.69B + 3.59 + 1.94D - 0.52AB - 0.20AC - 1.44AD - 1.18BC - 0.51BD - 0.44CD - 3.41A2 - 2.13B2 + 1.01C2 - 1.43D2 \quad (3)$$

In equation (3), Y is the biodiesel yield, A, B, C and D represent the temperature (°C), time (h), catalyst loading (wt.%) and methanol to oil molar ratio, respectively. According to the regression, all four factors A, B, C and D exhibited positive effect towards the biodiesel yield where increasing the temperature, time, catalyst loading and methanol to oil molar ratio will increase the biodiesel yield in tandem. According to the results obtained from ANOVA shown in table 2, the F value of the regression model is obtained at 13.43 and the small prob > F value of less than 0.0001 and this had indicated that the model is highly significant at 95% confidence level. The terms of A (temperature), B (time), C (catalyst loading), D (methanol to oil molar ratio), A^2, B^2 and D^2 had significant effect on the biodiesel yield. It was noted that the R-squared (R^2) of the model is 0.9260, indicating 92.6% of the variation can be explained by the model.

Table 2. Results for analysis of variance of quadratic regression

Source	Sum of Squares	Degree of freedom	Mean Square	F Value	p-value Prob > F
Model	2135.68	14	152.55	13.43	< 0.0001
A: Temperature	389.49	1	389.49	34.28	< 0.0001
B: Time	776.33	1	776.33	68.32	< 0.0001
C: Catalyst Loading	309.76	1	309.76	27.26	0.0001
D: Molar Ratio	90.77	1	90.77	7.99	0.0128
AB	4.40	1	4.40	0.39	0.5429
AC	0.61	1	0.61	0.05	0.8199
AD	33.35	1	33.35	2.93	0.1073
BC	22.11	1	22.11	1.95	0.1833
BD	4.14	1	4.14	0.36	0.5550
CD	3.12	1	3.12	0.27	0.6079
A^2	318.51	1	318.51	28.03	< 0.0001
B^2	124.31	1	124.31	10.94	0.0048
C^2	28.24	1	28.24	2.49	0.1358
D^2	56.01	1	56.01	4.93	0.0422
Residual	170.44	15	11.36		
Lack of Fit	142.12	10	14.21	2.51	0.1608
Pure Error	28.32	5	5.66		
Mean	71.64954			R^2	0.926091

3.3. Effect of parameters

Figure 3(a) shows the response surface plot of temperature, methanol to oil molar ratio and FAME yield at constant time of 5.8 h and 9.6 wt.% catalyst loading. Its shows that the FAME yield increased with the increase of both temperature and methanol to oil molar ratio. The increment of FAME yield with temperature is more significant than methanol to oil molar ratio. Increasing temperature helps in driving the reaction equilibrium towards the product side and also to promote protonation of the acid catalyst in order for reaction to occur [4, 18]. At lower methanol to oil molar ratio, the increment of

FAME yield with the temperature was more obvious compared to higher methanol to oil molar ratio. At 13:1 molar ratio, the FAME yield increased from about 66 to 76.5% when the temperature increased from 60 to 100 °C while at methanol to oil molar ratio of 29:1, FAME yield increased with the temperature up to 90 °C.FAME yield decreased slightly at higher temperature of 100 °C. On the other hand, at lower temperature, the effect of methanol to oil molar on the FAME yield was more obvious at higher temperature. At 60 °C, the FAME yield increased with the methanol to oil ratio from 13:1 to 29:1 while at 100 °C, the FAME yield slightly increased from molar ratio of 13:1 and reached the optimum point at 21:1 before started to decrease. Several researches had also investigated optimum methanol to oil molar ratio at around 20:1 [7, 8, 10]. Increasing the methanol to oil molar ratio can drive the reversible reaction towards the product side but excess addition of methanol would lead to dilution of the reaction system and reduced the collision rate of oil molecules and the catalyst particles resulting in lower products formed [4, 10].

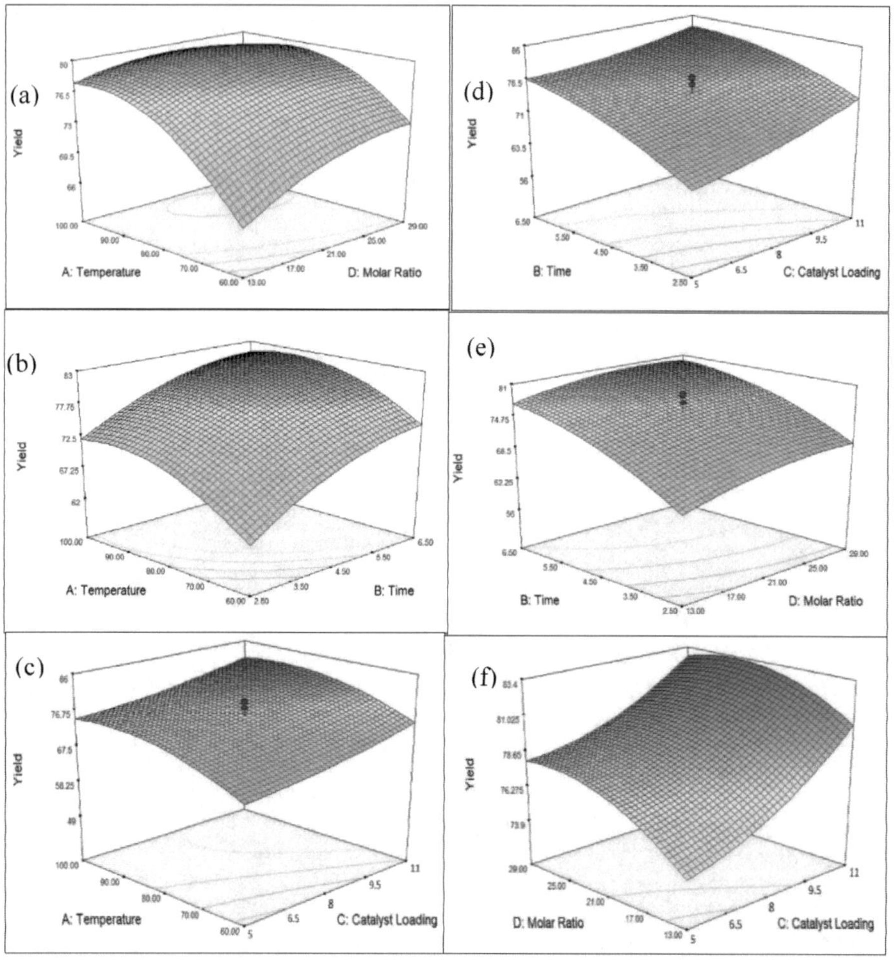

Figure 3. Response surface plot of (a) temperature (°C) and methanol to oil molar ratio (b) temperature (°C) and time (h) (c) temperature (°C) and catalyst loading (wt.%) (d) time (h) and catalyst loading (wt.%) (e) time (h) and methanol to oil molar ratio (f) catalyst loading (wt.%) and methanol to oil molar ratio on esterification reaction.

Figure 3(b) shows the response surface plot of temperature, time and FAME yield at constant methanol to oil molar ratio of 21:1 and 9.3 wt.% catalyst loading. The reaction time exhibited similar trend with the temperature whereby the FAME yield increased with reaction time. Longer reaction time is usually required to ensure a good mass transfer of catalyst in the reactants so that reaction has sufficient time to occur on the surface of catalyst [4, 19]. It shows that the FAME yield was the highest at 90 °C for all reaction time. On the other hand, it also shows that the FAME yield was always the highest at the reaction time of 6.5 h regardless of the change in temperature. In the other words, the change of temperature did not affect reaction time and vice versa. This may be due to the low interaction between these two variables. The highest FAME yield was only achieved when the temperature was at 90 °C at reaction of 6.5 h. The interaction effect of catalyst loading and temperature at time 4.5 h and 21:1 molar ratio on the FAME yield is shown in the surface response plot in figure 3(c). It shows a similar trend with figure 3(b) where the optimum temperature was at about 90 °C at all catalyst loading showing that these two are independent variables with low interaction effect [5]. The optimum temperature of 90 °C was in accordance with those reported in several literatures [4, 20].

Figure 3(d) shows the surface response plot of time, catalyst loading and FAME yield at constant temperature of 80 °C and methanol to oil molar ratio 21:1. Increment of reaction time will increase the FAME yield and the trend will not be affected by catalyst loading. On the other hand, at 2.5 h reaction time, the increment of catalyst loading will lead to the increase of FAME yield. However, when the reaction time was pro-longed to 6.5 h, increasing catalyst loading was less effective compared with increasing the FAME yield. In addition, the surface response plot of time, methanol to oil molar ratio and FAME yield at constant temperature of 80 °C and 9.3 wt.% catalyst loading was shown in figure 3(e). Overall, reaction and methanol to oil molar ratio had very less interaction as reported by Alejandro et al. [13]. It shows that the effect of methanol to oil molar ratio was more significant at short reaction time than long reaction time. At long reaction time, FAME yield only increased slightly with molar ratio. On the other hand, the effect of time on FAME yield was more obvious at lower molar ratio that the FAME yield only slightly increased with time when molar ratio was high. The effect of catalyst loading and molar ratio on the FAME yield at constant temperature of 80 °C and 5.8 h of reaction time is shown in figure 3(f). At constant temperature and time, catalyst loading had very obvious effect towards the FAME yield whereby the FAME yield increased drastically with catalyst loading. Catalyst plays an important role in lowering the activation energy of reaction, thus, increasing the amount of catalyst helps to promote the occurrence of reaction [19]. From the results obtained, methanol to oil molar ratio had lesser effect on the FAME yield. This outcome was similar with the findings reported by Dharma et al. [12]. At catalyst loading of 0.5 wt.%, the biodiesel was maximum at methanol to oil molar ratio of 26:1. However, maximum yield can be obtained at methanol to oil molar ratio of about 22:1 when the catalyst loading was increased to 11 wt.%.

3.4. Optimisation of esterification reaction

It was predicted that the maximum FAME yield of 85.94% can be achieved at the optimum condition of reaction temperature 89.24 °C, reaction time of 6.48 h, 21.94:1 methanol to oil molar ratio and 11wt.% of catalyst loading. Three experiment runs were conducted at the corresponding optimum condition and obtained an average yield of 83.48% which was close to the predicted value. The percentage error of the actual yield from the theoretical yield was only 2.86% which was considered low.

4. Conclusion

Esterification of PFAD and methanol was successfully carried out by using corncob derived carbon based heterogeneous acid catalyst. Catalyst was synthesised through sulfonation by arylation of 4-BDS with corncob derived activated carbon. RSM was carried out to study the effect of temperature, time, methanol to oil molar ratio and catalyst loading to the FAME yield of esterification reaction by using CCD. The optimum condition obtained from RSM study was at temperature 89.24 °C, reaction time of 6.48 h, 21.94:1 methanol to oil molar ratio and 11 wt.%. A highest FAME yield of 83.48%

was obtained at the optimum reaction conditions. As a whole, esterification reaction of PFAD catalysed by corncob derived heterogeneous acid catalyst has a high potential for further development in the future biodiesel industry which fulfils the sustainable framework by adding value into the waste materials.

Acknowledgements
The authors would like to acknowledge the funding given by Universiti Tunku Abdul Rahman (UTARRF) for this project and also the scholarship funding to Ms. Tang Zo Ee.

References
[1] British P 2017 *BP Statistical Review of World Energy June 2017* vol 67 p 2

[2] United-State EIA 2017 *Monthly Energy Review November 2017* (US: Energy Information Administration) vol 1 p 31

[3] Dehkhoda A M and Ellis N 2013 Biochar-based catalyst for simultaneous reactions of esterification and transesterification *Catal. Today* **207** 86-92

[4] Zhou Y, Niu S and Li J 2016 Activity of the carbon-based heterogeneous acid catalyst derived from bamboo in esterification of oleic acid with ethanol *Energy Convers. Manag.* **114** 188-96

[5] Ali C H, Qureshi A S, Mbadinga S M, Liu J-F, Yang S-Z and Mu B-Z 2017 Biodiesel production from waste cooking oil using onsite produced purified lipase from Pseudomonas aeruginosa FW_SH-1: Central composite design approach *Renew. Energy* **109** 93-100

[6] Ngaosuwan K, Goodwin Jr J G and Prasertdham P 2016 A green sulfonated carbon-based catalyst derived from coffee residue for esterification *Renew. Energy* **86** 262-69

[7] Konwar L J, Mäki-Arvela P, Salminen E, Kumar N, Thakur A J, Mikkola J-P and Deka D 2015 Towards carbon efficient biorefining: Multifunctional mesoporous solid acids obtained from biodiesel production wastes for biomass conversion *Appl. Catal. B Environ.* **176–177** 20-35

[8] Zhang F, Fang Z and Wang Y-T 2015 Biodiesel production direct from high acid value oil with a novel magnetic carbonaceous acid *Appl. Energ.* **155** 637-47

[9] Fu X, Li D, Chen J, Zhang Y, Huang W, Zhu Y, Yang J and Zhang C 2013 A microalgae residue based carbon solid acid catalyst for biodiesel production *Bioresour. Technol.* **146** 767-70

[10] Ezebor F, Khairuddean M, Abdullah A Z and Boey P L 2014 Oil palm trunk and sugarcane bagasse derived heterogeneous acid catalysts for production of fatty acid methyl esters *Energy* **70** 493-503

[11] Shariff A, Aziz S, Ismiza Ismail N and Abdullah N 2016 Corn Cob as a Potential Feedstock for Slow Pyrolysis of Biomass *J. Phys. Sci.* **27** 123–37

[12] Dharma S, Masjuki H H, Ong H C, Sebayang A H, Silitonga A S, Kusumo F and Mahlia T M I 2016 Optimization of biodiesel production process for mixed Jatropha curcas–Ceiba pentandra biodiesel using response surface methodology *Energy Convers. Manag.* **115** 178-90

[13] Regalado A, Juan M-M, Peralta E, Estrada V C, Gerardo M-V, Cordero M and Luis G Z 2017 Biodiesel Production from Jatropha oaxacana Oil by Reactive Vacuum Distillation: Optimization by Response Surface Methodology *Key Eng. Mat.***737** 385-92

[14] Dev Anand M, Vijay Ananth S, Jackson D and Prabhu N 2016 Investigation of Biodiesel Production from High Free Fatty Acid through RSM. *Indian Journal of Sci Technol* **9** 1-11

[15] Janaun J, Abang S and Anwar A M 2013 *Synthesis and characterization of carbon-based catalyst derived from seaweed* (Langkawi: IEEE) **1** 19-22

[16] Alves C, Franca A and Oliveira L 2013 Evaluation of an Adsorbent Based on Agricultural Waste (Corn Cobs) for Removal of Tyrosine and Phenylalanine from Aqueous Solutions **2013** 1-8

[17] Lokman I M, Rashid U and Taufiq-Yap Y H 2015 Production of biodiesel from palm fatty acid distillate using sulfonated-glucose solid acid catalyst: Characterization and optimization *Chinese J. Chem. Eng.* **23** 1857-64

[18] Ayodele O O and Dawodu F A 2014 Production of biodiesel from Calophyllum inophyllum oil using a cellulose-derived catalyst *Biomass Bioenerg.* **70** 239-48

[19] Dawodu F A, Ayodele O, Xin J, Zhang S and Yan D 2014 Effective conversion of non-edible oil with high free fatty acid into biodiesel by sulphonated carbon catalyst *Appl. Energ.* **114** 819-26

[20] Liu W-J, Tian K, Jiang H and Yu H-Q 2013 Facile synthesis of highly efficient and recyclable magnetic solid acid from biomass waste *Sci. Rep.* **3** 2419

IOP Conf. Series: Materials Science and Engineering **458** (2018) 012083 doi:10.1088/1757-899X/458/1/012083

Application of Smart Radiotracer Data Analysis (SRDA) Software for Industrial Application

N Othman[1*], M Ibrahim[2], N Yussup[2], N H Yussof[1] and M A Syafiq Yunos[1]

[1]Industrial Technology Division, Malaysian Nuclear, 43000 Kajang, Selangor, Malaysia
[2]Support Technical Division, Malaysian Nuclear, 43000 Kajang, Selangor, Malaysia

Email: noraishah@nuclearmalaysia.gov.my

Abstract. Radiotracers technique uses a radioactive source and scintillation detectors to accurately detect plant production anomalies and mechanical damage in the plant. The greatest benefit of radiotracer technology over the conventional methods is that the investigation can be carried out on-stream and without disrupting the operating process of the plant. The residence time distribution (RTD) is one of the important parameters that provide hydrodynamics of the reactor as well as process plant anomalies. Nevertheless, the data is obtained through multichannel analyzer and the analysis of RTD is carried out later and not on site. This is because the lengthy data treatments should be conducted to the counts before obtaining the RTD curve. This data treatment process is cumbersome and time consuming, Thus, this proposal is to describe the application of developed smart radiotracer data analysis (SRDA) software by Nuclear Malaysia so that it can accelerate industrial analysis. In this study, RTD in mixing process has been carried out using SRDA and instant result such as percentage of dead zone has been determined and optimized. The optimized parameters for mixing process are 100 rpm speed of impeller, clearance of 100mm and U_{tip} is 0.5m/s which result in 1.1% dead zone.

1. Introduction

Radiotracer technology (RT) has helped industries in the process optimization, improvement of yield and increase the process efficiency by identifying the mechanical problems and process irregularities in their respective plants. Although RT has been widely used for past 50 years, its credibility is remained unimpeded. The most significant and salient benefit of RT is the ability to be conducted without shutting down the plant and no interruption to the operating process, which is the most concern to the plant owner. Although scheduled shut down is compulsory for maintenance, the unnecessary shutdown of plant is not welcomed since the cost is high. RT is widely used globally in oil and gas as well as petrochemical industries, waste water treatment plant industries, mineral and mining processing plant industries. In this paper, the acquisition of data from radiotracer analysis is highlighted. Previously, the signals, which derived from radioactive signals (photons) are converted to electrical signal using photo multiplier tube (PMTs) when radiation strikes the scintillator in the NaI detector. The signals, in term of counts per second (cps) are recorded in the desktop and uploaded in Microsoft Excel or .dat file. Residence Time Distribution (RTD) is one of the most important parameter in chemical engineering and process. RTD is used to assess the hydrodynamics of the process and the behaviour of the flow field as well as process plant anomalies through RTD models. It also describes how long the particle stays inside the respective unit operations [1,2]. Nevertheless, in order to analyse RTD from the recorded raw data, it has to undergo several data treatment as describe in the next part. This is to ensure that only radioactive data is diagnosed in order to simulate the RTD model so that the actual reactor behaviour or process can be described well. In order to do so, the experimental data is plotted as cps vs time prior transforming to E(t) curve by using appropriate equations as described in the Research Methodology Section (2.0). Besides, Mean Residence Time (MRT) is also calculated in order to quantify the deviation of experimental model with the ideal model and identify the type of plant anomalies such as the presence of dead zone, stagnant zone or channelling and short-circuiting. The aforementioned steps are cumbersome, step by step and time consuming. Thus, Plant assessment Technology Group (PAT) has developed Smart Radiotracer Data Analysis Software (SRDA) to accelerate the plant diagnostic and troubleshooting. The objectives of the paper are to develop the on-line data acquisition monitoring for industrial application as well as to model the RTD for investigations of process plant behaviour.

Radiotracer Technology (RT) is using radioactive source, particularly gamma emitting source for industrial application. This is because, high energy is required to penetrate the thick and solid vessel in the plant for plant diagnostic and troubleshooting. Radiotracer technology is the only technique which utilizes small scale tracing injection in the system and yet able to provide significant and comprehensive information and flow field inside the vessel or reactor under study [3,4]. It is also the most trusted technology in providing correct information such as plant anomalies and mechanical failures of the opaque operating system or vessel. RTD study has been developed by Danckwerts (1953) and have been implemented by numerous researchers to identify possible malfunctions, pipe leakages, dead zone, mitigations and blockages in side pipelines as well as to estimate mixing homogeneity [5]. Besides RTD, RT is also used to measure the flow rates of liquid or gas as well as to calibrate the flow meter, verify leakage inside the heat exchanger system as well as to tracer the connectivity between injection well and production well in the oilfield application [6,7]. RT has the capability which outsmart conventional techniques such as in-situ detection, higher detection sensitivity, prompt respond, higher accuracy and reliability as well as availability and easy access of radioactive source especially Nuclear Malaysia has its own research reactor (RR) to produce radiotracer for industrial application [8,9,10].

International Atomic Energy Agency (IAEA) has established and recommended RTD software for the usage of its member states. Six models have been recommended to characterize the RTD model derived from radioactive based experiments. The RTD model will be fitted into the experimental RTD curve and the most optimized model is chosen based upon minimum value of Root Mean square (RMS) [2]. Figure 1 shows the fundamentals of a tracer experimental setup as described by Furman et al. (2011) [11]. There are at least two scintillation sodium iodide (NaI) detectors used for RTD investigation which are installed at the inlet and the outlet of the reactor or vessel respectively. The inlet detector is used to ensure that the tracer is entering the system correctly whereas the second detector is to record the gathered information inside the system. The peaks describe the transmitted radiation is received by the detector accordingly.

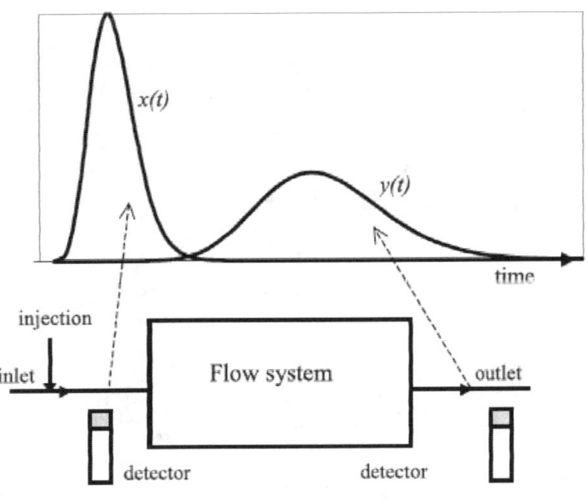

Figure 1. The principle of radiotracer experiment by Furman et al. 2011

The RTD curve of a radiotracer experiment is considered measurable after treatment of the raw data. The treatment of the data involves background correction, radioactive decay correction, starting point correction, filtering, and data extrapolation [12]. Besides, RTD is a fundamental parameter in reactor design because it can provide information on how long the substrate has been in the reactor and it can help characterise the extent of the deviation in the reactor behaviour from ideal. Besides, RTD can potentially be used to compare two different equipment designs and operating conditions. At presence, the data is obtained through multichannel analyzer and the analysis of RTD and MRT are carried out later and not on site. This is because the lengthy data treatments should be conducted to the radiotracer counts before obtaining the RTD curve. Moreira et al. (2007) used DTSpro Software for that purpose which enable the preliminary of radiotracer experiment [13]. Nevertheless, prior to do that the toggle should be used in the personal computer and since the establishment was done in year 2000, the toggle is consider obsolete and not compatible with most new data acquisition system Another option for data treatment software for radiotracer application was developed by Czech Technical University of Prague (1996). The software is not user friendly, complicated and MS DOS based [2]. Thus, this paper is to embark with the new development of smart radiotracer analysis so that it can be further utilized and accelerate any industrial applications analysis.

2. Methodology

2.1. Data treatment

Figure 2 shows the radiotracer experimental setup where the sodium iodide (NaI) scintillation detectors are located at the inlet and outlet of the mixing tank respectively. These detectors are connected to Data Acquisition System (DAS) that attached to PC for data tracking and monitoring. Water with flow rate of 5litre per minute (5lpm) is pumped throughout the mixer to ensure that continuous flow is achieved. The initial parameters are set with 100 rpm, clearance of impeller to floor is 100 mm and U_{tip} is 0.5m/s. The ^{99m}Tc radiotracer is injected once the stability of flow is achieved and the parameters have set beforehand. ^{99m}Tc is eluted from molybdenum generator and emits gamma rays with energy of 0.104MeV and half-life of 6 hours. The data plotted in the graph shows the ability of NaI detectors to detect emitted gamma rays (radioactive signals) from ^{99m}Tc. Nevertheless, the radioactive signals should be treated and the parasitic interference should be removed from the raw data.

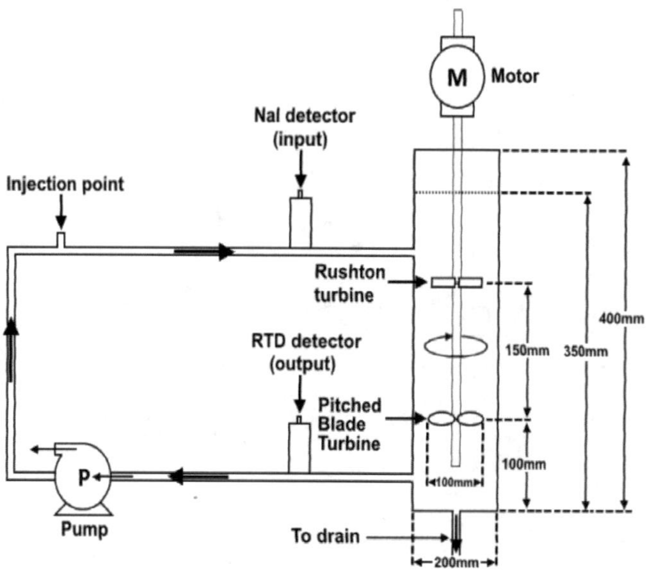

Figure 2. Schematic diagram of mixing vessel rig for radiotracer experimentation

Therefore, LabView and Mathlab application have been used to develop Smart Radiotracer Data Analysis (SRDA). The purpose is to eliminate the noises and obtain the actual radioactive representation. The treatment involves background correction, radioactive decay correction, starting point correction, filtering, and data extrapolation. All these treatments should be carried out prior Residence Time Distribution (RTD) determination. The steps of data treatment are as follows:

1. Background correction: This step is used to remove the existing background radiation

$$C_c = C_m - B_r \quad (1)$$

where C_c is the corrected radiotracer concentration, C_m is the measured radiotracer concentration, and B_r is the background radiation level.

2. Radioactive decay correction: Use the following equation to compensate for this decay:

$$C_r(t)=C_m(t)e^{\lambda t} \quad (2)$$

where $C_m(t)$ is the measured concentration as a function of time, $C_r(t)$ is the real concentration as a function of time, and λ is called the decay constant, and it is represented by the following equation:

$$\lambda=\ln(2)/t^{1/2} \quad (3)$$

3. Starting point correction: This step is used to make the starting point equal to zero to remove any counts before the injection.
4. Filtering: This step is used to eliminate the fluctuations resulting from counting statistical noise or electronic noise. *Cftool* from *Matlab* is used for filtering and smoothing purposes.
5. Extrapolation: This step is required to make the tracer concentration rates go back to zero after the end of the data acquisition. This step can be performed with the following equation:

$$C_e(t)=C(t)e^{-t} \quad (4)$$

2.2. Residence Time Distribution and Mean Residence Time (MRT)
In order to assess the mixing process operation and condition, the estimation of residence time distribution (RTD) is carried out. This parameter is enabled to identify the type of process anomalies such as presence of dead zone, stagnant zone as well as channelling and short-circuiting inside the mixing column. Hence, the treated data will undergo data transformation in which the following equation should be applied:

$$E(t)= Ci(t)/\int_0^{\infty} C(t)dt \quad (5)$$

The curve obtained is the RTD curve which represent the hydrodynamics of liquid behavior as well as the time taken of radiotracer in the reactor and the plotted graph is E(t) vs time (s).The MRT is the averaged data of RTD and is used to estimate or detect quantitatively the percentage of dead zone, channelling or fouling present in the reactor. The data from RTD will be transformed into MRT using following formula:

$$\tau = \int_0^{\infty} tE(t)dt \quad (6)$$

Therefore, SRDA will substitute the former technique in acquiring and analyzing radiotracer data in simple and quick manner.

3. Results and Discussions

Figure 3 shows the GUI of developed SRDA which comprises of data acquisition of radioisotopes (^{99m}Tc) which can be monitored online and RTD Analysis. The beauty of this developed software is, it can convert the raw data curve from experiment into well –curve treated data automatically. Figure 4 shows the actual plot of raw data prior data treatment is carried out. It can be observed clearly that the signal to noise ratio (SNR) is higher which influence the actual radioactive signal emitted from ^{99m}Tc.

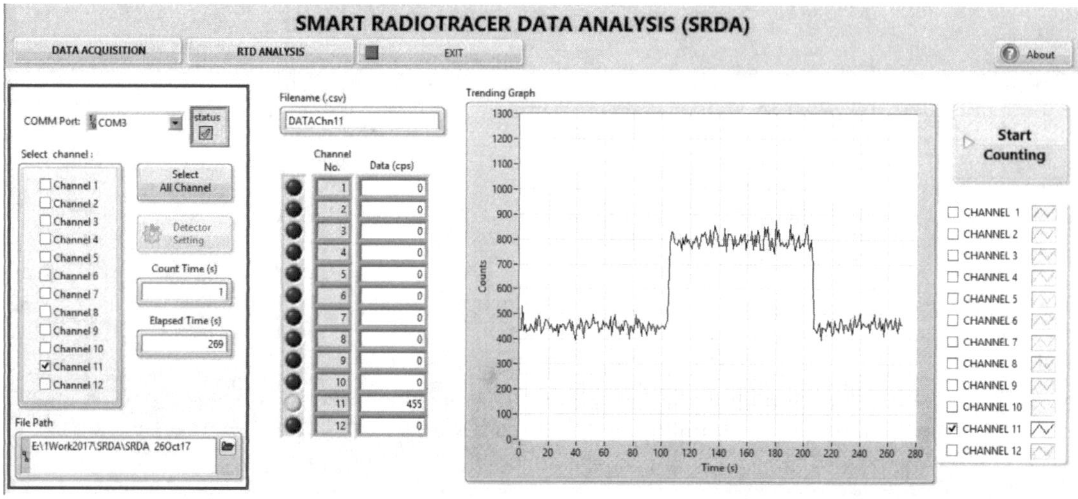

Figure 3. Online monitoring after radioactive injection into the system

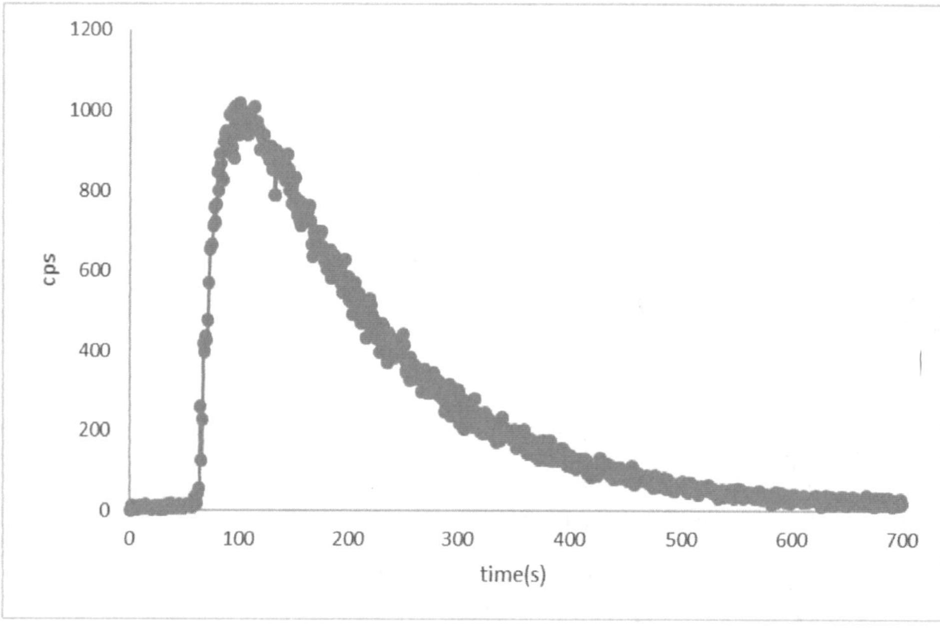

Figure 4. Plotted graph of raw data

By undergoing all data treatment steps as aforementioned, the step by step data treatment can be seen in Figure 5.Unlike conventional practice that is time consuming in which each of the data treatment should be calculated manually as described in Equations 1-6. It can be observed that parasitic noises are very imminent in the first measured data, and the treatments are followed consecutively from background correction, decay correction, zero point correction, filtered data and extrapolation respectively in order to obtain radioactive signals which are further developed for RTD analysis.

IOP Conf. Series: Materials Science and Engineering **458** (2018) 012083 doi:10.1088/1757-899X/458/1/012083

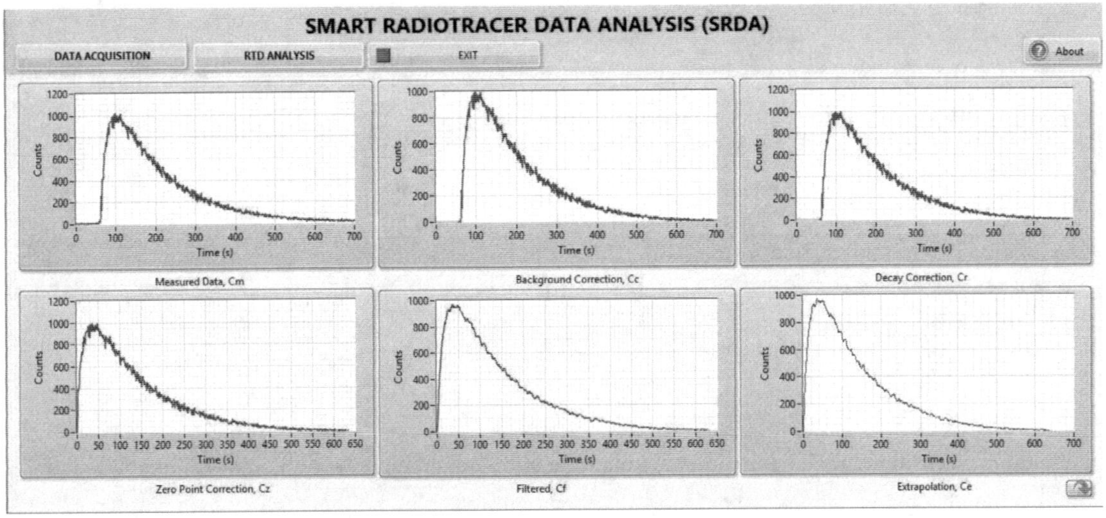

Figure 5. Automatic data treatment analysis

The final data treatment step, which is extrapolation has produced treated data that free from noises and can represent radioactive data accurately and precisely. The final counts (cps) vs time curve is then undergoing another analysis for determination of RTD which adopts Equation 5.

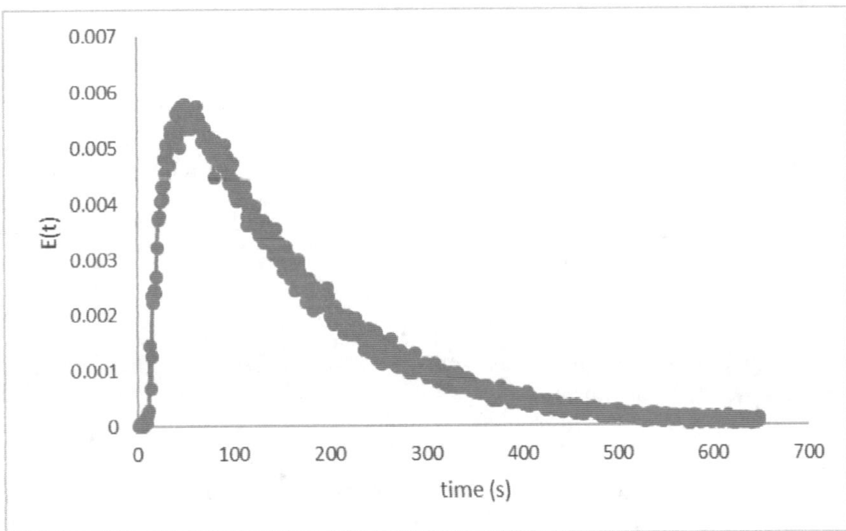

Figure 6. RTD determination using Eq. 5

Figure 6 shows the manually plotted curve of RTD and the value of MRT using Equation 6 is 133.60 s whereas Figure 7 provides the hassle free RTD curve and automatic calculated MRT which is 134s.

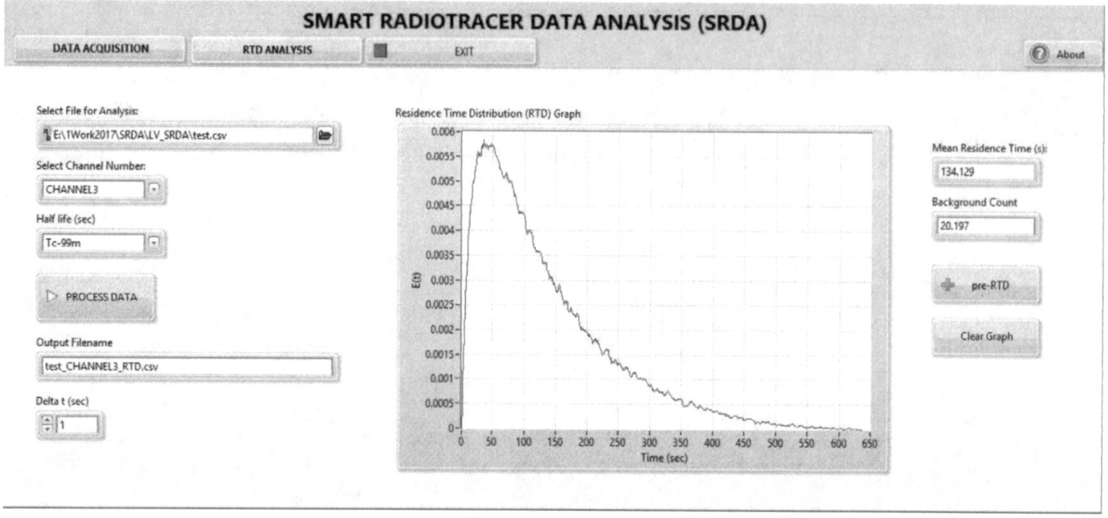

Figure 7. RTD Measurement and MRT Determination

Nevertheless, in order to determine quantitatively the percentage of dead zone or stagnant zone in the processing mixer column, the calculation of theoretical MRT should be carried out and the equation is as follows:

The theoretical MRT is the ratio of the vessel volume, V, to the volumetric flow rate, Q:

$$\mathrm{MRT}_{theory} = \frac{V}{Q} \tag{7}$$

The measured value of active volume, V by adopting $\pi r^2 h$ is 0.011m3 or 11L where r=d/2; 200mm/2 and h=350mm. The volumetric flow rate is 5 l/m or 0.083l/s. Therefore, theoretical MRT for this study is 132.5s. Thus, the relative difference between theoretical and experimental residence time or dead zone is:

$$V_{dead}\,(\%) = \frac{\mathrm{MRT}_{theory} - \tau}{\mathrm{MRT}_{theory}}\, x\, 100 \tag{8}$$

Hence, the percentage of dead zone in this case study is 1.1% which shows the existing parameters are sufficient to retain the process optimization

4. Conclusions

The development of Smart Radiotracer Data Analysis (SRDA) as an easy, fast and simple data analysis has been using as tool in assisting radiotracer engineer especially Plant Assessment Technology group in determining process anomalies efficiently and effectively in industrial plant in Malaysia.

Acknowledgments

The authors would like to express their special thanks to Plant Assessment Technology (PAT) staff for assisting us throughout this study. Not to forget International Atomic Energy Agency (IAEA) who provided CRP fund and Nuclear Malaysia in making the research fruitful.

References

[1] Othman N and Kamarudin S K 2014 *Sci. World J.* 768604
[2] International Atomic Energy Agency 2008 *Radiotracer residence time distribution method for industrial and environmental applications* (Vienna: IAEA)
[3] Arghya D, Raj Kumar G, Sunil G and Harish Jagat P 2017 *Appl. Radiat. Isot.* **130**
[4] Sarkar M, Sangal V K, Sharma V K, Samantray J, Bhunia H, Bajpai P K, Kumar A, Naithani, A.K and Pant H. 2017 Appl. Radiat. Isot. 130 270-275
[5] Danckwerts **P V** 1953 *Chem. Engng Sci.* **2** 1–13
[6] Sheoran M, Goswami S, Pant H J, Biswal J, Sharma V K, Chandra A, Bhunia H, Bajpai P K, Rao S M, and Dash A 2016 Appl. Radiat. Isot. 111 10-7
[7] Othman N, Mohamad Hassan N P, Yahya R, Meor Adnan A K, Hassan H and Mahmood A 2017 Malays. J. Anal. Sci. 21 445-451
[8] Sharma V K, Pant H J, Tandon D and Garg M O 2016 *Appl. Radiat. Isot.* **107** 57-63
[9] Pant H J, Thyn J, Zitny R and Bhatt B C 2001 *Appl. Radiat. Isot.* **54** 1-10
[10] Pant H J, Yelgoankar V N 2002 *Appl. Radiat. Isot.* **57** 319–325
[11] Furman L and Stegowski Z 2011 *Chem. Eng. Process.* **50** 300-304
[12] Kasban H, Zahran O, Arafa H, El-Kordy M, Elaraby S M and El-Samie F.A 2010 Appl. Radiat.Isot. 68 1049–1056
[13] Moreira R M, Pinto M F, Mesnier R, Lecler J P 2007 Appl. Radiat. Isot. 65 419–427

Characterization of Chicken Manure from Manjung Region

Gobind Singh[1], **M. R. Shamsuddin**[1,2*], **Aqsha**[1,2] and **S. W. Lim**[3]

[1] Chemical Engineering Department, Universiti Teknologi PETRONAS, 32610, Seri Iskandar, Perak, Malaysia.

[2] Centre for Biofuel and Biochemical Research (CBBR), Institute for Sustainable Living, Universiti Teknologi PETRONAS, 32610, Seri Iskandar, Perak, Malaysia.

[3] Dindings Poultry Development Centre Sdn Bhd, Batu 10 Persiaran Segari, Jalan Damar Laut, 32200 Segari, Lumut, Perak, Malaysia.

*mrashids@utp.edu.my

Abstract. Poultry business is a growing industry in Manjung Region, associated with high loading of daily manure generation. The manure contains food source that attract flies to lay eggs, which create an endemic health and societal problem. Chicken manure however can be a potential resource for value-added products such as biogas and bio fertilizer. As a start, manure was characterized for its contents such as using CHNS, TGA, AAS, XRF and ICP-OES. Key highlights include protein content of 34.5% and 27% as well as total organics of 45% and 51% for fresh and composted manure respectively. Stored energy was determined as 6905 J/g for fresh manure and 8239 J/g for composted manure. These and other parameters determined are comparable to literature values on chicken manure from various countries. The availability of energy and protein in the manure provides an avenue to convert them into another form of energy such as biogas. As the food source is digested, the fly issue could potentially be alleviated.

1. Introduction

Manjung district, a south-western part of the state of Perak in Malaysia is known for its major tourist attraction of Pangkor Island. The district alone supplies about 10% of the national chicken produce. At typical steady-state chicken population of 14 million, Manjung's poultry generates approximately 1400 tons of chicken dropping daily for disposal. Currently, the biomass is sold raw at RM 3 per 20 kg for agricultural use, thus resulting in widespread flies' problem across the region. Two commonly employed farming practices are open and closed farming. Both systems are ridden with flies due to the large daily manure generation with the open system being significantly detrimental.

Generally, flies are attracted to decomposed organics such as food waste and manure to feed and lay eggs [1]. When the infested manure is used as fertilizer in agricultural land, under suitable humidity the eggs hatch into flies that created an endemic health and societal problem. One harvesting cycle of broilers (chickens bred for meat) takes about 40 days, which is also when the manure is collected to be sold as raw fertilizer. Apart from fertilizer, the manure is a potential biomass for biogas production [2-5].

Prior to any treatment, the fresh manure was characterized in terms of moisture content, carbon and nitrogen content as well as macro and micronutrients. Analysis techniques used include CHNS, atomic absorption spectroscopy (AAS), inductive coupled plasma optical emission spectrometry (ICP-OES), and Bomb Calorimetry.

This paper discusses experimental findings on manure characterization from Manjung region in comparison to literature values obtained from various countries. The established values could serve as a baseline values for researchers to further enrich the nutritional content or to modify the parameters for specific premix formulations of biomass for biogas production.

2. Material and methods

2.1. Material

Fresh and composted chicken manure were obtained from Manjung's poultry closed farm. The fresh manure was collected after 40 days of a harvesting cycle while the composted manure was supplied by Dindings Poultry Sdn. Bhd.

2.2. Methods

2.2.1. Determination of moisture content

Moisture content was analysed in triplicate using a HX-240 Moisture Analyser by Mettler Toledo.

2.2.2. Determination of pH value

The pH value of chicken manure was determined in triplicate using a pH meter of Eutech pH 700 at weight ratio of 1:10 (manure: water) [6].

2.2.3. Determination of carbon and nitrogen content

Carbon and nitrogen content were analysed using a Perkin Elmer 2400 series CHNS analyser on 1 mg manure sample.

2.2.4. Determination of Protein, Macro and Micro Nutrients

Theoretical protein content was calculated by multiplying nitrogen content with 6.25 [7]. Microelements of K^+, Na^+, Ca^{2+}, Mg^{2+} were determined using two methods- AAS and ICP-OES for comparison. For both analyses, 0.1g of manure sample was added into 1g of concentrated nitric acid (HNO_3) and 5g of distilled water. The mixture was transferred into ETHOS One Microwave Digester at 170°C (500 W) for 50 minutes. Later, the sample was diluted to 1% (volume basis) concentration prior to analysis using AAS and ICP-OES.

2.2.5. Thermogravimetric Analysis (TGA)

5 mg manure was weighed and transferred to microbalance of a TGA Q50 analyser. The system was purged with nitrogen prior to thermal treatment under gas mixture of air at a flow rate of 20 $ml.min^{-1}$. The temperature range used was 25 - 900°C at a heating rate of 20°$C.min^{-1}$. The sample then was held at 900°C for 1 hour. Thermal degradation data was extracted to plot the percentage of weight loss over temperature [8].

3. Result and discussion

3.1. Characterization of chicken manure

Table 1 summarizes the characterization results in comparison to literature findings. Fresh and composted manure were determined to have high protein and carbohydrate at 34.5% and 31% at the highest. These are food sources for flies that could explain the rampant regional flies problem. Fresh manure contains 34.5% protein and 21% carbohydrate as compared to composted manure at 27.13% and 31% respectively. Both values reflect the poultry practice of overfeeding the chicken in order to meet minimum body weight of 2 kg prior to harvesting in 40 days. The chickens therefore do not have sufficient time to digest the food which are predominantly starch from mixed grains. Protein and carbohydrate content signify stored energy in the manure that could be potential for energy conversion into biogas.

In terms of C and N content, it is not expected for composted manure to have higher C and less N. In fact, the opposite trends typically prevail in various work [9]. Low C/N ratio (less than 20) signifies nitrogen volatilization in the form of ammonia and odours which is enhanced by high temperature and basic pH [10]. The pH recorded for composted chicken manure is 7.79 as compared to 6.10 for fresh chicken manure. This shows that composted manure is aged significantly to reach the classic pH of matured compost at about neutral. However, at C/N value of 7.19, the ratio is far less than normally recorded for matured compost in the range of 25 to 30. Compost with such C/N ratio signifies balance degradation process between carbon and nitrogen sources [11]. These results show that the composted manure has not reached maturation stage and could still attract flies based on its protein and carbohydrate leftovers. In general, the C and N content determined in this work differ greatly to the literature references, attributed to the chicken diet and farming practices worldwide. This is evidence by equally large variations between the literature values as well. This trend continues for macro and microelements in the manure as can be seen for K^+, Na^+, Ca^{2+} and Mg^{2+}.

Table 1. Elemental analysis of the chicken manure

Parameters	Experimental		Literature Review			
	Fresh Chicken Manure	Composted Chicken Manure	[12]	[13]	[14]	[15]
Total C	21.12%	31.20%	-	44.70 %	407.4 g/kg	22.70%
Total N	5.52%	4.34%	-	24.30 %	55.7 g/kg	2.40%
C/N Ratio	3.83	7.19	-	1.84	7.31	9.5
Moisture	39.73%	28.26%	-	-	-	-
Protein	34.5%	27.13%	-	-	-	-
K^+	537.03 ppm^a	574.25 ppm^a	588.9 ppm	0.70^c %	14469 ppm	-
Na^+	92.35 ppm^a	486.59 ppm^a	193.2 ppm	-	-	375.8 ppm^b
Ca^{2+}	1633 ppm^b	811 ppm^b	1958 ppm^a	5.60^c %	1480 ppm	1503 ppm^b
Mg^{2+}	101.8 ppm^b	206 ppm^b	75.6 ppm^a	0.30^c %	192 ppm	436.7 ppm^b
pH	6.10	7.79	7.29	-	7.7	6.94
Energy Content (J/g)	6905^d	8239^d	-	-	-	-

^{a} AAS analysis
^{b} ICP-OES
^{c} XRD
^{d} Bomb calorimeter

The elemental concentration increased after composting process except for Ca^{2+}. The values are comparable to literature findings suggesting that the natural chicken digestive system able to render the food into a uniform quality of manure. The only difference is on the protein (N) and carbohydrate (C) leftovers that are heavily dependent on the feeding practice. Cations are essential supplements that impact soil chemical properties which eventually affect plant development [7]. Fresh manure has lower Na^+ content while composted manure has a low concentration of Mg^{2+} relative to other elements, similar trend is represented in the literatures. Sodium (Na^+) is a micronutrient for plant while the rest of the cations are macronutrients. Based on low Na^+ alone, fresh manure could be a better plant's growth promoter as the element is only required in small amount, but other qualities should also be considered when choosing a suitable fertilizer.

During the decomposition of organic matter, nitrogen usually experiences two different stages of mineralization and immobilization. The latter results in nitrogen decomposition into plant accessible forms such as NH_4^+ and NO_3^- (via nitrification). Immobilization of nitrogen occurs when the accessible nitrogen species are taken up by microorganisms, preventing them from being accessible by plants. The extent of nitrogen immobilization is related to the bio-chemical composition of the manure. C/N ratio for composted chicken manure is significantly higher than fresh chicken manure. High C/N ratio in compost will promote nitrogen immobilization. The immobilized nitrogen will be available to plants after the microorganisms die and the nitrogen is released [16]. The results obtained in this study suggest that chicken manure can be a decent source of organic carbon which is essential in enhancing soil quality while the N plays a critical part in plant development [12]. However, without an analysis of common fertilizer quality benchmark of N, P and K content; both fresh and composted manure examined cannot be deemed suitable as an all-rounder fertilizer. Nevertheless, there are generous amount of literature to support the claim especially on chicken manure role to promote plant's growth [12].

3.2. Thermogravimetric analysis

TGA profiles of fresh and composted chicken manure are shown in Figure 1. The mass loss curves of chicken manure over temperature were summarized into Table 2. The percentage weight was calculated at temperature intervals from 160°C to 900°C, corresponding to percentage loss of water, surface organics, bound organics and low volatility organics [17].

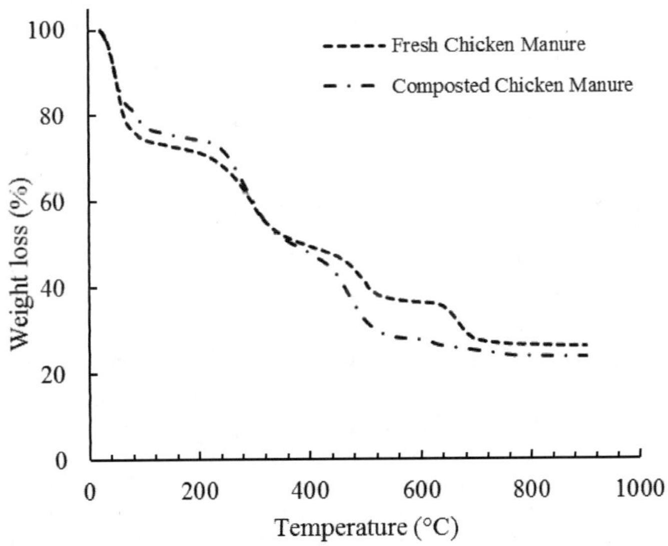

Figure 1. TGA curve for Fresh and Composted chicken manure

Table 2. Summary of TGA for fresh and composted chicken manure

Types of chicken manure	TGA composition				Total Organics	Ash
	0-160 (°C)	160-400 (°C)	400-600 (°C)	600-900 (°C)		
	A	B	C	D		
Fresh Chicken Manure (%)	27.37	22.97	13.12	10.24	45.11	26.20
Composted Chicken Manure (%)	24.60	27.20	20.36	4.00	50.70	23.74

[A] Water;
[B] Surface organics;
[C] Hydroxyl water and intercalated organics;
[D] Low volatility organics/compounds

From room temperature to 160°C, the weight loss of 24–28% indicates the moisture content of the manure where fresh manure has about 3% higher moisture than its composted version. The weight loss from 160°C to 400°C is mainly from thermal degradation of cellulose, hemicelluloses, protein, starch and microbial cell walls [18]. Composted manure has higher concentration of total organics and lower ash content compared to the fresh manure. The reason being that as the water mass is reduced, the other components are concentrated.

At 400°C to 600°C, about 13% of mass is further reduced for fresh manure, indicating the content of bound organics that are trapped within the sample lattice and heat-resistant organics such as lignin. Lignin is a recalcitrant biopolymer that resists degradation. Lignin degradation is suggested to occur between 300°C and 500°C [8]. Composted chicken manure did not exhibit this degradation curve between 400–600 °C, possibly due to thermal or chemical treatment used during the composting process to speed up the compost generation. The former could be ruled out as the sample contained moisture and surface organics.

The last region which is 600–900°C, the weight loss rate became steady and close to zero. The weight that loss between 600°C to 900°C corresponds to low volatility organics. These are particulate matters of compounds derived from the atmospheric transformation of organic species [19]. Residual mass of 23–27% after 900°C represents the ash and fixed carbon content. Composted manure contains less ash due to pre-treatment process employed such as screening and drying. Whereas fresh manure contains grits or sand particles from food source.

4. Conclusion

Fresh and composted manure samples contain high amount protein and carbohydrates which are a food source that attract flies. At 27% protein and 31% carbohydrate, the compost is far from maturation as also evidences by C/N ratio of 7.19. The elemental analysis revealed that both fresh and composted manure could be potential as fertilizer however more analysis in terms on N, P and K values should be conducted. Variation in characterization values between experiment and literature is mainly due to different broiler diets in various countries. Besides being sold at RM 3 per 20 kg, the manure has high potential for biogas generation via anaerobic digestion and biofertilizer production by aerobic composting. Future work will examine the effectiveness of both treatment processes to convert the stored energy into biogas and the compost should alleviate the flies problem. This adds values to what has traditionally been considered as waste.

Acknowledgments
This work was supported by Yayasan Universiti Teknologi PETRONAS grant (0153AA-H33). The author also would like to acknowledge Centre for Biofuel and Biochemical Research (CBBR) of UTP for the technical assistance on TGA, Bomb Calorimeter, TKN Analyser, ICP-OES, AAS, CHNS, and Moisture Analyser.

References

1. L. Beard, R. and D. Sands, *Factors Affecting Degradation of Poultry Manure by Flies*. Vol. 2. 1973. 801-806.
2. Fantozzi, F. and C. Buratti, *Biogas production from different substrates in an experimental Continuously Stirred Tank Reactor anaerobic digester.* Bioresour Technol, 2009. **100**(23): p. 5783-5789.
3. Bujoczek, G., et al., *High Solid Anaerobic Digestion of Chicken Manure.* J Agr Eng Res, 2000. **76**(1): p. 51-60.
4. Li, Y., et al., *Biogas production from co-digestion of corn stover and chicken manure under anaerobic wet, hemi-solid, and solid state conditions.* Bioresour Technol, 2013. **149**: p. 406-412.
5. Kolodynskij, V. and P. Baltrenas, *Experimental Research of Biogas Yield and Quality Produced from Chicken Manure*. 2017.
6. Tiquia, S.M. and N.F.Y. Tam, *Characterization and composting of poultry litter in forced-aeration piles.* Process Biochem, 2002. **37**(8): p. 869-880.
7. Lanyasunya, T., et al., *Factors Limiting Use of Poultry Manure as Protein Supplement for Dairy Cattle on Smallholder Farms in Kenya*. Vol. 5. 2006.
8. Várhegyi, G., et al., *Thermogravimetric Study of Biomass Pyrolysis Kinetics. A Distributed Activation Energy Model with Prediction Tests.* Energy & Fuels, 2011. **25**(1): p. 24-32.
9. Gao, M., et al., *Evaluation of stability and maturity during forced-aeration composting of chicken manure and sawdust at different C/N ratios.* Chemosphere, 2010. **78**(5): p. 614-619.
10. Diaz, L.F., M. de Bertoldi, and W. Bidlingmaier, *Compost Science and Technology*. 2011: Elsevier Science.
11. Shamsuddin, R.M., A. Borhan, and W.K. Lim, *Humic acid batteries derived from vermicomposts at different C/N ratios.* IOP Conf Ser Mater Sci Eng, 2017. **206**(1): p. 012067.
12. Dikinya, O. and N. Mufwanzala, *Chicken manure-enhanced soil fertility and productivity: Effects of application rates*. Vol. 1. 2010. 46-54.
13. Lee, J., et al., *Enhancement of energy recovery from chicken manure by pyrolysis in carbon dioxide.* J Clean Prod, 2017. **164**: p. 146-152.
14. Faridullah, et al., *Characterization of trace elements in chicken and duck litter ash.* Waste Manag, 2009. **29**(1): p. 265-271.
15. Ravindran, B., et al., *Assessment of nutrient quality, heavy metals and phytotoxic properties of chicken manure on selected commercial vegetable crops.* Heliyon, 2017. **3**(12): p. e00493.
16. Al-Bataina, B.B., T.M. Young, and E. Ranieri, *Effects of compost age on the release of nutrients.* Int Soil Water Conserv Res, 2016. **4**(3): p. 230-236.
17. Shamsuddin, R.M., C.J.R. Verbeek, and M.C. Lay, *Producing protein intercalated bentonite — Equilibrium, kinetics and physical properties of gelatin–bentonite system.* Appl Clay Sci, 2014. **87**: p. 52-60.
18. Kim, S.-S. and F.A. Agblevor, *Pyrolysis characteristics and kinetics of chicken litter.* Waste Manag, 2007. **27**(1): p. 135-140.
19. Kroll, J.H. and J.H. Seinfeld, *Chemistry of secondary organic aerosol: Formation and evolution of low-volatility organics in the atmosphere.* Atmos Environ, 2008. **42**(16): p. 3593-3624.

An In-Situ Thermogravimetric Study of Pyrolysis of Rice Hull with Alkali Catalyst of $CaCO_3$

D K W Gan[1], B L F Chin[1], A C M Loy[2,3], S Yusup[2,3] , M N Acda[4] , P Unrean[5] , E Rianawati[6], Z A Jawad[1] and R J Lee[1]

[1]Department of Chemical Engineering, Faculty of Engineering and Science, Curtin University Malaysia, CDT 250, 98009 Miri Sarawak, Malaysia.
[2]Biomass Processing Lab, Centre for Biofuel and Biochemical Research, Institute of Sustainable Living, Universiti Teknologi PETRONAS, Bandar Seri Iskandar, Tronoh 31750, Malaysia.
[3]Department of Chemical Engineering, Universiti Teknologi PETRONAS, Bandar Seri Iskandar, Tronoh 31750, Malaysia.
[4]Department of Forest Products and Paper Science, University of the Philippines Los Baños, College, Laguna 4031, Philippines.
[5]National Center for Genetic Engineering and Biotechnology (BIOTEC) Pathum Thani, Thailand.
[6]Resilience Development Initiative, Jl. Imperial 2, No. 52 Bandung 40135, Indonesia.

bridgidchin@curtin.edu.my

Abstract: Pyrolysis of rice hull (RH) with the presence of $CaCO_3$ catalyst was carried out in this study to understand the effect of alkali catalyst in the thermal degradation behaviour and evaluate the kinetic parameter of rice hull for bio-oil or syngas production. Five different heating rates of the pyrolysis experiments at 10, 20, 30, 50, and 100 $Kmin^{-1}$ were carried out in thermogravimetric analysis (TGA) equipment. Model fitting kinetic Coats Redfern integral method was applied in this study to estimate the activation energy (E_A) and pre-exponential (A) value of catalytic pyrolysis in RH. The results showed that the maximum degradation increased from 6.69 to 52.67 wt% min^{-1} as heating rates increases from 10 to 100 $Kmin^{-1}$. Besides that, the E_A of the catalytic pyrolysis for RH using $CaCO_3$ catalyst 60.86 $kJmol^{-1}$ which is lower than other similar pyrolysis reaction reported in literature i.e. 77.4 kJ/mol. Meanwhile, the A value for the catalytic pyrolysis for RH using $CaCO_3$ catalyst was 4.68×10^{10} min^{-1} which is significantly higher than 1.1×10^{6} min^{-1} as reported in literature for non-catalytic pyrolysis of rice husk.

1. Introduction

Recently, the world's energy consumption is mainly generated from fossil fuels and analysts have warned about the depletion of this non-renewable resources in the future. On the other hand, the environment awareness has increased due to greenhouse gas effect has restraint on fuel combustion emission, thus renewable energy has attracted plenty of attention in recent years [1].Biomass (e.g. rice hull, coconut shell, microalgae) is one of the promising renewable sources since it is cheap, biodegradable and environmentally friendly [2]. The world production of biomass was estimated to be 146 billion metric tons per year [3]. The biomass energy source can be attained through thermochemical conversion method such as pyrolysis [4], gasification [5] and direct combustion [6].

Among all the thermochemical conversion, pyrolysis is considered one of most effective method to convert the biomass to energy since it could reduce the possibility of unwanted oxidation in the pyrolyzer and enhance the liquid (bio-oil) and gaseous (syngas) production [7]. Pyrolysis process is a thermal decomposition of materials in an inert atmosphere under non-isothermal conditions. Hence, kinetic study is fundamental for investigating the thermal degradation behavior of biomass to promote the product yield and obtain the energy recovery.

In recent studies, there are plenty of research studies on the catalysts implementation to the pyrolysis process to reduce the energy input and enhance the syngas production [8]. Zhang et al. (2013) [9] has reported that cracking with the presence of catalyst can increase liquid yield and reduce operation cost. There are three types of catalysts composition for biomass gasification process such as alkali metal (reduce tar content and methane content), dolomite (removal of hydrocarbons), and nickel (removal of hydrocarbons and gas composition adjustment) [10]. There are numerous discussions carried out by the research communities on the alkali minerals exploitation. Giudicianni et al. (2018) [11] has reported the addition of alkali ions in raw xylon pyrolysis produced high solid residue and Chen et al. (2018) [12] reported that pyrolysis of RH treated with alkali able to increase the yield of aromatics.

Kinetic parameters such as activation energy (E_A) and pre-exponential value (A) that obtained in catalytic pyrolysis can be used as a guidelines to design, optimize, and scale up the industrial plant level for this thermochemical process [13]. Chin et al. (2014) [14] had reported that thermogravimetric analysis (TGA) is one of the most convenient and widely used method to understand the thermal degradation behavior of biomass and determine the kinetic parameters in pyrolysis process. Meanwhile, Coats-Redfern integral is one of the non-isothermal model fitted method to describe kinetic parameters of non-catalytic pyrolysis and catalytic pyrolysis of biomass. Balasundrum et al. (2017) [8] has reported the kinetic study of catalytic pyrolysis of biomass for possible bio-oil production using Coats-Redfern integral method, and Naqvi et al. (2015) [15] has reported the kinetic study of catalytic pyrolysis of paddy husk using Coats-Redfern integral method.

To our best knowledge, most of the catalysts reported in previous study are transition element [16], acid catalysts [17] and mesoporous materials [18]. Meanwhile, there are only a few studies described the effect of alkali catalysts in catalytic pyrolysis of biomass. In our previous work, we had investigated the non-catalytic pyrolysis of rice husk using Kissinger-Akahira-Sunose method [19]. Therefore, the aim of this study is to investigate the effect of alkali metal of $CaCO_3$ catalyst in thermal degradation of catalytic pyrolysis of rice hull and determine the activation energy (E_A) and pre-exponential factor (A) using Coats Redfern integral method and compared with the E_A obtained in non-catalytic pyrolysis of rice hull that had been reported in previous literature.

2. Experimental

2.1 Material preparation

RH was obtained from Malaysia rice mill. The RH was grounded and sieved to a particle less than 710 μm. The ultimate analysis of the RH was estimated using LECO CHNS-932 elemental analyzer and proximate analysis was analyzed using BS EN 14774-3:2009, BS EN 15148:2009 and BS EN 14775:2009, respectively [20]. Alkali metal catalyst of $CaCO_3$ was collected from Calrock Sdn. Bhd., Malaysia. The elemental analysis of the $CaCO_3$ catalyst was estimated using X-ray fluorescence (XRF) S8 Tiger Bruker.

2.2 Experimental procedure

In the present study, an approximate of 5.0 mg of RH was placed on a ceramic crucible with 0.5 mg of $CaCO_3$ catalyst which is 10% of the total weight of the RH. The experiments were conducted using

TGA under inert environment of nitrogen (N_2) gas. Before the experiments was conducted, a 100mL/min flowrate of N_2 was introduced into the TGA system to remove unwanted gaseous. This could minimize the unwanted gaseous presence in the system that could cause coke formation on the catalysts and oxidation on the samples. Subsequently, all the samples were heated from temperature 323 to 1173 K at five different heating rates of 10, 20, 30, 50, and 100 $Kmin^{-1}$ respectively. All the experiments were carried more than 3 times to ensure the accuracy of the data obtained.

2.3 Kinetic study

In this study, Coats-Redfern method was applied to study the kinetics of the catalytic pyrolysis of RH using MATLAB software R2015b. Coats-Redfern method uses the integral form of the non-isothermal rate law derived from Eq. (1) and integrate into Eq. (2).

$$\frac{d\alpha}{dT} = \frac{A}{\beta} e^{-\frac{E_A}{RT}} f(\alpha) \tag{1}$$

$$g(\alpha) = \int_0^{\alpha} \frac{d\alpha}{f(\alpha)} = \frac{A}{\beta} \int_{T_o}^{T} e^{-\frac{E_A}{RT}} dT \tag{2}$$

It utilized the asymptotic series expansion for approximating the temperature integral, *p(x)*, Eq. (3) and derived Eq. (4) shown in below:

$$\ln p(x) = \frac{g(\alpha)\beta R}{AE_A} \tag{3}$$

$$\ln \frac{g(\alpha)}{T^2} = \ln\left(\frac{AR}{\beta E_A}\left[1 - \left(\frac{2RT_{exp}}{E_A}\right)\right]\right) - \frac{E_A}{RT} \tag{4}$$

where T_{exp} is the mean experiment temperature. In this study, twenty integral functions, $g(\alpha)$ represents different reaction models and the expression of $g(\alpha)$ were collected from previous studies tabulated in Table 1 [10]. By plotting the $\left(\ln \frac{g(\alpha)}{T^2}\right)$ versus $\left(\frac{1}{T}\right)$ graph, the E_A and A can be obtained from the slope and intercept respectively. According previous studies, first order reaction is widely used as integral function model to determine the kinetic parameters [21].

Table 1: Reaction model functions and its mechanism [8, 22].

Model	Model name	Differential form, $f(a) = \frac{1}{k}\frac{da}{dt}$	Integral form, $g(a) = kt$
	Nucleation model		
1	Power law (P2)	$2\alpha^{1/2}$	$\alpha^{1/2}$
2	Power law (P3)	$3\alpha^{2/3}$	$\alpha^{1/3}$
3	Power law (P4)	$4\alpha^{3/4}$	$\alpha^{1/4}$
4	Avarami-Erofe'ev (A2)	$2(1-\alpha)[-\ln(1-\alpha)]^{1/2}$	$[-\ln(1-\alpha)]^{1/2}$
5	Avarami-Erofe'ev (A3)	$3(1-\alpha)[-\ln(1-\alpha)]^{2/3}$	$[-\ln(1-\alpha)]^{1/3}$
6	Avarami-Erofe'ev (A4)	$4(1-\alpha)[-\ln(1-\alpha)]^{3/4}$	$[-\ln(1-\alpha)]^{1/4}$
7	Prout-Tompkins (B1)	$a(1-\alpha)$	$\ln[a/(1-\alpha)]$
	Geometrical Contraction model		
8	Contracting area(R2)	$2(1-\alpha)^{1/2}$	$[1-(1-\alpha)^{1/2}]$
9	Contracting volume (R3)	$3(1-\alpha)^{2/3}$	$[1-(1-\alpha)^{1/3}]$
	Diffusion models		
10	1-D diffusion (D1)	$1/2\,\alpha$	α^2
11	2-D diffusion (D2)	$[-\ln(1-\alpha)]^{-1}$	$[(1-\alpha)\ln(1-\alpha)]+\alpha$
12	3-D diffusion-Jander eqn. (D3)	$\frac{3(1-\alpha)^{2/3}}{2(1-(1-\alpha)^{1/3})}$	$[1-(1-\alpha)^{1/3}]^2$
13	3-D diffusion Anti-Jander eqn.		$[(1+\alpha)^{1/3}-1]^2$
14	3-D diffusion Zhuralev, Lesokin, Tempelman equation		$[1/(1-\alpha)^{1/3}-1]^2$
15	Ginstling-Brounshtein (D4)	$(3/2\,(1-\alpha)^{-1/3}-1)$	$1-(2\alpha/3)-(1-\alpha)^{2/3}$
	Reaction-order models		
16	Zero-order (F0/R1)	1	α
17	Chemical reaction (n = 1)	$(1-\alpha)^{-2}$	$(1-\alpha)^{-1}$
18	First-order (F1)	$(1-\alpha)$	$-\ln(1-\alpha)$
19	Second-order (F2)	$(1-\alpha)^2$	$(1-\alpha)^{-1}-1$
20	Third-order (F3)	$(1-\alpha)^3$	$0.5((1-\alpha)^{-2}-1)$

3 Results and Discussion

3.1 Ultimate and proximate analysis of rice hull

The ultimate and proximate analysis for RH was shown in Table 2. The carbon, hydrogen, nitrogen, sulphur and oxygen content of the RH were estimated to be 39.65 wt%, 5.21 wt%, 1.20 wt%, 0.02 wt% and 53.92 wt % in dry and ash free basis respectively. From the proximate analysis, it showed that RH contained a high volatile content (69.20 wt%) and low ash content (1.23 wt%) can increase the bio-oil production which indicated it is suitable for fuel production [23].

Table 2: Characteristic of Rice Hull.

Ultimate analysis (wt %)[a]	
Carbon	39.65
Hydrogen	5.21
Nitrogen	1.20
Sulphur	0.02
Oxygen	53.92
Proximate analysis (wt %)[b]	
Moisture content	4.29
Volatile matter	69.20
Fixed carbon[a]	19.45
Ash	1.23

[a] *dry and ash-free* basis
[b] dry basis

3.2 Chemical Characterisation for Alkali Catalyst $CaCO_3$

The elemental analysis of alkali metal catalyst of $CaCO_3$ was listed in Table 3. The magnesium, calcium, iron, silicon, and other content of the alkali metal catalyst of $CaCO_3$ were estimated to be 1.05 wt%, 91.20 wt%, 2.63 wt%, 4.67 wt%, and 0.45 wt% respectively. It can be noticed that calcium content has the highest weightage in $CaCO_3$ which indicates the purity of $CaCO_3$. Sutton et al. (2001) [10] has reported that higher purity of Ca in $CaCO_3$ may increase in the reactivity of decomposition.

Table 3: Chemical characterisation of $CaCO_3$.

Elemental analysis (wt %)	
Magnesium	1.05
Calcium	91.20
Iron	2.63
Silicon	4.67
Others	0.45

3.3 Thermal Degradation Behavior of RH

Figure 1(a) and 1(b) show the thermogravimetric (TG) analysis in wt% and derivative thermogravimetric (DTG) analysis in wt% min^{-1} for catalytic pyrolysis of RH under five different heating rates of 10, 20, 30, 50, and 100 $Kmin^{-1}$ respectively. It was observed that the curves in the TG profile for the catalytic pyrolysis of RH showed similar trend of weight loss and almost identical in all heating rates studied. The TG-DTG curves were divided into three phases of thermal degradation: vaporization of moisture developed at temperature less than 450 K (Phase I), followed by Phase II which is the devolatilization phase of volatile matter at temperature range of 450 to 800 K and Phase III which is the lignin decomposition of biomass at temperature range of 800 to 1173 K. Moreover, there is a small peak can be observed in Fig.1(b) at temperature around 1023 to 1123 K was associated to carbonization of $CaCO_3$ as shown in Eq. (5).

$$CaCO_3 \leftrightarrow CaO + CO_2 \quad (5)$$

From Figure 1 (a) and 1(b), it can be observed that the mass loss curves shifted to the right as heating rates increased from 10 to 100 Kmin^{-1}. Also, it can be noticed there was lateral shift in temperature of the maximum degradation peak as heating rate increased. These right lateral shift occurrences were due to the combined effect of heat transfer at different heating rates and the heat conductive property of the biomass particle [24]. The maximum degradation rate has rose from 6.69 wt% min^{-1} to 59.37 wt% min^{-1} for the heating rate of 10 to 100 Kmin^{-1}. The rise of the maximum degradation peak showed that heating rate may affect the degradation rate of pyrolysis process. The residue produced from the catalytic pyrolysis of RH has increased compare to raw RH studies reported by Ahiduzzaman et al. (2015) [25]. This occurrence was good agreement with the studies reported by Giudicianni et al. (2018) [11] where the elimination of tar has converted to char.

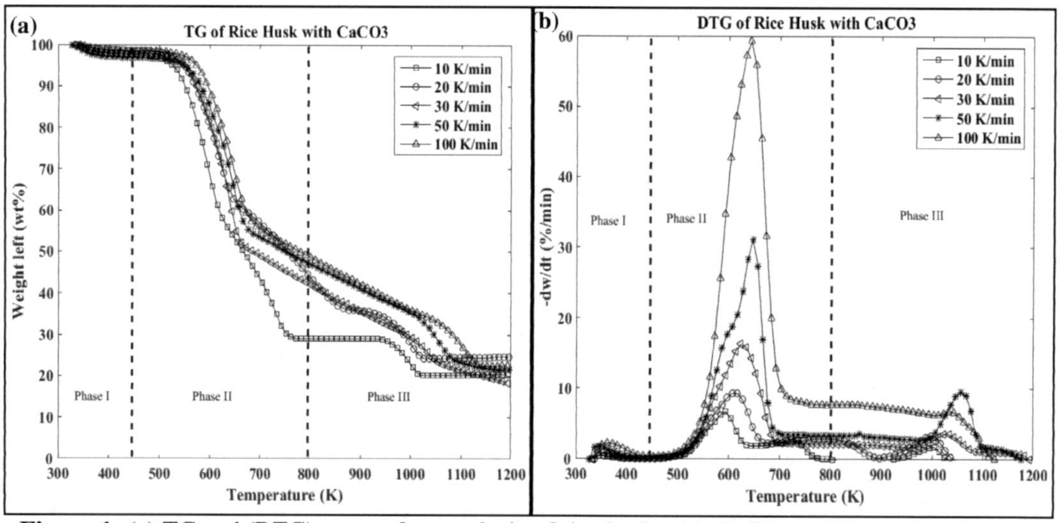

Figure 1: (a) TG and (DTG) curves for pyrolysis of rice husk with $CaCO_3$ at different heating rates.

3.4 Determination of kinetic parameters from Coats-Redfern method

The kinetic parameters of pyrolysis of RH with the presence of alkali metal catalyst of $CaCO_3$ were estimated using Coats-Redfern integral method. The E_A and A for different heating rates of 10, 20, 30, 50, and 100 Kmin^{-1} were tabulated in Table 4. The predicted E_A values ranged between 59.08 and 65.80 kJmol^{-1} while the average E_A value attained was 60.86 kJmol^{-1}. The calculated pre-exponential values range from 2.7 $\times 10^9$ to 189.4 $\times 10^9$ min^{-1}.

The comparison of E_A from pyrolysis of different biomass using Coats-Redfern integral method were shown in Table 5. From the previous literature, the E_A and A values of non-catalytic RH using Coats-Redfern kinetic model was 77.4 kJ mol^{-1} and 1.1 $\times 10^6$ min^{-1} respectively [25]. It is found that the E_A value in the present study of catalytic RH was reported 21.36 % lower compared to non-catalytic pyrolysis of RH, which explained the addition of catalyst has the effect to enhance the degradation rate of biomass. The reduced of E_A has the effect on the plant design specification, optimization process and the operation cost [26].

Table 4: Kinetics determined with Coats-Redfern equation for pyrolysis of rice husk with alkali metal catalyst of $CaCO_3$.

Heating rate, $Kmin^{-1}$	Fitted equation	R^2	T_{ave} (K)	E_A ($kJmol^{-1}$)	A (min^{-1})
10	y = -7106x - 1.715	0.9858	555.0	59.08	5.52×10^9
20	y = -7444x - 1.654	0.9879	562.5	61.89	1.88×10^{10}
30	y = -7367x - 1.896	0.9964	566.3	61.25	17.5×10^9
50	y = -6770x - 3.056	0.9979	571.1	56.29	2.70×10^9
100	y = -7914x - 1.477	0.9923	577.5	65.80	1.89×10^{11}
Average		0.9921		60.86	4.68×10^{10}

Table 5: Comparison of EA using Coats-Redfern method from previous studies.

Biomass	Heating rate ($Kmin^{-1}$)	Activation energy ($kJmol^{-1}$)	Pre-exponential factor (min^{-1})	Reference
RH	10 – 100	60.86	4.68×10^{10}	Present study
RH	20	77.4	1.1×10^6	[25]
Coal	10	92.6	3.31×10^{23}	[27]
Coal	5 – 30	83.46–93.41	$3.94 \times 10^5 - 5.44 \times 10^5$	[28]

4. Conclusions

The study of the thermal degradation behavior of catalytic pyrolysis for rice hull was carried out using thermogravimetric analysis (TGA) equipment. The Coats-Redfern integral model was chosen to perform the kinetic study and determine the kinetic parameters of RH with the presence of alkali metal of $CaCO_3$ catalyst in the pyrolysis process from heating rate of 10 to 100 $Kmin^{-1}$. Coats-Redfern is a single step method which aimed to extract a single value of E_A for an overall process and compare the catalytic result to the non-catalytic result at given heating rate. The calculated activation energy, E_A for catalytic pyrolysis of RH was 60.86 $kJmol^{-1}$. It is found that the heating rate of the pyrolysis process and the presence of alkali metal catalyst of $CaCO_3$ have strong relation to the degradation rate. The higher the heating rate and the addition of catalyst to the pyrolysis process may increase the rate of degradation of pyrolysis process. High degradation rate can minimize char formation and increase bio-oil and syngas production.

Acknowledgements

This research were supported by the International Foundation for Science, Stockholm, Sweden, through a grant to Dr Bridgid Chin Lai Fui on the project 'Catalytic fast pyrolysis of rice husk for syngas production' (Research Grant Agreement No. J-1-C6035-1) and also on the project 'Structure, Morphology and Properties Relationship of Well-Defined Architecture Cellulose Nanowhisker Reinforced Chitosan-g-Polylactic acid Nanocomposites via Atomic Transfer Radical Polymerization (ATRP) in an Ionic Liquid' (MOHE Ref. No: FRGS/1/2016/TK05/CURTIN/03/1). Furthermore, the authors would like to acknowledge the technical support from biomass processing lab, University Teknologi PETRONAS (UTP), Malaysia

Reference

[1] López D E, Goodwin J G, Bruce D A, and Lotero E 2005: 'Transesterification of triacetin with methanol on solid acid and base catalysts' *Appl. Catalysis A: General* **295** pp. 97-105

[2] McKendry P 2002: 'Energy production from biomass (part 1): overview of biomass' *Bioresour. Technol.* **83**

[3] Balat M, and Ayar G 2005: 'Biomass Energy in the World, Use of Biomass and Potential Trends' *Energy Sources* **27** pp. 931-940

[4] Lin Y, Liao Y, Yu Z, Fang S, and Ma X 2017: 'A study on co-pyrolysis of bagasse and sewage sludge using TG-FTIR and Py-GC/MS' *Energy Conver. Manag.* **151** pp. 190-198

[5] Shahbaz M, Yusup S, Inayat A, Patrick D O, Pratama A, and Ammar M 2017: 'Optimization of hydrogen and syngas production from PKS gasification by using coal bottom ash' *Bioresour. Technol.* **241** pp. 284-295

[6] Quispe I, Navia R, and Kahhat R 2017: 'Energy potential from rice husk through direct combustion and fast pyrolysis: A review' *Waste Manag.* **59** pp. 200-210

[7] Chin B L F, Yusup S, Al Shoaibi A, Kannan P, Srinivasakannan C, and Sulaiman S A 2014: 'Kinetic studies of co-pyrolysis of rubber seed shell with high density polyethylene' *Energy Convers. Manag.* **87** pp. 746-753

[8] Balasundram V, Ibrahim N, Kasmani R M, Hamid M K A, Isha R, Hasbullah H, and Ali R R 2017: 'Thermogravimetric catalytic pyrolysis and kinetic studies of coconut copra and rice husk for possible maximum production of pyrolysis oil' *J. Clean. Prod.* **167** pp. 218-228

[9] Zhang L., Liu R., Yin R., Mei Y. 2013: 'Upgrading of bio-oil from biomass fast pyrolysis in China: A review. *Renew. Sust. Energy Rev.* **24** pp. 66-72

[10] Sutton D., Kelleher B., Ross J.R.H. 2001: 'Review of literature on catalysts for biomass gasification' *Fuel Process. Technol.* **73** pp. 155-173

[11] Giudicianni P., Gargiulo V., Grottola C.M., Alfè M., Ragucci R. 2018 'Effect of alkali metal ions presence on the products of xylan steam assisted slow pyrolysis' *Fuel* **216** pp. 36-43.

[12] Chen H., Cheng H., Zhou F., Chen K., Qiao K., Lu X., Ouyang P., Fu J. 2018 'Catalytic fast pyrolysis of rice straw to aromatic compounds over hierarchical HZSM-5 produced by alkali treatment and metal-modification' *J. Anal. Appl. Pyrolysis* **131** pp. 76-84.

[13] Özsin G, and Pütün A E 2017: 'Insights into pyrolysis and co-pyrolysis of biomass and polystyrene: Thermochemical behaviors, kinetics and evolved gas analysis' *Energy Convers. Manag.* **149** pp. 675-685

[14] Chin B L F, Yusup S, Al Shoaibi A, Kannan P, Srinivasakannan C, and Sulaiman S A 2014: 'Comparative studies on catalytic and non-catalytic co-gasification of rubber seed shell and high density polyethylene mixtures' *J. Clean. Prod.* **70** pp. 303-314

[15] Naqvi S R, Uemura Y, Yusup S, Sugiura Y, and Nishiyama N 2015: 'In situ catalytic fast pyrolysis of paddy husk pyrolysis vapors over MCM-22 and ITQ-2 zeolites' *J. Anal. Appl. Pyrolysis* **114** pp. 32-39

[16] Guan G, Chen G, Kasai Y, Lim E W C, Hao X, Kaewpanha M, Abuliti A, Fushimi C, and Tsutsumi A 2012: 'Catalytic steam reforming of biomass tar over iron- or nickel-based catalyst supported on calcined scallop shell' *Appl. Catalysis B: Environ.* **115-116** pp. 159-168

[17] Mihalcik D J, Mullen C A, and Boateng A A 2011: 'Screening acidic zeolites for catalytic fast pyrolysis of biomass and its components' *J. Anal. Appl. Pyrolysis* **92** pp. 224-232

[18] López A, de Marco I, Caballero B M, Laresgoiti M F, Adrados A, and Aranzabal A 2011: 'Catalytic pyrolysis of plastic wastes with two different types of catalysts: ZSM-5 zeolite and Red Mud' *Appl. Catalysis B: Environ.***104** pp. 211-219

[19] Lim A C R, Chin B L F, Jawad Z A, and Hii K L 2016: 'Kinetic Analysis of Rice Husk Pyrolysis Using Kissinger-Akahira-Sunose (KAS) Method' *Procedia Eng.* **148** pp. 1247-1251

[20] Abu Bakar M S, and Titiloye J O 2013: 'Catalytic pyrolysis of rice husk for bio-oil production' *J. Anal. Appl. Pyrolysis* **103** pp. 362-368

[21] Mabuda A I, Mamphweli N S, and Meyer E L 2016: 'Model free kinetic analysis of biomass/sorbent blends for gasification purposes' *Renew. Sust. Energy Rev.* **53** pp. 1656-1664

[22] Khawam A, and Flanagan D R 2006: 'Basics and applications of solid-state kinetics: A pharmaceutical perspective' *J. Pharm. Sci.* **95** pp. 472-498

[23] Zhang S, Dong Q, Zhang L, and Xiong Y 2015: 'High quality syngas production from microwave pyrolysis of rice husk with char-supported metallic catalysts' *Bioresour. Technol.* **191** pp. 17-23

[24] Khan Z, Yusup S, Ahmad M M, Uemura Y, Chok V S, Rashid U, and Inayat A 2011: 'Kinetic Study on Palm Oil Waste Decomposition' *Biofuel's Eng. Proc. Technol.*

[25] Ahiduzzaman M, and Islam A K M S 2015: 'Thermo-gravimetric and Kinetic Analysis of Different Varieties of Rice Husk' *Procedia Eng.* **105** pp. 646-651

[26] Loy, A.C.M., Gan, K.W.D., Yusup, S., Chin B.L.F., Lam, M.K., M. Shahbaz, Unrean, P., Acda, M.N. 2018a.. Thermogravimetric kinetic modelling of in-situ catalytic pyrolysis conversion of rice husk to bioenergy using rice hull ash catalyst. *Bioresour. Technol.* **261**, pp. 213-222.

[27] Ebrahimi-Kahrizsangi R, and Abbasi M H 2008: 'Evaluation of reliability of Coats-Redfern method for kinetic analysis of non-isothermal TGA' *Trans. Nonferrous Metals Soc.China* **18** pp. 217-221

[28] Liu J, Ma J, Luo L, Zhang H, and Jiang X 2017: 'Pyrolysis of superfine pulverized coal. Part 5. Thermogravimetric analysis' *Energy Convers. Manage.* **154** pp. 491-502

Effect of the Density and the Refractive Index on the Toxicity of Imidazolium Ionic Liquids: A study of Ionic Interactions

Dzulkarnain Zaini[1,*], Muhammad Ishaq Khan[1], Azmi Mohd Shariff[1], Muhammad Athar[1], Muhammad Yasir Shamim[1]

[1] Centre for Advanced Process Safety (CAPS), Chemical Engineering Department, Universiti Teknologi Petronas, 32610 Bandar Seri Iskandar, Perak Darul Ridzuan, Malaysia.

* dzulkarnain.zaini@utp.edu.my

Abstract. Ionic Liquids (ILs) has got a significant attraction in the academia as well as in industries because of the salient characteristics and flexibility in tune-ability. The ionic nature and a huge number of possible combinations of cations, anions and functional groups make ILs potential candidates for industrially suitable materials. However, the toxic nature of ILs is still a big challenge. To overcome this problem, knowledge of the interactive mechanism of the ionic species of ILs is desired. In current work, we have investigated the ionic interaction of ILs by studying the effect of intensive properties of ILs "density" and "refractive index" of imidazolium NTf2 and PF6 ILs. Result produced in the current provided relationship of toxicity of ILs to their refractive indices and densities. The findings of current work are helpful to understand the mechanism of the toxic behavior of ILs.

1. Introduction

Ionic liquids (ILs) were thought to be green chemicals in the start of their invention because of their noticeable properties, for example, low vapor pressure, very less volatility, ionic nature, non-flammability, high stability[1]. The low vapor pressure of ILs makes them feasible for much industrial application where energy is considered as an economical barrier [2]. Nonflammability of ILs is a very important property particularly, from the safety perspective. Stability of ILs is related to the optimization and performance of the chemicals and the stability is considered as one of the major factors in the selection of chemicals for the adsorption, absorption and n catalysis. However, the investigation of their toxicological properties demonstrated that ILs cannot be placed into the category of green chemicals[3]. To understand the roots of the toxic behavior of ILs is a challenging task or the researchers.

In the past, the focus was on the synthesis of ILs to replace conventional organic and inorganic chemical in their respective application [4]. In recent years, the focus of the research is shifted towards the investigations of the properties responsible for their detrimental nature towards the environment.

An extensive research has already been done on an exploration of physicochemical properties of ILs[5]. Then properties of ILs provide a brief information about the interaction of the component of ILs with other compounds and with the components of itself. Cation-anion interaction determines the properties of ILs [6]. Different combination of cation and anion lead to synthesize of a large number of ILs. Hence, the properties of ILs are totally depending on the structure of ions. The physical, chemical properties, as well as toxic properties, are governed by ionic interactions.

The molecular interactions between the components of ILs define its physical and chemical properties as well as toxicological properties. Detailed study of the ionic interaction of molecules provides structural information which postulates different features of the ILs[7]. A schematic diagram is shown below in Figure1 which explains the mechanism of toxicity of ILs.

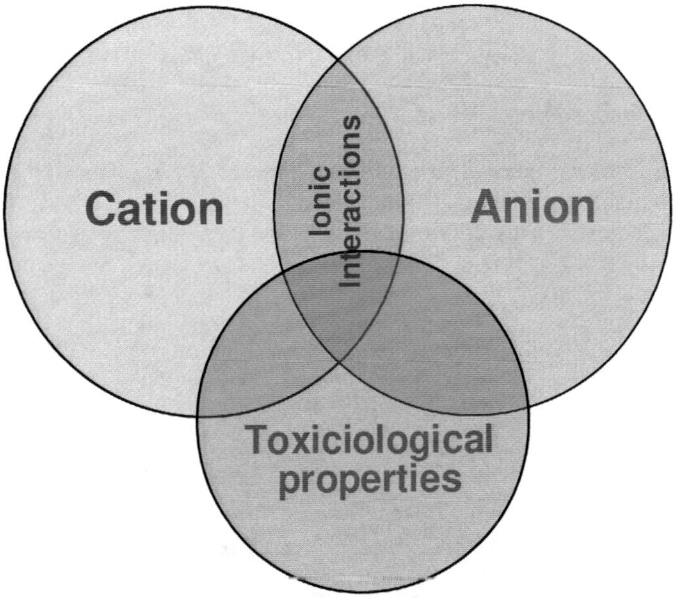

Figure 1 Interactions within the ILs

The discrete knowledge of the mechanism of combining components of ILs for example cations, anions, functional groups, alkyl chain, can define the toxic characteristics. A number of physical properties had been investigated in experimental studies by the researchers[8]. Some of the physical properties which are considered important are density, polarity, refractive index important where the applications of ILs are conferenced [9-11].

2. Methodology

2.1. Materials

In the current research, we have selected on two parameters i.e. the density and the refractive index of imidazolium bis(trifluoromethylsulfonyl)imide (NTf2) and hexafluorophosphate(PF6) ILs. The density and refractive index data were taken from a published work by Tariq et al.[11]. The toxicity data of the ILs are taken from the work done by Ventura et al, Ghanem et al, and Luis et al, [12-14] presented in Table 1.

Table 1. Toxicity of Imidazolium NTf_2 ILS towards different Bacterial species

Ionic Liquids	Density	Refractive index	EC50(mmol/L)
[EMIM][NTf_2]	1.5147 [11]	1.42251 [11]	0.844 [12]
[BIMI][NTf_2]	1.436 [11]	1.42653 [11]	0.339 [12]
[HMIM][NTf_2]	1.371 [11]	1.42954 [11]	0.051 [12]
[OMIM][NTf_2]	1.3189 [11]	1.43256 [11]	0.016 [13]
[BMIM][PF6]	1.3637 [11]	1.40844 [11]	1.174898 [14]
[HMIM][PF6]	1.2924 [11]	1.41648 [11]	0.138038 [14]
[OMIM][PF6]	1.234 [11]	1.42351 [11]	0.006761 [14]

The data presented in Table 1 is used in current method to understand and validate the ionic interactions of ILs to observe the effects of intensive properties of ILs to the toxicity.

2.2. Procedures

The current research work was carried out based on the interaction behavior of the intensive properties of ILs and their relevance was formed with the toxicity of ILs. Toxicity is the function of intensive properties of ILs, especially density and refractive index. To explain this dependency, the flow of the methodology is presented in Figure 2.

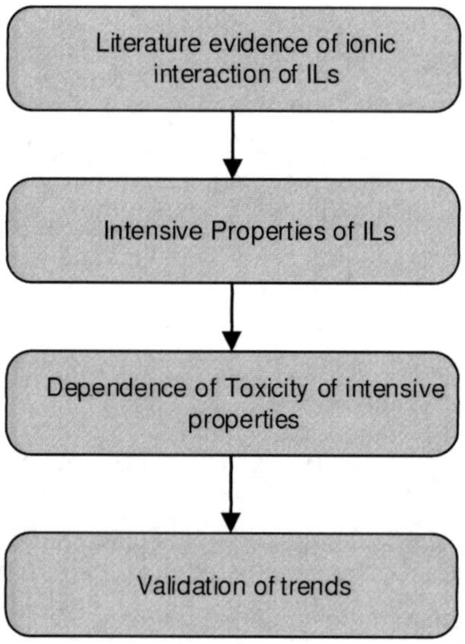

Figure 2 Methodological flow of interaction of intensive properties of IL with toxicity.

3. Results and Discussions

From the intensive properties of ILs especially refractive index and the density, the toxic behavior of ILs was studied. It was observed that with an increase in the refractive index, the density was reduced and the toxicity was increased. It was also noticeable the as the alkyl chain of ILs increased, the toxicity increased which was in line with previous studies done by many researchers [15]. Hence the increase in toxicity of ILs with altering refractive index and density helped to explore the ionic interaction in ILs.

IOP Conf. Series: Materials Science and Engineering **458** (2018) 012086 doi:10.1088/1757-899X/458/1/012086

3.1. Effect of the density on the toxicity of ILs

The length of alkyl chain of ILs is one of the major structural property of ILs which had been proved to affect their toxic behavior [16]. In current work, we noticed that with increased alkyl chain length, the reported toxicity was also increased. However, we found that increasing the alkyl chain actually alters the interaction amongst the constituents of imidazolium ILs. For example, when the density of imidazolium ILs was increased, we noticed that toxicity was decreased. Graphical illustration of results of toxicity of imidazolium ILs is presented in Figure 3.

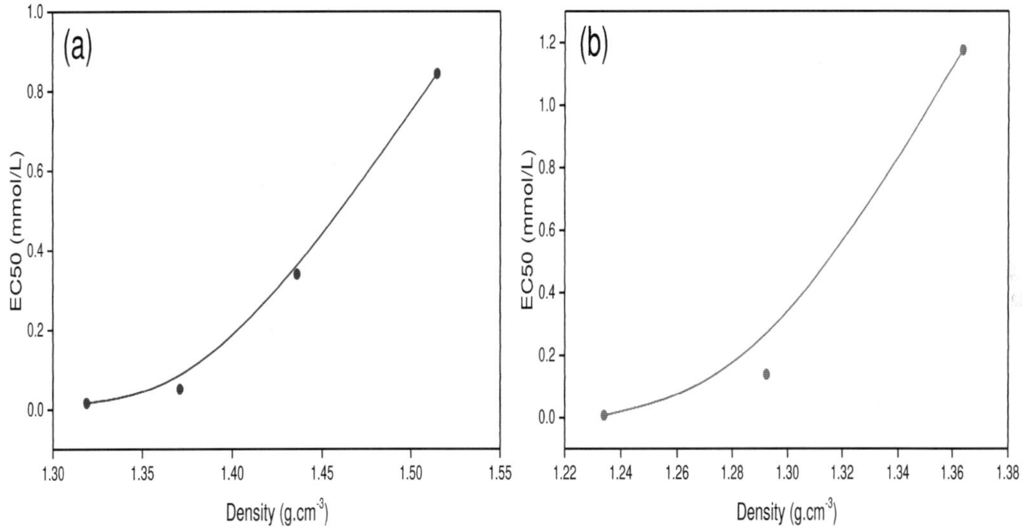

Figure 3. Toxicity of imidazolium ILs with (a) NTf2 and (b) PF_6 Vs Density.

Figure 3 showed that the EC50 of Imidazolium ILs increased linearly with density and EC50 of the ILs towards *Vibrio fischeri*. Increased density indicated that it was difficult for ILs to enter into the living organism because of the steric hindrance which may affect interaction with lipid membrane. Trends of increasing toxicity with increased density for imidazolium NTf2 ILs in Figure 3 (a) were validated and presented in Figure 3(b).

3.2. Effect of the refractive index on the toxicity of ILs

The graph was plotted between refractive indices of imidazolium NTf2 ILs and their toxicity. Figure 4(a) showed an inverse relationship between the refractive index and EC50 of imidazoliumNTf2 ILs, which specifies a linearly increase in toxicity with refractive index. The trends were validated in Figure 4(b). Refractive indices incorporate with molar volumes, which provide insights into the molecular interactions. However, the refractive index may vary at different temperatures but the toxicological investigations are carried out on room temperature ILs.

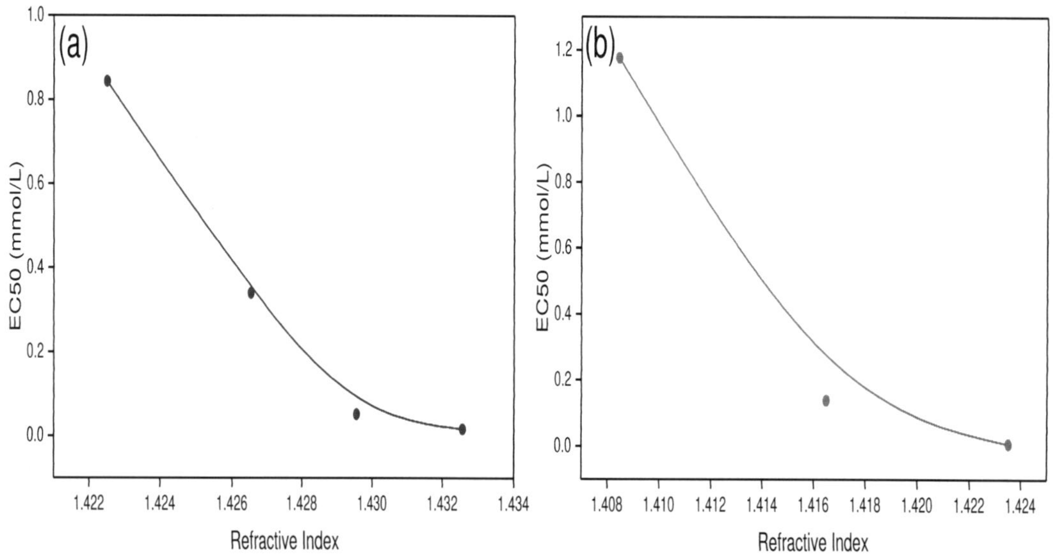

Figure 4. Toxicity of imidazolium ILs with (a) NTf_2 and (b) PF_6 Vs refractive index

Refractive index is the property of chemicals which can provide solubility properties by free volume analysis. The solubility of ILs in water is one of the major root cause of the toxicity of ILs towards the living organism when exposed. Hence, the refractive index is of great importance in the understand and assessment of the toxicological behavior of ILs.

4. Conclusion

In current work, we have investigated the interactive mechanism of ILs and studied the effects of density and refractive index on the toxicity of ILs. One of the basic root cause of toxic effects of ILs is water solubility property which makes it possible to mix ILs into the water and enter into the living organism through a cell membrane. We have highlighted the effect of density that the toxicity increase with decreasing the density and vice versa. It is difficult for denser ILs to enter into the cell wall of the cell membrane. In addition, the toxicity increased with increase in the refractive index of imidazolium ILs studied in current work. The refractive index incorporates the free volume of ILs which ultimately favor the solubility of ILs into the water. Hence the refractive index is linearly related with the toxicity of ILs. We have studied the interactive mechanism of two parameters which helped to understand the root cause of the toxic behavior of ILs. There are a number of other structural parameters can be studied to find the knowledge of toxicity on border level. In future, QSPR modeling can be done with a huge data on intensive properties and toxicities of ILs to different organisms of ecosystems.

Acknowledgment

The authors would like to gratitude Universiti Teknologi PETRONAS, Malaysia for providing research facilities and YUTP funding (0153AA-E19) to make this research feasible.

References

[1] Holbrey, J.D., Turner, M.B., and Rogers, R.D., Selection of ionic liquids for green chemical applications, in, ACS Publications, 2003.
[2] Bier, M. and Dietrich, S., Vapour pressure of ionic liquids, *Mol. Phys.* **108** 2010, 211-214.
[3] Khan, M.I., Zaini, D., Shariff, A.M., and Moniruzzaman, M., Framework for Ecotoxicological Risk Assessment of Ionic Liquids, *Proc. Engg.* **148** 2016, 1141-1148.
[4] de los Ríos, A.P., Irabien, A., Hollmann, F., and Fernández, F.J.H., Ionic liquids: Green solvents for chemical processing, *J. Chem.* **2013** 2013.
[5] Deng, Y., *Physico-chemical properties and environmental impact of ionic liquids*, PhD Thesis, Université Blaise Pascal-Clermont-Ferrand II, 2011.
[6] Fernandes, A.M., Rocha, M.A., Freire, M.G., Marrucho, I.M., Coutinho, J.A., and Santos, L.M., Evaluation of cation– anion interaction strength in ionic liquids, *The Journal of Physical Chemistry B* **115** 2011, 4033-4041.
[7] Barycki, M., Sosnowska, A., and Puzyn, T., Which structural features stand behind micelization of ionic liquids? Quantitative Structure-Property Relationship studies, *J. Colloid Interface Sci.* **487** 2017, 475-483.
[8] Stark, A. and Seddon, K.R., Ionic Liquids, in: *Kirk-Othmer Encyclopedia of Chemical Technology*, John Wiley & Sons, Inc., 2000.
[9] Ab Rani, M., Brandt, A., Crowhurst, L., Dolan, A., Hassan, N.H., Hallett, J., Hunt, P., Lui, M., Niedermeyer, H., and Perez-Arlandis, J., Erratum: Understanding the polarity of ionic liquids (Physical Chemistry Chemical Physics (2011) DOI: 10.1039/c1cp21262a), *PCCP* **13** 2011, 21653.
[10] Singh, T. and Kumar, A., Static dielectric constant of room temperature ionic liquids: internal pressure and cohesive energy density approach, *The Journal of Physical Chemistry B* **112** 2008, 12968-12972.
[11] Tariq, M., Forte, P., Gomes, M.C., Lopes, J.C., and Rebelo, L., Densities and refractive indices of imidazolium-and phosphonium-based ionic liquids: Effect of temperature, alkyl chain length, and anion, *The Journal of Chemical Thermodynamics* **41** 2009, 790-798.
[12] Ventura, S.P., Gonçalves, A.M., Sintra, T., Pereira, J.L., Gonçalves, F., and Coutinho, J.A., Designing ionic liquids: the chemical structure role in the toxicity, *Ecotoxicology* **22** 2013, 1-12.
[13] Ghanem, O.B., Mutalib, M.A., El-Harbawi, M., Gonfa, G., Kait, C.F., Alitheen, N.B.M., and Leveque, J.-M., Effect of imidazolium-based ionic liquids on bacterial growth inhibition investigated via experimental and QSAR modelling studies, *J. Hazard. Mater.* **297** 2015, 198-206.
[14] Luis, P., Garea, A., and Irabien, A., Quantitative structure–activity relationships (QSARs) to estimate ionic liquids ecotoxicity EC 50 (Vibrio fischeri), *J. Mol. Liq.* **152** 2010, 28-33.
[15] Matzke, M., Stolte, S., Thiele, K., Juffernholz, T., Arning, J., Ranke, J., Welz-Biermann, U., and Jastorff, B., The influence of anion species on the toxicity of 1-alkyl-3-methylimidazolium ionic liquids observed in an (eco)toxicological test battery, *Green Chemistry* **9** 2007, 1198-1207.
[16] Zhang, C., Zhu, L., Wang, J., Wang, J., Zhou, T., Xu, Y., and Cheng, C., The acute toxic effects of imidazolium-based ionic liquids with different alkyl-chain lengths and anions on zebrafish (Danio rerio), *Ecotoxicol. Environ. Saf.* **140** 2017, 235-240.

AUTHOR INDEX

AUTHOR INDEX

AUTHOR INDEX

AUTHOR INDEX